FUNDAMENTALS OF THE THEORY OF METALS

A. A. ABRIKOSOV

TRANSLATED FROM THE RUSSIAN BY
ARTAVAZ BEKNAZAROV

DOVER PUBLICATIONS
GARDEN CITY, NEW YORK

Copyright

Bibliographical Note

This Dover edition, first published in 2017, is an unabridged republication of the work originally published by North-Holland, Amsterdam, in 1988.

Library of Congress Cataloging-in-Publication Data

Names: Abrikosov, A. A. (Alekseæi Alekseevich), author. | Beknazarov, Artavaz, translator.
Title: Fundamentals of the theory of metals / A.A. Abrikosov ; translated from the Russian by Artavaz Beknazarov.
Other titles: Osnovy teorii metallov. English
Description: Dover edition. | Garden City, New York : Dover Publications 2017. | "This Dover edition, first published in 2017, is an unabridged republication of the work originally published by North-Holland, Amsterdam, in 1988"—Title page verso. | Includes bibliographical references and index.
Identifiers: LCCN 2017012583| ISBN 9780486819013 | ISBN 0486819019
Subjects: LCSH: Superconductors. | Superconductivity. | Metals.
Classification: LCC QC176.8.E4 A2613 2017 | DDC 530.4/1—dc23
LC record available at https://lccn.loc.gov/2017012583

Manufactured in the United States of America
81901905 2023
www.doverpublications.com

Dedicated to the memory of Lev Aslamazov

Preface

This book has grown out of courses and lectures given by the author at Moscow State University, the Moscow Engineering Physics Institute, the Moscow Institute of Steel and Alloys, and also at Delhi University (India) and Lausanne University (Switzerland). The book is intended for undergraduate students majoring in physics and also for graduate and postgraduate students, research workers and teachers. The presentation assumes a knowledge of quantum mechanics and quantum statistics. No special knowledge of metals that goes beyond a course in general physics is needed.

Part I is devoted to normal metals and is basically a thoroughly revised version of the author's book *An Introduction to the Theory of Normal Metals* originally published in Russian in 1972 in the USSR (also published in English in the USA and in German in the GDR). This part covers the following topics: the electronic spectra of metals; electrical and thermal conductivities; galvanomagnetic and thermoelectrical phenomena; the behavior of metals in high-frequency fields; sound absorption; and Fermi-liquid phenomena. Part I also deals with the present-day concept of the energy spectra of electrons in metals and with methods that make it possible to study the behavior of metals in dc and ac fields. It shows the reader how the energy spectrum can be deduced with the aid of the results of such investigations.

Much attention is paid to new findings in the theory of metals: quantum interference phenomena and the problem of localization of electrons by a random potential, nonlinear effects in the interaction of electrons with an electromagnetic field and with sound, the Kondo effect, etc.

Part II is concerned with the theory of superconductivity. Today we can no longer speak of the "phenomenon of superconductivity". As a matter of fact, this is a novel area of solid-state physics which embraces a large variety of phenomena. Many of these phenomena have found practical application not only in physical instrumentation but also in power engineering, transportation, medicine, electronics, and computational mathematics. The physics of superconductors and, in particular, the theory of superconductivity, has been presented in a large number of books. Most of these books were published many years ago and have become obsolete. At the same time, practically all of these books are either too specialized, covering only parts of the field, or too elementary in exposition, and therefore do not meet the requirements of the study of superconductivity by physicists and engineers interested in a wide range of allied disciplines.

The presentation of the theory of superconductivity is no doubt a more formidable task than that of the theory of normal metals. The point is that the theory of superconductivity covers many areas, which all employ their own specific techniques. These methods are, as a rule, highly sophisticated and are not accessible without special background knowledge.

To cope with these difficulties, I have chosen a compromise. The book offers a detailed description of two fundamental methods: the microscopic theory based on the Bogoliubov method, and the Ginzburg–Landau theory, which allows one to describe the behavior of superconductors close to the critical temperature. These methods provide the basis for a quantitative treatment of many important properties of superconductors: thermodynamic characteristics, linear electrodynamics, some problems of the kinetics, the theory of critical properties of thin films and wires, type-II superconductivity, paraconductivity, theory of the tunnel junction, the Josephson effect, etc. But there exist other phenomena of considerable physical interest which call for highly refined techniques and cumbersome calculations. In such cases a qualitative treatment is given, accompanied by simple estimates. I hope this book will be useful as a guide for readers who wish to become acquainted with more detailed information on the various problems of the theory of superconductivity.

In concluding this preface, I wish to express my thanks to my colleagues, especially Yu.V. Sharvin, V.F. Gantmakher, M.I. Kaganov, A.I. Larkin, G.M. Eliashberg, Yu.N. Ovchinnikov, A.G.Aronov, B.L. Altshuler, D.E. Khmelnitskii, L.A. Falkovskii, V.L. Gurevich, V.V. Shmidt, L.N. Bulaevskii, B.I. Ivlev, V.I. Kozub, R.I. Gurzhi, A.M. Finkelshtein, and N.B. Kopnin, for helpful stimulating discussions on various sections of the book and for valuable recommendations aimed at achieving a balanced presentation of complicated subjects. May all of them find here the expression of my gratitude.

Last but not least, I would like to acknowledge the sustained assistance rendered by the late Professor Lev G. Aslamazov during the writing of this book. With a feeling of deep sorrow at the tragic and untimely loss of his presence I give him here the tribute of my profound gratitude.

Professor Lev Aslamazov was a prominent specialist in the theory of superconductivity. In collaboration with A.I. Larkin, he has discovered the phenomenon of paraconductivity and investigated the Josephson effect at specified current. Lev Aslamazov has worked out a theory of Josephson junctions with a normal-metal and a semiconductor barrier, the high-frequency stimulation of superconductivity in bridges, and many other phenomena. The reader will repeatedly encounter his name in the second part of the book.

His work will not be forgotten and his distinctive personality will long remain a living force for all those who knew him and enjoyed the great pleasure of working with him.

Moscow, USSR
1987

A.A. Abrikosov

Contents

Part II. Superconducting Metals

Part I. Normal metals

1

*An electron in a periodic crystal lattice**

1.1. General properties

It is well known that metals are good conductors of electricity. This is because the outer electronic shells of atoms that make up a metal overlap to a considerable extent. Therefore, the electrons in these shells (called valence electrons) move easily from atom to atom, so that one cannot say to which atom they really belong. This collectivization of outer electrons leads to the generation of the large binding energy of metals and accounts for their specific mechanical properties.

As for the inner electronic shells, because of the small degree of overlap they may be regarded approximately the same as in isolated atoms.

Thus, a metal is a crystalline lattice made up of positive ions, into which are "poured" collectivized electrons of the valence shells. They are also called conduction or "free" electrons. In acutal fact, these electrons strongly interact with one another and with the lattice ions, the potential energy of these interactions being of the order of the kinetic energy of electrons.

The construction of the theory of such a system seems at first sight quite impossible. However, there actually exists at present a sufficiently rigorous description of most of the interesting phenomena that occur in metals. This is associated with two circumstances. First, the behaviour of a system of strongly interacting electrons (or of an electron liquid) is in many respects analogous to that of a system of noninteracting particles (i.e., a gas) in a certain external field, which is the averaged field of the lattice ions and the other electrons. Second, although this field is difficult to calculate exactly, one can deduce much from the fact that the averaged field displays the symmetry properties of the crystal lattice, in particular periodicity. We will start therefore from the study of the auxiliary problem of the behavior of an electron in a periodic field.

Let us consider an electron moving in an external field with a potential energy $U(\boldsymbol{r})$. The function $U(\boldsymbol{r})$ is periodic, i.e.,

$$U(\boldsymbol{r}+\boldsymbol{a}_n)=U(\boldsymbol{r}), \tag{1.1}$$

* The material presented in this chapter can be found practically in all books devoted to the theory of metals, and we could have simply referred the reader to the literature (for example, Peierls 1955). However, since these conceptions form the basis for what follows, we consider it useful to outline them here.

where $\boldsymbol{a}_n$ is an arbitrary lattice period. As is known, the vector $\boldsymbol{a}_n$ can always be represented as a linear combination of basis vectors $\boldsymbol{a}_i$:

$$\boldsymbol{a}_n = n_1\boldsymbol{a}_1 + n_2\boldsymbol{a}_2 + n_3\boldsymbol{a}_3, \tag{1.2}$$

where n_i are positive or negative integers or zeros.

The Schrödinger equation for an electron is

$$-\frac{\hbar^2}{2m}\nabla^2\psi(\boldsymbol{r}) + U(\boldsymbol{r})\,\psi(\boldsymbol{r}) = \varepsilon\psi(\boldsymbol{r}). \tag{1.3}$$

It is not difficult to see that $\psi(\boldsymbol{r}+\boldsymbol{a}_n)$ is also a solution of this equation, with the same eigenvalue ε. Therefore, if the electron level ε is nondegenerate, i.e. has a single eigenfunction ψ, then we must have

$$\psi(\boldsymbol{r}+\boldsymbol{a}_n) = C\psi(\boldsymbol{r}), \tag{1.4}$$

where C is a constant.

But if the level ε is degenerate, i.e., has several eigenfunctions ψ_ν, we may write

$$\psi_\mu(\boldsymbol{r}+\boldsymbol{a}_n) = \sum_\nu C_{\mu\nu}\psi_\nu(\boldsymbol{r}). \tag{1.5}$$

Since the functions ψ_μ form an orthogonal and normalized set, i.e.

$$\int \psi_\mu^*(\boldsymbol{r})\,\psi_\nu(\boldsymbol{r})\,\mathrm{d}V = \delta_{\mu\nu}, \tag{1.6}$$

it follows that by shifting the integration variable $\boldsymbol{r}$ by $\boldsymbol{a}_n$ and using formula (1.5) we obtain

$$\sum_\eta C_{\mu\eta}^* C_{\nu\eta} = \delta_{\mu\nu}. \tag{1.7}$$

Hence, $C_{\mu\nu}$ is a unitary matrix, i.e.,

$$C^+ = C^{-1}. \tag{1.8}$$

But such a matrix can be diagonalized. In other words, certain linear combinations of the functions ψ_ν exhibit the property (1.4). The normalization condition here gives

$$|C|^2 = 1. \tag{1.9}$$

Thus, we may write

$$C = e^{i\varphi(a_n)} \tag{1.10}$$

where φ is a real function of the displacement $\boldsymbol{a}_n$.

Let us now consider two successive displacements: $\boldsymbol{a}$ and $\boldsymbol{a}'$. In the first displacement the function ψ is multiplied by $C(\boldsymbol{a})$ and in the second by $C(\boldsymbol{a}')$. At the same time, the two successive displacements are equivalent to a single displacement by $\boldsymbol{a}+\boldsymbol{a}'$. Here the function ψ must simply be multiplied by $C(\boldsymbol{a}+\boldsymbol{a}')$. Hence,

$$C(\boldsymbol{a}+\boldsymbol{a}') = C(\boldsymbol{a})\,C(\boldsymbol{a}'). \tag{1.11}$$

From this it follows that the function φ in formula (1.10) must be a linear function of $\boldsymbol{a}_n$:

$$\varphi(\boldsymbol{a}_n) = \frac{\boldsymbol{p}\boldsymbol{a}_n}{\hbar}, \tag{1.12}$$

where $\boldsymbol{p}$ is a vector coefficient.

It is easy to see that this vector has been defined ambiguously. Namely, if to $\boldsymbol{p}$ we add the vector $\hbar\boldsymbol{K}$, which satisfies the condition $\boldsymbol{K}\boldsymbol{a}_n = 2\pi m$ for any lattice period $\boldsymbol{a}_n$ (where m is an integer), we will obtain the same coefficients $C(\boldsymbol{a}_n)$. The equations $\boldsymbol{K}\boldsymbol{a}_n = 2\pi m$ are satisfied by an infinite system of vectors, all of which may be written in the following form:

$$\boldsymbol{K} = q_1\boldsymbol{K}_1 + q_2\boldsymbol{K}_2 + q_3\boldsymbol{K}_3. \tag{1.13}$$

Here q_i are integers and $\boldsymbol{K}_i$ are the smallest noncoplanar vectors exhibiting the property $\boldsymbol{K}\boldsymbol{a}_n = 2\pi m$. From formula (1.2) it follows that this condition must be satisfied for the basis vectors $\boldsymbol{a}_i$. It will then be satisfied for any period $\boldsymbol{a}_n$. From this it is easy to obtain the vectors K_i:

$$\boldsymbol{K}_1 = \frac{2\pi[\boldsymbol{a}_2\boldsymbol{a}_3]}{(\boldsymbol{a}_1[\boldsymbol{a}_2\boldsymbol{a}_3])}, \qquad \boldsymbol{K}_2 = \frac{2\pi[\boldsymbol{a}_3\boldsymbol{a}_1]}{(\boldsymbol{a}_1[\boldsymbol{a}_2\boldsymbol{a}_3])}, \qquad \boldsymbol{K}_3 = \frac{2\pi[\boldsymbol{a}_1\boldsymbol{a}_2]}{(\boldsymbol{a}_1[\boldsymbol{a}_2\boldsymbol{a}_3])}. \tag{1.14}$$

Thus, we see that the vectors $\boldsymbol{K}_i$ are equal to 2π multiplied by the reciprocal heights of the unit cell. Taking $\boldsymbol{K}_1$, $\boldsymbol{K}_2$ and $\boldsymbol{K}_3$ as basis vectors, we can construct the so-called reciprocal lattice. Hence, the reciprocal lattice is wholly determined by the translational properties of the crystal under consideration (by the vectors $\boldsymbol{a}_i$), i.e., by its Bravais lattice, and has the same symmetry properties. But, as is known, there may exist various Bravais lattices with the same symmetry. The relationship between the Bravais lattice and the reciprocal lattice is as follows: if the Bravais lattice is body-centered, then the reciprocal lattice is face-centred, and vice versa; to a base-centered Bravais lattice there corresponds a base-centered reciprocal lattice.

The electron energy ε depends on the vector $\boldsymbol{p}$. Since $\boldsymbol{p}$ and $\boldsymbol{p} + \hbar\boldsymbol{K}$ are physically equivalent, it follows that the energy $\varepsilon(\boldsymbol{p})$ must evidently be a periodic function with periods $\hbar\boldsymbol{K}_i$. To each value of $\boldsymbol{p}$ there may, generally speaking, correspond several energy levels $\varepsilon_l(\boldsymbol{p})$ and each of these functions is periodic in the reciprocal lattice.

The wave function describing the movement of an electron in the periodic field and having the property

$$\psi(\boldsymbol{r}+\boldsymbol{a}_n) = e^{i\boldsymbol{p}\boldsymbol{a}_n/\hbar}\psi(\boldsymbol{r})$$

may be represented as

$$\psi(\boldsymbol{r}) = e^{i\boldsymbol{p}\boldsymbol{r}/\hbar}u(\boldsymbol{r}), \tag{1.15}$$

where $u(\boldsymbol{r})$ is a periodic function:

$$u(\boldsymbol{r}+\boldsymbol{a}_n) = u(\boldsymbol{r}).$$

Formula (1.15) is known as the Bloch theorem. The wave function ψ in the form (1.15) resembles a plane wave describing the motion of a free particle, but here the wave is modulated by a periodic function. Therefore, the vector $\boldsymbol{p}$, which is analogous to the momentum, is not in fact the momentum of a particle in the ordinary sense of the word. It is called the quasimomentum of the electron.

Since the vectors $\boldsymbol{p}$ and $\boldsymbol{p}+\hbar\boldsymbol{K}$ are physically equivalent, for the sake of uniqueness we may consider only one unit cell of the reciprocal lattice. The volume of the region of the unique determination of $\boldsymbol{p}$ is given by

$$\left(\frac{2\pi\hbar}{v}\right)^3([\boldsymbol{a}_2\boldsymbol{a}_3][[\boldsymbol{a}_3\boldsymbol{a}_1][\boldsymbol{a}_1\boldsymbol{a}_2]])=\frac{(2\pi\hbar)^3}{v},$$

where $v=(\boldsymbol{a}_1[\boldsymbol{a}_2\boldsymbol{a}_3])$ is the volume of the unit cell of the principal lattice.

In order to obtain the solution of the Schrödinger equation, one has to know the boundary conditions. However, in an infinitely large volume the successive states will be infinitely close to one another. We are actually interested only in the density of states, i.e., the number of states per energy interval or given volume element in quasimomentum space. The density of states is independent of the particular form of the boundary conditions, and therefore it is easier to determine by assuming the simplest conditions.

Assuming that the metal specimen under consideration has the shape of a rectangular parallelepiped, we specify periodic boundary conditions:

$$\psi(x+L_1, y, z)=\psi(x, y+L_2, z)=\psi(x, y, z+L_3)$$
$$=\psi(x, y, z). \tag{1.16}$$

Assuming that each of the dimensions L_1, L_2 and L_3 contains an integer number of periods in its direction, we obtain:

$$e^{ip_xL_1/\hbar}=e^{ip_yL_2/\hbar}=e^{ip_zL_3/\hbar}=1$$

from which it follows that

$$p_x=2\pi\hbar n_x/L_1, \qquad p_y=2\pi\hbar n_y/L_2, \qquad p_z=2\pi\hbar n_z/L_3, \tag{1.17}$$

where n_x, n_y and n_z are integers.

Thus, the vector $\boldsymbol{p}$ proves to be a discrete variable. But if the lengths L_1, L_2 and L_3 are very large, then the summation over the states may be replaced by an integration. To do this, we have to know the number of states in a given volume of $\boldsymbol{p}$-space. From eq. (1.17) we find

$$\Delta n_x\Delta n_y\Delta n_z=\frac{L_1L_2L_3\Delta p_x\Delta p_y\Delta p_z}{(2\pi\hbar)^3},$$

so that the number of states in the interval $\mathrm{d}^3\boldsymbol{p}=\mathrm{d}p_x\,\mathrm{d}p_y\,\mathrm{d}p_z$ is equal to

$$V\frac{\mathrm{d}^3\boldsymbol{p}}{(2\pi\hbar)^3}$$

where $V = L_1 L_2 L_3$ is the volume of the sample. This means that the density of states in $\boldsymbol{p}$-space is

$$\frac{V}{(2\pi\hbar)^3}. \tag{1.18}$$

As has already been pointed out, the region of the unique determination of $\boldsymbol{p}$ is the unit cell of the reciprocal lattice with a volume $(2\pi\hbar)^3/v$. Therefore, the total number of various values of $\boldsymbol{p}$ is equal to

$$\left[\frac{(2\pi\hbar)^3}{v}\right] \cdot \left[\frac{V}{(2\pi\hbar)^3}\right] = N,$$

where N is the number of unit cells in the sample under consideration. It must also be kept in mind that the electron has a spin $s = \frac{1}{2}$, whose projection on a certain axis may have two values, $s_z = \pm\frac{1}{2}$. This doubles the number of states. Thus, it turns out that to each of the functions $\varepsilon_l(\boldsymbol{p})$ there correspond $2N$ various states.

The functions $\varepsilon_l(\boldsymbol{p})$ are periodic in the reciprocal lattice and naturally oscillate between the maximal and minimal values. Hence, for each number l we obtain "bands" of allowed energy values. These bands may be separated by "energy gaps" (i.e., energy values unattainable for electrons), but they may also overlap.

Let us consider some general properties of the functions $\varepsilon_l(\boldsymbol{p})$. The complete Schrödinger equation has the form

$$i\hbar \frac{\partial \psi}{\partial t} = H\psi.$$

We will now turn to the complex-conjugate equation and perform the transformation $t \to -t$. Here we obtain

$$i\hbar \frac{\partial \psi^*(-t, \boldsymbol{r})}{\partial t} = H^* \psi(-t, \boldsymbol{r}),$$

that is, the same Schrödinger equation with a Hamiltonian H^*. But H is a Hermitian operator, i.e., the eigenfunctions and eigenvalues of the operators H and H^* are the same. From this it follows that if $\psi_{lp}(\boldsymbol{r}, t) = \exp[-i\varepsilon_l(p)t/\hbar]\, \psi_{lp}(\boldsymbol{r})$ is an eigenfunction of H, then the function $\psi^*_{lp}(\boldsymbol{r}, -t)$ is also an eigenfunction of H. Upon displacement of $\boldsymbol{r}$ by a period $\boldsymbol{a}$ the function ψ_{lp} acquires a factor $\exp(i\boldsymbol{pa}/\hbar)$ and the function $\psi^*_{lp}(\boldsymbol{r}, -t)$ acquires a factor $\exp(-i\boldsymbol{pa}/\hbar)$. It then follows that $\varepsilon_l(\boldsymbol{p}) = \varepsilon_l(-\boldsymbol{p})$.

We have so far used the unit cell of the reciprocal lattice as the region of the unique determination of the quasimomentum $\boldsymbol{p}$. But it is more convenient to define this region in a different way. Of course, it must have a volume equal to the volume of the unit cell of the reciprocal lattice and, besides, it must not contain points differing by a period of the reciprocal lattice or more. We will define it as follows.

Let us draw from some reciprocal lattice point all $\boldsymbol{K}$-vectors that connect it with the other lattice points. Then, we draw planes perpendicular to each of these vectors and dividing them in half. These planes will cut out a certain figure in the space of the reciprocal lattice which has the shape of a polyhedron. It is not difficult to see that such a polyhedron possesses all the required properties and may therefore be taken as the region of specification of the quasimomentum $\boldsymbol{p}$. It is called the Brillouin zone. Figure 1 shows examples of Brillouin zones for the face-centred cubic (a) and body-centered cubic (b) lattices.

As a rule, the crystal lattices of metals exhibit high symmetry. This gives rise to certain properties of the function $\varepsilon_l(\boldsymbol{p})$. Suppose, for example, that the symmetry plane perpendicular to the axis p_x passes through the point $\boldsymbol{p}=0$. If there exist faces of the Brillouin zone perpendicular to the p_x axis, then $\varepsilon_l(\boldsymbol{p})$ as functions of p_x must have extrema on these faces. Indeed, let us single out the points of these faces, $\boldsymbol{p}_1$ and $\boldsymbol{p}_2$, which are symmetric with respect to the symmetry plane (fig. 2). They differ by a reciprocal lattice period (multiplied by $\hbar$). Therefore, at these points

$$\varepsilon_l(\boldsymbol{p}_1)=\varepsilon_l(\boldsymbol{p}_2), \qquad \frac{\partial\varepsilon_l(\boldsymbol{p}_1)}{\partial p_x}=\frac{\partial\varepsilon_l(\boldsymbol{p}_2)}{\partial p_x}.$$

But by virtue of the symmetry with respect to the $p_x=0$ plane we have

$$\frac{\partial\varepsilon_l(\boldsymbol{p}_1)}{\partial p_x}=-\frac{\partial\varepsilon_l(\boldsymbol{p}_2)}{\partial p_x}.$$

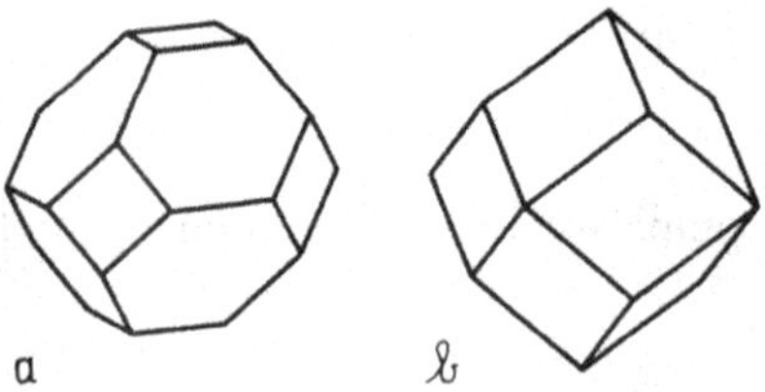

Fig. 1.

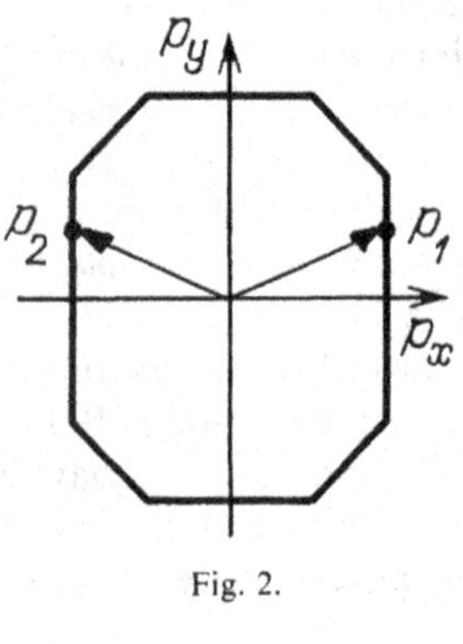

Fig. 2.

Hence,

$$\frac{\partial \varepsilon_l(\boldsymbol{p}_1)}{\partial p_x} = 0.$$

In an analogous way we obtain in this case

$$\left(\frac{\partial \varepsilon_l}{\partial p_x}\right)_{p_x=0} = 0.$$

Thus, we arrive at the conclusion that for symmetrical lattices, as a rule, there are extrema of the functions $\varepsilon_l(\boldsymbol{p})$ in the center of the Brillouin zone or at its boundaries.

The conclusions concerning the electron energy as a function of the quasimomentum are illustrated in fig. 3, which refers to the one-dimensional case. Evidently, the Brillouin zone here is the segment $-\pi\hbar/a < p < \pi\hbar/a$, where a is the period of a linear chain.

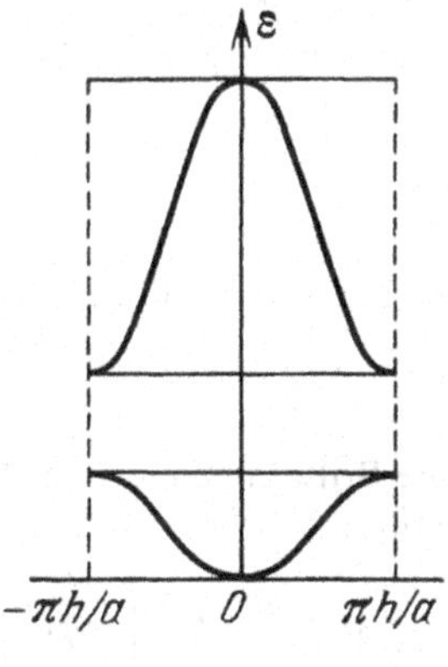

Fig. 3.

1.2. *The strong-coupling approximation*

For the function $\varepsilon_l(\boldsymbol{p})$ to be calculated exactly, use is made of rather complicated methods (see ch. 14). To illustrate the general properties of the functions $\varepsilon_l(\boldsymbol{p})$, we will consider in the following the two simplest techniques, although they are not very efficient for the exact determination of the functions $\varepsilon_l(\boldsymbol{p})$ in real metals.

We begin with the so-called strong-coupling method and for the sake of simplicity consider first a one-dimensional metal, i.e., a linear chain of atoms. We assume that the electronic shells overlap little and that, in the zeroth approximation, each electron belongs to one atom. The overlap of the shells is regarded as a perturbation.

The potential energy of the electron in the field of all the ions has the form

$$V(x)=\sum_n U(x-na) \tag{1.19}$$

and the Schrödinger equation is

$$-\frac{\hbar^2}{2m}\frac{d^2\psi}{dx^2}+\sum_n U(x-na)\,\psi(x)=\varepsilon\psi(x). \tag{1.20}$$

Let the following Bloch functions be the exact solutions of eq. (1.20):

$$\psi_p(x)=e^{ipx/\hbar}u_p(x)$$

and the corresponding eigenvalues be equal to $\varepsilon(p)$. We now form so-called Wannier functions from the functions ψ_p:

$$w_n(x)=N^{-1/2}\sum_p e^{-ipna/\hbar}\,\psi_p(x), \tag{1.21}$$

where N is the number of atoms in the chain, and the sum over p is limited by a one-dimensional Brillouin zone: $-\pi\hbar/a\leqslant p\leqslant \pi\hbar/a$. The inverse transformation has the form

$$\psi_p(x)=N^{-1/2}\sum_n e^{ipna/\hbar}w_n(x). \tag{1.22}$$

Indeed, substituting formula (1.21) into eq. (1.22), we have

$$\psi_p(x)=N^{-1}\sum_{n,p'} e^{i(p-p')na/\hbar}\psi_{p'}(x)=\psi_p(x)$$

(since p and p' are limited by the Brillouin zone). The functions $w_n(x)$ with different n are orthogonal. As a matter of fact,

$$\begin{aligned}\int w_\eta^*(x)\,w_m(x)\,dx &= N^{-1}\sum_{p,p'} e^{i(pn-p'm)a/\hbar}\int \psi_p^*(x)\,\psi_{p'}(x)\,dx\\ &=N^{-1}\sum_p e^{ip(n-m)a/\hbar}\\ &=N^{-1}\frac{L}{2\pi\hbar}\int_{-\pi\hbar/a}^{\pi\hbar/a} e^{ip(n-m)a/\hbar}\,dp\\ &=\left\{\begin{matrix}\dfrac{\sin[\pi(n-m)]}{\pi(n-m)}=0, & n\neq m\\ 1 & n=m\end{matrix}\right\}=\delta_{nm},\end{aligned} \tag{1.23}$$

where $L=Na$ is the length of the chain.

From the definition (1.21) it is seen that the function $w_n(x)$ is nonzero only near the nth ion. Indeed, if the Bloch function were simply a plane wave, without being modulated by the function $u_p(x)$, then w_n would be equal to $\delta(x-na)$. This property allows one to use a small degree of overlap of the shells. From the definition (1.21) it also follows that, because of the periodicity of $u_p(x)$, all the functions w_n are in fact the same function $w_0(x) \equiv w(x)$ with shifted arguments:

$$w_n(x) = w(x - na).$$

Since ψ_p satisfies the Schrödinger equation, it follows that substitution of expression (1.22) into eq. (1.20) yields

$$\sum_n \left(-\frac{\hbar^2}{2m}\frac{d^2}{dx^2} + U(x - na) \right) e^{ipna/\hbar} w_n(x) + \sum_n h(x)\, e^{ipna/\hbar} w_n(x)$$
$$= \varepsilon(p) \sum_n e^{ipna/\hbar} w_n(x), \tag{1.24}$$

where $h(x) = V(x) - U(x - na)$. The term with $h(x)$, which contains only the products $U(x - ma)\, w_n(x)$ with $n \neq m$, is small, since the function $w_n(x)$ is nonzero only near $x = na$.

In the zeroth approximation, neglecting this term, we see that the function $w(x)$, equal to the wave function of the electron in an isolated atom, satisfies eq. (1.24), i.e.,

$$w^{(0)} = \varphi(x). \tag{1.25}$$

Here it is evident that

$$\varepsilon^{(0)}(p) = \varepsilon_0,$$

where ε_0 is the corresponding level of the isolated atom.

We will now consider the following approximation. Assuming that $w = w^{(0)} + w^{(1)}$, we find

$$\sum_n \left(-\frac{\hbar^2}{2m}\frac{d^2}{dx^2} + U(x - na) - \varepsilon_0 \right) w^{(1)}\, e^{ipna/\hbar}$$
$$= [\varepsilon(p) - \varepsilon_0] \sum_n e^{ipna/\hbar} w_n^{(0)}(x) - \sum_n h(x)\, e^{ipna/\hbar} w_n^{(0)}(x).$$

This is a linear equation for $w^{(1)}$ with a right-hand side. According to the general rule, such an equation has a solution only if the right-hand side is orthogonal to the solution of the corresponding reduced equation with the same boundary conditions. These conditions consist of the vanishing of w at $\pm\infty$. From this it follows that the corresponding solution of the reduced equation is just $w(x)$ in the zeroth approximation, i.e., $\varphi(x)$. From the orthogonality condition we deduce that

$$\varepsilon(p) - \varepsilon_0 = \frac{\sum_n h(n)\, e^{ipna/\hbar}}{\sum_n I(n)\, e^{ipna/\hbar}}, \tag{1.26}$$

where

$$h(n)=\int \varphi^*(x)\, h(x)\, \varphi(x-na)\, \mathrm{d}x, \tag{1.27}$$

$$I(n)=\int \varphi^*(x)\, \varphi(x-na)\, \mathrm{d}x. \tag{1.28}$$

The atomic function φ can be chosen real. Then we evidently have $h(n)=h(-n)$ and $I(n)=I(-n)$.

The two functions $h(n)$ and $I(n)$ fall off very rapidly with increasing n, since the overlap is assumed to be small. We will therefore take into account only the first terms. The quantity $h(0)$ is of order $U(a)$, $h(1)$ of order $U(a)\varphi(a)/\varphi(0)$, and $I(0)=1$, $I(1)$ is of order $\varphi(a)/\varphi(0)$. Thus, we obtain

$$\varepsilon-\varepsilon_0=h(0)+2[h(1)-h(0)\, I(1)]\cos\left(\frac{pa}{\hbar}\right), \tag{1.29}$$

whence it follows that ε is a periodic function of p with a period $2\pi\hbar/a$, i.e., with the period of the inverse reciprocal chain.

Computations in three dimensions are more complicated. Additional difficulties arise if there are more than one atom per unit cell, so that these atoms are symmetrical under some symmetry transformations which are different from the displacement by a period. Moreover, the atomic state may correspond to a nonzero rotational momentum, and the corresponding level ε_0 will be degenerate. In the simplest case, when there is one atom per unit cell and this atom has only one s-electron, we find

$$\varepsilon-\varepsilon_0=\frac{\sum\limits_{n} h(\boldsymbol{n})\, \mathrm{e}^{\mathrm{i}pa_n/\hbar}}{\sum\limits_{n} I(\boldsymbol{n})\, \mathrm{e}^{\mathrm{i}pa_n/\hbar}}. \tag{1.30}$$

Let us consider the face-centred cubic (fcc) lattice. If we take into account only the nearest neighbors, then

$$\boldsymbol{a}_n=(\pm\tfrac{1}{2}a, \pm\tfrac{1}{2}a, \pm\tfrac{1}{2}a),$$

where a is the face of the cube. Here

$$\varepsilon-\varepsilon_0=h(0)+8[h(1)-h(0)\, I(1)] \times\cos\left(\frac{p_x a}{2\hbar}\right)\cos\left(\frac{p_y a}{2\hbar}\right)\cos\left(\frac{p_z a}{2\hbar}\right). \tag{1.31}$$

The coefficient $8[h(1)-h(0)\, I(1)]$ determines the width of the zone.

The physical significance of the results obtained consists in that the discrete levels of the isolated atoms expand into narrow zones, whose width depends on the degree of overlap or, more exactly, on the matrix element corresponding to the transfer of the electron to the neighboring atom. The formulas derived by this method are valid if the overlap of the atomic shells is small, i.e., for the inner shells. Therefore, some bands of the transition and rare-earth metals can be found using this procedure.

Another application of these formulas consists in that the corresponding functions $\varepsilon_l(\boldsymbol{p})$ have correct symmetry properties and may serve as the basis for empirical formulas representing experimental data on the electronic spectra even in a general case.

1.3. The model of weakly bound electrons

Let us now consider the opposite case, when the electrons in a metal are almost free and their interaction with the crystal is weak, so that perturbation theory may be applied. As before, let us first consider the one-dimensional model.

The normalized wave function of a free electron has the form

$$L^{-1/2}\,\mathrm{e}^{\mathrm{i}px/\hbar}, \tag{1.32}$$

where L is the length of the chain (it is convenient to take a finite chain and to set up a periodic boundary condition). The energy is equal to

$$\varepsilon^{(0)}(p)=\frac{p^2}{2m}. \tag{1.33}$$

The electron is acted on by a potential $U(x)$. In view of the fact that the potential is periodic, it can be expanded in a Fourier series:

$$U(x)=\sum_n U_n\,\mathrm{e}^{2\pi\mathrm{i}nx/a}. \tag{1.34}$$

Here $2\pi n/a$ are the periods of the reciprocal lattice for the one-dimensional case. The matrix elements of $U(x)$ with respect to the functions (1.32) are equal to

$$U(p,p')=L^{-1}\int U(x)\,\mathrm{e}^{-\mathrm{i}(p-p')x/\hbar}\,\mathrm{d}x.$$

They are evidently different from zero if

$$p-p'=2\pi\frac{n\hbar}{a}. \tag{1.35}$$

Here

$$U(p,p')=U_n. \tag{1.36}$$

The first-order correction to the energy $\varepsilon^{(0)}(p)$ is $\varepsilon^{(1)}(p)=U(p,p')=U_0$. This is a constant which changes only the origin of the energy scale. Therefore, we shall consider the second-order correction:

$$\varepsilon^{(2)}(p)=\sum_{n\neq 0}\frac{|U_n|^2}{\varepsilon^{(0)}(p)-\varepsilon^{(0)}(p-2\pi n\hbar/a)}. \tag{1.37}$$

The condition for the validity of perturbation theory is the smallness of this correction as compared to U_0. But obviously, this requirement is not fulfilled when the denominator becomes small. The latter can occur in reality. If p approaches $\pi n\hbar/a$ with any n, then in the corresponding term of the sum the value of $p' = p - 2\pi n\hbar/a$ approaches $-\pi n\hbar/a$ and, hence, $\varepsilon^{(0)}(p) \to \varepsilon^{(0)}(p')$. Thus, for such values of p ordinary perturbation theory is inapplicable. This is accounted for by the fact that the two states, p and p', have the same energy. Hence, this level is degenerate and we have to use perturbation theory for degenerate states.

Suppose that the sought-for wave function has the form

$$\psi = A_1\psi_1 + A_2\psi_2,$$

where ψ_1 corresponds to the first state and ψ_2 to the second. Substituting this expression into the Schrödinger equation $H\psi = \varepsilon\psi$, we find

$$A_1(\varepsilon_1 - \varepsilon)\psi_1 + U(A_1\psi_1 + A_2\psi_2) + A_2(\varepsilon_2 - \varepsilon)\psi_2 = 0,$$

where $\varepsilon_1 = p^2/2m$ and $\varepsilon_2 = p'^2/2m$. Multiplying by ψ_1^* and then by ψ_2^*, integrating and using the orthogonality of ψ_1 and ψ_2, we obtain

$$A_1(\varepsilon_1 - \varepsilon + U_0) + U_n A = 0, \qquad U_n^* A_1 + (\varepsilon_2 - \varepsilon + U_0)A_2 = 0.$$

The eigenvalues are found from the vanishing condition of the determinant of this homogenous linear system (the constant U_0 is incorporated into ε):

$$\varepsilon^2 - (\varepsilon_1 + \varepsilon_2)\varepsilon + \varepsilon_1\varepsilon_2 - |U_n|^2 = 0.$$

This equation has two solutions:

$$\varepsilon = (\tfrac{1}{2})(\varepsilon_1 + \varepsilon_2) \pm [(\tfrac{1}{4})(\varepsilon_1 - \varepsilon_2)^2 + |U_n|^2]^{1/2}. \tag{1.38}$$

The choice of the correct solution is dictated by the requirement that ε must be equal to $\varepsilon^{(0)}$ far from the "dangerous" momentum value. It is easy to see that near $p = \pi n\hbar/a$ we have to take the minus sign at $p < \pi n\hbar/a$ and the plus sign from the side $p > \pi n\hbar/a$. This means that at $p = \pi n\hbar/a$ the function $\varepsilon(p)$ undergoes a jump equal to $2|U_n|$ (fig. 4).

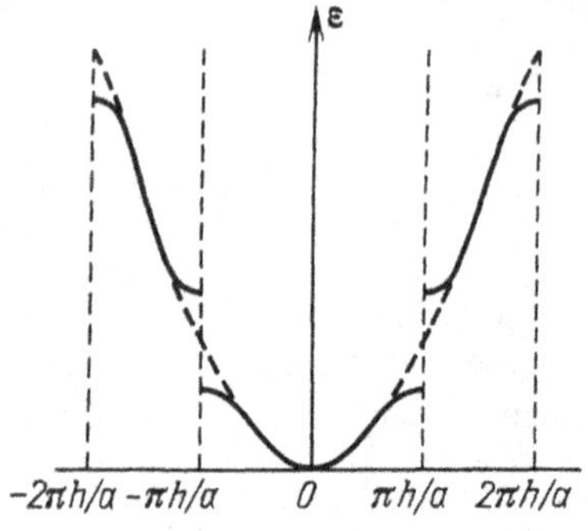

Fig. 4.

The quantity p which we have used so far is the momentum of a particle. But, as a matter of fact, an electron moving in the lattice is characterized by a quasimomentum. We can now pass to the quasimomentum by subtracting from p on each portion the corresponding period of the reciprocal lattice (multiplied by $\hbar$), so that the difference lies within the Brillouin zone. In such a manner we pass over from fig. 4 to fig. 3, which is a plot of the energy against the quasimomentum and not the momentum. Thus, there again arises the picture of energy bands separated by gaps.

In the three-dimensional case the periodicity of the potential $U(\boldsymbol{r})$ makes it possible to represent it in the form of a three-dimensional Fourier series:

$$U(\boldsymbol{r}) = \sum_{\boldsymbol{K}} U_{\boldsymbol{K}} \, e^{i\boldsymbol{K}\boldsymbol{r}}, \tag{1.39}$$

where $\boldsymbol{K}$ are the periods of the reciprocal lattice. Perturbation theory becomes invalid for those values of $\boldsymbol{p}$ for which

$$\varepsilon^{(0)}(\boldsymbol{p}) = \varepsilon^{(0)}(\boldsymbol{p} - \hbar\boldsymbol{K}). \tag{1.40}$$

Assuming that $\varepsilon^{(0)} = p^2/2m$, we obtain

$$\hbar \frac{\boldsymbol{K}\boldsymbol{p}}{m} = \frac{\hbar^2 K^2}{2m}$$

or

$$p \cos\theta = \tfrac{1}{2}\hbar K, \tag{1.41}$$

where θ is the angle between $\boldsymbol{p}$ and $\boldsymbol{K}$. But this is the equation of the plane in momentum space, which is perpendicular to the vector K and intersects it as a distance $\frac{1}{2}\hbar K$ from the coordinate origin. Just as in the one-dimensional case, the energy experiences a jump in this plane. If $\boldsymbol{K}$ is the smallest period in the corresponding direction, then this plane is simply the boundary of the Brillouin zone. The fact that the jump in the energy occurs here is one of the causes for the Brillouin zone being the most convenient region of determination of the quasimomentum.

If eq. (1.40) is satisfied only for one period $\boldsymbol{K}$ and not for two or three periods simultaneously, i.e., if we need not consider the intersection of two or three planes (1.41), then the energy is determined analogously to the one-dimensional case (1.38):

$$\begin{aligned}\varepsilon(\boldsymbol{p}) &= (\tfrac{1}{2})[\varepsilon^{(0)}(\boldsymbol{p}) + \varepsilon^{(0)}(\boldsymbol{p} - \hbar\boldsymbol{K})] \\ &\quad \pm [(\tfrac{1}{4})(\varepsilon^{(0)}(\boldsymbol{p}) - \varepsilon^{(0)}(\boldsymbol{p} - \hbar\boldsymbol{K}))^2 + |U_{\boldsymbol{K}}|^2]^{1/2}.\end{aligned} \tag{1.42}$$

Let us choose the axis p_z along $\hbar\boldsymbol{K}$ and introduce a new variable, $p_{z1} = p_z - \frac{1}{2}\hbar K$. Substituting $\varepsilon^{(0)} = p^2/2m$ into (1.42) and using the new variables, we find that

$$\varepsilon(\boldsymbol{p}) = \frac{[p_\perp^2 + p_{z1}^2 + (\frac{1}{2}\hbar \boldsymbol{K})^2]}{2m} \pm \left[\left(\frac{p_{z1}\hbar K}{2m}\right)^2 + |U_{\boldsymbol{K}}|^2\right]^{1/2}. \tag{1.43}$$

Let us consider the surfaces of constant energy in $\boldsymbol{p}$-space and find out how they intersect with the $p_{z1}=0$ plane. At point $p_{z1}=0$, $p_\perp=0$ the energy is given by

$$\varepsilon(0)=\frac{(\frac{1}{2}\hbar K)^2}{2m}\pm|U_{\boldsymbol{K}}|.$$

These are the energy values at which the surface $\varepsilon(p)=\text{const.}$ passes through the point $p_{z1}=p_\perp=0$. From this we may conclude that at

$$\varepsilon<\frac{(\frac{1}{2}\hbar K)^2}{2m}-|U_{\boldsymbol{K}}|$$

the surface $\varepsilon=\text{const.}$ does not intersect the boundary and looks like that in fig. 5a. If the energy lies in the interval

$$\frac{(\frac{1}{2}\hbar K)^2}{2m}-|U_{\boldsymbol{K}}|<\varepsilon<\frac{(\frac{1}{2}\hbar K)^2}{2m}+|U_{\boldsymbol{K}}|$$

the surface $\varepsilon=\text{const.}$ near the $p_{z1}=0$ plane looks like the one sketched in fig. 5b. And, finally, at

$$\varepsilon>\frac{(\frac{1}{2}\hbar K)^2}{2m}+|U_{\boldsymbol{K}}|$$

we have the situation depicted in fig. 5c.

The formation of zones in the weak coupling case has the following physical meaning. Let us substitute into eq. (1.41) $p=2\pi\hbar/\lambda$, where λ is the de Broglie wavelength, $K=nK_{\min}=2\pi n/d$, where $K_{\min}$ is the smallest period of the reciprocal lattice in the corresponding direction and $d=2\pi/K_{\min}$ is the distance between the successive crystal planes. Here we obtain

$$2d\cos\theta=n\lambda.$$

This formula, which is known as the Bragg formula, corresponds in this case to the coherent reflection of a plane wave, which describes the motion of an electron, from the crystal planes. As soon as this condition is satisfied the "free" electron moving in the lattice is strongly reflected from the crystal planes and its wave function undergoes a considerable change. The energy spectrum in this case corresponds to the condition that the component of the electron velocity $\boldsymbol{v}=\partial\varepsilon/\partial\boldsymbol{p}$ (section 3.1), which is perpendicular to the crystal plane, vanishes.

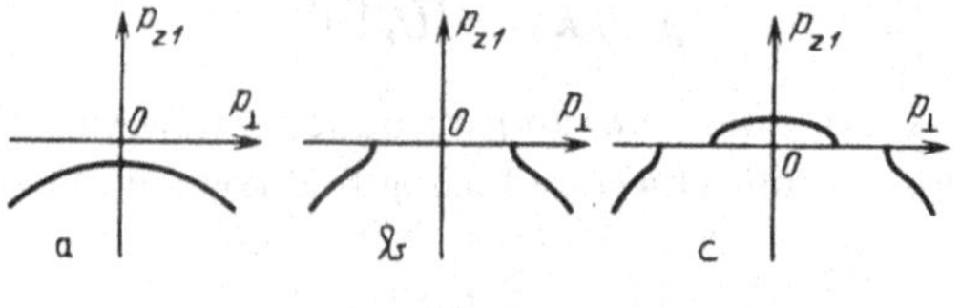

Fig. 5.

2 The electron Fermi liquid

2.1. The concept of quasiparticles

We have so far dealt with the behavior of one electron in the averaged field of the lattice and other electrons. Now we shall consider a real system of interacting electrons or an electron liquid. The behavior of such a system can be understood on the basis of the general concept proposed by Landau (1941) concerning the energy spectra of condensed quantum systems and the Landau theory of a Fermi liquid.

It is easier to illustrate the general Landau approach by considering as an example a vibrating crystal lattice. If the vibrations are small, the potential energy of the interaction of the lattice atoms may be expanded in powers of the displacements of atoms $\boldsymbol{u}$. The term of first order in the displacements is absent, since to the equilibrium position there corresponds the minimum of the potential energy. Thus, retaining only second-order terms, we have

$$U = U_0 + \tfrac{1}{2} \sum_{\substack{\boldsymbol{n},\boldsymbol{n}' \\ j,j' \\ \alpha,\alpha'}} A^{\alpha,\alpha'}_{\boldsymbol{n}j,\boldsymbol{n}'j'} u^{\alpha}_{\boldsymbol{n}j} u^{\alpha'}_{\boldsymbol{n}'j'}. \tag{2.1}$$

The lattice periods are $\boldsymbol{a}_{\boldsymbol{n}}$. The index j stands for the number of the atom in the unit cell $\boldsymbol{n}$. The index α corresponds to the projection of the displacement vector $\boldsymbol{u}$; $A^{\alpha,\alpha'}_{\boldsymbol{n}j,\boldsymbol{n}'j'}$ are the expansion coefficients.

Expression (2.1) is none other than the energy of a system of coupled oscillators. As is known, the quadratic form (2.1) can be diagonalized by means of a linear transformation of the oscillator coordinates, the vectors $\boldsymbol{u}_{\boldsymbol{n}j}$ in this particular case, following which we obtain a system of noninteracting linear oscillators. The energy will in this case be the sum of the energies of individual oscillators.

Since the study of lattice vibrations goes beyond the scope of our treatment*, we shall give only the results of such an approach. The solution of the equations of motion gives the following expression for displacements:

$$\boldsymbol{u}_{\boldsymbol{n}j} = \sum_{\boldsymbol{k},s} c(\boldsymbol{k}, s) \exp[\mathrm{i}\boldsymbol{k}\boldsymbol{a}_{\boldsymbol{n}} - \mathrm{i}\omega(\boldsymbol{k}, s)t]\boldsymbol{e}_j(\boldsymbol{k}, s). \tag{2.2}$$

Each set $\boldsymbol{k}$, s corresponds to one independent oscillator.

* More detailed information about lattice vibrations can be found in the book by Peierls (1955).

At the same time formula (2.2) is a superposition of plane waves propagating throughout the crystal. The wave vector $\boldsymbol{k}$ has the same properties as $\boldsymbol{p}/\hbar$ (where $\boldsymbol{p}$ is the quasimomentum). The index s denotes the type of wave, and the unit vector of polarization $\boldsymbol{e}_j$ defines how various atoms oscillate in a single unit cell. If the unit cell contains z atoms, then the index s takes on $3z$ various values. The vibration frequency ω depends on $\boldsymbol{k}$ and s.

Formula (2.2) is reminiscent of the wave function of free particles:

$$\exp\left(\mathrm{i}\frac{\boldsymbol{pr}}{h}-\mathrm{i}\frac{\varepsilon t}{\hbar}\right).$$

The role of the momentum $\boldsymbol{p}$ is played by $\hbar\boldsymbol{k}$ and that of the energy by $\hbar\omega$. Using this, we can introduce a new physical picture. Usually we deal with real particles, whose free motion is described by a plane wave. In this particular case we will treat expression (2.2) as the wave function of certain fictitious particles, which we call "quasiparticles". Since this notion is universal, the quasiparticles that correspond to lattice vibrations are specifically called "phonons". The origin of this term is associated with the fact that the quasiparticles in question have the same relation to elastic waves propagating in the lattice (i.e., to sound) as light quanta have to electromagnetic waves. Thus, it may be said briefly that usually in quantum mechanics waves describe the motion of particles and here particles are introduced for the description of waves.

The meaning of the description by means of quasiparticles becomes clearer if we consider the energy of a vibrating crystal. The energy levels are expressed by the formula for a system of noninteracting oscillators:

$$E-U_0=\sum_{\boldsymbol{k},s}\hbar\omega(\boldsymbol{k},s)\,(n(\boldsymbol{k},s)+\tfrac{1}{2}). \tag{2.3}$$

The numbers n are either equal to zero or positive integers. Let us write expression (2.3) as the sum of two terms:

$$E-U_0=\tfrac{1}{2}\sum_{\boldsymbol{k},s}\hbar\omega(\boldsymbol{k},s)+\sum_{\boldsymbol{k},s}\hbar\omega(\boldsymbol{k},s)\,n(\boldsymbol{k},s). \tag{2.4}$$

The first term corresponds to the lowest value of the energy and describes the ground state of the system. This is the energy of the so-called zero-point vibrations. The fact that the atoms of the crystal lattice must be vibrating even in the ground state is associated with the quantum uncertainty principle. According to this principle, a particle cannot be at rest in the equilibrium position, since in such a case it would have simultaneously a certain coordinate and a certain momentum.

In an excited state the numbers $n(\boldsymbol{k},s)$ are different from zero. Formula (2.4) corresponds in this case to a system of independent particles with energies $\hbar\omega(\boldsymbol{k},s)$. Since the numbers $n(\boldsymbol{k},s)$ can take on any positive integer values, it follows that any number of phonons may be in the same state. This means that they obey Bose statistics.

The concept of phonons is valid as long as the vibrational amplitude is small compared to the lattice period. Otherwise, one has to take into account the terms in the expansion of the potential energy U in higher powers of the displacement, and the total energy can no longer be expressed by formula (2.3). However, this occurs only near the melting point.

According to the idea put forward by Landau, any homogeneous system composed of a large number of particles has low-lying excited states of the same type as the vibrating lattice. Namely, the properties of any system may be described in terms of the quasiparticle model. Quasiparticles may have either an integer ($n\hbar$) or a half-integer ($(n+\frac{1}{2})\hbar$) spin, i.e., they may be either Bose particles or Fermi particles. The statistics of quasiparticles is not related uniquely to the statistics of the particles that make up the system. For example, as we have seen, phonons always obey Bose statistics, irrespective of the spin of the atoms that make up the lattice. The energy of quasiparticles is a function of their momentum. This dependence $\varepsilon(p)$ is the main characteristic of the low-lying excited states.

2.2. Quasiparticles in an isotropic Fermi liquid

Electrons have spin $\frac{1}{2}\hbar$. In view of this, the electron liquid is a so-called Fermi liquid. What are the properties of quasiparticles in such a liquid? According to the Landau hypothesis (1956), the energy spectrum of such a liquid is very similar to the spectrum of an ideal Fermi gas. The validity of this hypothesis was later rigorously proved. We do not give this proof here because it by far exceeds in complexity the level of this book*.

So, we begin with the ideal gas. The equilibrium distribution function is the well-known Fermi function**:

$$f=[e^{(\varepsilon-\mu)/T}+1]^{-1}$$

Here $\varepsilon=p^2/2m$ and μ is the chemical potential. At $T=0$ we have $f=1$ if $\varepsilon<\mu(0)$, and $f=0$ if $\varepsilon>\mu(0)$ (fig. 6, solid line). The quantity $\mu(0)$ is called the Fermi level. Introducing the Fermi momentum p_0 in accordance with the formula $\mu(0)=p_0^2/2m$, we find that at $T=0$ all the states contained in a sphere of radius $p=p_0$ (the Fermi sphere) in momentum space are occupied, all the external states being free. This is a consequence of the Pauli exclusion principle - only one particle may be in each of the states, and in this particular case at $T=0$ the lower states are occupied. The occupied volume in the phase space of momenta, coordinates and spins divided by $(2\pi\hbar)^3$ must be equal to the number of particles. The volume in momentum space

* The proof can be found in the book by Abrikosov et al. (1962).

** Here and henceforth the temperature is determined in energy units. To convert to degrees, it must be divided by the Boltzmann constant $k_B=1.38\times10^{-16}$ erg/K.

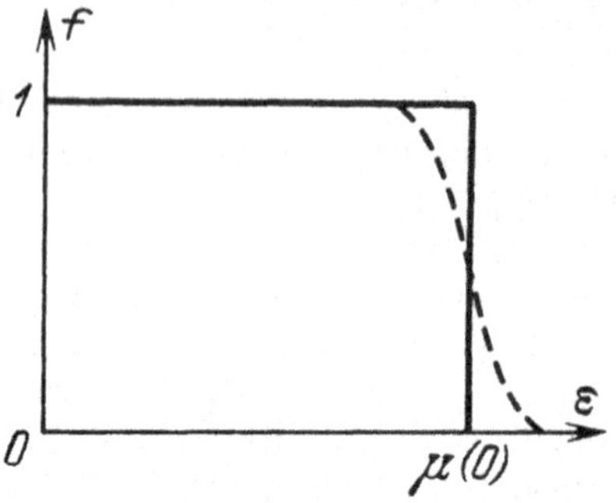

Fig. 6.

is the volume of the Fermi sphere. The possibility of the occurrence of two values for the spin projection is given by the factor 2. In view of this, we obtain:

$$N = 2(\tfrac{4}{3}\pi)p_0^3 V(2\pi\hbar)^{-3} \quad \text{or} \quad p_0 = \hbar(3\pi^2 N/V)^{1/3}. \tag{2.5}$$

We now turn to $T \neq 0$. The distribution function in this case is given by the dashed curve in fig. 6. The width of the "smeared-out" region is of the order of T. This is associated with the fact that some of the particles, after having received an extra energy of order T, escape from the Fermi sphere. The equilibrium state at $T \neq 0$ and, in general, any excited state can be generated from the state at $T = 0$ by way of successive displacements of particles from the interior of the Fermi sphere to the outside. Each such act results in a particle outside the Fermi sphere and a free site or an "antiparticle" inside it. These particles and antiparticles represent in this case the quasiparticles of the excited state*. Their energy must be counted from the Fermi level $\mu(0)$. Quasiparticles of particle-type have momenta larger than p_0 and their energy is given by

$$\xi_p(p) = \frac{p^2}{2m} - \frac{p_0^2}{2m}.$$

If $p - p_0 \ll p_0$, then

$$\xi_p \approx v(p - p_0),$$

where $v_0 = p_0/m$ is the velocity at the Fermi sphere. On the other hand, quasiparticles of antiparticle-type have momenta smaller than p_0 and their energy must be counted in the opposite direction:

$$\xi_a(p) = \frac{p_0^2}{2m} - \frac{p^2}{2m}$$

* The "antiparticles" constitute a full analogy to anti-particles in the theory of elementary particles (for example, the positron). The term "holes" frequently used for such quasiparticles is unjustified in our opinion because this term is used to denote another object: vacant sites in an unfilled band (Section 2.3).

or

$$\xi_a(p) \approx v(p_0 - p)$$

if $p_0 - p \ll p_0$. Such counting-off of the energy is due to the fact that the creation of antiparticles in the depth of the Fermi distribution requires a larger consumption of energy than at the Fermi surface.

According to the Landau hypothesis, the spectrum of quasiparticles in an isotropic Fermi liquid with a strong interaction between the particles is constructed in the same way as for the ideal gas. This means that there exists a certain value of p_0, which is connected, in accordance with the Landau theory, with the density of particles by the same relation as in the case of the ideal gas (formula 2.5). There are two types of quasiparticles: "particles" with $p > p_0$ and "antiparticles" with $p < p_0$. Their energies for the case $|p - p_0| \ll p_0$ are, respectively, equal to

$$\xi_p \approx v(p - p_0), \quad \xi_a \approx v(p_0 - p). \tag{2.6}$$

However, in this case v is simply a certain unknown coefficient, which has the dimension of velocity. Instead of v we may introduce another coefficient with the aid of the following relation:

$$v = p_0/m^*. \tag{2.7}$$

The constant m^* with the dimension of mass is called the effective mass*.

As has been pointed out above, these assumptions of the spectrum have been verified in a rigorous but rather complicated manner, but we can offer a simpler reasoning. If the state corresponding to the presence of a quasiparticle is not a true stationary state of the Fermi liquid, it must attenuate with time due to transitions to other states. The corresponding wave function will thus have the form

$$\exp\left[-i\frac{\xi(p)t}{\hbar} - \frac{\gamma(p)t}{\hbar}\right]. \tag{2.8}$$

We may meaningfully speak of quasiparticles only in those cases when $\gamma \ll |\xi|$. We have therefore to estimate γ. Evidently, it is proportional to the probability of the transition of the state under consideration to other states.

Let us first define this probability for a weakly interacting gas. If there is a particle 1 outside the Fermi distribution, then the process of first order in the interaction will be as follows (fig. 7). Particle 1 interacts with particle 2 inside the Fermi sphere, following which the two particles pass over to states 1′ and 2′ outside the Fermi sphere. Because or the Pauli principle this is the only possibility. The law of momentum conservation requires that

$$\boldsymbol{p}_1 + \boldsymbol{p}_2 = \boldsymbol{p}_1' + \boldsymbol{p}_2'$$

* In various books and articles the term "effective mass" is used for rather diverse quantities with the dimension of the mass. We shall use this term only for the isotropic spectrum.

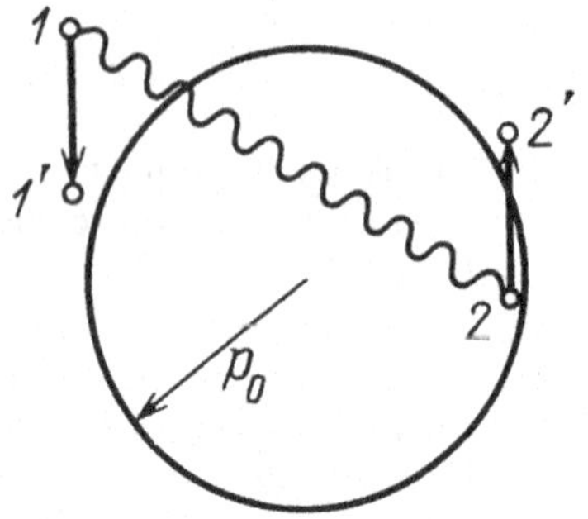

Fig. 7.

and, in accordance with what has been said above,

$$p_1 > p_0, \qquad p_2 < p_0, \qquad p_1' > p_0, \qquad p_2' > p_0.$$

The momentum conservation is shown graphically in fig. 8. The planes $(\boldsymbol{p}_1, \boldsymbol{p}_2)$ and $(\boldsymbol{p}_1', \boldsymbol{p}_2')$ do not coincide, generally speaking, and in fig. 8 they are simply superposed by rotation.

The scattering probability is given to within a constant by

$$W \propto \int \delta(\varepsilon_1 + \varepsilon_2 - \varepsilon_1' - \varepsilon_2')\, \mathrm{d}^3\boldsymbol{p}_2\, \mathrm{d}^3\boldsymbol{p}_1'.$$

The integration is carried out only over $\boldsymbol{p}_2$ and $\boldsymbol{p}_1'$, since $\boldsymbol{p}_2'$ is determined by the law of momentum conservation. The angle between the vectors $\boldsymbol{p}_1'$ and $\boldsymbol{p}_2'$ is actually specified by the law of energy conservation. The integration over this angle eliminates the δ-function. It now remains only to integrate over the absolute values of the vectors.

Suppose that p_1 is close to p_0. Then, all the remaining momenta will also be close to p_0 in absolute value, and, consequently, in fig. 8 they will make nearly equal angles with the horizontal line (with the sum $\boldsymbol{p}_1 + \boldsymbol{p}_2$). Hence, from the relationship between the projections on this axis we can write the relation between the absolute

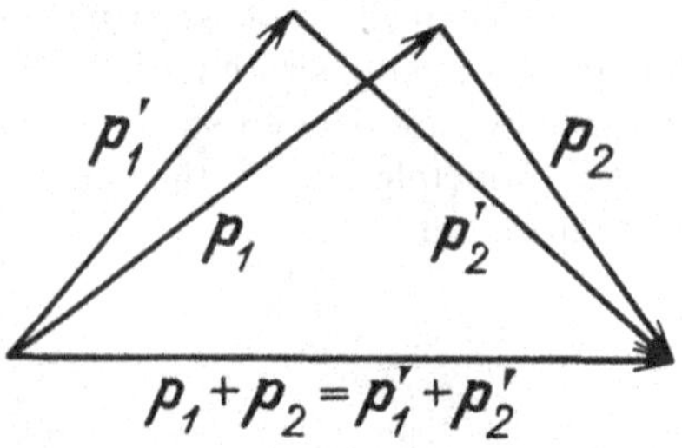

Fig. 8

values: $p_1' \approx p_1 + p_2 - p_2'$. Since $p_2' > p_0$, it follows that we have $p_1' < p_1 + p_2 - p_0$. But at the same time $p_1' > p_0$, from which it follows that $p_1 + p_2 - p_0 > p_0$ or $p_2 > 2p_0 - p_1$. The upper limit for p_2 is p_0. Thus, we have

$$0 > p_2 - p_0 > p_0 - p_1, \qquad 0 < p_1' - p_0 < (p_1 - p_0) + (p_2 - p_0).$$

On integration we obtain

$$\int \mathrm{d}p_2 \, \mathrm{d}p_1' = \tfrac{1}{2}(p_1 - p_0)^2.$$

Hence,

$$\gamma \propto W \propto (p_1 - p_0)^2. \tag{2.9}$$

The complete formula for γ can be obtained from dimensionality considerations. It must be proportional to the square of the interaction constant and, according to the above calculation, to the quantity $(p - p_0)^2$. Following this, we have to introduce a further factor made up of p_0, m and $\hbar$ in such a manner that the result has the dimension of energy.

Let us now turn to a strongly interacting system. As the interaction constant increases, other processes involving a larger number of particles may, in principle, become important, but it can be shown that the probabilities of such processes will contain higher powers of $p - p_0$. Hence, at $|p - p_0| \ll p_0$ the process in question will nevertheless predominate, i.e., we will again have $\gamma \propto (p - p_0)^2$. As for the other factors, we have to take into account the fact that in the liquid, whose volume is determined by the forces of particle interaction rather than by the walls of the container, the density is always such that the average kinetic energy of the particles and their potential energy of interaction will be approximately equal. It means that there is only one energy scale - the Fermi energy $\mu(0)$ or $p_0 v$. From this it follows that the quantity with the dimension of energy, which is proportional to $(p - p_0)^2$, must be equal to

$$\gamma = \alpha \frac{\xi^2}{\mu} \tag{2.10}$$

where $\alpha \sim 1$.

As has been said above, the quasiparticle concept is valid at $\gamma \ll |\xi|$. This really occurs near the Fermi level, i.e., at $|\xi| \ll \mu$. This justifies the above assumption of the spectrum in the vicinity of the Fermi level, i.e., for quasiparticles with small energies ξ.

If we deal with an equilibrium Fermi liquid at $T \neq 0$, the quasiparticles in it always have an energy $\xi \approx T$. The attenuation γ will be of order T^2/μ. It follows from this that the description of the liquid in terms of quasiparticles will be valid only as long as $T \ll \mu$.

The quantity μ can be estimated using the gas model. For electrons in a metal the value of $\hbar/p_0$ (the de Broglie wavelength) is of the order of interatomic spacings, i.e., 10^{-8} cm and, hence, $p_0 \sim 10^{-19}$ g · cm/s. It then follows that $p_0^2/2m \sim 1$–10 eV or, dividing by Boltzmann's constant, we find

$$T \ll T_0 \sim 10^4\text{–}10^5 \text{ K}. \tag{2.11}$$

This condition shows that the quasiparticle picture can really be applied to solid metals at all temperatures because T_0 is appreciably higher than the melting point in all cases.

Formulas (2.6) for the energy spectrum of quasiparticles may be written in a unified manner in the following form:

$$\varepsilon_{\text{qp}} = |\xi|, \qquad \xi = \varepsilon(\boldsymbol{p}) - \mu. \tag{2.6$'$}$$

Here it must be kept in mind that the "antiparticles" have a charge which is opposite to the charge of the "particles". However, we may introduce another, more familiar object. Let us visualize an ideal Fermi gas with a density N/V, which is composed of particles with a mass m^*. The spectrum of quasi-particles of such a gas is the same as in the case of the Fermi liquid. Therefore, such an ideal gas may describe the properties of a real interacting system. However, one has to bear in mind that the properties of the gas model that depend on the particles located far from the Fermi level do not correspond to a real Fermi liquid. In what follows, depending on the convenience, we will make use of both pictures: the gas model or quasiparticles with the spectrum (2.6′).

2.3. The anisotropic Fermi liquid

All the results described above refer to an isotropic Fermi liquid. In order to ascertain the meaning of the electronic spectra of metals, we first "turn off" the interaction of electrons or, more precisely, we will consider a gas composed of noninteracting electrons placed in an averaged periodic field. The states of one particle in such a field have been considered in the preceding chapter. It has been shown that the energy levels form bands separated by forbidden portions (energy gaps). Each band has $2N$ states, where N is the number of unit cells in the specimen.

If there are many noninteracting particles, they are distributed in some way between these states. At $T = 0$ (and in metals practically at all temperatures below the melting point) all the lower states will be occupied up to a certain maximum level (the Fermi energy) and all the higher states will be empty. There are two possibilities here.

(1) The Fermi level coincides with the upper edge of one of the bands, so that some of the bands are completely filled and the others are quite empty. In this case, not too strong an electric field cannot produce electric current. This follows from the fact that in the equilibrium state to each electron with momentum $\boldsymbol{p}$ there corresponds another one with momentum $-\boldsymbol{p}$ since $\varepsilon(\boldsymbol{p})$ is an even function of the momentum. Therefore, there is no current in the equilibrium state. In order to produce current, it is necessary to redistribute the electrons between these states.

But, because of the presence of an energy gap the redistribution of electrons requires a finite change in the energy. A low electric field cannot bring about such a change in energy, which is why the substance in question will be an insulator and not a metal.

(2) The Fermi level is in the middle of one of the bands. Such a band is called the conduction band (there may also be several such bands). It is only partially occupied. Of course, without electric field there is no current. However, the redistribution of electrons between states which is required for producing a current, can be realized at the expense of an infinitely small change in the energy; an arbitrarily weak electric field can produce a current. This situation corresponds to a metal. Semiconductors belong to the first case, but there the energy gap between the occupied and unoccupied states is small (this may be associated with the appearance or extra electronic levels upon incorporation of impurity atoms). Therefore, at not too low temperatures their properties are reminiscent of those of metals. We shall limit ourselves to true metals.

It is easy to see that when the number of electrons per unit cell is odd, then at least one of the bands must be partially occupied (recall that each band contains $2N$ states). However, even in the case where the number of electrons per unit cell is even the substance may be a metal since in a real three-dimensional case the bands may overlap. There will be several partially occupied bands in this case.

The Fermi level for the "gas in the lattice" is specified by the condition $\varepsilon(\boldsymbol{p}) = \mu$. In momentum space this equation describes a surface known as the Fermi surface. The symmetry of this surface is determined by the symmetry of the crystal. Here we can also define quasiparticles of "particle"-type with a momentum outside the Fermi surface and quasiparticles of "antiparticle"-type with momenta inside the Fermi surface.

The Fermi surface may have a rather complicated form in the general case. Two examples are given in figs. 9 and 10.

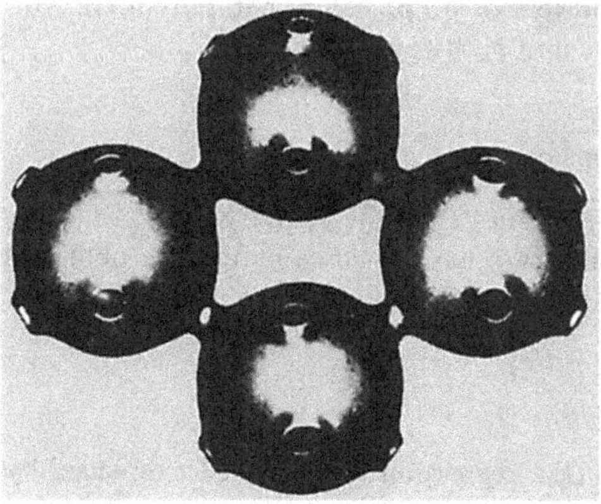

Fig. 9.

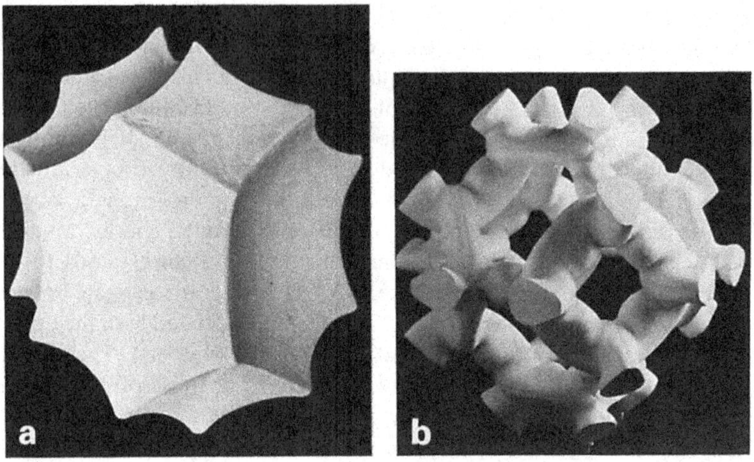

Fig. 10.

Figure 9 shows the Fermi surface of gold, where there is only one conduction band (one "sphere" corresponds to the Brillouin zone). Figure 10 presents the Fermi surface of lead, in which there are two conduction bands. The surface in fig. 10a encloses the energy region $\varepsilon < \mu$, and the interior of the tubes in fig. 10b corresponds to energies $\varepsilon > \mu$. The surfaces sketched in figs. 9 and 10 have been obtained experimentally.

But there are two cases where the Fermi surface turns out to be very simple. One case is an almost empty band. The number of electrons is very small and at $T = 0$ they occupy the lowest states. This means that they will be located near the minimum of the function $\varepsilon(\boldsymbol{p})$. If this minimum corresponds to the point $\boldsymbol{p}_{\mathrm{m}}$, the energy may be expanded in powers of $\boldsymbol{p} - \boldsymbol{p}_{\mathrm{m}}$. Here naturally there will be no linear terms. If the crystal is cubic and $\boldsymbol{p}_{\mathrm{m}} = 0$, we have

$$\varepsilon(\boldsymbol{p}) = \varepsilon_0 + \frac{p^2}{2m^*}, \tag{2.12}$$

where m^* is a coefficient called the effective mass. In a more general case, where the symmetry is arbitrary but $\boldsymbol{p}_{\mathrm{m}} = 0$ as before, we obtain instead of p^2 a positive quadratic form. After transformation to the principal axes it looks like

$$\varepsilon(\boldsymbol{p}) = \varepsilon_0 + \tfrac{1}{2}\left(\frac{p_x^2}{m_1} + \frac{p_y^2}{m_2} + \frac{p_z^2}{m_3}\right). \tag{2.13}$$

If $\boldsymbol{p}_{\mathrm{m}} \neq 0$, then in this expression all p_i must be replaced by $p_i - p_{\mathrm{m}i}$. However, in this case there are several equivalent energy minima, and the corresponding vectors $\boldsymbol{p}_{\mathrm{m}}$ form a "star" which has the symmetry of the crystal.

A very similar picture is observed in the case of an almost occupied band. Since the occupied band does not participate in the electric current, it may be assumed that the current is produced by "holes", i.e., empty sites left in the band. Evidently, they will be concentrated near the energy maxima.

If the metal has cubic symmetry and the energy maximum corresponds to $p_{\mathrm{m}} = 0$, then the electron energy near the maximum has the form

$$\varepsilon(\boldsymbol{p}) = \varepsilon_0 - \frac{p^2}{2m^*}. \tag{2.14}$$

It is inconvenient to use this formula because the effective mass of electrons proves negative. Passing over to holes the signs of the energy and charge change. Hence, the holes have an energy $-\varepsilon(\boldsymbol{p})$ and a charge $-e$. Their mass is positive. The generalization to the noncubic case or $\boldsymbol{p}_{\mathrm{m}} \neq 0$ is trivial. The Fermi surface in all the cases considered is an ellipsoid or consists of several ellipsoids (if $\boldsymbol{p}_{\mathrm{m}} \neq 0$).

Note that metals with a small number of electrons or holes in the conduction band are in fact seldom encountered. As has been pointed out above, if the number of electrons per unit cell is odd, then some of the bands may be occupied, but there will be, at least, one partially occupied band. If there is really one partially occupied band, it must contain N electrons. Since the entire Brillouin zone corresponds to $2N$ states, it follows that the volume inside the Fermi surface* must be equal to half the volume of the Brillouin zone. In order to produce a small number of electrons and holes it is required that the number of electrons per unit cell be even and the two upper bands intersect slightly. In this case, part of the electrons will go from the upper occupied band (the so-called valence band) to the lower unoccupied band (the conduction band) and there will be a small number of electrons in one band and the same number of holes in the other. This situation occurs in "semimetals": Bi, Sb, As, graphite.

Moreover, it may happen that the metal contains several partially occupied bands and in one of them the number of electrons is small or, conversely, the number of holes is small.

We have so far been concerned with the gas model consisting of noninteracting particles placed in a periodic field. This model in fact describes the properties of quasiparticles in a real metal in the same manner as the model of the isotropic ideal gas describes the properties of quasiparticles in an isotropic Fermi liquid. However, it is necessary to keep in mind that only those properties of the gas model correspond to reality which depend only on the particles near the Fermi surface.

In this connection, there may arise doubts as to whether our reasoning concerning the occupation of the bands has relation to reality (they are, in fact, also associated with "deep" particles). As a matter of fact, there exists a theorem due to Luttinger (1960a, b), which is an extension of the Landau formula (2.5) for the boundary

* If the Fermi surface approaches the boundaries of the Brillouin zone, the occupied volume is bounded by the Fermi surface and by the boundaries of the Brillouin zone.

Fermi momentum p_0. According to this Luttinger theorem, the density of electrons is given by

$$\frac{N_e}{V}=\frac{2qN}{V}+\frac{2V_F}{(2\pi\hbar)^3}, \tag{2.15}$$

where q is an integer and V is the volume inside the Fermi surface. This formula is in full accord with our reasoning about the occupation of the bands. The first term corresponds to occupied bands and the second to the conduction band.

2.4. Electronic heat capacity

We will now derive the expression for the electronic heat capacity of metals. The electron liquid is described by the model of a gas of particles exhibiting the properties of individual electrons in a periodic field. For the sake of simplicity, we will call these particles "electrons", but we should, of course, remember the difference between these "electrons" and the true electrons that make up the Fermi liquid. The energy of such a Fermi gas is given by the formula

$$E=2\int \varepsilon(\boldsymbol{p}) f d^3\boldsymbol{p}\frac{V}{(2\pi\hbar)^3}, \tag{2.16}$$

where $2V/(2\pi\hbar)^3$ is the density of states in quasimomentum space (the number 2 arises from the two directions of the spin) and f is the Fermi distribution function defined by

$$f=[\mathrm{e}^{(\varepsilon-\mu)/T}+1]^{-1}. \tag{2.17}$$

The integration over the quasimomentum is bounded by the Brillouin zone. If there are several partially occupied bands, then the summation must be carried out over all of them.

Differentiating (2.16) with respect to T, we obtain the heat capacity of a unit volume:

$$C=V^{-1}\left(\frac{\partial E}{\partial T}\right)_V=2\int \varepsilon(\boldsymbol{p})\frac{\partial f}{\partial T}\frac{\mathrm{d}^3\boldsymbol{p}}{(2\pi\hbar)^3}. \tag{2.18}$$

The derivative $\partial f/\partial T$ is given by

$$\frac{\partial f}{\partial T}=T^{-1}\frac{\mathrm{e}^{(\varepsilon-\mu)/T}}{(\mathrm{e}^{(\varepsilon-\mu)/T}+1)^2}\left(\frac{\varepsilon-\mu}{T}+\frac{\mathrm{d}\mu}{\mathrm{d}T}\right).$$

The derivative $\partial f/\partial \varepsilon$ has the form

$$\frac{\partial f}{\partial \varepsilon}=-T^{-1}\frac{\mathrm{e}^{(\varepsilon-\mu)/T}}{(\mathrm{e}^{(\varepsilon-\mu)/T}+1)^2}. \tag{2.19}$$

Hence, we may write

$$\frac{\partial f}{\partial T} = -\frac{\partial f}{\partial \varepsilon}\left(\frac{\varepsilon - \mu}{T} + \frac{d\mu}{dT}\right). \tag{2.20}$$

The chemical potential μ is found from the condition of conservation of the number of particles (recall that the number of particles of the gas model is equal to the number of particles of the liquid). So, we have

$$\frac{\partial}{\partial T}\left(\frac{N}{V}\right) = 2\int \frac{\partial f}{\partial T} d^3p \frac{1}{(2\pi\hbar)^3} = 0.$$

Substituting eq. (2.20) into eq. (2.18) and into the last condition, we obtain

$$C = -2\int \varepsilon \frac{\partial f}{\partial \varepsilon}\left(\frac{\varepsilon - \mu}{T} + \frac{d\mu}{dT}\right)\frac{d^3p}{(2\pi\hbar)^3}, \tag{2.21}$$

$$\frac{\partial}{\partial T}\frac{N}{V} = -2\int \frac{\partial f}{\partial \varepsilon}\left(\frac{\varepsilon - \mu}{T} + \frac{d\mu}{dT}\right)\frac{d^3p}{(2\pi\hbar)^3} = 0. \tag{2.22}$$

These two formulas contain $\partial f/\partial\varepsilon$. But while $T \ll \mu(0)$, which, as has been said above, is always valid for metals, $\partial f/\partial\varepsilon$ is nonzero only in a small region of energies of order T near the Fermi level $\mu(0)$ (see fig. 6). This proves that the description of electrons in terms of the gas model is really valid for calculation of the heat capacity.

The integrals in momentum space can be transformed as follows. Let us consider a constant-energy surface in momentum space, $\varepsilon(\boldsymbol{p}) = \text{const}$. Then, the integration over $d^3\boldsymbol{p}$ can be broken into an integration over this surface and an integration over $d\varepsilon$. If dS is an element of the constant-energy surface, then $d^3\boldsymbol{p} = dS\, dp_n$, where dp_n implies the integration over the normal to the element dS. But $dp_n = d\varepsilon/|\partial\varepsilon/\partial\boldsymbol{p}|$, where $\partial\varepsilon/\partial\boldsymbol{p}$ is the gradient of $\varepsilon(\boldsymbol{p})$ in momentum space. Applying the notation $v = |\partial\varepsilon/\partial\boldsymbol{p}|$, we obtain

$$\int \cdots d^3\boldsymbol{p} = \int \cdots d\varepsilon \int \frac{dS}{v}. \tag{2.23}$$

Since the distribution function depends only on energy, the integration over dS can be made independently. We use the notation

$$2(2\pi\hbar)^{-3}\int \frac{dS}{v} = \nu(\varepsilon). \tag{2.24}$$

This quantity is called the density of states.

To calculate (2.21) and (2.22), we employ the following method of integration. Since $\partial f/\partial\varepsilon$ is nonzero only near $\varepsilon = \mu$, we can expand all the other quantities in the integrand in powers of $\varepsilon - \mu$. This means that in an integral of the type $\int F(\varepsilon)\,(\partial f/\partial\varepsilon)\, d\varepsilon$ the function $F(\varepsilon)$ may be represented in the form

$$F(\varepsilon) = F(\mu) + (\varepsilon - \mu)F'(\mu) + \tfrac{1}{2}(\varepsilon - \mu)^2 F''(\mu) + \cdots.$$

Of course, this can be done only if the function F varies not too rapidly in the neighborhood of $\varepsilon=\mu$.

Formula (2.19) may be put in the following form:

$$\frac{\partial f}{\partial \varepsilon}=-(4T)^{-1}\cosh^{-2}\left(\frac{\varepsilon-\mu}{2T}\right)$$

and, because of the rapid decrease of this function with increasing distance from $\varepsilon=\mu$, the limits of integration over $z=\varepsilon-\mu$ may be taken $-\infty, \infty$. Since this function is even, the odd terms of the expansion of $F(\varepsilon)$ will give zero upon integration. The remaining integrals are

$$\int \frac{\partial f}{\partial \varepsilon}\,\mathrm{d}\varepsilon=-1,$$

$$\int (\varepsilon-\mu)^2\frac{\partial f}{\partial \varepsilon}\,\mathrm{d}\varepsilon=-(4T)^{-1}\int_{-\infty}^{\infty}\frac{z^2\,\mathrm{d}z}{\cosh^2(z/2T)}=-\frac{\pi^2T^2}{3}.$$

Thus, we have

$$\int F(\varepsilon)\frac{\partial f}{\partial \varepsilon}\,\mathrm{d}\varepsilon\approx -F(\mu)-\tfrac{1}{6}\pi^2T^2F''(\mu). \tag{2.25}$$

Applying this rule to relations (2.21) and (2.22), we obtain

$$C=\mu\nu(\mu)\frac{\mathrm{d}\mu}{\mathrm{d}T}+\tfrac{1}{3}\pi^2T\frac{\mathrm{d}}{\mathrm{d}\mu}(\mu\nu(\mu)),$$

$$\frac{\mathrm{d}}{\mathrm{d}T}\left(\frac{N}{V}\right)=\nu(\mu)\frac{\mathrm{d}\mu}{\mathrm{d}T}+\tfrac{1}{3}\pi^2T\frac{\mathrm{d}\nu(\mu)}{\mathrm{d}\mu}=0.$$

From the last condition we find

$$\frac{\mathrm{d}\mu}{\mathrm{d}T}=-\tfrac{1}{3}\pi^2T\frac{\nu'(\mu)}{\nu(\mu)}, \tag{2.26}$$

from which it follows that

$$\mu(T)=\mu(0)-\tfrac{1}{6}\pi^2T^2\frac{\nu'(\mu)}{\nu(\mu)}, \tag{2.27}$$

that is, the temperature correction to $\mu(0)$ is of relative order $(T/\mu)^2$.

Substitution of (2.26) into the expression for the heat capacity yields

$$C=\tfrac{1}{3}\pi^2T\nu(\mu). \tag{2.28}$$

The physical meaning of this formula is simple. Only electrons near the Fermi level participate in the thermal excitation of the system. The number of such electrons is of the order of the product T times the density of states $\nu(\mu)$. Each electron makes a contribution of the order of unity (or k_B, Boltzmann's constant in conventional units) to the heat capacity. This gives an expression which coincides in order of magnitude with (2.28). In the derivation of relation (2.28) we did not need concrete information about the Fermi surface. Hence, in all metals the heat capacity is proportional to the absolute temperature.

A quantitative estimation of C can be made for the isotropic model of a metal. For this case

$$\nu(\mu)=\frac{2}{(2\pi\hbar)^3}\int\frac{\mathrm{d}S}{v}=\frac{2}{(2\pi\hbar)^3}\frac{4\pi p_0^2}{p_0/m^*}=\frac{p_0 m^*}{\pi^2\hbar^3}, \tag{2.29}$$

where m^* is the effective mass. Hence,

$$C=\frac{p_0 m^*}{3\hbar^3}T. \tag{2.30}$$

The effective mass m^* can vary strongly from metal to metal but p_0 is of the same order of magnitude for all metals (with the exception of semimetals), namely $\hbar\pi/a$ (where a is the lattice period, which is known to differ little for different substances). For ordinary metals m^* is of the order of the mass of a free electron, but for transition metals, where the inner shells are not filled, the mass m^* is different. Its value can be determined from the relation $p_0/m^*=\partial\varepsilon/\partial p\approx\Delta\varepsilon/\Delta p$, where $\Delta\varepsilon$ is the band width and Δp is of the order of the period of the reciprocal lattice $\hbar\pi/a$. The inner shells overlap little and, hence, $\Delta\varepsilon$ (see eq. 1.29) is small and m^* is large. This fact is responsible for the heat capacity of transition metals being much higher (by 20 times in some cases) than that of the nontransition metals.

In experiments, of course, the total heat capacity is measured rather than the electronic heat capacity. However, as is known, at low temperatures the part of the heat capacity associated with lattice vibrations depends on temperature by a cubic law (see, for example, Peierls 1955). Therefore the total heat capacity may be represented in the following form:

$$C_{\mathrm{tot}}=AT+BT^3.$$

Having constructed a plot of C_{tot}/T versus T^2 (fig. 11), we obtain at low temperatures a straight line, from which we can determine the two coefficients A and B.

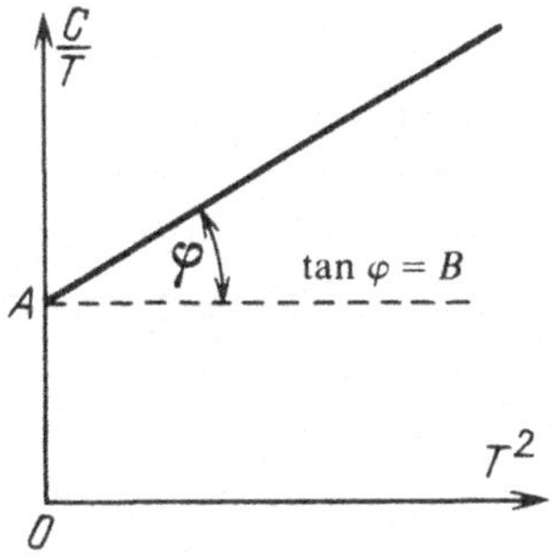

Fig. 11.

3

Electrical and thermal conductivity

3.1. The electron as a wave packet

So far we have described electrons* by means of the Bloch solutions of the Schrödinger equation $\psi = e^{ipr/\hbar}u(r)$.

However, these functions are useless for the study of transport phenomena, since they correspond to a certain definite value of the quasimomentum $\boldsymbol{p}$, whereas the coordinate remains completely undefined. This shortcoming can be removed by generating a wave packet of the Bloch states. It is still desirable here to define most precisely the quasimomentum $\boldsymbol{p}$. To do this, let us take the sum of Bloch waves over a small interval $\Delta \boldsymbol{p}$ (the contraction of this interval is limited by the accuracy of the determination of the coordinates). From the uncertainty principle we have

$$\Delta p\,\Delta r \sim \hbar.$$

Since the electrons are scattered and, hence, have a finite mean free path l, the uncertainty of the coordinates must be less than l. The upper allowable limit of Δp is obviously the particle's momentum itself. Thus, the following condition must be satisfied:

$$p \gg \Delta p \sim \frac{\hbar}{\Delta r} \gg \frac{\hbar}{l} \tag{3.1}$$

or

$$\lambda = \frac{2\pi\hbar}{p} \ll l.$$

It is only on this condition that the description in terms of the wave packet makes sense. In our case, $p \approx p_0$, and p_0 is of order $\hbar/a$, i.e., $\lambda \approx a$, where a is the interatomic distance. Hence, the condition (3.1) is satisfied if $l \gg a$ and difficulties arise only when l is of order a (section 4.8).

* Here and henceforth, "electrons" is taken to mean the particles of the gas model of noninteracting particles in a periodic field. In some cases, it will be more convenient to pass directly to quasiparticles (which are the same for the gas model and the true electron liquid) with the energy spectrum $\varepsilon_{\mathrm{qp}} = |\xi|$. This will be specified.

Another condition, which has already been mentioned, is associated with the attenuation of the wave function of quasiparticles. It had the form

$$|\xi| \gg \gamma,$$

where γ is the damping coefficient. It is connected with the average lifetime of the quasiparticles by the relation $\gamma \sim \hbar/\tau$. In section 2.2 we have found γ for the interaction of the quasiparticles themselves. As a matter of fact, there are other processes that limit the lifetime of quasiparticles (Ch. 4). In this case, the processes that lead to a change in energy by an amount of order ξ are of importance. They include the interaction of quasiparticles with one another, but the scattering from impurities, for example, does not belong in this category because it occurs elastically, without a change in energy. Therefore, the "energy relaxation" time τ_ξ may be different from the time $\tau = l/v$ for the relaxation of the momentum. Hence, the second condition for the validity of the quasiparticle model has the form

$$|\xi| \gg \frac{\hbar}{\tau_\xi}.$$

Usually, ξ is always of the order of the temperature T. Thus, we must have

$$T \gg \frac{\hbar}{\tau_\xi}. \tag{3.2}$$

In order to find the velocity of the electron described by the wave packet, let us consider the following integral:

$$I = \int_{\boldsymbol{p}^{(0)}+\Delta \boldsymbol{p}} \exp\left(\mathrm{i}\frac{\boldsymbol{pr}}{\hbar} - \mathrm{i}\frac{\varepsilon(\boldsymbol{p})\, t}{\hbar} \right) u_p(\boldsymbol{r})\, \mathrm{d}^3\boldsymbol{p}$$

taken over the vicinity of the radius $\Delta \boldsymbol{p}$ of a certain quasimomentum $\boldsymbol{p}^{(0)}$. At $\Delta p \ll p^{(0)}$ this integral is approximately equal to

$$I = u_{\boldsymbol{p}^{(0)}}(\mathbf{r}) \exp\left(\mathrm{i}\frac{\boldsymbol{p}^{(0)}\boldsymbol{r}}{\hbar} - \mathrm{i}\frac{\varepsilon(\boldsymbol{p}^{(0)})\, t}{\hbar} \right) \int_{\Delta p} \exp\left(\mathrm{i}\frac{\delta\boldsymbol{p}\left(\boldsymbol{r} - \frac{\partial \varepsilon}{\partial \boldsymbol{p}} \right) t}{\hbar} \right) \mathrm{d}^3\delta\boldsymbol{p}.$$

In the last expression the integral is of order

$$\Delta p^3 \qquad \text{if} \quad \left| \boldsymbol{r} - \frac{\partial \varepsilon}{\partial \boldsymbol{p}} t \right| \ll \frac{\hbar}{\Delta p}$$

$$\frac{\Delta p \hbar^2}{\left| \boldsymbol{r} - \frac{\partial \varepsilon}{\partial \boldsymbol{p}} t \right|^2} \qquad \text{if} \quad \left| \boldsymbol{r} - \frac{\partial \varepsilon}{\partial \boldsymbol{p}} t \right| \gg \frac{\hbar}{\Delta p}.$$

This means that

$$\boldsymbol{r} \approx \frac{\partial \varepsilon}{\partial \boldsymbol{p}} t,$$

that is, the wave packet moves with a velocity*

$$\boldsymbol{v} = \frac{\partial \varepsilon}{\partial \boldsymbol{p}}. \tag{3.3}$$

This is the so-called group velocity of the wave packet.

Let the electron move under the influence of an external electric field $\boldsymbol{E}$. The work done by the external field per unit time is equal to $\boldsymbol{v}e\boldsymbol{E}$. But it must be equal to the change in the electron energy

$$\frac{\partial \varepsilon}{\partial t} = \left(\frac{\partial \varepsilon}{\partial \boldsymbol{p}}\right) \frac{\mathrm{d}\boldsymbol{p}}{\mathrm{d}t} = \boldsymbol{v} \frac{\mathrm{d}\boldsymbol{p}}{\mathrm{d}t}.$$

From this we find

$$\frac{\mathrm{d}\boldsymbol{p}}{\mathrm{d}t} = e\boldsymbol{E}. \tag{3.4}$$

At first sight this is the Newton equation of motion of a free particle. However, $\boldsymbol{p}$ is not the momentum but the quasimomentum. An increase in $\boldsymbol{p}$ does not imply that the electron is necessarily accelerated. From fig. 12 it is seen that the velocity $\partial \varepsilon / \partial \boldsymbol{p}$ increases only in region A since it is only in this region that the electron is accelerated, while in region B the electron is slowed down and near the boundary of the Brillouin zone $\hbar \boldsymbol{K}/2$ its velocity vanishes. This corresponds to the reflection that occurs when the electron attains the maximum permissible energy.

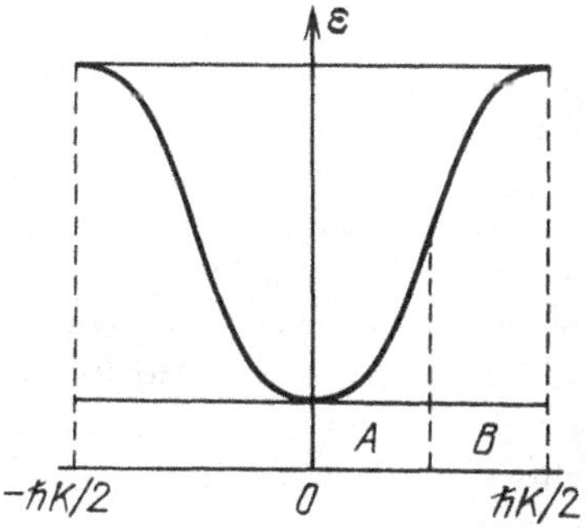

Fig. 12.

* Note that, although the wave packet is made up of the particle's wave functions in a periodic field, it moves with constant velocity as though "unaware" of the presence of the periodic field. This circumstance is a consequence of the quantum nature of electrons.

However, in practice such a situation never takes place. Because of scattering the momentum of the electron cannot be changed by a large amount under the action of the field. The acceleration of electrons occurs only on the mean free path, and electric fields that can be produced in metals are never strong enough to alter the quasimomentum of the electron significantly during the mean free time. In view of this, the electrons actually always remain in the vicinity of the Fermi surface.

3.2. *The kinetic equation*

The description in terms of a wave packet with restrictions (3.1) and (3.2) allows one to treat electrons as classical particles which have simultaneously a coordinate and a quasimomentum. Let us introduce a nonequilibrium distribution function f in the phase space $\boldsymbol{p}$, $\boldsymbol{r}$. The total change of f with time is expressed by a derivative $\mathrm{d}f/\mathrm{d}t$. This change occurs as a result of the collisions of electrons with other electrons and phonons and also as a result of scattering by crystal imperfections (impurity atoms, structural defects). Therefore, we can write

$$\frac{\mathrm{d}f}{\mathrm{d}t} = I(f), \tag{3.5}$$

where $I(f)$ is the so-called collision integral.

Relation (3.5) is known as the kinetic equation. It will be used below for the study of transport phenomena in metals. The left-hand side of eq. (3.5) may be written in the form

$$\frac{\mathrm{d}f}{\mathrm{d}t} = \frac{\partial f}{\partial t} + \frac{\partial f}{\partial \boldsymbol{r}}\frac{\mathrm{d}\boldsymbol{r}}{\mathrm{d}t} + \frac{\partial f}{\partial \boldsymbol{p}}\frac{\mathrm{d}\boldsymbol{p}}{\mathrm{d}t}.$$

Replacing $\mathrm{d}\boldsymbol{r}/\mathrm{d}t$ by $\boldsymbol{v}$ and $\mathrm{d}\boldsymbol{p}/\mathrm{d}t$ in a accordance with (3.4), we obtain

$$\frac{\mathrm{d}f}{\mathrm{d}t} = \frac{\partial f}{\partial t} + \boldsymbol{v}\frac{\partial f}{\partial \boldsymbol{r}} + e\boldsymbol{E}\frac{\partial f}{\partial \boldsymbol{p}}. \tag{3.6}$$

The collision integral depends on particular scattering processes. We start with the scattering by impurities (or other defects). We shall employ the Born approximation, assuming that the interaction with an impurity is weak. Although this is not so in actual fact, it can be shown, however, that in the case of a strong interaction the Born scattering amplitude must be replaced by an amplitude which is also constant. This is the only difference*.

The scattering probability of an electron in the field of impurity centers in the Born approximation is given by

$$\frac{2\pi}{\hbar}\int |V_{p'p}|^2 \delta(\varepsilon(\boldsymbol{p}) - \varepsilon(\boldsymbol{p}'))\frac{\mathrm{d}^3 p'}{(2\pi\hbar)^3}\, V, \tag{3.7}$$

* This is not exactly so if the impurity is a transition metal and impurity atoms have spin (section 4.6).

where $V_{p'p}$ is the matrix element of the energy of interaction of an electron with all the impurity atoms that occupy the positions $\boldsymbol{R}_1, \boldsymbol{R}_2, \dots$; $V(\boldsymbol{r}) = \sum_i v(\boldsymbol{r} - \boldsymbol{R}_i)$.

The matrix element in the Bloch functions

$$V^{-1/2}\, \mathrm{e}^{\mathrm{i}\boldsymbol{pr}/\hbar} u_{\boldsymbol{p}}(\boldsymbol{r})$$

(the factor $V^{-1/2}$ is introduced for normalization; the functions $u_{\boldsymbol{p}}(\boldsymbol{r})$ are normalized to unity) has the form

$$\begin{aligned} V_{p'p} &= V^{-1} \sum_i v(\boldsymbol{r} - \boldsymbol{R}_i)\, \mathrm{e}^{-\mathrm{i}\boldsymbol{p}'\boldsymbol{r}/\hbar}\, \mathrm{e}^{\mathrm{i}\boldsymbol{pr}/\hbar} u^*_{\boldsymbol{p}'}(\boldsymbol{r})\, u_{\boldsymbol{p}}(\boldsymbol{r})\, dV \\ &= V^{-1} \sum_i \mathrm{e}^{-\mathrm{i}(\boldsymbol{p}'-\boldsymbol{p})\boldsymbol{R}_i/\hbar} \int v(\boldsymbol{r})\, u^*_{\boldsymbol{p}'}(\boldsymbol{r})\, u_{\boldsymbol{p}}(\boldsymbol{r})\, \mathrm{e}^{-\mathrm{i}(\boldsymbol{p}'-\boldsymbol{p})\boldsymbol{r}/\hbar}\, \mathrm{d}V \\ &= V^{-1} \sum_i \mathrm{e}^{-\mathrm{i}(\boldsymbol{p}'-\boldsymbol{p})\boldsymbol{R}_i/\hbar}\, v_{p'p}. \end{aligned}$$

For the sake of simplicity, we assumed all the impurity atoms to be identical and to occupy equivalent positions in crystal unit cells, although, of course, the choice of these unit cells in itself is arbitrary. The scattering probability is given by

$$\frac{2\pi}{\hbar} V^{-2} \int |v_{p'p}|^2 \sum_{i,k} \exp\left(-\mathrm{i}\frac{(\boldsymbol{p}'-\boldsymbol{p})(\boldsymbol{R}_i - \boldsymbol{R}_k)}{\hbar}\right) \delta(\varepsilon(\boldsymbol{p}) - \varepsilon(\boldsymbol{p}')) \frac{\mathrm{d}^3\boldsymbol{p}'}{(2\pi\hbar)^3} V.$$

If the concentration of impurity atoms is small, they are distributed randomly and the average distance between them is much larger than the interatomic spacing. In this case, the scattering probability may be averaged over the positions of the impurity atoms. All the terms in

$$\sum_{i,k} \exp\left(-\mathrm{i}\frac{(\boldsymbol{p}'-\boldsymbol{p})(\boldsymbol{R}_i - \boldsymbol{R}_k)}{\hbar}\right),$$

while being rapidly oscillating functions, will vanish, except terms with $i = k$. Hence, we have

$$\overline{\sum_{i,k} \exp\left(-\mathrm{i}\frac{(\boldsymbol{p}'-\boldsymbol{p})(\boldsymbol{R}_i - \boldsymbol{R}_k)}{\hbar}\right)} = N_{\mathrm{i}},$$

where N_{i} is the number of impurity atoms. In view of this, the scattering probability becomes equal to

$$\frac{2\pi}{\hbar} \frac{N_{\mathrm{i}}}{V} \int |v_{p'p}|^2 \delta(\varepsilon(\boldsymbol{p}) - \varepsilon(\boldsymbol{p}')) \frac{\mathrm{d}^3\boldsymbol{p}'}{(2\pi\hbar)^3}. \tag{3.8}$$

All the above reasonings refer to an individual electron. Let us now consider a system of electrons described by a distribution function. Due to the scattering processes, the number of electrons with momenta in a given volume element of momentum space in the neighborhood of $\boldsymbol{p}$ may increase ($\boldsymbol{p}' \to \boldsymbol{p}$) or decrease ($\boldsymbol{p} \to \boldsymbol{p}'$). Each of the processes can take place only if there is an electron in the initial state and the final state is not occupied (since the Pauli principle does not permit two electrons to be in the same state). Therefore, if we deal with a system of electrons, then each individual scattering probability must be multiplied by the distribution function f for the initial state and by $1-f$ for the final state. Having done so, we finally obtain

$$I(f)=\frac{2\pi}{\hbar}n_i\int|v_{p'p}|^2\{f(\boldsymbol{p}')[1-f(\boldsymbol{p})]-f(\boldsymbol{p})[1-f(\boldsymbol{p}')]\}$$

$$\times\,\delta(\varepsilon(\boldsymbol{p})-\varepsilon(\boldsymbol{p}'))\frac{d^3p'}{(2\pi\hbar)^3}$$

$$=\frac{2\pi}{\hbar}n_i\int|v_{p'p}|^2[f(\boldsymbol{p}')-f(\boldsymbol{p})]\delta(\varepsilon(\boldsymbol{p})-\varepsilon(\boldsymbol{p}'))\frac{d^3p'}{(2\pi\hbar)^3},\tag{3.9}$$

where $n_i=N_i/V$. Singling out, as in section 2.4, the integration over the energies, we have

$$I(f)=\frac{2\pi}{\hbar}n_i\int|v_{p'p}|^2[f(\boldsymbol{p}')-f(\boldsymbol{p})]\frac{dS'}{v(\boldsymbol{p}')(2\pi\hbar)^3},\tag{3.10}$$

where the integral is taken over the constant-energy surface $\varepsilon(\boldsymbol{p}')=\varepsilon(\boldsymbol{p})$.

If we substitute into $I(f)$ the equilibrium distribution function f_0, which depends on the energy alone, then $I(f_0)$ will turn to zero. In a case where the external electric field and the temperature gradient are sufficiently small, the departure from the equilibrium will not be large and we may write

$$f=f_0+f_1,$$

where f_1 is a small correction: $|f_1|\ll f_0$. Only f_1 will be included in $I(f)$.

A further simplification of the expression for $I(f)$ may be made in the isotropic model. In this case, the momenta $\boldsymbol{p}$ and $\boldsymbol{p}'$ are equal in magnitude, the constant-energy surface is a sphere, the integration over dS' changes to an integration over the solid angle and, finally, $v_{p'p}$ depends only on the angle between $\boldsymbol{p}'$ and $\boldsymbol{p}$, which we shall denote by θ. We now introduce the quantity

$$W(\theta)=\frac{\pi}{\hbar}n_i|v(\theta)|^2\nu(\varepsilon).$$

The collision integral is written in the following form:

$$I(f)=\int W(\theta)[f_1(\boldsymbol{p}')-f_1(\boldsymbol{p})]\frac{d\Omega}{4\pi}.$$

Suppose that the violation of the equilibrium is associated with the presence of a weak electric field (the situation is the same in the presence of a small temperature gradient)*. The correction to the distribution function must be a scalar and in a first approximation it must depend linearly on $\boldsymbol{E}$. The only possible form is

$$f_1(\boldsymbol{p}) = \boldsymbol{pE}\eta(\varepsilon).$$

Substituting into the collision integral, we obtain

$$I(f) = pE\eta(\varepsilon)\int W(\theta)(\cos(\widehat{\boldsymbol{p}', \boldsymbol{E}}) - \cos(\widehat{\boldsymbol{p}, \boldsymbol{E}}))\frac{\mathrm{d}\Omega}{4\pi}.$$

The quantity $\cos(\widehat{\boldsymbol{p}', \boldsymbol{E}})$ may be transformed as follows, We choose the direction of the polar axis z along $\boldsymbol{p}$. Then, $\boldsymbol{p}'\boldsymbol{E} = p'_z E_z + \boldsymbol{p}'_\perp \boldsymbol{E}_\perp$, i.e.,

$$\cos(\widehat{\boldsymbol{p}', \boldsymbol{E}}) = \cos(\widehat{\boldsymbol{p}', \boldsymbol{p}})\cos(\widehat{\boldsymbol{E}, \boldsymbol{p}}) + \sin(\widehat{\boldsymbol{p}', \boldsymbol{p}})\sin(\widehat{\boldsymbol{E}, \boldsymbol{p}})\cos\phi_{p'E}$$

(where $\phi_{p'E}$ is the angle between the projections $\boldsymbol{p}'$ and $\boldsymbol{E}$ onto the plane perpendicular to $\boldsymbol{p}$). Upon integration over $\phi_{p'E}$ the second term vanishes. Since the angle $(\widehat{\boldsymbol{p}', \boldsymbol{p}})$ was denoted by θ above, substituting $\cos(\widehat{\boldsymbol{p}'\boldsymbol{E}})$ into $I(f)$ gives

$$I(f) = (\boldsymbol{pE})\eta(\varepsilon)\int W(\theta)(\cos\theta - 1)\frac{\mathrm{d}\Omega}{4\pi}.$$

Comparing this with the expected form of f_1, we see that

$$I(f) = -\frac{f_1}{\tau},$$

where

$$\tau^{-1} = \int W(\theta)(1 - \cos\theta)\frac{\mathrm{d}\Omega}{4\pi}. \tag{3.11}$$

The quantity τ plays the role of the average time between collisions.

Thus, in the case under consideration the kinetic equation has the form

$$\frac{\partial f}{\partial t} + \boldsymbol{v}\frac{\partial f}{\partial \boldsymbol{r}} + e\boldsymbol{E}\frac{\partial f}{\partial \boldsymbol{p}} = -\frac{(f - f_0)}{\tau}. \tag{3.12}$$

This form of the kinetic equation is often used for the study of various processes. However, one must keep in mind the assumptions that were made in its derivation. These assumptions are: (a) the isotropic model and (b) elastic scattering of electrons (i.e., the energy does not change upon scattering). The first of these assumptions does not correspond at all to a real metal, and the second does not apply for all scattering processes. Nevertheless, we will use eq. (3.12) in some cases to obtain simple estimates or to illustrate the calculation procedures. We begin with two examples of application of eq. (3.12).

* The simplification given below is suitable only for linear problems.

3.3. Electrical conductivity

Suppose that in a metal there is a homogeneous and constant electric field $\boldsymbol{E}$. We are to find the current that is produced. In the kinetic equation (3.12) the first two terms vanish and we obtain

$$e\boldsymbol{E}\frac{\partial f}{\partial \boldsymbol{p}} = -\frac{(f-f_0)}{\tau}. \tag{3.13}$$

Assuming the field to be weak, we substitute $f = f_0 + f_1$, where $f_1 \ll f_0$. Then, eq. (3.13) takes the form

$$e\boldsymbol{E}\frac{\partial f_0}{\partial \boldsymbol{p}} = -\frac{f_1}{\tau}. \tag{3.14}$$

Since f_0 depends only on the energy, we may write

$$\frac{\partial f_0}{\partial \boldsymbol{p}} = \frac{\partial f_0}{\partial \varepsilon}\frac{\partial \varepsilon}{\partial \boldsymbol{p}} = \boldsymbol{v}\frac{\partial f_0}{\partial \varepsilon}.$$

From eq. (3.14) we find

$$f_1 = -e\boldsymbol{E}\boldsymbol{v}\tau\frac{\partial f_0}{\partial \varepsilon}. \tag{3.15}$$

The electric current is expressed through the distribution function in the following manner:

$$\boldsymbol{j} = 2e\int \boldsymbol{v}f\frac{\mathrm{d}^3p}{(2\pi\hbar)^3}. \tag{3.16}$$

If we substitute the function $f_0(\varepsilon)$, the integral will vanish (this is because ε is an even function and $\boldsymbol{v}$ an odd function of $\boldsymbol{p}$). Therefore, f_1 must be substituted as f. We then find

$$\boldsymbol{j} = -e^2\int \boldsymbol{v}(\boldsymbol{v}\boldsymbol{E})\tau\frac{\partial f_0}{\partial \varepsilon}\nu(\varepsilon)\,\mathrm{d}\varepsilon\frac{\mathrm{d}\Omega}{4\pi},$$

where, as before, we have introduced the density of states $\nu(\varepsilon)$. This integral contains the derivative $\partial f_0/\partial\varepsilon$, which is not equal to zero only in the vicinity of $\varepsilon = \mu$. The rule of taking integrals containing $\partial f_0/\partial\varepsilon$ was found in section 2.4 (formula 2.25). To a first approximation, $\partial f_0/\partial\varepsilon$ may be treated as a δ-function: $\partial f_0/\partial\varepsilon \approx -\delta(\varepsilon-\mu)$. Using this circumstance and integrating over the angles, we obtain

$$\boldsymbol{j} = \tfrac{1}{3}e^2\boldsymbol{E}[v^2\tau\nu(\varepsilon)]_{\varepsilon=\mu}.$$

the conductivity σ is defined as the proportionality factor between $\boldsymbol{j}$ and $\boldsymbol{E}$: $\boldsymbol{j} = \sigma\boldsymbol{E}$.

Thus, we find

$$\sigma = \tfrac{1}{3}e^2[v^2\tau\nu(\varepsilon)]_\mu. \tag{3.17}$$

3.4. *Thermal conductivity*

In contrast to the electric current, i.e., the transfer of electric charge, energy transfer involves not only electrons but also phonons, and therefore the thermal conductivity has two parts, an electron and a phonon part. We shall here consider only the electronic thermal conductivity. Usually, it far exceeds the phonon thermal conductivity. The only exception is a metal containing a large amount of impurities and at very low temperatures. However, in this case too, we can single out the electron thermal conductivity, using its behavior in a magnetic field (ch. 6).

There is one significant difference between the mechanisms of charge and energy transfer. The charge is transferred by real electrons. Since the density of particles of the gas model ("electrons") is equal to the density of real electrons, we may consider the electric current that arises in the gas model under the action of the field, as has been done in the preceding section. As for the energy transfer, it is achieved by quasiparticles with the energy spectrum $\varepsilon_{\text{qp}} = |\xi|$ and therefore we will use the quasiparticle model here*.

First of all, we have to change the form of the kinetic equation slightly. Since the charge of "antiparticles" is opposite to that of "particles", instead of e we must write e sign ξ, where sign $\xi = \xi/|\xi|$. Further, the velocity of quasiparticles is

$$\frac{\partial|\varepsilon|}{\partial \boldsymbol{p}} = \frac{\partial|\xi|}{\partial \xi}\frac{\partial \xi}{\partial \boldsymbol{p}} = \boldsymbol{v} \operatorname{sign} \xi,$$

where $\boldsymbol{v} = \partial\varepsilon/\partial\boldsymbol{p} = \boldsymbol{p}/m^*$. Finally, the equilibrium distribution function for quasiparticles is equal to $f_0 = [e^{|\xi|/T}+1]^{-1}$ (the number of quasiparticles is not specified but is determined by the equilibrium condition, which is $\partial\Phi/\partial N_{\text{qp}} = \mu_{\text{qp}} = 0$). Taking all this into account, we can write the kinetic equation for quasiparticles in the form

$$\frac{\partial f}{\partial t} + \boldsymbol{v} \operatorname{sign} \xi \frac{\partial f}{\partial \boldsymbol{r}} + e\boldsymbol{E} \operatorname{sign} \xi \frac{\partial f}{\partial \boldsymbol{p}} = -\frac{(f-f_0)}{\tau} \tag{3.18}$$

(τ for quasiparticles coincides with τ for particles because it includes the square of the charge).

* In principle, in this case too we can use the gas model since the quasiparticles of the gas model and of the real liquid coincide. However, a direct calculation performed in terms of quasiparticles is simpler in this particular case, the more so because with such a procedure we can easily establish the relationship with the calculation of thermal conductivity in superconductors (section 19.1).

Let us now turn to the calculation of the thermal conductivity. If there is a temperature gradient, then only the first term will vanish in eq. (3.18). We will assume the temperature gradient to be weak and substitute $f=f_0+f_1$. Since the temperature depends on coordinates, the left-hand side of the kinetic equation will read

$$\boldsymbol{v}\ \mathrm{sign}\ \xi \frac{\partial f_0}{\partial \boldsymbol{r}} = \frac{\partial f_0}{\partial T}\ \mathrm{sign}\ \xi\ (\boldsymbol{v}\nabla T)$$

$$= -\frac{|\xi|}{T}\frac{\partial f_0}{\partial |\xi|}\ \mathrm{sign}\ \xi\ (\boldsymbol{v}\nabla T) = \frac{\xi}{T}\frac{\partial f_0}{\partial |\xi|}(\boldsymbol{v}\nabla T).$$

Here we have made the following substitution:

$$\frac{\partial f_0}{\partial T} = -\frac{|\xi|}{T}\frac{\partial f_0}{\partial |\xi|}.$$

The energy flux is expressed by

$$\boldsymbol{q} = 2\int |\xi| \frac{\partial |\xi|}{\partial \boldsymbol{p}} f \frac{\mathrm{d}^3\boldsymbol{p}}{(2\pi\hbar)^3} = 2\int \xi \boldsymbol{v} f_1 \frac{\mathrm{d}^3\boldsymbol{p}}{(2\pi\hbar)^3}. \tag{3.20}$$

From eq. (3.18) it follows that

$$f_1 = \tau \frac{\xi}{T}\frac{\partial f_0}{\partial |\xi|}(\boldsymbol{v}\nabla T). \tag{3.21}$$

Substituting f_1 into eq. (3.20) gives

$$\boldsymbol{q} = \int \boldsymbol{v}(\boldsymbol{v}\nabla T)\tau\nu(\varepsilon)\frac{\xi^2}{T}\frac{\partial f_0}{\partial |\xi|}\mathrm{d}\xi \frac{\mathrm{d}\Omega}{4\pi}. \tag{3.20'}$$

Since

$$\frac{\partial f_0}{\partial |\xi|} = -(4T)^{-1}\cosh^{-2}\frac{|\xi|}{2T} = -(4T)^{-1}\cosh^{-2}\frac{\xi}{2T},$$

it follows that the integral over ξ is the same as that evaluated in section 2.4. Substituting its value, we find that

$$\boldsymbol{q} = -\tfrac{1}{9}\pi^2 T\nabla T(\nu v^2\tau)_\mu. \tag{3.22}$$

Thermal conductivity is defined as the proportionality factor between the heat flux and the temperature gradient:

$$\boldsymbol{q} = -\varkappa\nabla T. \tag{3.23}$$

From expression (3.22) for $\boldsymbol{q}$ we find

$$\varkappa = \tfrac{1}{9}\pi^2 T(\nu v^2\tau)_\mu. \tag{3.24}$$

We should make two reservations concerning this calculation. First of all, the nonuniformity of temperature gives rise to inhomogeneity of μ and, hence, to a change in the spectrum of quasiparticles along the conductor. However, because of the smallness of the change of μ with temperature (section 2.4, formula 2.27), taking account of this effect leads to small corrections to $\varkappa$, of the order of $(T/\mu)^2$. Further, the presence of a temperature gradient leads to a simultaneous appearance in the metal of a certain electric field. This field is determined from the condition of absence of an electric current. But it is not difficult to show (section 6.1) that taking account of this circumstance also gives rise to corrections of order $(T/\mu)^2$. Taking such corrections into account does not correspond to the accuracy of the calculation made.

Comparing expression (3.24) with formula (3.17) for τ, we obtain

$$\frac{\varkappa}{\sigma T} = \frac{\pi^2}{3e^2}. \tag{3.25}$$

The constant on the right-hand side depends only on the electronic charge and contains no characteristics of the metal. It is called the Lorentz constant. Formula (3.25) is known as the Wiedemann–Franz law. Here it should again be emphasized that this relation was derived from the kinetic equation for the isotropic model under the assumption of elastic collisions.

One can ask whether this is more general than the kinetic equation (3.12) (or 3.18). Let us consider the case of an anisotropic metal; the collisions will, however, be assumed to be elastic. To do this, let us return to expression (3.10). Since $I(f_0)$ vanishes, it follows, as has been said above, that only $I(f_1)$ is left, i.e., generally speaking, a certain linear integral operator acting on f_1. Substituting f_0 into the left-hand side of the kinetic equation (in the form of 3.18), we arrive at an expression proportional to $\partial f_0/\partial|\xi|$. Hence, it may be presumed that f_1 is proportional to $\partial f_0/\partial|\xi|$. This is confirmed since the integral in (3.10) is taken along the constant-energy surface, and $\partial f_0/\partial|\xi|$ (and other energy-dependent factors) may be taken outside the integral sign. For this reason and also from the considerations of parity with respect to $\boldsymbol{p}$ it follows that the function f_1 must have the form

$$f_1 = \frac{\partial f_0}{\partial|\xi|}\left(e\boldsymbol{E}\boldsymbol{v} - \boldsymbol{v}\nabla T\frac{\xi}{T}\right)\zeta(\boldsymbol{p})$$

where ζ is an even function of the quasimomentum, which varies slowly in the neighborhood of $|\boldsymbol{p}| = p_0$.

In an anisotropic metal the connection between the current and the electric field will be expressed in the form $j_i = \sum_{i,k} \sigma_{ik} E_k$, i.e., instead of a single quantity σ we shall have a tensor σ_{ik}. The same is valid for thermal conductivity. On the basis of what has been said about the function f_1 we may conclude that the Wiedemann–Franz law will be satisfied for ratios of the corresponding components of σ_{ik} and $\varkappa_{ik}$. But if the collisions are not elastic, the Wiedemann–Franz law is not fulfilled.

3.5. The concept of a mean free path

The expressions for electrical and thermal conductivity, which have been derived from the kinetic equation, can be obtained in a simpler way in the gas model using the concept of a mean free path.

Let us define the mean free path as the average distance travelled by an electron between collisions. If τ is the time between collisions, then the mean free path $l = v\tau$ (v is the average velocity).

Let us first find the diffusion coefficient. Let the density of particles n_e be inhomogeneous along the x axis. This gives rise to a flux of particles along this axis. The diffusion flux, i.e., the number of particles passing across 1 cm^2 of this plane in 1 s is given by

$$j_D = \int v \cos\theta \frac{d\Omega}{4\pi},$$

where θ is the angle between the direction of the particle velocity and the x axis. But since a distance equal to the mean free path is travelled by the particles without undergoing collisions, $n_e\,(x - l\cos\theta)$ rather than $n_e(x)$ must be substituted into this formula. Assuming the gradient n_e to be small, we can expand this quantity in l. As a result, we obtain

$$j_D = -\frac{\partial n_e}{\partial x}\, lv\tfrac{1}{2}\int_{-1}^{1} \cos^2\theta \, d\cos\theta = -\tfrac{1}{3}lv\frac{\partial n_e}{\partial x}.$$

The proportionality factor between the particle flux and the density gradient is just what we call the diffusion coefficient:

$$j_D = -D\frac{\partial n_e}{\partial x}. \tag{3.26}$$

Hence,

$$D = \tfrac{1}{3}lv. \tag{3.27}$$

Electrical conduction is none other than the diffusion of electrons under the influence of an external force $F = eE$. We will write the electric current in the form of a diffusion flux of charge

$$j = ej_D = -eD\frac{\partial n_e}{\partial x}.$$

Substituting here

$$n_e = 2\int f(\varepsilon - e\phi)\frac{d^3p}{(2\pi\hbar)^3},$$

where ϕ is the electric field potential ($E = -\partial\phi/\partial x$), we obtain

$$j = 2e^2 D \int \frac{\partial f}{\partial \varepsilon}\frac{\partial \phi}{\partial x}\frac{d^3 p}{(2\pi\hbar)^3} = e^2 D\nu(\mu)E.$$

On the other hand, $j = \sigma E$, and hence*

$$\sigma = e^2 D\nu(\mu). \tag{3.28}$$

Substituting the formula for D (3.27) and taking into account that $l = v\tau$, we arrive at formula (3.17) for electrical conductivity.

In an analogous manner we can find the expression for thermal conductivity. The heat flux is expressed in terms of the energy density $\mathscr{E}/V$ by the following relation

$$q = \tfrac{1}{2}\int_{-1}^{1} \frac{\mathscr{E}}{V} v \cos\theta \, d\cos\theta.$$

But the energy density depends on the coordinates due to the presence of a temperature gradient. Here again, as before, we must substitute

$$\mathscr{E}[T(x - l\cos\theta)] \approx \mathscr{E}(T) - \frac{\partial \mathscr{E}}{\partial T}\frac{\partial T}{\partial x} l \cos\theta.$$

The quantity $C = V^{-1}\partial\mathscr{E}/\partial T$ is the heat capacity per unit volume. In this connection we obtain

$$q = -\tfrac{1}{3} C l v \frac{\partial T}{\partial x}.$$

Substituting here $l = v\tau$ and formula (2.27) for the heat capacity, we have

$$\varkappa = \tfrac{1}{3} C l v = \tfrac{1}{9}\pi^2 T v^2 \tau \nu(\mu),$$

which coincides with (3.34).

3.6. *Electrical and thermal conductivity in a gas of free electrons*

For a gas of free electrons formula (3.17) becomes very simple. In this model,

$$\nu(\mu) = \frac{\partial}{\partial \mu}\frac{2V_F}{(2\pi\hbar)^3}, \tag{3.29}$$

where V_F is the volume of the Fermi sphere. Since $\mu = p_0^2/2m$ and v_F is proportional to p_0^3, it thus follows that V_F is proportional to $\mu^{3/2}$. Hence, we have

$$\frac{\partial V_F}{\partial \mu} = \frac{3}{2}\frac{V_F}{\mu}$$

* The relationship between the diffusion coefficient and the coefficient that relates the particle flux to the force acting on the system is called the Einstein relation.

and

$$\nu(\mu) = \tfrac{3}{2}\frac{2}{(2\pi\hbar)^{-3}}\frac{V_F}{\mu}.$$

But the number of electrons per unit volume is

$$n_e = \frac{2V_F}{(2\pi\hbar)^3}.$$

Thus,

$$\nu(\mu) = \tfrac{3}{2}\frac{n_e}{\mu}.$$

The electrical conductivity is given by

$$\sigma = \tfrac{1}{3}e^2v^2\tau\nu(\mu) = \tfrac{2}{3}\frac{e^2\mu\tau\nu(\mu)}{m}$$

(since $\mu = \frac{1}{2}mv^2$). Substituting here $\nu(\mu)$, we obtain

$$\sigma = \frac{n_e e^2\tau}{m}. \tag{3.30}$$

From the Wiedemann–Franz formula (3.25) we have

$$\varkappa = \tfrac{1}{3}\pi^2\frac{n_e\tau T}{m}.$$

Although these formulas for σ and $\varkappa$ do not apply to real metals, they are nonetheless very useful for estimations.

4 Scattering processes

4.1. Scattering by impurities

For real metals the functions $\varepsilon(\boldsymbol{p})$ are very complicated. This refers even to a larger extent to the scattering amplitudes. Therefore, it is a very difficult matter to obtain exact numerical values of electrical and thermal conductivities. It is much easier to find the temperature dependences and the orders of magnitude of these coefficients. To do this, in most cases we need not solve the kinetic equation and the mean-free-path concept will just suffice. We begin with scattering by impurities. In section 3.2. we have already dealt with this process with the object of deriving a kinetic equation in the form (3.12) and we could have obtained the corresponding estimates for σ and $\varkappa$ from the formulas found in that section. However, we will not do this, and in the same way as we treat the other scattering mechanisms, we will obtain the values of σ and κ from qualitative but more spectacular considerations.

Let us define the effective cross section for scattering of electrons by impurities. For the impurity atom potential acting on the electron various assumptions can be made. We may assume either that the atom is neutral and behaves as a hard sphere with atomic dimensions or (and this is more probable) that the impurity atom is ionized. However, even in this case we may apply the hard-sphere model. As an illustration, we will now consider the screening of the impurity ion potential by electrons.

Suppose that an ion has been introduced into the lattice. If n_e' is the change in the electron density, then the Maxwell equation that determines the variation of the electrical potential has the form

$$\Delta\varphi = 4\pi e n_e' \tag{4.1}$$

(the presence of an ion at point $r=0$ implies that we are to choose a solution that has the form $-Ze/r$ for $r\to 0$). In the presence of the potential φ the energy ε is replaced by $\varepsilon - e\varphi$. But ε enters into the Fermi distribution in the combination $\varepsilon-\mu$, where μ is the chemical potential. Thus, it may be said that μ is replaced by $\mu + e\varphi$. Hence,

$$n_e' = n_e(\mu + e\varphi) - n_e(\mu) \approx e\varphi \frac{\partial n_e}{\partial \mu}.$$

The Maxwell equation assumes the form

$$\Delta\varphi - 4\pi e^2\varphi\frac{\partial n_e}{\partial\mu} = 0. \tag{4.2}$$

The solution of this equation is

$$\varphi = -Ze\frac{e^{-\varkappa_D r}}{r}, \tag{4.3}$$

where $-Ze$ is the charge of the ion and $\varkappa_D = r_D^{-1}$ is the inverse Debye radius equal to

$$\varkappa_D = \left(4\pi e^2\frac{\partial n_e}{\partial\mu}\right)^{1/2}. \tag{4.4}$$

Thus, the Coulomb potential of the impurity ion is screened by electrons at distances larger than the Debye radius $r_D = \varkappa_D^{-1}$. In order of magnitude we have

$$\varkappa_D \sim \left(e^2\frac{p_0^3}{\hbar^3}\frac{m}{p_0^2}\right)^{1/2} \sim \frac{p_0}{\hbar}\left(\frac{e^2 m}{p_0\hbar}\right)^{1/2} \sim \frac{p_0}{\hbar}\left(\frac{e^2}{\hbar v}\right)^{1/2}. \tag{4.5}$$

The quantity $e^2/\hbar v$ is dimensionless. In order to estimate it, we have to take into account that the atoms in the metal are bound due mainly to conduction electrons. And this means that the potential and kinetic energies must be of the same order of magnitude. The ratio of the average potential electron energy $e^2/\bar{r}$ to its kinetic energy $p_0^2/2m$ is equal to

$$\frac{e^2/\bar{r}}{p_0^2/m} \sim \frac{e^2 p_0/\hbar}{p_0^2/m} \sim \frac{e^2}{\hbar v} \sim 1. \tag{4.6}$$

Here use has been made of the fact that the average separation between the electrons $\bar{r}$ is of order $\hbar/p_0$. From relation (4.6) it follows that the quantity $e^2/\hbar v$ in eq. (4.5) is of the order of unity and that

$$r_D \sim \frac{\hbar}{p_0} \sim a \sim 10^{-8}\ \text{cm}.$$

This means that the impurity ion potential is screened at atomic distances. Hence, even in this case we can use the hard-sphere model with atomic dimension.

The derivation given here is not quite correct. The variation of the electron density is, in fact, connected with the potential by the following nonlocal relation:

$$n_e'(\boldsymbol{r}) = e\int K(\boldsymbol{r}-\boldsymbol{r}')\varphi(\boldsymbol{r}')\,\mathrm{d}V'.$$

For the isotropic model

$$K(\boldsymbol{r}) = K(r) = -\nu(\mu)\frac{p_0^3}{\pi\hbar^3}\left[\left(\frac{2p_0 r}{\hbar}\right)^{-3}\cos\left(\frac{2p_0 r}{\hbar}\right) - \left(\frac{2p_0 r}{\hbar}\right)^{-4}\sin\left(\frac{2p_0 r}{\hbar}\right)\right].$$

According to this formula (which is derived in the same way as formula 21.40; see section 21.3), under the action of the potential at a given point the change in electron density falls off mainly by a r^{-3} law, undergoing oscillations with a period $\pi\hbar/p_0$ (Friedel oscillations; Friedel 1952, 1954, 1958). Such a behavior of the electron density is a consequence of the Fermi degeneracy. The derivation of formula (4.3) is valid for a Boltzmann gas.

However, for our purposes, it is sufficient to use the potential averaged over distances larger than $\hbar/p_0$; this again leads us to the model of a hard-sphere with atomic dimensions.

Let us now consider the scattering of electrons by hard spheres. If $Q_{\rm eff}$ is the effective scattering cross section ($\sim 10^{-16}\,\mathrm{cm}^2$) and the electron moves by a distance of 1 cm, then on this path it will collide with impurity atoms in a volume of $1\,\mathrm{cm}^3 \cdot Q_{\rm eff}$. If $n_i = N_i/V$ is the density of impurity atoms, then the total number of collisions will be

$$n_i \cdot 1 \cdot Q_{\rm eff}.$$

Hence, the average distance travelled by the electron without collisions (in other words, the mean free path) is

$$l \sim (n_i Q_{\rm eff})^{-1}. \tag{4.7}$$

To evaluate the order of magnitude of the conductivity, we use formula (3.30) derived for a free electron gas:

$$\sigma \sim \frac{n_e e^2 \tau}{m} \sim \frac{n_e e^2 l}{mv} \sim \frac{n_e e^2 l}{p_0} \sim \frac{n_e e^2}{p_0 n_i Q_{\rm eff}}.$$

But the number of conduction electrons in ordinary metals has the same order of magnitude as the number of atoms, therefore $n_i/n_e \sim c_i$ is the atomic concentration of impurities. Recalling also that $e^2 \sim \hbar v$, we obtain

$$\sigma \sim \frac{v\hbar}{p_0 c_i Q_{\rm eff}} \sim \frac{\hbar}{m c_i Q_{\rm eff}} \sim \frac{10^{-27}\,\mathrm{g\,cm^2\,s^{-1}}}{10^{-27}\,\mathrm{g} \cdot c_i Q_{\rm eff}} \sim \frac{10^{16}}{c_i}\,\mathrm{s}^{-1}. \tag{4.8}$$

Thus, the conductivity is independent of the temperature.

From the Wiedemann-Franz law (3.25) and (4.8) we obtain

$$\varkappa \sim 10^4 T[\mathrm{K}] c_i^{-1} \frac{\mathrm{erg}}{\mathrm{cm\,s\,K}}. \tag{4.9}$$

4.2. *Scattering of electrons by electrons*

Another mechanism is the scattering of electrons by electrons. We have already considered the mutual scattering of Fermi particles in section 2.2. It has been found that the scattering probability is proportional to T^2; hence, the collision time τ must be proportional to T^{-2}. One needs only to introduce a coefficient so that τ has the dimension of time. Since the potential energy of interaction of electrons is of the order of their kinetic energy, the only form that τ may have is

$$\tau \sim \frac{\hbar\mu}{T^2}.$$

Thus, ($e^2 \sim \hbar v$),

$$\sigma \sim \frac{n_e e^2 \hbar\mu}{mT^2} \sim \frac{n_e \hbar^2 p_0 \mu}{m^2 T^2} \sim \left(\frac{\mu}{T}\right)^2 \frac{n_e \hbar^2 p_0}{\mu m^2}$$

$$\sim \left(\frac{\mu}{T}\right)^2 \frac{n_e \hbar^2}{p_0 m} \sim \left(\frac{\mu}{T}\right)^2 \left(\frac{p_0}{\hbar}\right)^3 \frac{\hbar^2}{p_0 m} \sim \left(\frac{\mu}{T}\right)^2 \frac{\hbar}{m} \left(\frac{p_0}{\hbar}\right)^2.$$

Since $\hbar/p_0 \sim 10^{-8}$ cm, we obtain

$$\sigma \sim 10^{16} (\mu/T)^2 \ \mathrm{s}^{-1} \tag{4.10}$$

(here T is taken in energy units). Thus, the conductivity varies as T^{-2}.

A remark should be made at this point concerning the nature of collisions. Collisions between electrons cannot be regarded as elastic because the changes in the energy may be of the order of T. The variations of the momentum are also considerable, of the order of p_0 (the Coulomb interaction of electrons is screened in the same way as the impurity potential at interatomic spacings; the variation of the momentum upon scattering is of order $\hbar/r_D \sim p_0$). But we may nevertheless employ the mean-free-path concept since each collision is effective in the sense of both the variation of the momentum and the variation of energy (and therefore τ appearing in various formulas has the same order of magnitude).

Because of this, the Wiedemann–Franz formula (3.25) is also applicable by order of magnitude. From this formula we obtain for thermal conductivity:

$$\kappa \sim \frac{10^{12}}{T\,[\mathrm{K}]} \frac{\mathrm{erg}}{\mathrm{cm\,s\,K}}. \tag{4.11}$$

4.3. *Scattering by lattice vibrations*

As has been pointed out in section 3.1, in an ideal periodic lattice the electron is moving as a free particle. However, the lattice is not ideal even at $T = 0$, because the atoms experience so-called zero-point vibrations (formula 2.4). Nonetheless, at $T = 0$ no electrical resistance is present. Indeed, the scattering process corresponds to the transfer of the momentum and energy from the electron to lattice vibrations or vice versa. The excited state of the lattice may be described in terms of the picture of a gas of vibrational quanta, i.e., phonons. Thus, the scattering process consists of the emission or absorption of phonons by the electron.

At $T=0$ there are no phonons, so there is nothing to be absorbed. At the same time, the electron liquid is in the ground state and cannot emit phonons - there is no energy available for this.* At $T\neq 0$ all these processes will, of course, take place. Since, as before, we can find only the temperature dependences and orders of magnitude of the quantities σ and κ, we shall always confine ourselves to simple estimations.

We begin with the quantization of lattice vibrations. The kinetic energy of the lattice is given by

$$E_K=\tfrac{1}{2}\sum_i M\dot{u}_i^2$$

where $\boldsymbol{u}_i$ is the displacement of the ith atom, M is its mass. For long-wavelength vibrations, instead of discrete $\boldsymbol{u}_i$, we may introduce a smooth function $\boldsymbol{u}(\boldsymbol{r})$ and write the kinetic energy in the form

$$E_K=\tfrac{1}{2}Mn\int \dot{u}^2(\boldsymbol{r})\,\mathrm{d}V,$$

where n is the number of ions in a unit volume. Note that here the replacement $\boldsymbol{u}_i\to\boldsymbol{u}(\boldsymbol{r})$ is, strictly speaking, justifiable only for long waves, but it is always suitable for estimates, and therefore we shall employ it. The commutation relations between the momenta of ions $M\dot{u}$ and their coordinates $\boldsymbol{u}$ are as follows:

$$M[\dot{u}_{i'}^{\alpha},u_{i'}^{\alpha'}]=-\mathrm{i}\hbar\delta_{ii'}\delta_{\alpha\alpha'}$$

(α is the projection of the displacement vector). Passing over from $\boldsymbol{u}_i$ to $\boldsymbol{u}(\boldsymbol{r})$, we have from the last relation**:

$$nM[\dot{u}^{\alpha}(\boldsymbol{r},t),u^{\alpha'}(\boldsymbol{r}',t)]=-\mathrm{i}\hbar\delta_{\alpha\alpha'}\delta(\boldsymbol{r}-\boldsymbol{r}').$$

We shall consider a single mode of lattice vibrations and expand the operator $u(\boldsymbol{r})$ in a Fourier integral. Since $u(\boldsymbol{r})$ is a real operator, it follows that

$$u(\boldsymbol{r})=V^{-1/2}\sum_{\boldsymbol{k}}(u_{\boldsymbol{k}}\,\mathrm{e}^{\mathrm{i}\boldsymbol{kr}-\mathrm{i}\omega(\boldsymbol{k})t}+\overset{+}{u}_{\boldsymbol{k}}\,\mathrm{e}^{-\mathrm{i}\boldsymbol{kr}+\mathrm{i}\omega(\boldsymbol{k})t}).$$

Substituting this expression into the formula for kinetic energy and taking the time average, we obtain

$$\bar{E}_K=\tfrac{1}{2}nM\sum_{\boldsymbol{k}}(u_{\boldsymbol{k}}\overset{+}{u}_{\boldsymbol{k}}+\overset{+}{u}_{\boldsymbol{k}}u_{\boldsymbol{k}})\omega^2(\boldsymbol{k}).$$

From the general commutation relation we find

$$[u_{\boldsymbol{k}},\overset{+}{u}_{\boldsymbol{k}'}]=\frac{\hbar\delta_{\boldsymbol{kk}'}}{2nM\omega(\boldsymbol{k})}.$$

* In fact, the electrons receive energy from the electric field, but consideration of this phenomenon leads to small corrections to the current, which are nonlinear in the field.

** The integral of the commutator with $\boldsymbol{u}(\boldsymbol{r}')$ over $\mathrm{d}V'$ multiplied by n is equivalent to the sum of commutators with discrete $\boldsymbol{u}_{i'}$, with respect to i'. In either case we have $-\mathrm{i}\hbar\delta_{\alpha\alpha'}$ on the right-hand side.

These commutation relations can be satisfied if it is assumed that the operator u_k has matrix elements equal to

$$(u_k)_{n_k}^{n_k-1} = \sqrt{n_k}\left(\frac{\hbar}{2nM\omega(\boldsymbol{k})}\right)^{1/2},$$

$$(u_k^+)_{n_k-1}^{n_k} = (u_k)_{n_k}^{n_k-1}, \tag{4.12}$$

where n_k are integers. Indeed,

$$[u_k, u_k^+]_{n_k}^{n_k} = (u_k)_{n_k+1}^{n_k}(u_k^+)_{n_k}^{n_k+1} - (u_k^+)_{n_k-1}^{n_k}(u_k)_{n_k}^{n_k-1}$$

$$= \left(\frac{\hbar}{2nM\omega(\boldsymbol{k})}\right)(n_k + 1 - n_k) = \frac{\hbar}{2nM\omega(\boldsymbol{k})}.$$

In harmonic vibrations the average kinetic energy is equal to the average potential energy. Therefore, the total energy can be found as the doubled average kinetic energy $E = 2\bar{E}_K$. Substituting formula (4.12) into $\bar{E}_K$, we find

$$E = \sum_k \hbar\omega(\boldsymbol{k})\,(n_k + \tfrac{1}{2}).$$

Thus, n_k is nothing but the number of phonons. According to (4.12), the operator u_k^+ has matrix elements corresponding to an increase in the number of phonons by one (i.e., emission of a phonon) and u_k has matrix elements corresponding to a decrease in the number of phonons by one (absorption of a phonon).

Let us now consider the interaction of electrons with lattice vibrations. If the lattice is distorted, polarization arises. The polarization charge per unit volume is div $\boldsymbol{P}$, where $\boldsymbol{P}$ is the polarization vector, i.e., the dipole moment per unit volume. The interaction energy of electrons with the polarization charge is given by

$$e\int Q(\boldsymbol{r}-\boldsymbol{r}')\,\mathrm{div}\,\mathbf{P}(\boldsymbol{r}')\,\mathrm{d}V'.$$

Without screening the integral kernel Q would be equal to $|\boldsymbol{r}-\boldsymbol{r}'|^{-1}$. Screening, however, makes the interaction short-range. If the atomic distances are assumed to be small, Q may be represented in the form of a δ-function: $Q(\boldsymbol{r}-\boldsymbol{r}') \approx a^2\delta(\boldsymbol{r}-\boldsymbol{r}')$, where a is of the order of atomic distances (i.e., $a \sim 10^{-8}$ cm $\sim \hbar/p_0$). The polarization vector $\boldsymbol{P}(\boldsymbol{r})$ is equal in order of magnitude to $ne\boldsymbol{u}$, where n is the density of atoms, which has the same order of magnitude as the density of electrons and $\boldsymbol{u}$ is the displacement vector. Since the interaction includes div $\boldsymbol{P}$, the Fourier component will contain $\mathrm{i}ku_k$ (or $-\mathrm{i}ku_k^+$). But for large wavelengths $\omega = sk$, where s is the velocity of sound. Thus, the Fourier component of the interaction will be of order

$$V_k \sim V^{-1/2}\mathrm{i}e^2a^2n\frac{\omega}{s}u_k.$$

Since the interaction is proportional to $\boldsymbol{u}$, it follows that in the first order of perturbation theory there may be processes that correspond to the matrix elements

u_k and u_k^+ (4.12), i.e., the emission or absorption of one phonon. For example, for the emission we have

$$V_{p-\hbar k,p} \sim -\mathrm{i}V^{-1/2}e^2a^2n\left(\frac{\hbar(n_k+1)}{nM\omega}\right)^{1/2}\frac{\omega}{s}$$

$$= -\mathrm{i}V^{-1/2}\frac{na^2e^2}{s}\left(\frac{\hbar\omega(n_k+1)}{nM}\right)^{1/2}$$

$$= -\mathrm{i}V^{-1/2}na^3\frac{e^2}{a}\left(\frac{\hbar\omega(n_k+1)}{n}\right)^{1/2}\frac{1}{s\sqrt{M}}.$$

But $na^3 \sim 1$, $e^2/a \sim p_0^2/m$ and the velocity of sound satisfies the relation $s\sqrt{M} \sim v\sqrt{m}$ (where M is the mass of the ion, m is the mass of the electron and v its velocity)*. Using this, we transform the matrix element of the interaction:

$$V_{p-\hbar k,p} \sim -\mathrm{i}V^{-1/2}\frac{p_0^2}{mv\sqrt{m}}\left(\frac{\hbar\omega(n_k+1)}{n}\right)^{1/2}$$

$$\sim -\mathrm{i}V^{-1/2}p_0\left(\frac{\hbar\omega(n_k+1)}{nm}\right)^{1/2}. \tag{4.13}$$

Now note that the function $u(\boldsymbol{r})$ in fact has a physical meaning only at points $\boldsymbol{r} = \boldsymbol{a}_n$, occupied by the atoms. The Fourier expansion $u(\boldsymbol{r})$ at these points will include the factors $e^{i\boldsymbol{k}\boldsymbol{a}_n}$. But these do not change upon variation of $\boldsymbol{k}$ by a reciprocal lattice period. Hence, any $\boldsymbol{k}$ and $\boldsymbol{K}$ are physically equivalent, and the function $\omega(\boldsymbol{k})$, like $\varepsilon(\boldsymbol{p})$, must be periodic in the reciprocal lattice. Hence, it varies within finite limits.

For small k (long wavelengths) we have sound and $\omega = sk$. This relation gives the correct order of magnitude for large k as well. The maximum physical value of k corresponds to the face of the Brillouin zone, i.e., is equal in order of magnitude to $K_{\min} \sim \pi/a$. Therefore, the limiting frequency of phonons is of order

$$\omega_D \sim sK_{\min} \sim \frac{s\pi}{a.}.$$

This limit is called the Debye frequency. For metals it usually corresponds to a few hundred degrees. Since $\hbar/a \sim p_0$ is the Fermi momentum of electrons, we may also write:

$$\hbar\omega_D \sim sp_0. \tag{4.14}$$

* That this is so can be seen, for example, from the formula for the oscillator frequency $\omega = \sqrt{\alpha/M}$, where α is the elastic coefficient. Since the Coulomb energies in a metal are of the order of the electron kinetic energy, it follows that with momenta of order p_0 the difference between $\hbar\omega$ and the electron energies consists in the factor $(m/M)^{1/2}$. But $\boldsymbol{v} = \partial\varepsilon/\partial\boldsymbol{p}$ and $\boldsymbol{s} = \partial\omega/\partial\boldsymbol{k}$, from which it follows that this difference extends to the velocities.

For the number of phonons n_k we can use the Bose distribution:

$$n_k = (e^{\hbar\omega/T} - 1)^{-1}, \tag{4.15}$$

with the exception of special cases, which will be discussed in section 4.4.

We shall now consider two limiting cases: high and low temperatures.

(1) $T \gg \hbar\omega_D$. At high temperatures the most probable is the emission and absorption of phonons with high energies of order $\hbar\omega_D$. But $\hbar\omega_D \ll T$, which is why from the Bose distribution we have

$$n_k \approx \frac{T}{\hbar\omega_D}.$$

Since $n_k \gg 1$, it follows that $n_k + 1 \approx n_k$.

The total scattering probability with the emission of a phonon is equal to

$$W = \frac{2\pi}{\hbar}\int |V_{p-\hbar k,p}|^2 \delta(\varepsilon(p) - \varepsilon(p - \hbar k) - \hbar\omega(k))\, V\frac{d^3 k}{(2\pi)^3}. \tag{4.16}$$

An analogous formula can be written for absorption. The integration of the δ-function over d^3k gives a factor of order*

$$\frac{K^3_{\min}}{\mu} \sim \frac{p_0^3/\hbar^3}{p_0^2/m} \sim \frac{p_0 m}{\hbar^3}.$$

Substituting the expressions for $V_{p-\hbar k,p}$ and n_k, we obtain

$$W \sim \frac{p_0^2}{\hbar}\frac{\hbar\omega_D}{mn}\frac{T}{\hbar\omega_D}\frac{p_0 m}{\hbar^3} \sim \frac{T}{\hbar}$$

(we have replaced n by $p_0^3/\hbar^3$). Thus,**

$$\tau \sim W^{-1} \sim \frac{\hbar}{T}. \tag{4.17}$$

Substituting this into formula (3.34) for the conductivity, we get

$$\sigma \sim \frac{n_e e^2 \tau}{m} \sim \frac{ne^2\hbar}{mT} \sim \frac{n\hbar^2 p_0}{m^2 T} \sim \frac{\mu}{T}\frac{n\hbar^2}{p_0 m} \sim \frac{\mu}{T}\frac{p_0^3\hbar^2}{\hbar^3 p_0 m}$$

$$\sim \frac{\mu}{T}\left(\frac{p_0}{\hbar}\right)^2 \frac{\hbar}{m} \sim 10^{16}\frac{\mu}{T}\,s^{-1}. \tag{4.18}$$

Thus, at high temperatures the conductivity varies according to a T^{-1} law.

* In fact, instead of (4.16) we have to deal with a collision integral that includes the distribution functions for electrons and phonons. Together with the δ-function they provide $k \sim K_{\min}$. An analogous situation occurs with expression (4.20), where $\hbar k \sim T/s$.

** The result obtained is dubious from the standpoint of the applicability of the quasiparticle concept (condition 3.2). However, this condition includes the time of energy relaxation τ_ε. At $T \gg \hbar\omega_D$ the electron energy varies only slightly in each interaction with a phonon. At $T \ll \hbar\omega_D$, as we shall see below, $\tau_\varepsilon = \tau_t \gg \hbar/T$. Therefore, only the region of $T \sim \hbar\omega_D$ is suspicious. However, in real metals the electron-phonon interaction constant contains a small numerical coefficient, of the order of several tenths. In τ^{-1} it enters squared. It may therefore be asserted that in the region of $T \sim \hbar\omega_D$ condition (3.2) still holds.

Since $\hbar\omega_D \ll T$, the electron energy varies little in collisions, and the collisions may be considered elastic. In view of this, the Wiedemann–Franz law holds here; it gives

$$\kappa = \text{const.} \sim 10^8 \frac{\text{erg}}{(\text{cm s K})}. \tag{4.19}$$

(2) $T \ll \hbar\omega_D$. The most significant role here is played by phonons with an energy $\hbar\omega \sim T$. The electron energy is therefore changed substantially in each collision. Since the momentum of phonons is equal on the average to $\hbar\omega/s \sim T/s$ and the momentum of the electron is of order p_0, it follows that the electron momentum is changed in each collision by a relative amount $T/(sp_0) \sim T/(\hbar\omega_D) \ll 1$. Thus, this case is in a certain sense opposite to the elastic collisions considered earlier, where the energy varied little and the momentum could change strongly. Therefore, the rules that we have used until now cannot be applied here mechanically.

The mean-free-path concept may be used if we introduce different τ for electrical and thermal conductivities. Since in each collision the energy varies by an amount of order T, it follows that for thermal conductivity each collision is effective. The appropriate value of τ is found as W^{-1}, where W is defined by formula (4.16). The δ-function appearing in this formula may be written in the form*

$$\delta\left(\frac{p^2}{2m} - \frac{(\boldsymbol{p} - \hbar\boldsymbol{k})^2}{2m} - \hbar\omega(\boldsymbol{k})\right) = \delta\left(\frac{\hbar\boldsymbol{pk}}{m} - \frac{\hbar^2 k^2}{2m} - \hbar\omega(\boldsymbol{k})\right)$$
$$= \delta\left(\cos\theta - \frac{\hbar k}{2p} - \frac{m\omega}{pk}\right)\frac{m}{\hbar pk},$$

where θ is the angle between $\boldsymbol{p}$ and $\boldsymbol{k}$. According to what was said above about the phonon momentum $\hbar k/p_0 \ll 1$. We have also $m\omega/pk \sim ms/p_0 \sim s/v \ll 1$.

The integral in (4.16) is taken over the magnitude of k and the angles. It is important here that the value of $\cos\theta$ at which the argument of the δ-function vanishes is very small, i.e., it definitely lies within the limits of variation of $\cos\theta$**. Therefore, the δ-function gives unity on integration over the angles. So, we have

$$W \sim \hbar^{-1} \frac{p_0^2 \hbar\omega}{mn} \frac{m}{\hbar p_0 k} k^3 \sim h^{-1} \frac{p_0^2 \hbar\omega}{mn} \frac{ms}{p_0 \hbar\omega}\left(\frac{\omega}{s}\right)^3$$
$$\sim \hbar^{-1} \frac{p_0^2 T}{m(p_0/\hbar)^3} \frac{ms}{p_0 T} \frac{T^3}{\hbar^3 s^3} \sim \frac{T}{h}\left(\frac{T}{\hbar\omega_D}\right)^2. \tag{4.20}$$

Here we have the relations $\hbar\omega \sim T$, $\omega \sim sk$ and $sp_0 \sim \hbar\omega_D$. The time $\tau_t \sim W^{-1}$ may be used in the expression for the thermal conductivity.

* Recall that $\delta(ax) = a^{-1}\delta(x)$.

** From this it also follows that the momentum of the phonon is almost perpendicular to that of the electron.

We shall now turn to electrical conductivity. When we derived the collision integral in the form (3.12), we actually took advantage only of the fact that the electron momentum changes little in magnitude during collisions. This is also valid in this particular case and, hence, formula (3.11) also holds. The scattering angle θ is equal in order of magnitude to the ratio of the phonon momentum to the electron momentum:

$$\theta \sim \hbar k/p_0 \sim T/\hbar\omega_{\mathrm{D}} \ll 1.$$

Expanding $1-\cos\theta$ in (3.11) in θ, we obtain the following expression for τ_e, which enters into the electrical conductivity:

$$\tau_e^{-1} \sim \overline{W(\theta)\theta^2} \sim \frac{T}{\hbar}\left(\frac{T}{\hbar\omega_{\mathrm{D}}}\right)^4.$$

Since the electrical conductivity is proportional to τ_e, we obtain the following expression through the comparison of τ_e with (4.17),

$$\sigma \sim \sigma(T \gg \hbar\omega_{\mathrm{D}})\left(\frac{\hbar\omega_{\mathrm{D}}}{T}\right)^4 \sim 10^{16}\frac{\mu}{T}\left(\frac{\hbar\omega_{\mathrm{D}}}{T}\right)^4 s^{-1}. \tag{4.21}$$

Thus, the electrical conductivity at low temperatures varies according to the law T^{-5} (the Bloch law).

For thermal conductivity we make use of $\tau_t = W^{-1}$. Here we have

$$\begin{aligned}\varkappa &\sim Clv \sim Cv^2\tau_t \sim \frac{p_0 m T}{\hbar^3}\left(\frac{p_0}{m}\right)^2 \frac{\hbar}{T}\left(\frac{\hbar\omega_{\mathrm{D}}}{T}\right)^2 \\ &\sim \varkappa(T \gg \hbar\omega_{\mathrm{D}})\left(\frac{\hbar\omega_{\mathrm{D}}}{T}\right)^2 \sim 10^8\left(\frac{T_{\mathrm{D}}}{T}\right)^2 \frac{\mathrm{erg}}{\mathrm{cm\,s\,K}}.\end{aligned} \tag{4.22}$$

Here $T_{\mathrm{D}} = \hbar\omega_{\mathrm{D}}$ is expressed in the same units as T.

Comparing the expressions for electrical and thermal conductivities, we obtain

$$\frac{\varkappa}{\sigma} \sim \frac{T}{e^2}\left(\frac{T}{\hbar\omega_{\mathrm{D}}}\right)^2. \tag{4.23}$$

The Wiedemann–Franz law is not satisfied.

4.4. *Umklapp processes*

We have so far been concerned with the main scattering processes and have determined electrical and thermal conductivities. In particular, we have considered the resistance due to the scattering of electrons by electrons. However, in reality it is not quite obvious how electron–electron collisons can lead to the appearance of resistance. If there is a flow of electrons, then the slowing-down of this flow implies a change of the total momentum of electrons. But if the total momentum is conserved in collisions, then such collisions cannot cause the slowing-down of the electron flow. Here it must be recalled that the quasimomentum considered is not in fact the true momentum. In particular, the quasimomentum is defined only to within $\hbar\boldsymbol{K}$, where $\boldsymbol{K}$ is any period of the reciprocal lattice. In view of this, electron–electron collisions are possible, in which the total momentum is not conserved.

We shall consider in more detail the collison integral for electron–electron scattering. Let the scattering probability for two electrons be $W(\boldsymbol{p}_1\boldsymbol{p}_2, \boldsymbol{p}_1'\boldsymbol{p}_2')$. The collision integral then contains two terms. One of them describes the escape of the electron from a given volume element of the momentum space and the other describes the entry of the electron. Each of the terms must contain factors taking account of the presence of electrons in the initial states and of their absence in the final states. Thus,

$$I(f) = -\int W(\boldsymbol{p}_1\boldsymbol{p}_2, \boldsymbol{p}_1'\boldsymbol{p}_2') \times \{f(\boldsymbol{p}_1)f(\boldsymbol{p}_2)[1-f(\boldsymbol{p}_1')][1-f(\boldsymbol{p}_2')] - f(\boldsymbol{p}_1')f(\boldsymbol{p}_2')[1-f(\boldsymbol{p}_1)][1-f(\boldsymbol{p}_2)]\} \times \delta(\varepsilon_1+\varepsilon_2-\varepsilon_1'-\varepsilon_2')\frac{\mathrm{d}^3\boldsymbol{p}_2\,\mathrm{d}^3\boldsymbol{p}_1'}{(2\pi\hbar)^6}.$$

Here we integrate over the momenta of the second electron $\boldsymbol{p}_2$ and of one of the electrons after scattering, $\boldsymbol{p}_1'$. As for $\boldsymbol{p}_2'$, this momentum is determined by the conservation law to within $\hbar\boldsymbol{K}$:

$$\boldsymbol{p}_1+\boldsymbol{p}_2=\boldsymbol{p}_1'+\boldsymbol{p}_2'+\hbar\boldsymbol{K}.$$

The vector $\hbar\boldsymbol{K}$ is arbitrary, and it is the summation over $\hbar\boldsymbol{K}$ that is meant in the collision integral, provided all the vectors, $\boldsymbol{p}_1$, $\boldsymbol{p}_2$, $\boldsymbol{p}_1'$ and $\boldsymbol{p}_2'$ do not go beyond the Brillouin zone.

Let us define the distribution function in the form

$$f=f_0-\psi\frac{\partial f_0}{\partial\varepsilon}.$$

The derivative $\partial f_0/\partial\varepsilon$ may be written as

$$\frac{\partial f_0}{\partial\varepsilon}=-T^{-1}\,\mathrm{e}^{(\varepsilon-\mu)/T}[\mathrm{e}^{(\varepsilon-\mu)/T}+1]^2=-\frac{f_0(1-f_0)}{T}.$$

In equilibrium f does not change. Therefore, $\mathrm{d}f_0/\mathrm{d}t=I(f_0)=0$. This means that the combination of the distribution functions enclosed in braces in the collision integral vanishes if the function f_0 is substituted for f. We shall now determine the terms of first order in ψ in $I(f)$. The number of such terms is 4: one with $\psi(\boldsymbol{p}_1)$, the second with $\psi(\boldsymbol{p}_2)$, etc. The term with $\psi(\boldsymbol{p}_1)\equiv\psi_1$ has the form

$$-\frac{\psi_1}{T}\{f_1(1-f_1)f_2(1-f_1')(1-f_2')+f_1'f_2'f_1(1-f_1)(1-f_2)\},$$

where the functions f in braces are equilibrium functions. But, recalling that the combination involved in the collision integral is equal to zero at $f=f_0$, we can transform the last expression to the following form:

$$\frac{\psi_1}{T} f_1 f_2 (1-f_1')(1-f_2').$$

The terms with $\psi(\boldsymbol{p}_2)$, $\psi(\boldsymbol{p}_1')$ and $\psi(\boldsymbol{p}_2')$ can be transformed in an analogous manner. As a result, it turns out that all these terms have the same coefficient consisting of the functions f_0, and the combination enclosed in braces in the collision integral assumes the form

$$f_1 f_2 (1-f_1')(1-f_2') \frac{\psi_1+\psi_2-\psi_1'-\psi_2'}{T}.$$

Let us now suppose that ψ has the form

$$\psi = cp_x, \qquad c = \text{const.}$$

For momentum-conserving processes the collision integral vanishes because in this case

$$p_{x1}+p_{x2}-p_{x1}'-p_{x2}'=0.$$

In the absence of an electric field the function $f=f_0-(\partial f_0/\partial \varepsilon)\psi$ has the property $\mathrm{d}f/\mathrm{d}t=0$. Thus, if we take into account only momentum-conserving processes this function is the solution of the kinetic equation without an electric field. But at the same time this function, when substituted into the expression for the electric current

$$\boldsymbol{j} = 2e\int \boldsymbol{v} f \frac{\mathrm{d}^3\boldsymbol{p}}{(2\pi\hbar)^3},$$

gives a finite value. This implies the presence of current in the absence of an electric field or, in other words, an infinite conductivity.

As a matter of fact, we must also take into account the processes in which

$$\boldsymbol{p}_1+\boldsymbol{p}_2-\boldsymbol{p}_1'-\boldsymbol{p}_2'=\hbar\boldsymbol{K}\neq 0.$$

These processes are known as Peierls Umklapp processes. It is just such processes that will give a finite resistance. As has been pointed out above, all the vectors $\boldsymbol{p}_i$ must lie within the Brillouin zone. Moreover, as we have seen in section 2.2, if the momentum $\boldsymbol{p}_1$ is near the Fermi surface, then all the other momenta will also lie in the vicinity of the Fermi surface. Thus, for a finite resistance to be produced, it is necessary that*

$$\max(\boldsymbol{p}_1+\boldsymbol{p}_2-\boldsymbol{p}_1'-\boldsymbol{p}_2') = 4\max p_0(\boldsymbol{n}) > \hbar|\boldsymbol{K}_{\min}|. \tag{4.24}$$

* More exactly, $\max(4p_0(\boldsymbol{n})-K_{\min}(\boldsymbol{n}))>0$, where $\boldsymbol{n}$ are possible directions of $\boldsymbol{K}$.

Evidently, this condition is satisfied for the case where the Fermi surface extends to the boundary of the Brillouin zone (fig. 5a, b). However, in alkali metals the Fermi surface never extends to the boundaries of the Brillouin zone. Nevertheless, even in this case the condition (4.24) is satisfied. The point is that the Fermi surface of any of the alkali metals is very similar to a sphere, and since there is one valence electron per atom in such metals, the volume of the Fermi sphere is equal to half the volume of the Brillouin zone (this is shown schematically in fig. 13a). Therefore, the Fermi sphere has a radius which is, for certain, larger than $\frac{1}{4}$ of the smallest period of the reciprocal lattice. This is quite enough for the condition (4.24) to be satisfied.

In metals, such as bismuth, the Fermi surface consists of several small closed regions (valleys). But their centers are at a distance of $\frac{1}{2}\hbar\boldsymbol{K}$ (fig. 13b) and this is sufficient for the fulfillment of the condition (4.24) (on intervalley transitions). It thus follows that everything is all right for electron-electron collisions.

The situation with electron-phonon collisions is, however, different. In section 3.4 we have used the equilibrium phonon distribution function. This is permissible if there exists an independent mechanism that sets up equilibrium in the phonon gas, say, the scattering of phonons by impurities or their mutual scattering. But if the impurity concentration is low, the first of these processes will be inefficient. As regards the second, it can set an equilibrium, like the mutual scattering of electrons, only due to Umklapp processes. At low temperatures the phonon momenta are small and therefore the condition (4.24) is surely not satisfied for phonon-phonon collisions. Thus, in a pure metal at low temperatures collisions with electrons constitute the only significant mechanism of phonon relaxation. But in such a case we are not in a position to substitute the equilibrium phonon function but, instead, we have to find it from the kinetic equation.

The collision integral for phonons in electron-phonon collisions is given by

$$I = \int W[f_1(1-f_2)(1+n_k) - (1-f_1)f_2 n_k] \\ \times \delta(\varepsilon_1 - \varepsilon_2 - \hbar\omega)\,\frac{\mathrm{d}^3 p}{(2\pi\hbar)^3}.$$

The first term corresponds to the emission of a phonon and the second to its absorption. The integration goes over all electron momenta. We will propose a distribution function for phonons in the form

$$n_k = n_{k0} + n_{k0}(1+n_{k0})\,\frac{\varphi}{T}.$$

For electrons we assume the form that has already been used:

$$f = f_0 + f_0(1-f_0)\,\frac{\psi}{T}.$$

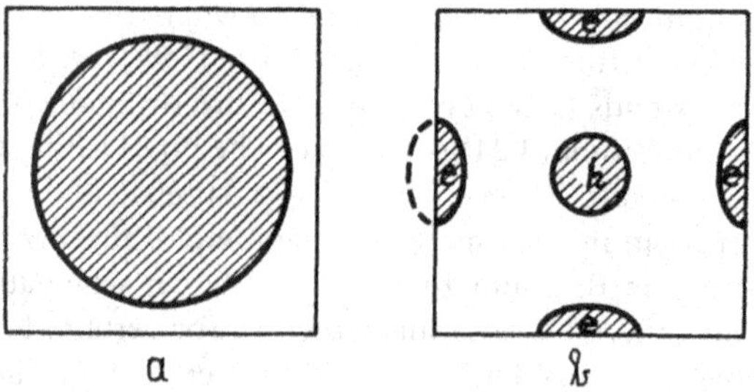

Fig. 13.

To first order in ψ and φ the collision integral takes the following form:

$$I \approx \int Wf_{10}(1-f_{20})(n_{k0}+1)(\psi_1-\psi_2-\varphi)\frac{d^3p}{(2\pi\hbar)^3}.$$

If we substitute here $\psi = cp_x$, $\varphi = c\hbar k_x$, the processes with momentum conservation will give zero. Incidentally, this also refers to the collision integral entering into the kinetic equation for electrons, and also to the phonon–phonon collision integral. Thus, this solution satisfies all kinetic equations (without Umklapps) and gives undamped electric current in the absence of an electric field, i.e., an infinite conductivity. The physical meaning of the obtained result is as follows. Although the momentum of the electron system is not conserved, the total momentum of electrons and phonons is conserved. In view of this, a joint unattenuated motion of both systems is possible, i.e., an electric current accompanied by a "phonon wind".

Thus, we have again to take into account Umklapp processes. But, as has already been noted, such processes are impossible for phonon–phonon collisions at low temperatures. For electron–phonon processes we have the following conservation law:

$$\boldsymbol{p}_1-\boldsymbol{p}_2-\hbar\boldsymbol{k}=\hbar\boldsymbol{K}.$$

At high temperatures the phonon momentum is of order p_0 and even for alkali metals Umklapp processes are possible. However, at low temperatures the momentum $\hbar\boldsymbol{k}$ is very small and may therefore be neglected. Instead of (4.24), we obtain the condition

$$2\max p_0(\boldsymbol{n}) > \hbar K_{\min}, \qquad (4.25)$$

which is not fulfilled in alkali metals (see fig. 13a) and, in general, in those cases where the Fermi surface does not extend to the face of the Brillouin zone.

In view of this, the emission and absorption of phonons with $\hbar\omega \sim T$ does not alter the total momentum of electrons. An efficient process is the interaction with phonons having momenta of order $\hbar K_{\min}$, i.e., energies of $T_0 \sim \hbar\omega_D$. But the number of such phonons is proportional to $e^{-T_0/T}$. It would appear that the exponential must be directly inserted into the expression for the resistance at $T < T_0$. However, the actual situation is more complicated*.

The change in the state of the electron that emits and absorbs phonons may be visualized as a diffusive motion of a point representing an electron of given momentum on the Fermi surface. For further reasoning it will be convenient to consider not the Brillouin zone alone but the entire reciprocal lattice. If the Fermi surface extends to the face of the Brillouin zone, then it is "open" in the entire reciprocal lattice, i.e., it extends throughout the reciprocal lattice (see fig. 15, section 5.1). The "diffusion" of the electron on such a surface also involves Umklapp processes. According to section 4.3, the corresponding mean free time is

$$\tau_e \sim [\overline{W\theta^2}]^{-1} \sim \omega_D^{-1} \left(\frac{\hbar\omega_D}{T}\right)^5.$$

But if the Fermi surface is "closed", i.e., does not extend to the face of the Brillouin zone, then when it is continued throughout the entire reciprocal lattice it becomes a set of periodically repeating closed portions.

Scattering with Unklapp implies a jump from one portion to another. The probability of such a jump is not difficult to determine. If it is assumed that the jump occurs to a distance which is still smaller than $\hbar K_{min}$, then the probability of the jump is of order

$$\tau_u^{-1} \sim \omega_D \left(\frac{T_0}{\hbar\omega_D}\right)^3 e^{-T_0/T} \frac{T}{\hbar\omega_D}.$$

Here T_0 is the energy if substantial phonons; the factor ω_D corresponds to τ^{-1} at T, $T_0 \to \hbar\omega_D$; the factor $(T_0/\hbar\omega_D)^3$ arises from the statistical weight k^3 of phonons with momenta T_0/s and, finally, the last factor corresponds to the fraction of time that an electron, while diffusing along a closed portion of the Fermi surface, spends near another portion, to which it can jump. This fraction is proportional to the ratio of the area of the dangerous portion (S_0) to the area of the entire Fermi surface (S_F), taking into account that the dangerous portion is determined by a distance not larger than $\delta p \sim T/s$, we obtain $S_0/S_F \sim T/(\hbar\omega_D)$.

Thus, the total mean free time is made up of the time during which the electron is diffusing along a closed portion and the time required for a jump with an Umklapp:

$$\tau_e' = \tau_e + \tau_u.$$

The longest of these times is the most significant. From the condition $\tau_u \sim \tau_e$ we find that $(\hbar\omega_D/T_0)x^4 \sim e^x$ or $x = 4 \ln x + \ln (\hbar\omega_D/T_0)$, where $x = T_0/T$. Assuming that $\ln(\hbar\omega_D/T_0) \sim 1$, we have $x \approx 7$. Hence, it may be stated that even for alkali metals the switchover to an exponential dependence of the resistance on temperature

* The reasonings given below have been put forward by R.N. Gurzhi. For complete calculations, see Gurzhi (1964) and Gurzhi and Kopeliovich (1973).

occurs smoothly over a relatively wide range of temperatures lower than T_0 ($T \sim 10$ K for potassium, and $T \sim 20$ K for sodium).

The picture outlined above occurs only for very pure metals. But if the dominant relaxation of the momentum is accomplished not by Umklapp processes but by other mechanisms, say, by scattering from defects or from the boundaries of the specimen, then a rather complicated situation arises associated with the interference of various scattering mechanisms. This is mainly caused by the fact that the electron-phonon scattering without Umklapps, which takes a time τ_e, while not leading to the relaxation of the total momentum of electrons, makes the electron liquid viscous. As a result, the motion of electrons reminds of the Poiseuille flow of a viscous liquid (for details, see Gurzhi 1964; Gurzhi and Kopeliovich 1973).

We have derived above the expressions for electrical and thermal conductivity, each time regarding a single scattering mechanism. In fact, all the mechanisms are operating simultaneously. We may take the simplest assumption that all scattering processes occur independently and that the corresponding probabilities add up. This means that not the conductivities (electrical and thermal) but the inverse quantities, i.e., the resistivities, are additive. Therefore, for the total electrical resistivity at low temperatures we obtain the following law (with the exception of alkali metals)*:

$$\rho = c + aT^2 + bT^5, \tag{4.26}$$

where c, a and b are constants.

At high temperatures electron-phonon scattering is dominant, and therefore

$$\rho = AT. \tag{4.27}$$

For the electronic thermal conductivity at low temperatures we obtain in an analogous way:

$$\varkappa^{-1} = \mathrm{d}T^{-1} + fT + gT^2. \tag{4.28}$$

At high temperatures $\kappa = \text{const}$.

As a matter of fact, various scattering mechanisms exert an effect on one another (interference) and this may occasionally lead to substantial deviations of the temperature dependence of the resistivity at low temperatures from the simple formula (4.26) (Kagan and Zhernov 1971). As a result of phonon scattering and of the anisotropy of the true phonon spectrum, the nonequilibrium part of the distribution function even in a cubic crystal does not have the form $\boldsymbol{Ep}\eta(\varepsilon)$ but depends on the reciprocal lattice periods. The anisotropic part η is of the same order of magnitude as the isotropic part; this brings about an increase in the numerical value of the resistivity. However, the scattering by impurities "mixes" electrons with different momenta and leads to a drastic decrease in the anisotropic part of the distribution

* In section 11.5 we will show that at the lowest temperatures the quantum correction to the conductivity, which depends on temperature according to the law $\delta\sigma \sim T^{1/2}$ (accordingly, $\delta\rho \sim -T^{1/2}$), becomes important.

function. This causes effects like a nonlinear dependence of ρ on the impurity concentration, dependence of the coefficient of T^5 on concentration at low temperatures, etc. These effects must occur (and they are actually observed) at small impurity concentrations in a temperature region where the impurity and phonon parts are of the same of order of magnitude. In very pure ($c_i < 10^{-7}$) and very impure ($c_i \sim 1$) metals these effects are not observed.

4.5. "Isotopic" scattering

If a metal is free from impurities and the temperature is lowered, we shall have an ever increasing conductivity. Experiments with tin have shown, however, that there exists a limit below which the resistivity does not fall off. It has also been found that the residual resistivity is independent of the size of the sample. It is evident that in an ideal crystal at $T=0$ the electrons can be scattered at the boundaries, but this effect must be dependent on the size of the crystal. It has been presumed in this connection that in this particular case the factor responsible for the resistivity is the isotopic composition (Pomeranchuk 1958).

What changes arise in a crystal if atoms with the same electronic structure but different masses reside at the lattice point? Since the potential energy depends on the electronic structure alone, it follows that this does not change. However, the kinetic energy of the ions will be different. The total kinetic energy of the system of ions may be separated into the average kinetic energy and the excess part depending on the difference in masses. We write this term as the sum over all the atoms:

$$\tfrac{1}{2}\sum_i (M_i - \bar{M})\dot{\boldsymbol{u}}_i^2. \tag{4.29}$$

This term will be regarded as a perturbation because the difference in mass is small. It must be pointed out that, while perturbation theory problems are usually concerned with part of the potential energy as a perturbation, in this case the role of a perturbation is played by part of the kinetic energy.

The displacement $\boldsymbol{u}$ has matrix elements with a change in the number of phonons by unity: $n_k \to n_k \pm 1$. Since expression (4.29) contains $\dot{\boldsymbol{u}}_i^2$, the number of phonons with this perturbation changes by a factor of 2 or does not change at all. The scattering of an electron may correspond, for example, to the process sketched in fig. 14. In this figure the vertices $\mathrm{V^I}$ correspond to electron-phonon interactions, and the vertex $\mathrm{V^{II}}$ corresponds to a phonon-phonon interaction. In the process presented in fig. 14, the number of phonons at the vertex $\mathrm{V^{II}}$ does not change. Other processes, involving creation or annihilation of two phonons at this vertex, could also be considered. They differ from fig. 14 by a change in the sequence of the processes and by the direction of the phonon lines and are of the same order of magnitude. Therefore, we shall consider only the process shown in fig. 14.

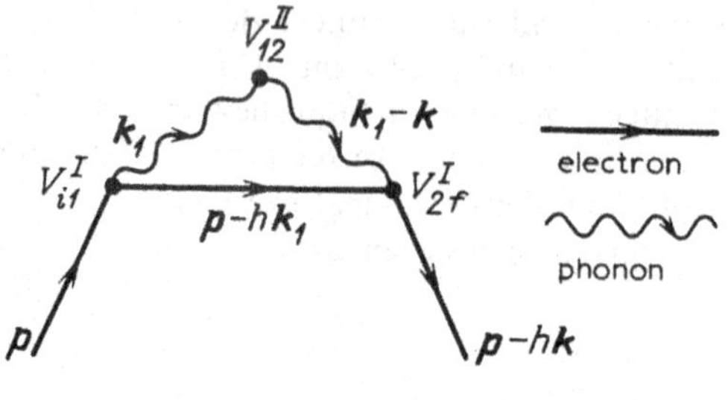

Fig. 14.

The new perturbation is given by

$$V^{-1}\sum_i \sum_{k_1} u_{k_1}\omega(\boldsymbol{k}_1)\, e^{ik_1 R_i}$$
$$\times \sum_{\boldsymbol{k}} u^+_{k_1-k}\omega(\boldsymbol{k}_1-\boldsymbol{k})\, e^{i(k-k_1)R_i}(M_i-\bar{M}) \tag{4.30}$$

(alternatively with $u^+_k \rightleftarrows u_k$). The scattering probability has the usual form (i stands for the initial state and f for the final state):

$$\frac{2\pi}{\hbar}\int |V_{fi}|^2\delta(\varepsilon_i-\varepsilon_f)\frac{d^3\boldsymbol{k}}{(2\pi)^3}. \tag{4.31}$$

The process under consideration involves three elementary interactions and, hence, the third order of perturbation theory must be taken:

$$V_{fi} = \sum_{1,2}\frac{V^{I}_{f2}V^{II}_{21}V^{I}_{1i}}{(\varepsilon_i-\varepsilon_1)(\varepsilon_i-\varepsilon_2)}. \tag{4.32}$$

Let us first consider the factor in $|V_{fi}|^2$, in which the summation goes over $\boldsymbol{R}_i$:

$$\sum_{i,k} e^{ik(\boldsymbol{R}_i-\boldsymbol{R}_k)}(M_i-\bar{M})(M_k-\bar{M}). \tag{4.33}$$

Since the distribution of isotopes is random, we may average (4.33) over their positions. Here, in view of the fact that the real values of k are of order a^{-1} (a is the interatomic distance), the terms with $\boldsymbol{R}_i \neq \boldsymbol{R}_k$ rapidly oscillate and vanish on averaging. Therefore, only the terms with $\boldsymbol{R}_i = \boldsymbol{R}_k$ are important; after averaging over the positions of the isotopes and summation over the lattice points, they give

$$N\overline{(M-\bar{M})^2} \tag{4.34}$$

where N is the number of atoms.

Expression (4.30) contains also the following factor:

$$u_{k_1}\omega(\boldsymbol{k}_1)\, u^+_{k_1-k}\omega(\boldsymbol{k}_1-\boldsymbol{k}).$$

In the process we are concerned with, the momenta of phonons will be of order p_0. Substituting the matrix elements that have been found above and recalling that at $T=0$ there are no phonons, i.e., $n_k=0$, $n_l+1=1$, we obtain

$$u_{k_1}\omega(\boldsymbol{k}_1)\,u^{+}_{k_1-k}\omega(\boldsymbol{k}_1-\boldsymbol{k})\sim\frac{\hbar\omega_{\mathrm{D}}}{Mn}. \tag{4.35}$$

We shall now find the power of the crystal volume V. It enters into the matrix elements and into the sums over the states ($\sum_k\to V\int \mathrm{d}^3\boldsymbol{k}/(2\pi)^3$). Since the matrix elements V^1 occur twice and, besides, V_{if} is squared, it follows that we have the following factor in the scattering probability:

$$NV^{-4}\frac{V\,\mathrm{d}^3\boldsymbol{k}_1}{(2\pi)^3}\,\frac{V\,\mathrm{d}^3\boldsymbol{k}_2}{(2\pi)^3}\,\frac{V\,\mathrm{d}^3\boldsymbol{k}}{(2\pi)^3}=n\frac{\mathrm{d}^3\boldsymbol{k}_1\,\mathrm{d}^3\boldsymbol{k}_2\,\mathrm{d}^3\boldsymbol{k}}{(2\pi)^3}.$$

The integration in the probability is carried out over three momenta. This is because the phonon $\boldsymbol{k}_1$ is virtual. The summation over $\boldsymbol{k}_1$ is made in the amplitude which is squared. This leads to two summations. Then, we are to carry out a summation over all possible finite momenta of the electrons.

Using (4.32), (4.34) and (4.35), we can now write the scattering probability in the form

$$\begin{aligned}W\sim\hbar^{-1}\int\, & n\overline{(M-\bar{M})^2}\left(\frac{\hbar\omega_{\mathrm{D}}}{Mn}\right)^2 p_0^4\left[\frac{\hbar\omega_{\mathrm{D}}}{mn}\right]^2\\ &\times[\varepsilon(\boldsymbol{p})-\varepsilon(\boldsymbol{p}-\hbar\boldsymbol{k}_1)-\hbar\omega(\boldsymbol{k}_1)]^{-1}\\ &\times[\varepsilon(\boldsymbol{p})-\varepsilon(\boldsymbol{p}-\hbar\boldsymbol{k}_1)-\hbar\omega(\boldsymbol{k}_1-\boldsymbol{k})]^{-1}\\ &\times[\varepsilon(\boldsymbol{p})-\varepsilon(\boldsymbol{p}-\hbar\boldsymbol{k}_2)-\hbar\omega(\boldsymbol{k}_2)]^{-1}\\ &\times[\varepsilon(\boldsymbol{p})-\varepsilon(\boldsymbol{p}-\hbar\boldsymbol{k}_2)-\hbar\omega(\boldsymbol{k}_2-\boldsymbol{k})]^{-1}\\ &\times\delta(\varepsilon(\boldsymbol{p})-\varepsilon(\boldsymbol{p}-\hbar\boldsymbol{k}))\frac{\mathrm{d}^3\boldsymbol{k}_1\,\mathrm{d}^3\boldsymbol{k}_2\,\mathrm{d}^3\boldsymbol{k}}{(2\pi)^9}.\end{aligned}$$

Let us now consider the factors in the denominator, for example:

$$\varepsilon(\boldsymbol{p})-\varepsilon(\boldsymbol{p}-\hbar\boldsymbol{k}_1)-\hbar\omega(\boldsymbol{k}_1)=\frac{\hbar\boldsymbol{p}\boldsymbol{k}_1}{m}-\frac{\hbar^2k_1^2}{2m}-\hbar\omega(\boldsymbol{k}_1).$$

The essential values of $\hbar k_1$ are of order p_0. Therefore this denominator varies within the limits from $\hbar\omega_{\mathrm{D}}$ to $\mu\sim p_0^2/m$. A low value of the denominator can be obtained if we choose the corresponding angle between $\boldsymbol{p}$ and $\boldsymbol{k}_1$. Of course, this reduces the domain of integration over the angles and gives a factor of order $\hbar\omega_{\mathrm{D}}/\mu$. But each difference $\varepsilon(\boldsymbol{p})-\varepsilon(\boldsymbol{p}-\hbar\boldsymbol{k}_1)$ occurs twice in the denominator. Therefore, confining ourselves to the small domain in the angular integrals, we nevertheless have a gain.

Thus, with this remark taken into account, we find

$$W\sim\hbar^{-1}n\overline{(M-\bar{M})^2}\left(\frac{\hbar\omega_{\mathrm{D}}}{Mn}\right)^2 p_0^4\left(\frac{\hbar\omega_{\mathrm{D}}}{mn}\right)^2\frac{p_0^9}{\hbar^9}\mu^{-1}\left(\frac{\hbar\omega_{\mathrm{D}}}{\mu}\right)^2(\hbar\omega_{\mathrm{D}})^{-4}$$

(the factor $p_0^9/\hbar^9$ arises from three $d^3\boldsymbol{k}/(2\pi)^3$, μ^{-1} from the δ-function in the energies, and $(\hbar\omega_D/\mu)^2$ from the angular integrals). The resultant expression can be transformed to the form

$$W \sim \frac{\overline{(M-\bar{M})^2}}{M^2}\left(\frac{\hbar\omega_D}{\mu}\right)^2 \frac{p_0^2}{m\hbar}.$$

Since $\hbar\omega_D \sim sp_0$, $\mu \sim vp_0$, it follows that $\hbar\omega_D/\mu \sim s/v \sim (m/M)^{1/2}$. From this we find

$$W \sim \frac{\overline{(M-\bar{M})^2}}{M^2} \frac{m}{M} \frac{\mu}{\hbar}. \tag{4.36}$$

The mean free path is of the following order of magnitude:

$$l \sim v\tau \sim \frac{v\hbar}{\mu} \frac{M^2}{\overline{(M-\bar{M})^2}} \frac{M}{m} \sim 10^{-8} \frac{M^2}{\overline{(M-\bar{M})^2}} \frac{M}{m} \text{ cm}. \tag{4.37}$$

For tin $\overline{(M-\bar{M})^2}/\bar{M}^2 \sim 10^{-3}$ and, hence,

$$l \sim 10^{-8} \times 10^3 \times 10^5 \sim 1 \text{ cm}.$$

The conductivity is of the order of

$$\sigma \sim \frac{n_e e^2 \tau}{m} \sim 10^{16} \frac{M}{m} \frac{\bar{M}^2}{\overline{(M-\bar{M})^2}} \text{ s}^{-1}. \tag{4.38}$$

The value of the mean free path for tin has been confirmed experimentally (Zernov and Sharvin 1959). The result obtained here shows that one cannot get rid of the finite residual resistivity in a metal if it has not been purified isotopically.

4.6. *The Kondo effect*

The scattering mechanisms considered above lead us to conclude that when the temperature is lowered the resistivity falls off or (at low temperatures) remains constant. However, it has been repeatedly observed in experiments that the resistivity of very pure gold, silver and copper passes in some cases through a minimum with the temperature being lowered and then increases with decreasing temperature. This phenomenon, which had remained puzzling for a long time, was first explained by Kondo (1964); it has come to be known as the Kondo effect. The cause of this effect is the presence in a metal of impurity atoms with unfilled inner shells which have a nonzero spin. Such magnetic impurities may be Mn, Fe, Cr, Co, Ce, Y and other transition and rare-earth metals. The interaction energy of an electron with such atoms contains, apart from the usual term $\sum_i v(\boldsymbol{r}-\boldsymbol{R}_i)$, a term dependent on the electron spin σ and the impurity spin $\boldsymbol{S}$. We shall write it in the form

$$V_S = -\frac{J}{n}\sum_i \sigma \boldsymbol{S}_i \delta(\boldsymbol{r}-\boldsymbol{R}_i). \tag{4.39}$$

Here $\boldsymbol{\sigma S} = \sigma^x S^x + \sigma^y S^y + \sigma^z S^z$; σ^k are the Pauli matrices. Since the spin $\boldsymbol{S}$ arises from the inner shells of the atoms, the interaction is regarded as a point interaction*.

Expression (4.39) corresponds to a physical scattering process, in which the electron spin can flip with a simultaneous change in the orientation of the impurity spin. When an electron is scattered from an ordinary atom, its spin retains its orientation. The coefficient J has the dimension of energy ($n = N/V$ is introduced for normalization). As a rule, J is several times lower than the electron energies (i.e., μ), whereas the ordinary spin-independent interaction of an electron with an impurity atom is of order μ. But since the two interactions do not interfere in the scattering probability, they may be considered separately.

The scattering amplitude of the interaction (4.39) is proportional to

$$-\frac{J}{n}(\boldsymbol{\sigma S})_{\sigma'\sigma},$$

where σ' corresponds to the final and σ to the intial orientation of the electron spin. But since it is only the electron momentum and not the electron spin that is important for the conductivity, the scattering probability must be summed over all the final orientations of the spin and averaged over the initial orientations. This yields

$$\tfrac{1}{2}\left(\frac{J}{n}\right)^2 \sum_{\sigma\sigma'} (\boldsymbol{\sigma S})_{\sigma\sigma'}(\boldsymbol{\sigma S})_{\sigma'\sigma} = \tfrac{1}{2}\left(\frac{J}{n}\right)^2 \sum_{\sigma} (\boldsymbol{\sigma S}\cdot\boldsymbol{\sigma S})_{\sigma\sigma}$$
$$= \left(\frac{J}{n}\right)^2 S(S+1).$$

Thus, the spin part of the interaction, to a first approximation, makes a contribution to the resistance of the order of J^2S^2/μ^2 with respect to the ordinary interaction with the same impurities. It would appear that this is a small effect, which may be neglected. It has been pointed out in section 3.2 that the ordinary interaction is not weak and therefore, strictly speaking, the Born approximation cannot be applied to it. However, a detailed calculation shows that in order to obtain a correct answer it is sufficient to replace the Born scattering amplitude by the true amplitude. But since both are independent of the electron energy, are of the same order of magnitude and are not easy to calculate for a real metal, it is clear that the refinement of the Born approximation does not change the picture at all for the ordinary interaction.

The situation is different for the spin part of the interaction. Although it is smaller than the spinless part and the application of the Born approximation seems to be justified in this case, there is nevertheless a significant difference. The point is that the correction to the Born approximation appears to be dependent on the energy of the electron (more exactly, on $\xi = \varepsilon - \mu$), which eventually leads to the temperature dependence of the impurity part of the resistance.

* In real transition and rare-earth metals with degenerate d- or f-levels this statement is incorrect. However, consideration of the angular dependence of scattering has a qualitative effect on the result only at the lowest temperatures, in which case perturbation theory is inapplicable (section 13.7).

We shall now consider the scattering of an electron in a metal at a given atom in the second Born approximation and, which is very important, we shall take into account the presence of other electrons described by the Fermi distribution function. Two processes are possible.

(1) The initial electron, which is in the state $\boldsymbol{p}\sigma$, goes over first to the state $\boldsymbol{p}_1\sigma_1$ and then to the final state $\boldsymbol{p}'\sigma'$. A sum is taken over the intermediate states, in which case it is required that these states be unoccupied. To provide this, a factor $1-f(\boldsymbol{p}_1)$ is introduced into the sum. Thus, for this variant we have

$$\left(\frac{J}{n}\right)^2 \sum_{\sigma_1} \frac{(\boldsymbol{\sigma S})_{\sigma'\sigma_1}(\boldsymbol{\sigma S})_{\sigma_1\sigma}(1-f(\boldsymbol{p}_1))}{\varepsilon(\boldsymbol{p})-\varepsilon(\boldsymbol{p}_1)} \frac{\mathrm{d}^3 \boldsymbol{p}_1}{(2\pi\hbar)^3}. \tag{4.40}$$

(2) One of the electrons in the occupied states, say $\boldsymbol{p}_1\sigma_1$, goes over to the state $\boldsymbol{p}'\sigma'$, and only after this does the electron in the state $\boldsymbol{p}\sigma$ go over to the free state $\boldsymbol{p}_1\sigma_1$. Here, conversely, the state $\boldsymbol{p}_1\sigma_1$ must be occupied and therefore one has to introduce a factor $f(\boldsymbol{p}_1)$. As a result, we obtain

$$-\left(\frac{J}{n}\right)^2 \sum_{\sigma_1} \int \frac{(\boldsymbol{\sigma S})_{\sigma_1\sigma}(\boldsymbol{\sigma S})_{\sigma'\sigma_1} f(\boldsymbol{p}_1)}{\varepsilon(\boldsymbol{p}_1)-\varepsilon(\boldsymbol{p}')} \frac{\mathrm{d}^3 \boldsymbol{p}_1}{(2\pi\hbar)^3}. \tag{4.41}$$

The minus sign is associated with the antisymmetry of the wave functions of electrons with respect to the permutation of the particles (see section 16.4).

Because of the elasticity of the process $\varepsilon(\boldsymbol{p}')=\varepsilon(\boldsymbol{p})$. Therefore the denominators in formulas (4.40) and (4.41) differ only in sign. The numerators of both expressions contain spin factors, which may be written as

$$(\boldsymbol{\sigma S})(\boldsymbol{\sigma S}), \qquad \sum_{i,k} \sigma^i \sigma^k S^k S^i.$$

The quantity S^k cannot be permuted with S^i because these are operators. Thus, the spin factors for the two variants are different. Using the properties of the Pauli matrices, we get

$$\begin{aligned} &(\boldsymbol{\sigma S})(\boldsymbol{\sigma S}) = S(S+1) - \boldsymbol{\sigma S}, \\ &\sum_{i,k} \sigma^i \sigma^k S^k S^i = S(S+1) + \boldsymbol{\sigma S}. \end{aligned} \tag{4.42}$$

Substituting this into eqs. (4.40) and (4.41), we have in the sum

$$\left(\frac{J}{n}\right)^2 \int \left\{ \frac{S(S+1)\delta_{\sigma'\sigma}}{\varepsilon(\boldsymbol{p})-\varepsilon(\boldsymbol{p}_1)} + \frac{2f(\boldsymbol{p}_1)-1}{\varepsilon(\boldsymbol{p})-\varepsilon(\boldsymbol{p}_1)} (\boldsymbol{\sigma S})_{\sigma'\sigma} \right\} \frac{\mathrm{d}^3 \boldsymbol{p}_1}{(2\pi\hbar)^3}.$$

Since $f(\boldsymbol{p}_1)$ depends on $\xi(\boldsymbol{p}_1)=\varepsilon(\boldsymbol{p}_1)-\mu$, we introduce the variables ξ and ξ_1 instead of $\varepsilon(\boldsymbol{p})$ and $\varepsilon(\boldsymbol{p}_1)$. Passing to the integral over the energies, we obtain

$$\left(\frac{J}{n}\right)^2 \int \left\{ \frac{S(S+1)}{\xi-\xi_1}\delta_{\sigma'\sigma} + \frac{2f(\xi_1)-1}{\xi-\xi_1} (\boldsymbol{\sigma S})_{\sigma'\sigma} \right\} \frac{\nu(\varepsilon)}{2} \mathrm{d}\xi_1. \tag{4.43}$$

The density of states $\nu(\varepsilon)$ depends only slightly on ε in the vicinity of $\varepsilon = \mu$. It may therefore be replaced by $\nu(\mu)$. The first term in the integral upon integration over ξ gives a value of order ξ/μ and since we are interested in the electrons near the Fermi boundary, this quantity may be neglected.

As for the second term, it has a factor

$$2f(\xi_1) - 1 = -\tanh\left(\frac{\xi_1}{2T}\right).$$

which is antisymmetric with respect to ξ_1. In view of this, the integral over ξ_1 may be transformed to the form

$$\int_{-\mu}^{\mu} \frac{2f(\xi_1)-1}{\xi - \xi_1}\,\mathrm{d}\xi_1 = \int_0^{\mu} [2f(\xi_1)-1]\left[\frac{1}{\xi-\xi_1} - \frac{1}{\xi+\xi_1}\right]\mathrm{d}\xi_1$$
$$= \int_0^{\mu} [2f(\xi_1)-1]\frac{2\xi_1\,\mathrm{d}\xi_1}{\xi^2 - \xi_1^2}$$

(the limits of the integral are of order $\pm\mu$ but, as we shall see immediately, their precise value is unimportant). If $\xi_1 \gg |\xi|$, then ξ^2 in the denominator of the integrand may be neglected. If we also have $\xi_1 \gg T$, then $2f(\xi_1) - 1 \approx -1$; the integral becomes logarithmic. But in a logarithmic integral it is sufficient to know the limits by order of magnitude only. A more precise determination of the limits gives a correction of the order of unity to the large logarithm. Neglecting such corrections, we may write for the last integral:

$$2\ln\left(\frac{\mu}{\max(|\xi|, T)}\right).$$

Thus, the second Born approximation to the scattering amplitude associated with spin interaction is given by

$$\left(\frac{J}{n}\right)^2 \nu(\mu)\,(\boldsymbol{\sigma}\mathbf{S})_{\sigma'\sigma} \ln\left(\frac{\mu}{\max(|\xi|, T)}\right).$$

The first approximation has the form

$$-\frac{J}{n}(\boldsymbol{\sigma}\mathbf{S})_{\sigma'\sigma}.$$

Hence, on the whole we can write for this amplitude

$$-\frac{J}{n}(\boldsymbol{\sigma}\mathbf{S})_{\sigma'\sigma}\left[1 - \frac{J}{n}\nu(\mu)\ln\left(\frac{\mu}{\max(|\xi|, T)}\right)\right]. \tag{4.44}$$

In order of magnitude

$$\frac{\nu(\mu)}{n} \sim \frac{\nu(\mu)}{n_e} \sim \mu^{-1}.$$

Thus, the correction has a relative order

$$\frac{J}{\mu}\ln\left(\frac{\mu}{\max(|\xi|, T)}\right).$$

The sign of the correction to the scattering amplitude depends on the sign of J. The scattering amplitude decreases if $J>0$ and increase if $J<0$. This result has a physical explanation. In the second order of perturbation theory the effects of the spatial correlation of the electron and impurity spins become important. If $J>0$, the interaction (4.39) has a ferromagnetic sign and tends to align the spins parallel to each other. At the same time, in a spin interaction of the type (4.39) the total electron plus impurity spin remains constant. Bit if the total spin is equal to the maximal value $S+\frac{1}{2}$, the electron cannot be scattered with a spin flip. Hence, the correlation of spins leads to a suppression of spin-flip scattering and hence to a decrease in the scattering amplitude. Conversely, at $J<0$ there is a tendency towards an antiparallel orientation of the electron and impurity spins. With such an orientation the total spin will be $S-\frac{1}{2}$ and processes with a flip of the electron spin ($S_z=S$, $s_z=-\frac{1}{2}\rightarrow S_z=S-1$, $s_z=\frac{1}{2}$) are quite possible ($\mathbf{s}=\frac{1}{2}\boldsymbol{\sigma}$ is the electron spin). This leads to an increase in the scattering amplitude at $J<0$.

We shall now turn to electrical resistance. Since electrons with $|\xi|\sim T$ are most important, we may write, with the accuracy adopted, $\ln(\mu/T)$. And since the resistance includes the square of the scattering amplitude, we may write:

$$\rho=\rho_v+\rho_J^{(0)}\left[1-2\frac{J}{n}\nu(\mu)\ln\left(\frac{\mu}{T}\right)\right] \tag{4.45}$$

where the first term originates from the ordinary (potential) interaction and the second from the spin interaction, with $\rho_J^{(0)}$ being the result in the first Born approximation. From this formula it is seen that at $J<0$ the resistance increases with decreasing temperature by a logarithmic law. This is consistent with experiment (Alekseevskii and Gaidukov 1956, 1957). The ratio of the temperature-dependent term to the main part of the residual resistance is of order

$$\left(\frac{J}{\mu}\right)^3\ln\left(\frac{\mu}{T}\right).$$

Besides this part, which increases (at $J<0$) with decreasing temperature, there is the part considered above, which falls off as the temperature is lowered. This leads to the appearance of a minimum in the total resistance. Usually the most significant decreasing term is associated with phonons and, hence, we may write on the whole:

$$\rho=\rho_v+c_{\mathrm{m}}a\ln\left(\frac{\mu}{T}\right)+bT^5, \tag{4.46}$$

where we separated the dependence on the atomic concentration of the magnetic impurity (c_{m}). Differentiating with respect to temperature, we find the position of the minimum:

$$T_{\min}=\left(\frac{c_{\mathrm{m}}a}{5b}\right)^{1/5}. \tag{4.47}$$

The temperature of the minimum is proportional to $c_{\mathrm{m}}^{1/5}$, i.e., it depends rather weakly on c_{m}.

From the low-temperature side the electrical resistance ceases to increase usually in the region where the interaction between the spins of the impurity atoms becomes significant. In a case where the concentration of magnetic impurities is small, no direct exchange interaction of the impurity spins takes place. However, due to the interaction with the spins of conduction electrons, there is an indirect exchange interaction of the impurity spins transmitted by conduction electrons. This interaction will be discussed in detail in section 21.3; its energy is expressed by formula (21.40). The oscillating dependence of this interaction on the separation between spins is characteristic. Since the impurity atoms are arranged randomly, such an interaction cannot lead to any regular ordering of the impurity spins. However, at temperatures below

$$\Theta\sim n_{\mathrm{m}}\left(\frac{J}{n}\right)^{2}\nu(\mu)\sim c_{\mathrm{m}}\frac{J^{2}}{\mu} \tag{4.48}$$

the spins are "frozen" with a random orientation, forming a so-called "spin glass"*.

Because of the orientation of the impurity spins being fixed, spin-flip scattering of electrons is no longer possible (recall that the total electron plus impurity spin is conserved). The logarithm in formula (4.45), having attained the value $\ln(\mu/\Theta)$, no longer increases. But the cessation of spin-flip scattering processes leads simultaneously to a decrease in the coefficient $\rho_J^{(0)}$ in formula (4.45), so that at $T\approx\Theta$ the $\rho(T)$ curve does not simply evolve to a plateau but has a small maximum.

The temperature Θ (4.48) at which the spins are "frozen" is proportional to the impurity concentration. Hence, by lowering the concentration it is possible to bring Θ to as small a value as desirable, even below the temperature at which the logarithmic term in formula (4.45) becomes large. This occurs at a temperature given by

$$T_{\mathrm{K}}\sim\mu\exp\left[-\frac{n}{|J|\,\nu(\mu)}\right] \tag{4.49}$$

(the Kondo temperature)**. In the case of $T\leqslant T_{\mathrm{K}}$ the approximation under which formula (4.45) was derived is no longer sufficient and one has to find a more precise

* The quantity Θ is obtained from (21.40) by substituting $r\sim n_{\mathrm{m}}^{-1/3}$, i.e., the mean spacing between the spins.

** The Kondo temperature (4.49) was deduced from formula (4.45) which was derived in the logarithmic approximation. A more precise result is: $T_{\mathrm{K}}\sim(J\mu)^{1/2}\exp[-n/(|J|\,\nu(\mu))]$ (Wilson 1974, 1975).

expression. We shall give here only some of the results. The summation of the dominant correction terms that contain the same power of $\ln(\mu/T)$ as J leads to the following result (Abrikosov 1965; Suhl 1965):

$$\rho = \rho_v + \rho_J^{(0)}\left[1 + \frac{J}{n}\nu(\mu)\ln(\mu/T)\right]^{-2}.$$

This formula will quite suffice if $J>0$. However, when $J<0$, at a certain temperature ρ turns to infinity. The reason for this increase is the tendency towards the formation of a bound complex of an impurity and electrons, so that the impurity spin is screened by the electron spins. Complete screening occurs only at $T=0$. According to quantum mechanics, the formation of a bound state always results in resonance scattering of particles of the corresponding energy (in this particular case, the resonance occurs at $\xi=0$). It is important to note that the resonance is a "collective effect", i.e., its existence is due to the entire electronic system (this follows even from the fact that the Kondo correction to the scattering amplitude depends on the electron distribution function).

As ξ, $T\to 0$ the effective interaction of an electron with an impurity becomes strong, which makes perturbation theory inapplicable. A detailed analysis shows that the spin part of the resistance, having reached a value of order ρ_v (we mean that part of ρ_v which is associated with magnetic impurities and which is proportional to c_m) no longer increases and tends towards a finite limit as $T\to 0$. Some of the properties of metals with magnetic impurities at low temperatures will be considered in section 13.7 on the basis of the theory of the Fermi liquid (Nozières 1974).

A further specific feature of the electrical resistance, which is associated with magnetic impurities, consists in that the resistance may fall off when the magnetic field is turned on.

In the absence of a magnetic impurity the action of the magnetic field on the resistance is associated with the curving of the electronic orbits. This effect will be discussed in Ch. 5. It always leads to an increase in electrical resistance, and in low fields

$$\Delta\rho_1 \sim \rho(\Omega\tau)^2, \tag{4.50}$$

where $\Omega \sim eH/(mc)$ is the Larmor frequency and τ is the collision time.

In the presence of magnetic impurities the magnetic field, apart from curving the electronic trajectories, polarizes the spins of the impurity atoms. The direction of the spin becomes fixed and, just as in the case of the formation of spin glass, the possibility of spin–flip scattering of electrons disappears. The total scattering probability decreases here and, hence, the resistivity also falls off. At $S\beta H \ll T$ (β is the Bohr magneton and $gS\beta$ is the magnetic moment of the impurity; $g\approx 2$)

$$\Delta\rho_2 \sim -\rho_J^{(0)}\left(\frac{gS\beta H}{T}\right)^2. \tag{4.51}$$

But if $S\beta H \gg T$, then

$$\Delta\rho_2 \sim -\rho_J^{(0)}.$$

The estimate of $\Delta\rho_1$ and $\Delta\rho_2$ shows that the total change in resistivity may be either positive or negative, depending on the particular characteristics of the sample. In particular, $\Delta\rho_1$ can be diminished by introducing an additional amount of nonmagnetic impurities. In such a case τ will decrease and so will $\Delta\rho_1$, where $\Delta\rho_2$ is not affected by ordinary impurities.

5

Galvanomagnetic properties of metals

5.1. The kinetic equation in the presence of a magnetic field

The dynamics of electrons in the presence of an electric and a magnetic field is determined to a considerable extent by the topology of its Fermi surface. Until now we have usually limited ourselves to the Brillouin zone. Now it will be more convenient to deal with the entire reciprocal lattice (in the same way as we did at the end of section 4.4). The energy of electrons is in this case a periodic function of the quasi-momentum. The same is true for any surface $\varepsilon(\boldsymbol{p}) = \text{const.}$, in particular for the Fermi surface. All Fermi surfaces may be divided into two groups.

(1). Closed surfaces. In the space of the reciprocal lattice we have a periodically recurring closed surface. If we are dealing with a single Brillouin zone, then this is the case where the Fermi surface does not reach the zone boundary*.

(2) Open surfaces. These are surfaces that extend throughout the entire reciprocal lattice. If a single Brillouin zone is considered, then evidently in this case the surface reaches the zone boundary. Examples for the two-dimensional case are given in fig. 15a, b. Some types of open surfaces in the three-dimensional case are shown in fig. 16a, b, c (cf. the experimental data on Au, fig. 9 and Pb, fig. 10).

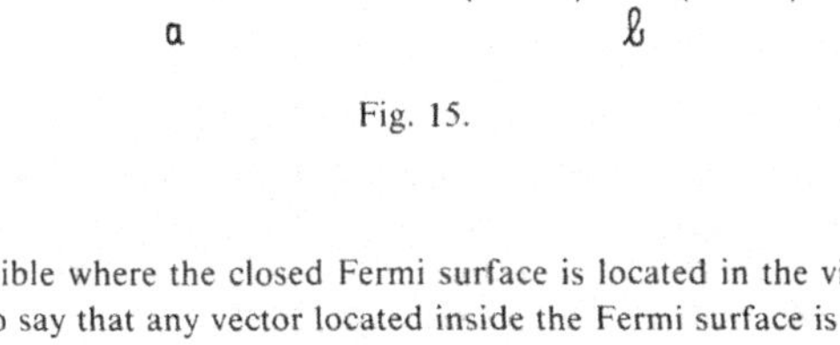

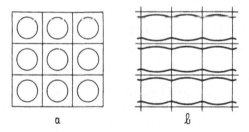

Fig. 15.

* Since cases are possible where the closed Fermi surface is located in the vicinity of the boundary, it will be more correct to say that any vector located inside the Fermi surface is smaller than the period of the reciprocal lattice.

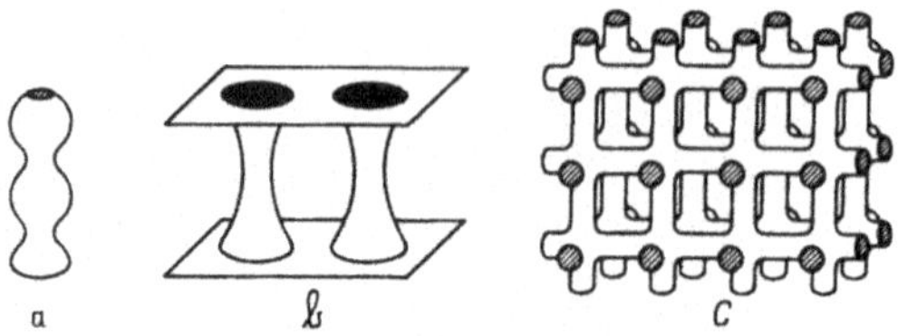

Fig. 16.

Let us consider an electron moving in a magnetic field. If we ignore quantum effects*, its equation of motion will have the form

$$\frac{\mathrm{d}\boldsymbol{p}}{\mathrm{d}t}=\frac{e}{c}[\boldsymbol{vH}]=\boldsymbol{F}. \tag{5.1}$$

The change in energy per second is given by

$$\frac{\mathrm{d}\varepsilon}{\mathrm{d}t}=\frac{\partial\varepsilon}{\partial\boldsymbol{p}}\frac{\mathrm{d}\boldsymbol{p}}{\mathrm{d}t}=\boldsymbol{vF}=0.$$

Hence,

$$\varepsilon=\text{const.} \tag{5.2}$$

From eq. (5.1) it also follows that

$$p_z=\text{const.} \tag{5.3}$$

if the magnetic field is directed along the z axis. Thus, the electron is moving in such a manner that ε and p_z are conserved.

In momentum space this is a trajectory that results from the intersection of the $\varepsilon(\boldsymbol{p})=\text{const.}$ surface with the plane $p_z=\text{const.}$ But, as is known, the most important role is played by the electrons in the vicinity of the Fermi surface, i.e., with $\varepsilon\approx\mu$. Hence, the possible trajectories of the electrons in momentum space are the intersections of the Fermi surface by the planes $p_z=\text{const.}$ If the Fermi surface is closed, then all its cross sections are closed contours, but if the Fermi surface is open, its cross sections may be either closed or open, i.e., they may extend throughout the entire reciprocal lattice.

* It will be shown in ch. 10 that the magnetic field leads to a quantization of the energy levels of the electron (Landau quantization). Therefore, in contrast to the equation $\dot{\boldsymbol{p}}=e\boldsymbol{E}$ in an electric field, eq. (5.1) is not exact. However, in view of the fact that in ordinary metals and in experimentally accessible magnetic fields $\beta H\ll\mu$ (β is the Bohr magneton), the quantization of the levels, except for the special cases considered in chapters 10–12, leads to small corrections which may be neglected to a first approximation.

We shall now write formula (5.1) in terms of its projections:

$$\frac{dp_x}{dt}=\frac{e}{c}v_yH, \qquad \frac{dp_y}{dt}=-\frac{e}{c}v_xH. \tag{5.4}$$

Squaring both equations and adding up, we obtain

$$\frac{dp_x^2+dp_y^2}{dt^2}=\left(\frac{e}{c}\right)^2H^2(v_x^2+v_y^2).$$

But $(dp_x^2+dp_y^2)^{1/2}=dl$ is a length element of the electron trajectory in momentum space. Hence,

$$\frac{dl}{dt}=\frac{e}{c}hv_\perp, \qquad v_\perp=(v_x^2+v_y^2)^{1/2}.$$

To put it another way,

$$dt=\frac{c}{eH}\frac{dl}{v_\perp}, \qquad t=\frac{c}{eH}\int\frac{dl}{v_\perp}. \tag{5.5}$$

If the electron trajectory is closed, then this integral may be taken over the entire contour, in which case we shall evidently obtain the period of motion along the contour:

$$T=\frac{c}{eH}\oint\frac{dl}{v_\perp}. \tag{5.6}$$

Let us take the plane $p_z=$ const. The area in this plane is represented by the integral

$$S=\int dp_x\,dp_y.$$

Instead of taking such an integral, we can plot the $\varepsilon=$ const. curves in the $p_z=$ const. plane and integrate along these contours and along the normal to them. The ring formed by the two contours with ε differing by $d\varepsilon$ has in this case a width given by

$$\frac{d\varepsilon}{|\partial\varepsilon/\partial \boldsymbol{p}_\perp|}=\frac{d\varepsilon}{v_\perp}.$$

The area of the ring is equal to $d\varepsilon\oint dl/v_\perp$ (the integral is taken along the contour) and the entire area in the $p_z=$ const. plane is represented by an integral:

$$S=\int d\varepsilon\oint\frac{dl}{v_\perp}.$$

Comparing with (5.6), we find

$$T=\frac{c}{eH}\frac{\partial S}{\partial\varepsilon}. \tag{5.7}$$

We now introduce the so-called cyclotron mass*:

$$m^* = \frac{1}{2\pi}\frac{\partial S}{\partial \varepsilon}. \tag{5.8}$$

The period is expressed in terms of the cyclotron mass as follows:

$$T = \frac{2\pi c m^*}{eH}. \tag{5.9}$$

The angular frequency $\Omega = 2\pi/T = eH/cm^*$ is called the Larmor or cyclotron frequency. The cyclotron mass can be determined only for closed orbits. If the orbit $\varepsilon = \text{const.}$ surrounds the lower-energy region, then m^* is positive. This is the case where the carriers are "electrons". This also includes a case with a small number of electrons in the band concentrated near its bottom. Conversely, if the contour surrounds the higher-energy region, m^* will be negative. A special case is an almost filled band. In this case, it is natural to pass over to the concept of "holes", which are carriers with a positive mass and a charge opposite to the electronic charge (section 2.3).

For free electrons

$$S = \pi p_\perp^2 = \pi(p^2 - p_z^2) = \pi(2m\varepsilon - p_z^2)$$

and, hence,

$$\frac{1}{2\pi}\frac{\partial S}{\partial \varepsilon} = m$$

is the mass of a free electron.

It is not difficult to pass from the motion of the electron in momentum space to the motion in coordinate space. Equation (5.1) may be rewritten in the form

$$\mathrm{d}\boldsymbol{p} = \frac{e}{c}[\mathrm{d}\boldsymbol{r}\boldsymbol{H}]. \tag{5.10}$$

It then follows that the projection of the electron's trajectory onto a plane perpendicular to H repeats, in fact, the trajectory in momentum space with the replacement of the coordinates $x \to -(c/eH)p_y$, $y \to (c/eH)p_x$. Moreover, the motion in the direction z with a velocity $v_z = \partial\varepsilon/\partial p_z$ is retained. If the intersection of the Fermi surface with the plane $p_z = \text{const.}$ is closed, then the trajectory in coordinate space will have the shape of a helix with the axis along H. But if the trajectory is open in momentum space, it will then be open in the (x, y) plane as well.

* The cyclotron mass on the Fermi surface for the isotropic model coincides with the "effective mass" (see relation 2.7). In order to avoid misunderstanding, recall that the concept of "effective mass" was used in this book only for the isotropic model. In the anisotropic case m^* will always denote the cyclotron mass.

In the presence of a magnetic field it is convenient to introduce, instead of p_x and p_y, two new variables: the energy ε and the "time of motion along the trajectory" given by

$$t_1 = \frac{c}{eH}\int \frac{\mathrm{d}l}{v_\perp}.$$

It should be kept in mind that in this case this is not the true time but a certain function of p_x and p_y connected with them by eqs. (5.4). According to what has just been said, we have

$$\int \mathrm{d}p_x\,\mathrm{d}p_y = \int \mathrm{d}\varepsilon \frac{\mathrm{d}l}{v_\perp}.$$

But since $(c/eH)\,\mathrm{d}l/v_\perp = \mathrm{d}t_1$, it follows that in the new variables the integrals over momentum space assume the form

$$\frac{2}{(2\pi\hbar)^3}\int \mathrm{d}p_x\,\mathrm{d}p_y\,\mathrm{d}p_z = \frac{2}{(2\pi\hbar)^3}\frac{eH}{c}\int \mathrm{d}p_z\,\mathrm{d}t_1\,\mathrm{d}\varepsilon. \tag{5.11}$$

The kinetic equation in the presence of constant electric and magnetic fields may be written as follows:

$$\frac{\partial f}{\partial t_1}\dot{t}_1 + \frac{\partial f}{\partial p_z}\dot{p}_z + \frac{\partial f}{\partial \varepsilon}\dot{\varepsilon} = I(f). \tag{5.12}$$

For $\dot{\varepsilon}$ we have

$$\dot{\varepsilon} = \frac{\partial \varepsilon}{\partial t} = \frac{\partial \varepsilon}{\partial \boldsymbol{p}}\frac{\mathrm{d}\boldsymbol{p}}{\mathrm{d}t} = \boldsymbol{v}\frac{\mathrm{d}\boldsymbol{p}}{\mathrm{d}t}.$$

In the presence of electric and magnetic fields

$$\frac{\mathrm{d}\boldsymbol{p}}{\mathrm{d}t} = \frac{e}{c}[\boldsymbol{vH}] + e\boldsymbol{E}, \tag{5.13}$$

which is why

$$\dot{\varepsilon} = e\boldsymbol{vE}, \tag{5.14}$$

$$\dot{p}_z = eE_z. \tag{5.15}$$

The variable t_1 is determined from eqs. (5.4), which differ from eq. (5.13) by the absence of an electric field. But in metals for not too weak magnetic fields $(v/c)H$ is always larger than E*. Therefore, the difference between t_1 and t is small and

* For example, the magnetic field of the Earth (H_0) is equal to about 0.5 Oe. If we assume that $E = (v/c)H_0$, then at room temperature, according to formula (4.18), $j = \sigma E \sim 10^6$–10^7 A/cm^2. Such a current density in a normal metal would lead to an immediate vaporization of the conductor.

$dt_1/dt \approx 1$. Thus, we obtain

$$\frac{\partial f}{\partial t_1} + \frac{\partial f}{\partial p_z} eE_z + \frac{\partial f}{\partial \varepsilon} e\boldsymbol{v}\boldsymbol{E} = I(f). \tag{5.16}$$

We shall assume f to be given by

$$f = f_0 - \frac{\partial f_0}{\partial \varepsilon} \psi. \tag{5.17}$$

Since ε, p_z and t_1 are independent variables we must consider f_0 to be independent of p_z. By virtue of this, substitution of eq. (5.17) into eq. (5.16) gives in a lower order in ψ:

$$\frac{\partial \psi}{\partial t_1} - I(\psi) = e\boldsymbol{v}\boldsymbol{E}. \tag{5.18}$$

This is a general equation, which will be solved in each particular case separately.

Let us now consider the boundary conditions with respect to t_1. If the electron's trajectory is closed, then evidently the function ψ must periodically depend on t_1. But if the trajectory is open, the function ψ need not necessarily be periodic, but it must be finite everywhere. These conditions provide a unique solution of eq. (5.18).

5.2. Galvanomagnetic phenomena in a weak magnetic field

We shall first consider a weak magnetic field. In this case, the scattering processes are important and we can obtain only the order of magnitude of the effect. But in order to demonstrate the solution of the kinetic equation (5.18) on a simple example, let us consider the isotropic model with $I(\psi) = -\psi/\tau$. The kinetic equation assumes the form

$$\frac{\partial \psi}{\partial t_1} + \frac{\psi}{\tau} = e\boldsymbol{v}(t_1)\boldsymbol{E}. \tag{5.19}$$

The solution of this equation is

$$\psi = \int_c^{t_1} e\boldsymbol{v}(t_2)\boldsymbol{E}\, \mathrm{e}^{-(t_1 - t_2)/\tau}\, dt_2. \tag{5.20}$$

The constant c must be determined from the boundary conditions. We shall assume the function ψ to be periodic, since in the isotropic case all the orbits are closed (the constant-energy surfaces are spheres). Here, in eq. (5.20) we must put $c = -\infty$. Indeed,

$$\psi(t_1 + T) = \int_{-\infty}^{t_1 + T} e\boldsymbol{v}(t_2)\boldsymbol{E}\, \mathrm{e}^{-(t_1 + T - t_2)/\tau}\, dt_2$$

$$=\int_{-\infty}^{t_1} e\boldsymbol{v}(t_2+T)\boldsymbol{E}\,e^{-(t_1-t_2)/\tau}\,dt_2=\psi(t_1)$$

since $\boldsymbol{v}(t_2+T)=\boldsymbol{v}(t_2)$ is a periodic function.

According to eq. (5.11), (5.17) and (5.20), the electric current is given by

$$j_\alpha=\frac{2e}{(2\pi h)^3}\int f v_\alpha\,d^3\boldsymbol{p}=-\frac{2He^2}{(2\pi\hbar)^3c}\int d\varepsilon\frac{\partial f_0}{\partial\varepsilon}\int dp_z\,dt_1\,v_\alpha\psi$$

$$=\frac{2He^3}{(2\pi\hbar)^3}\int_{-p_0}^{p_0}dp_z\int_0^T dt_1\,v_\alpha(t_1)\int_{-\infty}^{t_1}v_\beta(t_2)\,e^{-(t_1-t_2)/\tau}E_\beta\,dt_2. \quad (5.21)$$

In the last integral all the momenta lie at the Fermi surface.

Let $\boldsymbol{E}\perp\boldsymbol{H}$. The subscripts α and β run over the values of x, y. Equations (5.4) for t_1 have the solution

$$v_x=v_\perp\cos\Omega t_1,\qquad v_y=-v_\perp\sin\Omega t_1,$$

where $p_{x,y}=m^*v_{x,y}$, $\Omega=eH/m^*c$. From this we find

$$\begin{Bmatrix}j_x\\j_y\end{Bmatrix}=\frac{2He^3}{(2\pi\hbar)^3c}\int_{-p_0}^{p_0}v_\perp^2\,dp_z\int_0^T dt_1\begin{Bmatrix}\cos\Omega t_1\\-\sin\Omega t_1\end{Bmatrix}$$

$$\times\int_{-\infty}^{t_1}(E_x\cos\Omega t_2-E_y\sin\Omega t_2)\,e^{-(t_1-t_2)/\tau}\,dt_2.$$

Further,

$$\int_{-\infty}^{t_1}e^{t_2/\tau}\,e^{i\Omega t_2}\,dt_2=\frac{e^{t_1/\tau+i\Omega t_1}}{\tau^{-1}+i\Omega}=\frac{e^{t_1/\tau+i\Omega t_1}}{\tau^{-2}+\Omega^2}\times(\tau^{-1}-i\Omega)$$

$$=\frac{e^{t_1/\tau}}{\tau^{-2}+\Omega^2}[(\tau^{-1}\cos\Omega t_1+\Omega\sin\Omega t_1)$$

$$+i(\tau^{-1}\sin\Omega t_1-\Omega\cos\Omega t_1)].$$

The real and imaginary parts of this formula define the integrals over dt_2 with $\cos\Omega t_2$ and $\sin\Omega t_2$. Thus, we obtain:

$$\begin{Bmatrix}j_x\\j_y\end{Bmatrix}=\frac{2He^3}{(2\pi\hbar)^3c}\int_{-p_0}^{p_0}\frac{dp_z v_\perp^2}{\Omega^2+\tau^{-2}}\int_0^T dt_1\begin{Bmatrix}\cos\Omega t_1\\-\sin\Omega t_1\end{Bmatrix}$$

$$\times[E_x(\tau^{-1}\cos\Omega t_1+\Omega\sin\Omega t_1)-E_y(\tau^{-1}\sin\Omega t_1-\Omega\cos\Omega t_1)].$$

The integration over the total period gives

$$\int_0^T\cos^2\Omega t_1\,dt_1=\int_0^T\sin^2\Omega t_1\,dt_1=\tfrac{1}{2}T,$$

$$\int_0^T\cos\Omega t_1\sin\Omega t_1\,dt_1=0.$$

Therefore,

$$\begin{Bmatrix} j_x \\ j_y \end{Bmatrix} = \frac{2He^3}{(2\pi\hbar)^3 c}\,\frac{T}{2}\times\frac{1}{\Omega^2+\tau^{-2}}\begin{Bmatrix} \tau^{-1}E_x+\Omega E_y \\ -\Omega E_x+\tau^{-1}E_y \end{Bmatrix}\int_{-p_0}^{p_0} v_\perp^2\,\mathrm{d}p_z .$$

The last integral is easy to evaluate. Since $v_\perp = p_\perp/m^*$, we have

$$\int_{-p_0}^{p_0} v_\perp^2\,\mathrm{d}p_z = m^{*-2}\int_{-p_0}^{p_0} p_\perp^2\,\mathrm{d}p_z = m^{*-2}\int_{-p_0}^{p_0}(p_0^2-p_z^2)\,\mathrm{d}p_z = \frac{4}{3}\frac{p_0^3}{m^{*2}} .$$

Thus, we find

$$\begin{Bmatrix} j_x \\ j_y \end{Bmatrix} = \frac{2He^3}{(2\pi\hbar)^3 c}\,\frac{4}{3}\frac{p_0^3}{m^{*2}}\,\frac{2\pi c m^*}{2He}\,\frac{1}{\Omega^2+\tau^{-2}}\begin{Bmatrix} \tau^{-1}E_x+\Omega E_y \\ -\Omega E_x+\tau^{-1}E_y \end{Bmatrix}$$

or

$$\begin{Bmatrix} j_x \\ j_y \end{Bmatrix} = \frac{n_e e^2}{m^*}\,\frac{1}{\Omega^2+\tau^{-2}}\begin{Bmatrix} \tau^{-1}E_x+\Omega E_y \\ -\Omega E_x+\tau^{-1}E_y \end{Bmatrix}, \tag{5.22}$$

where $n_e = p_0^3/(3\pi^2\hbar^3)$ is the density of electrons in the isotropic model.

If the experiment is set up so that the current flows only in the x-direction, then $j_y = 0$. From (5.22) we obtain in such a case:

$$E_y = \Omega\tau E_x, \qquad j_x = \frac{n_e e^2\tau}{m^*}E_x . \tag{5.23}$$

Thus, in this model the magnetic field does not alter the conductivity. However, it leads to the appearance of an electric field in the direction y perpendicular to the current and the magnetic field. This is the so-called Hall effect. The Hall constant is defined as the ratio

$$R = \frac{E_y}{Hj_x} = \frac{eH}{m^*c}\,\frac{\tau E_x m^*}{n_e e^2\tau H E_x} = (n_e ec)^{-1} . \tag{5.24}$$

Thus, for the isotropic model the Hall coefficient depends only on the number of electrons in a metal. Real metals exhibit neither this property nor a conductivity independent of magnetic field. In a weak field the Hall constant depends on the scattering probability in a complicated way, and the resistance varies with the magnetic field. Since $\boldsymbol{H}$ is a vector, it is evident that the variation of the resistance in low fields will depend quadratically on $\boldsymbol{H}$.

The physical factor responsible for this dependence is the curving of the electron trajectories in the magnetic field (this has been discussed in section 4.6). As a result, the effective mean free path decreases. It is easy to find out that the correction to the resistance is of order*

$$\Delta\rho \sim \rho\left(\frac{l}{r_L}\right)^2 ,$$

* The difference in length between the arc and the chord is equal to $2r_L\varphi - 2r_L\sin\varphi \approx \frac{1}{3}r_L\varphi^3$, where 2φ is the angular size of the arc and r_L is the radius of the arc. If we assume that $r_L\varphi \sim l$, then the difference in length is of the order of $l(l/r_L)^2$, i.e., the relative order of the correction is $(l/r_L)^2$.

where l is the mean free path and $r_L \sim v/\Omega \sim cp_0/eH$ is the Larmour radius of electron precession in the magnetic field; $l \ll r_L$. Since $l \sim v\tau$, we also find that

$$\Delta\rho \sim \rho(\Omega\tau)^2. \tag{5.25}$$

Formula (5.25) describes the quadratic dependence of $\Delta\rho$ on H and specifies the order of magnitude of the effect in low fields. In the next section we shall see that in the opposite limiting case of high fields ($\Omega\tau \gg 1$) we can also find the dependence of ρ on H and the order of magnitude of ρ. The situation is much worse with moderate fields ($\Omega\tau \sim 1$). For an anisotropic metal, in this region it is extremely difficult to find the dependence of ρ on the field. In view of this, use is sometimes made of the so-called Kohler rule, which is the similarity law for $\Delta\rho(H)$. Although this rule can be derived, strictly speaking, only for the isotropic model, it like any similarity law describes experimental data better than the detailed formulas for the initial model. Examples of an analogous situation are the law of corresponding states and the van der Waals equation for a gas-liquid phase transition.

Kohler's rule is based on the idea of the presence of a universal mean free path, which is independent of the electron momentum. The resistance is inversely proportional to this path length. As the magntic field increases the role of the mean free path is gradually taken over by the Larmor radius. In view of this, it may be assumed that $\rho(H, T)/\rho(0, T)$ depends only on the ratio l/r_L. But since $r_L \propto H^{-1}$ and $l \propto [\rho(0, T)]^{-1}$, we may consider $\rho(H, T)/\rho(0, T)$ to be dependent only on the combination $H/\rho(0, T)$. To normalize this variable, it is multiplied by $\rho(0, 300\text{ K})$. If the metal is sufficiently pure, then $\rho(0, 300\text{ K})$ is determined by phonon scattering alone and does not depend on the impurities, i.e., $\rho(0, 300\text{ K})$ may be looked upon as a characteristic of a given metal. Subtracting unity from $\rho(H, T)/\rho(0, T)$, we get

$$\frac{\Delta\rho}{\rho(0, T)} = f\left(\frac{\rho(0, 300\text{ K})H}{\rho(0, T)}\right).$$

This is what we call Kohler's rule.

It should again be emphasized that this rule is not strictly valid for an anisotropic metal. If the anistropy plays a significant role, for example, in the presence of open trajectories, or in the case of a magnetic breakdown (section 10.7), strong deviations from Kohler's rule arise. Moreover, it will evidently break down if the dependence of the resistance on the field is due not only to the curving of the electron trajectories but to other factors as well, as in the presence of magnetic impurities (section 4.6) or when the quantum correction to the resistance becomes significant (section 11.3).

5.3. *Galvanomagnetic phenomena in a strong magnetic field. Closed trajectories*

We shall now consider another limiting case, $\Omega\tau \gg 1$, i.e., a strong magnetic field (Lifshitz et al. 1956, 1957). Here we assume an arbitrary energy spectrum. We introduce a small parameter $\gamma=(\Omega\tau)^{-1}$ and find the solution of eq. (5.18) using the method of successive approximations. A point of importance here is that the collision integral $I(\psi)$ is linear in ψ. Let

$$\psi=\sum_k \psi_k, \tag{5.26}$$

where $\psi_k \sim \gamma^k$. In the zeroth approximation from (5.18) we obtain

$$\frac{\partial \psi_0}{\partial t_1}=0. \tag{5.27}$$

In the first approximation*

$$\frac{\partial \psi_1}{\partial t_1}-I(\psi_0)=e\boldsymbol{v}\boldsymbol{E}. \tag{5.28}$$

In the second approximation

$$\frac{\partial \psi_2}{\partial t_1}-I(\psi_1)=0. \tag{5.29}$$

The subsequent approximations are analogous to the last equation.

From eq. (5.27) we obtain $\psi_0=C_0$, where C_0 is a constant. Thus, ψ_0 does not depend on t_1. From (5.28) we find

$$\psi_1=\int_0^{t_1}[I(C_0)+e\boldsymbol{v}(t_2)\boldsymbol{E}]\,\mathrm{d}t_2+C_1. \tag{5.30}$$

All the other equations yield

$$\psi_k=\int_0^{t_1} I(\psi_{k-1})\,\mathrm{d}t_2+C_k, \quad k=2,3,\ldots. \tag{5.31}$$

The first term in eqs. (5.30) and (5.31) is denoted by $\tilde{\psi}_k$. The constants C_k can be determined by averaging eqs. (5.28), (5.29), ... The finiteness of ψ gives in all cases

$$\overline{\frac{\partial \chi}{\partial t_1}}=\lim_{T_1\to\infty}\frac{1}{T1}\int_0^{T_1}\frac{\partial \psi}{\partial t_1}\,\mathrm{d}t_1=\lim_{T_1\to\infty}\frac{\psi(T_1)-\psi(0)}{T_1}=0.$$

Therefore, we obtain

$$-\overline{I(\psi_0)}=e\boldsymbol{E}\bar{\boldsymbol{v}}, \qquad \overline{I(\psi_k)}=0, \quad k=1,2,\ldots.$$

Using the notation $\psi_k=\tilde{\psi}_k+C_k$ and denoting $\overline{I(\text{const.})}$ as I_0, we find

$$-I_0(C_0)=e\boldsymbol{E}\bar{\boldsymbol{v}}, \tag{5.32}$$

$$-I_0(C_k)=\overline{I(\tilde{\psi}_k)}, \qquad k=1,2,\ldots. \tag{5.33}$$

The last two equations determine the constants C_k.

* It is convenient to add the right-hand side of eq. (5.18) to the first-approximation equation. In this case, ψ_0 is determined by the electric field (see eq. 5.32) and the subsequent approximations are expressed in terms of ψ_0.

Let us consider the case of closed orbits. Here v_x and v_y are periodic and $\bar{v}_x = \bar{v}_y = 0$. From eq. (5.32) we conclude that C_0 depends only on $\bar{v}_z E_z$. Hence, the terms proportional to E_x and E_y may be contained only in ψ_1. Finding v_x and v_y from eqs. (5.4), we obtain

$$\psi_1 = \int_0^{T_1} \frac{c}{H}\left(E_y \frac{dp_x}{dt_2} - E_x \frac{dp_y}{dt_2}\right) dt_2 + \varphi(t_1)E_z + C_1$$

$$= C_1 + \frac{c}{H}[E_y(p_x - p_x(0)) - E_x p_y] + \varphi(t_1)E_z, \tag{5.34}$$

where

$$\varphi(t_1)E_z = \int_0^{t_1} I(C_0)\, dt_2 + e\int_0^{t_1} v_z(t_2)\, dt_2 E_z$$

and the origin for measuring t_1 is chosen so that $p_y(0) = 0$.

Let us determine, for example, the current component j_x. We find in a first approximation that

$$j_x = \frac{2He^2}{(2\pi\hbar)^3 c}\int dp_z \int_0^T v_x(t_1)\,\psi_1\, dt_1$$

$$= -\frac{2e}{(2\pi\hbar)^3}\int dp_z \int_0^T \frac{dp_y}{dt_1}\psi(t_1)\, dt_1$$

$$= -\frac{2e}{(2\pi\hbar)^3}\int dp_z\left[\left[\int_0^T \frac{dp_y}{dt_1}[p_x - p_x(0)]E_y \frac{c}{H}\, dt_1\right.\right.$$

$$\left. - \int_0^T \frac{dp_y}{dt_1} p_y E_x \frac{c}{H}\, dt_1 + \int_0^T \frac{dp_y}{dt_1} C_1\, dt_1 + \int_0^T \varphi(t_1)\frac{dp_y}{dt_1} E_z\, dt_1\right]$$

(here use has been made of the equation $v_x = -(c/eH)\, dp_y/dt_1$ and of formula (5.34). The second and third of these integrals and also the part of the first integral with $p_x(0)$ are equal to zero because of the periodicity of $p_y : p_y(T) = p_y(0)$. Hence, the xy-component of the conductivity tensor is given

$$\sigma_{xy} = -\frac{2ec}{(2\pi\hbar)^3 H}\int dp_z \int_0^T p_x \frac{dp_y}{dt_1}\, dt_1$$

$$= -\frac{2ec}{(2\pi\hbar)^3 H}\int dp_z \oint p_x\, dp_y$$

$$= \frac{2ec}{(2\pi\hbar)^3 H}\int dp_z [S_e(p_z, \mu) - S_h(p_z, \mu)]. \tag{5.35}$$

Here S_e and S_h are, respectively, the areas of the cross-sections of the Fermi surface by the plane $p_z = \text{const.}$ in those cases where the region of lower energies (electrons,

S_e) or the region of higher energies (holes, S_h) is inside the contour. The sign for holes (S_h) is changed because their motion along the orbit takes place in the opposite direction to the electrons. The total sign can be easily obtained by comparing with the isotropic case ($p_x = p_\perp \cos \Omega t_1$, $p_y = -p_\perp \sin \Omega t_1$). Integration over dp_z gives

$$\sigma_{xy} = \left(\frac{ec}{H}\right)(n_e - n_h), \tag{5.36}$$

where n_e is the electron density and n_h is the hole density ($n = 2V_F/(2\pi\hbar)^3$, where V_F is the volume of the momentum space enclosed inside the corresponding portion of the Fermi surface). Here we assume that these numbers are not equal. Formula (5.36) holds also when the Fermi surface is open, but in a given direction of the field all its intersections with the planes $p_z = \text{const.}$ are closed. Here V_F means the volume bounded by the Fermi surface and by the faces of the Brollouin zone.

A further conclusion that can be drawn from the form of j_x is the absence of terms linear in γ in the component σ_{xx} of the conductivity tensor. Hence, σ_{xx} will be at least of second order in γ. The presence of such terms can be demonstrated with the aid of the equation of the next approximation. The same also refers to σ_{yy}. From the expression for j_x it follows that σ_{xz} is first order in γ. It can be shown that σ_{yz} is first order in γ and σ_{zz} is zero order. All the coefficients that are linear in γ exhibit the property of $\sigma_{ik} = -\sigma_{ki}$. This is a consequence of the Onsager symmetry principle of the kinetic coefficients: $\sigma_{ik}(\boldsymbol{H}) = \sigma_{ki}(-\boldsymbol{H})$ (ch. 6).

Thus, the complete conductivity tensor may be written in the form

$$\sigma_{ik} = \begin{pmatrix} \gamma^2 a_{xx} & \gamma a_{xy} & \gamma a_{xz} \\ \gamma a_{yx} & \gamma^2 a_{yy} & \gamma a_{yz} \\ \gamma a_{zx} & \gamma a_{zy} & a_{zz} \end{pmatrix}. \tag{5.37}$$

From this we can obtain the resistivity tensor $\rho_{ik} = \sigma_{ik}^{-1}$. Retaining the terms of lower order in γ, we have

$$\rho_{ik} = \begin{pmatrix} b_{xx} & \gamma^{-1} b_{xy} & b_{xz} \\ \gamma^{-1} b_{yx} & b_{yy} & b_{yz} \\ b_{zx} & b_{zy} & b_{zz} \end{pmatrix}, \tag{5.38}$$

where

$$b_{xx} = \frac{a_{yy}a_{zz} - a_{zy}a_{yz}}{a_{xy}a_{yx}a_{zz}}, \qquad b_{xy} = a_{yx}^{-1},$$

$$b_{yy} = \frac{a_{xx}a_{zz} - a_{xz}a_{zx}}{a_{xy}a_{yx}a_{zz}}, \qquad b_{yx} = a_{xy}^{-1},$$

$$b_{zz} = \frac{1}{a_{zz}}, \qquad b_{xz} = \frac{a_{zy}}{a_{xy}a_{zz}}, \qquad b_{yz} = \frac{-a_{zx}}{a_{yx}a_{zz}}. \tag{5.39}$$

Defining, as usual, the Hall constant

$$R = \frac{E_y}{j_x H} = \frac{\rho_{yx}}{H} = \frac{1}{\sigma_{xy} H}$$

we obtain

$$R = [(n_e - n_h)ec]^{-1}. \tag{5.40}$$

Thus, in a strong magnetic field the Hall constant is determined indeed by the numbers of electrons and holes. The components ρ_{xx} and ρ_{yy} are independent of H. This means that in high fields the resistance across the field has the tendency to saturate*.

Until now we have been concerned with the case $n_e \neq n_h$. However, as has been said above, each metal containing an even number of electrons per unit cell must have the same number of electrons and holes. In this case, the linear term in the component σ_{xy} cancels out and the latter becomes quadratic in γ. Taking this into account, we obtain

$$\sigma_{ik} = \begin{pmatrix} \gamma^2 a_{xx} & \gamma^2 a_{xy} & \gamma a_{xz} \\ \gamma^2 a_{yx} & \gamma^2 a_{yy} & \gamma a_{yz} \\ \gamma a_{zx} & \gamma a_{zy} & a_{zz} \end{pmatrix}. \tag{5.41}$$

The inverse tensor has the form

$$\rho_{ik} = \begin{pmatrix} \gamma^{-2} b_{xx} & \gamma^{-2} b_{xy} & \gamma^{-1} b_{xz} \\ \gamma^{-2} b_{yx} & \gamma^{-2} b_{yy} & \gamma^{-1} b_{yz} \\ \gamma^{-1} b_{zx} & \gamma^{-1} b_{zy} & b_{zz} \end{pmatrix}. \tag{5.42}$$

Thus, in this case the resistivity across the field increases quadratically with H.

Usually experiments on the Hall effect determine the difference in the Hall fields for different signs of $\boldsymbol{H}$:

$$R = \frac{E_y(\boldsymbol{H}) - E_y(-\boldsymbol{H})}{2Hj_x} = \frac{\rho_{yx}(\boldsymbol{H}) - \rho_{yx}(-\boldsymbol{H})}{2H}.$$

Evidently, for the case $n_e = n_h$, where $\rho_{yx} \propto H^2$, this difference is equal to zero. However, the quantity E_y itself is not constant, but increases with the field as H^2.

5.4. Galvanomagnetic phenomena in a strong magnetic field and the topology of open Fermi surfaces

We shall now turn to open Fermi surfaces (Lifshitz and Peschanskii 1958). The trajectories of an electron determined by the equations $\varepsilon = \text{const.}$, $p_z = \text{const.}$ may

* A slight linear increase in magnetoresistance at $\Omega\tau \gg 1$, which is observed in alkali metals (the Fermi surface is close to a sphere), is accounted for quite satisfactorily by the warping of the current lines in real inhomogenous samples (Sampsell and Garland 1976).

in this case be open or closed. We shall consider several examples that are most frequently encountered in metals.

(1) Suppose that the constant-energy surface is topologically equivalent to a cylinder (fig. 16a). Then, if the magnetic field is perpendicular to the axis of the cylinder, the trajectory in momentum space will be open. This case is illustrated in fig. 17a. However, if the magnetic field is not perpendicular to the axis of the cylinder, the orbits will be closed. Therefore, in the case of a cylindrical surface open trajectories occur when the direction of the magnetic field lies in a plane perpendicular to the axis of the cylinder.

(2) Let us consider a case where the Fermi surface is something similar to a lattice composed of tubes (fig. 16a*). If the magnetic field is directed along the normal to one of the planes, only closed trajectories will be obtained (fig. 17b). If now the magnetic field is tilted, then within a limited range of angles near such a normal trajectories will be formed that are open in one direction. For instance, an increase in the angle θ in fig. 18 will lead to a continuous change of the trajectories

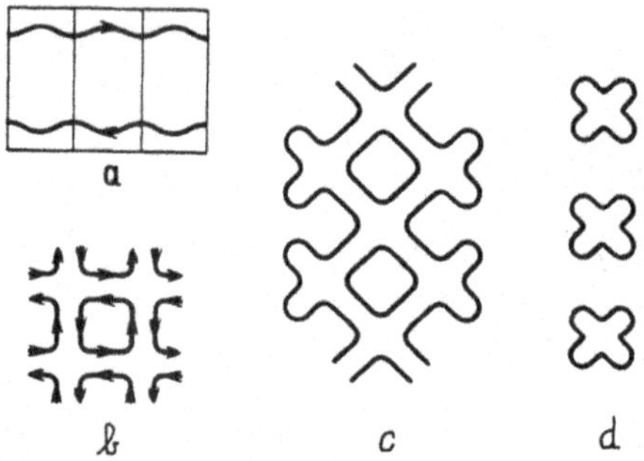

Fig. 17.

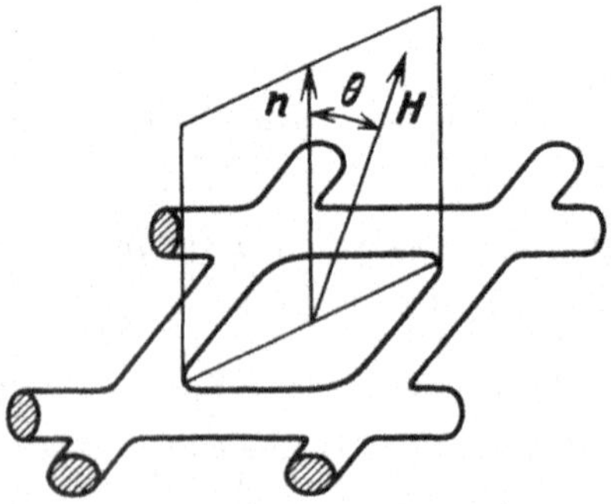

Fig. 18.

* The case shown in fig. 16b occurs more seldom.

from the type shown in fig. 17c to the one depicted in fig. 17d. However, if the magnetic field rotates and lies in one of the basal planes all the time, it will be always perpendicular to the axis of one of the tubes, and this implies the presence of open trajectories.

The topological features indicated can be depicted with the aid of a stereographic projection (fig. 19). The center of the circle corresponds to the normal in fig. 18 and the boundary of the circle corresponds to $\theta = 90°$. The hatched portions are regions where open trajectories are encountered. The solid lines from the center to the boundary and the circle $\theta = 90°$ correspond to open trajectories that arise when the field is tilted in one of the principal planes*. The dots in the center and on the circle correspond to the directions where there are only closed trajectories; the dashed lines delineate the regions of trajectories which are closed but extend throughout many periods in the reciprocal lattice.

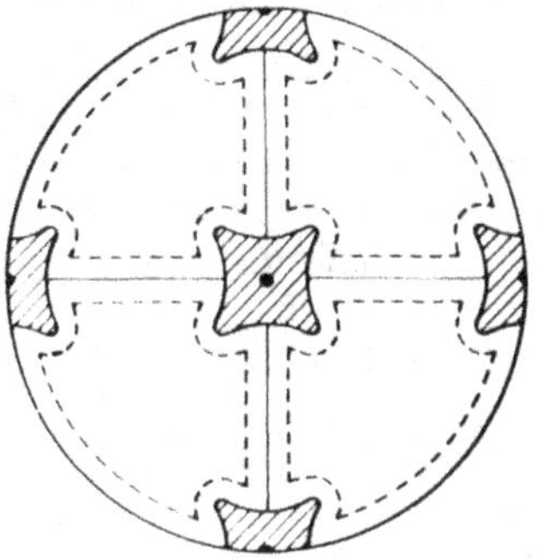

Fig. 19.

Let us consider the simplest variant – a corrugated cylinder in a case where the magnetic field is perpendicular to the axis of the cylinder (fig. 20). The location of the axes is indicated in the figure (the axis of the cylinder is along p_x). Let us consider the equation of motion (5.4). From fig. 20 it follows that in the direction p_x the electron can move to infinity, but in the direction p_y the motion is bounded. Averaging eqs. (5.4), we conclude that $\bar{v}_x = 0$ and $\bar{v}_y \neq 0$. The latter is seen from the fact that

$$\bar{v}_y = \frac{c}{eH} \lim_{T_1 \to \infty} \frac{p_x(T_1) - p_x(0)}{T_1}$$

* In fact, in metals that have an orthorhombic, tetragonal or cubic lattice there are additional directions of openess lying in the planes that make rational angles with the basal planes. This leads to the appearance of additional segments (fig. 19) extending from the centers along the radii to a finite length (Alexeyevskii et al. 1960). We shall not dwell on this problem because our purpose is to classify the specific features of the components ρ_{ik} rather than to study particular metals.

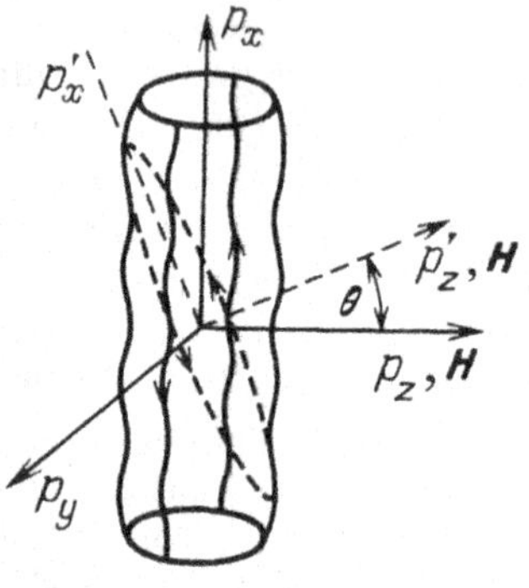

Fig. 20.

and the electron can move to infinity in the direction p_x. Thus, in contrast to the case of closed trajectories, where $\bar{v}_x = v_y = 0$, here only $\bar{v}_x = 0$. Therefore, from (5.32) we may conclude that $\psi_0 = C_0$ depends not only on E_z but on E_y as well. Hence the component σ_{yy} no longer has a small factor, just as the σ_{zz} component. Of course, the off-diagonal components $\sigma_{yz} = \sigma_{zy}$ will not be small either in this case. The remaining components of σ_{ik} have the same order of magnitude as in the case of closed orbits.

Thus, the conductivity tensor has the form

$$\sigma_{ik} = \begin{pmatrix} \gamma^2 a_{xx} & \gamma a_{xy} & \gamma a_{xz} \\ \gamma^2 a_{yx} & a_{yy} & a_{yz} \\ \gamma a_{zx} & a_{zy} & a_{zz} \end{pmatrix}. \tag{5.43}$$

From this we find the resistivity tensor:

$$\rho_{ik} = \begin{pmatrix} \gamma^{-2} b_{xx} & \gamma^{-1} b_{xy} & \gamma^{-1} b_{xz} \\ \gamma^{-1} b_{yx} & b_{yy} & b_{yz} \\ \gamma^{-1} b_{zx} & b_{zy} & b_{zz} \end{pmatrix}. \tag{5.44}$$

Thus, in this case $\rho_{xx} \propto H^2$. Comparing ρ_{xy} in eq. (5.44) with the expression (5.38) for the case of closed orbits, we notice that this component depends linearly on H in both cases, but it can be shown that the corresponding constants b_{xy} do not coincide with one another.

It is now of interest to find out how the dependence of the galvanomagnetic coefficients on the magnetic field is changed if we pass over from closed to open trajectories by gradually changing the direction of the magnetic field. To do this, we consider a case where the magnetic field is tilted at a small angle to the direction perpendicular to the axis of the cylinder (fig. 20, dashed axes). The length of the electron trajectory evidently increases with decreasing angle θ in proportion to θ^{-1}. Hence, the period of motion along that trajectory increases proportionately to θ^{-1}

in accordance with formula (5.6) ($v_\perp$ is always of the order of v) and, hence, the frequency of rotation along such an orbit is

$$\Omega \sim \Omega_0 \theta, \tag{5.45}$$

where $\Omega_0 \sim eH/mc$ is the ordinary cyclotron frequency. With a sufficiently small angle θ the frequency Ω becomes so small that the condition $\Omega\tau \gg 1$ breaks down. Evidently, it is in this case that the crossover from σ_{ik} of type (5.37) to σ_{ik} of type (5.43) occurs. Thus, the corresponding angle is $\theta \sim (\Omega_0\tau)^{-1} \sim \gamma \ll 1$.

The tensor $\rho_{ik} = \sigma_{ik}^{-1}$ at these angles varies smoothly from the form (5.38) to the form (5.44). In particular, the component ρ_{xx} is proportional to H^2 in a narrow angle range $\theta \sim \gamma \sim (\Omega_0\tau)^{-1}$, and is constant at all other angles. Hence, the dependence of the resistivity on the angle has the form shown in fig. 21a.

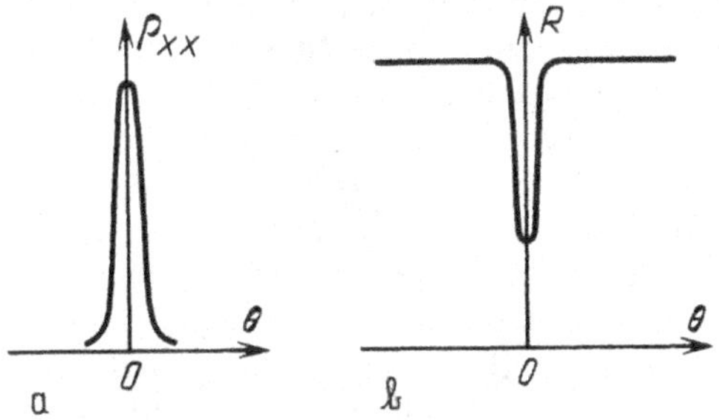

Fig. 21.

From expression (5.44) we can also draw a conclusion concerning the Hall coefficient. The component ρ_{xy} is linear in H for both large and small angles, but the proportionality factors are different in the two cases. The crossover between them occurs in a narrow angle range $\theta \sim \gamma$. A more detailed analysis shows that the Hall coefficient must be lower in the region of small angles (fig. 21b). If the field increases, the width of the minimum decreases, but its depth, as well as the height of its wings, is not changed. Note also that for a Fermi surface of the type shown in fig. 16a the Hall coefficient in strong fields is independent of the angles (except for very small angles θ) and is equal to $[(n_e - n_h)ec]^{-1}$, where n_e and n_h are determined by the volumes in momentum space bounded by the Fermi surface and by the faces of the Brillouin zone (fig. 22). This has been discussed in section 5.3.

If the topology of the Fermi surface corresponds to fig. 16c, the angular dependence of ρ_{xx} in strong fields may have singular points of several different types. We shall consider the stereographic projection in fig. 19. If we go from one of the four sectors to another, crossing the diameter, this will give a singularity of the type shown in fig. 21a, since these diameters bounded by dashed lines are the directions perpendicular to the cylindrical portions of the Fermi surface.

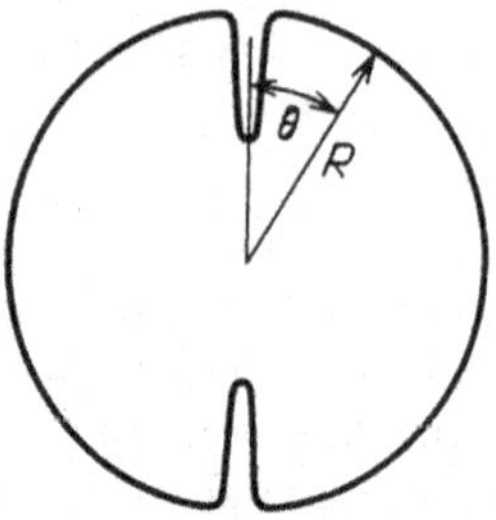

Fig. 22.

Another singularity can be obtained by crossing the boundary of a hatched region. The hatched regions are the field directions for which the trajectory is open in one direction. Hence, the resistance in this region must increase as H^2, while in the unhatched region it remains constant. This type of singularity is shown in fig. 23a.

There are also special directions of the field marked by dots on the stereographic projection in fig. 19. In these directions there are only closed trajectories and therefore ρ_{xx} remains constant. But these dots are surrounded by hatched regions where $\rho_{xx} \sim H^2$, i.e., ρ_{xx} is large. This type of singularity is sketched in fig. 23b.

Thus, we see that the study of the resistance of a monocrystalline sample in a strong magnetic field enables one to determine the topology of the Fermi surface, depending on the direction of the field and its magnitude. Of course, this method provides no information about the closed portions of the Fermi surface. It can, however, be performed relatively simply in practice and has been actually applied for the study of the Fermi surface in many metals. It should be noted that in most cases the Fermi surface has open portions.

Figure 24 shows, as an example, the angular diagram for ρ_{xx}, the result obtained for gold (Justi 1940).

A limitation of this method is the condition $\Omega\tau \gg 1$. This condition may be rewritten in the following form: $vT \ll v\tau$ or $r_L \ll l$, where r_L is the Larmor radius and l is the mean free path. One way to satisfy this condition is to increase the field (r_L will decrease in this case) and another is to purify the metal and to lower the temperature

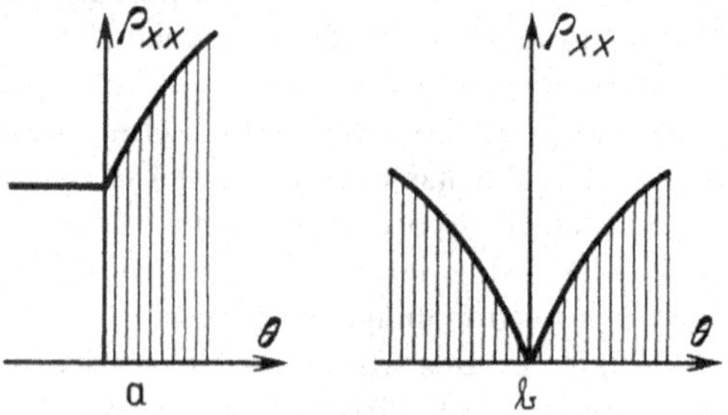

Fig. 23.

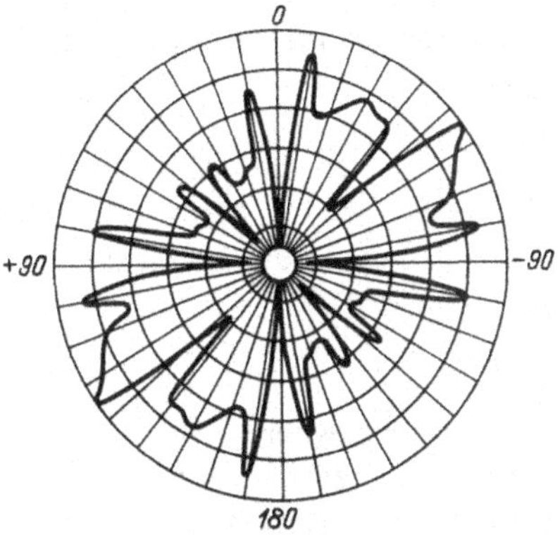

Fig. 24.

(l will increase). The Larmor radius for fields of the order of 10^4 Oe is of order $r_L \sim cp_0/eH \sim 10^{-3}$ cm. In the region of the residual resistance

$$l \sim (n_i Q_{eff})^{-1} \sim (10^{22} c_i 10^{-16})^{-1} \sim \frac{10^{-6}}{c_i}.$$

Hence, the condition can be satisfied if the atomic concentration of the impurities is less than 10^{-3}; this can be easily realized in practice.

5.5. *The magnetoresistance of polycrystals*

We have so far been concerned with the properties of single crystals. If the sample is a polycrystal, it is necessary to average over crystallite orientations. In the presence of open trajectories the high anisotropy of the conductivity tensor in a strong magnetic field leads to very peculiar dependences $\rho(H)$ (Dreizin and Dykhne 1972). We shall outline the theory only for a single case - a slightly corrugated cylinder. The reader interested in other possible cases (a strongly corrugated cylinder, a space grid) is referred to the original paper.

In section 3.5 we have found the connection between the diffusion coefficient and the conductivity (3.28). In the case under consideration the diffusion coefficient is simpler to find. Suppose that the electron density varies in space and with time. These variations are interrelated by a continuity equation expressing a conservation law for the number of electrons:

$$\frac{\partial n_e}{\partial t} + \operatorname{div} j_D = 0. \tag{5.46}$$

Substituting here the expression for the diffusion flux (3.26), we obtain the well-known diffusion equation:

$$\frac{\partial n_e}{\partial t} = D\Delta n_e.$$

According to this equation, the mean distance traveled by an electron per unit time is given by the formula

$$\overline{r^2(t)} \sim Dt. \tag{5.47}$$

Of course, this holds only for $t \gg \tau$, where τ is the collision time. Formula (5.47) will be used here to find the coefficient D.

Let an electron be moving for some period of time t. The square of its displacement in a direction perpendicular to the magnetic field in crystallites of a certain orientation is $D(\theta)\,\mathrm{d}t$; here θ characterizes the orientation (fig. 20) and $\mathrm{d}t$ characterizes the time which the electron spends in crystallites with orientations in the interval from θ to $\theta + \mathrm{d}\theta$. This time is the product of t and the probability of encountering such a crystallite. With a sufficiently uniform distribution of crystallite orientations the probability is proportional to the solid angle, i.e., $2\pi \sin(\frac{1}{2}\pi - \theta)\,\mathrm{d}\theta \sim \mathrm{d}\theta$. Thus, $\mathrm{d}t \sim t\,\mathrm{d}\theta$. For the transverse conductivity in "normal" crystallites with $\theta \gg \gamma \sim (\Omega_0\tau)^{-1}$ we have the formula $\sigma \sim \sigma_0(\Omega\tau)^{-2}$ (section 5.4). At small angles $1 \gg \theta \gg \gamma$, $\Omega \sim \Omega_0\theta$ (see eq. 5.45). Substituting this into σ, we obtain $\sigma \sim \sigma_0\gamma^2/\theta^2$. The same also refers to D_0, i.e.,

$$D(\theta) \sim \frac{D_0\gamma^2}{\theta^2}. \tag{5.48}$$

Since this is valid up to angles $\theta \sim \gamma$, the total displacement is equal to

$$\overline{r^2(t)} \sim t\int_\gamma^{\pi/2} D_0\frac{\gamma^2}{\theta^2}\,\mathrm{d}\theta \sim tD_0\gamma. \tag{5.49}$$

However, the derivation made here is not quite accurate. It does not take into account the fact that for crystallites with very small θ the diffusion coefficient is very high, i.e., the diffusion in the transverse direction occurs rapidly. It may turn out that the displacement of electrons in such crystallites is limited not by the time of observation but simply by their boundaries. We denote the corresponding critical angle by θ_0 and assume that $\theta_0 \gg \gamma$ (this will be proved below). Then,

$$\overline{r^2(t)} = \overline{(r^2)}_1 + \overline{(r^2)}_2,$$

where $\overline{(r^2)}_1$ is the total displacement in crystallites with $\theta > \theta_0$ and $\overline{(r^2)}_2$ is the displacement in crystallites with $\theta < \theta_0$. For the former we may use the integral in (5.49), taking θ_0 rather than γ as the lower limit. Here we have

$$D_1 \sim \frac{(r^2)_1}{t} \sim D_0\frac{\gamma^2}{\theta_0}.$$

The contribution to D coming from crystallites with $\theta < \theta_0$ must be treated in a different manner. The movement of an electron may be visualized as follows. Since the transverse diffusion in nonsingular crystallites is small, the electron is diffusing mainly along the magnetic field. This occurs until it encounters a singular crystallite (with $\theta < \theta_0$). The electron is displaced in the transverse direction to a length corresponding to the crystallite size (d), whereupon the situation repeats itself. This process has a probabilistic nature and therefore the motion along the field may be treated as a "mean free path" and the encounter with a singular crystallite as "scattering". The "mean free path" has the order

$$\mathscr{L} \sim (nd^2)^{-1}$$

where n is the number of singular crystallites per unit volume and d^2 plays the role of the "effective cross section". The quantity nd^3 is simply the volume fraction occupied by crystallites, in other words, it is the orientation probability $\theta < \theta_0$. But this probability is of order θ_0 and, hence, $nd^3 \sim \theta_0$ and

$$\mathscr{L} \sim \frac{d}{\theta_0}.$$

Since the diffusion coefficient along the magnetic field is of order D_0 (see eq. 5.43), the flight time for distance $\mathscr{L}$ is given by

$$\tau_1 \sim \frac{\mathscr{L}^2}{D_0} \sim \left(\frac{d}{\theta_0}\right)^2 \frac{1}{D_0}.$$

During this time the displacement in the transverse direction corresponds to the crystallite size, i.e., d. Hence, for singular crystallites the transverse diffusion coefficient is

$$D_2 \sim \frac{d^2}{\tau_1} \sim D_0\theta_0^2.$$

Thus, the total diffusion coefficient is given by

$$D \sim D_1 + D_2 \sim D_0\left(\theta_0^2 + \frac{\gamma^2}{\theta_0}\right). \tag{5.50}$$

It now remains to define θ_0. Since this is the angle at which the displacement in a single crystallite becomes of the order of d, the crossover from the $\theta > \theta_0$ regime to the regime at $\theta < \theta_0$ must occur in a continuous way. Hence, θ_0 is determined from the condition that the two terms in (5.50) be of the same order, i.e.,

$$\theta_0 \sim \gamma^{2/3}. \tag{5.51}$$

This estimate can also be obtained in a different way. During the time τ_1 the electron is moving along the entire length $\mathscr{L}$ and spends, on average, an equal time at any portion. Hence, in a singular crystallite it spends the time $\tau_1 d/\mathscr{L}$. During this time its motion in the transverse direction will be of order

$$\left(\frac{D(\theta_0)\tau_1 d}{\mathscr{L}}\right)^{1/2} \sim \left(\frac{D_0\gamma^2}{\theta_0^2}\frac{\mathscr{L}^2}{D_0}\frac{d}{\mathscr{L}}\right)^{1/2}$$
$$\sim \left(\frac{\gamma^2 \mathscr{L} d}{\theta_0^2}\right)^{1/2} \sim d\gamma\theta_0^{-3/2}$$

[here $D(\theta)$ is expressed by formula (5.48), $\mathscr{L} \sim d/\theta_0$]. The angle θ_0 is deduced from the condition that this displacement be of order d, from which we again arrive at formula (5.51).

If we substitute this formula into (5.50), we obtain $D \sim D_0\gamma^{4/3}$ or

$$\sigma_\perp \sim \sigma_0\gamma^{4/3}. \tag{5.52}$$

As before, we have to take into account the fact that $\rho_\perp$ rather than $\sigma_\perp$ is determined in experiments. Taking the component of the inverse tensor, we obtain

$$\rho_\perp \approx \frac{\sigma_\perp}{\sigma_\perp^2 - \sigma_{xy}\sigma_{yx}}.$$

If $n_e \neq n_h$, then $\sigma_{xy} = -\sigma_{yx} = (n_e - n_h)ec/H$. Here $\sigma_{xy} \gg \sigma_\perp$ and we have

$$\rho_\perp \approx \sigma_\perp/\sigma_{xy}^2 \sim \rho_0\gamma^{-2/3}. \tag{5.53}$$

But if $n_e = n_h$, then $\sigma_{xy} = \sigma_{yx} \sim \sigma_0\gamma^2 \ll \sigma_\perp$. In this case,

$$\rho_\perp \approx \sigma_\perp^{-1} \sim \rho_0\gamma^{-4/3}. \tag{5.54}$$

Thus, it turns out that in high fields the transverse resistance of polycrystalline samples of metals with a Fermi surface of the type of a corrugated cylinder must increase like $H^{2/3}$ at $n_e \neq n_h$ and like $H^{4/3}$ at $n_e = n_h$. It is interesting to note that neglect of the different diffusion regime at $\theta < \theta_0$ leads to formula (5.49), i.e., to $\sigma_\perp \sim \sigma_0\gamma$. Here, $\sigma_{xy} \leqslant \gamma$ and in both cases, $n_e \neq n_h$ and $n_e = n_h$, we obtain $\rho_\perp \sim \rho_0\gamma^{-1}$, i.e., $\rho_\perp \propto H$. This law can be distinguished from the true laws only in very high fields, such that $(\Omega\tau)^{1/3} \gg 1$.

A linear or nearly linear increase in the resistance with the field has been observed in experiments. This result was first obtained directly in a polycrystalline sample (Kapitza 1929). The same dependence was obtained by averaging the angular resistance diagram found for a single crystal (Alekseevskii and Gaidukov 1958). The latter method, however, does not agree with what happens in polycrystals, because it ignores the above-described singularities of the motion of electrons in the case of the high conductivity anisotropy. For the case of the corrugated cylinder considered this procedure necessarily gives a linear dependence $\rho_\perp(H)$. Indeed, the main contribution to $\bar{\rho}_\perp$ comes from the angles $\theta \leqslant \gamma$, where $\rho_{xx} \sim \gamma^{-2}$. Hence,

$$\bar{\rho}_\perp \sim \int \rho_{xx}(\theta)\,\mathrm{d}\theta \sim \frac{\rho_0}{\gamma^2}\gamma \sim \rho_0/\gamma.$$

In conclusion, it should be noted that in the derivation carried out above it was assumed that each crystallite has a tensor σ_{ik} characteristic of a single crystal with an appropriate orientation. This is possible only if the crystallite size is much larger than the Larmor radius. This condition is simultaneously the minimum requirement for the fulfillment of the inequality $\Omega\tau \gg 1$ because the motion of an electron is sharply altered in going from one crystallite to another, and therefore the mean free path cannot be smaller than the crystallite size.

6

Thermoelectric and thermomagnetic phenomena

6.1. Thermoelectric phenomena

In chapter 3 we have considered the conductivity, i.e., the electric current that arises in a metal under the action of an electric field, and thermal conductivity, i.e., the energy flux produced under the influence of a temperature gradient.

We shall now be concerned with phenomena that take place in the simultaneous presence of an electric current and a temperature gradient. For the sake of simplicity, we begin with the isotropic case. If a small temperature gradient is produced in a metal and simultaneously an electric field is applied to it, an electric current and a heat flux will arise:

$$\boldsymbol{j} = \sigma \boldsymbol{E} + \beta \nabla T, \tag{6.1}$$

$$\boldsymbol{q} = \gamma \boldsymbol{E} + \zeta \nabla T. \tag{6.2}$$

There exists a relationship between the coefficients β and γ, which follows from Onsager's principle of symmetry of the kinetic coefficients. This principle (for the derivation, see, for example, Landau and Lifshitz 1976) consists in the following. Suppose that certain generalized forces $X_1, \ldots, X_n$ are applied to the system. Under the influence of these forces generalized currents $J_1, \ldots, J_n$ arise in the system, which are connected with the forces by the following relation:

$$J_i = -\sum_k \gamma_{ik} X_k. \tag{6.3}$$

Suppose also that the forces and currents are determined so that the variation of the entropy of the system with time could be written in the form

$$\dot{S} = -\sum_i J_i X_i. \tag{6.4}$$

In this case, according to Onsager's principle, the matrix of the coefficients γ_{ik} must be symmetric, i.e.,

$$\gamma_{ik} = \gamma_{ki}. \tag{6.5}$$

In the case under consideration, the variation of the entropy with time results from two factors. First, due to the heat flux an amount of heat equal to div $\boldsymbol{q}$ is released from the unit volume per unit time. Second, due to the work done by the electric field, an amount of energy equal to $\boldsymbol{jE}$ is dissipated in each unit volume per unit time. Hence, we obtain

$$\dot{S} = -\int \frac{\operatorname{div} \boldsymbol{q}}{T} \,\mathrm{d}V + \int \frac{\boldsymbol{jE}}{T} \,\mathrm{d}V.$$

Taking the first integral by parts, we have

$$\dot{S} = \int \boldsymbol{q} \nabla(T^{-1}) \,\mathrm{d}V + \int \frac{\boldsymbol{jE}}{T} \,\mathrm{d}V. \tag{6.6}$$

If we assume $\boldsymbol{j}$ and $\boldsymbol{q}$ to be generalized currents, then, according to (6.4), the role of generalized forces will be played by $-\boldsymbol{E}/T$ and $-\nabla(T^{-1}) = T^{-2}\nabla T$. Writing relations (6.1) and (6.2) in the form

$$\boldsymbol{j} = -\sigma T \frac{-\boldsymbol{E}}{T} + \beta T^2 \frac{\nabla T}{T^2}, \qquad \boldsymbol{q} = -\gamma T \frac{-\boldsymbol{E}}{T} + \zeta T^2 \frac{\nabla T}{T^2}$$

we obtain from (6.5)

$$\gamma = -\beta T. \tag{6.7}$$

The electric current is always easier to monitor in experiments than the electric field inside the metal. In view of this, we solve the system (6.1) and (6.2) for $\boldsymbol{E}$. This gives

$$\boldsymbol{E} = \rho \boldsymbol{j} + Q \nabla T, \tag{6.8}$$

$$\boldsymbol{q} = \Pi \boldsymbol{j} - \varkappa \nabla T, \tag{6.9}$$

where

$$\rho = \sigma^{-1}, \qquad Q = -\frac{\beta}{\sigma}, \qquad \Pi = \frac{\gamma}{\sigma}, \qquad \varkappa = \gamma \frac{\beta}{\sigma} - \zeta.$$

By virtue of relations (6.7) we have

$$\Pi = QT, \qquad \varkappa = -\left(T \frac{\beta^2}{\sigma} + \zeta \right). \tag{6.10}$$

For metals the first term in parentheses is always much smaller than the second (see formula 6.18 below). It is for this reason that in section 3.4 we calculated the coefficient $-\zeta$ instead of $\varkappa$.

From eq. (6.8) it follows that in a nonuniformly heated conductor in the absence of current there arises an electric field

$$\boldsymbol{E} = Q \nabla T.$$

If we write $\boldsymbol{E} = -\nabla \varphi$, then it follows that

$$\frac{-\mathrm{d}\varphi}{\mathrm{d}t} = Q. \tag{6.11}$$

This is the thermal emf per unit increment in temperature. It is called the thermopower.

We shall now consider the thermocouple. Schematically, this is a loop of two dissimilar metals welded together. Here, naturally, the question arises of the boundary conditions at the junction of two metals. Since the Fermi boundaries in two different metals differ, it follows that when an electrical contact is set up the electrons will start to flow from one metal to the other. Therefore, in the boundary region an electrical double layer is produced. The thickness of this layer is of the order of interatomic distances, i.e., it is vanishingly small on the scales we are dealing with. In this layer a jump of the electrical potential occurs, which equilibrates the Fermi boundaries of the two metals being in contact with each other. This jump is called the contact potential.

Let us find the electromotive force of a thermocouple made of metals a and b in the absence of current, i.e., when the loop is disrupted at some point, say, in metal a (fig. 25). It is assumed that the temperature of the terminals formed is the same (T_0). The temperatures of the junctions are assumed to be equal to T_1 and T_2. From eq. (6.8) at $j = 0$ we have

$$V_{\text{therm}} = \oint Q \frac{\mathrm{d}T}{\mathrm{d}l} \mathrm{d}l$$
$$= \int_{T_0}^{T_1} Q_\mathrm{a} \,\mathrm{d}T + \int_{T_1}^{T_2} Q_\mathrm{b} \,\mathrm{d}T + \int_{T_2}^{T_0} Q_\mathrm{a} \,\mathrm{d}T = \int_{T_1}^{T_2} (Q_\mathrm{b} - Q_\mathrm{a}) \,\mathrm{d}T. \tag{6.12}$$

Thus, the thermal emf is expressed in terms of the difference in thermopower between the two metals.

Let us consider the so-called Peltier effect. Suppose that an electric current passes through the junction of two dissimilar metals (fig. 26). We assume that the temperature along the entire combined conductor is constant. According to (6.9), the

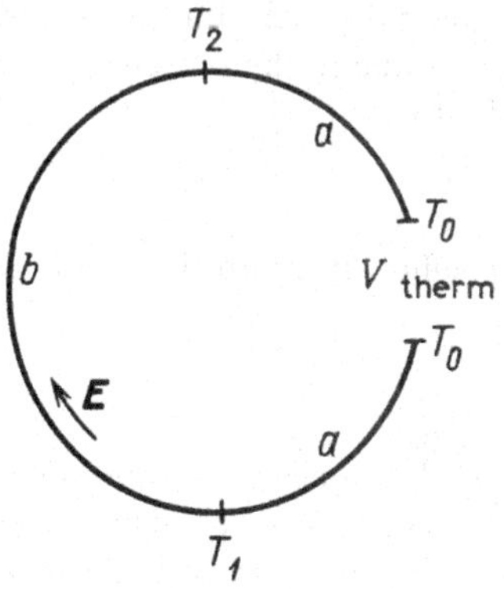

Fig. 25.

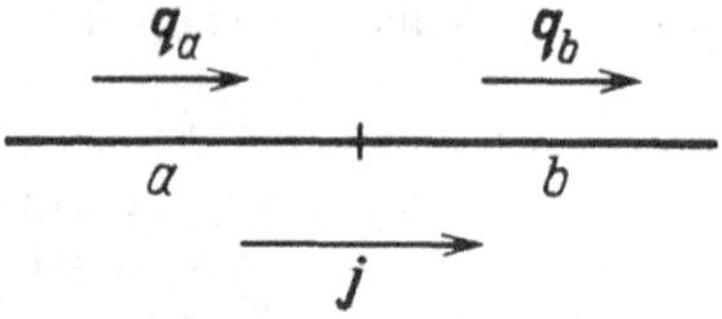

Fig. 26.

presence of a current leads to the generation of a heat flux. But, because of the constant Π being different in dissimilar metals, the heat flux in the two metals will be different. In view of this, excess heat will be released or absorbed at the junction. The quantity of this heat will be equal to

$$W_{ab} = q_a - q_b = (\Pi_a - \Pi_b)j. \tag{6.13}$$

The quantity Π is known as the Peltier coefficient. Note that the sign of the effect, i.e., whether intake or output of heat occurs at the junction, depends on the direction of the current. According to eq. (6.10), a knowledge of the thermopower is sufficient for this effect to be determined.

Finally, let us consider the Thomson effect, which consists in that in a nonuniformly heated conductor the amount of heat released is different from that generated at constant temperature. As has been pointed out, the amount of heat released in a unit volume per unit time is equal to

$$W = \boldsymbol{jE} - \operatorname{div} \boldsymbol{q}.$$

Substituting here formulas (6.8) and (6.9) and assuming that the entire nonuniformity results from the temperature gradient alone, we get

$$W = \rho j^2 + jQ\nabla T - j\frac{\mathrm{d}\Pi}{\mathrm{d}t} T\nabla T. \tag{6.14}$$

The first term in this expression is the usual Joule heat and the second, the Thompson heat, is the extra heat released due to the presence of a temperature gradient. The second term is written in the form

$$W_T = -\mu_T j\nabla T, \tag{6.15}$$

where μ_T is the Thomson coefficient. From (6.14) and (6.10) we obtain

$$\mu_T = -Q + \frac{\mathrm{d}\Pi}{\mathrm{d}T} = T\frac{\mathrm{d}Q}{\mathrm{d}T}. \tag{6.16}$$

Thomson heat, in contrast to Joule heat, is proportional to j, i.e., it changes sign when the direction of the current is reversed. This makes it possible to separate it from Joule heat. It will suffice to carry out two experiments with opposite directions of the current.

It thus turns out that all thermoelectric phenomena in the isotropic case are described by a single coefficient, namely the thermopower Q.

Let us now find this coefficient for the isotropic model of a metal under the assumption of elastic collisions. Just as in section 3.4, we make use of the quasiparticle model and apply the kinetic equation in the form (3.18). Substituting $f=f_0+f_1$, where $f_1 \ll f_0$, we obtain

$$\frac{\partial f_0}{\partial|\xi|}\left(e\boldsymbol{E}\boldsymbol{v}-\frac{\xi}{T}\boldsymbol{v}\nabla T\right)=-\frac{f_1}{\tau}, \tag{6.17}$$

whence we find for the electric current

$$\begin{aligned} j &= 2\int (e\,\mathrm{sign}\,\xi)\cdot(\boldsymbol{v}\,\mathrm{sign}\,\xi) f \frac{\mathrm{d}^3 p}{(2\pi\hbar)^3} = 2e\int \boldsymbol{v} f \frac{\mathrm{d}^3 p}{(2\pi\hbar)^3} \\ &= -e^2 \int \boldsymbol{v}(\boldsymbol{v}\boldsymbol{E})\,\tau\nu(\varepsilon)\frac{\partial f_0}{\partial|\xi|}\,\mathrm{d}\xi\frac{\mathrm{d}\Omega}{4\pi} \\ &\quad + e\int \boldsymbol{v}(\boldsymbol{v}\nabla T)\frac{\xi}{T}\tau\nu(\varepsilon)\frac{\partial f_0}{\partial|\xi|}\,\mathrm{d}\xi\frac{\mathrm{d}\Omega}{4\pi}. \end{aligned} \tag{6.18}$$

We are interested in the second term in eq. (6.18). After integrating over the angles there remains the integral over ξ. To a first approximation, the integrand is odd in ξ and the integral turns to zero. It is for this reason that we have to expand the factor at $\partial f_0/\partial|\xi|$ in ξ:

$$(\nu v^2\tau)_\xi \approx (\nu v^2\tau)_\mu + \xi\frac{\mathrm{d}}{\mathrm{d}\mu}(\nu v^2\tau)_\mu.$$

After this the integral over ξ assumes the same form as eq. (3.20′). Taking this integral, we find the coefficient β (see eq. 6.1):

$$\beta = -\tfrac{1}{9}\pi^2 eT\frac{\mathrm{d}}{\mathrm{d}\mu}(v^2\tau\nu)_\mu. \tag{6.19}$$

Using the relation $Q=-\beta/\sigma$, we find

$$Q=\frac{\pi^2 T}{3e}\frac{\mathrm{d}}{\mathrm{d}\mu}[\ln(v^2\tau\nu)_\mu]. \tag{6.20}$$

If we deal with free electrons scattered from impurities, then $\tau=l/v$, where l is independent of μ (section 4.1) and $v\propto\sqrt{\mu}$. The density of states $\nu=p_0 m/(\pi^2\hbar^3)$ is proportional to $\sqrt{\mu}$. It follows that

$$\frac{\mathrm{d}}{\mathrm{d}\mu}[\ln(v^2\tau\nu)_\mu]=\mu^{-1}.$$

Thus, for free electrons

$$Q=\frac{\pi^2}{3e}\frac{T}{\mu}\sim 10^{-8}\,T\ \mathrm{V/K} \tag{6.21}$$

(where T is measured in K). This estimate holds practically in all cases. It is justified experimentally for pure nonmagnetic metals. From formulas (6.10), (6.12) and (6.16) we get

$$\Pi \propto T^2, \qquad V_{\text{therm}} \propto T_2^2 - T_1^2, \qquad \mu_T \propto T. \tag{6.22}$$

The situation may change substantially in the presence of magnetic impurities (section 4.6) (Kondo 1965). We shall not give a lengthy and complicated analysis of this question here. We only note that, because of the interference of the amplitude of ordinary scattering and that of the spin interaction between the electron and the impurity $(J/n)\boldsymbol{\sigma S}$, terms appear in the scattering probability which vary rapidly in the vicinity of the Fermi surface. Moreover, it turns out that such terms are odd in ξ. Under these conditions the usual rule of taking integrals with the Fermi function (2.25) does not apply. As a result of all this, the small factor T/μ cancels out in the second term of (6.18).

The overall result for this case under the assumption of isotropy of the scattering amplitude may be represented in the following form:

$$Q \sim e^{-1}\left[\frac{\nu(\mu)J}{n}\right]^2 \frac{JVS(S+1)}{V^2+J^2S(S+1)}, \tag{6.23}$$

where V is the amplitude corresponding to ordinary electron scattering mechanisms, i.e., to the interaction with impurities, which is independent of the spin, and to electron–electron and electron–phonon interactions. In all cases $J \ll V$, so that $Q \sim \nu^2 J^3/(en^2 V)$. As the temperature is lowered the value of V decreases until it reaches a limit corresponding to the interaction of electrons with impurities. This leads to an increase in the absolute value of Q with decreasing temperature. If the metal contains only magnetic impurities, then in the limit J/V is determined only by the ratio of the amplitudes of the spin and nonspin interactions, which is usually of the order of 0.2. Since $\nu V/n \sim 1$, the limiting value of Q is found to be of the order of 10^{-6} V/K. The sign of Q is determined, according to (6.23), by the sign of the product VJ. (Usually, in this region $Q<0$, so that when the temperature is lowered Q changes sign.) Hence, at a sufficiently low temperature Q no longer depends on temperature. It should be noted here that the quantity Q itself is here independent of the impurity concentration. Only the temperature at which Q begins to deviate from the law (6.21) depends on the impurity concentration.

The constancy of Q persists until the temperature drops to the Kondo temperature (4.48), where the thermopower begins to fall off rapidly in magnitude, presumably reaching zero at $T=0$.

We have so far dealt with the isotropic model of the metal. In the general case of an arbitrary crystal, instead of (6.8) we obtain expressions with tensor coefficients:

$$E_i = \sum_k \rho_{ik} j_k + \sum_k Q_{ik} \frac{\partial T}{\partial x_k}, \qquad q_i = \sum_k \Pi_{ik} j_k - \sum_k \varkappa_{ik} \frac{\partial T}{\partial x_k}. \tag{6.24}$$

From the symmetry principle for the kinetic coefficients we have

$$\rho_{ik} = \rho_{ki}, \qquad \varkappa_{ik} = \varkappa_{ki}, \qquad \Pi_{ik} = TQ_{ki}. \tag{6.25}$$

It follows that in the general case the matrix coefficients Π_{ik} and Q_{ik} are nonsymmetric.

6.2. Thermomagnetic phenomena in a weak field

If a magnetic field is applied to a metal, eqs. (6.24) retain their form, but the coefficients become dependent on the magnetic field. Since the derivation of the symmetry principle for the kinetic coefficients (see Landau and Lifshitz 1976) makes use of the symmetry of the equations of mechanics with respect to the change of the sign of time, it must be kept in mind that with such a replacement the magnetic field changes sign. Hence, in this case, instead of relations (6.26) we obtain

$$\rho_{ik}(\boldsymbol{H}) = \rho_{ki}(-\boldsymbol{H}), \qquad \varkappa_{ik}(\boldsymbol{H}) = \varkappa_{ki}(-\boldsymbol{H}), \qquad \Pi_{ik}(\boldsymbol{H}) = TQ_{ki}(-\boldsymbol{H}). \tag{6.26}$$

This property of the matrix ρ_{ik} (or $\sigma_{ik} = \rho_{ik}^{-1}$) has already been mentioned in section 5.3.

Thus, according to eq. (6.24), the properties of a metal in the presence of a temperature gradient and an electric field are characterized by 36 coefficients that obey 15 various relations (6.26). Here, all the 21 remaining independent coefficients are complex functions of the magnitude and direction of the magnetic field. It is clear that a large variety of phenomena arise here, which are difficult to classify. In view of this, one is usually interested in what happens in the presence of a weak magnetic field ($\Omega\tau \ll 1$) in an isotropic metal. In practice, the results of such an investigation are applicable either to polycrystalline samples or to alkali metals, whose energy spectrum is almost isotropic.

The magnetic field introduces anisotropy even in the case of an isotropic metal because relations (6.24) remain tensor relations. However, if we retain linear terms in $\boldsymbol{H}$, we can make use of the fact that $\boldsymbol{H}$ is an axial vector, whereas $\boldsymbol{E}$, ∇T, $\boldsymbol{q}$ and $\boldsymbol{j}$ are polar vectors. It follows from this that the overall change will reduce to adding terms proportional to $[\boldsymbol{Hj}]$ and $[\boldsymbol{H}\nabla T]$ to the right-hand sides of eqs. (6.8) and (6.9). Thus, we have

$$\boldsymbol{E} = \rho\boldsymbol{j} + R[\boldsymbol{Hj}] + Q\nabla T + N[\boldsymbol{H}\nabla T], \tag{6.27}$$

$$\boldsymbol{q} = \Pi\boldsymbol{j} + B[\boldsymbol{Hj}] - \varkappa\nabla T + L[\boldsymbol{H}\nabla T]. \tag{6.28}$$

From the principle of symmetry of the kinetic coefficients it follows that $B = NT$. The relative order of the additional terms in these equations is $\Omega\tau$, say, $NH/Q \sim \Omega\tau$.

Equations (6.27) and (6.28) describe a number of effects. For definiteness, suppose that the field $\boldsymbol{H}$ is directed along the z axis, and the principal direction of current flow is x.

Evidently, the second term in (6.27) corresponds to the Hall effect. But, apart from the ordinary Hall effect, which is observed with the condition $\partial T/\partial y = 0$, one can also study the so-called adiabatic Hall effect on the condition that $q_y = 0$. Here, from eqs. (6.27) and (6.28) we obtain

$$E_y = \left(R + \frac{QB}{\varkappa}\right) H j_x. \tag{6.29}$$

Further, let us suppose that there is a current j_x but $\partial T/\partial x = 0$. In this case the y-component of eq. (6.28) at $q_y = 0$, $j_y = 0$ yields the generation of a temperature gradient in the direction y, with

$$\frac{\partial T}{\partial y} = \frac{B}{\varkappa} H j_x. \tag{6.30}$$

This phenomenon is known as the Ettingshausen effect and the quantity $B/\varkappa$ is the corresponding Ettingshausen coefficient.

Now we specify the following conditions. Suppose that the current $j_x = 0$ and a finite temperature gradient exists, $\partial T/\partial x$. In this case a temperature gradient arises also along the axis y. According to (6.28), at $j_x = j_y = q_y = 0$ one has

$$\frac{\partial T}{\partial y} = \frac{L}{\varkappa} H \frac{\partial T}{\partial x}. \tag{6.31}$$

This phenomenon is called the Righi–Leduc effect. The corresponding coefficient is $L/\varkappa$.

The temperature gradient along the x-axis is capable, like the current j_x, of producing an emf along the y-axis. From eq. (6.27) under the conditions $j_x = j_y = \partial T/\partial y = 0$ we find

$$E_y = NH \frac{\partial T}{\partial x}. \tag{6.32}$$

This phenomenon is known as the Nernst effect. But here, instead of the condition $\partial T/\partial y = 0$, we may put another condition, namely $q_y = 0$. In such a case we have the adiabatic Nernst effect. From eqs. (6.27) and (6.28) it follows that this effect is expressed by

$$E_y = \left(N + \frac{QL}{\varkappa}\right) H \frac{\partial T}{\partial x}. \tag{6.33}$$

There are also numerous effects of the order of $(\Omega\tau)^2$, such as, for example, magnetoresistance. However, these effects are all corrections to coefficients that also exist in the absence of a magnetic field, and therefore we shall not consider them.

6.3. *Thermal conductivity and thermoelectric effects in a strong magnetic field*

In sections 5.3 and 5.4 we have found the tensors σ_{ik} and $\rho_{ik} = \sigma_{ik}^{-1}$ for the case $\Omega\tau \gg 1$. Let us now calculate the other tensor coefficients in eqs. (6.24) in a strong magnetic field (Azbel' et al. 1957; Bychkov et al. 1959). Since for the condition $\Omega\tau \gg 1$ to be realized a large mean free path is required, it will be natural to consider the case of very low temperatures, where the scattering of electrons is governed by impurities. But in such a case the collisions are elastic, i.e., the energy of the electron is conserved in collisions.

The kinetic equation for quasiparticles in the case under consideration becomes

$$\frac{\partial f_1}{\partial t_1} - I(f_1) = \left[-e\boldsymbol{v}\boldsymbol{E} + \frac{\xi}{T}\boldsymbol{v}\nabla T\right]\frac{\partial f_0}{\partial|\xi|}. \tag{6.34}$$

It is natural to seek the solution of this equation in the form

$$f_1 = -\sum_i eE_i\psi_i\frac{\partial f_0}{\partial|\xi|} + \sum_i \frac{\partial T}{\partial x_i}\varphi_i\frac{\xi}{T}\frac{\partial f_0}{\partial|\xi|}. \tag{6.35}$$

Since the energy is conserved in collisions, after substitution of expression (6.35) into the collision integral the terms that depend only on the energy are taken outside the integral sign. It then turns out that ψ_i and φ_i satisfy the same equations of the following form:

$$\frac{\partial\psi_i}{\partial t_1} - I_i(\psi) = v_i. \tag{6.36}$$

Considering that the boundary conditions for the two functions are also the same, we come to the conclusion that

$$\varphi_i = \psi_i. \tag{6.37}$$

The expressions for the current and the energy flux have the form

$$\begin{aligned} j_i &= \frac{2e^2H}{(2\pi\hbar)^3c}\int v_i f\,\mathrm{d}\xi\,\mathrm{d}t_1\,\mathrm{d}p_z \\ &= -\frac{2e^3H}{(2\pi\hbar)^3c}\sum_k\int\frac{\partial f_0}{\partial|\xi|}\mathrm{d}\xi\int v_i\psi_k E_k\,\mathrm{d}t_1\,\mathrm{d}p_z \\ &\quad + \frac{2e^2H}{(2\pi\hbar)^3cT}\int\frac{\partial f_0}{\partial|\xi|}\xi\,\mathrm{d}\xi\sum_k\int v_i\varphi_k\frac{\partial T}{\partial x_k}\mathrm{d}t_1\,\mathrm{d}p_z, \end{aligned} \tag{6.38}$$

$$\begin{aligned} q_i &= \frac{2eH}{(2\pi\hbar)^3c}\int \xi v_i f\,\mathrm{d}\xi\,\mathrm{d}t_1\,\mathrm{d}p_z \\ &= -\frac{2e^2H}{(2\pi\hbar)^3c}\int\frac{\partial f_0}{\partial|\xi|}\xi\,\mathrm{d}\xi\sum_k v_i\psi_k E_k\,\mathrm{d}t_1\,\mathrm{d}p_z \\ &\quad + \frac{2eH}{(2\pi\hbar)^3c}\int\frac{\partial f_0}{\partial|\xi|}\xi^2\,\mathrm{d}\xi\sum_k v_i\varphi_k\frac{\partial T}{\partial x_k}\mathrm{d}t_1\,\mathrm{d}p_z. \end{aligned} \tag{6.39}$$

These relations are tensor generalizations of eqs. (6.1) and (6.2). We now write them in the form

$$j_i = \sum_k \sigma_{ik} E_k + \sum_k \beta_{ik} \frac{\partial T}{\partial x_k}, \tag{6.40}$$

$$q_i = \sum_k \gamma_{ik} E_k + \sum_k \zeta_{ik} \frac{\partial T}{\partial x_k}. \tag{6.41}$$

As for the coefficients in these equations, two remarks may be made. First, the symmetry principle for the kinetic coefficients requires that

$$\gamma_{ik}(\boldsymbol{H}) = -T\beta_{ki}(-\boldsymbol{H}). \tag{6.42}$$

This is a generalization of relation (6.7). However, formulas (6.38) and (6.39) with account taken of (6.37) yield something more, namely

$$\gamma_{ik}(\boldsymbol{H}) = -T\beta_{ik}(\boldsymbol{H}).$$

Hence, the two coefficients satisfy the conditions

$$\beta_{ik}(\boldsymbol{H}) = \beta_{ki}(-\boldsymbol{H}), \qquad \gamma_{ik}(\boldsymbol{H}) = \gamma_{ki}(-\boldsymbol{H}), \tag{6.43}$$

that is, the same conditions satisfied by σ_{ik} and ζ_{ik}. This is a consequence of the assumption of the elasticity of collisions.

Further, let us examine the last term in eq. (6.39). We take the integral over ξ using the rule (2.25). We thus obtain

$$\zeta_{ik} = -\frac{2eH}{(2\pi\hbar)^3 c} \frac{\pi^2 T}{3} \left[\int v_i \varphi_k \, \mathrm{d}t_1 \, \mathrm{d}p_z \right]_\mu .$$

On the other hand, from eq. (6.38) we have

$$\sigma_{ik} = \frac{2e^3 H}{(2\pi\hbar)^3 c} \left[\int v_i \psi_k \, \mathrm{d}t_1 \, \mathrm{d}p_z \right]_\mu .$$

Taking into account eq. (6.37), we obtain

$$\zeta_{ik} = -\frac{\pi^2 T}{3e^2} \sigma_{ik}. \tag{6.44}$$

Now we shall express, using eqs. (6.40) and (6.41), $\boldsymbol{E}$ and $\boldsymbol{q}$ in terms of $\boldsymbol{j}$ and ∇T. We obtain eqs. (6.24), where

$$\rho_{ik} = \sigma_{ik}^{-1}, \qquad Q_{ik} = -\sum_l \rho_{il} \beta_{lk},$$

$$\Pi_{ik} = \sum_l \gamma_{il} \rho_{lk}, \qquad \varkappa_{ik} = \sum_{l,m} \gamma_{il} \beta_{lm} \beta_{mk} - \zeta_{ik}. \tag{6.45}$$

From the form of the integral over ξ in the coefficients β_{ik} (6.40) and (6.38) it follows that they will be proportional to T; hence, on the basis of (6.42) $\gamma_{ik} \propto T^2$. In view of this, the first term in expression (6.45) for the thermal conductivity is small as compared to the second (this has been noted in section 6.1). It then follows that $\varkappa_{ik} = -\zeta_{ik}$, and from (6.44) we have

$$\varkappa_{ik} = \frac{\pi^2 T}{3e^2}\sigma_{ik}. \tag{6.46}$$

Thus, it turns out that the Wiedemann–Franz law for the case of elastic collisions holds also in the presence of a strong magnetic field.

This implies that the thermal conductivity displays the same features as σ_{ik}. These have been described in section 5.3. Note, in particular, that the components $\varkappa_{xx}$ and $\varkappa_{yy}$ (the z axis is directed along the magnetic field) for all directions of the magnetic field, with which the trajectories are closed, decrease with the field as H^{-2}. This makes it possible to separate the phonon thermal conductivity, which is independent of the magnetic field. Thus, it proves possible to separate the two parts of the total thermal conductivity of a metal.

Further, the component $\varkappa_{xy}$ for a metal with different numbers of electrons and holes equals in accordance with eqs. (6.36) and (6.46)

$$\varkappa_{xy} = \frac{\pi^2 Tc}{3eH}(n_e - n_h). \tag{6.47}$$

This means that the difference in the number of electrons and holes can be found not only from the Hall effect but from the Righi–Leduc effect as well ($\varkappa_{xy}$ correspond to $-LH$ in eq. 6.31).

Let us now turn to the thermopower tensor Q_{ik}. Applying the rule (2.25) after taking the integral over ξ we obtain (see eqs. 6.40 and 6.38)

$$\beta_{ik} = -\frac{2e^2 H\pi^2 T}{3(2\pi\hbar)^3 c}\frac{\partial}{\partial\mu}\left[\int v_i\varphi_k \, dt_1 \, dp_z\right]_\mu.$$

Comparing with the expression for σ_{ik} and taking into account that $\psi_k = \varphi_k$, we find

$$\beta_{ik} = -\frac{\pi^2 T}{3e}\frac{\partial}{\partial\mu}\sigma_{ik}(\mu). \tag{6.48}$$

It thus follows that the dependence of the coefficients β_{ik} on H must in general be the same as the dependence of the coefficients σ_{ik}. An exception is the case of closed Fermi surfaces at $n_e = n_h$ since in this case, generally speaking,

$$\frac{\partial}{\partial\mu}[n_e(\mu) - n_h(\mu)] \neq 0.$$

Let us now find the tensor Q_{ik} for the main cases.

(1) Closed Fermi surfaces at $n_e \neq n_h$. The tensor β_{ik} has the following form on the basis of eqs. (6.48) and (5.37):

$$\beta_{ik} = \begin{pmatrix} \gamma^2 c_{xx} & \gamma c_{xy} & \gamma c_{xz} \\ -\gamma c_{xy} & \gamma^2 c_{yy} & \gamma c_{yz} \\ -\gamma c_{xz} & -\gamma c_{yz} & c_{zz} \end{pmatrix}. \tag{6.49}$$

Multiplying by $-\rho_{ik}$ (5.38), we obtain

$$Q_{ik} = \begin{pmatrix} \nu_{xx} & \gamma\nu_{xy} & \nu_{xz} \\ \gamma\nu_{yx} & \nu_{yy} & \nu_{yz} \\ \gamma\nu_{zx} & \gamma\nu_{zy} & \nu_{zz} \end{pmatrix}. \tag{6.50}$$

Here $\nu_{xx} = \nu_{yy} = b_{xy}c_{xy}$. Considering that $b_{xy} = a_{yx}^{-1} = -a_{xy}^{-1}$ (see eq. 5.37) and using eqs. (6.48) and (5.36), we get

$$Q_{xx} = Q_{yy} = \frac{\pi^2 T}{3e} \frac{\partial}{\partial\mu} \ln[n_e(\mu) - n_h(\mu)]. \tag{6.51}$$

Thus, the transverse thermopower is isotropic, does not depend on collisions and is determined by the energy spectrum alone. From it we can deduce a new characteristic of the spectrum, namely $(\partial/\partial\mu)\ln[n_e(\mu) - n_h(\mu)]$, which cannot be found in any other way.

(2) Closed Fermi surfaces at $n_e = n_h$. The tensor β_{ik} retains its previous form (6.49), but the tensor ρ_{ik} is expressed by formula (5.42). According to eq. (6.45), we obtain

$$Q_{ik} = \begin{pmatrix} \gamma^{-1}\nu_{xx} & \gamma^{-1}\nu_{xy} & \gamma^{-1}\nu_{xz} \\ \gamma^{-1}\nu_{yx} & \gamma^{-1}\nu_{yy} & \gamma^{-1}\nu_{yz} \\ \nu_{zx} & \nu_{zy} & \nu_{zz} \end{pmatrix}. \tag{6.52}$$

In this case $Q_{xx} = -Q_{yy}$, but their particular form depends on the collision integral. Note that the first two lines of the rows of the tensor Q_{ik} increase in proportion to H, i.e., the thermopower of such metals in high fields may be very high.

(3) Fermi surface open in the direction p_x. According to eqs. (6.48) and (5.43) we have in this case

$$\beta_{ik} = \begin{pmatrix} \gamma^2 c_{xx} & \gamma c_{xy} & \gamma c_{xz} \\ -\gamma c_{xy} & c_{yy} & c_{yz} \\ -\gamma c_{xz} & c_{yz} & c_{zz} \end{pmatrix}. \tag{6.53}$$

The tensor ρ_{ik} is given by formulas (5.44). From (6.45) we obtain

$$Q_{ik} = \begin{pmatrix} \nu_{xx} & \gamma^{-1}\nu_{xy} & \gamma^{-1}\nu_{xz} \\ \gamma\nu_{yx} & \nu_{yy} & \nu_{yz} \\ \gamma\nu_{zx} & \nu_{zy} & \nu_{zz} \end{pmatrix}. \tag{6.54}$$

Thus, in this particular case the components Q_{xy} and Q_{xz} strongly increase and, on the contrary, the components Q_{yx} and Q_{zx} decrease with the field. Just as for the tensor ρ_{ik}, the crossover from the laws (6.50) to the laws (6.54) occurs in a narrow region of directions of the magnetic field of the order of $\theta \sim (\Omega\tau)^{-1}$ from the dangerous direction, in which open trajectories appear.

It should be stressed once more that many of the derivations made in this section become invalid if the collisions are not quite elastic, i.e., the scattering of electrons involves not only impurities but phonons as well.

6.4. *Thermopower and Lifshitz transitions*

It has been demonstrated in the previous sections that thermoelectric and thermomagnetic phenomena can yield additional information about the electronic spectrum and scattering processes. Here we will show that the measurement of thermopower is a rather sensitive detection method for a special type of electronic transitions called Lifshitz transitions or second-and-half order transitions (Lifshitz 1960).

At the end of section 1.3 we have dealt with the types of intersection of the constant-energy surface with the face of the Brillouin zone in the model of weakly coupled electrons. In a general case, the formulas of weak-coupling theory do not apply, but qualitatively they give a good description of the behavior of the energy spectrum in the vicinity of the face of the Brillouin zone. Let us visualize that by means of an external influence, say isotropic compression, uniaxial deformation, or a gradual change of the composition we can alter the relative position of the Fermi surface and of the face of the Brillouin zone. This may entail changes in the topology of the Fermi surface shown in fig. 5: the formation of a "neck" or a new portion of the surface. It is rather obvious that such changes in the topology will be accompanied by certain singularities of the thermodynamic and kinetic characteristics.

Let us find the singularity in the total energy in the weak-coupling model in going from fig. 5a to fig. 5b. The variable parameter chosen is the chemical potential μ. The number of states corresponding to the interior of the Fermi surface is given by

$$n(\mu) = \frac{2}{(2\pi\hbar)^3} \int 2\pi p_\perp \, dp_\perp \, dp_{z1} = \frac{1}{4\pi^2\hbar^3} \int dp_\perp^2 \, dp_{z1}. \tag{6.55}$$

The integral over $dp_\perp^2$ at given μ and p_{z1} is simply equal to $p_\perp^2$, which is expressed by formula (1.43), i.e.,

$$p_\perp^2 = 2m\mu - (\tfrac{1}{2}\hbar K)^2 - p_{z1}^2 + [(p_{z1}\hbar K)^2 + (2m|U_K|^2)]^{1/2}. \tag{6.56}$$

The integral over p_{z1} is taken in the case depicted in fig. 5a from some finite value $p_{z1}^{(0)}$ to a certain upper limit denoted by P, and in the case of fig. 5b it is taken from 0 to P.

The value of $p_{z1}^{(0)}$ is determined from formula (6.56) at $p_\perp^2 = 0$. We shall be interested here in small values of p_{z1} and of $\Delta\mu = \mu - \mu_c$, where μ_c given by

$$\mu_c = \left(\frac{\frac{1}{2}\hbar K}{2m}\right)^2 - |U_K| \tag{6.57}$$

is the value of μ at which tangency occurs. Putting $\Delta\mu < 0$ and expanding the right-hand side of formula (6.56) in p_{z1}^2, we have

$$\begin{aligned} p_\perp^2 &\approx 2m\Delta\mu - p_{z1}^2 + \frac{(p_{z1}\hbar K)^2}{4m|U_K|} \\ &= 2m\Delta\mu + \frac{p_{z1}^2[\frac{1}{2}(\hbar K)^2 - 2m|U_K|]}{2m|U_K|}. \end{aligned}$$

Setting $p_\perp^2 = 0$, we obtain

$$p_{z1}^{(0)} = 2m|U_K|^{1/2}|\Delta\mu|^{1/2}[\tfrac{1}{2}(\hbar K)^2 - 2m|U_K|]^{-1/2}. \tag{6.58}$$

The main part of the integral in (6.55) corresponds to large p_{z1}. But we are interested here in the singular part, which depends on $\Delta\mu$; it is determined by the value of the integral at the lower limit. Thus, we have at $\Delta\mu < 0$ ($P \sim \hbar/a$ is the conditional upper limit):

$$\begin{aligned} n(\mu) &= \frac{1}{4\pi^2\hbar^3}\int_{p_{z1}^{(0)}}^{P}\left\{2m\Delta\mu + \frac{p_{z1}^2[\frac{1}{2}(\hbar K)^2 - 2m|U_K|]}{2m|U_K|}\right\}\mathrm{d}p_{z1} \\ &= n_{\text{reg}} - \frac{1}{4\pi^2\hbar^3}\left\{2m\Delta\mu p_{z1}^{(0)} - \tfrac{1}{3}p_{z1}^{(0)3}\frac{[\frac{1}{2}(\hbar K)^2 - 2m|U_K|]}{2m|U_K|}\right\} \\ &= n_{\text{reg}} + \frac{1}{6\pi^2\hbar^3}(2m)^2|U_K|^{1/2}[\tfrac{1}{2}(\hbar K)^2 - 2m|U_K|]^{-1/2}|\Delta\mu|^{3/2}. \end{aligned} \tag{6.59}$$

In case $\Delta\mu > 0$ there is only the regular part of $n(\mu)$.

An increment in the total energy can be found as $-\int^{\mu} n(\mu_1)\,\mathrm{d}\mu_1$. Evidently, the singular part exists at $\mu < \mu_c$ and is proportional to $|\mu - \mu_c|^{5/2}$, where μ_c is given by formula (6.57). Suppose, for the sake of definiteness, that the variable external factor is isotropic compression (p is the pressure). In this case, $\Delta\mu = \mu - \mu_c$ is proportional to $\Delta p = p - p_c$ and the singular part of the total energy is proportional to $|\Delta p|^{5/2}$ on one side of the transition point. This resembles a second-order phase transition. If we assume that $T = \text{const.}$ and vary the pressure, then on one side of the transition point in the thermodynamic potential there will appear an additional term proportional to $(\Delta p)^2$ (within the framework of Landau theory; see Appendix 2); this term is responsible for a jump in the compression bulk modulus:

$$\alpha = -V^{-1}\left(\frac{\partial V}{\partial p}\right)_{T_c} = -V^{-1}\left(\frac{\partial^2 \Phi}{\partial p^2}\right)_{T_c}.$$

Since in our case the increment of the energy is proportional to $|\Delta p|^{5/2}$, such a transition is sometimes called a second-and-half order transition.

It should, however, be borne in mind that there is a fundamental difference between the Lifshitz transition and the second-order phase transition. Whereas the second-order phase transition is associated with a change in symmetry and occurs along the entire curve in the (T, p) plane, the Lifshitz transition takes place, strictly speaking, only at $T=0$, and at finite temperatures the singularity is smoothed out. In real metals, because of the smallness of T/μ there remains a rather marked trace from the singularity of the Lifshitz transition; levelling-off occurs only in a small neighborhood of the transition point. Note that the transition case in question, which is associated with the disruption of the "neck" of the Fermi surface at the face of the Brillouin zone, is not the only one possible. Another example is the transition from fig. 5b to fig. 5c, i.e., the formation of a new portion of the Fermi surface. In principle, such topological changes are possible not only at the Brillouin zone boundary but inside the zone as well. But in all cases the singularity in the neighborhood of the transition is of the same nature as in the example considered above.

The question now arises of the detection of Lifshitz transitions. Evidently, the most efficient method is to measure a physical quantity that contains a sufficiently large derivative of the function $n(\mu)$ and therefore tends to infinity at the transition point. Such a quantity is the thermopower.

Differentiating (6.59) with respect to μ, we obtain $\nu(\mu)$:

$$\nu(\mu)=\frac{\partial n(\mu)}{\partial \mu}=\nu_{\text{reg}}-\frac{1}{4\pi^2\hbar^3}(2m)^2|U_{\text{K}}|^{1/2} \times[\tfrac{1}{2}(\hbar K)^2-2m|U_{\text{K}}|]^{-1/2}|\Delta\mu|^{1/2}. \tag{6.60}$$

According to eq. (6.20), the thermal emf is proportional to $\sigma'(\mu)/\sigma(\mu)$, where σ is the conductivity. However, when calculating σ one cannot use the isotropic model because of the sharp anisotropy of the spectrum (1.43). At first glance it seems that the singularity disappears altogether because at the point $p_{z1}=0$, $p_{\perp}=0$ the components of the velocity v_z and $v_{\perp}$ vanish. Actually, the situation is somewhat more complicated.

The major contribution to the thermopower comes from electrons with momenta far from the point $p_{z1}=0$, $p_{\perp}=0$. However, due to scattering processes they change momentum and get "virtually" into the neighborhood of the singular point. Therefore, the scattering probability as well as $\tau=W^{-1}$, acquires a correction proportional to $|\Delta\mu|^{1/2}$. As a matter of fact, this arises from $\nu(\mu)$ not for the initial states of electrons but for the virtual states which arise upon scattering. Therefore, when differentiating in formula (6.20), it may be assumed that $v^2\nu(\mu)$ are constants of the usual order, and in τ one has to take into account a correction of the same relative order as the relative value of the correction to $\nu(\mu)$ but with an opposite sign. As a result, we find

$$Q_{\text{sing}}\propto\frac{\tau'(\mu)}{\tau(\mu)}\propto-\frac{\nu'(\mu)}{\nu(\mu)}\propto-|\Delta\mu|^{-1/2}. \tag{6.61}$$

Note that this singularity occurs at $\Delta\mu < 0$ and is absent at $\Delta\mu > 0$, i.e., is unsymmetric. In other types of Lifshitz transitions (the formation of a neck with a decrease in the energy; the formation of a new piece with increase or decrease in the energy) a singularity appears always from the side corresponding to a larger number of pieces of the Fermi surface; note that in all cases $Q_{\rm sing} \propto -|\Delta\mu|^{-1/2}$. As regards the regular part of Q, it may have either sign. For example, when μ is near the bottom of the band, $Q > 0$, and when μ is near the top of the band, $Q < 0$. In view of this, a change in sign of the thermopower is, in principle, possible in the vicinity of a singularity.

Experimental investigations of the thermopower in Lifshitz transitions in lithium–magnesium alloys of varying composition and in bismuth–antimony alloys subjected to anisotropic deformation (Yegorov and Fedorov 1982, Yegorov et al. 1984), have revealed sharp, somewhat asymmetric maxima, which decrease rapidly with increasing temperature. The factor responsible for the smoothing out of the singularity and for a decrease in asymmetry is the scattering of electrons by defects and the finite temperature (Varlamov and Pantsulaya 1985). The decrease in the amplitude of the maxima with rising temperature is a consequence of the changeover from impurity scattering to phonon scattering, in which the momentum is changed little; because of this, electrons with momenta far from the singular point have a lesser change to get into it (Abrikosov and Pantsulaya 1986).

However, at low temperatures the maxima of the termpower are quite distinctly pronounced and presumably provide the simplest method of determining Lifshitz transitions.

7 Metals in a high-frequency electromagnetic field. Cyclotron resonance

7.1. The normal skin effect

Let us consider a metal placed in a high-frequency electromagnetic field. We shall be concerned with a case where use is made of the ordinary expression for the current $\boldsymbol{j} = \sigma \boldsymbol{E}$, and for simplicity we assume that the metal is isotropic. The Maxwell equations have the form

$$\operatorname{rot} \boldsymbol{E} = -c^{-1} \frac{\partial \boldsymbol{H}}{\partial t}, \tag{7.1}$$

$$\operatorname{rot} \boldsymbol{H} = \frac{4\pi}{c} \boldsymbol{j} = \frac{4\pi}{c} \sigma \boldsymbol{E}. \tag{7.2}$$

Suppose the metal occupies a half-space $x > 0$ and that the wave falls normally on its surface, the vector $\boldsymbol{E}$ being directed along the axis y and $\boldsymbol{H}$ along the axis z (fig. 27).

We shall seek a solution proportional to $e^{ikx - i\omega t}$. Substituting into (7.1) and (7.2) yields

$$ikE_y = \frac{i\omega}{c} H_z, \qquad -ikH_z = \frac{4\pi}{c} \sigma E_y.$$

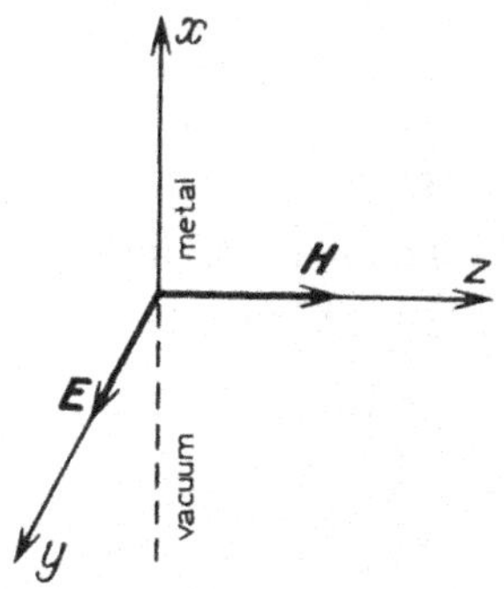

Fig. 27.

Combining these equations, we obtain the value of the wave vector:

$$k^2 = \frac{4\pi i\omega\sigma}{c^2}.$$

Hence,

$$k = \left(\frac{4\pi i\omega\sigma}{c^2}\right)^{1/2} = \left(\frac{2\pi\omega\sigma}{c^2}\right)^{1/2}(1+i) = k_1 + ik_2. \tag{7.3}$$

Substituting eq. (7.3) into $e^{ikx - i\omega t}$, we see that the electromagnetic field decreases inside the metal by the law $e^{-k_2 x}$, i.e., the field penetrates only into a relatively thin surface layer of the metal. This phenomenon is known as the skin effect. Let us define the skin penetration depth as follows:

$$\delta = k_2^{-1} = \left[\frac{c^2}{2\pi\omega\sigma}\right]^{1/2}. \tag{7.4}$$

In order to characterize the properties of the metal in a high-frequency field, we introduce the so-called impedance or surface resistance Z. It is defined as the ratio of the electric field on the metal surface to the current density integrated over the thickness of the metal:

$$Z = \frac{E_y(0)}{\int_0^\infty j_y(x)\,dx}. \tag{7.5}$$

The impedance Z is a complex quantity and is written in the form $R - iX$, where R is the resistance and X is the reactance. The quantities R and X can be determined from the change in the amplitude and phase of the wave reflected from the metal surface. The resistance R determines the energy loss of the electromagnetic wave upon reflection and can be found from the production of heat in the metal when it is place in a high-frequency field.

Relation (7.5) may be transformed to a different form. From eq. (7.2) we find

$$Z = R - iX = \frac{E_y(0)}{-(c/4\pi)H_z|_0^\infty} = \frac{4\pi}{c}\frac{E_y(0)}{H_z(0)}. \tag{7.6}$$

From eq. (7.1) we now have

$$Z = \frac{4\pi}{c}\frac{E_y(0)}{H_z(0)} = \frac{4\pi}{c^2}\frac{\omega}{k}. \tag{7.7}$$

Formulas (7.6) and (7.7) are valid for any relationship between $\boldsymbol{j}$ and $\boldsymbol{E}$. In the case $\boldsymbol{j} = \sigma\boldsymbol{E}$, substituting (7.3) into (7.7), we have

$$Z = \left[\frac{4\pi\omega}{i\sigma c^2}\right]^{1/2} = \left[\frac{2\pi\omega}{\sigma c^2}\right]^{1/2}(1-i).$$

Thus, in this case

$$R = X = \left[\frac{2\pi\omega}{\sigma c^2}\right]^{1/2}. \tag{7.8}$$

This formula describes the so-called "normal" skin effect.

7.2. *The anomalous skin effect. Inefficiency concept*

Let us examine formula (7.4) for the skin penetration depth. For pure metals at low temperatures the conductivity σ becomes large and δ decreases. The value of δ can be lowered even more strongly by increasing the frequency ω. At the same time, as the temperature is lowered the mean free path l increases. At sufficiently low temperatures and high frequencies there may arise a situation where δ becomes less than l. In this case, the simple connection between the current and the field, $\boldsymbol{j} = \sigma \boldsymbol{E}$, is no longer valid because this formula has been derived from the kinetic equation under the assumption of a homogeneous field. But if the field varies significantly at distances comparable to or even less than the mean free path, it cannot be considered homogeneous.

Let us turn to this case known as the anomalous skin effect (London 1940). We shall consider the physical picture in a limiting situation where $l \gg \delta$. The electrons can move parallel to the metal surface or at a large angle to it. The electrons moving at a large angle spend little time in the electric field and therefore practically do not interact with it. Those electrons which are moving parallel to the surface or at a small angle are accelerated by the electric field and therefore take from the wave a fraction of its energy.

In the isotropic case the number of electrons with momenta directed within a certain solid angle is proportional to that angle. Essential electrons are those which are moving within the skin layer throughout their entire mean free path. For these electrons (Fig. 28) $\mathrm{d}\theta \sim \delta / l$ and, hence,

$$\mathrm{d}\Omega \sim 2\pi \sin\theta \, \mathrm{d}\theta \sim 2\pi \, \mathrm{d}\theta \sim \frac{2\pi\delta}{l} \qquad (\theta \approx \tfrac{1}{2}\pi).$$

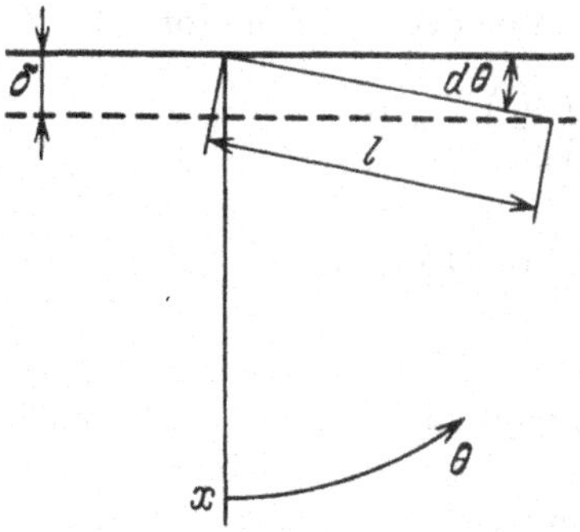

Fig. 28.

Thus, the effective electron density is

$$n_{\text{eff}} \sim n_e \frac{d\Omega}{4\pi} \sim \frac{n_e \delta}{l}.$$

The conductivity is proportional to the number of electrons. Therefore, the effective conductivity compared to the ordinary conductivity contains a factor of order δ / l. We assume the effective conductivity to have the form

$$\sigma_{\text{eff}} = \frac{ia\sigma}{kl}, \tag{7.9}$$

where k is the wave vector; a is a real coefficient of the order of unity. This notation is possible since $\delta = k_2^{-1}$ and $k_1 \sim k_2$. However, formula (7.9) actually reflects the fact that in a rigorous theory at $|kl| \gg 1$ the quantity i/k begins to play the role of l (see the derivation of formula 7.23). In this connection, formula (7.9) makes it possible to deduce the correct relationship between the real and the imaginary part of the impedance; this formula is the basis for the so-called "inefficiency concept".

We use the expression

$$k = \left(\frac{4\pi i \omega \sigma}{c^2}\right)^{1/2}$$

and substitute (7.9) instead of σ into it. As a result, we obtain an equation for k. Solving it, we find

$$k = \left[\frac{4\pi\omega a \sigma}{c^2 l}\right]^{1/3} e^{i\pi/3}. \tag{7.10}$$

The penetration depth of the field is found as the inverse value of the imaginary part k. Hence,

$$\delta = \left(\frac{c^2 l}{4\pi\omega a \sigma}\right)^{1/3} \frac{1}{\sin \frac{1}{3}\pi}. \tag{7.11}$$

The surface impedance can be deduced from formula (7.7):

$$Z = \frac{4\pi\omega}{c^2 k} = \left(\frac{4\pi\omega}{c^2}\right)^{2/3} \left(\frac{l}{a\sigma}\right)^{1/3} e^{-i\pi/3}$$

$$= \left(\frac{2}{a}\right)^{1/3} \left(\frac{\pi\omega}{c^2}\right)^{2/3} \left(\frac{l}{\sigma}\right)^{1/3} (1 - i\sqrt{3}). \tag{7.12}$$

Thus, we have obtained the following results: (a) Z is proportional to $\omega^{2/3}$; (b) $X = \sqrt{3}R$; (c) the conductivity enters into eqs. (7.11) and (7.12) only in the combination σ / l; but since $\sigma \sim n_e e^2 \tau / m \sim n_e e^2 l / p_0$, it follows that $\sigma / l \sim n_e e^2 / p_0$ (this relation is independent of the temperature and is determined by the electronic spectrum alone).

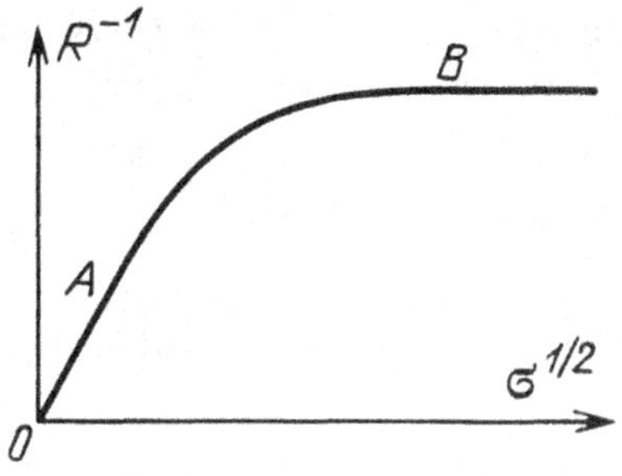

Fig. 29.

If the results obtained are depicted as a plot of R^{-1} versus $\sigma^{1/2}$, the plot must have the form shown in fig. 29, where region A corresponds to the normal skin effect and region B to the anomalous skin effect. In fact, the normal skin effect occurs in the region of low conductivity (high temperatures), where $R \propto \sigma^{-1/2}$, and in the region of large σ one has the anomalous skin effect and R does not depend on σ. This plot has been confirmed experimentally.

Let us find the restriction on the frequency at which the anomalous skin effect can be observed. From the condition $\delta \ll l$, formula (7.4) and the expression for the conductivity $\sigma = n_e e^2 l / p_0$ we find

$$\omega > \frac{c^2 p_0}{2\pi n_e e^2 l^3}.$$

Assuming that $n_e \sim 10^{22}$ cm^{-3} and $p_0 = \hbar(3\pi^2 n_e)^{1/3} \sim 10^{-19}$ g . cm/s (this holds for a good metal), we obtain

$$\omega > 10^{-2}(l[\mathrm{cm}])^{-3}\ \mathrm{s}^{-1}$$

If $l \sim 10^{-3}$ cm (this is the residual resistance region), then $\omega > 10^7$ s^{-1}.

7.3. The anomalous skin effect. Solution of the kinetic equation

It has been pointed out in the preceding section that the surface impedance in the region of the anomalous skin effect contains the ratio σ/l, which depends only on the parameters of the electronic spectrum and not on the scattering processes. Therefore, we may hope that the study of the anomalous skin effect will make it possible to extract information about the electronic spectrum. To this end, we shall present here a rigorous derivation of the surface impedance on the basis of the kinetic equation (Reuter and Sondheimer 1948).

From formulas (7.1) and (7.2) we obtain

$$-\mathrm{rot}\,\mathrm{rot}\,\boldsymbol{E} = -\mathrm{grad}\,\mathrm{div}\,\boldsymbol{E} + \nabla^2 \boldsymbol{E} = -\frac{4\pi}{c^2}\mathrm{i}\omega \boldsymbol{j}.$$

But div $\boldsymbol{E}=0$, since it is assumed that the vector $\boldsymbol{E}$ has only y- and z-components, which depend on x alone; so we may write

$$\frac{\mathrm{d}^2 E_\alpha}{\mathrm{d}x^2}=-\frac{4\pi \mathrm{i}\omega}{c^2} j_\alpha \tag{7.13}$$

(we mean here that all the quantities are proportional to $\mathrm{e}^{-\mathrm{i}\omega t}$). In order to determine the current j_α, we must have a distribution function, which we shall seek in the form

$$f=f_0-\left(\frac{\partial f_0}{\partial \varepsilon}\right)\psi.$$

Using this expression, we find

$$j_\alpha=2e\int v_\alpha f\frac{\mathrm{d}^3\boldsymbol{p}}{(2\pi\hbar)^3}=\frac{2e}{(2\pi\hbar)^3}\oint v_\alpha\psi\frac{\mathrm{d}S}{v}. \tag{7.14}$$

Here use has been made of the transformation of the integral over the momenta (2.23) and of the fact that $\partial f_0/\partial\varepsilon=\delta(\varepsilon-\mu)$. The integration in (7.14) is made along the Fermi surface.

We shall write the kinetic equation in the form (3.12). Of course, the form of the collision integral

$$I(f)=-\frac{(f-f_0)}{\tau}$$

is, generally speaking, incorrect for an anisotropic metal. Nevertheless, we may utilize it. This is associated, in the first place, with the fact that in the limiting anomalous case the result will not depend on τ at all. Moreover, an interesting feature of the case under consideration is that the collision integral really has such a form, but with τ dependent on p (see the end of this section). Substituting the expected form of the f-function, we obtain from (3.12):

$$-\mathrm{i}\omega\psi+v_x\frac{\partial\psi}{\partial x}+\frac{\psi}{\tau}=eE_\alpha v_\alpha. \tag{7.15}$$

Here account has been taken of the fact that $\boldsymbol{E}$ has only y- and z-components and therefore the indices α take only these values. If the index is repeated, summation is meant (i.e., $E_\alpha v_\alpha=E_y v_y+E_z v_z$). Equation (7.15) expresses ψ in terms of $\boldsymbol{E}(x)$. Solving it, we can find $\boldsymbol{j}(\boldsymbol{E})$ and substitute it into the Maxwell equation (7.13), whence we can determine $\boldsymbol{E}$ and, eventually, the impedance.

Dividing eq. (7.15) by v_x and denoting $(\tau^{-1}-\mathrm{i}\omega)/v_x$ by L, we obtain

$$\frac{\partial\psi}{\partial x}+L\psi=\frac{eE_\alpha v_\alpha}{v_x}.$$

The solution of this equation is

$$\psi = e^{-Lx} \int_c^x eE_\alpha(x_1) \frac{v_\alpha}{v_x} e^{Lx_1}\, dx_1, \tag{7.16}$$

where c is a certain constant. In order to find this constant, we have to introduce boundary conditions. First of all, it is clear that with $x \to \infty$

$$\psi \to 0.$$

Another boundary condition depends on the type of the metal surface. As is known, on the metal surface there is a potential jump which prevents the escape of electrons from the metal. This potential jump reflects electrons into the bulk of the metal.

If the surface is ideal, the electrons are reflected in a specular way. In an isotropic metal this would imply that p_y and p_z do not vary and that p_x changes sign. In the anisotropic case under consideration the situation is more complicated. From the homogeneity of the problem in the directions y and z it again follows that p_y and p_z remain unchanged, but this time, instead of the change of the sign of p_x we have to require that the energy remain constant.

In an isotropic metal, the waves with p_x and $-p_x$, which refer to the same energy, are added up so that the wave function vanishes at $x=0$ (the potential wall). In the case under consideration, we also have a superposition of the waves. One of them has $v_x = \partial\varepsilon/\partial p_x > 0$ and the other has $v_x < 0$. But at $\varepsilon(p_x, p_y, p_z) = \varepsilon(p'_x, p_y, p_z)$, generally speaking, $p'_x \neq -p_x$. This gives rise to serious complications in the general case.

However, in the limiting case of the anomalous skin effect, which is dealt with here, only those electrons are important which move almost parallel to the surface, i.e., those which have $v_x \approx 0$ (section 7.2; see also the further derivation). In a small neighborhood of such momenta the energy of the electron has the form

$$\varepsilon = a(p_y, p_z)\,[p_x - b(p_y, p_z)]^2 + d(p_y, p_z)$$

(at $p_x = b(p_y, p_z)$, $\partial\varepsilon/\partial p_x = 0$). On specular reflection the condition of invariability of p_y and p_z implies that $v_y \approx \partial d/\partial p_y$ and $v_z = \partial d/\partial p_z$ remain unaltered. The invariability of the energy in going from $v_x > 0$ to $v_x < 0$ means that $p_x - b(p_y, p_z)$ changes to $-p_x + b(p_y, p_z)$ and, hence, $v_x = 2a(p_y, p_z)\,(p_x - b(p_y, p_z))$ changes to $-v_x$. Thus, under the conditions of the limiting anomalous skin effect upon specular reflection we may write the boundary condition at $x=0$ in the form

(Case I) $\psi(v_x, v_y, v_z) = \psi(-v_x, v_y, v_z)$.

Usually, even the best metal surface has irregularities with sizes of the order of the atomic dimensions. As a result, part of the electrons are reflected from the surface in a diffusive manner, i.e., they do not retain the memory of their initial state. Therefore, apart from specular reflection, it will be reasonable to consider the opposite limiting case too - purely diffuse reflection. Here for $x=0$, $v_x>0$ we may put

(Case II)* $\psi=0$.

The boundary condition $\psi\to 0$ with $x\to\infty$ gives the constant c in (7.16) at $v_x<0$. Indeed, since here Re $L<0$ in the exponent e^{-Lx}, we are to choose $c=\infty$. From this it follows that

$$\psi_{v_x<0}=\int_{\infty}^{x} eE_\alpha \frac{v_\alpha}{v_x} e^{L(x_1-x)}\,dx_1. \tag{7.17}$$

This is valid for both cases. If $v_x>0$, then e^{-Lx} itself vanishes with $x\to\infty$. In view of this, the constant c must be determined from the condition at $x=0$ and the result will naturally be different in the two cases.

Case I. For $v_x>0$ in the case of specular reflection we obtain at $x=0$

$$\int_c^0 eE_\alpha(x_1)\frac{v_\alpha}{v_x} e^{Lx_1}\,dx_1=-\int_{\infty}^{0} eE_\alpha(x_1)\frac{v_\alpha}{v_x} e^{-Lx_1}\,dx_1$$

since if $v_x>0$ then $-v_x<0$ (it should be kept in mind that L is proportional to v_x^{-1}). Thus, from (7.16) we obtain

$$\begin{aligned}\psi_{v_x>0}&=\int_0^x eE_\alpha(x_1)\frac{v_\alpha}{v_x} e^{L(x_1-x)}\,dx_1+\int_c^0 eE_\alpha(x_1)\frac{v_\alpha}{v_x} e^{L(x_1-x)}\,dx_1\\ &=\int_0^x eE_\alpha(x_1)\frac{v_\alpha}{v_x} e^{L(x_1-x)}\,dx_1+\int_0^{\infty} eE_\alpha(x_1)\frac{v_\alpha}{v_x} e^{-L(x_1+x)}\,dx_1.\end{aligned} \tag{7.18}$$

Substituting eqs. (7.17) and (7.18) into the expression for the current (7.14) and denoting v_α/v as n_α, we obtain

$$j_\alpha=\frac{2e^2}{(2\pi\hbar)^3}\left\{\int_{n_x>0} dS\frac{n_\alpha n_\beta}{n_x}\right.$$
$$\times\left[\int_0^x E_\beta(x_1)\,e^{-L(x-x_1)}\,dx_1+\int_0^{\infty} E_\beta\, e^{-L(x+x_1)}\,dx_1\right]$$

* This condition, which holds for the given geometry of the problem, is not universal for any form of the diffusely reflecting boundary. The most general form of the boundary condition for diffuse reflection suitable for a boundary of an arbitrary form is $\psi(x=0, v_x>0)=\text{const.}$, which implies that the distribution function for reflected electrons is independent of the direction of reflection. The value of this constant is determined from the condition of the absence of current along the normal to the boundary, i.e., $j_x=0$ at $x=0$, which must be fulfilled for any boundary. In the case of a planar boundary under consideration for the limiting anomalous skin effect $\delta\ll l$, whence we have $\psi(x=0, v_x>0)=0$.

$$-\int_{n_x<0} dS\frac{n_\alpha n_\beta}{n_x}\int_x^\infty E_\beta(x_1)\,e^{-L(x-x_1)}\,dx_1\bigg\}.$$

The last two integrals may be combined if the last one is brought to integration over the region of $n_x>0$. To do this, we change the signs of all the momenta in the integral; all the velocities also change sign. As a result, we have

$$j_\alpha=\frac{2e^2}{(2\pi\hbar)^3}\int_{n_x>0} dS\frac{n_\alpha n_\beta}{n_x}\left[\int_0^x E_\beta(x_1)\,e^{-L(x-x_1)}\,dx_1\right.$$
$$\left.+\int_0^\infty E_\beta(x_1)\,e^{-L(x+x_1)}\,dx_1+\int_x^\infty E_\beta(x_1)\,e^{L(x-x_1)}\,dx_1\right].$$

Integration over x_1 occurs, of course, over the region inside the metal. But if it is formally assumed that the electric field is symmetrically extended outside the metal, then in the second integral we may replace x_1 by $-x_1$, in which case we have

$$\int_{-\infty}^0 E_\beta(x_1)\,e^{-L(x-x_1)}\,dx_1.$$

The sum of the first two integrals in the expression for $\boldsymbol{j}$ will then give

$$\int_{-\infty}^x E_\beta(x_1)\,e^{-L(x-x_1)}\,dx_1.$$

Combining this integral with the last term in the expression for $\boldsymbol{j}$, we obtain

$$j_\alpha(x)=\frac{2e^2}{(2\pi\hbar)^3}\int_{n_x>0} dS\frac{n_\alpha n_\beta}{n_x}\int_{-\infty}^\infty E_\beta(x_1)\,e^{-L|x-x_1|}\,dx_1. \tag{7.19}$$

Case II. Let us now consider the diffuse scattering of electrons from the boundary. Applying the corresponding boundary condition to eq. (7.16) at $v_x>0$, we obtain $c=0$. Substituting this and eq. (7.17) into eq. (7.14) gives

$$j_\alpha-\frac{2e^2}{(2\pi\hbar)^3}\left[\int_{n_x>0} dS\frac{n_\alpha n_\beta}{n_x}\int_0^x E_\beta(x_1)\,e^{-L(x-x_1)}\,dx_1\right.$$
$$\left.-\int_{n_x<0} dS\frac{n_\alpha n_\beta}{n_x}\int_x^\infty E_\beta(x_1)\,e^{-L(x-x_1)}\,dx_1\right].$$

If now we pass over in the second integral from the integration over $n_x<0$ to the integration over $n_x>0$, we obtain

$$j_\alpha=\frac{2e^2}{(2\pi\hbar)^3}\int_{n_x>0} dS\frac{n_\alpha n_\beta}{n_x}\left[\int_0^x E_\beta(x_1)\,e^{-L(x-x_1)}\,dx_1\right.$$
$$\left.+\int_x^\infty E_\beta(x_1)\,e^{L(x-x_1)}\,dx_1\right].$$

We can again combine these two terms, as a result of which we have

$$j_\alpha = \frac{2e^2}{(2\pi\hbar)^3} \int_{n_x>0} \mathrm{d}S \frac{n_\alpha n_\beta}{n_x} \int_0^\infty E_\beta(x_1)\, \mathrm{e}^{-L|x-x_1|}\, \mathrm{d}x_1 . \tag{7.20}$$

The difference between this formula for j_α and formula (7.19) in the specular reflection case is that in the latter case the limits of integration over x_1 were $-\infty$ and ∞, and in (7.20) they are from 0 to ∞. Thus, we obtain

$$j_\alpha = \begin{cases} \displaystyle\int_{-\infty}^{\infty} K_{\alpha\beta}(x-x_1)\, E_\beta(x_1)\, \mathrm{d}x_1, & \text{specular case,} \\ \displaystyle\int_0^{\infty} K_{\alpha\beta}(x-x_1)\, E_\beta(x_1)\, \mathrm{d}x_1, & \text{diffuse case,} \end{cases} \tag{7.21}$$

where the kernel $K_{\alpha\beta}$ is given by

$$K_{\alpha\beta}(x) = \frac{2e^2}{(2\pi\hbar)^3} \int_{n_x>0} \mathrm{d}S \frac{n_\alpha n_\beta}{n_x} \exp\left(-\frac{|x|}{l^* n_x}\right). \tag{7.22}$$

(we have introduced the notation $l^{*-1} = (\tau^{-1} - \mathrm{i}\omega)/v$).

Let us consider the Fourier transform of the kernel $K_{\alpha\beta}$:

$$K_{\alpha\beta}(k) = \int_{-\infty}^{\infty} K_{\alpha\beta}(x)\, \mathrm{e}^{-\mathrm{i}kx}\, \mathrm{d}x.$$

We have

$$\int_{-\infty}^{\infty} \mathrm{e}^{-\mathrm{i}kx-\gamma|x|} = 2\int_0^\infty \mathrm{e}^{-\gamma x} \cos kx\, \mathrm{d}x = \frac{2\gamma}{\gamma^2 + k^2}$$

if $\mathrm{Re}\, \gamma > 0$. Thus, we obtain

$$K_{\alpha\beta}(k) = \frac{4e^2}{(2\pi\hbar)^3} \int_{n_x>0} \mathrm{d}S \frac{n_\alpha n_\beta}{n_x} \frac{l^* n_x}{1 + (l^* n_x k)^2}. \tag{7.23}$$

This expression simplifies significantly for the limiting anomalous skin effect, i.e., when $k \sim \delta^{-1} \gg l^{-1}$.

First of all, it should be noted that it is more convenient to pass from the integral over the Fermi surface to the integral over the angles that characterize the direction of the normal to that surface (i.e., the velocity $\boldsymbol{v} = \partial\varepsilon/\partial\boldsymbol{p}$). Let us consider a surface element $\mathrm{d}S$ shown in fig. 30. Let the principal radii of curvature be R_1 and R_2. Obviously, the area of the surface element is $R_1 R_2\, \mathrm{d}\theta_1\, \mathrm{d}\theta_2$. Now we choose the polar system of coordinates θ, φ for the direction of the normal so that the plane of the angle θ_1 coincides with the plane $\varphi = \text{const.}$, i.e., $\mathrm{d}\theta_1 = \mathrm{d}\theta$. Then, evidently, $\mathrm{d}\theta_2 = \sin\theta\, \mathrm{d}\varphi$. It then follows that

$$\mathrm{d}S = R_1 R_2\, \mathrm{d}\Omega = \frac{\mathrm{d}\Omega}{K(\theta, \varphi)}, \tag{7.24}$$

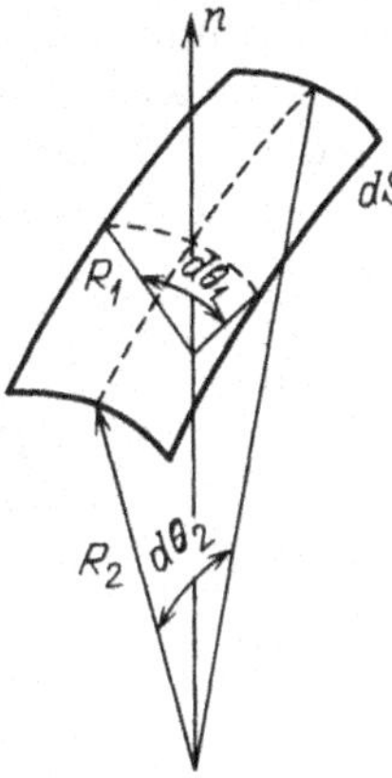

Fig. 30.

where K is the so-called Gaussian curvature of the surface; it is equal to the product of the inverse principal radii of curvature, $K=(R_1R_2)^{-1}$, at a point of the surface where the direction of the normal is θ, φ.

So, we now have

$$K_{\alpha\beta}=\frac{4e^2}{(2\pi\hbar)^3}\int_0^{2\pi}\mathrm{d}\varphi\int_0^{\pi/2}\frac{l^*}{1+(l^*k\cos\theta)^2}\frac{n_\alpha n_\beta}{K(\theta),\varphi)}\sin\theta\,\mathrm{d}\theta. \tag{7.25}$$

The upper limit of the integral over θ is $\frac{1}{2}\pi$ (from the condition $n_x>0$). If $kl^*\gg 1$, then only small values of $\cos\theta$ are important. Therefore, we can immediately make the replacement

$$K(\theta,\varphi)\to K(\varphi)\equiv K(\tfrac{1}{2}\pi,\varphi).$$

The remaining integral over $\cos\theta$ becomes equal to

$$\int_0^1\frac{l^*\mathrm{d}x}{1+(l^*kx)^2}=\frac{1}{|k|}\int_0^{l^*|k|}\frac{\mathrm{d}y}{1+y^2}\approx\frac{1}{|k|}\int_0^\infty\frac{\mathrm{d}y}{1+y^2}=\frac{\pi}{2|k|}.$$

Thus, for the case $l^*k\gg 1$ we obtain

$$K_{\alpha\beta}(k)=\frac{4e^2}{(2\pi\hbar)^3}\int_0^{2\pi}\mathrm{d}\varphi\frac{n_\alpha n_\beta}{K(\varphi)}\frac{\pi}{2|k|}=\frac{e^2B_{\alpha\beta}}{4\pi^2\hbar^3|k|}, \tag{7.26}$$

where

$$B_{\alpha\beta}=\int_0^{2\pi}\frac{n_\alpha n_\beta}{K(\varphi)}\mathrm{d}\varphi \tag{7.27}$$

is the tensor in the (y,z) plane.

Note that only the values of $\cos\theta\ll 1$, i.e., $v_x\approx 0$, are important in the integral. This circumstance corresponds to the "inefficiency concept"; it has been used here in the formulation of the boundary condition for specular reflection.

The subsequent calculations are performed in different ways in the two cases. The calculations for specular reflection are much simpler than those for diffuse reflection. But the results obtained in both cases differ little, so we shall limit ourselves to calculations for the specular case, and give only the final result for diffuse reflection.

In the derivation of formula (7.19) we assumed the electric field to be symmetrically extended to the region of $x<0$. But this means that at $x=0$ the derivative $\mathrm{d}E/\mathrm{d}x$ must undergo a jump (fig. 31) from $-E'(0)$ to $E'(0)$, i.e., the second derivative $\mathrm{d}^2E/\mathrm{d}x^2$ must behave as a δ-function at the $x=0$ boundary. To take this into account, we write eq. (7.13) in the form

$$\frac{\mathrm{d}^2E_\alpha}{\mathrm{d}x^2}-2E'_\alpha(0)\,\delta(x)=-\frac{4\pi\mathrm{i}\omega}{\mathrm{c}^2}\int_{-\infty}^{\infty}K_{\alpha\beta}(x-x_1)\,E_\beta(x_1)\,\mathrm{d}x_1. \tag{7.28}$$

Taking the Fourier transform of this equation, we get

$$-k^2E_{k\alpha}-2E'_\alpha(0)=-\frac{\mathrm{i}\omega e^2}{c^2\pi\hbar^3}B_{\alpha\beta}\,|k|^{-1}E_{k\beta}. \tag{7.29}$$

This is a system of two equations ($\alpha, \beta=y, z$). We may choose the axes in the (y, z) plane as we wish. Let us choose them along the principal axes of the tensor $B_{\alpha\beta}$. We obtain two independent equations: one for E_y with B_1 and the other for E_z with B_2. Since they are identical in form, we shall consider one of them:

$$-k^2E_k-2E'(0)=-\frac{\mathrm{i}\omega\mathrm{e}^2}{\mathrm{c}^2\pi\hbar^3}\,B\,|k|^{-1}E_k. \tag{7.30}$$

Solving it for E_k, we obtain

$$E_k=-2E'(0)\left(k^2-\frac{\mathrm{i}b}{|k|}\right)^{-1}, \tag{7.31}$$

where

$$b=\frac{\omega e^2B}{c^2\pi\hbar^3}. \tag{7.32}$$

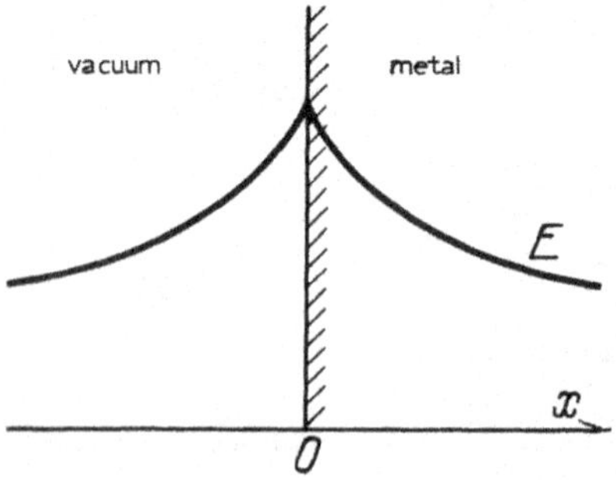

Fig. 31.

Since E_k is symmetric with respect to k, then, passing over to $E(x)$, we find

$$E(x)=\int_{-\infty}^{\infty}\frac{dk}{2\pi}E_k\,e^{ikx}=2\int_0^{\infty}\frac{dk}{2\pi}E_k\cos kx$$
$$=-\frac{2E'(0)}{\pi}\int_0^{\infty}\frac{\cos kx}{k^2-ib/k}\,dk=-\frac{2E'(0)}{\pi}\int_0^{\infty}\frac{k\cos kx}{k^3-ib}\,dk. \tag{7.33}$$

The field at the surface is given by

$$E(0)=-\frac{2E'(0)}{\pi}\int_0^{\infty}\frac{k\,dk}{k^3-ib}.$$

This integral can be calculated analytically and it is equal to

$$\int_0^{\infty}\frac{k\,dk}{k^3-ib}=\frac{2\pi}{b^{1/3}}\frac{e^{i\pi/6}}{3^{3/2}}.$$

Thus,

$$\frac{E(0)}{E'(0)}=-4\cdot 3^{-3/2}\,e^{i\pi/6}b^{-1/3}.$$

According to (7.6), the surface impedance is given by

$$Z=\frac{4\pi}{c}\frac{E_y(0)}{H_z(0)}.$$

But, on the other hand,

$$\frac{dE_y}{dx}=\frac{i\omega}{c}H_z$$

and, hence, $H_z(0)=cE_y'(0)/(i\omega)$. We therefore have

$$Z=\frac{4\pi i\omega}{c^2}\frac{E_y(0)}{E_y'(0)}. \tag{7.34}$$

Substituting $E(0)/E'(0)$, we obtain

$$Z=\frac{8\pi}{3^{3/2}}(1-i\sqrt{3})\omega c^{-2}b^{-1/3}.$$

Finally, substituting here the value of b (7.32), we find

$$Z=\frac{8\pi^{4/3}}{3^{3/2}}\,\omega^{2/3}e^{-2/3}c^{-4/3}\hbar B^{-1/3}(1-i\sqrt{3}). \tag{7.35}$$

Thus, Z is a tensor in the (y, z) plane, whose principal values are expressed in terms of the principal values of the tensor $B_{\alpha\beta}$. This result was obtained for the specular case.

The solution for the diffuse case is more complicated, and we shall not give it here; the result is a formula for Z, which is similar to (7.35), but it has the coefficient $\sqrt{3}$ instead of $8/3^{3/2}$ in (7.35). This means that the ratio of the Z's in both cases is $\frac{9}{8}$. This difference does not mean much in comparing theory with experiment because independent evaluations of the absolute value of Z differ by no less than 10 percent. Hence, it is impossible to deduce the type of reflection from the high-frequency impedance*.

Note that the collision time did not appear in eq. (7.35). The tensor $B_{\alpha\beta}$ depends only on the characteristics of the Fermi surface. Thus, the measurement of the impedance may be used to find the Fermi surface of real metals. As a matter of fact, this method was one of the first procedures used for this purpose.

The result (7.35) coincides with the result that has been found in the previous section by means of the "inefficiency" concept. However, the microscopic derivation using the kinetic equation made it possible to determine the unknown constant and phase factor.

Since the collision time was not included into the result, it is not important in which form the collision integral is chosen. However, it can be shown (Azbel' and Kaner 1956) that the form adopted was accurate in the limiting anomalous case. Indeed, for scattering by impurities we have

$$I(f) = \int W(\boldsymbol{p}, \boldsymbol{p}')\,[f(\boldsymbol{p}') - f(\boldsymbol{p})]\,\frac{\mathrm{d}S'}{v'(2\pi\hbar)^3},$$

where the integration goes over the surface $\varepsilon(\boldsymbol{p}') = \varepsilon(\boldsymbol{p}) = \text{const}$. Inserting $f = f_0 - (\partial f_0/\partial\varepsilon)\psi$ yields

$$I(f) = \frac{\partial f_0}{\partial\varepsilon}\int W(\boldsymbol{p}, \boldsymbol{p}')[\psi(\boldsymbol{p}) - \psi(\boldsymbol{p}')]\,\frac{\mathrm{d}S'}{v'(2\pi\hbar)^3},$$

where this time $\boldsymbol{p}$ and $\boldsymbol{p}'$ lie at the Fermi surface.

The quantity ψ as a function of $\boldsymbol{p}$ is close to $\delta(n_x)$. This can be seen from expression (7.23) for the kernel $K_{\alpha\beta}$, which determines the current. If we substitute $\psi \sim \delta(n_x)$ into the collision integral, the term with $\psi(\boldsymbol{p})$ will retain the character of the δ-function, and in the term with $\psi(\boldsymbol{p}')$ the δ-function will disappear upon integration. Hence, this term will become smooth and unimportant for the determination of the δ-shaped function ψ. Denoting $\int W(\boldsymbol{p}, \boldsymbol{p}')\mathrm{d}S'/v'(2\pi\hbar)^3$ by $\tau^{-1}(\boldsymbol{p})$, we obtain the collision integral with the time $\tau(\boldsymbol{p})$. Since $\tau(\boldsymbol{p})$ has no singularities at

* A more detailed calculation, in which account is taken of the surface structure (Fal'kovskii 1971 and 1979) shows that in the case of the limiting anomalous skin effect the reflection of electrons from the surface is practically specular. This result has a simple physical interpretation. If an electron falls onto the boundary at an angle θ_1, the de Broglie wavelength corresponding to its motion along the normal to the surface is of the order of $\hbar/(p_0\theta_1)$. If there is an irregularity of atomic dimensions, it gives rise to an uncertainty of the wave phase on the surface of order $a/\lambda \sim \theta_1$. In the case of the anomalous skin effect the relevant values are $\theta_1 \sim \delta/l \ll 1$. With this accuracy the boundary may be considered specular.

$n_x = 0$, it follows that in the integral (7.25) the quantity l^* may be regarded as being independent of θ, and it drops out upon integration over θ. It can be shown in an analogous manner that the adopted form of the collision integral applies for any scattering mechanism.

7.4. *Cyclotron resonance*

We have so far been concerned with metals in a high-frequency field or in a strong magnetic field. We shall now consider a case where there are both fields present. The most interesting phenomenon here is the so-called cyclotron resonance.

Suppose a metal occupies a half-space $x > 0$ and the magnetic field is parallel to the metal surface and is directed along the axis z perpendicular to the plane of the figure (fig. 32). It is assumed that the Fermi surface is closed. It is also assumed that the field is sufficiently strong, so that the Larmor radius is much smaller than the mean free path. In this case, the electrons will move along helical trajectories with the axis along the z axis (a circular projection onto the (x, y) plane is shown for the sake of simplicity; in fact, we shall consider the general case). If the temperature is sufficiently low and the frequency of the alternating field is high, the skin depth δ is much less than r_L. In this case, some of the electrons can return several times into the skin layer, although they will spend most of the time outside that layer.

The electric field in the skin layer varies with time and if the rotation frequency of the electron is equal to the frequency of the applied field, then each time the electron enters the skin layer the electric field will push it in one direction. Of course, the same will happen in a case where the frequency of the electromagnetic field is an integer multiple of the Larmor frequency, $\omega = n\Omega$, since the variation of the field during the time the electron is outside the skin layer is unimportant.

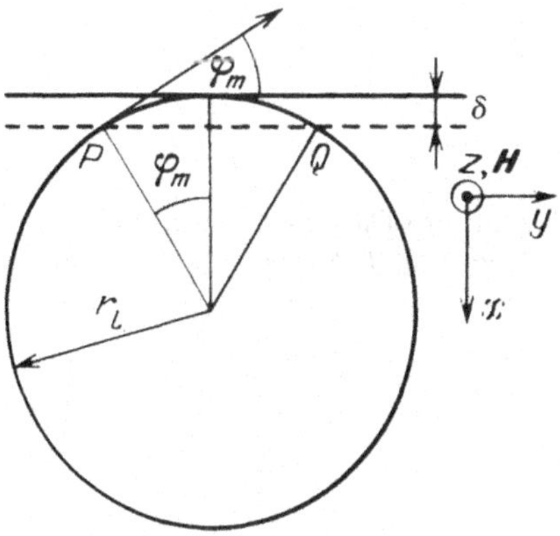

Fig. 32.

From the foregoing it can be seen that if the resonance conditions are satisfied, the electrons will effectively absorb the energy of the electromagnetic field. This phenomenon is very similar to what occurs in cyclotrons - accelerators of elementary particles - and therefore it has been called cyclotron resonance (Azbel' and Kaner 1956). Since this phenomenon depends on the electronic orbits, the corresponding experiments may be expected to yield some information about the electronic spectrum.

We shall not give the detailed theory of this phenomenon here. It will be simpler to employ the inefficiency concept (section 7.2). We start with the electrons that are responsible for cyclotron resonance, although, as we shall see below, their contribution to the total current is relatively small.

Let us write a time-dependent kinetic equation:

$$\frac{\partial f}{\partial t}+\frac{\partial f}{\partial t_1}+\boldsymbol{v}\frac{\partial f}{\partial \boldsymbol{r}}+eE_z\frac{\partial f}{\partial p_z}+e\boldsymbol{v}\boldsymbol{E}\frac{\partial f}{\partial \varepsilon}=-\frac{f-f_0}{\tau}. \tag{7.36}$$

The use of τ is justifiable for the same reasons as in the absence of a magnetic field (section 7.3). As always, we seek the distribution function in the form

$$f=f_0-\frac{\partial f_0}{\partial \varepsilon}\psi,$$

and assuming that ψ depends on time according to the law $\mathrm{e}^{-\mathrm{i}\omega t}$, we obtain

$$(-i\omega+\tau^{-1})\psi+\frac{\partial \psi}{\partial t_1}+\boldsymbol{v}\frac{\partial \psi}{\partial \boldsymbol{r}}=e\boldsymbol{v}\boldsymbol{E}. \tag{7.37}$$

This equation can be solved by using the method of characteristics. Let us write:

$$\mathrm{d}t_1=\frac{\mathrm{d}x}{v_x}=\frac{\mathrm{d}y}{v_y}=\frac{\mathrm{d}z}{v_z}$$

$$=\frac{\mathrm{d}\psi}{e\boldsymbol{E}(\boldsymbol{r})\,\boldsymbol{v}-(-\mathrm{i}\omega+\tau^{-1})\psi}. \tag{7.38}$$

The first equations give

$$\boldsymbol{r}(t_1)-\boldsymbol{r}(t_{1'})=\int_{t_1'}^{t_1}\boldsymbol{v}(t_3)\,\mathrm{d}t_3. \tag{7.39}$$

This corresponds to the motion of an electron along a trajectory. The coordinate $\boldsymbol{r}(t_1)$ is inserted into $\boldsymbol{E}(\boldsymbol{r})$, so that $\boldsymbol{E}$ becomes a function of t_1. Then we solve the equation

$$\frac{\partial \psi}{\partial t_1}+(-\mathrm{i}\omega+\tau^{-1})\psi=e\boldsymbol{v}(t_1)\,\boldsymbol{E}(\boldsymbol{r}(t_1)).$$

The solution found is

$$\psi(t_1)=\int_{\mathrm{c}}^{t_1}e\boldsymbol{v}(t_2)\,\boldsymbol{E}(\boldsymbol{r}(t_2))\exp[(-\mathrm{i}\omega+\tau^{-1})(t_2-t_1)]\,\mathrm{d}t_2. \tag{7.40}$$

If the trajectory does not touch the surface, the lower limit of the integral must be taken to be equal to $-\infty$. As we have seen in section 5.2, the quantity ψ will then be periodic. As for the trajectories that touch the surface, they make no contribution to the resonance, and therefore we shall not consider them for the moment.

Moreover, one should take into account the fact that we are dealing with ψ at a given point $\boldsymbol{r}$. But, since the electron moves, according to eq. (7.39), along a trajectory $\boldsymbol{r}(t_1)$, we must choose a trajectory such that at time t_1' the electron is at point $\boldsymbol{r}$. From (7.39) we obtain

$$\boldsymbol{r}(t_1) = \boldsymbol{r} + \int_{t_1'}^{t_1} \boldsymbol{v}(t_3)\, \mathrm{d}t_3 . \tag{7.41}$$

This is exactly the argument $\boldsymbol{r}(t_2)$ of $\boldsymbol{E}(\boldsymbol{r}(t_2))$ in eq. (7.40).

Substitution of eq. (7.40) into the expression for the current gives

$$j_\alpha(\boldsymbol{r}, \omega) = \frac{2e^2}{(2\pi\hbar)^3} \oint \frac{\mathrm{d}S}{v} v_\alpha \times \int_{-\infty}^{t_1} v_\beta(t_2) E_\beta(\boldsymbol{r}(t_2)) \exp[(t_2 - t_1)(\tau^{-1} - \mathrm{i}\omega)]\, \mathrm{d}t_2 \tag{7.42}$$

(where $\alpha, \beta = y, z$). Instead of an exact evaluation of the integral in (7.42) we shall find the result by resorting to simple physical reasonings.

Let us consider the electronic orbit in fig. 32. In this figure it is assumed that $\delta \ll r_\mathrm{L}$; φ_m represents the maximum angle corresponding to the electron that touches the metal surface. From fig. 32 it follows that

$$\delta = r_\mathrm{L} - r_\mathrm{L} \cos \varphi_\mathrm{m} \quad \text{or} \quad r_\mathrm{L}\varphi_\mathrm{m}^2 \sim \delta .$$

Hence,

$$\varphi_\mathrm{m} \sim \left(\frac{\delta}{r_\mathrm{L}}\right)^{1/2} . \tag{7.43}$$

The length of the portion of the orbit inside the skin layer $\mathrm{PQ} \approx 2r_\mathrm{L}\varphi_\mathrm{m}$, and, hence, the time spent by the electron in the skin layer is equal to $2r_\mathrm{L}(\varphi_\mathrm{m}/v_y)$.

If the electron does not touch the surface, then in (7.42) one has to integrate from $-\infty$ to t_1. Up to time t_1 there are many short intervals when the electron enters the skin layer. Each of these time intervals is very short, and in the integration over a single interval we shall replace the integrand by its average value. Only these time intervals make a contribution to the integral over t_2, since the field $\boldsymbol{E}$ is equal to zero when the electron is moving outside the skin layer. However, since before entering the skin layer the electron moves during the entire period T, it follows that each such interval adds in the integral (7.42) a phase factor $\exp[-T(\tau^{-1} - \mathrm{i}\omega)]$ as compared to the next interval. Introducing the notation $w = T(\tau^{-1} - \mathrm{i}\omega)$, we obtain after integration over $\mathrm{d}t_2$ the sum

$$\frac{2r_\mathrm{L}\varphi_\mathrm{m}}{v_y}(1 + \mathrm{e}^{-w} + \mathrm{e}^{-2w} + \cdots = \frac{2r_\mathrm{L}\varphi_\mathrm{m}}{v_y}(1 - \mathrm{e}^{-w})^{-1} . \tag{7.44}$$

It will be shown below that the factor $(1-e^{-w})^{-1}$ gives rise to the resonance. The integral over dS may be rewritten as

$$\int dS \to \int \frac{d\Omega}{K(\theta, \varphi)},$$

where K is the Gaussian curvature of the surface at a point where the direction of the normal to it is defined by the angles θ, φ (section 7.3).

Let us now apply the inefficiency concept. Since only those electrons are important here which are moving in the skin depth parallel to the surface, the integration over $d\theta$ gives a value of order $v_{x,\max}/v$. But from fig. 32 it follows that $v_{x,\max} = v_y \varphi_m$. Thus, the integration over $d\theta$ yields $v_y \varphi_m/(vK(\varphi))$, where $K(\varphi) \equiv K(\frac{1}{2}\pi, \varphi)$. Substituting this into the integral (7.42), we find

$$j_\alpha \sim \frac{2e^2}{(2\pi\hbar)^3} \int (1-e^{-w})^{-1} \frac{2r_L\varphi_m}{v_y} \frac{v_y \varphi_m}{vK(\varphi)} v_\alpha v_\beta E_\beta \frac{d\varphi}{v}.$$

Inserting $\varphi_m \sim (\delta/r_L)^{1/2}$ from (7.43), we obtain the effective conductivity:

$$\sigma_{r\alpha\beta} \sim \frac{2e^2}{(2\pi\hbar)^3} \int_0^{2\pi} (1-e^{-w})^{-1} \frac{n_\alpha n_\beta}{K(\varphi)} d\varphi \cdot \delta. \tag{7.45}$$

In the absence of a magnetic field $w \to \infty$ and the effective conductivity is

$$\sigma_r \sim \frac{e^2 p_0^2 \delta}{\hbar^3} \sim \frac{\sigma\delta}{l},$$

where

$$\sigma \sim \frac{n_e e^2 \tau}{m} \sim n_e e^2 l/p_0 \sim e^2 p_0^2 l/\hbar^3$$

is the ordinary conductivity. This quantity corresponds to the anomalous skin effect without field (cf. eq. 7.9, $k \sim \delta^{-1}$).

Let us now consider the contribution to the current from trajectories that touch the surface. The most important of these are the so-called "hopping trajectories" shown in fig. 33 (Fal'kovskii 1981). This is associated with the circumstance mentioned above: the electrons that fall onto the surface at a small angle are reflected from it mainly specularly (see the footnote on p. 128). Such electrons always move within the skin depth and introduce a major contribution to the current. Let us estimate the current by applying the inefficiency concept*.

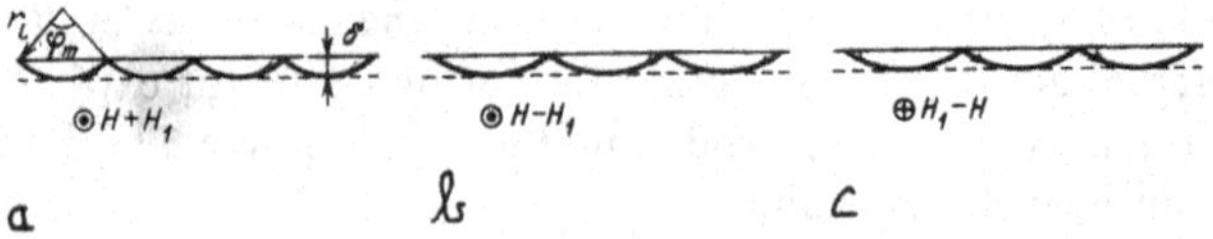

Fig. 33.

* The reasonings given below have been put forward by Fal'kovskii (1983).

For the electrons to stay within the skin layer, it is required that the angle in fig. 33 be smaller than $\varphi_m \sim (\delta/r_L)^{1/2}$. The effective number of electrons will be of the order of $n_e \varphi_m$. The effective conductivity varies accordingly. However, no account is taken here of the fact that in each collision with the metal surface there is a small probability of diffuse reflection. If we denote this probability by q, then the formula for the effective conduction of electrons moving along the hopping trajectories may be written as

$$\sigma_s \sim \frac{\sigma \varphi_m}{(1+\zeta q)}, \tag{7.46}$$

where ζ is the number of collisions with the surface along the mean free path. Since the length of a single portion between reflections is of order $r_L \varphi_L \sim (\delta r_L)^{1/2}$, we have $\zeta \sim l/(\delta r_L)^{1/2}$.

A strict microscopic treatment shows that if the surface irregularities have atomic dimensions, then $q \sim \varphi_m$ (this corresponds to the estimate in the footnote on p. 128). It follows that

$$\zeta q \sim \frac{l}{(\delta r_L)^{1/2}} \cdot \left(\frac{\delta}{r_L}\right)^{1/2} \sim l/r_L \gg 1.$$

In this case the second term in parenthesis in formula (7.46) is considerably larger than the first, and therefore use may be made of the expression

$$\sigma_s \sim \frac{\sigma}{\zeta} \sim \frac{\sigma(\delta r_L)^{1/2}}{l}. \tag{7.47}$$

Comparing this expression with the expression for the effective conductivity of resonance electrons $\sigma_r \sim \sigma\delta/l$, we see that $\sigma_s \gg \sigma_r$ (with the possible exception of the immediate vicinity of the resonances with small n; see below).

We then proceed analogously to the calculation of the impedance in the case of the anomalous skin effect, the only difference being that here we use δ instead of k^{-1}. Let us write the total effective conductivity in the form:

$$\sigma_{eff} \sim \sigma \frac{(\delta r_L)^{1/2}}{l}(1+\alpha),$$

where $\alpha = \sigma_r/\sigma_s \ll 1$; we then insert it into expression (7.4) for δ. We obtain the equation for δ, whose solution is

$$\delta \sim \left(\frac{c^2 l}{\omega \sigma}\right)^{2/5} r_L^{-1/5}(1+\alpha)^{-2/5}. \tag{7.48}$$

Inserting this formula (7.7) for the impedance, we have

$$Z \sim c^{-2}\frac{\omega}{k} \sim c^{-2}\omega\delta \sim \left(\frac{\omega}{c^2}\right)^{3/5}\left(\frac{\sigma}{l}\right)^{-2/5} r_L^{-1/5}(1-\tfrac{2}{5}\alpha). \tag{7.49}$$

The relative value of the resonance part of the impedance is of the order of $\alpha = \sigma_r/\sigma_s$. Substituting the corresponding formulae and expressing δ with the aid of (7.48), we find

$$\frac{Z_r}{Z} \sim \frac{\sigma_r}{\sigma_s} \sim \left(\frac{\delta}{r_L}\right)^{1/2} \sim \left(\frac{c^2 l}{\omega\sigma}\right)^{1/5} r_L^{-3/5} \ll 1. \tag{7.50}$$

We will now consider the resonance part Z_r in more detail. It is proportional to the conductivity of resonance electrons (7.45). This formula contains the tensor

$$B_{\alpha\beta} = \int_0^{2\pi} \frac{n_\alpha n_\beta}{K(\varphi)} \frac{d\varphi}{1 - e^{-w(\varphi)}}. \tag{7.51}$$

The denominator of the integrand contains

$$1 - e^{-w} = 1 - \exp\left(\frac{2\pi i\omega}{\Omega} - \frac{2\pi}{\Omega\tau}\right).$$

If $\Omega\tau \gg 1$, then the factor $\exp[-2\pi/(\Omega\tau)]$ is unimportant, and for $\omega = n\Omega$ we obtain the resonance. It should, however, be noted that Ω depends on φ and upon integration over φ the singularity of the integral may vanish. Therefore, the integral must be analyzed more thoroughly.

As has been shown in section 5.1, $\Omega = eH/(m^*c)$, where m^* is the cyclotron mass given by

$$m^* = (2\pi)^{-1}\left(\frac{\partial S(p_z, \varepsilon)}{\partial \varepsilon}\right)_{\varepsilon=\mu}.$$

The contribution to the current comes only from those points at the Fermi surface where the velocity lies in the plane of the metal surface, i.e., $v_x = 0$. This is illustrated in fig. 34. This figure shows a closed Fermi surface. The dashed line connects the points where $v_x = 0$. The vertical flat contours are the intersections of the Fermi surface with the $p_z = $ const. planes.

If the dependence of the energy on the momentum were quadratic, i.e., $\varepsilon = \alpha_{ik} p_i p_k$, then, obviously, the quantity m^* would be the same for all p_z values. In the general case, this is not so, of course, and m^* depends on p_z, varying within certain limits.

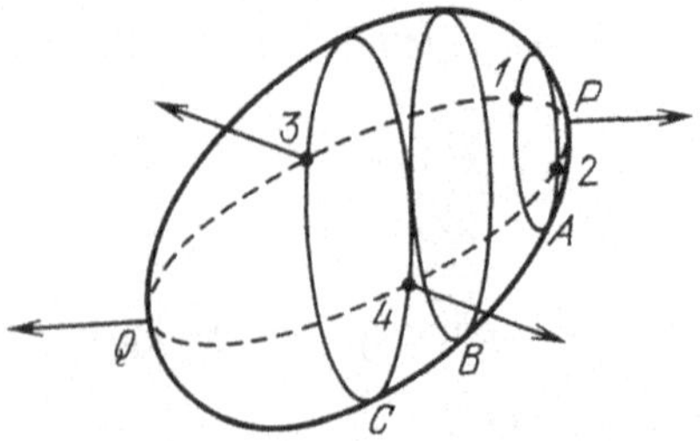

Fig. 34.

Suppose that the frequency and magnetic field are such that the electrons of a certain cross section produce a resonance. If we slightly change the field or frequency, the electrons of the neighboring cross section will correspond to a resonance. From this it can be seen that if electrons with any intermediate mass m^* are in resonance, this cannot give rise to a singularity in the impedance.

The situation is different if electrons with an extremal (maximum or minimum) value of m^* are in resonance. In this case, when the field is altered to one side, other electrons prove to be in resonance, but when the field is altered to the other side, the resonance will disappear altogether. Here, the impedance should be expected to exhibit a singularity. Hence, the resonance condition is

$$\omega = n\Omega_{\text{extr}}, \qquad \Omega_{\text{extr}} = \frac{eH}{m^*_{\text{extr}}c}. \tag{7.52}$$

Let us examine this problem in more detail. Usually, in experiments the frequency ω is kept constant and the magnetic field is varied. Suppose that we are near the resonance corresponding to $H = H_n$. Of importance in the integral (7.51) are obviously the values of φ near φ_0, corresponding to m^*_{extr}. In view of this, Ω may be represented approximately in the following form:

$$\Omega = \frac{\omega}{n}[1 + \Delta + a(\varphi - \varphi_0)^2],$$

where $\Delta = (H - H_n)H_n$, $a \sim 1$. If the mass at φ_0 is minimal, then $a < 0$; but if the mass is at a maximum, then $a > 0$. The quantity $w = 2\pi/(\Omega\tau) - 2\pi i\omega/\Omega$, which enters into (7.50), is close to $-2\pi i n$. Suppose that $\Omega\tau \gg 1$. Expanding $1 - e^{-w}$ in Δ, $(\varphi - \varphi_0)^2$ and $(\Omega\tau)^{-1}$, we find

$$1 - e^{-w} \approx w + 2\pi i n \approx \frac{2\pi n}{\omega\tau} + 2\pi i n\Delta$$
$$+ 2\pi i n a(\varphi - \varphi_0)^2.$$

Since the integral in (7.51) converges rapidly near $\varphi = \varphi_0$, all the quantities that vary slowly in this neighborhood may be replaced by their values at $\varphi = \varphi_0$, and the limits of integration over $x = \varphi - \varphi_0$ may be regarded as infinite. Thus, we obtain

$$B_{\alpha\beta} = \frac{n_\alpha(\varphi_0)\, n_\beta(\varphi_0)}{K(\varphi_0)} \int_{-\infty}^{\infty} \frac{dx}{2\pi i n a(x^2 + \Delta/a - i/a\omega\tau)}. \tag{7.53}$$

A point of interest here is the following. In fact, the extremal mass refers to a given cross section $p_z = \text{const.}$, which intersects the contour $v_x = 0$ at two points (except the limiting point; see below), corresponding to two angles, φ_{01} and φ_{02}. Therefore, obviously one should consider the neighborhood of both these angles. The integrals will here be the same and the only replacement required involves the factor in front of the integral in (7.53): we have to take the sum for φ_{01} and φ_{02}:

$$B^{(0)}_{\alpha\beta} = \frac{n_\alpha(\varphi_{01})\, n_\beta(\varphi_{01})}{K(\varphi_{01})} + \frac{n_\alpha(\varphi_{02})\, n_\beta(\varphi_{02})}{K(\varphi_{02})}. \tag{7.54}$$

Consider the case where $|\Delta| \gg (\omega\tau)^{-1}$. If $a > 0$, we have

$$B_{\alpha\beta} = B^{(0)}_{\alpha\beta} \times \begin{cases} -\mathrm{i}(2n\sqrt{\Delta a})^{-1}, & \Delta > 0, \\ (2n\sqrt{|\Delta| a})^{-1}, & \Delta < 0. \end{cases} \tag{7.55}$$

But if $a < 0$, we obtain

$$B_{\alpha\beta} = B^{(0)}_{\alpha\beta} \times \begin{cases} (2n\sqrt{\Delta |a|})^{-1}, & \Delta < 0, \\ \mathrm{i}(2n\sqrt{|\Delta||a|})^{-1}, & \Delta < 0. \end{cases} \tag{7.56}$$

From the foregoing it is seen that the sign has a substantial effect on the result, i.e., the maximum and minimum cyclotron masses give quite different types of effects in the impedance.

The maximum value in the resonance corresponds to

$$\Delta \sim \frac{1}{\omega\tau}, \quad \text{i.e.,} \quad B_{\alpha\beta} \sim B^{(0)}_{\alpha\beta} \frac{(\omega\tau)^{1/2}}{n}.$$

If $n \sim 1$, then $\omega \sim \Omega$, i.e.,

$$B_{\alpha\beta} \sim B^{(0)}_{\alpha\beta}(\omega\tau)^{1/2} \sim B^{(0)}_{\alpha\beta}\left(\frac{l}{r_L}\right)^{1/2}.$$

Since σ_r is proportional to $B_{\alpha\beta}$, then, according to (7.50), Z_r/Z may reach a value of order $(l/r_L)^{1/2} \cdot (\delta/r_L)^{1/2} \sim (l\delta)^{1/2}/r_L$. In principle, this value may even be larger, but in actual experiments it never reaches large values. As n increases the height of the maxima falls off rapidly.

For resonances with a large value of n one can no longer say that $2\pi/\Omega\tau \ll 1$ and it is better to use the approximation $e^{-w} \ll 1$. Then, the oscillatory part of the tensor $B_{\alpha\beta}$ corresponds to the integral

$$\tilde{B}_{\alpha\beta} = \int_0^{2\pi} \frac{n_\alpha n_\beta}{K(\varphi)} e^{-w}\, d\varphi$$
$$= \int_0^{2\pi} \frac{n_\alpha n_\beta}{K(\varphi)} \exp\left(-\frac{2\pi}{\Omega\tau}\right) \exp\left(\frac{2\pi \mathrm{i}\omega}{\Omega}\right) d\varphi.$$

Inserting $\Omega = \Omega_{\mathrm{extr}}[1 + a(\varphi - \varphi_0)^2]$ and assuming that $\omega\tau \gg 1$, we obtain an integral which can be evaluated by the method of steepest descent. As a result, we have

$$\tilde{B}_{\alpha\beta} = B^{(0)}_{\alpha\beta}\left(\frac{\Omega_{\mathrm{extr}}}{2\omega|a|}\right)^{1/2} \exp(\tfrac{1}{4}\mathrm{i}\pi \operatorname{sign} a) \exp\left(\frac{-2\pi}{\Omega_{\mathrm{extr}}\tau}\right) \exp\left(\frac{2\pi \mathrm{i}\omega}{\Omega_{\mathrm{extr}}}\right). \tag{7.57}$$

It follows that as n increases the set of resonance peaks is transformed to damped oscillations of the impedance.

Let us return to resonance condition (7.52). It is of interest to find the exact location of cross sections with extremal masses. If the Fermi surface has a center of symmetry, then the central cross section (C in fig. 34) will definitely have an extremal mass. There may also be fortuitous cross sections that display this property. And, finally, this property is exhibited by the points P and Q in fig. 34, which are called the elliptic limiting points (at these points the surface is tangent to the $p_z = \text{const.}$ plane). In order to prove this, consider a cross section which is very close to P, say A. This cross section intersects the $v_x = 0$ line at two points 1 and 2. These correspond to different values of φ, but belong to a single cross section and, hence, have the same value of $m^* = (2\pi)^{-1}\partial S/\partial\varepsilon$. Hence, a certain value of φ between these points corresponds to the extremum of $\partial S/\partial\varepsilon$. Obviously, this is the point P.

In order to calculate $\partial S/\partial\varepsilon$ at point P, consider again a cross section close to P. It will be an ellipse (fig. 35) with half-axes given by

$$a = (R_1^2 - (R_1 - \delta p_z)^2)^{1/2} \approx (2R_1\delta p_z)^{1/2}, \qquad b \approx (2R_2\delta p_z)^{1/2},$$

where R_1 and R_2 are two principal curvature radii of the Fermi surface. The cross-sectional area is given by

$$\delta S = \pi ab = 2\pi(R_1R_2)^{1/2}\delta p_z = 2\pi\delta p_z/\sqrt{K}. \tag{7.58}$$

At point P the velocity has the direction $\boldsymbol{H}$, i.e., $v_z = v = \partial\varepsilon/\partial p_z$; hence,

$$\delta p_z = \frac{\partial\varepsilon}{v} \quad \text{and} \quad \delta S = \frac{2\pi}{\sqrt{K}}\frac{\delta\varepsilon}{v}.$$

Since $m^* = (2\pi)^{-1}\partial S/\partial\varepsilon$, then for the elliptic limiting points we have

$$m^* = (v\sqrt{K})^{-1}. \tag{7.59}$$

Here, however, an uncertainty arises. If the velocity at point P is parallel to the magnetic field, then what is the Larmor radius? It is apparently equal to zero in

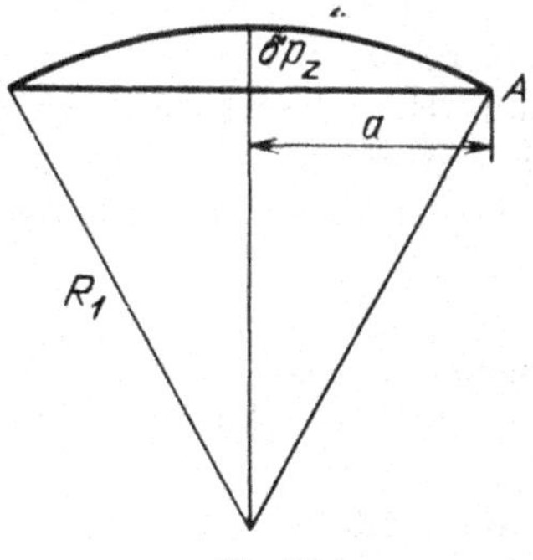

Fig. 35.

this case, while the phenomenon is entirely based on the assumption that $r_L \gg \delta$. As a matter of fact, there is no contradiction at all. Near the limiting point

$$\Omega = \Omega_0[1 + a(\varphi - \varphi_0)^2],$$

where $a \sim 1$. But in formula (7.50) $w = T\ (\tau^{-1}) - i\omega)$, i.e., it contains the term $2\pi/(\Omega\tau)$. This term leads to a smearing out of the resonance, which implies the participation in it of points adjacent to P. The size of this region is determined by the condition $(\omega/\Omega)(\varphi - \varphi_0)^2 \sim (\Omega\tau)^{-1}$ or by

$$|\varphi - \varphi_0| \sim (\omega\tau)^{-1/2}. \tag{7.60}$$

On the other hand, the Larmor radius is evidently of order $r_L \sim r_{L0}(\varphi - \varphi_0)$ (where r_{Lo} is a certain average value for a given surface). Thus, the Larmor radii that participate in the resonance are of order $r_L \sim r_{L0}(\omega\tau)^{-1/2}$ near the limiting points. The condition $r_L \gg \delta$ implies that

$$(\omega\tau)^{1/2} \ll r_{L0}/\delta. \tag{7.61}$$

Thus, although it is necessary that $\omega\tau \gg 1$ for cyclotron resonance to be observed, this quantity must not be too large. Otherwise, it is impossible to observe the resonance from the limiting points.

In an experimental investigation of cyclotron resonance it is easy to distinguish between different types of points with extremal masses. At the limiting points P and Q in fig. 34 the electron velocity is directed along the magnetic field $\boldsymbol{H}$. Since the interaction with the electric field is expressed by the quantity $e\boldsymbol{vE}$, the resonance at a limiting point must disappear when $\boldsymbol{E}$ is perpendicular to the constant magnetic field (recall that the wave is incident normally on the metal surface and $\boldsymbol{E}$ certainly lies in the plane of the surface).

If the extremal masses are connected with the central cross section C in fig. 34, the velocities at points 3 and 4 which are important in our problem are just opposite. Hence, one can find a certain direction of the electric field perpendicular to both velocities. In this case, the resonance disappears. However, the velocities here are not parallel to the constant magnetic field.

Finally, if the cross section with an extremal mass is fortuitous, the velocities at two points of intersection of the contour $v_x = 0$ with the contour $p_z = \text{const.}$ are directed in different directions. Here the resonance cannot disappear altogether.

As has been said above, in experiments it is easier to alter the magnetic field than to alter the frequency. Writing the resonance condition in the form

$$H^{-1} = ne/(m^* c\omega)$$

we see that the surface impedance must be periodic as a function of H^{-1} (two periods are distinctly seen in fig. 36) (Khaikin 1961). From the period of oscillations we can determine m^*.

Let us now consider certain conditions that must be fulfilled for the cyclotron resonance to be observed. First of all, we will find the possible limit of the tilting

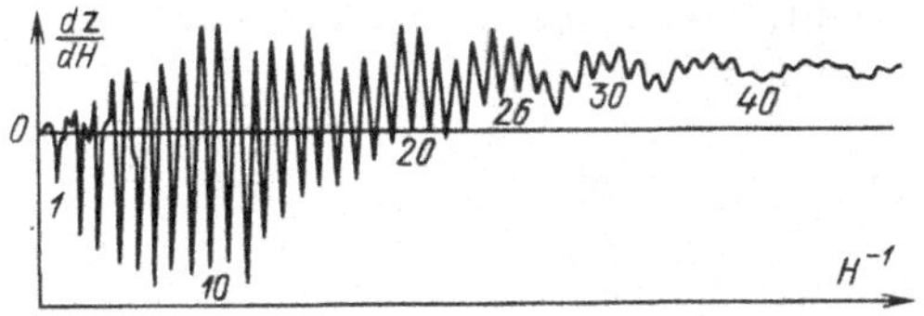

Fig. 36.

of the magnetic field to the metal surface. Let us inspect fig. 37a. For the resonance to arise, an electron must enter the skin layer at least twice. The period is equal to $2\pi r_L/v$. If the average velocity of motion along the field is of order v, the electron moves in the direction x with a velocity $v\psi$. During one period its displacement in the x direction will be $(2\pi r_L/v)v\psi$. It must be smaller than δ, i.e.,

$$\psi < \frac{\delta}{2\pi r_L}. \tag{7.62}$$

This condition is fulfilled for $\psi < 1°$ in the case of ordinary metals in fields of the order of 10^4 Oe.

At the limiting points $r_L/v \sim r_{Lo}/v$. Therefore, the condition (7.62) is valid for them as well.

The central cross sections are under special conditions. For them the average electron velocity in the direction of the magnetic field is equal to zero. Hence, the electron trajectory is closed rather than helical (fig. 37b). The condition for observation of the resonance in this case is not as restricting as in the general case. It can be deduced as follows. Although $\bar{v}_z = 0$ at the central cross section itself, the resonance in fact involves a certain neighborhood of the central cross section. This neighborhood is determined form the same considerations as in the case of a limiting point, i.e., it is expressed by formula (7.60). It is not difficult to see that the average velocity v_z at a certain $\varphi - \varphi_0$ will be of the order of $v(\varphi - \varphi_0)$, i.e., of order $v(\omega\tau)^{-1/2}$. Further, reasoning in the same way as in the derivation of formula (7.62), we find

$$\psi \ll \frac{\delta(\omega\tau)^{1/2}}{2\pi r_L}. \tag{7.63}$$

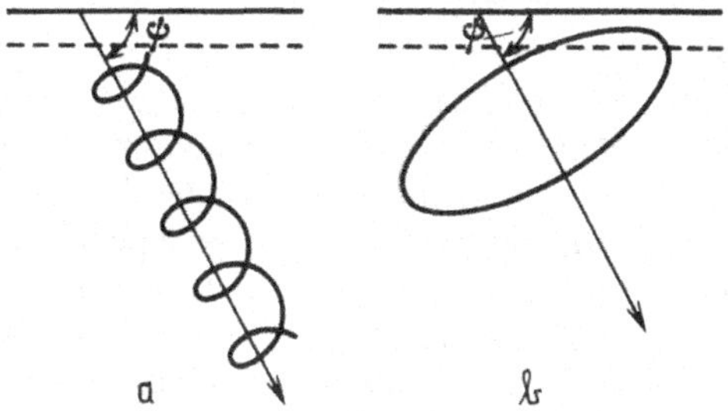

Fig. 37.

From formula (7.57) it follows that at large n the amplitude of oscillations of the impedance is proportional to

$$\exp\left(\frac{-2\pi}{\Omega\tau}\right)=\exp\left(\frac{-2\pi n}{\omega\tau}\right)=\exp\left(\frac{-n}{\nu\tau}\right), \tag{7.64}$$

where $\nu=\omega/2\pi$ is the cyclic frequency. If ω remains constant, then this factor falls off rapidly with increasing n, i.e., at a given $\nu\tau$ resonances can be observed with numbers which do not strongly exceed $\nu\tau$. This means that for the resonance to be observed, it is necessary to provide large τ values. For the frequency $\nu=10^{10}$ Hz the condition $n \gg 1$ gives $10^{10}\tau \sim 10^{10} l/v \sim 10^2 l \gg 1$ (since for most metals $v \sim p_0/m \sim 10^8$ cm/s). This means $l \gg 10^{-2}$ cm or the impurity concentration $c_i \ll 10^{-4}$.

Even with this condition the number of observed resonance peaks is limited. The point is that during the time interval when the electron is moving in the skin layer the electric field of the wave must practically remain constant. The length of this portion of the trajectory, as we have found out, is of order $(r_L\delta)^{1/2}$, and the electron moves along it with velocity v. Hence, the time taken by the electron to pass through the skin layer is of order $(r_L\delta)^{1/2}/v$, and the condition of invariance of the field is

$$\frac{(r_L\delta)^{1/2}}{v} \ll \nu^{-1}. \tag{7.65}$$

Considering that $r_L \sim v/\Omega = vn/\omega$, we obtain

$$n \ll \left(\frac{v}{\nu}\right)^2 \frac{\omega}{\delta v} \sim \frac{2\pi v}{\nu\delta}. \tag{7.66}$$

For $\nu \sim 10^{10}$ Hz, $\delta \sim 10^{-4}$ cm; from this we obtain $n < 10^3$, which is in agreement with experiment for the purest samples.

7.5. Nonlinear effects. Current states

Until now we have assumed the high-frequency field to be weak and, in fact, we have calculated the distribution of currents in a metal that are produced under the action of that field in first order with respect to the field. This kind of calculation is known as linear response theory. At the same time, the use of very pure metals and low temperatures has made it possible to observe nonlinear effects which depend on the intensity of the incident wave.

A large number of such phenomena exists. For example, when a wave is incident on a metal, in contrast to vacuum where the electric and magnetic fields of the wave are equal in amplitude, inside the metal the magnetic field is much stronger than the electric field. But the conductivity of pure metals depends on the magnetic field. This gives rise to a dependence of the surface impedance on the wave intensity and also leads to the generation of a wave with a doubled frequency, the appearance of a wave with a difference frequency when two waves of different frequency are incident on the metal surface simultaneously, etc.

The nonlinear effects manifest themselves especially strongly when a small isolated group of electrons is involved in this phenomenon. While being incapable of distorting the distribution function of all the electrons, the incident wave can exert an appreciable effect on a small group of electrons. An example is cyclotron resonance, in which the major role is played by a small group of electrons in the neighborhood of the intersection point of the planar cross section $p_z = \text{const.}$, which has an extremal cyclotron mass m^*, with the contour $v_x = 0$.

We shall consider here a very interesting nonlinear effect, namely the appearance of so-called "current states" (Dologopolov 1975). Suppose the problem is formulated in the same way as in the case of cyclotron resonance. However, this time we shall assume that the amplitude of the varying magnetic field H_1 is not small compared to the amplitude of the constant field H applied to the sample. In this case the motion of electrons will be substantially different, depending on the relative orientation of the fields $\boldsymbol{H}_1$ and $\boldsymbol{H}$.

It has been pointed out in the preceding section that if the skin depth is small compared to the Larmor frequency, the major contribution to the current comes from electrons that are moving along hopping trajectories (fig. 33). The shape of these trajectories depends significantly on the value of the Larmor radius. In the presence of a constant field an asymmetry arises between the currents in different half-periods, and consequently a rectified current appears at the surface. This current produces an increment of the constant magnetic field in the sample. The effect of "current states" consists of the possibility of a spontaneous appearance of a rectified current and of the corresponding constant field without application of any external constant field. The field and the current maintain each other. Of course, this may occur only at a sufficiently high amplitude of the incident high-frequency radiation.

We shall assume two conditions to be fulfilled for the hopping trajectories to occur. First, the number of jumps per mean free path must be large:

$$(\delta r_L)^{1/2} \ll l. \tag{7.67}$$

Second, the following condition must be satisfied:

$$\delta \ll r_L. \tag{7.68}$$

We shall assume, as before, that $\boldsymbol{E}$ and $\boldsymbol{H}$ are parallel to the sample surface and for the sake of simplicity we also assume that $\boldsymbol{E} \perp \boldsymbol{H}$ (E_y, H_z). Here one may become suspicious that even neglecting nonlinearity the current depends on the relative orientation of $\boldsymbol{E}$ and $\boldsymbol{H}$, i.e., is different in different half-periods. In fact, the electrons are turned in one direction by the magnetic field. Therefore, the electrons that move parallel to the surface parallel to $\boldsymbol{E}(\boldsymbol{vE} > 0)$, with a certain direction of $\boldsymbol{H}$ turn to the surface and form hopping trajectories and with another direction of $\boldsymbol{H}$ move into the bulk of the metal and escape from the skin layer. However, the current is produced not only due to acceleration of electrons moving in the direction $\boldsymbol{E}$ but also as a result of the deceleration of those which move in the opposite direction. For the latter the entire picture is opposite. It follows that the relative orientation of $\boldsymbol{E}$ and $\boldsymbol{H}$ is unimportant, and the the current depends only on the magnitude of $\boldsymbol{H}$.

As has been said above, the constant field $\boldsymbol{H}$ is made up of the external field and the field of the rectified current. Integrating the Maxwell equation rot $\boldsymbol{H}=(4\pi/c)\boldsymbol{j}$ and averaging over time, we obtain the following boundary condition:

$$\bar{H}_{\text{int}}-\bar{H}_{\text{ext}}=-\frac{4\pi}{c}\int_0^\infty \bar{j}\,\mathrm{d}x.$$

Here $\bar{H}_{\text{int}}$ is the average field in the bulk of the metal and $\bar{H}_{\text{ext}}$ is the average field outside the metal. Since the rectified current is short-circuited inside the metal, it does not produce a magnetic field outside the metal. Hence, $\bar{H}_{\text{ext}}=H_0$, $\bar{H}_{\text{int}}=H_0+\mathscr{H}$, where $\mathscr{H}$ is the field created by the rectified current. Thus, we have

$$\mathscr{H}=-\frac{4\pi}{c}\int_0^\infty \bar{j}\,\mathrm{d}x=f(H_0+\mathscr{H}). \tag{7.69}$$

This is an equation for $\mathscr{H}$. We put $H_0=0$; if eq. (7.69) has nonetheless nonzero solutions, this means spontaneous generation of a rectified current and a constant field which maintain each other.

To find the rectified current, we use formula (7.46) for the effective conductivity for hopping trajectories. The current $\bar{j}$ is found as the difference between the currents with the magnetic fields $|H_1-\mathscr{H}|$ and $H_1+\mathscr{H}$ (H_1 is the amplitude of the varying field). Since the current flows only in the skin layer, it follows that $\int \bar{j}\,\mathrm{d}x\sim\bar{j}\delta$.

Since $\bar{j}$ is a difference between currents, a question arises about its sign. This can be solved on the basis of the following reasoning. The external magnetic field causes curving of the electronic orbits, which in turn leads to the appearance in the bulk of the metal of an additional field, opposite to the external field (diamagnetism, see ch. 10). However, the current generated by electrons in hopping trajectories is opposite to the diamagnetic current from the bulk electrons and therefore always creates a paramagnetic effect, i.e., the magnetic field is parallel to the external field. If the constant field $H=H_0+\mathscr{H}$ is higher than the varying field H_1, then during the two half-periods the total field is directed in one direction (fig. 33a, b) and the average field is parallel to this field. But if $H_1>H$, the situation is different. In one half-period the field produced by the electrons in hopping trajectories is directed in one direction (fig. 33a) and in the other half-period it is directed in the opposite direction (fig. 33c). The time-averaged field is determined by the half-period during which the current j or, in other words, the effective conductivity, will be higher. This depends on the relationship between r_{L} and l. If $r_{\mathrm{L}}\gg l$, then, according to (7.46), $\sigma_{\text{eff}}\sim\sigma\varphi_{\mathrm{m}}\sim\sigma(\delta/r_{\mathrm{L}})^{1/2}$. In this case, σ_{eff} increases with increasing field, and it will be higher in that half-period when H_1 is parallel to H. In this case the averaged field created by the electrons of hopping trajectories, i.e., $\mathscr{H}$, is directed parallel to H. But if $r_{\mathrm{L}}\ll l$, then from (7.46) it follows that $\sigma_{\text{ef}}\sim\sigma(\delta r_{\mathrm{L}})^{1/2}/l$, i.e., the conductivity falls off with increasing field and it is larger in the half-period when H_1 is directed antiparallel to H. This leads to a change in the sign of the averaged field created by the electrons of hopping trajectories. However, in practice the amplitude of the incident electromagnetic wave is never so large that $r_L(H_1) \ll l$. For this reason, we shall assume that $r_L \gg l$.

For what follows it is very important which of the quantities should be considered specified. From the equation rot $\boldsymbol{E} = -c^{-1}\partial \boldsymbol{H}/\partial t$ we find that inside the metal

$$E \sim \frac{\delta\omega}{c} H_1 \sim \frac{\delta}{\lambda} H_1, \tag{7.70}$$

where $\lambda = c/\omega$ is the wavelength of the alternating field in vacuum. In the microwave region $\lambda \gg \delta$, i.e., in a metal $E \ll H_1$. From the boundary conditions at the metal surface it follows that H_1 is of the order of the amplitude of the incident wave. It is therefore natural to assume H_1 to have been specified.

We are interested here in the spontaneous generation of current states and therefore we put $H_0 = 0$ in eq. (7.69). Substituting $\sigma_{\text{eff}} = \sigma(\delta/r_L)^{1/2}$, we obtain

$$\begin{aligned} f(\mathcal{H}) &\sim \frac{\sigma}{c}\,\delta^{3/2}[r^{-1/2}(H_1 + \mathcal{H}) - r_L^{-1/2}(|H_1 - \mathcal{H}|)]E \\ &\sim \frac{\sigma\omega}{c^2}\,\delta^{5/2}[r_L^{-1/2}(H_1 + \mathcal{H}) - r_L^{-1/2}(|H_1 - \mathcal{H}|)]H_1. \end{aligned} \tag{7.71}$$

Let us consider the function $f(\mathcal{H})$ in the limiting cases. If $\mathcal{H} \ll H_1$, then we expand in $\mathcal{H}$:

$$f(\mathcal{H}) \sim \frac{\sigma\omega}{c^2}\,\delta^{5/2}\mathcal{H}H_1^{1/2}\left(\frac{e}{cp_0}\right)^{1/2}.$$

The skin penetration depth can be found from eq. (7.4). After substituting the expression for the effective conductivity* we obtain the equation for δ, whence we find

$$\delta \sim \left(\frac{c^2}{\omega\sigma}\right)^{2/5}\left(\frac{cp_0}{eH_1}\right)^{1/5}. \tag{7.72}$$

Inserting this expression into the preceding formula, we obtain

$$f(\mathcal{H}) \sim \mathcal{H}. \tag{7.73}$$

In the opposite limiting case we expand eq. (7.71) in H_1:

$$f(\mathcal{H}) \sim \frac{\sigma\omega}{c^2}\,\delta^{5/2}H_1^2\mathcal{H}^{-3/2}\left(\frac{e}{cp_0}\right)^{1/2}.$$

* In the qualitative derivation given above the quantity σ remains somewhat indefinite becuase of the difference in σ_{eff} at different half-periods. It seems to be reasonable to take the average value of σ_{eff}.

The formula for δ has the form (7.72) with H_1 being replaced by $\mathscr{H}$. Substituting it, we find

$$f(\mathscr{H}) \sim \frac{H_1^2}{\mathscr{H}}. \tag{7.74}$$

From this we cannot, of course, come to a definite conclusion concerning the possibility of spontaneous current states; to do this, the exact solution of the problem is required. However, the limiting formulas for $f(\mathscr{H})$, (7.73) and (7.74), show that a nonzero solution of eq. (7.69) at $H_0 = 0$ may exist in the region of $\mathscr{H} \sim H_1$. The necessary condition is the presence of hopping trajectories, ie., $(\delta r_L)^{1/2} \ll l$. Substituting δ in accordance with (7.72), we obtain

$$H_1 > H_c \sim \frac{cp_0}{e}\left(\frac{c^2}{\omega\sigma l^5}\right)^{1/3}. \tag{7.75}$$

The threshold value H_c falls off rapidly with increasing frequency and, especially, with increasing mean free path. For ordinary metals $\sigma \sim (ne^2/p_0)l \sim 10^{22} l$, i.e., $H_c \sim 5\omega^{-1/3} l^{-2}$ Oe. If we substitute here the actual experimental values: $\omega \sim 10^7\ \mathrm{s}^{-1}$, $l \sim 3 \times 10^{-2}$ cm, then we have $H_c \sim 20$ Oe, this being in accord with observations.

It has been experimentally found (Dolgopolov 1980) that when the "pumping" power exceeds a certain threshold, the metal jumps to a state with a rectified current and with the magnetic field created by it, i.e., to a magnetized state. This transition can be traced either form the appearance of an induction emf in the measuring coil surrounding the sample or from the hysteresis in any of the properties of the metal that depend on the magnetic field. For the threshold field H_c the law $\omega^{-1/3}$ corresponding to formula (7.75) is well observed. Current states have been detected in bismuth, tin, copper, indium, and molybdenum.

The transition described above is one of the examples of a "dynamical" phase transition. In contrast to ordinary phase transitions, we are speaking here of transitions between nonequilibrium states which occur with a change in the magnitude of "pumping", i.e., an external factor maintaining a state of nonequilibrium. Other examples of such transitions will be described in ch. 22.

8 Size effects

8.1. Cutoff of cyclotron resonance orbits

We will now consider certain phenomena that occur in thin metal films in varying and constant fields. There are many such phenomena and they are of different nature, but taken together they all form a special group of so-called size effects.

One of these effects is associated with cyclotron resonance (Kaner 1958). The geometry of the experiment is shown in fig. 38. Let us integrate one of the equations (5.4) over t:

$$\frac{\mathrm{d}p_y}{\mathrm{d}t} = -\frac{e}{c} v_x H$$

within the limits from $t = 0$ when the electron is at a minimum distance from the surface A (at this point, of course, $v_x = 0$), to the moment t' when the electron approaches the boundary B most closely (and again $v_x = 0$). Here we obtain

$$p_y(t') - p_y(0) = -\frac{e}{c} H \int_0^{t'} v_x \,\mathrm{d}t.$$

But the integral on the right-hand side of this equation is simply the displacement of the electron in the direction x. We shall call it the orbit diameter d. Hence,

$$p_y(t') - p_y(0) = -\frac{e}{c} Hd. \tag{8.1}$$

Suppose now that this is one of the orbits of cyclotron resonance. By virtue of relation (7.48) we have

$$p_y(t') - p_y(0) = -\frac{m^*\omega}{n} d_n \tag{8.2}$$

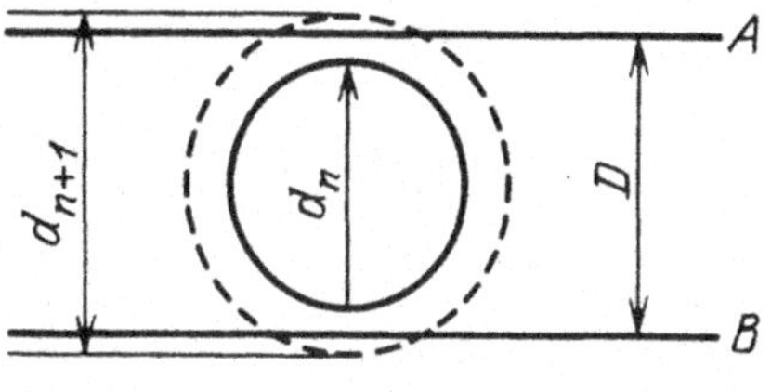

Fig. 38.

for any n. In the preceding section it has been found that each value of n corresponds to resonance. According to (8.2), the orbit diameter increases with increasing n. Therefore there exists a certain maximum n for which the diameter d_n is smaller than the thickness of the film D, whereas the diameter of the next order of magnitude, d_{n+1}, is larger than this thickness. The latter circumstance means that such electrons are reaching the boundary of the film and, hence, cannot give a resonance.

Instead of the plot shown in fig. 36, we obtain in this case the plot given in fig. 39 (Khaikin 1961)*. Enumerating the last observed peak and assuming that $d_n = D$ in (8.2), we find the value of the quantity $p_y(t') - p_y(0)$. If the resonance is associated with the central cross section of the Fermi surface, then, obviously, $t' = \frac{1}{2}T$, where T is a period and, moreover, $p_y(\frac{1}{2}T) = -p_y(0)$. In this case, from (8.2) we obtain

$$p_y(0) = \frac{m^* \omega}{2n} d_n. \tag{8.3}$$

Thus, this method enables one to measure the Fermi momentum. This is the p_y-coordinate of that point of the Fermi surface where the central cross section $p_z = 0$ intersects the $v_x = 0$ line. The accuracy of this measurement depends on the number n at which the resonance is cut off. Indeed, we obtain $d_n < D < d_{n+1}$. Hence, the maximum error is equal to n^{-1} and therefore conditions must be created for n to be as large as possible. According to (8.3), to do this we must have a high frequency ω. In pure tin crystals values have been obtained of $n \sim 25$–30.

In thin films one can also observe a special type of cyclotron resonance, which is of interest for the study of the energy spectrum, namely cyclotron resonance on

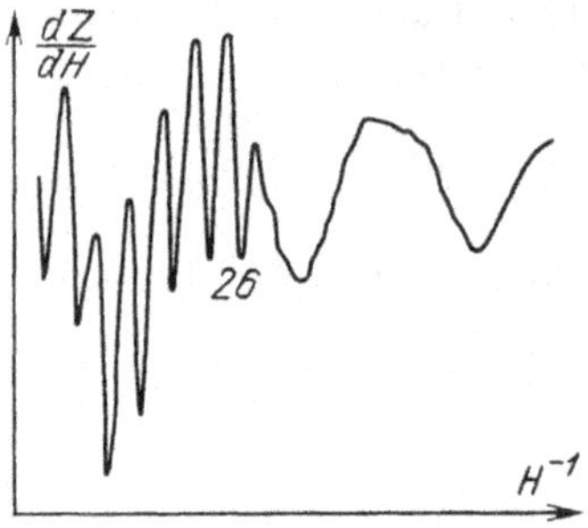

Fig. 39.

* The abrupt disappearance of the higher harmonic in fig. 39 is evidence of the diffuse reflection of electrons from the surface. This is quite understandable from the viewpoint of the reasoning given in the footnote on p. 128. If the nth orbit is the last one that can be placed in the sample, then the $(n+1)$st orbit approaches the surface at an angle of $\varphi \sim n^{-1/2}$. This gives rise to an uncertainty in the momentum of order $n^{-1/2}$ and, hence, to the same uncertainty in the Larmor radius. But for the resonance to be observable, the harmonics must not overlap, i.e., $\Delta r_L / r_L < n^{-1}$. Hence, the $(n+1)$st orbit cannot be a resonance orbit.

nonextremal orbits (Volodin et al. 1973). Suppose that we observe a cutoff of the $(n+1)$st orbit connected with the central cross section of the Fermi surface, which has the largest p_y. The other cross sections have lower values of p_y and, hence, smaller diameters of the orbit. By varying p_z, starting from the central cross section, we arrive at some other cross section, for which the following relation is valid:

$$p_y(t') - p_y(0) = -\frac{m^*\omega}{n+1} D.$$

At a given frequency this is a quite definite cross section. It is special in the sense that the cross sections closer to the central one cannot participate in the $(n+1)$st resonance, and the cross sections lying behind the given one can take part in it. Therefore, a singularity must appear in the impedance, at

$$H = m^* \frac{c\omega}{e(n+1)}.$$

An analogous formula with $n+2q$ corresponds to the next such cross section, etc.

This phenomenon has been observed in experiments (Volodin et al. 1973); it has been found that the amplitude of the corresponding maxima in the $\mathrm{d}R/\mathrm{d}H$ curve is one order of magnitude less than that of the maxima associated with the resonance at the central cross section. Nevertheless, the effect is observable and so far is the only effect that can be used for direct measurements of the dependence of the mass m^* on p_z. Note that, apart from the constant-mass case (i.e., a quadratic spectrum of the type of $\varepsilon = a_{ik} p_i p_k$) each such resonance has its own value of m^* and therefore there is no periodicity of the maxima with respect to H^{-1}, as in the case of the resonance at the extremal cross section.

8.2. Internal splashes of a high-frequency field in cyclotron resonance

As has been noted earlier, a high-frequency field penetrates the metal to a depth δ. However, in the presence of a magnetic field inside the metal layers are formed where the high-frequency field is also different from zero (Azbel' 1960).

Consider an electronic orbit that enters the skin layer of a metal (fig. 40). The length of the arc AB will be of order $r_L\varphi$ or $(r_L\delta)^{1/2}$. If the total current passing through the skin layer is J, the current density will be $j = J/\delta$. The electron moves along the orbit and therefore the component of its velocity along the surface, v_y, at any point C is $v \cos\theta$; the y-component of the total current will be proportional to $J\cos\theta$. This current will be distributed along the arc A′B′ of the same length as before. But now this arc has rotated and its projection onto the x axis is equal to $(r_L\delta)^{1/2} \sin\theta$. Therefore, the current density will now be $J\cos\theta/[(r_L\delta)^{1/2}\sin\theta]$, i.e., compared with the value in the skin layer it is reduced by a factor of $(\delta/r_L)^{1/2}$

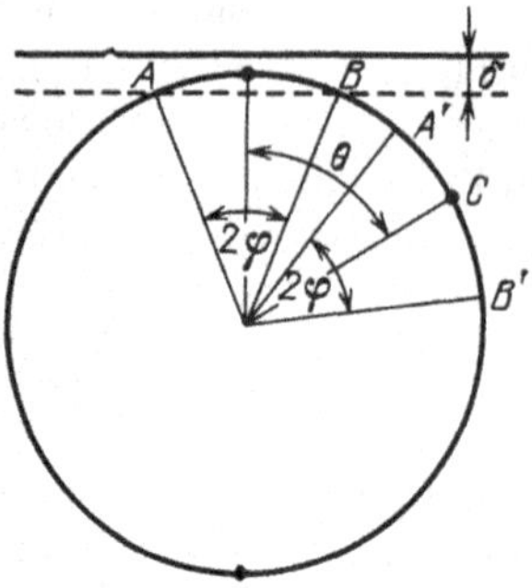

Fig. 40.

at average angles θ. However, when we arrive at point D of the orbit, we again obtain a high current density.

All this refers to a single orbit, but, actually, there are various orbits with different diameters. For free electrons they are equal to $cp_{\perp}/(eH) = (c/eH)\ (2m\varepsilon - p_z^2)^{1/2}$. Therefore, the true picture looks like that shown in fig. 41 and at any depth the density of "current-carrying" electrons is at least δ/r_{L} of the density in the skin layer.

This shortcoming can be overcome in the case of cyclotron resonance. As is known, the cyclotron resonance involves electrons with frequencies close to extremal values. We have the following condition:

$$|\omega - n\Omega| \sim \tau^{-1}.$$

Since Ω is close to the extremal value, it follows that

$$\Omega \approx \Omega_0 + \tfrac{1}{2}\frac{\partial^2\Omega}{\partial p_z^2}(\Delta p_z)^2$$

(here we do not consider the limiting points; for them $\delta\Omega \propto \Delta p_z$). In order of magnitude, $\partial^2\Omega/\partial p_z^2 \sim \Omega/p_0^2$ (p_0 is the average Fermi momentum). Therefore we

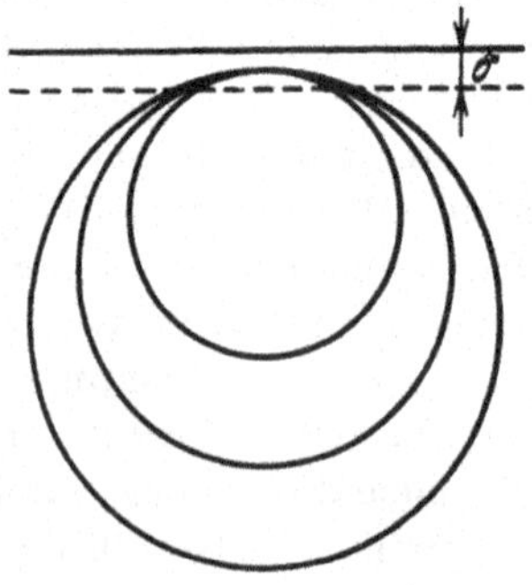

Fig. 41.

have

$$\frac{\omega}{p_0^2}(\Delta p_z)^2 \sim \tau^{-1} \quad \text{or} \quad \Delta p_z \sim \frac{p_0}{(\omega\tau)^{1/2}}.$$

Let us now examine this phenomenon in two different cases.

(1) Central cross section. Here, not only $\partial S/\partial \varepsilon$ is extremal but the orbit diameter also has an extremal value. Thus,

$$d = d_0 + \tfrac{1}{2}\frac{\partial^2 d}{\partial p_z^2}(\Delta p_z)^2.$$

Since $\partial^2 d/\partial p_z^2 \sim d_0/p_0^2$, we obtain

$$\Delta d \sim (\Delta p_z)^2 \frac{d_0}{p_0^2} \sim \frac{p_0^2}{\omega\tau}\frac{d_0}{p_0^2} \sim \frac{d_0}{\omega\tau}.$$

We also know that $d_0 \sim r_{\rm L} \sim v/\Omega \sim nv/\omega$, so that the spread of the orbit diameters will be $\Delta d \sim nv/(\omega^2\tau)$. For the electric field in the region AB to be transferred into a deeper point D in fig. 40. it is required that the condition $\Delta d < \delta$ be fulfilled, i.e.,

$$\omega\tau > \frac{d_0}{\delta} \sim \frac{r_{\rm L}}{\delta}. \tag{8.4}$$

We have seen above that for the resonance it is required that $\omega\tau \gg 1$. The condition (8.4) is much more stringent. However, it can nevertheless be fulfilled in practice.

(2) Noncentral cross section with an extremum $\partial S/\partial \varepsilon$. Here

$$\Delta d \sim \Delta p_z \frac{d_0}{p_0} \sim \frac{d_0}{(\omega\tau)^{1/2}}.$$

The condition $\Delta d < \delta$ now gives $(d_0 \sim r_{\rm L})$

$$\omega\tau > \left(\frac{r_{\rm L}}{\delta}\right)^2. \tag{8.5}$$

This condition is very difficult to realize in practice.

Thus, only the resonances at central cross sections can lead to the transfer of a high-frequency field into the bulk of the metal. In such a case, at a distance $d_{\rm extr}$ from the surface we obtain a layer of thickness δ where there will be a high-frequency field comparable in amplitude with the field present in the skin depth. This splash in the field will in turn lead to the appearance of a second one at a depth of $2d_{\rm extr}$, then a third at a depth of $3d_{\rm extr}$, etc. An exact calculation shows that the amplitudes of E and j inside the metal vary as shown in fig. 42.

The occurence of splashes in the field can be detected in thin films. One should expect the appearance of singularities in surface impedance when one of the additional "skin-layers" is found to be at the opposite surface of the film. However, this is not easy to attain. In fact, at a given frequency the resonance arises at certain

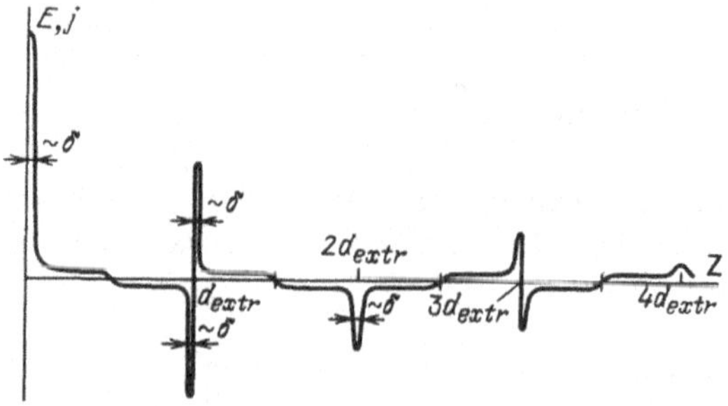

Fig. 42.

values of the magnetic field, i.e., with certain sizes of the resonance orbits. The diameters of such orbits differ, generally speaking, from the thickness of the film. For the splashes to be observed it is necessary either to vary the diameters of the resonance orbits by altering the frequency of the varying field and to "adjust" them to the film thickness or to vary the latter. Either of these is difficult to accomplish in experiments.

The same data on the electronic spectrum can, however, be obtained with the aid of nonresonance size effects by varying the magnetic field alone. They will be dealt with in the subsequent sections.

8.3. *Nonresonant size effect*

Equation (8.1) may also be applied to nonresonance orbits. If the diameter of the orbit is larger than the thickness of the film, the electron is diffusely scattered on the surface and does not remember its previous motion. If, however, the external field is increased, the diameter of the orbit will decrease and eventually become less than the film thickness. If this diameter was extremal, with an appropriate value of the field there must appear a singularity in the surface impedance (Gantmakher 1962). A detailed calculation shows that the derivative $\mathrm{d}Z/\mathrm{d}H$ has an extremum with this field. It should be noted that this effect is not associated with the resonance and therefore the position of the singularity is independent of the frequency of the varying field.

The nature of this phenomenon becomes understandable if we apply a reasoning analogous to the preceding one. For the extremal orbit

$$\Delta d \sim \tfrac{1}{2}\frac{\partial^2 d}{\partial p_z^2}(\Delta p_z)^2 \sim (\Delta p_z)^2\frac{d_0}{p_0^2}.$$

If $\Delta d \leqslant \delta$, then the high-frequency field is transferred into the bulk of the metal and an additional "skin-layer" appears at a depth d_{extr}. However, here we obtain

$$\Delta p_z < p_0\left(\frac{\delta}{r_{\mathrm{L}}}\right)^{1/2}.$$

Hence, only a small fraction of electrons (with orbits close to the extremal one) participate in this effect. Hence, the current density at a distance d_{extr} inside the metal is $(\delta/r_{\text{L}})^{1/2}$ of the current density in the skin layer. Nevertheless, it is greater than the current density at other distances from the skin layer because there it is at least a factor of δ/r_{L} smaller than that in the skin layer.

If the thickness of the film is sufficiently large, then the presence in the metal of a layer with a high-frequency field will lead to the formation of one more layer at a distance $2d_{\text{extr}}$, etc. If there exists several extremal diameters, then the splash of the field associated with one of these diameters will lead to the appearance of a second one associated with another d_{extr}. This means that the second splash of the field will occur at a depth $d^{(1)}_{\text{extr}} + d^{(2)}_{\text{extr}}$. In the general case any combinations of various d_{extr} may occur and the splashes will take place at all distances $\sum n_i d^{(i)}_{\text{extr}}$ (fig. 43). However, since with each following splash the current density will be reduced by a factor of $(\delta/r_{\text{L}})^{1/2}$, the current amplitude in every layer will be of order

$$\left(\frac{\delta}{r_{\text{L}}}\right)^{\sum_i \frac{1}{2} n_i}. \tag{8.6}$$

If one of these additional layers is found to be at the opposite surface of the film, the presence of a high-frequency current will result in the emission of an electromagnetic field. In other words, the film will have increased transparency for the electromagnetic wave incident on it. This manifests itself as a singularity of the surface impedance (the extremum of $\mathrm{d}Z/\mathrm{d}H$, fig. 44) at the corresponding value of the constant magnetic field. According to eq. (8.1) the condition for the appearance of a singularity in the surface impedance has the form

$$\sum_i n_i [p_y(t') - p_y(0)]^{(i)}_{\text{extr}} = -\frac{e}{c} HD. \tag{8.7}$$

The amplitude of the singularity depends on its order according to the law (8.6).

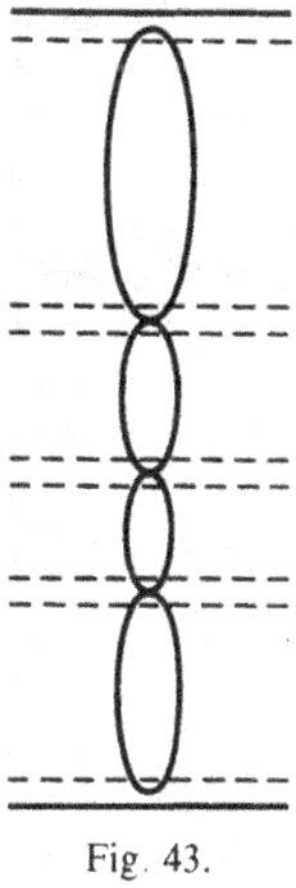

Fig. 43.

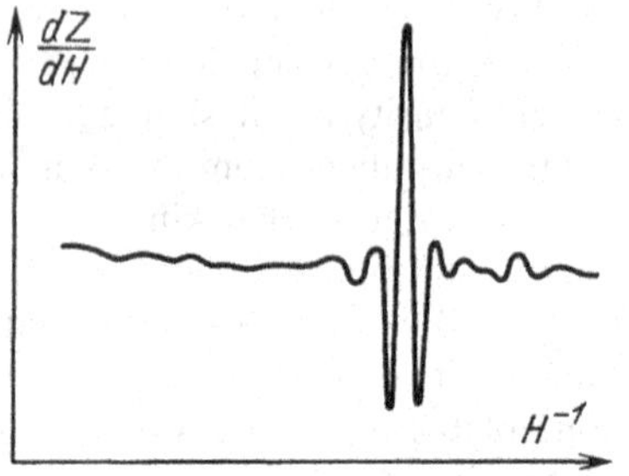

Fig. 44.

This method has proved most efficient for the direct measurement of the Fermi momentum. Of course, it applies only in a case where the mean free path is larger than the film thickness or, at least, is of the same order of magnitude.

8.4. Nonresonant size effect in a tilted field

There is one more size effect of the same type as the one considered in section 8.3, but its amplitude does not fall so rapidly with the order of the singularity. This effect occurs in a tilted field and is associated with the central cross section of the Fermi surface (Kaner 1963).

Let us consider fig. 45. For the central cross section the average velocty along the magnetic field $v_z = 0$. All the other electrons move along helical orbits and escape from the skin layer. Near the central cross section there is a group of electrons whose entire mean free path is within the skin layer. For these electrons, apparently

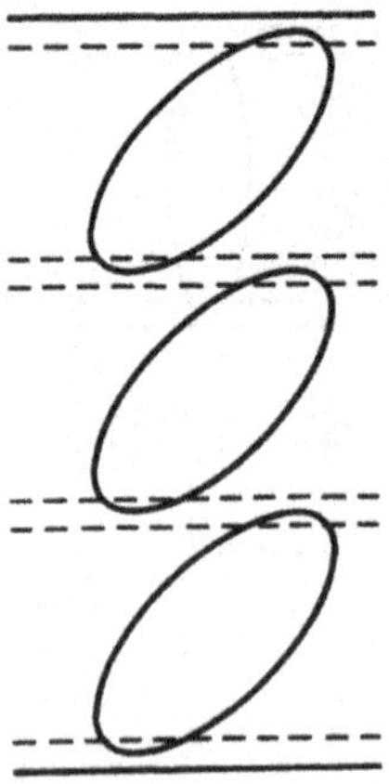

Fig. 45.

$v_z\tau \sin\psi < \delta$ or $v_z < \delta/(\tau \sin\psi)$. The relative number of these electrons is equal to $n/n_e \sim v_z/v \sim \delta/(\tau v \sin\psi)$. But because of their repeated return into the skin layer, their contribution to the conductivity is $\Omega\tau$ times larger than that of the corresponding number of ordinary electrons, which do not pass repeatedly into the skin layer. Therefore, the effective conductivity from these electrons is of order

$$\sigma\left(\frac{n}{n_e}\right)\Omega\tau \sim \frac{\sigma\delta\Omega}{v \sin\psi} \sim \frac{\sigma\delta}{r_L \sin\psi}.$$

In a case where

$$\sin\psi < \frac{\delta}{r_L}$$

the conductivity will be mainly determined by these electrons.

On the other hand, the spread of orbital diameters for electrons near the central cross section is of order

$$\Delta d \sim d_0\left(\frac{p_z}{p_0}\right)^2 \sim d_0\left(\frac{v_z}{v}\right)^2 \sim d_0\left(\frac{\delta}{l \sin\psi}\right)^2.$$

If this quantity is less than δ, then the skin layer will be reproduced in the bulk of the metal, i.e., there arises a splash in the field and in the current (fig. 45). Hence, it is necessary that

$$r_L\left(\frac{\delta}{l \sin\psi}\right)^2 < \delta$$

or in conjunction with the preceding condition

$$\frac{(\delta r_L)^{1/2}}{l} < \sin\psi < \frac{\delta}{r_L}. \tag{8.8}$$

If the sample is a film, just as in the previous case, then one can vary the magnetic field so as to create the situation that one of the splashes is found to occur at the opposite side of the film. This will give rise to a singularity in surface impedance (the extremum of $\mathrm{d}Z/\mathrm{d}H$). The location of the singularity is governed by the condition

$$np_y(0) = \frac{eH}{2c}\,\frac{D}{\cos\psi}. \tag{8.9}$$

In contrast to the size effect in a parallel field, the current in additional skin layers is not attenuated noticeably as compared with the current in the skin layer. Therefore, the amplitude of the singularities varies little with the number.

8.5. The Sondheimer effect

This effect can be observed both in static conductivity and in high-frequency impedance. We shall consider the case of static conductivity (Sondheimer 1950). Suppose we again have a film in a magnetic field directed at an angle ψ to the surface. The electron is moving along a helical trajectory (fig. 46). If $\bar{v}_z$ is the average velocity of the electron in the direction z, then its average velocity along the normal to the surface is

$$v_\zeta = \bar{v}_z \sin\psi.$$

Suppose that the electron is moving during a time equal to an integer number of periods

$$t = nT = n\frac{2\pi m^* c}{eH}.$$

It may happen that during this time the electron will just travel the distance from one surface of the film to the other. The corresponding condition has the form

$$v_\zeta t = n\frac{2\pi m^* c}{eH} v_\zeta = D,$$

whence

$$H = \frac{2\pi n c m^* v_\zeta}{eD} = \frac{2\pi n c m^* \bar{v}_z \sin\psi}{eD}.$$

It may therefore be expected that various properties of the metal will depend on the magnetic field. But the quantity $m^*\bar{v}_z$ depends on p_z. Again, the extremal values are the most important. The period is given by

$$\Delta H = \frac{2\pi c \sin\psi (m^*\bar{v}_z)_{\text{extr}}}{eD}. \tag{8.10}$$

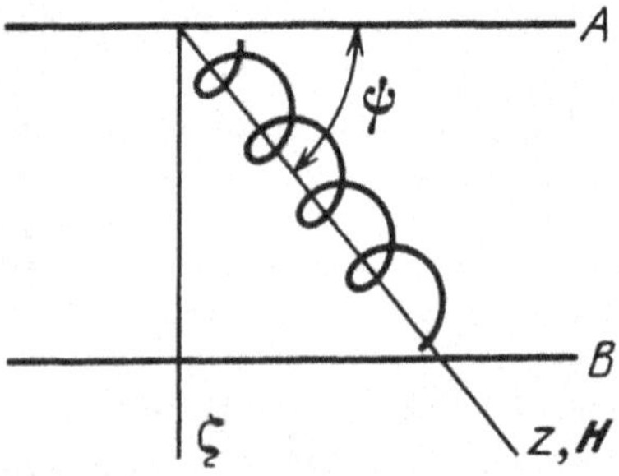

Fig. 46.

Consider now the quantity $m^*\bar{v}_z$. Since $m^* = (2\pi)^{-1}\partial S/\partial \varepsilon$ and $v_z = \partial\varepsilon/\partial p_z$, it follows that $m^*\bar{v}_z = (2\pi)^{-1}\partial S/\partial p_z$. Obviously, we have here the extremum of $\partial S/\partial p_z$ on the Fermi surface. Such an extremum may arise in a certain accidental cross section, in which

$$\partial^2 S/\partial p_z^2 = 0$$

(fig. 47, the line ab). In this case we have

$$\Delta H = \frac{c \sin\psi (\partial S/\partial p_z)_{\text{extr}}}{eD}. \tag{8.11}$$

Besides, the extremal values of $\partial S/\partial p_z$ correspond to the elliptic limiting points. Actually, these are simply the boundary values, and in a general case at the limiting points $\partial^2 S/\partial p_z^2 \neq 0$. In accordance with (7.58), at these points we have $\partial S/\partial p_z = 2\pi/\sqrt{K}$, where K is the Gaussian curvature. Hence, the period of oscillations is given by

$$\Delta H = \frac{2\pi c K^{-1/2} \sin\psi}{eD}. \tag{8.12}$$

Thus, it is possible now to find the Gaussian curvature at a limiting point experimentally. Comparing it with formula (7.59) for the cyclotron mass, we see that from the Sondheimer effect together with the cyclotron resonance we can determine the Fermi velocity at the limiting point. By varying the direction of the magnetic field, we can find this velocity for nearly the entire Fermi surface.

Let us determine the values of the oscillatory contribution to conductivity. For simplicity we confine ourselves to the case where the field is directed perpendicular to the surface.

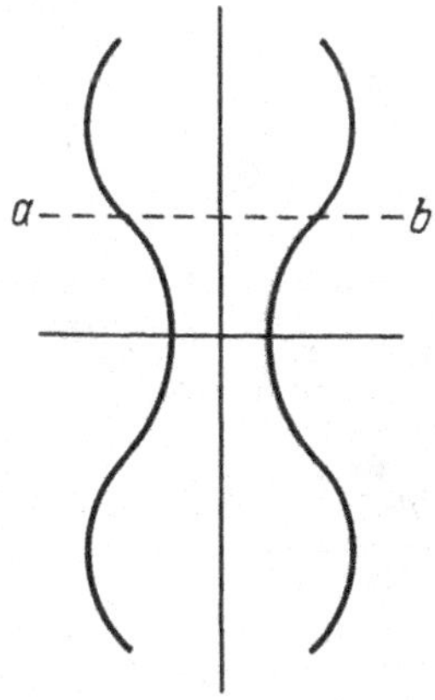

Fig. 47.

We start with the cross section for which $\partial^2 S/\partial p_z^2 = 0$. The effect involves electrons residing in a certain small neighborhood of this cross section, which is determined from the condition that two neighboring periods of oscillations must not merge, i.e., $\Delta n < 1$. According to the preceding reasoning, it is necessary here that

$$\Delta\left(\frac{D}{\bar{v}_z T}\right) < 1. \tag{8.13}$$

But the quantity $(\bar{v}_z T)^{-1}$ being proportional to $(\partial S/\partial p_z)^{-1}$ has an extremum in the case under discussion and, hence, we have

$$\Delta(\bar{v}_z T)^{-1} = (\tfrac{1}{2}) \frac{\partial^2}{\partial p_z^2}\left(\frac{1}{\bar{v}_z T}\right)(\Delta p_z)^2$$
$$\sim \left(\frac{\Delta p_z}{p_0}\right)^2 (\bar{v}_z T)^{-1} \sim \left(\frac{\Delta p_z}{p_0}\right)^2 r_L^{-1},$$

where r_L is the Larmor frequency. Thus, the interval Δp_z is determined by the condition

$$\Delta p_z \sim p_0 \left(\frac{r_L}{D}\right)^{1/2}.$$

It follows from the foregoing that the effective number of electrons taking part in the oscillatory component of the current is of order

$$n_{\text{eff}} \sim n_e \frac{\Delta p_z}{p_0} \sim n_e (r_L/D)^{1/2}.$$

The role of the time τ in the conductivity is evidently played in this case by the period of rotation, i.e., T. Moroever, we have to take into account the fact that the electrons in the plane of the film perform a finite motion and therefore the current arises only due to the fact that the final turn of the helix remains unclosed. This adds one more factor of order $n^{-1} \sim r_L/D$ to the oscillatory part of the conductivity*. Collecting all the terms, we find (Gurevich 1958)

$$\sigma_{\text{osc}} \sim n_{\text{eff}} e^2 \tau_{\text{eff}} m^{-1} \frac{r_L}{D} \sim \sigma(\Omega\tau)^{-1}(r_L/D)^{3/2}, \tag{8.14}$$

where $\Omega = 2\pi/T$. Thus, the amplitude of oscillations depends on H by the law $H^{-5/2}$.

We shall now perform an analogous estimate for a limiting point. In this case, $(v_z T)^{-1}$ has a boundary extremum and, hence,

$$\Delta(\bar{v}_z T)^{-1} = \left(\frac{\partial}{\partial p_z}\right)(v_z T)^{-1} \Delta p_z \sim \left(\frac{\Delta p_z}{p_0}\right) r_L^{-1}.$$

* We mean here the average conductivity equal to $J/(DE)$, where J is the total current passing through a film 1 cm wide.

According to the condition (8.13), we have**

$$\frac{\Delta p_z}{p_0} \sim \frac{r_L}{D}.$$

The effective number of electrons depends on Δp_z in a different manner from that in the preceding case. The area of the cross section located at a distance Δp_z from the reference point is proportional to Δp_z (see formula 7.58). Hence, the corresponding volume inside the Fermi surface is proportional to $(\Delta p_z)^2$, i.e.,

$$n_{\text{eff}} \sim n_e \left(\frac{\Delta p_z}{p_0}\right)^2 .$$

The effective collision time is of order T. Again, since the contribution to the current arises only from the incomplete final turn of the helix, a factor $n^{-1} \sim r_L/D$ appears. With this taken into account, we obtain (Gurevich 1958):

$$\sigma_{\text{osc}} \sim n_{\text{eff}} e^2 \tau_{\text{eff}} m^{-1} \frac{r_L}{D} \sim \sigma(\Omega\tau)^{-1} \left(\frac{r_L}{D}\right)^3 . \tag{8.15}$$

Thus, in this case $\sigma_{\text{osc}}/\sigma \propto H^{-4}$.

8.6. *Drift focusing of the high-frequency field*

From formulas (8.14) and (8.15) it follows that the Sondheimer effect introduces only a small correction to static conductivity. With the surface impedance we have an analogous situation. However, in the case of the high-frequency field the drift of electrons along the magnetic field causes one more, considerably stronger effect associated with the formation of splashes of the high-frequency field in the bulk of the metal (Gantmakher and Kaner 1963).

First of all, it should be noted that "effective" electrons must move in the skin layer parallel to the electric field. If they form an additional skin layer, they must move in the same direction there. Hence, the distance at which such a skin layer is formed must correspond to the drift of the electron along the magnetic field during an integer number of periods. Further, since the displacement along the field during a period is different for different electrons, the extremal displacements will obviously play the dominant role. Hence, the splashes of the high-frequency field will arise at distances from the surface equal to nu_{extr}, where u_{extr} is the extremal displacement during a period in the direction of the normal to the film. The strongest effect in

** Here $r_L \sim \bar{v}_z T$ denotes a value of the order of the ordinary Larmor radius $cp_0/(eH)$. We emphasize this circumstance because the true Larmor radius for electrons near a limiting point is of order $cp_\perp/eH$, where $p_\perp \sim (p_0 \Delta p_z)^{1/2}$ (see the derivation of formula 7.58).

impedance occurs if one of such skin layers is located at the opposite surface of the film, i.e., when

$$nu_{\text{extr}} = n(\bar{v}_z T)_{\text{extr}} \sin\psi$$
$$= n\frac{2\pi c}{eH}(m^*\bar{v}_z)_{\text{extr}} \sin\psi = D.$$

It is not difficult to see that this is exactly the same condition as in the case of the Sondheimer effect. However, this effect does not introduce small corrections but leads to appreciable changes in impedance. This circumstance makes it possible to use the effect for the study of electron characteristics. The most interesting is the fact that drift focusing may arise from the limiting points. Therefore, the corresponding results enable one to study the properties of electrons with a given momentum.

Let us find the conditions under which this effect can be observed (Gantmakher and Kaner 1963). The high-frequency field near the surface can be described as a superposition of plane waves with a continous set of wave vectors $\boldsymbol{k}$ having a width $\Delta k \sim \delta^{-1}$. The electrons, while moving along the helical trajectories, will interact with these waves, the strongest interaction being, apparently, that with harmonics for which $n\lambda = u_{\text{extr}}$, where u is the displacement of the electron during a period in the direction of propagation of the wave, i.e., of the normal to the film ζ. By virtue of this, such waves penetrate the metal to a large depth and their interference gives rise to splashes in the metal at depths of $\zeta = nu_{\text{extr}}$. This will, of course, occur only if $\delta \ll u_{\text{extr}}$.

For an electron to interact strongly with a wave, the electron must move, during its motion in the skin layer and at a depth of $\zeta = nu_{\text{extr}}$, along the front of the wave, i.e., here it is necessary that $v_\zeta = 0$. This means that a fraction of the electrons with $u(p_z) = u_{\text{extr}}$ must have $v_\zeta = 0$. At the limiting point itself at any finite angle ψ, $v_\zeta = v \sin\psi \neq 0$. But if the angle ψ is small, then even in a small neighborhood of this point electrons with $v_\zeta = 0$ may exist. The effective region near the limiting point is determined from the following considerations. During the mean free time τ electrons with different velocities will be apart a distance (along ζ) of order $[v_\zeta(\psi) - v_\zeta(0)]\tau = [v_z(\psi) - v_z(0)]\tau \sin\psi$. For the effect to be observable, this distance must be less than the skin depth δ. But since in the neighborhood of the limiting point $v_z(\psi) = v\cos\psi \approx v(1-\frac{1}{2}\psi^2)$, it follows that $v\psi^3\tau \ll \delta$ or $\psi^3 \ll \delta/l$. On the other hand, as has been said above, for the effect to be observed it is necessary that $\delta \ll u_{\text{extr}}$. Since $u_{\text{extr}} \sim \psi r_{\text{L}}$, we obtain $\psi \gg \delta/r_{\text{L}}$. Thus, the two conditions taken together yield

$$\left(\frac{\delta}{r_{\text{L}}}\right)^3 \ll \psi^3 \ll \frac{\delta}{l}. \tag{8.16}$$

The experimental results for tin (Gantmakher and Kaner 1965) are presented in fig. 48.

The effect of drift focusing of the high-frequency field also enables one to measure the mean free path of an electron with a given momentum and its temperature

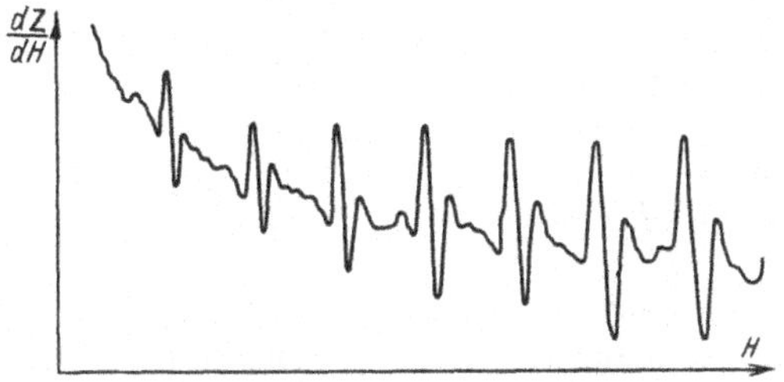

Fig. 48.

dependence (Gantmakher and Sharvin 1965). In order to see this, one has to take into account that the electron in the neighborhood of the limiting point is moving practically exactly in the direction z. Let the average mean free path be l_0. Then the number of electrons with a mean free path larger than l is proportional to $\exp(-l/l_0)$. Thus, the number of electrons that have reached the other surface of the film is proportional to $\exp[-D/(l_0 \sin \psi)]$. Obviously, the amplitude of the splashes must be proportional to this quantity. Therefore, the plot of $\ln A$ (where A is the amplitude) versus $D/\sin \psi$ must be a straight line (fig. 49), whose slope determines the mean free path. A straight line is actually obtained experimentally, which confirms the above considerations.

Thus, from what has been said above we can determine the mean free path as a function of the electron momentum and temperature. It should be noted that scattering even to a small angle will violate the condition of focusing of electrons. In view of this, l_0 is the mean free path with respect to a single collision (even with a phonon), i.e., this is the mean free path that enters into the thermal conductivity

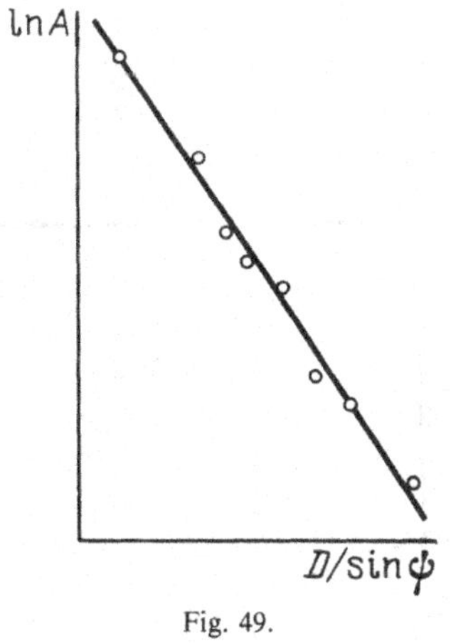

Fig. 49.

rather than the one that determines the electrical conductivity, where scattering to a small angle is ineffective. Naturally, experiment gives $l_0^{-1} = a + bT^3$.

8.7. Size effect on open trajectories

We shall now consider the size effect associated with open trajectories (Gantmakher and Kaner 1963). Let the magnetic field be parallel to the z axis and suppose there is an open trajectory in the direction p_x.

The equation for p_x is

$$\frac{dp_x}{dt} = \left(\frac{eH}{c}\right) v_y.$$

Integration over t within certain limits gives

$$p_x(t) - p_x(0) = \frac{eH}{c} \int_0^t v_y \, dt.$$

If the geometry is the same as in fig. 50, then we have

$$v_\zeta = v_y \sin\theta$$

$$\int_0^t v_\zeta \, dt = \frac{c}{eH}[p_x(t) - p_x(0)] \sin\theta.$$

The left-hand side represents the displacement of the electron in the direction ζ. Let it be equal to the thickness of the film. Further, let the magnetic field be such that the variation of the momentum is equal to one of the periods of the reciprocal lattice in the direction p_x. This gives

$$H = \sin\theta \frac{n\hbar Kc}{eD}, \tag{8.17}$$

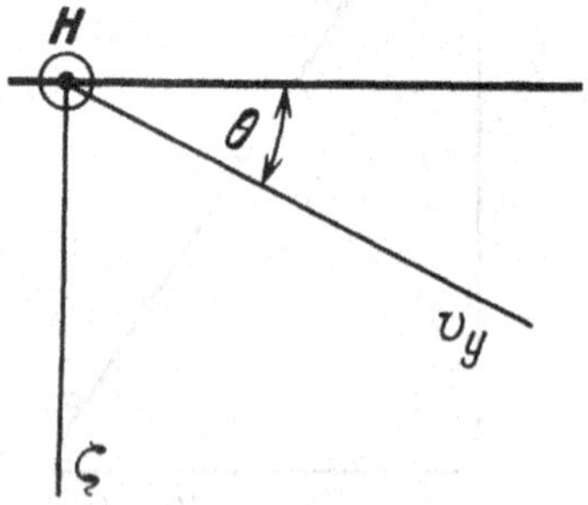

Fig. 50.

where K is the minimal period of the reciprocal lattice. Apparently, the effective conductivity will oscillate in this case with a period corresponding to eq. (8.17), i.e., with a period determined by the lattice period. This method can be used for measurement of the lattice periods or for the establishment of open Fermi surfaces.

The size effect on open trajectories exists in both static conductivity and in the high-frequency surface impedance. In the latter case it is associated with the formation of splashes of the field in the bulk of the metal and its amplitude is considerably larger than in the static case.

9

Propagation of electromagnetic waves in the presence of a magnetic field

9.1. Helicons in metals with unequal numbers of electrons and holes

Consider a metal in a high-frequency field. We have already concluded that an alternating field cannot get into the metal (except for the skin layer at the surface). However, in the preceding chapter we have seen that in the presence of a magnetic field under certain conditions a high-frequency field can also exist in the bulk of the metal, forming additional "skin layers" at certain distances from the surface. We shall show here that in the presence of a strong magnetic field there exist waves which can propagate in the metal without being damped significantly (Konstantinov and Perel' 1960). Numerous works have lately been devoted to this question. It has been found that there is a large variety of such waves and also of conditions under which they can propagate. We shall not consider all these waves here and shall only describe the theory of two modes of long-wavelength oscillations, which are the simplest for observations. More detailed information can be found in a review article by Kaner and Skobov (1968).

Let us write again the Maxwell equation

$$\operatorname{rot} \boldsymbol{H} = \frac{4\pi}{c} j, \qquad \operatorname{rot} \boldsymbol{E} = -c^{-1} \frac{\partial \boldsymbol{H}}{\partial t}. \tag{9.1}$$

Taking the rotation of the last equation and substituting into the first, we obtain

$$\operatorname{rot}\operatorname{rot} \boldsymbol{E} = \operatorname{grad}\operatorname{div} \boldsymbol{E} - \nabla^2 \boldsymbol{E} = -\frac{4\pi}{c^2}\frac{\partial \boldsymbol{j}}{\partial t}. \tag{9.2}$$

We will assume that the wavelength is large (the exact criterion will be obtained later). In this case, the electric field varies slowly in space and we may apply the conductivity tensor calculated for a homogeneous uniform electric field. Hence,

$$j_i = \sum_k \sigma_{ik} E_k.$$

We shall seek a solution of eq. (9.2) which is proportional to $e^{-i\omega t + ikr}$. From eq. (9.2) we obtain

$$\sum_k (k^2 \delta_{ik} - k_i k_k) E_k = \frac{4\pi i \omega}{c^2} \sum_k \sigma_{ik} E_k. \tag{9.3}$$

This is a homogeneous system of equations. It has solutions if the determinant vanishes. So,

$$\mathrm{Det}\left[k^2\delta_{ik} - k_i k_k - \frac{4\pi \mathrm{i}\omega}{c^2}\sigma_{ik} \right] = 0. \tag{9.4}$$

This condition gives the dispersion law $\omega(k)$.

However, before solving eq. (9.4) it is necessary to find the conductivity tensor. The kinetic equation used in ch. 7 has the form

$$\frac{\partial \psi}{\partial t} + \frac{\partial \psi}{\partial t_1} + \boldsymbol{v}\frac{\partial \psi}{\partial \boldsymbol{r}} + \frac{\psi}{\tau} = e\boldsymbol{v}\boldsymbol{E}.$$

Substituting $\psi \propto \mathrm{e}^{-\mathrm{i}\omega t + \mathrm{i}\boldsymbol{k}\boldsymbol{r}}$, we obtain

$$(-\mathrm{i}\omega + \tau^{-1})\psi + \mathrm{i}\boldsymbol{k}\boldsymbol{v}\psi + \frac{\partial \psi}{\partial t_1} = e\boldsymbol{v}\boldsymbol{E}.$$

The derivative $\partial\psi/\partial t_1$ has the order $\Omega\psi$, where $\Omega = eH/(m^*c)$ is the cyclotron frequency. The quantity vk/Ω is of order kr_{L} since $v/\Omega \sim r_{\mathrm{L}}$. Assuming that

$$kr_{\mathrm{L}} \ll 1 \tag{9.5}$$

we may drop the term with $\boldsymbol{kv}$, i.e., neglect the spatial variation of ψ. Moreover, if

$$|-\mathrm{i}\omega + \tau^{-1}| \ll \Omega \tag{9.6}$$

then we arrive at the kinetic equation which gives the static conductivity tensor in a strong magnetic field. Thus, assuming that conditions (9.5) and (9.6) are satisfied, we can apply the conductivity tensor found in section 5.3.

If the Fermi surface is closed and the number of holes is not equal to the number of electrons, we may use formula (5.37), where

$$\gamma \sim \frac{(\tau^{-1} - \mathrm{i}\omega)}{\Omega}$$

and

$$\sigma_{xy} = \gamma a_{xy} = ec\frac{(n_{\mathrm{e}} - n_{\mathrm{h}})}{H}. \tag{9.7}$$

Formula (5.37) was derived under the assumption that the field is directed along the z axis. The other axes are chosen so that the vector $\boldsymbol{k}$ is in the (y, z) plane, making an angle θ with the z axis (fig. 51). In the determinant (9.4) we retain only the terms of zeroth and first orders in γ. Here we get

$$\begin{vmatrix} k^2 & -4\pi \mathrm{i}\omega c^{-2}\sigma_{xy} & -4\pi \mathrm{i}\omega c^{-2}\sigma_{xz} \\ 4\pi \mathrm{i}\omega c^{-2}\sigma_{xy} & k_z^2 & -k_y k_z - 4\pi \mathrm{i}\omega c^{-2}\sigma_{yz} \\ 4\pi \mathrm{i}\omega c^{-2}\sigma_{xz} & -k_y k_z + 4\pi \mathrm{i}\omega c^{-2}\sigma_{yz} & k_y^2 - 4\pi \mathrm{i}\omega c^{-2}\sigma_{zz} \end{vmatrix} = 0. \tag{9.8}$$

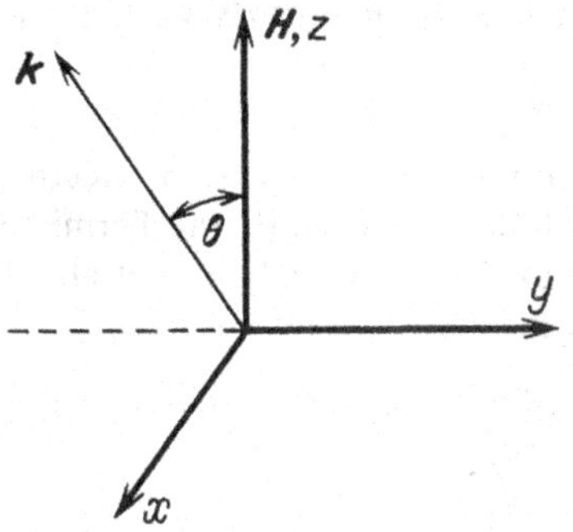

Fig. 51.

In accordance with (5.37), σ_{zz} is the largest component. If we assume that

$$k^2 \sim 4\pi\omega c^{-2}\sigma_{xy} \sim 4\pi\omega c^{-2}\sigma_{yz} \ll 4\pi\omega c^{-2}\sigma_{zz}$$

then from (9.8) we get

$$k^2k_z^2 + (4\pi \mathrm{i}\omega c^{-2})^2\sigma_{xy}^2 = 0$$

(the assumption is thus confirmed).

Inserting eq. (9.7) and $k_z = k \cos\theta$, we find

$$\omega = \frac{c^2k^2\cos\theta}{4\pi|\sigma_{xy}|} = \frac{cHk^2\cos\theta}{4\pi e|n_e - n_h|}. \tag{9.9}$$

We see that ω is proportional to H and k^2. The wave can propagate at any angle to H, except $\theta = \frac{1}{2}\pi$. Waves of this type are called helicons.

Let us see how the electric field varies in such a wave. From the last equation (9.3) [corresponding to the last row of the matrix (9.8)] we see that, because of the large value of σ_{zz} compared to the other components of σ_{ik},

$$E_z \ll E_x, E_y.$$

Passing to the first of eqs. (9.3), we obtain

$$k^2E_x - 4\pi \mathrm{i}\omega c^{-2}\sigma_{xy}E_y = 0.$$

Substituting σ_{xy} from (9.7) and ω from (9.9), we find

$$E_x = \mathrm{i}\cos\theta\, E_y. \tag{9.10}$$

This is an elliptically polarized wave. If $\theta = 0$, we have $E_x = \mathrm{i}E_y$. Such a wave is circularly polarized.

9.2. Magnetoplasmon waves in metals with equal numbers of electrons and holes

Consider now a metal with an even number of electrons per unit cell. Here the number of electrons and holes is the same; the Fermi surfaces are again assumed to be closed (Kaner and Skobov 1963; Khaikin et al. 1963). In the static case the conductivity tensor takes the form (5.41):

$$\sigma_{ik} = \begin{pmatrix} \gamma^2 a_{xx} & \gamma^2 a_{xy} & \gamma a_{xz} \\ \gamma^2 a_{yx} & \gamma^2 a_{yy} & \gamma a_{yz} \\ \gamma a_{zx} & \gamma a_{zy} & a_{zz} \end{pmatrix}.$$

Now let the frequency be so high that*

$$\tau^{-1} \ll \omega, \qquad kv \ll \omega \tag{9.11}$$

but not high enough for condition (9.6) or condition (9.5) to be violated. In this case, use may be made of the conductivity tensor, but it must be somewhat altered. Since the kinetic equation includes the combination $\tau^{-1} - \mathrm{i}\omega$, then in formula (9.11) we have to make the replacement $\tau^{-1} \rightarrow -\mathrm{i}\omega$. For example, in the static case with $n_e = n_h$

$$\sigma_{xy} \sim \frac{ecn}{H}(\Omega\tau)^{-1}.$$

Hence, in this case,

$$\sigma_{xy} = \frac{ecn}{H}\left(-\mathrm{i}\frac{\omega}{\Omega}\right) a_{12}, \tag{9.12}$$

with $a_{12} \sim 1$. All the other components of the conductivity tensor can be determined in the same manner. As a result, we find

$$\sigma_{ik} = \frac{ecn}{H} \begin{pmatrix} -\mathrm{i}\dfrac{\omega}{\Omega} a_1 & -\mathrm{i}\dfrac{\omega}{\Omega} a_{12} & a_{13} \\ -\mathrm{i}\dfrac{\omega}{\Omega} a_{12} & -\mathrm{i}\dfrac{\omega}{\Omega} a_2 & -a_{32} \\ -a_{13} & a_{32} & \mathrm{i}\dfrac{\Omega}{\omega} a_3 \end{pmatrix} \tag{9.13}$$

where all a_i and $a_{ik} \sim 1$.

Now we introduce two coordinate systems as shown in fig. 52, where ζ is directed along $\boldsymbol{k}$, the axis η lies in the $(\boldsymbol{H}, \boldsymbol{k})$ plane, and the axis x is common for the two

* Conditions (9.11) are associated with the fact that in the case under consideration the important components of the conductivity tensor are those of second order in γ. Because of conditions (9.11) these components have the form (9.12) rather than $(ecn/H)(kv/\Omega)$ or $(ecn/H)(\Omega\tau)^{-1}$.

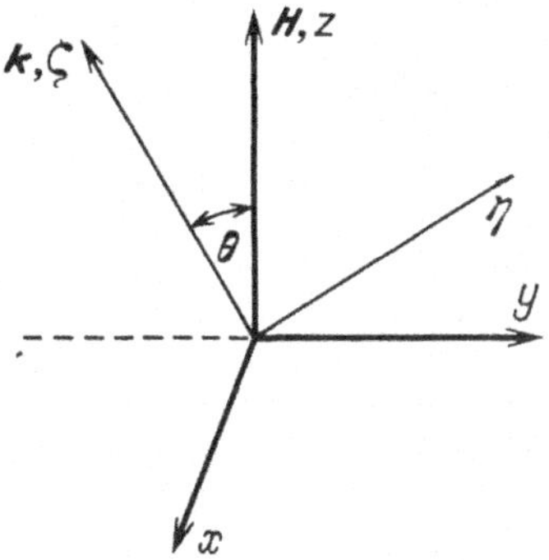

Fig. 52.

coordinate systems. Consider the third of eq. (9.3) in the coordinate system x, η, ζ. From it we obtain

$$E_\zeta = -(\sigma_{\zeta x}E_x + \sigma_{\zeta\eta}E_\eta)\sigma_{\zeta\zeta}.$$

If we substitute this into the other two equations of (9.3), we obtain two homogeneous equations for E_x and E_η. Equating the determinant of the system to zero, we have

$$\mathrm{Det}[k^2\sigma_{\alpha\beta} - 4\pi \mathrm{i}\omega c^{-2}\tilde{\sigma}_{\alpha\beta}] = 0. \tag{9.14}$$

The subscripts α and β may take the values of x and η.

$$\tilde{\sigma}_{\alpha\beta} = \sigma_{\alpha\beta} - \frac{\sigma_{\alpha\zeta}\sigma_{\zeta\beta}}{\sigma_{\zeta\zeta}}. \tag{9.15}$$

We transform the conductivity tensor (9.13) calculated in the system x, y, z to the system x, η, ζ. This is a simple, but rather tedious procedure. Retaining only the dominant terms in ω/Ω, we have

$$\tilde{\sigma}_{\alpha\beta} = -\mathrm{i}\frac{\omega}{\Omega}\frac{nec}{H}A_{\alpha\beta}, \tag{9.16}$$

where

$$A_{\alpha\beta} = \begin{pmatrix} a_1 + \dfrac{a_{13}^2}{a_3} & \left(a_{12} + \dfrac{a_{13}a_{32}}{a_3}\right)\sec\theta \\ \left(a_{12} + \dfrac{a_{13}a_{32}}{a_3}\right)\sec\theta & \left(a_2 + \dfrac{a_{23}^2}{a_3}\right)\sec^2\theta \end{pmatrix}. \tag{9.17}$$

In formula (9.16) we singled out the factor that determines the order of the quantity $\tilde{\sigma}_{\alpha\beta}$. There $n = n_e = n_h$, $\Omega = eH/mc$, where m is the mass of the free electron.

Inserting eq. (9.16) into eq. (9.14), we can write the latter in the form

$$\mathrm{Det}\left[\left(\frac{v_a k}{\omega}\right)^2\delta_{\alpha\beta} - A_{\alpha\beta}\right] = 0, \tag{9.18}$$

where $v_a^2 = c\Omega H/(4\pi ne)$, i.e., $v_a = H/(4\pi nm)^{1/2}$*. From (9.18) we find

$$(v_a k/\omega)^4 - (v_a k/\omega)^2 \operatorname{Sp} A_{\alpha\beta} + \operatorname{Det} A_{\alpha\beta} = 0$$

from which we have

$$\frac{v_a k}{\omega} = \{\tfrac{1}{2} \operatorname{Sp} A_{\alpha\beta} \pm [(\tfrac{1}{2} \operatorname{Sp} A_{\alpha\beta})^2 - \operatorname{Det} A_{\alpha\beta}]^{1/2}\}^{1/2}$$

or

$$\frac{\omega}{k} = v_a (\operatorname{Det} A_{\alpha\beta})^{-1/2} \{\tfrac{1}{2}(A_{xx} + A_{\eta\eta}) \mp [\tfrac{1}{4}(A_{xx} - A_{\eta\eta})^2 + A_{x\eta}^2]^{1/2}\}^{1/2}. \quad (9.19)$$

Thus there exist two waves with a linear dispersion law $\omega \propto k$. These waves are called magnetoplasmon waves. The velocities of both waves are proportional to H. If the magnetic field is directed along the symmetry axis of the crystal, then the tensor $A_{\alpha\beta}$ simplifies (all $a_{ik} = 0$ if $i \neq k$ and $a_1 = a_2$):

$$A_{\alpha\beta} = A_{xx} \begin{pmatrix} 1 & 0 \\ 0 & \sec^2\theta \end{pmatrix}. \quad (9.20)$$

Substituting this tensor into eq. (9.19) gives

$$\frac{\omega}{k} = 2^{-1/2} v_a A_{xx}^{-1/2} \cos\theta [(1 + \sec^2\theta) \pm (1 - \sec^2\theta)]^{1/2}.$$

Hence, for one wave

$$\frac{\omega_1}{k} = v_a A_{xx}^{-1/2} \quad (9.21)$$

and for the other

$$\frac{\omega_2}{k} = v_a A_{xx}^{-1/2} \cos\theta, \quad (9.22)$$

which gives

$$\frac{\omega_2}{\omega_1} = \cos\theta. \quad (9.23)$$

From eq. (9.22) it follows that when $\cos\theta = 0$ only one wave is left. This conclusion is not quite strict, because in the derivation of the tensor $A_{\alpha\beta}$ (9.17) we assumed that $\cos\theta \gg \omega/\Omega$. If the inverse inequality is valid, i.e., if θ is very close to $\frac{1}{2}\pi$, then the calculation has to be performed anew. Nevertheless, if we do this and then substitute the result into eq. (9.14), only one wave with a linear dispersion law will remain.

* An estimate of v_a can be obtained from the following simple reasoning. In the preceding section we have obtained $k^2 \sim 4\pi i\omega c^{-2}\sigma_{xy}$. If we substitute eq. (9.12) here and solve the resulting equation for ω, we shall obtain $\omega = vk$, where $v \sim H/(4\pi nm^*)^{1/2}$.

9.3. Experimental investigations

There are various experimental methods for the study of waves. One method (the standing wave technique) consists of the measurement of the high-frequency surface impedance of metal films as a function of the magnetic field. By varying the magnetic field at a given frequency, we alter the wavelength λ of oscillations propagating through the metal. At certain values of H an integer number of half-wavelengths will correspond to the thickness of the film ($D=\frac{1}{2}q\lambda$, where q is an integer). This is the condition of the existence of a standing wave. Thus, we can anticipate a periodic dependence of surface impedance on D/λ or Dk. In the first case considered above $\omega \propto Hk^2$. Hence, the surface impedance will be a periodic function of $H^{-1/2}$. If $H \sim 10^4$ Oe and $D \sim 1$ mm, the effect in ordinary metals can be observed at frequencies of the order of $\omega/2\pi = \nu \sim cH(D^2en_e) \sim 10^4$ Hz (i.e., at relatively low frequencies). Waves of this type have been studied experimentally in Na, Cu, Ag, Au, and Al and in many other metals.

Waves of the second type have been detected in bismuth, where the number of electrons is equal to the number of holes. In this case, the values of n and m^* are small ($n \sim 10^{17}$ cm^{-3}, $m^* \sim 10^{-2}$ m). Therefore, the following condition must be satisfied: $\nu \sim H(Dn^{1/2}m^{*1/2}) \sim 10^{10}$ Hz in fields of $H \sim 10^3$ Oe ($D \sim 1$ mm). Again a standing wave was observed, but the surface impedance was periodic as a function of H^{-1}. In this way both magnetoplasmon waves have been detected; in the case where the magnetic field was directed along the principal axis of the crystal, the frequencies satisfied relation (9.23).

Another method of studying waves is the so-called interference technique. This technique uses the interference of a wave incident on a metal film with a wave that has passed through it.

Apart from the dispersion law, i.e., the function $\omega(k)$, another effect of great interest is the damping of electromagnetic waves in a metal, especially in the region of frequencies and magnetic fields where resonance damping mechanisms not associated with electron collision (i.e., which exist also for $\tau \to \infty$) become important.

One of these damping mechanisms is known as Landau damping and it was first detected in the study of oscillations in a gas plasma. This mechanism consists in the following. In a magnetic field an electron moves along a helical trajectory. Its average velocity along the magnetic field is equal to v_z. We pass over to the coordinate system that moves with this velocity. The electromagnetic field, which was proportional to $\exp(i\boldsymbol{kr} - i\omega t)$ in the initial coordinate system, will be proportional to $\exp[i\boldsymbol{kr} - i(\omega - k_z v_z)t]$ in the new coordinate system. If

$$\omega = k_z v_z \tag{9.24}$$

then the electron will "see" this field as static. In this case, conditions are created for an appreciable pumping of the energy from the wave to the electrons.

Landau damping may be described in a somewhat different way. Figure 53 shows the motion of the electron and the propagation of the wave. The dashed lines

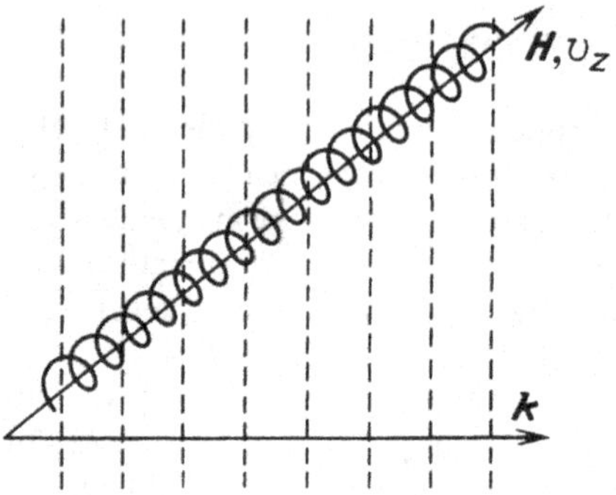

Fig. 53.

correspond to equal phase planes. Condition (9.24) is an expression of the fact that the electron is permanently moving in phase with the wave.

Taking into account that the electron velocities vary within the limits from $-v$ to v, where v is the maximal value of $\bar{v}_z$, which, as a rule, corresponds to the limiting point, we see that condition (9.24) can be satisfied for a certain group of electrons at sufficiently low frequencies, namely,

$$\omega < kv \cos \theta. \tag{9.25}$$

If the frequency is fixed and the field varies, then this condition specifies the lower limit of the magnetic fields. By measuring this field, we can find v, the velocity at the Fermi boundary at the limiting point. In experiment Landau damping is a very efficient absorption mechanism, so that, for example, in the method used to measure the impedance of films the experimental picture is as shown in fig. 54 (Khaikin and

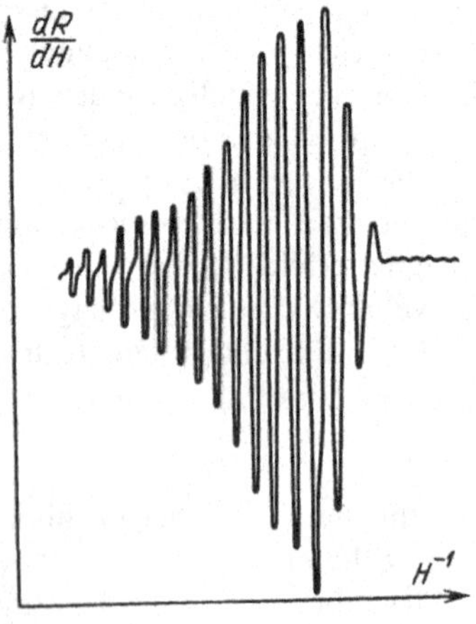

Fig. 54.

Edel'man 1965) [this figure refers to an even metal, bismuth; if the metal is odd, the picture is the same but in coordinates $Z(H^{-1/2})$].

Formula (9.25) with $\omega(k)$ in the form (9.9) or (9.19) may be used only in a case where condition (9.5) is fulfilled. This usually occurs in experiments with odd metals. Indeed, at $H \sim 10^4$ Oe, $\lambda = 2\pi/k \sim 0.1$ cm, $kr_L \sim 10^{-2}$. But in experiments with even metals the condition (9.5) may be violated. In fact, the criterion (9.25) is also valid at $kr_L \gtrsim 1$, although in this case we can no longer use the dispersion laws $\omega(k)$ in the form of (9.10) and (9.19).

In the case of $kr_L \sim 1$ one more interesting damping mechanism known as Doppler-shifted cyclotron resonance comes into play. This mechanism is associated with the fact that in the coordinate system moving with velocity $\bar{v}_z$ the electron moves along a closed contour with a cyclotron velocity $\Omega = eH/m^*c$. If the frequency of the electromagnetic field in the new frame of reference coincides with Ω conditions are created for cyclotron resonance. Thus, the corresponding condition is

$$\omega - k_z\bar{v}_z = \Omega. \tag{9.26}$$

In an odd metal, usually $\omega \ll \Omega$ (for example at $H \sim 10^4$ Oe, $\lambda = 2\pi/k \sim 0.1$ cm, $\omega \sim 10^4$ s^{-1}, $\Omega \sim 10^{11}$ s^{-1}). Hence, this mechanism begins to operate when

$$kv \cos\theta > \Omega. \tag{9.27}$$

In this case, the experimental picture is approximately as shown in fig. 54 [but only in coordinates $Z(H^{-1/2})$]. In an even metal, under usual experimental conditions $\omega \sim \Omega$ and therefore from (9.26) it follows that the damping region may, in principle, be bounded from two sides:

$$\omega - kv\cos\theta < \Omega < \omega + kv\cos\theta$$

Depending on the relation between the parameters of the metal and the experimental conditions, the left of these inequalities either may or may not give a real constraint on H^{-1}. In the first case, the experimental result is similar to that shown in fig. 55, and in the second it corresponds to fig. 54.

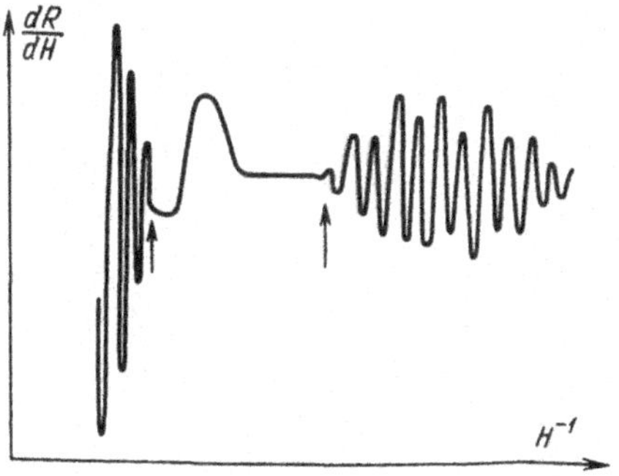

Fig. 55.

Formulas (9.27) and (9.28) refer to the region where condition (9.5) is violated and therefore formulas (9.9) and (9.19) are inapplicable. But since the true dispersion law is established in the same experiment, it can be used instead of the theoretical formula. This makes it possible to determine from the damping limit in the case of an odd metal the quantity $m^*v = K^{-1/2}$, where K is the Gaussian curvature at the limiting point, and in the case of an even metal both m^* and $v = m^{*-1}K^{-1/2}$ can be determined. Such determinations have made it possible to obtain more accurate parameters of the spectra in several metals.

10 Magnetic susceptibility and the de Haas–van Alphen effect

10.1. Pauli spin paramagnetism

Up until now we have considered the electron as a classical particle moving along a certain trajectory. Now we turn our attention to phenomena in which the quantum nature of electrons is important. We start with the behavior of electrons in a static magnetic field. Let us first consider a weak field.

Each electron has spin $\hbar \mathbf{s}$ and magnetic moment $2\beta\mathbf{s}$, where $\mathbf{s}$ may have a projection onto a certain direction equal to $\frac{1}{2}$ or $-\frac{1}{2}$, and $\beta = e\hbar/2mc$ is Bohr's magneton. Evidently, in the presence of an external magnetic field the magnetic moments of electrons are polarized and a net magnetic moment arises. This phenomenon is called spin paramagnetism or Pauli paramagnetism.

The corresponding magnetic susceptibility is easy to find. The interaction energy of the magnetic moment of an electron with the external field is equal to $-2\beta\mathbf{s}\mathbf{H}$. Therefore, if the electron spin is directed along the magnetic field, its energy will be $\varepsilon - \beta H$, and if the spin is in the opposite direction to the field, then it is $\varepsilon + \beta H$. But the Fermi distribution includes the energy in combination with the chemical potential $\varepsilon - \mu$. Thus, instead of considering the variation of the energy under the influence of the field, we may assume that the chemical potential is equal to $\mu + \beta H$ for electrons with a spin along the field and to $\mu - \beta H$ for electrons with a spin opposite to the field.

At $T = 0$, $H = 0$, the number of electrons per unit volume is

$$n = \int_0^{\mu} \nu(\varepsilon)\, \mathrm{d}\varepsilon.$$

In the case under consideration the number of electrons with different spin orientations will be different. It is not difficult to ascertain that they are expressed as follows:

$$n_+ = \tfrac{1}{2}\int_0^{\mu+\beta H} \nu(\varepsilon)\, \mathrm{d}\varepsilon, \qquad n_- = \tfrac{1}{2}\int_0^{\mu-\beta H} \nu(\varepsilon)\, \mathrm{d}\varepsilon.$$

The total magnetic moment per unit volume will be equal to $\beta(n_+ - n_-)$. Thus,

$$M = \beta(n_+ - n_-) = \tfrac{1}{2}\int_{\mu-\beta H}^{\mu+\beta H} \nu(\varepsilon)\, \mathrm{d}\varepsilon \approx \beta^2 H \nu(\mu) \tag{10.1}$$

if $\beta H \ll \mu$. Hence, the magnetic susceptibility is given by

$$\chi = \beta^2 \nu(\mu). \tag{10.2}$$

At first sight this formula seems to be correct from the standpoint of the ideas of the Fermi liquid too (section 2.2) since the integration in eq. (10.1) covers only the neighborhood of the Fermi surface. Actually, this is not so. In the Landau theory of the Fermi liquid it is shown that the energy spectrum itself changes under the influence of the variation of the distribution function, and in a number of cases this leads to a change in the results. In particular, this also refers to formula (10.2), which, in fact, determines only the order of magnitude of the effect. These questions will be treated in more detail in ch. 13.

Note also that we assumed $T=0$. But, as is always the case with electrons, the temperature corrections are of order $(T/\mu)^2$ and they may be neglected here.

10.2. Quantization of the levels of a free electron in a magnetic field

Consider the motion of a free electron in a constant magnetic field. Replacing the momentum operators $\boldsymbol{p} = -\mathrm{i}\hbar\nabla$ by $\boldsymbol{p} - (e/c)\boldsymbol{A}$ (where $\boldsymbol{A}$ is the vector potential), we write the Hamiltonian in the form

$$\mathcal{H} = \frac{1}{2m}\left(-\mathrm{i}\hbar\nabla - \frac{e}{c}\boldsymbol{A}\right)^2. \tag{10.3}$$

We did not include the spin term $(-2\beta \boldsymbol{s}\boldsymbol{H})$ because it gives only a displacement of the levels by an amount $\pm\beta H$. Suppose that the field H is directed along the z axis and take the vector potential in the form $A_y = Hx$, $A_x = A_z = 0$. The Schrödinger equation $\mathcal{H}\psi = \varepsilon\psi$ will take the form

$$-\frac{\hbar^2}{2m}\frac{\partial^2\psi}{\partial x^2} + \frac{1}{2m}\left(-\mathrm{i}\hbar\frac{\partial}{\partial y} - \frac{e}{c}Hx\right)^2\psi - \frac{\hbar^2}{2m}\frac{\partial^2\psi}{\partial z^2} = \varepsilon\psi. \tag{10.4}$$

Since this equation includes only the coordinate x, we may seek the solution in the form

$$\psi(x, y, z) = \mathrm{e}^{\mathrm{i}p_y y/\hbar}\,\mathrm{e}^{\mathrm{i}p_z z/\hbar}\,\psi(x). \tag{10.5}$$

Substituting this into eq. (10.4), we obtain

$$-\frac{\hbar^2}{2m}\frac{\mathrm{d}^2\psi(x)}{\mathrm{d}x^2} + \frac{1}{2m}\left(p_y - \frac{e}{c}Hx\right)^2\psi(x) + \frac{p_z^2}{2m}\psi(x) = \varepsilon\psi(x)$$

or

$$-\frac{\hbar^2}{2m}\frac{\mathrm{d}^2\psi(x)}{\mathrm{d}x^2} + \frac{e^2H^2}{2mc^2}\left(x - \frac{p_y c}{eH}\right)^2\psi(x) = \left(\varepsilon - \frac{p_z^2}{2m}\right)\psi(x). \tag{10.6}$$

Equation (10.6) is very similar to the equation for a one-dimensional quantum oscillator:

$$-\frac{\hbar^2}{2m}\frac{d^2\psi}{dx^2}-\tfrac{1}{2}kx^2\psi=E\psi. \tag{10.7}$$

The eigenvalues of this last equation are given by

$$E=\hbar\omega(n+\tfrac{1}{2}) \tag{10.8}$$

where $\omega=(k/m)^{1/2}$. Comparing eqs. (10.7) and (10.6), we see that they are identical if only we shift the point of equilibrium to a distance cp_y/eH from the origin. The role of ω is played by the quantity eH/mc and the quantity $\varepsilon - p_z^2/2m$ plays the role of the oscillator energy. Thus, we have

$$\varepsilon=\frac{p_z^2}{2m}+\beta H(2n+1), \tag{10.9}$$

where $\beta=e\hbar/2mc$. This is the formula for the so-called Landau levels (Landau 1930). Thus, instead of the continuous dependence of ε on p_y and p_x, we obtained discrete levels. This transformation is illustrated in fig. 56. Since the separation between Landau levels is proportional to H, each of them "collects" an interval of the continuous spectrum proportional to H. Hence, the Landau levels must be degenerate and this degeneracy must be proportional to H.

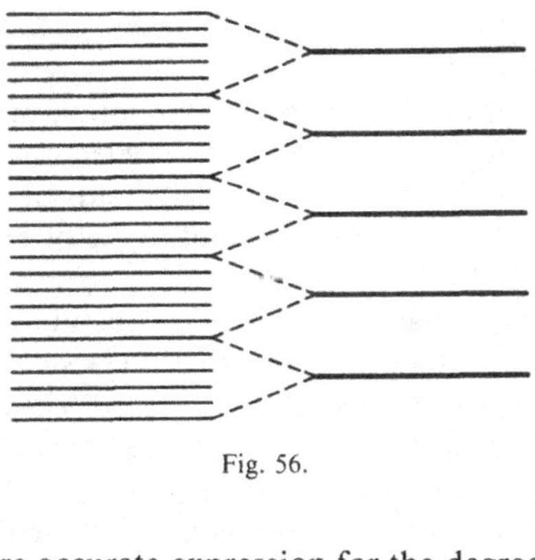

Fig. 56.

In order to find a more accurate expression for the degree of degeneracy of each level, let us consider, as has been done earlier, a rectangular box with edge lengths L_1, L_2 and L_3. The position of the center of the oscillator depends on p_y. This center must lie within the interval from 0 to L_1, so that

$$0<\frac{cp_y}{eH}<L_1 \quad \text{or} \quad 0<p_y<\frac{eHL_1}{c}.$$

The number of states in the intervals dp_z and dp_y is the same as for a free electron, i.e.,

$$dn_y = \frac{dp_y L_2}{2\pi\hbar}, \qquad dn_z = \frac{dp_z L_3}{2\pi\hbar}.$$

Hence, the number of states for the entire interval p_y is $L_1 L_2 eH/(2\pi c\hbar)$. Multiplying by dn_z, we obtain the complete expression:

$$\frac{eH}{c}\frac{V dp_z}{(2\pi\hbar)^2},$$

which gives the number of states corresponding to the interval dp_z and to a certain n in eq. (10.9). In order to find the density of states, we have to take into account the fact that the energy ε depends on p_z^2, i.e.,

$$\int_{-\infty}^{\infty} dp_z \varphi(\varepsilon) = 2\int_0^{\infty} \varphi(\varepsilon)\, d|p_z| = \int_0^{\infty} \varphi(\varepsilon)\frac{d|p_z|}{d\varepsilon}\, d\varepsilon,$$

where φ is an arbitrary energy function. Taking also into account the two-fold degeneracy associated with the spin, we obtain

$$\nu(\varepsilon) = \frac{4V}{(2\pi\hbar)^2}\frac{eH}{c}\frac{d|p_z|}{d\varepsilon}. \tag{10.10}$$

10.3. Landau diamagnetism

When a metal is placed in a magnetic field, the motion of electrons changes: they begin to move along helical trajectories. So, the electrons produce an additional magnetic field, which is opposite to the external field. In other words, the change of the orbital motion of electrons under the influence of external field gives rise to a diamagnetic effect.

It is not difficult to show that this effect is of quantum nature and is absent in the classical approximation. Indeed, according to classical mechanics, in the magnetic field the momentum of the electron is replaced by the difference $\boldsymbol{p} - (e/c)\boldsymbol{A}$, where $\boldsymbol{A}$ is the vector potential. However, since the thermodynamic quantities depend only on the energy of the electron and are integrals over all momenta, we may pass over to integration over $\boldsymbol{p} - (e/c)\boldsymbol{A}$. Here, the integrals will assume the same form as in the absence of a magnetic field.

In fact, diamagnetism occurs and results from the quantization of electron energy levels in a magnetic field.

Let us calculate the thermodynamic potential:

$$\Omega = -T\sum_i \ln(1 + e^{(\mu - \varepsilon_i)/T}), \tag{10.11}$$

where the sum is taken over all the states. This expression can be derived from the formula

$$\left(\frac{\partial \Omega}{\partial \mu}\right)_T = -N = -\sum_i n(\varepsilon_i),$$

where $n(\varepsilon_i)$ is the Fermi distribution. Introducing the density of states $\nu(\varepsilon)$ in accordance with eq. (10.10), we obtain

$$\Omega = -\frac{4VT}{(2\pi\hbar)^2}\frac{eH}{c}\int_0^\infty d\varepsilon \sum_n \frac{d|p_z|}{d\varepsilon}\ln(1+e^{(\mu-\varepsilon)/T}).$$

Integrating by parts and substituting

$$|p_z(\varepsilon, n)| = [2m(\varepsilon - \beta H(2n+1))]^{1/2} \tag{10.12}$$

from eq. (10.9), we find

$$\Omega = -\frac{2V}{(2\pi\hbar)^2}\frac{eH}{c}\int_0^\infty d\varepsilon \sum_n \frac{[2m(\varepsilon - \beta H(2n+1))]^{1/2}}{\exp[(\varepsilon-\mu)/T]+1}. \tag{10.13}$$

The summation over n is limited here by the condition $\varepsilon > \beta H(2n+1)$.

Expression (10.13) is valid for arbitrary temperature. But is $T \ll \mu$, we may put $T=0$, since the correction in the case of a small magnetic field will be of order $(T/\mu)^2$. Therefore, we replace the Fermi distribution by a step function and, altering the order of summation over n and integrating over ε, we obtain

$$\Omega = -\frac{4VeH}{(2\pi\hbar)^2 c}\sum_n \int_{\beta H(2n+1)}^{\mu} d\varepsilon[2m(\varepsilon - \beta H\,(2n+1))]^{1/2}$$
$$= \frac{8}{3}\frac{VeH(2m)^{1/2}}{(2\pi\hbar)^2 c}\sum_n [\mu - \beta H\,(2n+1)]^{3/2}\left(n < \frac{\mu}{2\beta H} - \frac{1}{2}\right).$$

At small H we may replace the summation by integration using the formula

$$\sum_{n=0}^{n_0} f(n) = \int_{-1/2}^{n_0+1/2} f(n)\,dn - \tfrac{1}{24}[f'(n_0+\tfrac{1}{2}) - f'(-\tfrac{1}{2})]. \tag{10.14}$$

As a result, we obtain

$$\Omega = -\frac{8}{15}\frac{Ve(2m)^{1/2}\mu^{5/2}}{\beta c(2\pi\hbar)^2} + \frac{1}{3}\frac{Ve\beta H^2(2m)^{1/2}\mu^{1/2}}{(2\pi\hbar)^2 c}.$$

Substituting $\beta = e\hbar/2mc$ and $\mu = p_0^2/2m$, we find

$$\Omega = -\frac{p_0^5 V}{15m\pi^2\hbar^3} + \frac{V\beta^2H^2}{6}\frac{p_0 m}{\pi^2\hbar^3}.$$

The magnetic moment per unit volume is defined as

$$M = -V^{-1}\frac{\partial \Omega}{\partial H} = -\tfrac{1}{3}\beta^2 H\frac{p_0 m}{\pi^2\hbar^3}.$$

Thus, the magnetic susceptibility is given by (Landau 1930)

$$\chi = -\tfrac{1}{3}\beta^2 \frac{p_0 m}{\pi^2 \hbar^3}. \tag{10.15}$$

From formula (2.28) it follows that $p_0 m/\pi^2\hbar^3$ is just $\nu(\varepsilon)$ for free electrons. Comparing eq. (10.15) with eq. (10.2), we find that the quantization of energy levels leads to a diamagnetic susceptibility, which for a gas of free electrons compensates only $\frac{1}{3}$ of the paramagnetic susceptibility associated with spins. Thus, the free electron gas will be definitely paramagnetic.

But, actually there exist many diamagnetic metals in nature. Qualitatively, this may be accounted for by the fact that the electronic energy spectrum of a metal differs from the spectrum of an ideal gas. Take, for example, the simplest case – the isotropic spectrum $\varepsilon = p^2/2m^*$ with effective mass m^*. In this case, the quantities β which enter into formulas (10.2) and (10.15) are different. β in formula (10.2) is the Bohr magneton and contains the mass of a free electron. Formula (10.15) contains β^* associated with the orbital motion of the electron, i.e, $e\hbar/(2m^*c)$. The density of states $\nu(\varepsilon)$ also contains m^*. From this we may conclude that

$$\frac{\chi_{\text{dia}}}{\chi_{\text{para}}} = \frac{1}{3}\left(\frac{m}{m^*}\right)^2.$$

This quantity may be larger than unity.

This is, of course, the simplest model. In actual fact, diamagnetism is difficult to calculate because the virtual transitions from the deep electronic states to the conduction band (polarization effects) also make a contribution to it.

10.4. Quasiclassical quantization of the energy levels for an arbitrary spectrum

We shall not consider formula (10.13) at high fields and finite temperatures since the most interesting consequences can be obtained not only for the free-electron model but for an arbitrary energy spectrum as well. Of course, in a general case it is impossible to determine the energy levels exactly. But if we compare the spacing between the levels βH with the chemical potential μ, we can see that for the usually employed fields of $\sim 10^4$ Oe

$$\frac{\beta H}{\mu} \sim 10^{-4}.$$

Since the quantum effects are associated with the discreteness of the energy levels, this estimate shows that they must be small, giving corrections of order $(\beta H/\mu)^m$, where the exponent m is different for different particular effects. Since $\mu \gg \beta H$, only higher levels with large values of n are important, and we can apply the quasiclassical Bohr quantization rules.

In the presence of a magnetic field the momentum operators $\boldsymbol{p} = -\mathrm{i}\hbar\nabla$ are replaced by

$$\boldsymbol{P} = \boldsymbol{p} - \frac{e}{c}\boldsymbol{A}$$

Taking $A_y = Hx$, $A_x = A_z = 0$, we find

$$P_x = -\mathrm{i}\hbar\frac{\partial}{\partial x}, \qquad P_y = -\mathrm{i}\hbar\frac{\partial}{\partial xy} - \frac{e}{c}Hx.$$

They satisfy the commutation relation

$$P_y P_x - P_x P_y = -\mathrm{i}\frac{\hbar e H}{c}. \tag{10.16}$$

Introducing the notation $Y = (c/eH)P_x$, we see that Y plays the role of the coordinate canonically conjugate to the momentum P_y:

$$P_y Y - Y P_y = -\mathrm{i}\hbar.$$

If we have a closed orbit, we can apply the Bohr quantization rule:

$$\oint P_y\, \mathrm{d}Y = 2\pi\hbar[n + \gamma(n)],$$

where $\gamma(n)$ $(0 < \gamma < 1)$ is a slowly varying function. Substituting the definition of Y in terms of P_x, we have

$$\oint P_y\, \mathrm{d}P_x = 2\pi\frac{e\hbar H}{c}[n + \gamma(n)].$$

But the integral on the left is an area bounded by an electron trajectory. As we know, the trajectory is determined by the conditions $p_z = \text{const.}$, $\varepsilon = \text{const.}$, i.e., it is the intersection of a constant-energy surface with the $p_z = \text{const.}$ plane. Thus, we obtain (Onsager 1952)

$$S(\varepsilon, p_z) = \left(2\pi\frac{e\hbar H}{c}\right)[n + \gamma(n)]. \tag{10.17}$$

For free electrons

$$S = \pi p_\perp^2 = \pi(p^2 - p_z^2) = \pi(2m\varepsilon - p_z^2).$$

Hence,

$$\varepsilon = \frac{p_z^2}{2m} + \frac{e\hbar H}{2mc}(2n + 2\gamma),$$

which coincides with eq. (10.9) for large values of n.

Formula (10.17) can also be derived easily from the correspondence principle. For closed orbits the cyclotron frequency is $\Omega = eH/m^*c$, where $m^* = (2\pi)^{-1}\partial S/\partial\varepsilon$. In the quasiclassical case, we must obtain an equidistant system of levels with spacings equal to $\hbar\Omega$. Hence,

$$\Delta\varepsilon = \hbar\Omega = \frac{eH\hbar}{c(2\pi)^{-1}\partial S/\partial\varepsilon}.$$

This formula may be written in the form

$$\frac{\partial S}{\partial\varepsilon}\Delta\varepsilon = \Delta S = 2\pi\frac{e\hbar H}{c}, \tag{10.18}$$

from which we obtain formula (10.17).

According to formula (5.10), the projection of the trajectory in coordinate space onto the (x, y) plane has a form similar to the trajectory in momentum space. The corresponding area is equal to $(c/eH)^2 S$. In accordance with formula (10.18), the variation of this area in going from the level n to $n+1$ is $2\pi c\hbar/eH$. Multiplying by H, we obtain the variation of the magnetic flux passing through this contour. It is found to be equal to

$$\Delta\Phi = 2\pi\frac{c\hbar}{e} = 2\Phi_0. \tag{10.19}$$

The quantity $\Phi_0 = \pi c\hbar/e$, which contains only universal constants, is called the magnetic flux quantum (F. London 1950)*. It is equal to 2.05×10^{-7} Oe cm^2. Thus, the electron trajectory changes in such a way that the magnetic flux passing through it changes by an even number of flux quanta (fig. 57). This rule is universal (see, for example, section 11.2).

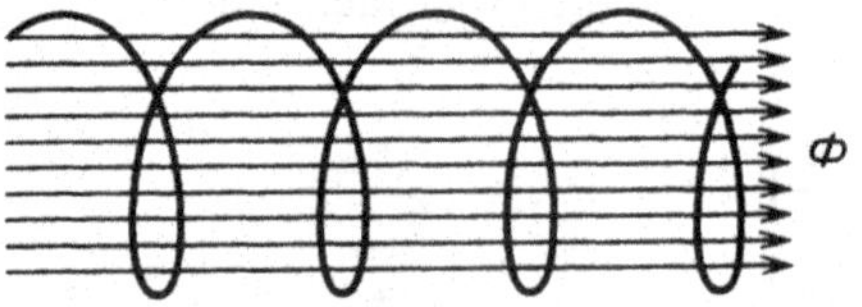

Fig. 57.

* For universality, we define a flux quantum as $\pi\hbar c/e$. This quantity plays a significant role in superconductors and in quantum corrections to the conductivity of normal metals (section 11.4). In his work on the theory of superconductivity F. London (1950) introduced a quantity twice as large since in 1950 it was not yet known that the superconducting current is transported by "Cooper pairs" with charge $2e$ (Part II). It is exactly this quantity ($2\Phi_0$) that determines $\Delta\Phi$ in a normal metal, but the introduction of two flux quanta, one for a normal metal and the other for a superconductor, seems to be unjustified.

Let us now find, in the quasiclassical case ($n \gg 1$), the number of states per level n and interval dp_z. Generally speaking, the number of states is always equal to

$$\frac{V}{(2\pi\hbar)^3}\,dp_z \iint dP_y\,dP_x.$$

The integral $\iint dP_y\,dP_x$ is the area in the $p_z = \text{const.}$ plane. In our particular case, this must be an area corresponding to an interval, say from n to $n+1$. But this is exactly the area ΔS defined by formula (10.18). Thus, the number of states is

$$\frac{eH}{c}\,\frac{V\,dp_z}{(2\pi\hbar)^2}.$$

This is the same expression as in the case of free electrons.

Note that all of the above derivation refers only to closed trajectories. If open trajectories are also possible for a certain direction of the field, the spectrum undergoes a considerable change and consists of continuous bands of finite width. This case will not be considered here.

10.5. The de Haas-van Alphen effect

In sections 10.1 and 10.3 we have found the magnetic susceptibility of a free electron gas. The approximation used there was the first correction to the thermodynamic potential from the magnetic field. Its order was $(\beta H/\mu)^2$. In this section we shall find the next correction. We shall see that it oscillates rapidly with magnetic field and therefore its derivative with respect to H (at sufficiently low temperatures) may exceed the monotonic part of the magnetic moment. This phenomenon was theoretically predicted by Landau (1930) and discovered experimentally by de Haas and van Alphen (1930).

We shall give here the calculation for an arbitrary energy spectrum (Lifshitz and Kosevich 1955). The expressions for the energy levels and the number of states have been derived in the preceding section.

The thermodynamic potential is written in the form (10.11), i.e., in this particular case,

$$\Omega = -\frac{VTeH}{(2\pi\hbar)^2 c}\sum_n\sum_\sigma \int \ln\left[1+\exp\left(\frac{\mu-\varepsilon_\sigma(n,p_z)}{T}\right)\right] dp_z. \tag{10.20}$$

The subscript σ denotes the projection of the spin, and $\varepsilon_\sigma = \varepsilon(n, p_z) \pm \beta H$ *. For

* In weak fields the effects caused by spin polarization and by the quantization of the orbits simply add up, which makes it possible to calculate them independently. This is exactly what has been done in sections 10.1 and 10.3. However, in calculating the oscillatory corrections such a division is no longer possible. It should also be mentioned that we disregard spin-orbit interaction here; this is justified for metals with Z not too large (see section 10.7). Otherwise, β should be replaced by some β' which depends on the orientation of $\boldsymbol{H}$ and which may exceed β significantly.

what follows we shall need Poisson's formula

$$\sum_{n_0}^{\infty} \varphi(n) = \int_a^{\infty} \varphi(n)\,\mathrm{d}n + \mathrm{Re} \sum_{k=1}^{\infty} \int_a^{\infty} \varphi(n)\,\mathrm{e}^{2\pi ikn}\,\mathrm{d}n, \tag{10.21}$$

where a lies between $n_0 - 1$ and n_0. The proof of this formula is as follows. It is not difficult to show that

$$\sum_{n=-\infty}^{\infty} \delta(x-n) = \sum_{k=-\infty}^{\infty} \mathrm{e}^{2\pi ikx}. \tag{10.22}$$

Actually, the right-hand side must be equal to zero everywhere, except for $x = n$. If $x \neq n$, then, going from k to $k + k_1$ (k and k_1 are integers), we get

$$\sum_{k=-\infty}^{\infty} \mathrm{e}^{2\pi ikx} = \mathrm{e}^{2\pi ik_1x} \sum_{k=-\infty}^{\infty} \mathrm{e}^{2\pi ikx}.$$

Since $\mathrm{e}^{2\pi ik_1x} \neq 1$, then $\sum_{k=-\infty}^{\infty} \mathrm{e}^{2\pi ikx} = 0$ for $x \neq n$. On the other hand, the right-hand side of (10.22) is a periodic function of x with a period of 1. We integrate the right-hand side of eq. (10.22) over the interval $-\delta$, δ and then put $\delta \to 0$. We find

$$\int_{-\delta}^{\delta} \sum_{k=-\infty}^{\infty} \mathrm{e}^{2\pi ikx}\,\mathrm{d}x = \sum_{k=-\infty}^{\infty} \frac{\mathrm{e}^{2\pi ik\delta} - \mathrm{e}^{-2\pi ik\delta}}{2\pi ik}$$

$$= \sum_{k=-\infty}^{\infty} \frac{\sin 2\pi k\delta}{\pi k} = \int_{-\infty}^{\infty} \frac{\sin 2\pi k\delta}{\pi k}\,\mathrm{d}k = 1.$$

The transformation of the sum to an integral is justified here if $\delta \to 0$ since only terms with large k are important.

Thus, we have shown that near $x = 0$ the right-hand side of eq. (10.22) behaves as the δ-function. By virtue of periodicity and also of the vanishing of this sum at any noninteger x, we obtain formula (10.22). Multiplying this formula by $\varphi(x)$ and integrating from a to ∞, where $n_0 - 1 < a < n_0$, we obtain Poisson's formula (10.21). The first term corresponds to $k = 0$ and the second term is composed of the terms with k and $-k$ combined.

Let us apply Poisson's formula (10.21) to the sum in eq. (10.20). It is split into two integrals. Only the second integral in (10.21) gives an oscillatory correction, and therefore we shall consider only this integral. Thus, we have

$$\tilde{\Omega} = -2\,\mathrm{Re} \sum_{\sigma,k} I_{k\sigma}, \tag{10.23}$$

where $\tilde{\Omega}$ denotes the oscillatory part of Ω and I_k is given by

$$I_{k\sigma} = \frac{VeHT}{(2\pi\hbar)^2 c} \int_a^{\infty} \mathrm{d}n \int_{-\infty}^{\infty} \mathrm{d}p_z \ln\left[1 + \exp\left(\frac{\mu_\sigma - \varepsilon(n, p_z)}{T}\right)\right] \mathrm{e}^{2\pi ikn}.$$

Here, we have replaced $\mu-\varepsilon_\sigma(n, p_z)$ by $\mu_\sigma-\varepsilon(n, p_z)$, where $\mu_\sigma=\mu+\beta H\sigma_z$. Instead of integrating over n we shall integrate over the energy. This yields

$$I_{k\sigma}=\frac{VTeH}{(2\pi\hbar)^2c}\int_0^\infty \mathrm{d}\varepsilon\,\ln\left[1+\exp\left(\frac{\mu_\sigma-\varepsilon}{T}\right)\right]$$
$$\times\int_{p_{z,\min}}^{p_{z,\max}}\frac{\partial n}{\partial\varepsilon}(p_z,\varepsilon)\,\mathrm{e}^{2\pi \mathrm{i}kn(p_z,\varepsilon)}\,\mathrm{d}p_z. \tag{10.24}$$

The lower limit of the integral over energies is taken to be equal to zero. But, actually what is essential here is the neighborhood of the Fermi surface alone and therefore the choice of the lower limit is unimportant. The limits $p_{z,\min}$ and $p_{z,\max}$ refer to a given value of ε.

The factor $\mathrm{e}^{2\pi\mathrm{i}kn}$ in the integral over $\mathrm{d}p_z$ is a radpily oscillating function. Suppose that n attains an extremal value n_m at some value of p_z, say, $p_z^{(\mathrm{m})}$. Then,

$$n(\varepsilon, p_z)=n_\mathrm{m}+\tfrac{1}{2}\left(\frac{\partial^2 n}{\partial p_z^2}\right)_{p_z^{(m)}}(p_z-p_z^{(\mathrm{m})})^2$$

(of course, n_m depends on ε here). Upon integration over p_z only these extrema are important. Near any of these extrema the integral over p_z becomes equal to

$$\left(\frac{\partial n}{\partial\varepsilon}\right)_\mathrm{m}\mathrm{e}^{2\pi\mathrm{i}kn_\mathrm{m}}\int_{-\infty}^{\infty}\exp\left[\mathrm{i}\pi k\left(\frac{\partial^2 n}{\partial p_z^2}\right)_\mathrm{m} z^2\right]\mathrm{d}z, \tag{10.25}$$

where $z=p_z-p_z^{(\mathrm{m})}$. If $(\partial^2 n/\partial p_z^2)_\mathrm{m}$ is positive, then

$$\int\exp\left[\mathrm{i}\pi k\left(\frac{\partial^2 n}{\partial p_z^2}\right)_\mathrm{m} z^2\right]\mathrm{d}z=\mathrm{e}^{\mathrm{i}\pi/4}\int_{-\infty}^{\infty}\exp\left[-\pi k\frac{\partial^2 n}{\partial p_z^2}y^2\right]\mathrm{d}y,$$

where we have put $z=\mathrm{e}^{\mathrm{i}\pi/4}y$. In the case of $(\partial^2 n/\partial p_z^2)_\mathrm{m}<0$ we set $z=\mathrm{e}^{-\mathrm{i}\pi/4}y$. Integrating, we obtain

$$\int\exp\left[\mathrm{i}\pi k\left(\frac{\partial^2 n}{\partial p_z^2}\right)z^2\right]\mathrm{d}z=\mathrm{e}^{\pm\mathrm{i}\pi/4}\left(k\left|\frac{\partial^2 n}{\partial p_z^2}\right|_\mathrm{m}\right)^{-1/2}. \tag{10.26}$$

Substituting eqs. (10.25) and (10.26) into eq. (10.24) leads to

$$I_{k\sigma}=\frac{VTeH}{(2\pi\hbar)^2c}\sum_\mathrm{m}\int_0^\infty \mathrm{d}\varepsilon\,\ln\left[1+\exp\left(\frac{\mu_\sigma-\varepsilon}{T}\right)\right]$$
$$\times\mathrm{e}^{\pm\mathrm{i}\pi/4}\left(\frac{\partial n}{\partial\varepsilon}\right)_\mathrm{m}\mathrm{e}^{2\pi\mathrm{i}kn_\mathrm{m}}\left(k\left|\frac{\partial^2 n}{\partial p_z^2}\right|_\mathrm{m}\right)^{-1/2}, \tag{10.27}$$

where the sum is taken over all the extrema of $n(p_z, \varepsilon)$.

Let us integrate this expression over ε. Here, we must take into account that only the logarithm and $\mathrm{e}^{2\pi\mathrm{i}kn_\mathrm{m}}$ are rapidly varying functions of ε, and therefore the remaining factors may be considered constant. After integrating by parts we have

$$I_{k\sigma}=\frac{VeH}{(2\pi\hbar)^2ck^{3/2}2\pi i}\sum_{\mathrm{m}} e^{\pm i\pi/4}\int_0^{\infty}\frac{d\varepsilon\, e^{2\pi ikn_{\mathrm{m}}}|\partial^2 n/\partial p_z^2|_{\mathrm{m}}^{-1/2}}{e^{(\varepsilon-\mu_\sigma)/T}+1}. \tag{10.28}$$

Here we have retained only the oscillatory term.

The most significant part of the integral arises from the neighborhood of $\varepsilon=\mu_\sigma$. It has to be taken into account that $e^{2\pi ikn_{\mathrm{m}}}$ is a rapidly oscillating function. But the quantity n_{m} itself varies smoothly and can be expanded in powers of $\varepsilon-\mu_\sigma$:

$$n_{\mathrm{m}}(\varepsilon)=n_{\mathrm{m}}(\mu_\sigma)+\left(\frac{\partial n_{\mathrm{m}}}{\partial\varepsilon}\right)_{\mu_\sigma}(\varepsilon-\mu_\sigma).$$

As a result, we have

$$\int_0^{\infty}\frac{e^{2\pi ikn_{\mathrm{m}}}}{e^{(\varepsilon-\mu_\sigma)/T}+1}d\varepsilon$$
$$=e^{2\pi ikn_{\mathrm{m}}(\mu_\sigma)}\int_{-\infty}^{\infty}\frac{\exp[2\pi ikx(\partial n_{\mathrm{m}}/\partial\varepsilon)_{\mu_\sigma}]}{e^{x/T}+1}dx,$$

where $x=\varepsilon-\mu_\sigma$ (the lower limit may be set equal to $-\infty$). Using the value of the integral

$$\int_{-\infty}^{\infty}\frac{e^{i\alpha y}}{e^y+1}dy=-\frac{i\pi}{\sinh\alpha\pi}$$

we obtain

$$I_{k\sigma}=-\frac{VeHT}{2(2\pi\hbar)^2ck^{3/2}}\sum_{\mathrm{m}} e^{\pm i\pi/4}\frac{e^{2\pi ikn_{\mathrm{m}}(\mu_\sigma)}|\partial^2 n/\partial p_z^2|_{\mathrm{m},\mu_\sigma}^{-1/2}}{\sinh[2\pi^2k(\partial n_{\mathrm{m}}/\partial\varepsilon)_{\mu_\sigma}T]}. \tag{10.29}$$

When substituting this expression into (10.23) we are to carry out the summation over the spin projections. Since only the factor $e^{2\pi ikn_{\mathrm{m}}(\mu_\sigma)}$ is rapidly oscillating, we may substitute $\mu_\sigma\approx\mu$ into all the remaining terms. Using the fact that $\beta H\ll\mu$, we may write:

$$\sum_\sigma e^{2\pi ikn_{\mathrm{m}}(\mu_\sigma)}=e^{2\pi ikn_{\mathrm{m}}(\mu)}\sum_\sigma\exp\left[2\pi ik\left(\frac{\partial n_{\mathrm{m}}}{\partial\varepsilon}\right)_\mu\beta H\sigma_z\right]$$
$$=2\,e^{2\pi ikn_{\mathrm{m}}(\mu)}\cos\left(2\pi k\left(\frac{\partial n_{\mathrm{m}}}{\partial\varepsilon}\right)_\mu\beta H\right). \tag{10.30}$$

According to eq. (10.17), we have

$$n_{\mathrm{m}}\approx\frac{cS_{\mathrm{m}}}{2\pi e\hbar H}. \tag{10.31}$$

Substituting eqs. (10.29), (10.30) and (10.31) into eq. (10.23), we finally obtain for the oscillatory part of the thermodynamic potential*:

$$\tilde{\Omega} = \frac{V}{2^{3/2}\pi^{7/2}\hbar^3}\left(\frac{e\hbar H}{c}\right)^{5/2} \sum_{\mathrm{m}} \left|\frac{\partial^2 S}{\partial p_z^2}\right|^{-1/2} [m^*(\mu, p_z^{(\mathrm{m})})]^{-1}$$

$$\times \sum_k \frac{\psi(k\lambda)}{k^{5/2}} \cos\left(k\frac{cS_{\mathrm{m}}}{e\hbar H} \pm \tfrac{1}{4}\pi\right) \cos\left(\pi k \frac{m^*}{m}\right), \tag{10.32}$$

where $\psi(z) = \dfrac{z}{\sinh z}$, $\lambda = \dfrac{2\pi^2 T c m^*}{e\hbar H}$, $m^* = \dfrac{1}{2\pi}\dfrac{\partial S}{\partial \varepsilon}$.

The sum over m implies the summation over the cross sections of the Fermi surface that have an extremal area.

Now let us examine the result. First of all, we shall estimate the role played by the oscillatory correction to the thermodynamic quantities. To this end, we find the ratio of Ω to Ω_0, which is the potential at $H = 0$. The order of magnitude of Ω_0 is

$$\Omega_0 \sim V n_e \mu \sim V \frac{p_0^3}{\hbar^3}\frac{p_0^2}{m} \sim \frac{V S^{5/2}}{m\hbar^3}.$$

From (10.32) at $\psi \sim 1$, $m^* \sim m$ we have

$$\tilde{\Omega} \sim \left(\frac{e\hbar H}{c}\right)^{5/2} \frac{V}{m\hbar^3}.$$

Hence,

$$\frac{\tilde{\Omega}}{\Omega_0} \sim \left(\frac{e\hbar H}{cS}\right)^{5/2} \sim \left(\frac{\beta H}{\mu}\right)^{5/2}.$$

This quantity is small. The nonoscillatory correction to Ω_0 from the magnetic field is of the order of $(\beta H/\mu)^2$. Thus, the oscillatory part is small even in comparison with this correction.

Let us now determine the role of the oscillating term in the magnetic moment. The oscillatory part of the magnetization is given by

$$\tilde{M} = -\frac{\partial \tilde{\Omega}}{\partial H}.$$

From eq. (10.32) we find

$$\tilde{M} = -2^{-3/2}\pi^{-7/2}\hbar^{-3}\left(\frac{e\hbar}{c}\right)^{3/2} H^{1/2} \sum_{\mathrm{m}} \left|\frac{\partial^2 S_{\mathrm{m}}}{\partial p_z^2}\right|^{-1/2} \frac{S_{\mathrm{m}}}{m^*}$$

$$\times \sum_{k=1}^{\infty} \frac{\psi(k\lambda)}{k^{3/2}} \sin\left[k\frac{cS_{\mathrm{m}}}{e\hbar H} \pm \tfrac{1}{4}\pi\right] \cos\left(\pi k \frac{m^*}{m}\right). \tag{10.33}$$

* For heavier metals, where spin-orbit effects become important, the bare mass in the last cosine should be replaced by some effective value (see footnote on p. 181).

For low temperatures $\psi \sim 1$. Then,

$$M \sim h^{-3}\left(\frac{e\hbar}{c}\right)^{3/2} H^{1/2}\frac{S}{m}.$$

The nonoscillatory part of the magnetic moment is of the order of

$$M_0 \sim \beta^2 \hbar^{-3} p_0 m H.$$

Thus, we obtain

$$\frac{\tilde{M}}{M_0} \sim \left(\frac{\mu}{\beta H}\right)^{1/2}. \qquad (10.34)$$

This means that at low temperatures the oscillatory part makes a greater contribution to the magnetic moment than the monotonic part. This is associated with the high frequency of oscillations, which strongly enhances the effect upon differentiation.

If the temperature is not too low, then in the sum over k in eq. (10.33) it will suffice to retain only one term, in which case we have

$$\tilde{M} \approx -\frac{2^{1/2}T}{\pi^{3/2}}\left(\frac{e\hbar}{c}\right)^{1/2} H^{-1/2}\sum_m S_m \left|\frac{\partial^2 S_m}{\partial p_z^2}\right|^{-1/2} \times \exp\left(-\frac{2\pi^2 Tcm^*}{e\hbar H}\right)\sin\left(\frac{cS_m}{e\hbar H} \pm \tfrac{1}{4}\pi\right)\cos\left(\frac{\pi m^*}{m}\right). \qquad (10.35)$$

Thus, the magnetic moment M is a periodic function of H^{-1} (fig. 58 shows a M/H plot for bismuth, Brandt and Ventsel' 1958). The period is given by

$$\Delta(H^{-1}) = 2\pi\frac{e\hbar}{cS_m}. \qquad (10.36)$$

At low temperatures we can no longer retain only one term in the sum over k (10.33). However, the periodicity of the moment as a function of H^{-1} with a period specified by eq. (10.36) is kept.

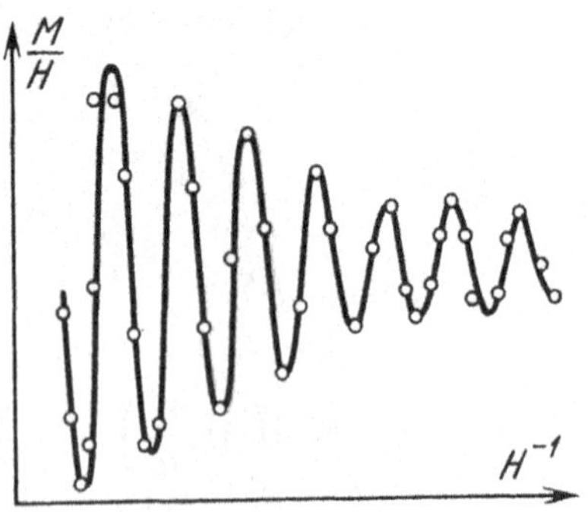

Fig. 58.

In the above derivation we ignored the scattering of electrons. The role of collisions is easy to estimate qualitatively. The collisions in this case have the same effect as the temperature – they smear out the sharp boundary of the Fermi distribution*. If the average collision time is τ, then the uncertainty in the electron energy is of the order of $\hbar/\tau$. This quantity plays the same role as the temperature. It may therefore be expected that taking account of collisions in (10.35) will lead to a new exponential factor. This is confirmed by a rigorous derivation. The collisions add the following factor to formula (10.35) (Dingle 1952):

$$\exp\left(\frac{-2\pi\, cm^*}{e\tau H}\right). \tag{10.37}$$

As a result of this lengthy calculation we have obtained the de Haas–Van Alphen oscillations. However, their physical origin can be explained rather easily if we consider the occupation of the electronic states. Let us examine fig. 59a. It shows the levels corresponding to different n (more precisely, each horizontal line signifies the bottom of the band $\varepsilon_n(p_z)$. The dashed line corresponds to the chemical potential μ. The lower levels are occupied and the upper ones are empty. Since $\mu \gg \beta H$, the number of levels n below μ is very large. Now suppose that the magnetic field becomes larger. Then, the separation between the levels increases and, finally, one of the lower levels will cross the Fermi boundary $\varepsilon = \mu$ (fig. 59b). The distribution of levels will then be very similar to the preceding one with the only difference that the number of levels below μ is now $n-1$ instead of n. But since n is very large, this difference is extremely small, which is why the new state must be expected to

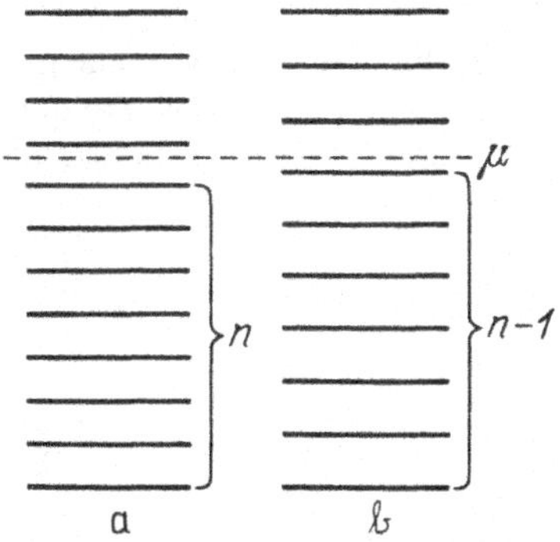

Fig. 59.

* As a matter of fact, the role of the temperature is different from that of collisions. Whereas a finite temperature leads to a redistribution of electrons among the energy levels, collisions lead to a change in the levels themselves (because of the nonconservation of the numbers n and p_z in collisions the classification of the levels is changed). In a case where the collision time is not too short, it may be assumed that the levels $\varepsilon(n, p_z)$ acquire a finite width of the order of $\hbar/\tau$. In the case under consideration this leads to the same effects as a finite temperature.

be almost equivalent to the old one. This implies periodic dependence on the magnetic field.

We can give a very simple derivation of oscillations of the magnetic moment if we use a two-dimensional metal model in a perpendicular magnetic field and it is assumed that $T=0$ (Peierls 1933). In this model the levels have the form $\varepsilon = \beta H\,(2n+1)$. Each level is degenerate and the corresponding number of states is equal to pSH ($p = e/(2\pi\hbar c) =$ const., S is the area of the two-dimensional metal).

If the total number of electrons is less than pSH, then they will all occupy the lower level with an energy βH. Hence, in this case the total energy will be

$$E = N_e \beta H$$

and the magnetic moment per unit area will be given by

$$M = -\frac{1}{S}\left(\frac{\partial E}{\partial H}\right) = -n_e\beta,$$

where N_e is the total number of electrons and $n_e = N_e/S$ is the electron density.

Suppose now that the magnetic field became lower, so that $2pH > n_e > pH$ or $n_e/2p < H < n_e/p$. In this case the electrons have to occupy not only the first level but also the second one with $n=1$. Therefore we have

$$\frac{E}{S} = \beta H pH + 3\beta H(n_e - pH) = 3\beta n_e H - 2\beta pH^2,$$

$$M = -3\beta n_e + 4\beta pH.$$

This means that at $H = n_e/p$ the magnetic moment jumps from the value $-n_e\beta$ to $n_e\beta$ and in smaller fields it changes linearly down to the value $-n_e\beta$ at $H = n_e/2p$.

Consider now the case $3pH > n_e > 2pH$ or $n_e/3p < H < n_e/2p$. This time the level with $n=2$ becomes involved, so that

$$\frac{E}{S} = \beta H\,pH + 3\beta H\,pH + (n_e - 2pH)5\beta H = 5\beta H n_e - 6\beta pH^2,$$

$$M = -5\beta n_e + 12\beta pH,$$

and so on.

If we construct a $M(H)$ plot, we shall obtain the picture shown in fig. 60. The jumps occur at $H = n_e/qp$ or $H^{-1} = qp/n_e$, where q is an integer. Of course, the two-dimensional model at $T=0$ does not reproduce many of the details of the true three-dimensional picture, but it provides a qualitative explanation of the effect.

Let us go back to formula (10.35) and find out what physical conditions must be realized for the effect to be observable. Suppose that the metal is sufficiently pure, so that the factor (10.37) arising from τ is equal to unity. (The expression in the

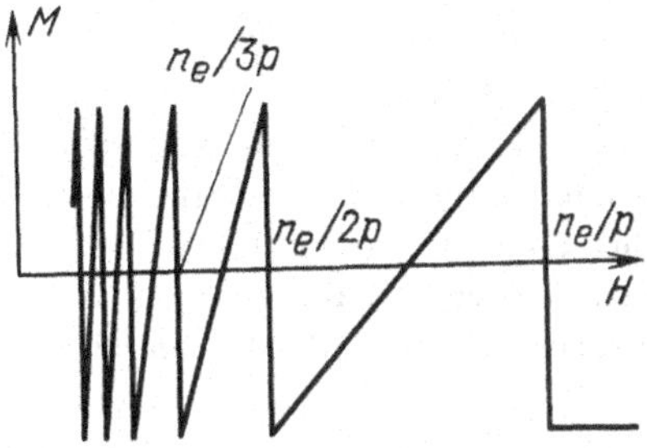

Fig. 60.

exponential is of the order of $2\pi/\Omega\tau$, where Ω is the cyclotron frequency, which is why we must have $\Omega\tau \gg 2\pi$.)

Consider the temperature exponential. In the exponent we have $2\pi^2 T/\beta^* H \approx 20T/\beta^* H$. If $m^* \sim m$, so that $\beta^* \sim \beta$ and $T \sim 1$ K, then for the exponential to be close to unity, it is required that the field $H \sim 5 \times 10^5$ Oe. Such fields can, in principle, be achieved, though it is not a simple matter. Further, for $\beta H \sim 20$ K the argument in the sine of (10.35) is of the order of $\mu/\beta H \sim 10^3$ since $\mu \sim (2\text{–}3) \times 10^4$ K. But the period of oscillations is 2π. Hence, the field has to be kept homogeneous and constant to within 1–0.1%. This adds to the difficulty of practical observation of the effect*.

However, in semimetals the situation is much simpler. If $m^* \sim 0.01\, m$, then $\beta^* \sim 100\beta$. Thus, at $T \sim 1$ K a field of the order of 10^4 Oe will suffice. The number of electrons in a semimetal is also small. Therefore, the Fermi energy is several hundredths of that in ordinary metals. Therefore, the argument in the sine appears to be of the order of unity, and the requirement of field uniformity will not be extremely stringent.

In "good" metals there also exist portions of the Fermi surface with small cross sections and small effective masses. These portions are difficult to examine by using the methods described in the preceding chapters because their influence on conductivity is negligible. But the de Haas–van Alphen oscillation from those portions will have a large amplitude and a large period. Of course, the extremal cross sectional area is not such a direct characteristic of the spectrum as the value of the Fermi momentum. But nonetheless the measurement of this quantity with various directions of the field practically allows one to reproduce the Fermi surface with good accuracy.

Historically, it happened so that the de Haas–van Alphen effect was the first phenomenon which made it possible to determine the energy spectrum, and so the energy spectrum of a semi-metal, namely, bismuth, was determined first (Shoenberg 1939).

* The requirement of field uniformity is not an insurmountable obstacle because in present-day setups it is possible to provide field uniformity to within 10^{-3}–10^{-4}%.

10.6. Diamagnetic domains

In the formula of the preceding section the external field H played the role of the acting field. Actually, of course, the field inside the metal differs from the external field. But as long as the magnetic susceptibility is low, this difference may be neglected. It is not difficult to see that the de Haas–van Alphen effect can really lead to a high magnetic susceptibility. Indeed, from (10.33) we obtain at $T \leqslant \beta H$:

$$\chi = \frac{\partial M}{\partial H} \sim \left(\frac{\mu}{\beta H}\right)^{3/2} \gg 1.$$

In view of this, one has to find out what field must be substituted into formulas (10.33) and (10.35) for M.

The electrons in metals move along the Larmor orbits. Therefore, the magnetic field produced by them is averaged over regions with a size of the order $r_L \sim cp_0/eH \sim 10^{-3}$ cm in a field $\sim 10^4$ Oe. But the average separation between the electrons is 10^{-8} cm. It is clear therefore that the field that acts on the electrons in a metal is an averaged field. As is known, the average value of the microscopic field is nothing else than the induction B. Hence, in the formula for the moment under conditions where χ becomes large, H must be replaced by B. Here, the dependence $B(H)$ or $M(H)$ is no longer given by formula (10.33) but, instead, it is found from the equation

$$H = B - 4\pi M(B). \tag{10.38}$$

This leads to a rather unexpected consequence. If the temperature is not too low, then $M(B) \ll B$ and the dependence $H(B)$ is as shown in fig. 61a. But when the temperature is lowered, the amplitude of oscillations increases and we eventually obtain the picture shown in fig. 61b. Here, there appear regions of H which correspond to different values of B. Such an ambiguity is evidence of an instability of the state in the same way as in the curve representing the van der Waals equation of state, $p(V)$ (fig. 62).

This analogy is not accidental. Just as in the case of the van der Waals equation, where by virtue of the thermodynamic inequality $(\partial p/\partial V)_T < 0$ the section of the curve between points 1 and 2 cannot be realized, here too we have a thermodynamic

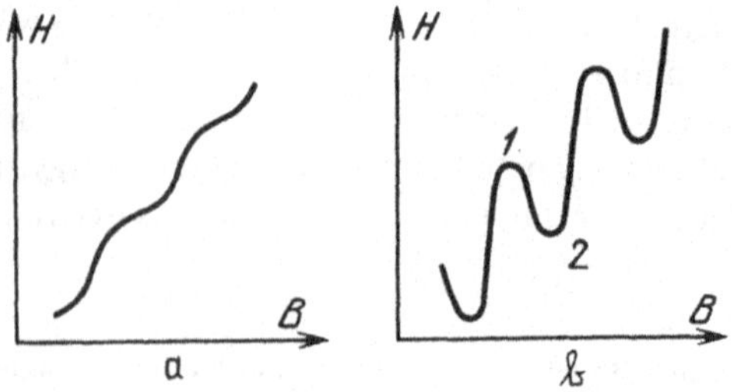

Fig. 61.

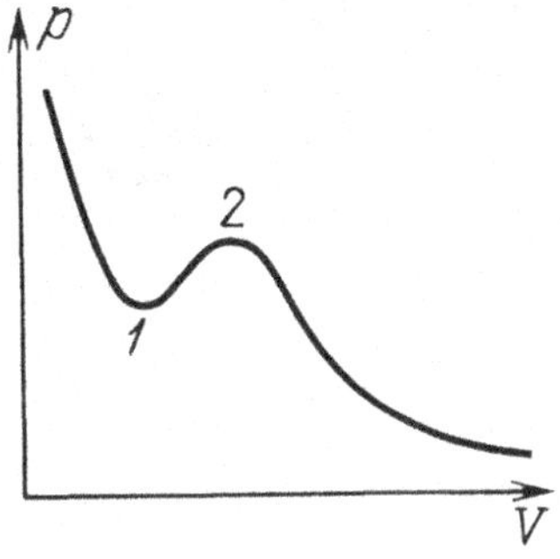

Fig. 62.

inequality, $\partial H/\partial B>0$, which forbids the section 1–2 in fig. 61b (Shoenberg 1962). The analogy goes further. Just as the van der Waals curve in fact describes a first-order phase transition from a gas into a liquid, the curve in fig. 61b describes successive phase transitions with a discontinuous change in the induction (Pippard 1963). Actually, the situation in the magnetic field is more complicated because of the demagnetization effect. But for the moment we assume that we have a cylindrical sample in a longitudinal field. With such geometry the analogy is fully retained.

Thus, the condition for jumps in the induction is the appearance of sections in the curve $H(B)$ curve where $\partial H/\partial B<0$, or

$$\chi=\frac{\partial M(B)}{\partial B}>\frac{1}{4\pi}. \tag{10.39}$$

A discontinuity in the induction occurs in a constant external field and is determined by the equality of the free energies in a given field, i.e., $\int B\,\mathrm{d}H$. This equality may be written as

$$\int_1^2 B\,\mathrm{d}H=H_0(B_2-B_1)-\int_1^2 H\,\mathrm{d}B=0, \tag{10.40}$$

which means the equality of the shaded areas in fig. 63. Thus, in this case the $B(H)$ curve consists of smooth sections and discontinuities and the $M(H)$ curve is as shown in fig. 64a. Let us consider noncylindrical geometry. Suppose that the sample has the shape of an arbitrary ellipsoid with a demagnetization factor n. The effective internal field is known to be $H_\mathrm{i}=H_0-4\pi nM$ (see Appendix III); at the same time, according to (10.38), $H_\mathrm{i}=B-4\pi M$. Thus, we have

$$H_0=B-4\pi(1-n)M. \tag{10.41}$$

Since in this case H_i plays the role of the field H, the dependence $H_\mathrm{i}(B)$ is given by the plot in fig. 61. Hence, for the "singular" portions we have $H_\mathrm{c}=H_0-4\pi nM$ or

$$M=\frac{H_0-H_\mathrm{c}}{4\pi n}, \tag{10.42}$$

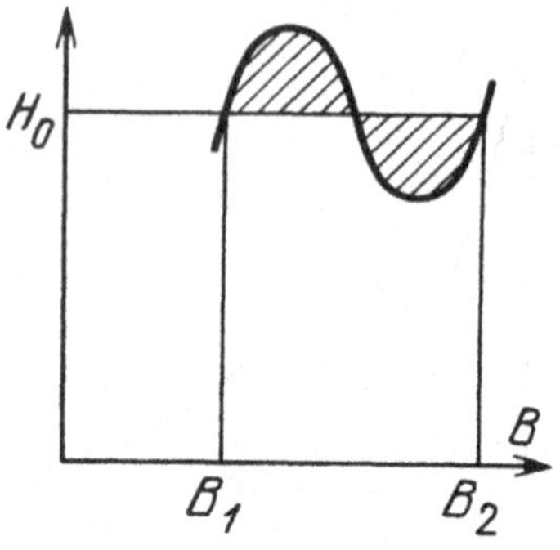

Fig. 63.

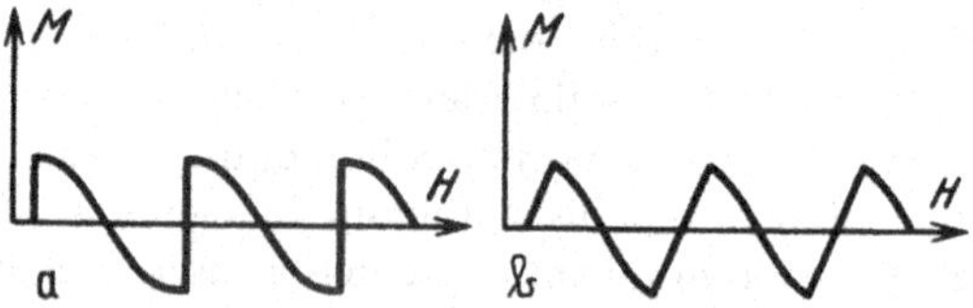

Fig. 64.

that is, a linear dependence instead of the discontinuities. This dependence is valid in a region where $B_1 < B < B_2$ or, according to eqs. (10.41) and (10.42),

$$nB_1 + (1-n)H_c < H < nB_2 + (1-n)H_c. \tag{10.43}$$

On the whole, the dependence $M(H)$ is shown in fig. 64b.

Now, however, a question arises: What is the "singular" region in this case? As a matter of fact, the sample may be either in a state with induction B_1 or in a state with induction B_2. And in the case $n \neq 0$ its induction assumes all intermediate values. The point is that the formal theory outlined above describes only the fully averaged values. The induction B, which appears in eq. (10.41), is the averaged value over the entire sample. Since only the values B_1 and B_2 are actually possible, it follows that the sample is split into alternating domains with the induction values B_1 and B_2 (Condon 1962). The ratio of the phase volumes is such that the average induction corresponds to B. The situation looks especially simple in the case of a planar plate perpendicular to the field ($n = 1$). In this case, the boundary condition of continuity of the normal component of the induction implies that $B = H_0$. Therefore, the region of separation into domains is $B_1 < H_0 < B_2$ and, hence, some domains are magnetized paramagnetically and the others diamagnetically.

The types of dependence of magnetization on field shown in fig. 64a, b have been obtained experimentally (Condon 1966). Moreover, it has been proved with the aid of the NMR technique (see section 21.4) that at low temperatures a silver sample splits into domains with different values of induction, the induction values remaining unchanged with the variation of the external field; only the phase concentration is changed (Condon and Walstedt 1968).

The above reasonings allow one to establish the presence of domains, the values of induction in the domains and the relative concentration of phases. However, the detailed magnetic structure itself remains indefinite. It is most natural to suppose that, as a result of the spatial homogeneity and symmetry of the problem, in a plate perpendicular to the external field there arises a periodic pattern of alternating domains which have the shape of planar layers parallel to the external field. In order to determine the period of this structure, one has to introduce a new concept - the surface energy at the boundary between layers.

Clearly the transition layer between the phases is in a special state. The induction in it varies from B_1 to B_2, i.e., it is inhomogeneous in space. This means that some additional energy must be associated with this layer. The whole picture described earlier is valid only in the case where this surface energy is positive; otherwise it would be energetically favorable if the number of interfaces increased infinitely, which would lead to a complete mixing of the phases. This would occur not only when $n \neq 0$ but for a cylinder in a longitudinal field as well. In this case the overall picture of the transition would be quite different. In principle, this can actually occur (Azbel' 1967) but the case of positive surface energy is encountered most frequently (Privorotskii 1967) and we are going to consider it here qualitatively.

Since thermodynamic equilibrium requires that the substance be either in a state with induction B_1 or in a state with induction B_2, any intermediate induction value must give an excess energy. This energy must vanish at $B = B_1$ and $B = B_2$. Hence, if B_1 and B_2 differ little from each other, the excess energy will be proportional to

$$(B_1 - B_2)^2 - (2B - B_1 - B_2)^2.$$

Throughout the entire transition layer this quantity is of the order of $(B_1 - B_2)^2$. Since the expression for the free energy includes $(4\pi)^{-1} \int H \, dB$, it follows that the energy per unit volume of the transition layer must be of the order of

$$\frac{(B_1 - B_2)^2}{4\pi}.$$

Multiplying this value by the thickness of the transition layer d, we obtain the energy per unit area of the boundary, i.e., the surface energy

$$\sigma \sim d \frac{(B_1 - B_2)^2}{4\pi}.$$

If $B_1 - B_2$ is of the order of the period of de Haas-Van Alphen oscillations, then the only parameter having the dimension of length is the Larmor radius r_L, so that $d \sim r_L$.

Take a plate of thickness D, the other dimensions being equal to unity. If the period of the structure is b, then the number of layers is equal to b^{-1} and the total area of all the boundaries is $2D/b$. Thus, the surface energy makes a contribution to the total energy equal to $2\sigma D/b$.

Let us now see how the layers emerge to the sample surface. This is shown in fig. 65. The continuity of the magnetic flux ($\operatorname{div} B = 0$) implies the continuity of magnetic lines of force. Far from the sample the lines are distributed with a uniform density, but inside the sample their density is different in different domains. On the other hand, a strong distortion of the lines of force gives rise to an extra energy. It follows that the layer boundaries must be curved in the vicinity of the surface, as shown in fig. 65. The linear dimensions of such a distortion will be of the order of the thickness of the layers, i.e., of the order of the period of the structure b, and the energy for these portions will evidently be of the order of $(B_1 - B_2)^2 b^2/4\pi$. Since the number of layers is b^{-1}, the total contribution of the distortions to the energy is of the order of $(B_1 - B_2)^2 b/4\pi$. Thus, the total energy is given by

$$C_1(B_1 - B_2)^2\left(\frac{dD}{b} + C_2 b\right)\Big/4\pi,$$

where C_1, $C_2 \sim 1$. Taking the minimum of this expression with respect to b, we find the period of the structure:

$$b = \left(\frac{dD}{C_2}\right)^{1/2},$$

that is, the dimensions of the domains increase with increasing plate thickness as $D^{1/2}$.

The formation of domains due to the de Haas–van Alphen effect is analogous in many respects to other cases of formation of magnetic domains with noncylindrical geometry, namely, in ferromagnetics and type I superconductors (the intermediate

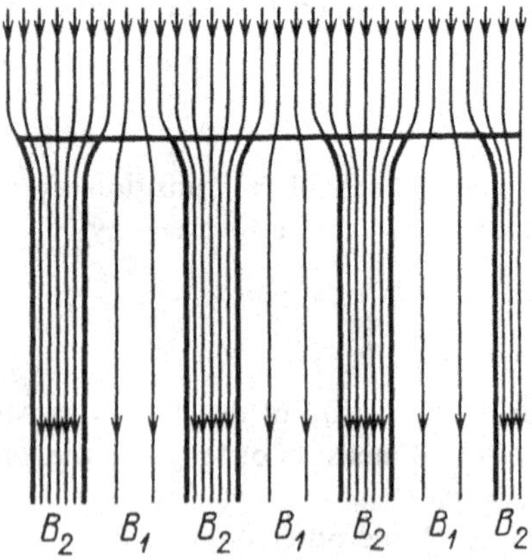

Fig. 65.

state; see section 15.3). In both these cases there is an equilibrium between two different phases and there is an extra surface energy at the interface. Just as in the case under consideration, the size of the domains is found to be proportional to $D^{1/2}$.

10.7. Magnetic breakdown

Up until now we have always compared the separation between the Landau levels βH with T and μ, but we have not compared this quantity with the width of the gap in the electronic energy spectrum, which we label ε_g. It is clear that if βH is of the order of ε_g (as a matter of fact, as we shall see below, even long before this happens), simple quasiclassical quantization is no longer applicable. Usually, the spacing between magnetic levels is of the order of 10^{-8} eV per 1 Oe, whereas the energies characterizing the band structure are of the order of 2–10 eV. But there are cases where the band splitting energy is much lower than 1 eV. A well-known case is that where the splitting is caused by the spin–orbit coupling of electrons.

What we mean here is that it is quite possible that the levels of an electron moving in the electrostatic potential of ions and of other electrons prove to be degenerate. Usually this occurs in those points of the space of the reciprocal lattice that have a high symmetry. It may happen that the degeneracy is lifted due to a peculiar relativistic effect called spin-orbit coupling. This arises from the fact that in a coordinate system moving together with the electron a small magnetic field is added to the electric field; it appears to be of order $(v/c)\nabla V$, where v is the electron velocity, c is the velocity of light and V is the periodic electrostatic potential of the lattice. This magnetic field interacts with the magnetic moment of the electron and may alter the electronic energy spectrum. The relative value of the correction to the energy is of order $(v/c)^2 f(Z)$, where $f(Z)$ is an increasing function of the atomic number. Since $v \sim 10^8$ cm/s, these energies are usually of the order of 10^{-3}–10^{-2} eV. Thus, this is a small effect, but it may turn out to be significant in the context of lifting the level degeneracy. For example, let us consider a case shown in fig. 66, where two bands are in contact with each other at a point corresponding to the energy maximum for one band and to the minimum for the other. In this case, spin–orbit coupling can lift the degeneracy, which will lead to the appearance of a small gap in the energy spectrum.

If the magnetic field acting on such a metal is small, the electrons will move along the corresponding quasiclassical orbits. But as the field increases a situation arises in which such small gaps become unimportant. The motion of the electrons will then occur along the orbits that airse with complete neglect of the small gaps. This phenomenon is known as magnetic breakdown (Cohen and Falicov 1961).

With intermediate values of the magnetic field the energy levels expand into energy bands, leading to certain peculiar effects (for example, to a change in the mode of electron scattering). But this refers only to the transition region itself and does not occur outside this region.

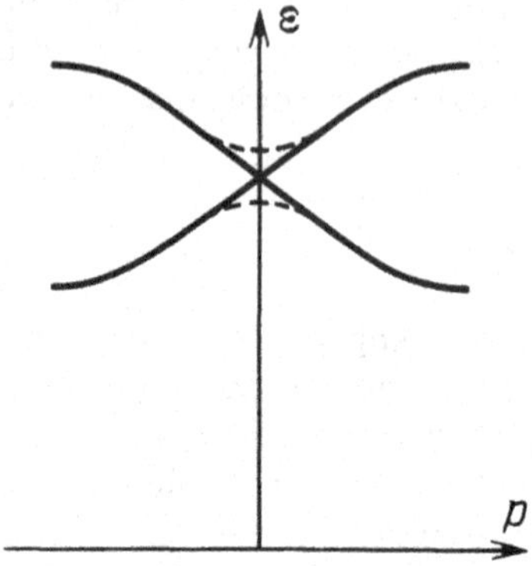

Fig. 66.

Consider, for example, a case where an almost free electron is placed in a lattice which is periodic in one direction (fig. 67a). In accordance with ch. 1, we are to consider only the Brillouin zone, while the part of the constant-energy surface that extends beyond the Brillouin zone must be interpreted as one belonging to the next energy band. From fig. 67b we see that at sufficiently high energies we obtain a warped cylinder in one band and a closed surface in the other. If the magnetic field is directed along the z axis, then in small fields we shall observe de Haas–van Alphen oscillations from the closed surfaces alone, and the resistance ρ_{xx} will be proportional to H^2, since this corresponds to the case of a warped cylinder in a perpendicular field (section 5.4). However, when the magnetic field is sufficiently high, magnetic breakdown will restore the original Fermi sphere for free electrons, and its equatorial cross section will give a period of de Haas–van Alphen oscillations. For the same reason, ρ_{xx} will tend to saturate.

Let us estimate the order of magnetic fields required for magnetic breakdown. Inspect fig. 68. This figure may be interpreted not only as a representation of the constant-energy surface but also as a representation of its intersection with the

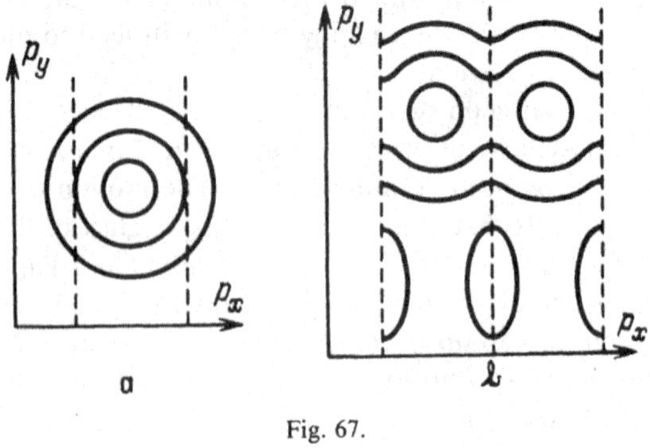

Fig. 67.

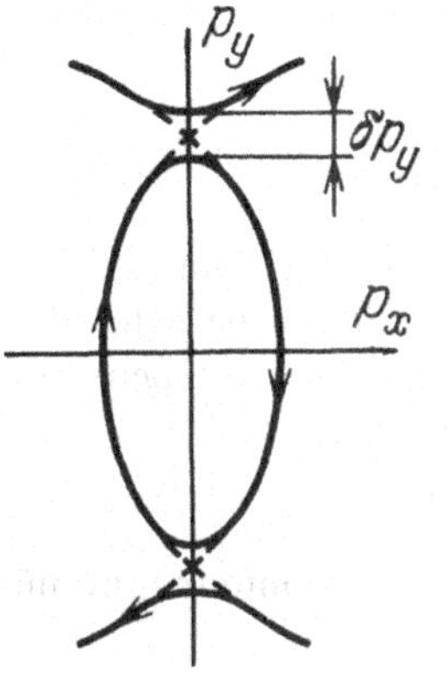

Fig. 68.

$p_z = \text{const.}$ plane, i.e., an electron trajectory. The dependence of the energy on the momentum for almost free electrons has been considered in section 1.3. According to formula (1.42), with an appropriate redefinition of the momenta we have the equation of the trajectories depicted in fig. 68:

$$\frac{1}{2m}\left[p_x^2+p_y^2+(\tfrac{1}{2}\hbar K)^2\right] \pm \left[\left(p_x\frac{\hbar K}{2m}\right)^2+|U_K|^2\right]^{1/2} = \varepsilon. \tag{10.44}$$

where $2U_K$ plays the role of the energy gap.

The shortest distance between the trajectories corresponds to $p_x = 0$ and is given by

$$\begin{aligned}\delta p_y &= \left\{2m\left[\varepsilon - \frac{1}{2m}(\tfrac{1}{2}\hbar K)^2 + |U_K|\right]\right\}^{1/2} \\ &\quad - \left\{2m\left[\varepsilon - \frac{1}{2m}(\tfrac{1}{2}\hbar K)^2 - |U_K|\right]\right\}^{1/2} \\ &\approx (2m)^{1/2}|U_K|\left[\varepsilon - \frac{1}{2m}(\tfrac{1}{2}\hbar K)^2\right]^{1/2} \sim \left(\frac{m}{\mu}\right)^{1/2}|U_K|.\end{aligned}$$

Here we assume that $U_K \ll \varepsilon - (2m)^{-1}(\tfrac{1}{2}\hbar K)^2 \sim \mu$. As has been pointed out in section 5.1 (formula 5.10), the trajectory in coordinate space is similar to the trajectory in momentum space. Therefore, by changing the scale we find from the magnitude of the "forbidden portion" in momentum space δp_y the size of the forbidden portion in coordinate space:

$$\delta x \sim \frac{c}{eH}\left(\frac{m}{\mu}\right)^{1/2} U_K. \tag{10.45}$$

This portion is a potential barrier that cannot be passed by a classical electron. But in quantum mechanics the probability exists of passing through the barrier with the aid of the so-called tunnelling effect. The probability of tunnelling is proportional to

$$W \sim \exp\left[\frac{-2}{\hbar}\,\mathrm{Im}(p_x)\,\delta x\right].$$

The imaginary part of p_x is obtained from eq. (10.44) in the forbidden portion in fig. 68. By order of magnitude it can be determined from the condition that the imaginary p_x under the root in (10.44) compensate the term $|U_K|^2$. Hence, we have

$$\mathrm{Im}(p_x) \sim \frac{U_K m}{(\hbar K)} \sim U_K\left(\frac{m}{\mu}\right)^{1/2}. \tag{10.46}$$

Substituting eqs. (10.45) and (10.46) into the formula for the tunneling probability, we find (Blount 1962):

$$W \sim \exp\left(-\alpha\frac{2cm}{ehH}\frac{U_K^2}{\mu}\right) = \exp\left(-\alpha\frac{U_K^2}{\beta H\mu}\right), \tag{10.47}$$

where $\alpha \sim 1$. Thus, it follows that for magnetic breakdown we must have $\beta H \sim U_K^2/\mu$ rather than $\beta H \sim U_K$, i.e., a much smaller quantity. For example, at $U_K \sim 10^{-2}$ eV, $\mu \sim 1$ eV a field of 10^4 Oe is sufficient.

As a matter of fact, the model under discussion is not realized directly in nature, and the conditions for observation of magnetic breakdown are encountered in rare cases. One example is magnesium. In high fields in magnesium a cross-sectional area larger than the cross section of the Brillouin zone has been observed in the de Haas–van Alphen effect (Cohen and Falicov 1961). This may be interpreted as follows. Magnesium has a close-packed hexagonal lattice (i.e., its reciprocal lattice has the symmetry of a regular hexagonal prism). Consider the central cross section of the Fermi surface which is perpendicular to the principal axis. The cross section of the Brillouin zone in this case has the shape of a regular hexagon (fig. 69).

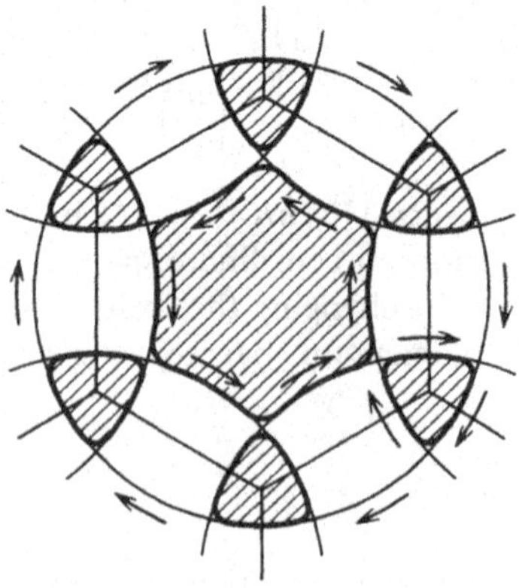

Fig. 69.

Let us sketch the electronic orbits in accordance with the free electron model (section 14.3). To this end, let us draw circles with their center located in the center of each hexagon. These circumferences intersect. As always, the intersection is actually removed and we obtain a closed contour in the middle and "triangular contours" at the edges of the hexagon. An essential point is that the lifting of the degeneracy occurs only due to the spin-orbit coupling. With small fields the periods of de Haas-van Alphen oscillations are determined by the shaded areas in fig. 69. But if the field is sufficiently high, the orbit of the free electron is restored and a cross section larger than the area of the hexagon is obtained.

11

Quantum effects in conductivity

11.1. The Shubnikov–de Haas effect

Oscillations of the same type as in the de Haas-van Alphen effect are observed in kinetic phenomena, for example, in the electrical and thermal conductivities. Oscillations of the conductivity (Shubnikov and de Haas 1930) are most convenient for experimental observations; therefore we shall consider this effect in detail. The kinetic equation that has been used so far is inapplicable in this case and the construction of a complete quantum theory of kinetic phenomena, because of its complexity, goes beyond the scope of this book*. In view of this, we shall find the order of magnitude of the oscillatory correction to the conductivity by using the fact that the major contribution to it comes from variation of the scattering probability (Adams and Holstein 1959).

We shall consider scattering from impurities and for the sake of simplicity we assume it to be isotropic. In section 3.2 we have found the scattering probability for the classical case,

$$W=\frac{\pi}{\hbar}n_i|v|^2\nu(\mu)$$

$$=\frac{2\pi}{\hbar}n_i|v|^2\int\delta(\varepsilon-\mu)\frac{\mathrm{d}^3p}{(2\pi h)^3}. \tag{11.1}$$

Let us extend formula (11.1) to the quantum case. The states here are characterized by the quantum numbers n and p_z. Instead of formula (11.1), we get

$$W=\frac{2\pi}{\hbar}n_i|v|^2\int\mathrm{d}p_z\sum_n\delta(\varepsilon(n,p_z)-\mu)\frac{eH}{c}\frac{1}{(2\pi\hbar)^2}. \tag{11.2}$$

For large n ($\mu\gg\beta H$) this formula turns into (11.1) because $\sum_n\to\int\mathrm{d}n\to\int\mathrm{d}p_z\,\mathrm{d}p_yc/(2\pi e\hbar H)$.

In considering ordinary impurity scattering we could have used formula (11.1) for any temperatures. However, in the quantum case the temperature, being much

* Readers who wish to be acquainted with the complete theory are referred to the books by Abrikosov, Gor'kov and Dzyaloshinskii (1962) and Lifshitz and Pitaevsky (1979) and an article by Abrikosov (1969).

lower than μ, is not necessarily low as compared to βH. In view of this, we make the replacement

$$\delta(\varepsilon-\mu)\rightarrow-\frac{\partial f}{\partial\varepsilon},$$

where f is the Fermi function. Thus, we finally write the scattering probability for electrons with given spin projection in the form*

$$W=\frac{2\pi}{\hbar}n_i|v|^2\frac{eH}{c}\frac{1}{(2\pi\hbar)^2}\cdot I, \tag{11.3}$$

where

$$I=-\sum_n\int \mathrm{d}p_z\frac{\partial f}{\partial\varepsilon}. \tag{11.4}$$

Let us apply, just as in section 10.5, Poisson's formula (10.21). We then have

$$I=-\int_a^\infty \mathrm{d}n\int \mathrm{d}p_z\frac{\partial f}{\partial\varepsilon}+2\,\mathrm{Re}\sum_{k=1} I_k, \tag{11.5}$$

where

$$I_k=-\int_{-\infty}^{\infty}\mathrm{d}p_z\int_a^\infty \mathrm{d}n\,\mathrm{e}^{2\pi ikn}\frac{\partial f}{\partial\varepsilon}. \tag{11.6}$$

It is not difficult to see that the first term in formula (11.5) gives the classical scattering probability (11.1).

We shall now find the quantity I_k in the same manner as we found $I_{k\sigma}$ in section 10.5. The integration over n is replaced by integration over ε. In this case a rapidly varying factor appears: $\exp[2\pi ikn(\varepsilon, p_z)]$. The exponent is expanded near the extremal points of n in $p_z(n=n_\mathrm{m})$ at given ε. As a result, we obtain the integral (10.25). Taking this integral, we have

$$I_k=-\sum_n\int_0^\infty \mathrm{d}\varepsilon\frac{\partial f}{\partial\varepsilon}\mathrm{e}^{\pm i\pi/4}\left(\frac{\partial n}{\partial\varepsilon}\right)_\mathrm{m}\mathrm{e}^{2\pi ikn_\mathrm{m}}\left(k\left|\frac{\partial^2 n}{\partial p_z^2}\right|_\mathrm{m}\right)^{-1/2}. \tag{11.7}$$

Here the rapidly varying functions of ε are the factors $\partial f/\partial\varepsilon$ and $\exp(2\pi ikn_\mathrm{m})$. Again, as before, putting

$$n_\mathrm{m}\approx n_\mathrm{m}(\mu_\sigma)+\left(\frac{\partial n_\mathrm{m}}{\partial\varepsilon}\right)_\mu(\varepsilon-\mu_\sigma),$$

* At first sight it seems that we are going to find the correction to the density of states and that such a correction will not affect the conductivity because $\sigma\sim e^2v^2\nu\tau$ and $\tau=W^{-1}$, i.e., is inversely proportional to ν (see formula 11.1). Actually, we calculate the correction not to ν but to the collision integral, i.e., τ^{-1}. Strict calculation shows that σ_{zz}, σ_{xx} and σ_{yy}, apart from the corrections to W, include quite different sums over n, so that no cancellation occurs. It turns out that the result found below gives an exact answer for σ_{zz} and determines the order of magnitude of the relative correction to σ_{xx} and σ_{yy}.

we obtain the integral

$$-\int \frac{\partial f}{\partial \varepsilon} e^{2\pi i k n_m} d\varepsilon$$

$$\approx (4T)^{-1} e^{2\pi i k n_m(\mu_\sigma)} \int_{-\infty}^{\infty} \cosh^{-2}\left(\frac{x}{2T}\right) \exp\left[2\pi i \frac{\partial n_m}{\partial \varepsilon} x\right] dx.$$

Using the formula

$$\int_{-\infty}^{\infty} \exp(i\alpha y) \cosh^{-2} y \, dy = \frac{-\pi\alpha}{\sinh \frac{1}{2}\pi\alpha}$$

and substituting all these results into eq. (11.7), we find

$$I_k = -\sum_m e^{\pm i\pi/4} \frac{\partial n_m}{\partial \varepsilon} \left|\frac{\partial^2 n}{\partial p_z^2}\right|_m^{-1/2} \times \sum_k \psi(k\lambda)\, k^{-1/2} \exp[2\pi i k n_m(\mu_\sigma)]. \tag{11.8}$$

Here we considered the scattering probability for electrons with a definite spin projection, i.e., $\mu_\sigma = \mu + \beta H \sigma_z$. However, the conductivity is determined by electrons with both spin projections, and the small corrections in a first approximation simply add up. In view of this, the resultant expression obtained for I_k can be averaged over the spin projections. This leads to the replacement

$$\exp[2\pi i k n_m(\mu_\sigma)] \to \exp[2\pi i k n_m(\mu)] \cos\left[2\pi k \left(\frac{\partial n}{\partial \varepsilon}\right)_m \beta H\right].$$

Substituting this into eq. (11.5), expressing n_m in terms of S_m with the aid of (10.31) and using the definition $m^* = (2\pi)^{-1}(\partial S/\partial \varepsilon)$, we find

$$I = \frac{c}{eH}(2\pi\hbar)^2 \frac{\nu(\mu)}{2} - 2(2\pi)^{1/2}\left(\frac{c}{\hbar e H}\right)^{1/2} \times \sum_m m_m^* \left|\frac{\partial^2 S_m}{\partial p_z^2}\right|^{-1/2} \times \sum_k k^{-1/2} \psi(k\lambda) \cos\left(\frac{c S_m k}{\hbar e H} \pm \tfrac{1}{4}\pi\right) \cos\left(\pi k \frac{m^*}{m}\right). \tag{11.9}$$

Substitution of I into eq. (11.3) yields

$$W = W_0\left[1 - \frac{2^{1/2}}{\pi^{3/2}\hbar^3}\left(\frac{e\hbar H}{c}\right)^{1/2} \nu^{-1}(\mu) \times \sum_m m_m^* \left|\frac{\partial^2 S_m}{\partial p_z^2}\right|^{-1/2} \times \sum_k k^{-1/2} \psi(k\lambda) \cos\left(\frac{c S_m k}{\hbar e H} \pm \tfrac{1}{4}\pi\right) \cos\left(\frac{\pi k m^*}{m}\right)\right]. \tag{11.10}$$

Since the conductivity is proportional to $\tau = W^{-1}$, it may be expected that the relative correction to the conductivity will be equal in order of magnitude to the value of the relative correction to W. A more detailed theory shows that if we write $W = W_0 (1-\tilde{\alpha})$, then $\sigma_{zz} = \sigma_{zz}^{(0)}(1+\tilde{\alpha})$. The same order of magnitude is found for the relative corrections to σ_{xx} and σ_{yy} [recall that in the case under consideration we assume $\Omega\tau \gg 1$, so that $\sigma_{xx}^{(0)}/\sigma_{zz}^{(0)} \sim (\Omega\tau)^{-2}$].

Thus, the relative correction to the conductivity at low temperatures is of order $(\beta H/\mu)^{1/2}$. If we compare the second term of formula (11.10) with expression (10.33) for M, we shall come to the conclusion that the relative correction to the conductivity may be written in the form

$$\frac{\Delta\tilde{\sigma}_{zz}}{\sigma_{zz}} \sim \frac{\Delta\tilde{\sigma}_{xx}}{\sigma_{xx}} \sim \frac{\Delta\tilde{\sigma}_{yy}}{\sigma_{yy}}$$

$$\sim \nu^{-1}(\mu)\sum_m \left(\frac{m_{\mathrm{m}}^* S_{\mathrm{m}}}{H}\right)^2 \frac{\partial \tilde{M}_{\mathrm{m}}}{\partial H} \tag{11.11}$$

where $\tilde{M}_{\mathrm{m}}$ is the term corresponding to one extremum of S_{m} in formula (10.33).

The oscillations of high-frequency surface impedance have a very similar form (Azbel' 1958). The relative order of these oscillations is the same as in static conductivity; hence, at $T=0$

$$\frac{\tilde{Z}}{Z(H)} \sim \left(\frac{e\hbar H}{cS}\right)^{1/2} \sim \left(\frac{\beta H}{\mu}\right)^{1/2}. \tag{11.12}$$

As has been said in the preceding chapter, because of the high oscillation frequency each differentiation with respect to H increases the relative amplitude of the oscillatory correction. Therefore, experiments in which $\partial X/\partial H$ or $\partial R/\partial H$ is determined are especially useful from the standpoint of quantum oscillations. Measurements of the impedance have an added advantage as compared with the de Haas-van Alphen effect. In the latter case, all the extremal cross sections participate in the magnetic moment and if there are several cross sections, it is not easy to distinguish between different types of oscillations. Impedance measurements are, in fact, concerned with the determination of the effective conductivity σ_{eff}, which is proportional to $v_\alpha v_\beta$. Therefore, by varying the polarization of the incident wave it is possible to suppress some of the oscillations. In particular, the oscillations from the central cross section are easily suppressed due to the fact that both essential points where the contour $v_x = 0$ intersects the central cross section (fig. 34) have opposite velocity directions. Taking $\boldsymbol{Ev} = 0$, we completely destroy the corresponding oscillations.

In concluding this section, it is worthwhile to note qualitative changes that arise in a tilted field. As has been said above, in this case one cannot use the quasiclassical approach described above. As a result of strict calculations (see Kaner and Skobov 1968), it turns out that the magnitude of oscillations in this case increases as compared to the case of the parallel field by the factor $k_z v/\max(\tau^{-1}, \omega)$, where k_z is the projection of the wave vector onto the magnetic field direction. Hence, we have

$$\frac{\tilde{Z}}{Z(H)} \sim \left(\frac{\beta H}{\mu}\right)^{1/2} \frac{k_z v}{\max(\tau^{-1}, \omega)}.$$

At sufficiently high fields and frequencies this ratio can exceed unity, i.e., giant oscillations arise. As a matter of fact, this phenomenon is similar to giant oscillations in sound absorption (section 12.7). If we substitute into the above estimate $\omega \sim 10^{10}\,\mathrm{s}^{-1}$, $k \sim \delta^{-1} \sim 10^4\,\mathrm{cm}^{-1}$, $H \sim 10^4$ Oe, then the ratio $\tilde{Z}/Z \in 0.1$–1, from which we see that giant oscillations of the impedance are quite realizable.

11.2. Cyclotron resonance on "hopping" orbits

While considering cyclotron resonance and current states we dealt with "hopping trajectories". Let us now examine this question in more detail.

In comparing the wavelength of an electron and the parameters of surface inhomogeneity, only the wavelength associated with the motion of electrons along the normal to the surface is important. The corresponding component of the momentum can decrease strongly if the electron moves at a small angle to the surface*. Here, conditions are created for specular reflection.

Suppose the metal is placed in a magnetic field parallel to its surface. The possible electron trajectories are of the type shown in fig. 70. If the angle at which the electron approaches the surface is small, the electron experiences specular reflection and the subsequent portion of the trajectory reproduces the preceding one, as a result of which "hopping trajectories" appear.

Along the hopping trajectory the motion of the electron along the normal to the surface is periodic and, hence, it is subject to quantization. Discrete levels arise, and transitions between them are possible, in which absorption of the energy of the high-frequency field occurs. As a result, in the region of small magnetic fields a peculiar resonance mechanism appears in the absorption of the energy of the high-frequency field (Khaikin 1960, 1968, Nee and Prange 1967).

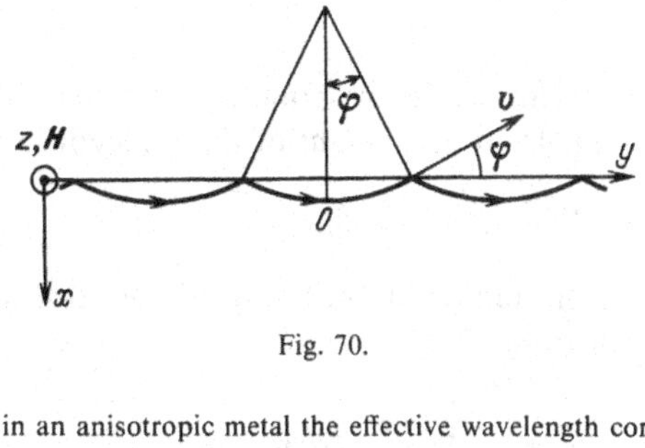

Fig. 70.

* It can be shown that in an anisotropic metal the effective wavelength corresponds to $\hbar/(p_x - p_{x0})$, where p_{x0} is the value of p_x at a point where $v_x = 0$.

Although the physical picture is the same here as in cyclotron resonance, there is one significant difference. In cyclotron resonance, due to the inequality $\hbar\Omega \ll \mu$ we are speaking of the transitions between very high quantum levels; therefore, the effect is classical, in point of fact. The resonance on hopping trajectories involves the lowest quantum levels, so that classical calculation of these levels is no longer possible.

In momentum space the motion along the trajectory in fig. 70 corresponds to the motion along the closed orbit which surrounds the shaded segment in fig. 71. According to formula (10.17), the shaded area is equal to $(2\pi e\hbar H/c)[n+\gamma(n)]$, where γ may depend on n and lies within the interval $0<\gamma<1$. The quasiclassical quantization rule provides no further information about the function γ. This was not important in the case of the quantization of ordinary trajectories because higher levels were involved, but here $n\sim 1$, so that knowledge of $\gamma(n)$ becomes important. In the problem under discussion the function $\gamma(n)$ can be determined for an arbitrary spectrum due to the fact that only a very small portion of the entire closed orbit is essential and it may be treated as part of a circle. It turns out that $\gamma(n)\approx -\frac{1}{4}$ for any n. The relative error in $n+\gamma(n)$ even for $n=1$ is less than 2% and as n increases it decreases as n^{-2}. We shall not give the details of this calculation here. We denote $n+\gamma(n)=n-\frac{1}{4}$ by n_1.

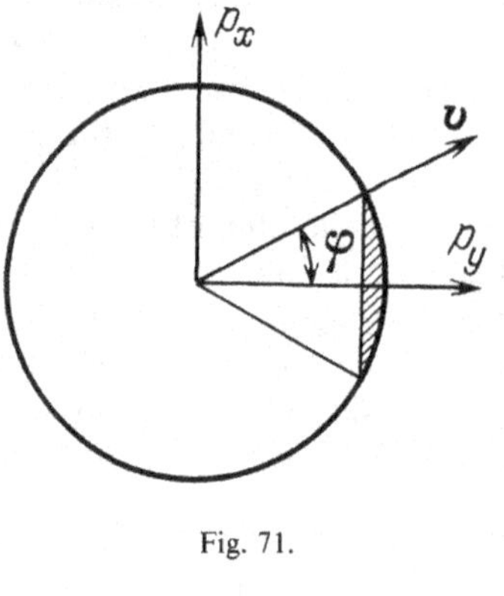

Fig. 71.

Since the angle φ is small, the corresponding portion of the trajectory will be approximated by a circle. Then, the height of the segment in fig. 70 is given by

$$x=R(1-\cos\varphi)\approx\tfrac{1}{2}R\varphi^2,$$

where $R=cp/eH$ (p is the radius of curvature of the orbit in momentum space). The area of the segment in fig. 70 is

$$\frac{2\varphi}{2\pi}\pi R^2-R^2\sin\varphi\cos\varphi\approx\tfrac{2}{3}R^2\varphi^3=\tfrac{4}{3}(2x^3R)^{1/2}.$$

But, according to formula (5.10), the area in momentum space is equal to the area in coordinate space multiplied by $(eH/c)^2$. Thus, in accordance with the rule of quantization, we have

$$S_n = \tfrac{4}{3}\left(\frac{eH}{c}\right)^2 (2x_n^3 R)^{1/2} = 2\pi \frac{e\hbar H}{c} n_1$$

from which we find

$$x_n = \tfrac{1}{2}(3\pi\hbar)^{2/3}\left(\frac{c}{eHp}\right)^{1/3} n_1^{2/3}. \tag{11.13}$$

Now it is not difficult to find the energy levels. Since the electron is acted on by the Lorentz force $(e/c)\,[\boldsymbol{vH}]$ and its velocity $\boldsymbol{v}$ is almost parallel to the boundary, it may be roughly assumed that the force is directed along the normal and equals $(e/c)v_yH$ in magnitude. In this case, we may introduce a potential $V(x)=(e/c)v_yHx$ $(x>0)$. The specular reflection from the boundary will be provided if the boundary is assumed to be an infinitely high potential wall. So we obtain the infinite potential well shown in fig. 72. Evidently, to each possible turning point x_n there corresponds the energy

$$\varepsilon_n = \frac{eHv_yx_n}{c}. \tag{11.14}$$

The resonance frequencies corresponding to the lines of absorption are equal to the differences ε_n divided by $\hbar$:

$$\omega_{nm} = \frac{\varepsilon_n - \varepsilon_m}{\hbar} = \frac{1}{2}\left(\frac{3\pi e}{c}\right)^{2/3}\left(\frac{H^2}{p\hbar}\right)^{1/3} v_y(n_1^{2/3} - m_1^{2/3}). \tag{11.15}$$

In fact, the frequency is specified and the magnetic field is varied. The resonance

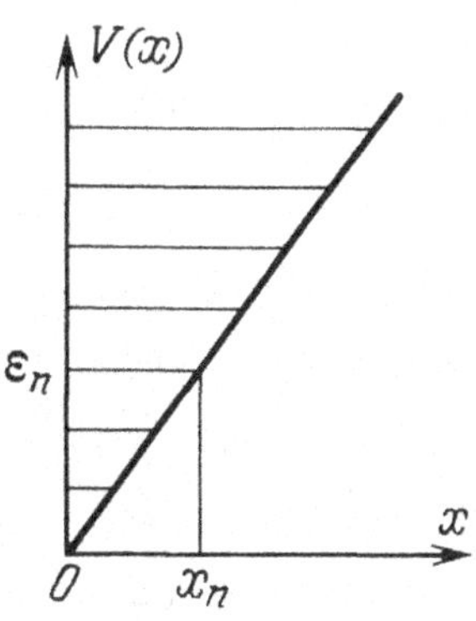

Fig. 72.

values of the field are as follows

$$H_{nm} = \frac{2^{3/2}\hbar^{1/2}}{3\pi}\frac{c}{e}\omega^{3/2}\left(\frac{p}{v_y^3}\right)^{1/2}(n_1^{2/3} - m_1^{2/3})^{-3/2}. \tag{11.16}$$

The factor $(p/v_y^3)^{1/2}$ (p is the radius of curvature of the orbit in momentum space) corresponds to a line on the Fermi surface at which $v_x = 0$. It depends on p_z. Evidently, extremal values of p/v_y^3 must manifest themselves in H_{nm}. But the effect will be much stronger if the parameter p/v_y^3 varies little along an entire portion of the Fermi surface. This occurs when the Fermi surface contains cylindrical portions. It is these portions that give the absorption maxima found experimentally in bismuth, tin and indium.

The resonance lines described by formula (11.16) may be classified according to series. Each series is characterized by a certain m_1 and all the lines of this series can be obtained from (11.16) by setting $n_1 = m_1 + 1, m_1 + 2, \ldots$. It is not difficult to see that each of the successive series will begin at a higher field than the preceding one, and all the lines of this series will be counted off to the side of a smaller field. Formula (11.16) formally allows one to obtain as large values of resonance fields as desired. But, in fact, only those orbits will be effective which are within the skin layer. As we shall see below, the numbers m_1 and n_1 corresponding to these orbits are relatively small.

We shall assume that n_1 is large and that $n_1 - m_1 = \Delta n \ll n_1$. Then, from eq. (11.16) we obtain

$$H \approx \frac{3^{1/2}}{\pi}\frac{\hbar^{1/2}c}{e}\omega^{3/2}\left(\frac{p}{v_y^3}\right)^{1/2} n^{1/2}(\Delta n)^{-3/2}.$$

Substituting H into eq. (11.13), we get

$$x_n \approx \tfrac{1}{2}(3^{1/2}\pi)\left(\frac{\hbar v_y}{p\omega}\right)^{1/2}(n\Delta n)^{1/2} < \delta.$$

The lowest value of Δn is 1 and, hence, we have

$$n < \frac{4}{3\pi^2}\frac{\delta^2 p\omega}{\hbar v_y} = n_{\lim}. \tag{11.17}$$

For $\omega \sim 10^{10}\ \mathrm{s}^{-1}$, $\delta \sim 10^{-4}$ cm, $p/\hbar \sim 10^8\ \mathrm{cm}^{-1}$ and $v_y \sim 10^8$ cm/s we find that n cannot exceed a few tens. At large values of n the intensity of the resonance peaks will drop sharply*.

* Moreover, at large n and m the resonances begin to merge since the difference between the neighboring ω_{nm} values becomes less than τ^{-1}.

The value of $\Delta n = 1$ corresponds to the beginning of the series. For large Δn we obtain the restriction

$$1 < \Delta n < \frac{n_{\text{lim}}}{n}, \tag{11.18}$$

where n_{lim} is the absolute limit of n.

Substituting $\Delta n = 1$ and $n = n_{\text{lim}}$ into the formula for H, we find

$$\Delta H < 1\text{–}10 \text{ Oe}.$$

Since $\delta \propto \omega^{-1/3}$, an increase in the frequency contributes little to the increase of permissible n, though, of course, the resonance values of H increase in this case.

It should be noted that since the electrons moving along hopping trajectories spend all the time in the skin layer, they are very efficient in absorbing electromagnetic waves, and the resonances associated with cylindrical portions of the Fermi surface give appreciable peaks in the total surface impedance.

11.3. Interference correction to the conductivity

It has been shown in section 3.1 that on the condition $p_0 l \gg \hbar$ the electrons may be treated as classical particles. This is what we have used in all cases, except for those where the quantum effects associated with the discreteness of the energy levels in the magnetic field become important. In the absence of a magnetic field the quantum corrections are small, and until recently they have not attracted attention at all. However, more thorough investigations have shown that these corrections depend strongly on temperature, frequency of the applied electric field and on a very small magnetic field, which is why they can be singled out against the background of the major effect.

Complete calculations of quantum corrections to the conductivity and other physical characteristics are performed by using methods of quantum field theory. However, there exist simple and elegant reasonings which make it possible to obtain these corrections in order of magnitude. An advantage of such an approach is the very clear physical picture of the origin of various effects*.

Let us first consider the correction that arises in the model of noninteracting electrons which are scattered from impurities. Suppose that the electron, while being scattered from impurities, moves from point A to point B. While doing so, it can travel along various paths (fig. 73). According to quantum mechanics, not the probabilities of these paths must be added but the corresponding amplitudes. If we

* The reasonings given below have been advanced mainly by Larkin and Khmel'nitskii (1982) and have been presented also by Al'tshuler in his thesis (1983), which we follow in this and subsequent sections.

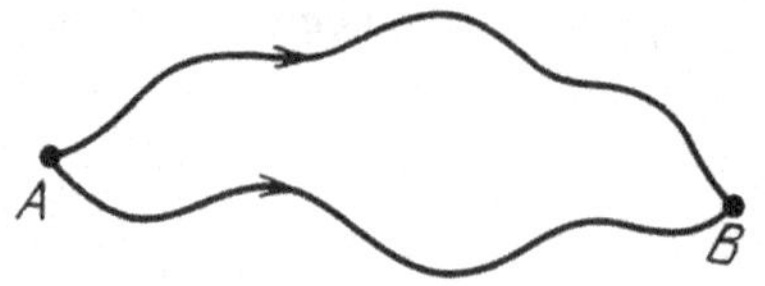

Fig. 73.

denote these amplitudes as A_i, the total probability of going from point A to point B is equal to the square of the modulus of the sum of all amplitudes, i.e.,

$$W = \left| \sum_i A_i \right|^2 = \sum_i |A_i|^2 + \sum_{i \neq j} A_i A_j^* . \tag{11.19}$$

The first term describes the sum of the probabilities of travelling each of the paths and the second term corresponds to the interference of various amplitudes. For most trajectories the interference is not important, since the lengths of these trajectories are significantly different and therefore the changes in the phase of the wave function

$$\Delta\varphi = \hbar^{-1} \int_A^B \boldsymbol{p}\, \mathrm{d}\boldsymbol{l} \tag{11.20}$$

differ strongly on these trajectories. In summing over all trajectories the interference term vanishes because of its oscillatory nature.

However, among all the trajectories there are special ones, for which the interference cannot be neglected. These are self-intersecting trajectories (fig. 74). Each such trajectory actually corresponds to two amplitudes which differ in the direction of traversal of the loop. Since with a change in the direction of motion in formula (11.20) $\boldsymbol{p}$ is replaced by $-\boldsymbol{p}$ and $\mathrm{d}\boldsymbol{l}$ is replaced by $-\mathrm{d}\boldsymbol{l}$, the quantity $\Delta\varphi$ for the loop remains unchanged here. Hence, the two amplitudes prove to be coherent (i.e., have identical phases), and so for them

$$|A_1 + A_2|^2 = |A_1|^2 + |A_2|^2 + A_1 A_2^* + A_2 A_1^* = 4|A_1|^2,$$

that is, twice as large as upon addition of the probabilities. The interference of scattering amplitudes is of quantum origin, whereas the addition of the probabilities corresponds to the classical description of the electron by means of the kinetic

Fig. 74.

equation. As we have seen above, interference effects are important for self-intersecting trajectories, and they lead to an increase in the total scattering probability, i.e., to an increase in resistivity or a decrease in conductivity.

Let us now estimate the order of magnitude of the interference correction to the conductivity. The relative value of the correction is determined by the probability of a self-intersecting trajectory. This probability would be equal to zero for a classical point, but for an electron the trajectory should be treated as a tube of finite thickness of the order of the wavelength $\lambda \sim \hbar/p_0$. Suppose that the electron is moving during a time t, which is much longer than the collision time τ. During this time, due to diffusion the electron can reach any point at a distance of $\sqrt{\overline{x^2}} \sim (Dt)^{1/2}$ (section 5.5, formula 5.47). In other words, a volume of order $(Dt)^{3/2}$ is accessible to it. In order for "self-intersection" to occur during a time interval dt it is required that the final point of the electron's path enter the volume element $v\,dt \cdot \lambda^2$. The probability of this event is equal to the volume ratio.

Since we are interested here in the total probability, we are to integrate over the time. We must take τ as the lower limit because only with larger times is the diffusion concept applicable. As for the upper limit, it must be taken into account that, apart from the scattering from impurities, there are inelastic electron-electron and electron-phonon interactions, which lead to phase relaxation and to the breakdown of amplitude coherence. This time is denoted here as τ_φ. Thus, we obtain*:

$$\frac{\Delta\sigma}{\sigma} \sim -\int_{\tau}^{\tau_\varphi} v\lambda^2 (Dt)^{-3/2}\,dt. \tag{11.21}$$

On integration we have

$$\frac{\Delta\sigma}{\sigma} \sim -v\lambda^2 D^{-3/2}(\tau^{-1/2} - \tau_\varphi^{-1/2}) \sim -\left(\frac{\lambda}{l}\right)^2 + \frac{\lambda^2}{(lL_\varphi)} \tag{11.22}$$

(here we have substituted $D \sim lv$ and $\tau \sim l/v$ and introduced the notation $L_\varphi = (D\tau_\varphi)^{1/2}$. Thus, the relative correction to conductivity is of order $(\lambda/l)^2 \sim (\hbar/p_0 l)^2$. If use is made of the estimate for the conductivity itself,

$$n_e e^2 \frac{\tau}{m} \sim n_e e^2 \frac{l}{p_0} \sim p_0^2 e^2 \frac{l}{\hbar^3}, \tag{11.23}$$

then from (11.22) we obtain (Gor'kov et al. 1979):

$$\Delta\sigma \sim -\frac{e^2}{\hbar l} + \frac{e^2}{\hbar L_\varphi}. \tag{11.24}$$

Although on the whole the correction is small and moreover the last term in (11.22) is small compared to the first, it is this last term which is of main interest since τ_φ depends strongly on temperature. As has been said, the phase relaxation

* The sign is determined by the fact that the interference increases the scattering probability and lowers the conductivity.

time τ_φ is determined by inelastic scattering. For interference effects to occur it is necessary that $\tau \ll \tau_\varphi$, i.e., low temperatures are required (region of residual resistivity). Inelastic processes are here the scattering of electrons from electrons and from phonons. In the first case $\tau_\varphi \sim \hbar\mu/T^2$. According to formula (11.22), the interference temperature correction is of order

$$\frac{\sigma(T)-\sigma(0)}{\sigma(0)} \sim \left(\frac{\lambda}{l}\right)^{3/2} \frac{T}{\mu}. \tag{11.25}$$

It is interesting to compare this with the temperature correction from electron-electron scattering. As has been noted in section 4.4, the resistivities add up. In view of this,

$$\frac{\Delta\sigma_{ee}(T)}{\sigma(0)} = -\frac{\rho_{ee}(T)}{\rho(0)} = -\frac{\sigma(0)}{\sigma_{ee}(T)},$$

where σ_{ee} is the conductivity defined in section 4.2. Substituting this value and (11.23) as $\sigma(0)$, we obtain

$$\frac{\Delta\sigma_{ee}(T)}{\sigma(0)} \sim -\frac{l}{\lambda}\left(\frac{T}{\mu}\right)^2.$$

Thus, at sufficiently low temperatures the interference correction becomes larger.

But if the time τ_φ is determined by the emission and absorption of phonons, it is of order (section 4.3):

$$\tau_\varphi \sim \frac{\hbar}{T}\left(\frac{\hbar\omega_D}{T}\right)^2.$$

In this case, from (11.22) we have

$$\frac{\sigma(T)-\sigma(0)}{\sigma(0)} \sim \left(\frac{\lambda}{l}\right)^{3/2}\left(\frac{T}{\mu}\right)^{1/2}\left(\frac{T}{\hbar\omega_D}\right). \tag{11.26}$$

The relationship between the two phase relaxation mechanisms is given by the ratio of the corresponding probabilities τ_φ^{-1}. It is easy to see that electron-electron scattering predominates at $T<(\hbar\omega_D)^2/\mu \sim 1$ K; at higher temperatures electron-phonon scattering becomes dominant.

Moreover, it should be noted that the interference temperature correction to conductivity is positive, i.e., the correction to the resistivity is negative. In other words, the resistivity falls off with temperature, just as in the Kondo effect. In spite of this interesting circumstance, we shall not consider this correction in more detail because, as we shall see in the next section, there exists a quantum temperature correction from electron-electron interaction, which exceeds those found above. The latter circumstance is valid for the case of a three-dimensional bulk specimen which we have dealt with so far. However, for a thin metal film or wire the situation is different. In view of this, we shall briefly consider the interference correction in this case too.

If the thickness b of a film or wire is small*, the particle has a chance to diffuse repeatedly from one wall to the other, and the probability of finding it at any point across the film or wire is the same. So, the volume of the intersection region $\lambda^2 v\,dt$ must be referred not to $(Dt)^{3/2}$ but to $(Dt)^{d/2}b^{3-d}$, where d is the dimension of the sample: $d=2$ for a film and $d=1$ for a wire. Thus, we have

$$\frac{\Delta\sigma}{\sigma} \sim -\int_\tau^{\tau_\varphi} \frac{v\lambda^2\,dt}{(Dt)^{d/2}b^{3-d}} \sim -\frac{v\lambda^2}{D}\times\begin{cases} b^{-1}\ln(\tau_\varphi/\tau), & d=2,\\ b^{-2}L_\varphi, & d=1.\end{cases} \tag{11.27}$$

If we find the product $\Delta\sigma\cdot b^{3-d}$ from (11.27), we obtain (Gor'kov et al. 1979):

$$\Delta\sigma_d = \Delta\sigma\cdot b^{3-d} \sim -\frac{e^2}{\hbar}\times\begin{cases}\ln(L_\varphi/l), & d=2,\\ L_\varphi, & d=1.\end{cases} \tag{11.28}$$

The quantity $\Delta\sigma_d$ introduced here corresponds to a correction to inverse total resistivity of a square film or a wire of unit length. Note that we are speaking of the macroscopic dimensions rather than of dimensions of the order of atomic dimensions (in the latter case, the increase of the self-intersection probability is obvious). The phase relaxation time in a system of small dimensions ($d=1$, 2) is determined by an electron-electron interaction with a small energy transfer (Al'tshuler et al. 1982). We shall not give the corresponding derivation here because of its complexity; we present only the main results. In the case of $d=2$ (a film) we have

$$L_\varphi \sim \frac{l}{\lambda}\left(\frac{\hbar v}{T}\right)^{1/2} b^{1/2}\ln\left(\frac{lb}{\lambda^2}\right), \tag{11.29}$$

$$\Delta\sigma_d \sim -\left(\frac{e^2}{\hbar}\right)\ln\left(\frac{b\hbar v}{T}\lambda^{-2}\right). \tag{11.30}$$

For $d=1$ (thin wire) we have

$$L_\varphi \sim \left(\frac{l}{\lambda}\right)^{2/3}\left(\frac{\hbar v}{T}\right)^{1/2} b^{2/3}, \tag{11.31}$$

$$\Delta\sigma_d \sim -\frac{e^2}{\hbar}\left(\frac{l}{\lambda}\right)^{2/3}\left(\frac{\hbar v}{T}\right)^{1/3} b^{2/3}. \tag{11.32}$$

* The criterion for the smallness b of the cross-sectional size varies, depending on how the problem is stated. Strictly speaking, for $d=1$, 2 in formulas (11.27) there is a term $-(\lambda/l)^2$, just as in (11.22) for the three-dimensional case; this is associated with the fact that for short times any sample behaves as a three-dimensional one. Therefore, if we are speaking of the conditions under which the behavior of the L_φ-dependent term of interest in the correction is changed, the criterion is $b<L_\varphi$. If we wish to know when we may drop the term $-(\lambda/l)^2$, then we find $l \ll b \ll (L_\varphi l)$ for the wire and $l\ll b\ll l\ln(L_\varphi/l)$ for the film. Strictly speaking, formulae (11.27) are valid only when the latter conditions are satisfied; however, the parts dependent on L_φ retain their form also under less stringent conditions $l\ll b\ll L_\varphi$.

Comparison of the interference corrections to the conductivity (11.30 and 11.32) with the quantum correction from the interaction of electrons (see below) shows that for $d=2$ both corrections are of the same order of magnitude, and for $d=1$ the interference correction is dominant.

If the conductivity in a high-frequency field of frequency $\omega > \tau_\varphi^{-1}$ is studied, the upper limit of integrals (11.21) and (11.27) is ω^{-1} because the main phase relaxation mechanism is the action of the external field (the corresponding time is of order $\omega^{-1} \ll \tau_\varphi$). If $\omega^{-1} \gg \tau$, then the formulas derived earlier retain their form, with L_φ replaced by L_ω:

$$L_\omega = \left(\frac{D}{\omega}\right)^{1/2}. \qquad (11.33)$$

11.4. Interference effects in a magnetic field

The behavior of the interference correction in a magnetic field is very interesting. In the presence of a field one should replace $\boldsymbol{p}$ by $\boldsymbol{p} - (e/c)\boldsymbol{A}$. With a change in the direction of going around a closed loop the momentum $\boldsymbol{p}$ changes sign, but the vector potential $\boldsymbol{A}$ retains its sign. As a result, a phase difference appears in the interfering amplitudes given by

$$\Delta\varphi_H = \frac{2e}{c\hbar}\oint \boldsymbol{A}\, \mathrm{d}\boldsymbol{l} = \frac{2e}{c\hbar}\int \mathrm{rot}\, \boldsymbol{A}\, \mathrm{d}\boldsymbol{S} = \frac{2e}{c\hbar}\Phi = 2\pi\frac{\Phi}{\Phi_0}, \qquad (11.34)$$

where Φ is the magnetic field flux across the loop and Φ_0 is the flux quantum (section 10.4). The appearance of a phase difference results in destruction of the interference, i.e., in a decrease of the resistivity (Al'tshuler et al. 1980, Al'tshuler 1983). The interference correction thus leads to negative magnetoresistance even in the absence of magnetic impurities (see end of section 4.6).

In order to estimate the magnitude of negative magnetoresistance, we introduce a time t_H defined as follows. If in formula (11.34) we use the diffusion length $(Dt)^{1/2}$ as the characteristic size of the loop, the magnetic field flux across the loop is $\Phi \sim HDt$. Let us define t_H so that in (11.34) $\Delta\varphi_H \sim 2\pi$. This gives

$$t_H \sim \frac{\Phi_0}{HD} \sim \frac{L_H^2}{D}. \qquad (11.35)$$

Here $L_H = (\hbar c/2eH)^{1/2}$ is the so-called magnetic length (it corresponds to the size of the Landau eigenfunction in the magnetic field, section 10.2). The essential fields are determined by the condition $\tau_H \sim \tau_\varphi$, i.e.,

$$H \sim \frac{\Phi_0}{(D\tau_\varphi)}. \qquad (11.36)$$

Substituting $D \sim lv \sim v^2\tau \sim \mu\tau/m$, $\Phi_0 \sim c\hbar/e$, we obtain

$$\Omega\tau \sim \hbar/(\tau_\varphi\mu) \ll 1, \qquad (11.37)$$

where $\Omega = eH/mc$ is the Larmor frequency; we are thus speaking here of fields much smaller than the classically strong fields for which $\Omega\tau \gg 1$.

The asymptotic formulas for the conductivity are derived from formulas (11.21) and (11.27) through the replacement of τ_φ by t_H if $t_H \ll \tau_\varphi$, i.e., H is larger than the value given by (11.36):

$$\sigma_d(H) - \sigma_d(0) \sim \frac{e^2}{\hbar} \begin{cases} \ln\left(\dfrac{eHD\tau_\varphi}{\hbar c}\right), & d=2, \\ \left(\dfrac{eH}{\hbar c}\right)^{1/2}, & d=3. \end{cases} \tag{11.38}$$

We have already pointed out that the interference effect of negative magnetoresistance manifests itself in classically weak fields, $\Omega\tau \ll 1$. Other peculiarities are the following: (a) in the three-dimensional case the effect is independent of the angle between the current and the field (in the two-dimensional case it was assumed that the field is directed across the film); (b) the effect may strongly depend on the temperature due to τ_φ (crossover to the asymptotics (11.38) and the asymptotics itself in the two-dimensional case); (c) because of the fact that the phase difference defined by (11.34) is governed by the magnetic field flux, i.e., by the size of the loop, it is clear that in the one-dimensional case (a wire thinner than L_φ) the effect is much weaker than in a bulk sample; in the two-dimensional case (a film thinner than L_φ), the effect must be small with the field directed in the plane of the film, i.e., there is a strong anisotropy, depending on the orientation of the field with respect to the film.

Consider now the resistance of a hollow thin-walled metal cylinder (fig. 75). Suppose that there is a magnetic field in the cavity parallel to the cylinder axis. This may be produced by a solenoid present entirely in the cavity, so that the magnetic field in the metal will be equal to zero. We assume that the mean free path of the electron l is much shorter than the perimeter of the cylinder L, so that no quantization

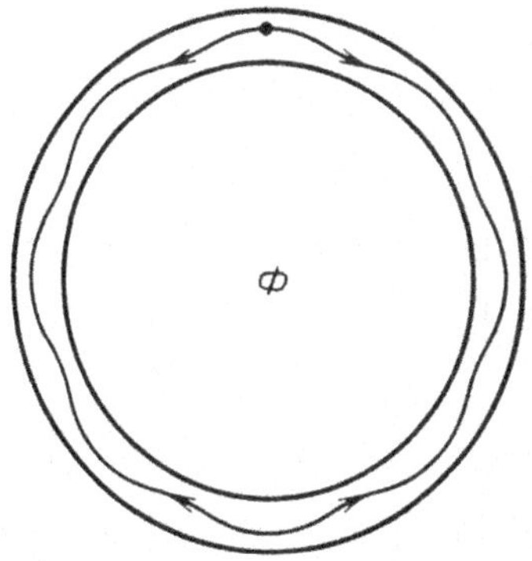

Fig. 75.

of the motion of the electron will occur. In this case, according to (11.34), the phase difference between de Broglie waves that travel along the circumference of the cylinder clockwise and anticlockwise will be equal to $\Delta\Phi = 2\pi\Phi/\Phi_0$. Therefore, because of the interference of the amplitudes the probability of the electron returning to a given point will oscillate with a period of $\Delta\Phi = \Phi_0$. Since the flux Φ is the same for all the trajectories traversing the cylinder, the total resistivity will be an oscillatory function of the flux with a period of $\Delta\Phi = \Phi_0$ (Al'tshuler et al. 1981). Of course, for the effect to be observed the phase coherence along the circumference of the cylinder must be conserved, i.e., $L \ll L_\varphi$.

At first glance this phenomenon seems to be paradoxical. If the magnetic field exists only in the cavity and does not penetrate the metal, then how can the electrons "be aware" of its existence? The cause is the quantum nature of the electron. Although the electron is moving inside the metal, the phase of its wave function depends on the vector potential, which is different from zero not only in the cavity but also in the bulk of the metal. We can put it differently: the phase of the wave function is determined by the field in the cavity. This property of the wave function was first predicted by Aharonov and Bohm (1959). The effect of the oscillations of the resistance of the hollow cylinder with magnetic field has been detected experimentally (D. Sharvin and Yu. Sharvin 1981; Yu. Sharvin 1984). Figure 76 shows the result for lithium at $T = 1.1$ K. The data of an exact theoretical calculation are shown by a dashed curve.

Quantum interference depends very significantly on the electron spin if there exist scattering mechanisms leading to the flip of the electron spin. We have already considered one mechanism of this kind - scattering from impurities (section 4.6). But even if no magnetic impurities are present, the electron spin is influenced by

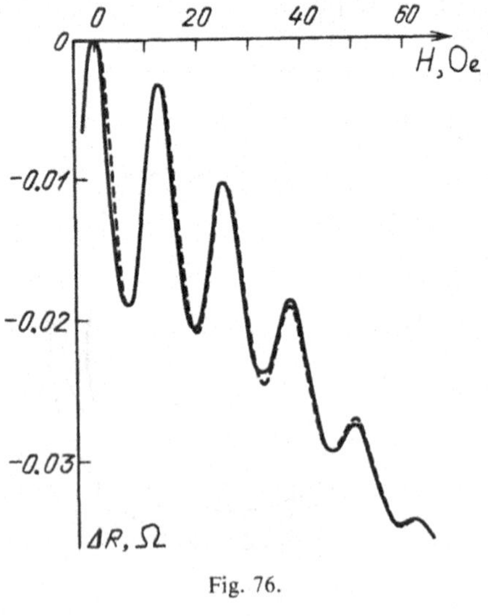

Fig. 76.

spin-orbital effects during ordinary scattering from nonmagnetic impurities. In section 10.7, which is devoted to magnetic breakdown, we have already considered the spin-orbit interaction of the electron with a periodic crystal lattice. The same effect occurs upon interaction of the electron with an impurity. It gives a correction to the scattering amplitude proportional to $[\boldsymbol{p}\times\boldsymbol{p}']\boldsymbol{s}f(Z)$, where $\boldsymbol{p}$ and $\boldsymbol{p}'$ are the electron momenta before and after scattering, $\boldsymbol{s}$ is the electron spin operator, and $f(Z)$ is the increasing function of the atomic number Z. In Part II we shall see that spin-orbit scattering gives rise to some interesting properties of superconductors (section 21.4). At this point we will show that it can have a substantial effect on the interference correction in a normal metal (Hikami et al. 1980).

Consider the interference correction to the scattering amplitude:

$$A_1^*A_2+A_2^*A_1=2\,\mathrm{Re}(A_1A_2^*).$$

For a wave function φ_σ in the initial state ($\sigma=\pm$ characterizes the projection of the electron spin) and a wavefunction $\varphi_{\sigma'}$ in the final state, the interference term may be written in the form

$$C=A_1A_2^*=\tfrac{1}{2}\sum_{\sigma\sigma'}\varphi_\sigma^{(1)}\varphi_{\sigma'}^{(2)}\varphi_{\sigma'}^{(1)*}\varphi_\sigma^{(2)*} \tag{11.39}$$

(here the sum is taken over the final states and the average is taken over the initial states). Now we shall proceed as follows. We shall assume that fig. 73, instead of two possible trajectories of a single particle, shows the trajectory of two particles moving simultaneously. The expression for C may then be written with the aid of bilinear combinations corresponding to the given total spin of the two particles and to its projection (see Landau and Lifshitz 1974). The total spin may be equal to 1 (with projections 1, 0, -1) or 0. The corresponding functions have the form

$$\Psi_0=2^{-1/2}(\varphi_+^{(1)}\varphi_-^{(2)}-\varphi_-^{(1)}\varphi_+^{(2)}),$$
$$\Psi_{1,1}=\varphi_+^{(1)}\varphi_+^{(2)},$$
$$\Psi_{1,-1}=\varphi_-^{(1)}\varphi_-^{(2)},$$
$$\Psi_{1,0}=2^{-1/2}(\varphi_+^{(1)}\varphi_-^{(2)}+\varphi_-^{(1)}\varphi_+^{(2)}). \tag{11.40}$$

After performing simple calculations we find from (11.39)

$$C=\tfrac{1}{2}\sum_{m=0,\pm1}|\Psi_{1,m}|^2-|\Psi_0|^2. \tag{11.41}$$

In the presence of spin-orbit scattering with a characteristic time $\tau_{so}\ll\tau_\varphi$ the states with total spin 1 and those with total spin 0 behave differently. The states $\Psi_{1,m}$ carry information about the electron spin and therefore they are damped with time τ_{so}. As regards the function Ψ_0, it is damped only with time τ_φ. Therefore, we obtain

$$\frac{\Delta\sigma}{\sigma}\sim-\int_\tau^{\tau_\varphi}\frac{\lambda^2 v\,\mathrm{d}t}{b^{3-d}(Dt)^{d/2}}\left(\tfrac{3}{2}\,\mathrm{e}^{-t/\tau_{so}}-\tfrac{1}{2}\right). \tag{11.42}$$

Integrating, we find

$$\frac{\Delta\sigma}{\sigma} \sim \lambda^2 v \begin{cases} D^{-3/2}(\frac{3}{2}\tau_{so}^{-1/2} - \frac{1}{2}\tau_{\varphi}^{-1/2} - \tau^{1/2}), & d=3, \\ (Db)^{-1}\left[-\frac{3}{2}\ln\left(\frac{\tau_{so}}{\tau}\right) + \frac{1}{2}\ln\left(\frac{\tau_{\varphi}}{\tau}\right)\right], & d=2, \\ D^{-1/2}b^{-2}(-\frac{3}{2}\tau_{so}^{1/2} + \frac{1}{2}\tau_{\varphi}^{1/2}), & d=1. \end{cases} \tag{11.43}$$

Since only $\tau_{\varphi} \sim T^{-p}$ depends on temperature, according to this formula we find that at $\tau_{so} \ll \tau_{\varphi}$ the temperature dependence of the resistivity changes sign: $\Delta\sigma$ falls off with increasing temperature.

The same refers to magnetoresistance. In weak fields the magnetoresistance becomes positive. Only when the magnetic field becomes so strong that the characteristic time t_H defined by formula (11.35) is less than τ_{so} or

$$H > \frac{\Phi_0}{D\tau_{so}} \tag{11.44}$$

does the magnetoresistance change sign and become negative. This has been observed in experiments with copper (Gershenson and Gubankov 1982) and magnesium (Bergmann 1982) films.

11.5. Quantum correction to the density of states and conductivity arising from the electron interaction

All the preceding discussion has been based on the picture of an electron in a self-consistent periodic field produced by lattice ions and other electrons. Such a picture proves invalid when the electrons approach one another to interatomic distances. In such a case, a direct Coulomb repulsion manifests itself, leading to electron–electron scattering (section 4.2). However, the Coulomb interaction also changes the energy spectrum.

As is known, the first-order correction of perturbation theory to the energy is simply a diagonal matrix element of the interaction energy. In section 2.2 we have dealt with the scattering of particles during a weak interaction (fig. 7). But we ignored one possibility there which does not correspond to true scattering but does provide a correction to the energy spectrum, namely: electrons 1 and 2 change positions in the interaction. This is the exchange energy defined by

$$\Delta\varepsilon = -\int_{|p-\hbar k|<p_0} g(\boldsymbol{k})\frac{d^3\boldsymbol{k}}{(2\pi\hbar)^3}. \tag{11.45}$$

Here $g(\boldsymbol{k})$ is the Fourier component of the interaction energy (Coulomb repulsion with screening at atomic distances); the minus sign is associated with the interchange of the electrons.

We will now calculate $\Delta\varepsilon$ for the three-dimensional case. For simplicity, we consider the unscreened Coulomb interaction and $p > p_0$. We have

$$\Delta\varepsilon = -4\pi e^2 \int_{|p-\hbar k|<p_0} k^{-2} \frac{d^3k}{(2\pi)^3} = -4\pi e^2 h^2 \int_{p'<p_0} (\boldsymbol{p}-\boldsymbol{p}') \frac{d^3\boldsymbol{p}'}{(2\pi\hbar)^3}$$

$$= -\frac{e^2}{\pi\hbar}\left(p_0 - \frac{p^2 - p_0^2}{2p} \ln \frac{p+p_0}{p-p_0}\right).$$

The first term is a constant and simply represents the renormalization of the chemical potential. The second term in the case of $p - p_0 \ll p_0$ has the form

$$\Delta\varepsilon = \frac{e^2}{\pi\hbar v} \xi \ln\left(\frac{2p_0 v}{\xi}\right). \tag{11.46}$$

In order to take into account the screening, one has to replace k^{-2} in the integral by $(k^2+\varkappa^2)^{-1}$, where $\varkappa = r_D^{-1}$ is the reciprocal Debye radius, which has already been introduced in section 4.1. Here, subtracting the constant, $\Delta\varepsilon$ will be expressed by formula (11.46) with replacement of $\ln(2p_0 v/\xi)$ by $\ln(2p_0/\hbar\varkappa)$ at $\hbar\varkappa \ll p_0$. A point of interest to us is that the exchange interaction leads to a positive correction of order $e^2/\hbar$ to the velocity of quasiparticles.

As a matter of fact, the interference of the states was ignored in this estimate*. Two states differing in energy by δ will be coherent during a time of order $\hbar/\delta$. If we are dealing with a point interaction (in our case it has an action radius of the order of interatomic distances, i.e., does not practically differ from a point interaction), then because of the interference the effective interaction constant will be increased so that the relative correction to the interaction constant will be of the order of the probability of two particles meeting at the same point (with accuracy λ) within the time interval $\hbar/|\xi|$. Since the quasiparticles execute a diffusive motion, the probability of finding them at one point is given by formula (11.27) with the upper limit $\hbar/|\xi|$:

$$\alpha_\xi = \int_\tau^{\hbar/|\xi|} \frac{v\lambda^2\, dt}{(Dt)^{d/2} b^{3-d}}. \tag{11.47}$$

Therefore, the effective interaction constant becomes

$$e_{\text{eff}}^2 \sim e^2(1+\alpha_\xi) \tag{11.48}$$

and the correction to the velocity at the Fermi boundary will be of order

$$\Delta v \sim \frac{e^2}{\hbar}\alpha_\xi. \tag{11.49}$$

* The physical reasoning outlined below and the estimate of the correction to the density of states have been given by A.G. Aronov.

Whatever dimensionality of the sample is ($d=1$, 2, 3), the density of states is proportional to v^{-1}. By replacing v by $v+\Delta v$, we obtain (Al'tshuler and Aronov 1979):

$$\frac{\Delta\nu}{\nu} \sim -\frac{e^2}{\hbar v}\alpha_\xi \sim \frac{e^2}{\hbar v}\begin{cases} \left(\frac{\lambda}{l}\right)^{3/2}\left(\frac{|\xi|}{\mu}\right)^{1/2}, & d=3, \\ -\frac{\lambda^2}{lb}\ln\left(\frac{\hbar}{|\xi|\tau}\right), & d=2, \\ -\frac{\lambda^{5/2}}{l^{1/2}b^2}\left(\frac{\mu}{|\xi|}\right)^{1/2}, & d=1. \end{cases} \tag{11.50}$$

Since the conductivity is proportional to ν, a correction appears in the conductivity as well. The relative value of the correction is of the same order of magnitude as $\Delta\nu/\nu$, with $|\xi|$ replaced by T. The quantity $e^2/\hbar v$ may be assumed unity in estimates. For the three-dimensional case we have

$$\frac{\sigma(T)-\sigma(0)}{\sigma(0)} \sim \left(\frac{\lambda}{l}\right)^{3/2}\left(\frac{T}{\mu}\right)^{1/2}. \tag{11.51}$$

Comparing with (11.25), we see that the quantum correction arising from electron-electron interaction proves to be larger than the interference correction. Note also that this correction is positive and increases with temperature as $T^{1/2}$. Due to this correction the resistivity decreases with temperature even in the absence of magnetic impurities.

The condition for the observation of this effect is $\hbar/T \gg \tau$ or

$$T \ll \frac{\hbar}{\tau}. \tag{11.52}$$

This inequality does not contradict condition (3.2) which makes it possible to introduce quasiparticles. The point is that condition (3.2) implies the time of inelastic scattering associated with the smearing-out of the wave packet over the energies, whereas formula (11.52) contains the time of elastic scattering from impurities at which the energy of the quasiparticle does not change.

From formulas (11.50) upon substitution $\Delta\sigma/\sigma \sim \Delta\nu/\nu$ and $|\xi| \sim T$ and also from formulas (11.30) and (11.32) it follows that at $d=2$ the interference correction is of the order of the correction from the interaction and at $d=1$ the interference correction dominates.

11.6. *Anderson localization. The metal–insulator transition*

We have so far dealt with cases where the quantum corrections are small. However, in the cases $d=2$ and $d=1$ the interference correction increases with decreasing temperature due to the increase of τ_φ (or $L_\varphi=(D\tau_\varphi)^{1/2}$) and, eventually, ceases to be small (see formula 11.28). Our reasoning then becomes invalid and the question arises of what will happen with the conductivity. Note that at $d=1$ or $d=2$ this question arises even when the condition $p_0 l/\hbar \gg 1$ or $l \gg \lambda$ is satisfied. In the three-dimensional case the interference correction never exceeds $(\lambda/l)^2$ in accordance with (11.22). In this case, we can put the question of how the conductivity will behave when the condition $l/\lambda \gg 1$ is violated.

Anderson (1958) has put forward arguments in favor of the conductivity of the sample to vanish at $T=0$ when the condition $l/\lambda \gg 1$ is violated, i.e., each electron will be localized in a certain region of the conductor. This phenomenon has come to be known as Anderson localization.

It became clear at a later time that in a purely one-dimensional metal, which is a chain of atoms with overlapping valence shells, the localization in the case $T=0$ arises with an arbitrarily small concentration of defects (Mott and Twose 1961). The next step was the prediction that the same is valid for an infinite wire of finite thickness (Thouless 1977). It was subsequently shown in a rigorous way that both in the first case (Berezinskii 1973) and in the second (Efetov 1983) the resistivity increases exponentially with increasing length of the sample. The characteristic radius of localization in the first case is the mean free path l and in the second $(b/\lambda)^2 l$ (b being the wire diameter). Probably, the exponential rise of the resistivity with increasing size occurs also for a metal film, although this statement has not yet been proved rigorously.

The origin of Anderson localization may be explained as follows. Imagine an airplane flying over mountains. If the airplane flies high, it does not encounter obstacles. As the airplane loses height it finds itself below the top of the mountains. If its course has been strictly predetermined (one-dimensional motion), it will eventually encounter a summit. But if the airplane can fly around the summit (two-dimensional motion), it may go down until it is below the mountain passes. After this it will be blocked. This is the situation in classical mechanics.

In quantum mechanics, the particles are described by a wave function, which, in principle, extends to an infinite distance. Therefore, even if the particle flies high above the potential relief, the mountains produce an "echo" in the form of reflected waves. The scattering from a random potential of impurities occurs elastically, i.e., the reflected waves correspond to the same energy as the principal wave. If we are dealing with a purely one-dimensional conductor, the Fermi surface reduces to two points $p=\pm p_0$. This means that the reflected waves have the same wavelength λ as the original one. All these waves interfere with one another, as a result of which the particle appears to be localized.

From the above reasoning it is seen that the factor responsible for localization is the interference of the incident and reflected waves. For localization to occur, one has to rule out completely the violation of the interference by inelastic processes. It is therefore necessary that the temperature be equal to zero.

Consider the interference corrections (11.21) and (11.27) to the conductivity at $T=0$. Evidently, in this case $\tau_\varphi \to \infty$ and the upper limit of the integrals will simply be determined by the size of the sample: $t_L \sim L^2/D$ (we assume all large sizes to be equal to L, i.e., we shall consider a cube with a side L at $d=3$ and a square film at $d=2$). In the ratio $\Delta\sigma/\sigma$, L_φ is replaced by L. Obviously, for $d=3$ at $L \gg l$ the interference correction is always of the order of $(\lambda/l)^2$; it becomes significant only at $l \sim \lambda$, which is practically unattainable in good metals but may occur in semi-metals and semiconductors. For $d=1, 2$ the correction increases with the size of the sample. Let us find the critical size at which $\Delta\sigma/\sigma \sim 1$. From (11.27) we have

$$L_c = \begin{cases} l \exp\left(\dfrac{bl}{\lambda^2}\right), & d=2, \\ l\left(\dfrac{b}{\lambda}\right)^2, & d=1. \end{cases} \tag{11.53}$$

As has been said, the existence of localization has been strictly proved for a one-dimensional chain of atoms and for a wire of finite thickness, and L_c coincides with the estimate (11.53) ($d=1$). Although no rigorous theory has been constructed for $d=2$ and $d=3$, one may make the rather probable supposition that, just as in the case of $d=1$, localization occurs when $\Delta\sigma/\sigma \sim 1$. This means that for a metal film ($d=2$) localization takes place at arbitrary impurity concentration and that the localization radius is expressed by formula (11.53) ($d=2$); for a three-dimensional metal localization is realized only at $l \sim \lambda$.

If the temperature is not too low, the interference effects give only a small correction to the conductivity; this has been discussed in sections 11.3 and 11.4. The theory concerned with the study of this correction is sometimes called "weak localization" theory and the correction itself is called the localization correction rather than the interference correction. But here we are interested in the case of strong localization, i.e., strictly speaking, $T=0$.

Consider a quantity which is the inverse of the total resistance for samples with $d=1, 2, 3$, for which all large sizes are the same; this quantity is called "conductance"*. The behavior of the conductance with increasing sizes can be ascertained on the basis of the scaling hypothesis (Abrahams et al. 1979) analogous to that used in the theory of second-order phase transitions (see Appendix II). According to this hypothesis, the conductance G is the only quantity that determines the behavior of the system when its size is changed. This condition is written in the following form:

$$G(qL) = f[q, G(L)]. \tag{11.54}$$

Relation (11.54) may be rewritten in differential form. We put $q = 1+\alpha$, where $\alpha \ll 1$. In the zeroth approximation we have

$$G(L) = f[1, G(L)].$$

* This name is used in order to distinguish it from the conductivity. Sometimes the dimensionless quantity G/G_0, where $G_0 = e^2/(\pi^2\hbar)$ (see below), is called the conductance.

To first order in α we obtain

$$\alpha L G'(L) = \alpha \left(\frac{\partial f}{\partial q}\right)_{q=1}.$$

We divide this equality by G and denote the function $(\partial f/\partial q)_{q=1}/G$ as $\beta(G)$. As a result, we have

$$\frac{\partial \ln G}{\partial \ln L} = \beta(G). \tag{11.55}$$

Let us now consider the behavior of the function $\beta(G)$ (called the Gell-Mann–Low function). In the case when G is large, the ordinary theory of conductivity is valid. Here,

$$G = \sigma \begin{cases} \dfrac{S_\perp}{L} = L, & d = 3, \\ \dfrac{bL_\perp}{L} = b, & d = 2, \\ \dfrac{b^2}{L}, & d = 1. \end{cases} \tag{11.56}$$

According to formula (11.55), in the zeroth approximation

$$\beta(G) \approx d - 2. \tag{11.57}$$

The interference corrections (11.21) and (11.27) obtained earlier can be used to derive the next approximation to the function $\beta(G)$ at large G. For $d = 3$ we get from eq. (11.21)

$$\ln G \approx \ln(\sigma L) + \frac{\Delta\sigma}{\sigma} \approx \ln(\sigma_1 L) + \frac{\lambda^2}{lL},$$

where $\sigma_1 = \sigma + \Delta\sigma$ $(L = \infty)$. From this relation it follows that

$$\beta(G) = \frac{\partial \ln G}{\partial \ln L} \approx 1 - \frac{\lambda^2}{lL}$$

$$\approx 1 - \frac{\lambda^2 \sigma}{lG} = 1 - \alpha_3 \frac{G_0}{G},$$

where

$$G_0 = \frac{e^2}{\pi^2 \hbar} \tag{11.58}$$

and $\alpha_3 \sim 1$ (it is convenient to introduce a single constant G_0 for all dimensions).

For $d = 2$ we find from eq. (11.27)

$$\ln G = \ln \sigma b + \frac{\Delta\sigma}{\sigma} = \ln \sigma b - \frac{\lambda^2}{bl} \ln\left(\frac{L}{l}\right),$$

which gives $\beta(G)=-\lambda^2/bl$. But according to the scaling idea, $\beta(G)$ may depend only on G. In zero order $G=b\sigma=(e^2/\hbar)bl/\lambda^2$. In view of this, we may write with the accuracy required:

$$\beta(G)=-\alpha_2\frac{G_0}{G},$$

where $\alpha_2\sim 1$.

Finally, for $d=1$ we obtain from eq. (11.27)

$$\ln G\approx\ln\left(\frac{\sigma b^2}{L}\right)+\frac{\Delta\sigma}{\sigma}\approx\ln\left(\frac{\sigma b^2}{L}\right)-\lambda^2\frac{L}{b^2l},$$

from which it follows that

$$\beta(G)\approx -1-\alpha_1\frac{G_0}{G}.$$

Thus, we have

$$\beta(G)\approx d-2-\alpha_d\frac{G_0}{G}, \tag{11.59}$$

where $\alpha_d\sim 1$.

Small G values imply localization. Therefore, the asymptotics at small G can be obtained on the basis of the assumption that

$$G\sim G_0\exp\left(\frac{-L}{L_c}\right) \tag{11.60}$$

(according to formulae 11.58 and 11.59, $G_0=e^2/\pi^2\hbar$ is the natural scale for G). From eq. (11.60) we find that

$$\beta(G)\approx\ln\left(\frac{G}{G_0}\right). \tag{11.61}$$

On the basis of the asymptotic formulae for large and small G values obtained we may presume the dependence $\beta(G)$ shown in fig. 77. It can be seen that the curves have substantially different forms for different dimensions. For $d=1$, 2 the function $\beta(G)$ is always negative. It follows that as the size increases G always falls off, i.e., there is localization.

In the three-dimensional case the situation is different. If we begin with a small G, then β is negative and G drops with increasing size. But if we take G from the right of the point of intersection $G=G_c$, then β is positive and G increases with increasing size, approaching the usual dependence $G\propto L$.

At the point $G=G_0$ itself (evidently, $G_c\sim G_0$), $\beta=0$, and in accordance with eq. (11.55), G is independent of the size. On the strength of this, the point G_c is called a fixed point. Since an infinitesimal departure from the fixed point leads to variation of G with size, i.e., to deviation of G from G_c, G_c is called an unstable fixed point.

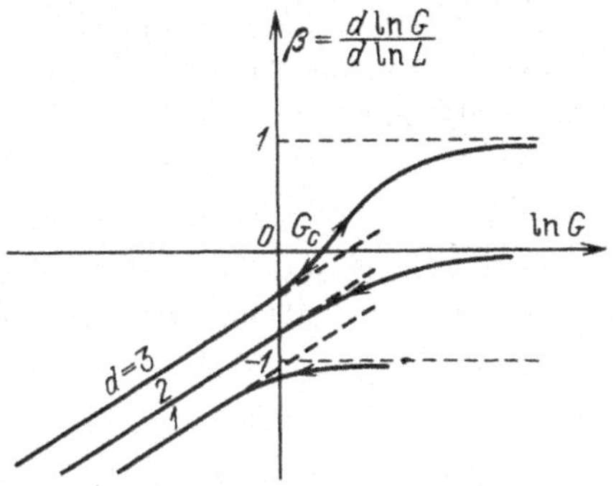

Fig. 77.

Let us make the simplest assumption that in the neighborhood of G_c

$$\beta(G) \approx \gamma(\ln G - \ln G_c) \approx \gamma\left(\frac{G}{G_c} - 1\right). \tag{11.62}$$

We assume the initial condition: $G = G^{(0)}$ at $L = L_0$ ($G^{(0)}$ being close to G_c). In this case,

$$G \approx G_c\left(\frac{G^{(0)}}{G_c}\right)^{(L/L_0)^\gamma}. \tag{11.63}$$

This formula reflects the tendencies described above: if $G^{(0)} > G_c$, then G increases with L; if $G^{(0)} < G_c$, then G decreases with L, and, finally, if $G^{(0)} = G_c$, the value of G does not change with L.

Formula (11.63) can be used to find the physical dependencies of the quantities in the neighborhood of the localization threshold. It will be reasonable to suppose that for sizes of the order of the mean free path l all the dependences are simple, i.e., for these sizes

$$\sigma \sim \frac{e^2 p_0^2 l}{\hbar^3}, \qquad G = \sigma L \approx \frac{e^2 p_0^2 l^2}{\hbar^3} \sim \frac{e^2}{\hbar}\left(\frac{l}{\lambda}\right)^2.$$

Therefore, in eq. (11.63) we take $L_0 \sim l$. Suppose there is a variable parameter x in the system. This may be the deformation or the atomic concentration of impurities. As a result, p_0 or l may vary. It may be supposed that $G^{(0)}$ has no singularity with respect to this parameter at $G^{(0)} = G_c$, i.e., we may write $G^{(0)} = G_c(1 + x)$. Substituting this into formula (11.63) at $L_0 = l$, we have

$$G = G_c(1+x)^{(L/l)^\gamma} \approx G_c \exp\left[x\left(\frac{L}{l}\right)^\gamma\right]. \tag{11.64}$$

The last transformation is valid only in those cases where $x \ll 1$ and $L/l \gg 1$.

If $x<0$, then as L increases G falls off exponentially. The characteristic length, or the localization radius, is found from the condition that magnitude of the exponent in eq. (11.64) be of the order of unity. Hence,

$$L_c \sim l|x|^{-1/\gamma}, \quad x<0. \tag{11.65}$$

If $x>0$, then $G \to \infty$ as the size increases. This may continue until $G \sim G_c \sim G_0$. There then occurs a matching with the usual Ohm law, i.e., $G = \sigma L$, where $\sigma = \text{const}$. So we get

$$\sigma \sim \frac{G_0}{L_c} \sim \left(\frac{G_0}{l}\right) x^{1/\gamma}, \quad x>0. \tag{11.66}$$

Thus, it turns out that the conductivity falls off by a power law upon approach to the localization threshold. From the other side of the threshold the substance is an insulator, the localization radius increasing infinitely by a power law with the same exponent γ^{-1}. Experimental investigations (for example, Hess et al. 1982, Brandt et al. 1981) support these conclusions, with $\gamma = 0.6 \pm 0.1$ (the localization radius can be determined from the dielectric constant: $\varepsilon_0 \propto L_c^2$).

In our discussion of localization we assumed, strictly speaking, that $T=0$ and that the conductivity is measured at zero frequency. As a matter of fact, the first condition cannot be realized, and the second complicates experiments considerably. In this connection one may ask: What is the range of validity of the formulas derived above? It is not difficult to see that for localization to be observed, the following inequalities must be satisfied:

$$L_c \ll L \ll L_\varphi, \ L_\omega, \tag{11.67}$$

that is, in any case

$$L_c \ll L_\varphi, \ L_\omega. \tag{11.68}$$

However, condition (11.67) cannot be realized in practice. In metals it is possible to make L less than L_φ, but there the condition of strong localization cannot be fulfilled. In semiconductors, for the condition (11.67) to be satisfied it is necessary either to use very small samples, which reduces the accuracy of measurements, or to attain extremely low temperatures. Experiments are therefore carried out on samples for which $L \gg L_\varphi$; the measurements are made at a finite frequency at different temperatures and the results are extrapolated to $T=0$ (or at zero frequency but at $T \neq 0$). It is in this way that formulas (11.65) and (11.66) have been verified.

However, this procedure is open to serious doubts. The possibility is not excluded that in going from $L \gg L_\varphi$ to $L < L_\varphi$ the behavior of the substance changes sharply and the extrapolation becomes illegitimate. There are certain grounds for such doubts. In the case of a purely one-dimensional model of the metal (a chain of atoms) many of the quantities can be calculated exactly, and here it turns out that the conductivity of an atomic chain of finite length at $T=0$, $\omega=0$ is not a self-averaging quantity, i.e., its average relative fluctuation does not drop but increases with length (Abrikosov and Ryzhkin 1978, Abrikosov 1981). We can state this in a different way. The probability of conductivity fluctuations is not described by the usual Gauss law; it has a much broader distribution function at which $\overline{\sigma^2} \neq (\bar{\sigma})^2$, $\rho = (\overline{\sigma^{-1}})$ differs from $(\bar{\sigma})^{-1}$, etc. This is because the averaging over "realizations of the random potential", i.e., over the location of impurities, is of quite a different nature than the thermodynamic averaging. In fact, it follows that measurements on different samples, even if they were prepared under identical conditions, must lead to quite different results.

It has recently been demonstrated (Al'tshuler 1985; see also section 11.7) that an analogous situation occurs in any dimensionality. This refers, however, only to lengths smaller than L_φ. At $L \gg L_\varphi$ the relative fluctuations fall off with size, i.e., self-averaging is restored. It is for this reason that one may suspect that the extrapolation through $L=L_\varphi$ is illegitimate.

On the other hand, if the conductivity is a non-self-averaging quantity, then all the reasonings based on the scaling hypothesis become pointless. Because of this, the possibility is not excluded that they just refer to those quantities which are obtained by extrapolation from the region of $L \gg L_\varphi$. This conclusion can be drawn from the agreement between formulas (11.65) and (11.66) and experiment (Hess et al. 1982, Brandt et al. 1981).

At a finite temperature or frequency no strict localization takes place. In samples of larger dimensions Ohm's law is fulfilled, and the relative fluctuations of conductivity decrease with increasing size. In calculations it may be assumed from the start that the sample is infinite. Of course, if the quantum correction becomes large (for $d=1$ or 2 this corresponds to $L_\varphi \gtrsim L_c$, and for $d=3$, to $l \lesssim \lambda$), the calculation of the conductivity is very complicated.

At present there are only a few results for a purely one-dimensional metal, i.e., a chain of atoms (see Abrikosov and Ryzhkin 1978) and for a quasi-one-dimensional metal, i.e., an assembly of chains with a low probability of jumps. For a three-dimensional metal the following experimental result has been reported (Gurvitch 1983): in metallic glasses at high temperatures, when the mean free path is of order $\lambda = \hbar/p_0$, the resistivity deviates from the law $\rho \propto T$ and saturates.

Up until now we have been concerned only with the scattering of electrons from impurities, and the electron-electron interaction has been assumed to be weak and to determine only τ_φ, with $\tau \ll \tau_\varphi$. This is valid at low temperatures and for sufficiently disordered metals. Let us consider another limiting case where the metal is pure and the dominant role is played by electron-electron interaction. In this case, a metal-insulator transition can also take place.

Consider an even metal and assume that, because of the variation of some external parameter, say pressure, the overlapping of the bands decreases. The electrons and holes may remain "free" as long as their kinetic energy is larger than or of the order of the potential energy of their Coulomb interaction, i.e., $p_0^2/m^* \gtrsim e^2/\varepsilon_0\bar{r}$, where ε_0

is the dielectric constant of the "medium", i.e., the part which is not associated with free carriers. This condition yields

$$\frac{e^2}{\varepsilon_0 \hbar v} \leqslant 1, \tag{11.69}$$

where $v = p_0/m^*$ (recall that $p_0 = \hbar(3\pi^2 n_e)^{1/3}$). If this condition is violated, the electrons and holes may combine into neutral complexes: Mott–Wannier excitons. An "exciton phase" appears, in which either there are no free carriers, i.e., it is an insulator, or the number of free carriers drastically decreases. In the latter case, a series of excitonic transitions may take place. The excitonic ttransitions may be either first or second order.

The question arises: Why does no excitonic transition occur in semi-metals, where the number of electrons and holes is small ($\sim 10^{-5}$ per atom in bismuth)? This is associated with the fact that the effective mass of semi-metals is very small ($m^* \sim 10^{-2}\, m$ in Bi), so that $v = p_0/m^* \sim 10^8$ cm/s, as in an ordinary metal*. Moreover, the dielectric constant of these metals is large ($\varepsilon_0 \sim 100$ in Bi), so that condition (11.69) is fulfilled with a large margin.

It was originally thought that when condition (11.69) is violated, a first-order transition occurs to the insulating phase, it was termed the Mott transition (Mott 1961, 1974, Mott and Davis 1971). Hence came the concept of "minimum metallic conductivity". It was thought that when the conductivity falls off due either to the number of carriers or the mean free path, it attains its minimum value and then turns to zero discontinuously. The minimum value was obtained from the condition (11.69), i.e., $e^2/(\varepsilon_0 \hbar v) \sim 1$, and the assumption that $l \sim \lambda \sim \hbar/p_0$. This gave $\sigma \sim n_e e^2 l/p_0 \sim p_0 e^2/\hbar^2 \sim e^2/\hbar r_B^*$, where $r_B^* = \hbar^2 \varepsilon_0/(m^* e^2)$ is the effective Bohr radius.

Actually, the scattering of electrons from impurities impedes an excitonic transition. Therefore, the concept that the conductivity governs the metal-insulator transitions has no grounds whatsoever. Excitonic transitions have been observed experimentally in pure substances, and metal-insulator transitions associated with Anderson localization, which has been described above, have been detected in strongly disordered doped semiconductors. In such transitions the conductivity gradually drops to zero according to eq. (11.66).

11.7. Mesoscopics

It has been pointed out in the preceding section that the properties of small samples with a size of $L \ll L_\varphi$ are found to be unusual in the sense that their kinetic characteristics are not self-averaging quantities. The probability distribution of the conductivity and other kinetic characteristics of such samples is not Gaussian with

* The energy spectrum of carriers in semi-metals is, in fact, close to a linear function, $\varepsilon = vp$, where $v \sim 10^8$ cm/s (Abrikosov and Fal'kovskii 1962).

a relative width proportional to $N^{-1/2}$, where N is the number of particles or the volume of the system, but in fact it is much broader. From this it follows first of all that the properties of samples prepared by standard procedures will really be different.

Here we refer not to microscopic objects consisting of a few atoms or molecules but to specimens containing many atoms, which may be expected to display macroscopic behavior. The samples usually considered are those with sizes within the range

$$l \ll L \ll L_\varphi . \tag{11.70}$$

The area of physics concerned with the study of such objects has come to be known as "mesoscopics". Estimates show that at $T < 0.1$ K the mesoscopic effects can be observed in samples with a size less than a micron (10^{-4} cm).

Actually, one cannot prepare such samples and their contacts with the current source in a standard way and therefore, at first sight, the problem seems to be purely academic. However, in reality the mesoscopic effects also manifest themselves in the unusual behavior of a single sample upon variation of the external parameters. This is associated with the fact that the mesoscopic effects result from quantum interference, which is characteristic of the given configuration of the scatterers. As we have found out above, the interference effects depend on the external parameters: the magnetic field and the temperature. The distinction between mesoscopic effects and those discussed earlier consists in that they disappear upon averaging over the configurations of the impurity.

Let us consider a simple example. It has been shown in section 11.4 that the resistivity of a hollow cylinder placed in a longitudinal magnetic field depends periodically on the field, the period being characterized by the fact that the magnetic field flux through the cavity varies by one flux quantum Φ_0. Such behavior is associated with the trajectories shown in fig. 75, i.e., those which go around the cylinder and return to the same point. Interference arises between the amplitudes of passing such trajectories in the clockwise and reverse directions.

Let us now consider a trajectory of a different type. Suppose that one trajectory emerges from point A and arrives at point B. Another trajectory starts from the same point A and also arrives at point B, but in doing so it passes round the cavity from the other side. The phase difference between these trajectories is given by

$$\hbar^{-1} \oint \boldsymbol{p}\, \mathrm{d}\boldsymbol{l} - \frac{e}{\hbar c} \oint \boldsymbol{A}\, \mathrm{d}\boldsymbol{l}, \tag{11.71}$$

where the integral is taken over a closed loop formed by the two trajectories. The second term, after being transformed, changes to $\pi\Phi/\Phi_0$, where $\Phi = \int \boldsymbol{B}\, \mathrm{d}\boldsymbol{S}$ is the magnetic field flux through the cavity. When Φ is replaced by $2\Phi_0$, this term changes by 2π. Hence, oscillations with a period of $2\Phi_0$ rather than Φ_0 may be expected to occur.

At this point, however, one should recall the first term in expression (11.71), which, in contrast to the first case discussed in section 11.4, does not cancel in the interference term. The averaging of $\exp(i\hbar^{-1}\oint \boldsymbol{p}\,d\boldsymbol{l})$ over various trajectories or, which is the same, over various scattered configurations, makes the corresponding term vanish. This is what happens when the experiment is carried out in a long thin-walled cylinder. If, however, instead of the cylinder, we use a thin ring, then such averaging does not take place and this means that, in addition to oscillations with a period Φ_0, oscillations must be observed with a period $2\Phi_0$ (Al'tshuler and Khmel'nitskii 1985). This has been observed in experiments (Kramer et al. 1984).

Apart from the oscillatory effect, which is associated with the interference of special trajectories, there are also other interference terms in the scattering probability, which are strongly dependent on the field and which lead to random fluctuations in the magnetoresistance curve. In contrast to the thermal "noise", these fluctuations are fully reproducible for a given sample, because they are determined only by a particular specified configuration of the scatterers. They are sometimes called "grass" in the regular (averaged) curve.

We shall consider this "grass" by using another example, which is more important from a practical point of view. Let us find fluctuations in the current-voltage characteristic of a wire with a size

$$l \ll b \ll L \ll L_\varphi \tag{11.72}$$

(l is the mean free path; b is the diameter of the wire; and L is the length of the wire). They can be evaluated from the following considerations (Larkin and Khmel'nitskii 1986). Interference is possible only between electrons having close energies. The permissible interval is determined from the condition that the phase difference be of the order of unity, i.e.,

$$\frac{1}{\hbar}[p(\varepsilon+\delta E)-p(\varepsilon)]\cdot X \sim \frac{1}{\hbar}\frac{dp}{d\varepsilon}\,\delta\varepsilon\cdot X \sim \frac{1}{\hbar v}\cdot X\delta\varepsilon \sim 1,$$

where X is the total distance travelled by the particle. In the case of diffusive motion $X = vt \sim v\cdot L^2/D$, where D is the diffusion coefficient. Hence,

$$\delta\varepsilon \sim \frac{\hbar v}{X} \sim \frac{\hbar D}{L^2}. \tag{11.73}$$

The scale of fluctuations of the current-voltage characteristic along the axis V is determined by the relation $eV_c \sim \delta\varepsilon$ or by

$$V_c \sim \frac{\hbar D}{eL^2}. \tag{11.74}$$

The amplitude of fluctuations with respect to the current is determined as follows. The current is produced by the electrons residing in a region of order eV near the Fermi boundary, where V is the applied potential difference. If $V \gg V_c$, then interference is possible, but only between the electrons within the intervals eV_c and, hence, the region eV breaks up into $(eV)/(eV_c)$ independent intervals. The interference correction to the current from one of such intervals is of order (R being the resistance)

$$\Delta\left(\frac{V_c}{R}\right)=\Delta\left[\frac{V_c}{L/\sigma b^2}\right]=\frac{V_c}{L}b^2\Delta\sigma\sim V_c\frac{e^2}{\hbar}$$

[here we have substituted formula (11.28) for $d=1$: $\Delta\sigma_d=\Delta\sigma b^2\sim(e^2/\hbar)L$]. But different intervals of order V_c are not intercorrelated, and the interference corrections to the current may have different signs. The overall effect can be determined by using the method for finding the total distance travelled by the particle in diffusive motion: $x\sim\sqrt{Dt}\sim\sqrt{lvt}\sim\sqrt{l\cdot X}$, where l is the mean free path covered in a single event (corresponding to V_c) and $X=vt$ is the total length of the curved path (corresponding to V). Thus, we obtain the amplitude I of fluctuations along the axis:

$$I_c\sim\frac{e^2}{\hbar}(VV_c)^{1/2}. \tag{11.75}$$

These fluctuations are added to the regular part corresponding to Ohm's law ($I=V/R_0$). Note that the ratio I_c/V_c is of order

$$\frac{e^2}{\hbar}\frac{(VV_c)^{1/2}}{V_c}\sim\frac{e^2}{\hbar}\left(\frac{V}{V_c}\right)^{1/2},$$

that is, it increases with increasing V. In the case of $I_c/V_c>R_0^{-1}$, i.e., $V>(\hbar/e^2R_0)^2V_c$, the current-voltage curve exhibits portions with a negative slope (N-shaped portions), which may lead to the formation of mobile domains with different magnitudes of the electric field (the Gunn effect).

The examples given above are not the only manifestations of mesoscopics. There are a number of other effects. All these effects tend to disappear as the temperature increases. This is associated not only with the decrease of L_φ but also with the variation of the distribution of electrons among the energies (Stone and Imry 1986). If the temperature exceeds $\delta\varepsilon$ (11.73), then, just as in the case of the potential difference, the entire temperature interval splits into independent intervals, in which the interference corrections may have different signs. The overall effect, like the preceding one, is proportional to $(T\delta\varepsilon)^{1/2}$. The average value is obtained by dividing by the interval T, i.e., is of order $(\delta\varepsilon/T)^{1/2}$ and decreases with temperature as $T^{-1/2}$. It is exactly such behavior that has been observed experimentally. The condition $\delta\varepsilon/T\gtrsim 1$ gives

$$L\lesssim L_T\sim\left(\frac{\hbar D}{T}\right)^{1/2}.$$

Thus, for mesoscopic effects to be observable, it is necessary not only that $L\lesssim L_\varphi$ but also that $L\lesssim L_T$. Note that the temperature smearing-out of the Fermi distribution is, in fact, equivalent to averaging over the configurations of the scatterers and therefore it has no influence on the effects discussed in sections 11.3 and 11.5.

As has been said above, the mesoscopic effects are actually observed in micron-size samples and at temperatures lower than 0.1 K and therefore, at first sight, they seem to be exotic and to have little relation with practice. It should, however, be kept in mind that at present there is a trend towards miniaturization in computer technology. In order to suppress thermal fluctuations and to reduce the amount of heat released, use is made of low temperatures. Mesoscopics imposes restrictions on this tendency, because it demonstrates that when certain limits are reached, it is impossible not only to produce standard cells but also to hope that, by way of selection, one will be able to find cells possessing desired properties (for example, a smooth current-voltage characteristic).

12

Absorption of sound in metals

12.1. The absorption coefficient in the absence of a magnetic field. Low frequencies

The absorption of sound in metals depends on the quantity $\omega\tau$ and also on the ratio of the wavelength to the mean free path λ/l or (in the case of $\lambda/l \ll 1$) on the ratio λ/δ, where δ is the depth of the skin layer at a frequency equal to the sound frequency. If $\omega\tau \gg 1$, we may speak of the emission and absorption of individual quanta. In the case where $\omega\tau \ll 1$ the sound wave plays the role of an external field acting on electrons. It is this case we will consider here first (Akhiezer et al. 1957).

Up until now we have studied the properties of electrons in a crystal at rest. Since the passage of a sound wave makes the crystal lattice vibrate, it is necessary to take into account the fact that the energy spectrum of electrons is specified in a coordinate system bound to the crystal lattice. We will use this coordinate system in the discussion below.

The energy of electrons in the field of a sound wave acquires the following correction:

$$\varepsilon(\boldsymbol{p}, \boldsymbol{r}, t) = \varepsilon(\boldsymbol{p}) + \lambda_{ik}(\boldsymbol{p})\, u_{ik}(\boldsymbol{r}, t), \tag{12.1}$$

where

$$u_{ik} = \tfrac{1}{2}\left(\frac{\partial u_i}{\partial x_k} + \frac{\partial u_k}{\partial x_i}\right) \tag{12.2}$$

is the strain tensor; $\boldsymbol{u}$ is the displacement vector for the medium at point $\boldsymbol{r}$; λ_{ik} is a tensor called the deformation potential. In principle, the acceleration of electrons should also have given an inertial term $m\boldsymbol{v}\,\partial\boldsymbol{u}/\partial t$ [the result of the expansion of the energy $\frac{1}{2}m(\boldsymbol{v}+\partial\boldsymbol{u}/\partial t)^2$ in $\boldsymbol{u}$]. But the order of magnitude of this term is $mvsku$ (s is the sound velocity and k is the wave vector). At the same time, the second term in eq. (12.1) is of order $u\mu k \sim mv^2uk$ ($\lambda_{ik} \sim \mu$) and, hence, is $v/s \sim 10^3$ times larger than the inertial term.

The kinetic equation has the form

$$\frac{\partial f}{\partial t} + \boldsymbol{v}\frac{\partial f}{\partial \boldsymbol{r}} + \dot{\boldsymbol{p}}\frac{\partial f}{\partial \boldsymbol{p}} = I(f). \tag{12.3}$$

Since in the most interesting cases, which will be considered in this chapter, the collisions play a minor role, we use the collision time approximation for $I(f)$, i.e.

$$I(f) = -\frac{f - f_0(\varepsilon)}{\tau}.$$

However, in this case f_0 is the equilibrium function in the coordinate system bound to be crystal lattice. This means that its energy is expressed by formula (12.1) and, moreover, the chemical potential is renormalized. Thus,

$$f_0(\varepsilon) \approx f_0(\varepsilon_0) + \frac{\partial f_0}{\partial \varepsilon}(\lambda_{ik} u_{ik} - \Delta\mu).$$

The correction $\Delta\mu$ is obtained from the condition of electron density conservation. The point is that, as has been pointed out earlier, any charge in a metal is screened at atomic distances. Since the sound wavelength is presumed to be much larger than the interatomic spacing, the space charge should be equal to zero. In other words, the electron density must not differ from the equilibrium density in the absence of sound. It thus follows that

$$\int \frac{\partial f_0}{\partial \varepsilon}(\lambda_{ik} u_{ik} - \Delta\mu) \frac{d^3 \boldsymbol{p}}{(2\pi\hbar)^3} = -\int (\lambda_{ik} u_{ik} - \Delta\mu) \frac{\mathrm{d}S}{(2\pi\hbar)^3 v} = 0,$$

or

$$\Delta\mu = \bar{\lambda}_{ik} u_{ik},$$

where $\bar{\lambda}_{ik}$ represents the average over the Fermi surface. Hence,

$$f_0(\varepsilon) = f_0(\varepsilon_0) + \frac{\partial f_0}{\partial \varepsilon} \Lambda_{ik} u_{ik}, \qquad (12.4)$$

where $\Lambda_{ik} = \lambda_{ik} - \bar{\lambda}_{ik}$.

The derivative $\dot{\boldsymbol{p}}$, which enters into the kinetic equation, is given by

$$\dot{\boldsymbol{p}} = -\frac{\partial \varepsilon}{\partial \boldsymbol{r}} + e\boldsymbol{E}.$$

The first term is associated with the variation of the energy spectrum in the field of the sound wave and the second with the electric fields produced by the sound. Because of the absence of a space charge the field $\boldsymbol{E}$ may be only of vortex origin, i.e. $\operatorname{div} \boldsymbol{E} = 0$. Thus, $\boldsymbol{E}$ is determined from the Maxwell equations

$$\operatorname{rot} \boldsymbol{E} = -\frac{1}{c}\frac{\partial \boldsymbol{H}}{\partial t}, \qquad \operatorname{rot} \boldsymbol{H} = \frac{4\pi}{c}\boldsymbol{j}.$$

Combining these equations and using the condition $\operatorname{div} \boldsymbol{E} = 0$, we get

$$\nabla^2 \boldsymbol{E} = \frac{4\pi}{c^2}\frac{\partial \boldsymbol{j}}{\partial t}. \qquad (12.5)$$

The kinetic equation (12.3) will be solved here by means of the same method used earlier. Substituting $f = f_0(\varepsilon) - (\partial f_0/\partial \varepsilon)\psi$, we obtain

$$\frac{\partial \psi}{\partial t} + \boldsymbol{v}\frac{\partial \psi}{\partial \boldsymbol{r}} + \frac{\psi}{\tau} = \Lambda_{ik}\dot{u}_{ik} + e\boldsymbol{v}\boldsymbol{E}. \tag{12.6}$$

The complete procedure for the solution of the problem is as follows. The kinetic equation gives ψ as a function of $\boldsymbol{E}$. From this we find $\boldsymbol{j}$ as a function of $\boldsymbol{E}$ and substituting it into eq. (12.5), we obtain the equation for the determination of the electric field.

As a matter of fact, it can be shown that the vortex electric field plays an important part in sound absorption only if $kl \gg 1$ and $k\delta \ll 1$, where $\delta \sim (cp_0/4\pi\omega n_e e^2)^{1/3}$ is the depth of the skin layer in the anomalous limit at a frequency ω and, in this case, gives a contribution to the absorption of sound of the same order of magnitude as that obtained without taking account of the term with $\boldsymbol{E}$ in eq. (12.6). The entire expression will depend on the tensor Λ_{ik}, which is very complicated for an anisotropic metal. Because of this complicated structure only the determination of the frequency dependence of the absorption coefficient, and of its order of magnitude, is meaningful (just as in the case of the electrical resistance discussed in ch. 4). But for such a problem to be solved, one needs not take into account the term with $\boldsymbol{E}$ in eq. (12.6), because under no conditions does it change the order of magnitude of the effect.

Let us assume that $u_{ik} = u_{ik}^0\,\mathrm{e}^{i\boldsymbol{kr}-i\omega t}$ and look for ψ, which is proportional to the same exponent. Substitution into eq. (12.6) (with $\boldsymbol{E} = 0$), gives*

$$\psi = \frac{-\mathrm{i}\omega\Lambda_{ik}u_{ik}}{\tau^{-1} + \mathrm{i}(\boldsymbol{vk} - \omega)}. \tag{12.7}$$

The absorption coefficient may be expressed directly via ψ. Let us turn to fig. 78. Let I be the energy flux incident on 1 cm^2 of surface, Q the amount of energy

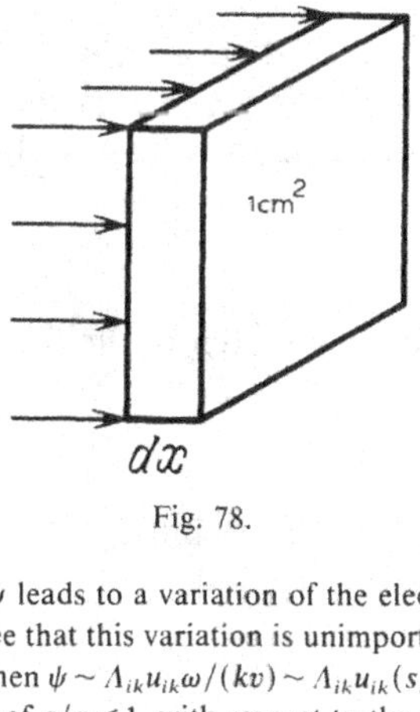

Fig. 78.

* The nonequilibrium correction ψ leads to a variation of the electron density, which is proportional to $\bar{\psi}$. It is not difficult, however, to see that this variation is unimportant. Indeed, in the case of $\tau^{-1} \gg kv$, $\bar{\psi} = 0$ due to $\bar{\Lambda}_{ik} = 0$. But if $kv \gg \tau^{-1}$, then $\psi \sim \Lambda_{ik}u_{ik}\omega/(kv) \sim \Lambda_{ik}u_{ik}(s/v)$. The correction to the absorption coefficient that arises is of the order of $s/v \ll 1$, with respect to the main effect.

absorbed in 1 cm^3/s. In this case, the volume shown in fig. 78 will absorb in one second an energy $Q\,\mathrm{d}x$ or a part $(Q/I)\,\mathrm{d}x$ of the incident energy. Designating Q/I by Γ, we may write the equation for the flux as

$$\mathrm{d}I = -\Gamma I\,\mathrm{d}x,$$

from which we get

$$I = I_0\,\mathrm{e}^{-\Gamma x}.$$

Thus, Γ is the absorption coefficient. According to what has just been said, we have

$$\Gamma = \frac{Q}{I}. \tag{12.8}$$

In harmonic vibrations the average kinetic energy is equal to the average potential energy. Hence, for the average total energy we have

$$\bar{E} = \bar{K} + \bar{U} = 2\bar{K}.$$

The kinetic energy per unit volume is $\frac{1}{2}\rho v^2$ ($\rho = MN/V$ is the mass density). Upon propagation of the wave, $v = v_0\cos(kx - \omega t)$. This yields

$$\bar{E} = \rho\overline{v^2} = \tfrac{1}{2}\rho v_0^2.$$

However, $v_0 = u_0\omega$, where u_0 is the displacement amplitude. Thus, the total average energy per unit volume is equal to $\frac{1}{2}\rho u_0^2\omega^2$. Multiplying by the sound velocity s, we find the flux

$$I = \tfrac{1}{2}\rho u_0^2\omega^2 s. \tag{12.9}$$

Let us now consider Q. This is the energy dissipated in unit volume per second. Hence, we have

$$Q = 2\int \dot{\varepsilon} f \frac{\mathrm{d}^3\boldsymbol{p}}{(2\pi\hbar)^3}. \tag{12.10}$$

This expression must be averaged over time. If we substitute here $f = f_0(\varepsilon) - (\partial f_0/\partial\varepsilon)\psi$, then the combination $f_0(\varepsilon)\dot{\varepsilon}$ is the time derivative of a certain function of ε; therefore, on averaging over time we have zero. Hence,

$$Q = -2\int \dot{\varepsilon}\frac{\partial f_0}{\partial\varepsilon}\psi\frac{\mathrm{d}^3\boldsymbol{p}}{(2\pi\hbar)^3}. \tag{12.11}$$

The kinetic equation (12.6) may be written in the form

$$\frac{\mathrm{d}\psi}{\mathrm{d}t} - \dot{\varepsilon} = -\frac{\psi}{\tau}.$$

Expressing $\dot{\varepsilon}$ in terms of ψ, substituting into (12.11) and recalling that the time derivative gives zero on averaging, we find

$$Q = -\frac{2}{\tau}\int \psi^2\frac{\partial f_0}{\partial\varepsilon}\frac{\mathrm{d}^3\boldsymbol{p}}{(2\pi\hbar)^3}. \tag{12.12}$$

The remaining expression is quadratic in ψ. Because of this, we cannot use the complex form for ψ but must take only the real part. Let, for example, $\psi = A\,e^{i(kx-\omega t+\varphi)}$, where A is real. Then,

$$\mathrm{Re}\,\psi = A\cos(kx-\omega t+\varphi)$$

and

$$\overline{(\mathrm{Re}\,\psi)^2} = \tfrac{1}{2}A^2, \quad \text{or} \quad \tfrac{1}{2}|\psi|^2.$$

Thus, we have

$$Q = -\int \frac{|\psi|^2}{\tau}\frac{\partial f_0}{\partial \varepsilon}\frac{d^3\boldsymbol{p}}{(2\pi\hbar)^3}. \tag{12.13}$$

We shall first consider long waves with $kl \ll 1$. Since $l \sim v\tau$, $vk \ll \tau^{-1}$, and the inequality $\omega\tau \ll 1$ was assumed from the beginning. Hence, from (12.7) we get

$$\psi = -i\omega\tau\Lambda_{ik}u_{ik}. \tag{12.14}$$

Therefore, in order of magnitude we have

$$Q \sim \frac{\omega^2\tau\mu^2k^2u_0^2p_0^3}{\hbar^3\mu} \sim \omega^2\tau\mu k^2u_0^2n_e$$

(where n_e is the electron density). Dividing by the flux (12.9), we obtain

$$\Gamma \sim \frac{\tau\mu k^2 n_e}{\rho s} \sim \frac{\omega^2\tau\mu n_e}{\rho s^3} \sim \frac{\omega^2\tau}{s}. \tag{12.15}$$

(Here it is taken into account that $s^2 \sim v^2 m/M$ and $\rho = Mn \sim Mn_e$.) Thus, in this case the absorption coefficient is proportional to ω^2 and to the mean free path $l(\tau \sim l/v)$.

Consider the opposite limiting case $kl \gg 1$. In accordance with eq. (12.7), we may write

$$\frac{|\psi|^2}{\tau} = \omega^2|\Lambda_{ik}u_{ik}^0|^2\frac{\tau^{-1}}{(\boldsymbol{kv}-\omega)^2+\tau^{-2}}.$$

The last factor is equal to $\pi\delta(\boldsymbol{kv}-\omega)$, to within terms of order $(kl)^{-1}$. To put it another way, the absorption is associated with those electrons for which $\boldsymbol{kv} \approx \omega$. But since $\omega = sk$, we obtain $v\cos\theta = s$. This means that the sound is absorbed by electrons that move in phase with the sound wave. This is nothing more than the Landau absorption mechanism, which has been discussed in section 9.3, although there we were dealing with the interaction of electrons with an electromagnetic wave. Since $s/v \ll 1$, the relevant electrons move almost perpendicular to $\boldsymbol{k}$ (fig. 79). Integration of the δ-function over $\cos\theta$ yields $(vk)^{-1}$. Thus, we obtain

$$Q \sim \frac{\omega^2\mu k^2u_0^2n_e}{vk},$$

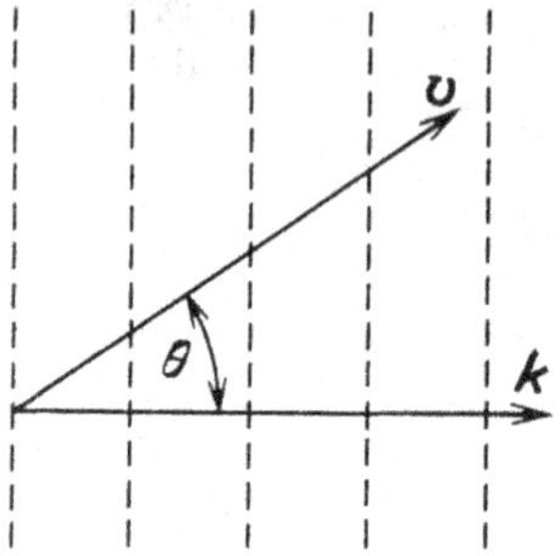

Fig. 79.

and

$$\Gamma \sim \frac{\omega \mu n_e}{\rho v s^2} \sim \frac{\omega}{v}. \tag{12.16}$$

Hence, the absorption coefficient no longer depends on l. Formally, in eq. (12.15), l was replaced by k^{-1}.

12.2. The absorption coefficient in the absence of a magnetic field. High frequencies

Let us now consider the opposite limiting case $\omega\tau \gg 1$ (Skobov 1962). Here, we have to deal with the emission and absorption of individual quanta. For the emission of a phonon we have the following conservation laws:

$$\boldsymbol{p}' = \boldsymbol{p} + \hbar \boldsymbol{k}, \qquad \varepsilon(\boldsymbol{p}') = \varepsilon(\boldsymbol{p}) + \hbar\omega.$$

Hence,

$$\varepsilon(\boldsymbol{p} + \hbar \boldsymbol{k}) = \varepsilon(\boldsymbol{p}) + \hbar\omega.$$

Since in a real situation k is much smaller than p_0, the energy can be expanded in $\boldsymbol{k}$:

$$\varepsilon(\boldsymbol{p} + \hbar \boldsymbol{k}) \approx \varepsilon(\boldsymbol{p}) + \frac{\partial \varepsilon}{\partial \boldsymbol{p}} \hbar k = \varepsilon(\boldsymbol{p}) + \hbar \boldsymbol{v}\boldsymbol{k}.$$

Hence, we have

$$\boldsymbol{v}\boldsymbol{k} = \omega. \tag{12.17}$$

This is the same condition as in the case $kl \gg 1$, $\omega\tau \ll 1$.

The Hamiltonian of an electron in an external periodic field is given by

$$\mathcal{H}' = \tfrac{1}{2}(U\,e^{-i\omega t} + U^{+}\,e^{i\omega t}).$$

If we deal with the interaction with a sound wave, then

$$U = U_0\,\mathrm{e}^{i\boldsymbol{kr}}, \qquad U_0 = \Lambda_{ik}(\boldsymbol{p})\,u^0_{ik}.$$

The probability of a transition under the influence of a perturbation periodic in time is

$$W = \frac{\pi}{2\hbar^2}\left[|U_{p'p}|^2\,\delta(\omega_{p'p}-\omega) + |U^+_{p'p}|^2\,\delta(\omega_{p'p}+\omega)\right], \tag{12.18}$$

where $\hbar\omega_{p'p} = \varepsilon(\boldsymbol{p}') - \varepsilon(\boldsymbol{p})$. The first term in eq. (12.18) corresponds to the absorption of a quantum, the second to the emission of a quantum. This probability must be multiplied by $f_0(\boldsymbol{p})\,[1-f_0(\boldsymbol{p}')]$, which is the probability that the state $\boldsymbol{p}$ is occupied and the state $\boldsymbol{p}'$ empty. If the result is multiplied by a change in the electron energy $\hbar\omega_{p'p}$, and summed over all the momenta $\boldsymbol{p}$ and $\boldsymbol{p}'$ and spin projections (which gives the factor 2 since the spin is conserved), we will then obtain the energy absorbed per second. By referring it to unit volume, we have

$$Q = V^{-1}\sum_{p,p'}\hbar\omega_{p'p}\frac{\pi}{\hbar^2}\left[|U_{p'p}|^2\delta(\omega_{p'p}-\omega)\right.$$
$$\left. + |U^+_{p'p}|^2\delta(\omega_{p'p}+\omega)\right] f_0(\varepsilon(\boldsymbol{p}))[1-f_0(\varepsilon(\boldsymbol{p}'))].$$

In the second term we make the replacement $U^+_{p'p} \to U^*_{pp'}$. While taking the sum over $\boldsymbol{p}$, $\boldsymbol{p}'$ in this term, we interchange the subscripts $\boldsymbol{p} \rightleftarrows \boldsymbol{p}'$. As a result, we get

$$Q = \frac{\pi\omega}{V}\sum_{p,p'}|U_{p'p}|^2\delta(\hbar\omega_{p'p}-\hbar\omega)[f_0(\varepsilon(\boldsymbol{p})) - f_0(\varepsilon(\boldsymbol{p}'))]. \tag{12.19}$$

The difference term in brackets may be expanded in $\varepsilon' - \varepsilon$, which gives

$$Q = -\frac{\pi\omega^2}{V}\sum_{p,p'}|U_{p'p}|^2\delta(\omega_{p'p}-\omega)\frac{\partial f_0}{\partial\varepsilon}. \tag{12.20}$$

Here $U_{p'p}$ has matrix elements only for transitions with $\boldsymbol{p}' = \boldsymbol{p} + \hbar\boldsymbol{k}$, which are equal to $\Lambda_{ik}u^0_{ik}$. Taking this circumstance into account, we have

$$Q = -\frac{\pi\omega^2}{V}\sum_{p}|\Lambda_{ik}u^0_{ik}|^2\delta(\boldsymbol{kv}-\omega)\frac{\partial f_0}{\partial\varepsilon}.$$

Passing from the sum over $\boldsymbol{p}$ to an integral ($\sum_p = [V/(2\pi\hbar)^3]\int \mathrm{d}^3\boldsymbol{p}$) and dividing by the energy flux, we obtain

$$\Gamma \sim \frac{\omega\mu n_e}{\rho v s^2} \sim \frac{\omega}{v}. \tag{12.21}$$

This formula, however, coincides with the result (12.16) for the case $\omega\tau \ll 1$ and $kl \gg 1$. It then follows that the true parameter that determines the mode of absorption is kl rather than $\omega\tau$. This occurs because ω and k enter into the kinetic equation in the combination $\boldsymbol{kv} - \omega$. And since $kv \gg \omega = sk$, the absorption is governed by the quantity $kv\tau \sim kl$ and not by $\omega\tau$.

The sound absorption at $kl \gg 1$ is the absorption of individual quanta and is provided by electrons moving in phase with the sound wave. In fact, in the opposite case $kl \ll 1$, the electrons, between collisions, move in the homogeneous field of the sound wave. The factor responsible for sound absorption in this case is the viscosity of the electron liquid.

12.3. Geometric resonance

We now turn to the treatment of sound absorption in the presence of a magnetic field. We begin with a case where the quantization of the energy levels of electrons in the magnetic field may be neglected, i.e., where the classical picture of electrons moving along trajectories is applicable. Evidently, this occurs when the field is not too strong, i.e. $\hbar\Omega \ll T$. The criterion of importance here is also

$$kr_L \gg 1.$$

However, at the same time we shall assume that

$$\Omega\tau \gg 1.$$

First of all, let us consider a case where the vectors $\boldsymbol{H}$ and $\boldsymbol{k}$ are perpendicular to each other. In this case, the phase of the sound wave does not change along the field direction, and therefore the displacement of an electron in this direction is unimportant for the absorption of sound. Consider the projection of the motion of the electron onto the (x, y) plane (fig. 80); the z axis is, as always, directed along $\boldsymbol{H}$. The trajectory is assumed to be closed. The dashed lines denote the planes of equal phase. The electron spends different time intervals near those planes, the longest interval when its velocity is parallel to the plane; this is in points 1 and 2 in fig. 80. The absorption is mainly determined by these points.

The difference between points 1 and 2 and other points of the orbit consists in that near points 1 and 2 the electron resides for a long time in a more or less constant field, while in other places of the orbit the sound-wave field acting on the electron

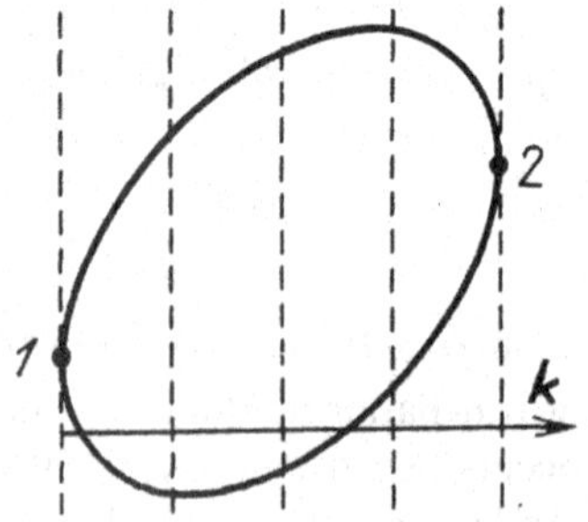

Fig. 80.

is rapidly changing. Of course, this is correct only when the number of wavelengths per diameter of the orbit is large, i.e. $kr_L \gg 1$.

Suppose the function ψ has an extremum for a certain phase difference of the sound field between the planes passing through points 1 and 2. In this case, the same difference will be preserved if the number of wavelengths accommodated in the diameter of the orbit is changed by an integral number, i.e.

$$\varphi = \boldsymbol{k} \int_{t_1}^{t_2} \boldsymbol{v}\, \mathrm{d}t$$

is changed by $2\pi n$. Since $\boldsymbol{k}$ is directed along the x axis, we find that

$$\frac{\mathrm{d}p_y}{\mathrm{d}t} = -\frac{e}{c} H v_x, \qquad \varphi = \frac{ck}{eH}\left(p_y^{(1)} - p_y^{(2)}\right).$$

Therefore, if $\varphi \gg 1$, the quantity ψ becomes an oscillatory function of H^{-1} with a period

$$\Delta \mathrm{H}^{-1} = \frac{2\pi e}{ck}\left(p_y^{(1)} - p_y^{(2)}\right)^{-1}. \tag{12.22}$$

Of course, the difference $p_y^{(1)} - p_y^{(2)}$ depends on p_z and the true oscillations correspond to the extrema of $p_y^{(1)} - p_y^{(2)}$. This effect is analogous to the nonresonant size effect in metal films (section 8.3). It is called geometric resonance (Bömmel 1955, Pippard 1957).

12.4. *Magnetoacoustic resonance phenomena*

We shall now consider a case where $\boldsymbol{k}$ is not perpendicular to $\boldsymbol{H}$. We begin with the case of a closed trajectory in momentum space. Here, in coordinate space the electron moves along the magnetic field in helical motion. At the same time, one must remember that only electrons with $\boldsymbol{kv} \approx 0$ play a role in absorption. Consider fig. 81. Suppose that a certain moment of time the electron has a velocity perpendicular to $\boldsymbol{k}$. Naturally, the same will occur after a period T. During this time the electron will be displaced along $\boldsymbol{H}$ over a distance $\bar{v}_z T$, and in the $\boldsymbol{k}$ direction it

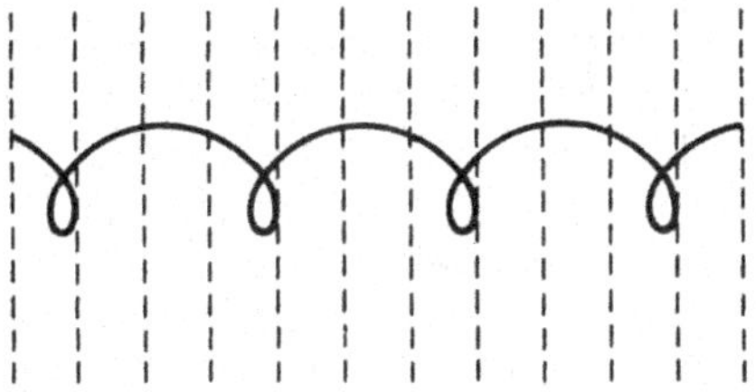

Fig. 81.

will move over a distance $\bar{v}_z T \cos\theta$ (θ is the angle between $\boldsymbol{k}$ and $\boldsymbol{H}$). Evidently, the maximal effect will be realized when the phase of the sound wave is changed by an amount $2\pi n$. Putting it another way,

$$k\bar{v}_z T \cos\theta = 2\pi n.$$

The most important role will evidently be played by extreme values of $\bar{v}_z T$. The density of states for such values increases strongly and one can anticipate a resonance effect in the absorption of sound. Substituting $T = 2\pi m^* c/eH$, we find the condition for the magnetic field,

$$H_n = \frac{kc}{en}(m^* \bar{v}_z)_{\text{extr}} \cos\theta.$$

Recalling

$$m^* = \frac{1}{2\pi}\frac{\partial S}{\partial \varepsilon}, \qquad v_z = \frac{\partial \varepsilon}{\partial p_z},$$

we get

$$H_n = \frac{kc\cos\theta}{2\pi ne}\left(\frac{\partial S}{\partial p_z}\right)_{\text{extr}}. \tag{12.23}$$

This effect is similar to the drift focusing of a high-frequency field (section 8.6), from which we can determine the same quantity, $(\partial S/\partial p_z)_{\text{extr}}$. It is obvious that $\partial S/\partial p_z$ may have an extremum only for a nonconvex Fermi surface (see fig. 47) or at the limiting point.

Apart from the points with $(m^* v_z)_{\text{extr}}$ at $\boldsymbol{k}$ not perpendicular to $\boldsymbol{H}(\boldsymbol{kH} \neq 0)$, the so-called boundary points play a special role. The point is that the condition $\boldsymbol{kv} = 0$, which is necessary for the absorption of a phonon, is fulfilled not for all electrons but only for a part of the electrons (fig. 82). If the contour $p_z = \text{const.}$ is traced out, the condition $\boldsymbol{kv} = 0$ can be seen to be fulfilled in the middle of the surface at two

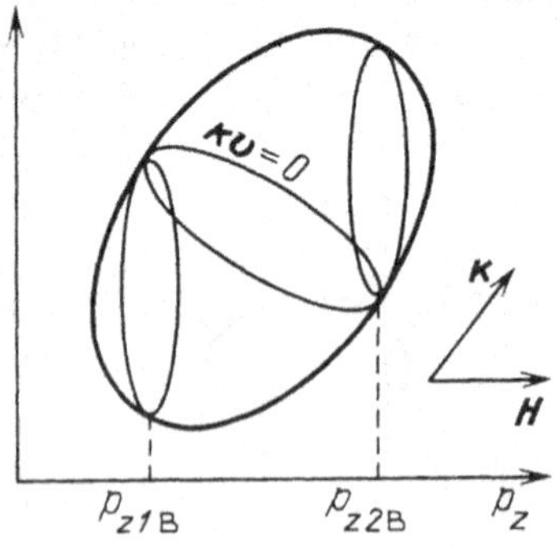

Fig. 82.

points, but as p_z varies to one or the other side these points merge into one, following which the condition $\boldsymbol{kv} = 0$ is no longer satisfied for any point of the contour. Thus, there is every ground for expecting a singularity in the absorption coefficient at

$$H_{1n} = \frac{kc \cos\theta}{2\pi ne}\left(\frac{\partial S}{\partial p_z}\right)_{\mathrm{b}}, \tag{12.24}$$

where the subscript b signifies that the derivative is taken at the boundary value of the momentum $p_z = p_{z,b}$.

Finally, let us consider the case of an open trajectory. Here, the electron performs a drift across the field. The characteristic time which replaces the Larmor period is the time during which the electron travels a period of the reciprocal lattice. If the openness is in the direction p_x, then, integrating the following equation over t,

$$\frac{\mathrm{d}p_x}{\mathrm{d}t} = \frac{eH}{c} v_y,$$

we find the resonance condition

$$\hbar K = \frac{eH}{c}\int v_y \,\mathrm{d}t = \frac{eH}{c}\cos\theta n\lambda.$$

where K is the smallest period of the reciprocal lattice; λ is the wavelength of sound; and θ is the angle between the directions $\boldsymbol{k}$ and y. In other words, the resonance values of H are given by

$$H_n = \frac{c\hbar Kk\cos\theta}{2\pi en}. \tag{12.25}$$

All these phenomena have come to be known as magnetoacoustic resonances (Kaner et al. 1961).

12.5. Quantitative theory of geometric resonance

We will consider the case of a closed Fermi surface. In the presence of a magnetic field eq. (12.5) acquires an additional term, $\partial\psi/\partial t_1$. Assuming $\boldsymbol{E} = 0$ and ψ depends on the coordinates and time according to the law $e^{i(\boldsymbol{kr}-\omega t)}$, we obtain

$$i(\boldsymbol{kv} - \omega)\psi + \frac{\mathrm{d}\psi}{\mathrm{d}t_1} + \frac{\psi}{\tau} = g, \tag{12.26}$$

where $g \equiv \Lambda_{ik}\dot{u}_{ik}$. Multiplying this equation by ψ^* and adding it to the complex-conjugate equation multiplied by ψ, we find

$$\frac{1}{2}\frac{\mathrm{d}|\psi|^2}{\mathrm{d}t_1} + \frac{|\psi|^2}{\tau} = \frac{1}{2}(g\psi^* + g^*\psi) = \mathrm{Re}(g^*\psi).$$

The absorption coefficients includes $\int d^3\boldsymbol{p}(|\psi|^2/\tau)(\partial f_0/\partial\varepsilon)$. Replacing the variables $dp_x\, dp_y \to (eH/c)\, dt_1\, d\varepsilon$ (see eq. 5.11), we see that upon integration over t_1 the term $d|\psi|^2/dt_1$ vanishes. Hence, we have

$$\int \frac{|\psi|^2}{\tau}\, dt_1 = \int \mathrm{Re}(g^*\psi)\, dt_1. \tag{12.27}$$

Below we assume that $\omega \ll \tau^{-1}$ and therefore ω is neglected. Solving eq. (12.26) in the same way as was done in section 5.2 and in ch. 7, we obtain

$$\psi(t_1) = \int_{-\infty}^{t_1} g(t_2) \exp\left\{\int_{t_1}^{t_2} [\mathrm{i}\boldsymbol{kv}(t_3) + \tau^{-1}]\, dt_3\right\} dt_2. \tag{12.28}$$

Now we split the integral $\int_{-\infty}^{t_1} \cdots dt_2$ into integrals over intervals of length T and in each such integral we make the replacement of the variables for the limits of integration to be the same for all the terms. For example,

$$\int_{t_1-T(n+1)}^{t_1-Tn} f(t_2)\, dt_2 = \int_{t_1}^{t_1+T} f[t_2 - T(n+1)]\, dt_2.$$

Since $g(t_2)$ is a periodic function of t_2, it does not change with such a replacement and, hence, the integrand in (12.28) after the replacement will be given by

$$g(t_2) \exp\left\{\int_{t_1}^{t_2-T(n+1)} [\mathrm{i}\boldsymbol{kv}(t_3) + \tau^{-1}]\, dt_3\right\}.$$

But since the integrand in the exponential is a periodic function of t_3, we obtain

$$g(t_2) \exp\left\{\int_{t_1}^{t_2} (\mathrm{i}\boldsymbol{kv} + \tau^{-1})\, dt_3\right\} \exp[-T(n+1)(\mathrm{i}\boldsymbol{k}\bar{\boldsymbol{v}} + \tau^{-1})].$$

Hence,

$$\psi(t_1) = \int_{t_1}^{t_1+T} g(t_2) \exp\left\{\int_{t_1}^{t_2} (\mathrm{i}\boldsymbol{kv} + \tau^{-1})\, dt_3\right\} dt_2$$
$$\times \sum_{n=0}^{\infty} \exp[-T(n+1)(\mathrm{i}\boldsymbol{k}\bar{\boldsymbol{v}} + \tau^{-1})].$$

Summing over n, we get

$$\psi(t_1) = \int_{t_1}^{t_1+T} g(t_2) \exp\left\{\int_{t_1}^{t_2} (\mathrm{i}\boldsymbol{kv} + \tau^{-1})\, dt_3\right\} dt_2 [\mathrm{e}^{T(\mathrm{i}\boldsymbol{k}\bar{\boldsymbol{v}} + \tau^{-1})} - 1]^{-1}. \tag{12.29}$$

The numerator in the integrand contains a large exponent. We apply to this integral the stationary-phase method. The stationary points of the exponent are determined by the condition

$$\left(\frac{\partial}{\partial t_2}\right) \int_{t_1}^{t_2} \mathrm{i}\boldsymbol{kv}(t_3)\, dt_3 = 0, \qquad \boldsymbol{kv}(t_2) = 0.$$

Let the points $t = t_\alpha$ be the solution of the equation $\boldsymbol{kv}(t_2) = 0$. Expanding in $t = t_2 - t_\alpha$ and extending the limits of integration over t from $-\infty$ to ∞, we get

$$\int_{t_1}^{t_1+T} g(t_2) \exp\left\{\int_{t_1}^{t_2} (\mathrm{i}\boldsymbol{kv} + \tau^{-1})\,\mathrm{d}t_3\right\} \mathrm{d}t_2$$

$$= \sum_{t_1 < t_\alpha < t_1+T} g_\alpha J_\alpha \exp\left\{\int_{t_1}^{t_\alpha} (\mathrm{i}\boldsymbol{kv} + \tau^{-1})\,\mathrm{d}t_3\right\}, \tag{12.30}$$

where $g_\alpha = g(t_\alpha)$ and

$$J_\alpha = \int_{-\infty}^{\infty} \mathrm{d}t[\tfrac{1}{2}\mathrm{i}\boldsymbol{kv}'_\alpha t^2 + \tfrac{1}{6}\mathrm{i}\boldsymbol{kv}''_\alpha t^3] \tag{12.31}$$

(the subscript α corresponds to the moment t_α).

Usually, in the stationary phase method only the first term, which is proportional to t_α^2, is retained. But here we should be careful because $\boldsymbol{kv}'_\alpha = 0$ at points $p_{z,b}$. In view of this, we also retain the term with t^3. If $\boldsymbol{kv}'_\alpha \neq 0$, we obtain

$$J_\alpha = \left|\frac{2\pi}{(\boldsymbol{kv}'_\alpha)}\right|^{1/2} \exp[\tfrac{1}{4}\pi\mathrm{i}\,\mathrm{sign}(\boldsymbol{kv}'_\alpha)] \tag{12.32}$$

(sign x signifies the sign function, i.e., 1 if $x > 0$ and -1 if $x < 0$). But if $\boldsymbol{kv}'_\alpha = 0$, then

$$J_\alpha = 3^{-1/2}\Gamma(\tfrac{1}{3}) \left|\frac{6}{\boldsymbol{kv}''_\alpha}\right|^{1/3}. \tag{12.33}$$

According to eqs. (12.8), (12.13), (12.27) and (5.11), the absorption coefficient is

$$\Gamma = \frac{eH}{c}\frac{1}{(2\pi\hbar)^3 I}$$

$$\times \mathrm{Re} \int \mathrm{d}p_z \int_0^T \mathrm{d}t_1\, g^*(t_1)$$

$$\sum_{t_1 \leq t_\alpha \leq t_1+T} g_\alpha J_\alpha \frac{\exp\{\int_{t_1}^{t_\alpha} (\mathrm{i}\boldsymbol{kv} + \tau^{-1})\,\mathrm{d}t_3\}}{\exp[T(\mathrm{i}\boldsymbol{k\bar{v}} + \tau^{-1})] - 1}.$$

The same stationary phase method will now be applied for evaluating the integral over t_1; as a result, we find

$$\Gamma = \frac{eH}{c}\frac{1}{(2\pi\hbar)^3 I}$$

$$\times \mathrm{Re} \int \mathrm{d}p_z \frac{\mathrm{d}p_z}{\exp[T(\mathrm{i}\boldsymbol{k\bar{v}} + \tau^{-1})] - 1}$$

$$\times \sum_{0 \leq t_\alpha,\, t_\beta < T} g_\alpha J_\alpha g_\beta^* J_\beta^* \exp\left\{\int_{t_\beta}^{t_\alpha} (\mathrm{i}\boldsymbol{kv} + \tau^{-1})\,\mathrm{d}t_3\right\}. \tag{12.34}$$

This is the general expression.

We shall now consider in particular the case of geometric resonance (Gurevich 1959). Here $\boldsymbol{k}\bar{\boldsymbol{v}}=0$ (since $\bar{\boldsymbol{v}}\|\boldsymbol{H}$ and $\boldsymbol{kH}=0$). On the electronic orbit there are two points where $\boldsymbol{kv}=0$. They are labelled here as t_1 and t_2 (see fig. 80). At these points $\boldsymbol{kv}'(t_{(1)})$ and $\boldsymbol{kv}'(t_{(2)})$ do not, in general, vanish, with $\boldsymbol{kv}'$ having different signs at these points. Hence, if we denote by t_1 the point at which $\boldsymbol{kv}'_{(1)}>0$, then certainly $\boldsymbol{kv}'_{(2)}<0$. At $\boldsymbol{k}\bar{\boldsymbol{v}}=0$ in formula (12.34) there remains $\mathrm{e}^{T/\tau}-1$ and since $\tau \gg T$, this expression is equal to T/τ. The sum over t_α and t_β includes terms with $t_\alpha=t_\beta=t_{(1)}$, $t_\alpha=t_\beta=t_{(2)}$ and two terms with $t_\alpha=t_{(1)}$, $t_\beta=t_{(2)}$; $t_\alpha=t_{(2)}$, $t_\beta=t_{(1)}$. From this we find

$$\Gamma=\frac{2\pi eH}{c(2\pi\hbar)^3 I}\int\frac{\tau}{T}\,\mathrm{d}p_z \times\left\{\left|\frac{g_{(1)}^2}{\boldsymbol{kv}'_{(1)}}\right|+\left|\frac{g_{(2)}^2}{\boldsymbol{kv}'_{(2)}}\right|+\frac{2|g_{(1)}g_{(2)}|}{[(\boldsymbol{kv}'_{(1)})(\boldsymbol{kv}'_{(2)})]^{1/2}}\sin\left(\int_{t_{(1)}}^{t_{(2)}}\boldsymbol{kv}(t_3)\,\mathrm{d}t_3\right)\right\} \tag{12.35}$$

(here we have again made use of the fact that $\tau\gg T$).

Let us choose the x axis along $\boldsymbol{k}$ and use the equation

$$\frac{\mathrm{d}p_y}{\mathrm{d}t}=-\frac{e}{c}\,v_x H.$$

So, we have

$$\int_{t_{(1)}}^{t_{(2)}}\boldsymbol{kv}(t_3)\,\mathrm{d}t_3=\frac{kc}{eH}\,(p_y^{(1)}-p_y^{(2)}), \tag{12.36}$$

that is, we have obtained the same relation as in section 12.3.

The first two terms in the braces in eq. (12.35) give a monotonic, nonoscillatory part of the absorption coefficient. By order of magnitude,

$$\Gamma_{\mathrm{mon}}\sim\frac{eH\tau p_0\Lambda^2\omega^2u_{ik}^2}{c\hbar^3 kvI}\sim\frac{eH\tau p_0\mu^2\omega^2k^2u^2}{c\hbar^3\rho u^2\omega^2 skv}$$

$$\sim\Omega\tau\frac{\omega\mu n_e}{\rho vs^2}\sim\Gamma_0\Omega\tau. \tag{12.37}$$

The quantity $\Gamma_0\sim\omega\mu n_e/(\rho vs^2)$ is the absorption coefficient in the absence of magnetic field.

The term in eq. (12.35) with a sine gives the oscillatory part of the absorption coefficient. Since the sine is a rapidly oscillating function of p_z, we apply the saddle point method upon integration over p_z. Writing $p_y^{(1)}-p_y^{(2)}=\Delta p_y$ and choosing the extremal values of this difference, we get

$$\Delta p_y=\Delta p_{y\mathrm{m}}+\tfrac{1}{2}\left(\frac{\partial^2\Delta p_y}{\partial p_z^2}\right)_{\mathrm{m}}(p_z-p_{z\mathrm{m}})^2.$$

After this the integration is carried out easily and we obtain

$$\left(\frac{2\pi eH}{ck}\right)^{1/2}\left|\frac{\partial^2\Delta p_y}{\partial p_z^2}\right|_{\mathrm{m}}^{-1/2}\sin\left(\frac{ck}{eH}\,\Delta p_{y\mathrm{m}}\pm\tfrac{1}{4}\pi\right),$$

where the plus sign is taken at $(\partial^2 \Delta p_y/\partial p_z^2)_m > 0$ and the minus sign in the opposite case. Upon integration over p_z all the other quantities in (12.35) must be regarded as constants (the values at $p_z = p_{zm}$).

Comparing the result for Γ_{osc} and Γ_{mon}, we obtain

$$\frac{\Gamma_{osc}}{\Gamma_{mon}} \sim (kr_L)^{-1/2} \sum_m \sin\left(\frac{ck}{eH} \Delta p_{ym} \pm \tfrac{1}{4}\pi\right). \tag{12.38}$$

Thus, geometric resonance, like the size effect, may be used as a method for determination of the Fermi surface.

Evidently, here we find the difference in the values of p_y at points of intersection of the contour $p_z = p_{zm}$ with the contour $\boldsymbol{kv} = 0$ at the Fermi surface (fig. 83), the contour $p_z = p_{zm}$ corresponding to the extremal value of Δp_y. For example, this may be the central cross section.

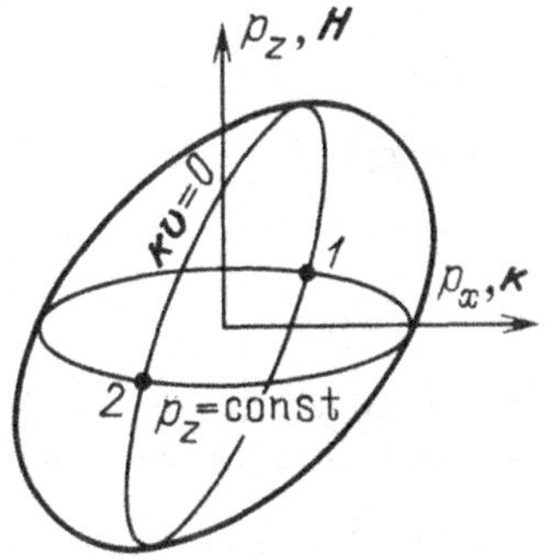

Fig. 83.

12.6. Quantitative theory of magnetoacoustic resonances

Magnetoacoustic resonances at $\boldsymbol{kH} \neq 0$ (Kaner et al. 1961) stem from the factor

$$C(p_z) = \frac{1}{\exp[T(\mathrm{i}\boldsymbol{k}\bar{\boldsymbol{v}} + \tau^{-1})] - 1}$$

in formula (12.34). This factor has narrow and sharp maxima as a function of p_z. If we fix ω and H, these maxima will then correspond to p_z values at which

$$T\boldsymbol{k}\bar{\boldsymbol{v}} = \frac{k_z c}{eH} \left| \frac{\partial S(p_{zn})}{\partial p_z} \right| = 2\pi n.$$

Generally speaking, these values do not correspond to the extrema of $\partial S/\partial p_z$. But we can choose ω or H such that this condition will be satisfied for $(\partial S/\partial p_z)_{extr}$.

Then, the maximum in $C(p_z)$ will be much broader and its contribution to absorption will be much larger.

In the neighborhood of such an extremum, for example of $p_z = p_s$, we may expand the expression in the exponent in $C(p_z)$ in $(T\boldsymbol{k}\bar{\boldsymbol{v}})_{\text{extr}} - 2\pi n$:

$$C(p_z) \approx \frac{1}{\exp[\gamma + \mathrm{i}\Delta + \frac{1}{2}\mathrm{i}q(p_z - p_s)^2] - 1}$$

where we have introduced the notation $\gamma = T/\tau$, $\Delta = (T\boldsymbol{k}\bar{\boldsymbol{v}})_{\text{extr}} - 2\pi n$, $q = [\partial^2(T\boldsymbol{k}\bar{\boldsymbol{v}})/\partial p_z^2]_{p_s}$. Because of the narrowness of the maxima of $C(p_z)$ we may regard all the other functions as constants at $p_z = p_s$. Substituting $C(p_z)$ into eq. (12.34) and denoting

$$\varphi_\alpha = \int_0^{t_\alpha} (\boldsymbol{k}\boldsymbol{v})\, \mathrm{d}t_3$$

we obtain

$$\Gamma = \frac{eH}{c} \frac{1}{(2\pi\hbar)^3 I} \left| \sum_{t_\alpha} g_\alpha J_\alpha \exp(\mathrm{i}\varphi_\alpha) \right|^2_{p_s}$$

$$\times \operatorname{Re} \int \frac{\mathrm{d}p_z}{\exp[\gamma + \mathrm{i}\Delta + \frac{1}{2}\mathrm{i}q(p_z - p_s)^2] - 1}$$

The exponential in the denominator may be expanded at sufficiently low Δ values. The integral then can be taken, as a result of which we obtain

$$\operatorname{Re} \int \cdots \mathrm{d}p_z = \frac{\pi}{|q|^{1/2}} \left[\frac{(\gamma^2 + \Delta^2)^{1/2} - \Delta \operatorname{sign} q}{\gamma^2 + \Delta^2} \right]^{1/2}. \tag{12.39}$$

It is this factor that determines the resonance lineshape. The maximum is reached at $\Delta = \gamma \operatorname{sign} q/\sqrt{3}$, i.e., it is shifted relative to $\Delta = 0$. The line is asymmetrical. The line-width with respect to Δ is of order γ or, in terms of the magnetic field, $\Delta H \sim H_n(k_z l)^{-1}$, i.e., the absolute linewidth increases with decreasing number $H_n = H_1/n$. At the maximum the factor in brackets is of order $\gamma^{-1/2}$ and, hence, $\operatorname{Re}\int \cdots \mathrm{d}p_z \sim p_0(n\gamma)^{-1/2}$. The total absorption coefficient will be of order

$$\Gamma_n \sim \Gamma_0(n\gamma)^{-1/2} \gg \Gamma_0, \tag{12.40}$$

where Γ_0 is the absorption coefficient at $H = 0$. Since $\gamma = T/\tau$, $T \propto H^{-1}$ and $H_n \propto n^{-1}$, then $\gamma n \propto n^2$, i.e., the height of the maxima decreases with increasing n as n^{-1}. On the whole, the picture is as shown in fig. 84.

Let us now examine the role of the points $p_{z,\mathrm{b}}$. For simplicity, we take a centrosymmetric convex Fermi surface. As has been pointed out above, at $|p_z| < p_{z,\mathrm{b}}$ there are two points t_α where $\boldsymbol{k}\boldsymbol{v}(t_\alpha) = 0$, and at $|p_z| > p_{z,\mathrm{b}}$ there are no such points at all. Evidently, since at $p_{z,\mathrm{b}}$ two points with the same value of $\boldsymbol{k}\boldsymbol{v}$ merge, then it follows that at this point $\boldsymbol{k}\boldsymbol{v}'(t) = 0$, i.e., we have to use formula (12.33).

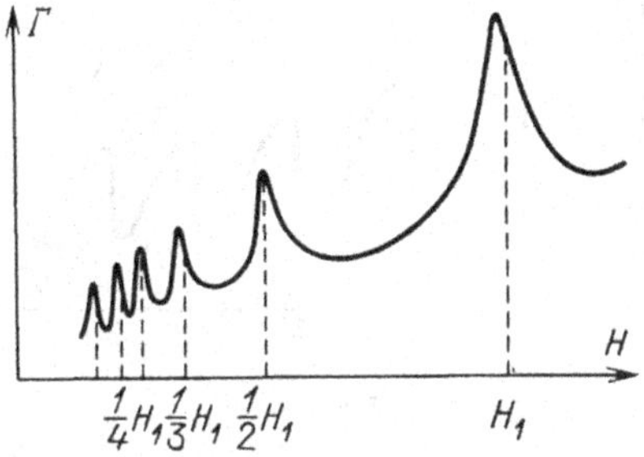

Fig. 84.

Consider the neighborhood of $p_{z,b}$. In the general formula (12.34) we may put $t_\alpha = t_\beta$ in the numerator and the factor $C(p_z)$ may be written in the form

$$C(p_z) \approx \frac{1}{\exp[\gamma + i\Delta + iq_1(p_z - p_{z,b})] - 1},$$

where this time

$$\Delta = T(k\bar{v})_b - 2\pi n, \qquad q_1 = \left(\frac{\partial(T(kv))}{\partial p_z}\right)_b.$$

The absorption coefficient has the form

$$\Gamma = \frac{eH}{cI}\frac{1}{(2\pi\hbar)^{-3}}|g_\alpha J_\alpha|^2_{p_{z,b}}$$

$$\times \mathrm{Re}\int^{p_{z,b}} dp_z \frac{dp_z}{\exp[\gamma + i\Delta + iq_1(p_z - p_{z,b})] - 1}. \qquad (12.41)$$

The integral appears to be equal to

$$\mathrm{Re}\int \cdots dp_z = \frac{\pi}{|q_1|}\left[\tfrac{1}{2} + \frac{\mathrm{sign}\ q_1}{\pi}\arctan\frac{\Delta}{\gamma}\right]. \qquad (12.42)$$

If Δ sign $q_1 > 0$ and $|\Delta| \gg \gamma$, then the term in brackets will be unity. But if Δ sign $q_1 < 0$, $|\Delta| \gg \gamma$, we obtain zero. Thus, the absorption coefficient undergoes jumps at $H = H_n = H_1/n$, where $H_1 = (k_z c/2\pi e)|\partial S/\partial p_z|_{p_{z,b}}$.

From eqs. (12.41), (12.42) and (12.33) we find the amplitude of the jumps:

$$\Delta\Gamma \sim \Gamma_0 n^{-2/3}, \qquad (12.43)$$

that is, the amplitude rises with increasing H. The picture, as a whole, is as shown in fig. 85 if $q_1 > 0$.

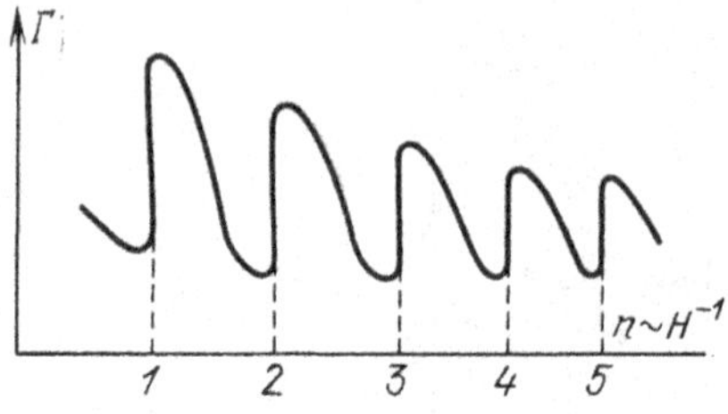

Fig. 85.

But if $q_1 < 0$, the jumps occur in the opposite direction: Γ increases discontinuously with increasing H.

Differentiating (12.42), we obtain the following formula:

$$\frac{\partial \Gamma}{\partial \ln H} \sim \frac{\Gamma_0 n^{1/3} \gamma}{\gamma^2 + \Delta^2}. \tag{12.44}$$

This is a curve with maxima at points $H = H_n$. The height of each maximum is

$$\frac{\mathrm{d}\Gamma}{\mathrm{d}H} \sim \frac{\Gamma_0 n^{1/3}}{\gamma H}, \tag{12.45}$$

that is, increases with decreasing H_n as $n^{1/3}$ (since $\gamma = T/\tau \propto n$).

As has been noted in section 8.5, the extremum of $T\boldsymbol{k}\bar{\boldsymbol{v}}$ corresponds, in particular, to the limiting point. Here, however, for the resonances to be observable, certain inequalities must be satisfied. According to fig. 81, for the limiting point to take part in magnetoacoustic resonance it is required that the contour $\boldsymbol{kv} = 0$ pass near it, and for this to be realized $\boldsymbol{k}$ must almost perpendicular to $\boldsymbol{H}$.

Let $\boldsymbol{k}$ lie in the (x, z) plane. We denote the angle between $\boldsymbol{k}$ and the z axis by φ. This angle must not be too small. This is because the displacement of an electron during a period in the direction k must be much larger than the wavelength. This leads to $v\varphi \gg \lambda$ or $\varphi \gg (kr_{\mathrm{L}})^{-1}$. On the other hand, at any finite φ the $\boldsymbol{kv} = 0$ contour no longer passes through the limiting point. It participates in the resonance only due to a small amount of smearing-out that results from the finiteness of its flight, i.e., from the presence of τ^{-1} in the exponent. The order of magnitude of this smearing-out is given by the relation

$$k_z(v - v_0) \sim \tau^{-1}.$$

But since $v_0 - v \sim v_0(1 - \cos\varphi) \sim v_0\varphi^2$ and $k_z \sim k\varphi$, we find $\varphi \ll (kl)^{-1/3}$. Thus, the condition for observation of magnetoacoustic resonance from the limiting point is

$$(kl)^{-1/3} \gg \varphi \gg (kr_{\mathrm{L}})^{-1}. \tag{12.46}$$

Note again the similarity of the entire situation with the effect of drift focusing of the high-frequency field (see condition 8.16).

In the case of a resonance at the limiting point we encounter a special situation where on one side there is a boundary extremum of $T(k\bar{v})$ and on the other side a boundary point arises where $\boldsymbol{k}\boldsymbol{v}'(t_\alpha)=0$. Therefore, this particular case needs special consideration (Kozub 1974)*. In line with the reasoning given above, not the limiting point itself but a certain neighborhood of this point contributes to the resonance. Throughout this region the points t_α and t_β do not coincide and therefore for J_α one may employ formula (12.32). On a contour with a specified p_z,

$$v_\perp \sim \frac{(p_0^2-p_z^2)^{1/2}}{m} \sim \frac{[p_0(p_0-p_z)]^{1/2}}{m}$$

(p_0 corresponds to the limiting point), i.e., $|J_\alpha|^2 \sim Tm/\{k[p_0(p_0-p_z)]^{1/2}\}$.

Hence, $|J_\alpha|^2$ depends strongly on p_z and therefore, instead of formula (12.41) we must write an analogous formula but with $|J_\alpha|^2$ under the integral sign. A simple calculation gives

$$\int \cdots |J_\alpha|^2 \, dp_z \sim \frac{Tm}{k(p_0|q_1|)^{1/2}} \left[\frac{(\Delta^2+\gamma^2)^{1/2} - \Delta \operatorname{sign} q_1}{\Delta^2+\gamma^2}\right]^{1/2}. \tag{12.47}$$

This formula is reminiscent of formula (12.42) in the sense that at $\Delta q_1>0$, $|\Delta| \gg \gamma$ we obtain a finite value and at $\Delta q_1<0$, $|\Delta|\gg\gamma$ we obtain zero, i.e., the absorption undergoes jumps. However, in the vicinity of the jumps, i.e., at $|\Delta|\ll\gamma$, formula (12.42) gives a value of order $Tm/[k(p_0q_1\gamma)^{1/2}]$, and in the case of an ordinary boundary point, according to (12.33) and (12.42), $|J_\alpha|^2\int\cdots dp_z \sim T^{4/3}/[(kv)^{2/3}q_1]$. Substituting the orders of magnitude, we find that the limiting point gives an amplitude $n^{-1/3}(kl)^{1/2}$ times larger than the ordinary boundary point. Comparing with formula (12.43), we see that before undergoing a jump the absorption coefficient attains a maximum which is of order

$$\Gamma_n \sim \frac{\Gamma_0 (kl)^{1/2}}{n}. \tag{12.48}$$

For not too large n this maximum by far exceeds the absorption coefficient in the absence of the field.

In conclusion, let us make certain that the use of formula (12.32) was legitimate. To this end it is required that $[kv'(t_\alpha)]^{1/2} \gg [kv''(t_\alpha)]^{1/3}$, i.e., $(kv_\perp/T)^{1/2} \gg (kv_\perp/T)^{1/3}$. Substituting here $v_\perp \sim [p_0(p_0-p_z)]^{1/2}/m$ and $p_0-p_z\sim\gamma/q_1$, we obtain the condition $kl\gg1$, which was assumed to have been fulfilled.

* The author is grateful to V. Kozub for the private communication of this unpublished result.

12.7. Nonlinear absorption of sound. Effect of the magnetic field

Just as in the case of the interaction of electrons with an electromagnetic field considered in section 12.5, nonlinear effects are of substantial interest in the absorp-absorption of sound. There are a number of such effects. They include the dependence of the absorption coefficient Γ on the sound intensity and the so-called electroacoustic effects. The latter consists of the generation of a constant current due to the conduction electrons being entrained by the travelling sound wave (or the appearance of an electric field when the circuit is broken). At low sound intensities the current is proportional to the intensity and at high intensities a more complicated dependence appears.

As has been pointed out in section 7.5, especially strong nonlinear effects arise under conditions where a small group of electrons play a significant role. The field of the sound wave, while being incapable of noticeably changing the entire distribution function, can distort it considerably for a small group of electrons. This situation occurs upon absorption of a short-wavelength sound ($kl \gg 1$) when only electrons in the vicinity of the contour at the Fermi surface on which $\boldsymbol{kv}=0$ interact with the sound wave. This interaction leads to a distortion of the momentum distribution of electrons, which in turn alters the absorption coefficient. This effect is called momentum nonlinearity (Gal'perin et al. 1972). We shall describe here only the qualitative theory of nonlinear absorption. The quantitative theory is described in original works (Gal'perin et al. 1972; Gal'perin and Kozub 1975).

Consider the effect of the wave on the motion of resonance particles. Due to the deformation interaction the sound wave produces an effective field which acts on electrons with a potential energy of $U=\Lambda_{ik}u_{ik}$. In this field the electrons break down into two groups: drifting electrons, which perform an infinite motion, and trapped electrons, whose trajectories are localized in potential wells of the effective field of the wave. The latter electrons perform a vibrational motion with a characteristic velocity given by

$$\tilde{v} \sim \left(\frac{U_0}{m}\right)^{1/2}$$

($U_0=\Lambda_{ik}u^0_{ik}$ is the amplitude of the potential) and a frequency

$$\omega_0 \sim \frac{\tilde{v}}{\lambda} \sim \tilde{v}k \sim \left(\frac{U_0}{m}\right)^{1/2} k. \tag{12.49}$$

However, each scattering event, while changing the direction of the velocity of the electron, removes it from the resonance group. Therefore, we may speak of trapping only when the mean free time is longer than the characteristic period of vibrations, i.e.,

$$\omega_0\tau \sim \left(\frac{U_0}{m}\right)^{1/2} k\tau \sim \left(\frac{U_0}{\mu}\right)^{1/2} kl \gtrsim 1, \tag{12.50}$$

where $\mu \sim mv^2$ is the Fermi energy. The appearance of a group of trapped particles implies a substantial distortion of the distribution in the resonance group and a change in the sound absorption coefficient brought about by the distortion.

The character of this change can be understood from very simple considerations. In the linear limit one may neglect the effect of the sound wave on the electron distribution (which is described by the equilibrium function f_0). If we pass to the coordinate associated with the wave, this distribution will be asymmetric with respect to velocities: the number of electrons "lagging" behind the wave will be larger than the number of those which "outrun" it. As a result, the wave imparts more energy to electrons than it receives from them, the absorption coefficient Γ_0 being proportional to the derivative $\partial f_0/\partial v_x$ (we chose the direction $\boldsymbol{k}$ as the x axis).

If a group of trapped electrons is formed, then during the time ω_0^{-1} the distribution over velocity v_x is "mixed", so that the preferred direction in the coordinate system associated with the wave is lost. If scattering were absent, the distribution in this coordinate system would become equilibrium and the energy absorption would stop. In the presence of scattering the absorption coefficient is nonzero and proportional to

$$\Gamma \sim \frac{\Gamma_0}{\omega_0 \tau}. \tag{12.51}$$

Since $\omega_0 \propto (\Lambda u_0)^{1/2} \propto I^{1/4}$, then Γ is proportional to $I^{-1/4}$ (I is the sound intensity)*.

Let us estimate the possibilities of observation of the effect. At present a sound intensity of $I \sim 1\ \mathrm{W/cm^2}$ is quite attainable. This corresponds to $u_{ik}^0 \sim ku_0 \sim (\omega/s) u_0 \sim (I/\rho s^3)^{1/2} \sim 10^{-5}$. Since $\Lambda \sim \mu \sim 10\ \mathrm{eV} \sim 10^{-11}$ erg, it follows that $\Lambda_{ik} u_{ik}^0 \sim 10^{-16}$ erg. Here $(U_0/\mu)^{1/2} \sim 3 \times 10^{-3}$ and, according to (12.50), $\omega_0 \tau \gtrsim$ if $kl \gtrsim 300$. If we take a sound frequency of $\nu \sim 10^9$ Hz, then $k \sim \nu/s \sim 10^4\ \mathrm{cm^{-1}}$ and a mean path of $l > 3 \times 10^{-2}$ cm is required, which is quite possible.

Let us now see how this absorption mechanism is influenced by the magnetic field (Gal'perin and Kozub 1975). As we shall see, even a very weak magnetic field leads to the suppression of the effect. In view of this, we shall not impose the condition $\Omega\tau \gtrsim 1$, which is necessary for the magnetic field to exert an appreciable influence on the absorption of sound in the linear theory. We assume that the field is not directed along the x axis (i.e., $\boldsymbol{k}$). Only those electrons interact effectively with the wave whose velocity is almost perpendicular to $\boldsymbol{k}$. Therefore, if $\boldsymbol{H}$ is not directed along x, the Lorentz force acting on electrons will contain a component along the axis x, of order $(e/c)\, vH$. This force brings about an increase in the velocity v_x, thereby leading to the withdrawal of electrons from the group of electrons trapped by the sound wave (as the velocity v_x increases their energy becomes higher than U_0). The removal of electrons is efficient if the Lorentz force is larger than or of the order of the force acting from the wave side. Hence, the field suppresses the

* With condition (12.50) fulfilled most of the electrons are trapped by the wave.

nonlinear absorption if $(e/c)vH \sim \partial U_0/\partial x \sim kU_0$ or

$$\left(\frac{U_0}{\mu}\right) kr_L \sim 1. \tag{12.52}$$

As has been found above, in the absence of a field the absorption occurs due to scattering events with an effective time τ. Let us now introduce the "magnetic" time τ_1, which is required for the Lorentz force to accelerate an electron to a velocity $\tilde{v}$. From the equation of motion we have

$$\frac{d}{dt} m\tilde{v} \sim \frac{m\tilde{v}}{\tau_1} \sim \frac{e}{c} vH.$$

Hence (see formula 12.49),

$$\tau_1 \sim \frac{\omega_0}{\Omega \mathrm{kv}}. \tag{12.53}$$

At $\tau_1 < \tau$ the removal of electrons from the group of trapped electrons due to the magnetic field is more important than that brought about by collisions. The condition $\tau_1 < \tau$ may be written in the form

$$\left(\frac{U_0}{\mu}\right)^{1/2} \frac{r_L}{l} < 1. \tag{12.54}$$

Since $U_0/\mu \ll 1$ and $r_L/l \sim (\Omega\tau)^{-1}$, condition (12.54) is much less stringent than $\Omega\tau > 1$ and can be fulfilled with rather small fields.

An estimate of the absorption coefficient in the nonlinear regime has the form

$$\Gamma \sim [\omega_0 \min(\tau_1, \tau)]^{-1} \Gamma_0. \tag{12.55}$$

If $\tau_1 < \tau$, the absorption coefficient is proportional to H and to $I^{-1/2}$. The absorption coefficient increases with the field until $\omega_0\tau_1 \gtrsim 1$, following which Γ approaches Γ_0, the value for linear absorption. The condition $\omega_0\tau_1 \sim 1$ coincides with eq. (12.52) after expressions (12.53) and (12.49) are substituted.

Let us evaluate the magnitude of the magnetic field at which it begins to exert a substantial effect on the absorption of sound. From the conditions $\omega_0\tau_1 > 1$ and $\tau_1 < \tau$ we have $\omega_0\tau > 1$. On the other hand, the condition $\tau_1 < \tau$ gives: $\omega_0 < \Omega\, kv\tau \sim \Omega kl$. Hence, in any case $\Omega kl\tau > 1$ or $(\Omega/v)kl^2 > 1$. But since $v/\Omega \sim r_L$, we obtain

$$r_L < kl^2. \tag{12.56}$$

If we take a sound frequency of the order 10^9 Hz, then $k \sim 10^4\ \mathrm{cm}^{-1}$. In a pure metal it is possible to make $l \sim 10^{-1}$ cm. As a result, we obtain $r_L < 10^2$ cm and $H > 10^{-1}$ Oe. Hence, in this case, the magnetic field of the Earth proves to be sufficient for the suppression of nonlinear absorption.

In conclusion, we note that throughout this section the motion of electrons has been described by the classical Newton equation. This is justified if the number of quantum levels inside the potential well of the sound wave is large. The appropriate condition has the following form:

$$\frac{1}{m}\left(\frac{\hbar}{\lambda}\right)^2 \sim \frac{(\hbar k)^2}{m} \ll U_0. \tag{12.57}$$

There is another possibility of suppressing the quantum effects, namely, when the smearing out of the levels due to collisions is larger than the spacing between them. The relevant condition has the form

$$\frac{(\hbar k)^2}{m} \ll \frac{\hbar}{\tau}. \tag{12.58}$$

When one of these conditions is fulfilled, the problem becomes classical.

The above theoretical results pertaining to the dependence of the absorption coefficient on the sound intensity and to the suppression of nonlinear absorption by a weak magnetic field have been confirmed by experimental data (see the review article by Gal'perin et al. 1979).

12.8. *Giant oscillations of sound absorption due to quantization of levels in a magnetic field*

Up until now, while speaking of the effect of the magnetic field on the absorption of sound, we confined ourselves to the classical theory. Let us now see, first qualitatively and then quantitatively, what happens when account is taken of the quantization of the levels in a magnetic field. Evidently, it manifests itself only at temperatures $T \ll \hbar\Omega$.

In the free-electron model the levels are described by the formula

$$\varepsilon = \hbar\Omega(n+\tfrac{1}{2}) + \frac{p_z^2}{2m}.$$

Upon absorption of a phonon the conservation laws give

$$p_z' = p_z + \hbar k_z,$$

$$\hbar\Omega(n'+\tfrac{1}{2}) + \frac{p_z'^2}{2m} = \hbar\Omega(n+\tfrac{1}{2}) + \frac{p_z^2}{2m} + \hbar\omega.$$

Assuming k_z to be small, we obtain from the last expression

$$\hbar\Omega(n'-n) + \frac{p_z \hbar k_z}{m} = \hbar\omega. \tag{12.59}$$

If $n \neq n'$, then the term $\hbar\Omega(n'-n)$ will be very large compared to the remaining ones, and therefore condition (12.59) can be fulfilled only at $n'=n$. From this we obtain

$$\frac{p_z \hbar k_z}{m} = \hbar\omega$$

or

$$p_z = \frac{m\omega}{k_z} = \frac{msk}{k_z} = \frac{ms}{\cos\theta}. \tag{12.60}$$

The value of p_z given by this relation depends on θ. By varying $\cos\theta$ we can increase or decrease the value of p_z at which the absorption takes place.

If we plot the energy as a function of p_z, we shall obtain a system of parabolas (fig. 86). If $T \ll \hbar\Omega \ll \mu$, then only those electrons will participate in absorption which lie in a band of width T in the vicinity of μ. The momenta of these electrons lie in the intervals 1, 2, 3 in fig. 86. Suppose the value of p_z that satisfies condition (12.60) is P. If P is located as shown in fig. 86, then this is a "forbidden region" for the absorption of sound, i.e., the absorption will be negligible. If now we begin to increase the field, all the parabolas will go upward. This will lead to the movement of the intervals 1, 2, and 3 along the axis p_z. Eventually, the point P will be found in one of these intervals. In this case, absorption is possible. The distinction between the "allowed" and "forbidden" absorption of sound is very large, and therefore these oscillations of Γ with the magnetic field have come to be known as giant quantum oscillations (Gurevich et al. 1961).

We shall now consider this effect from a quantitative point of view. The energy dissipated per unit time in 1 cm^3 is given by formula analogous to (12.18):

$$Q = \frac{\pi}{2hV} \sum_{a,a'} [f_0(\varepsilon_a) - f_0(\varepsilon_{a'})] \hbar\omega_{a'a} |U_{a'a}|^2 \delta(\hbar\omega_{a'a} - \hbar\omega). \tag{12.61}$$

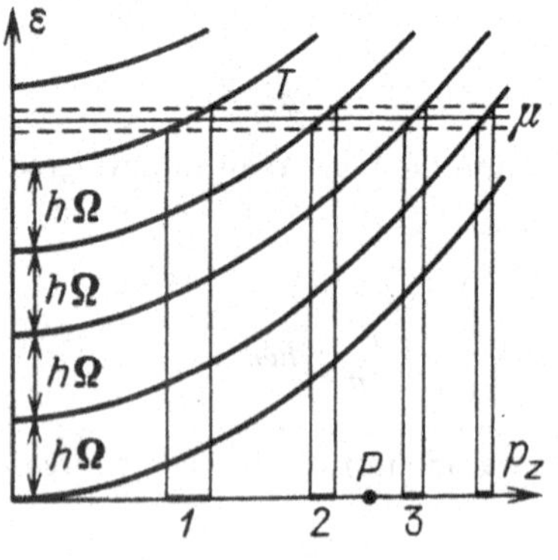

Fig. 86.

The subscripts a signify p_y, p_z and n *. If $\hbar\omega \ll T$, we can expand the distribution function in the energy difference:

$$\varepsilon_{a'} - \varepsilon_a \equiv \hbar\omega_{a'a} = \hbar\omega.$$

so we have

$$Q = -\frac{\pi}{2hV}\sum_{a,a'}\frac{\partial f_0}{\partial\varepsilon}(\hbar\omega)^2|U_{a'a}|^2\delta(\hbar\omega_{a'a} - \hbar\omega). \tag{12.62}$$

We had before

$$U = \Lambda_{ik}u_{ik}^0\,e^{i\boldsymbol{kr}}.$$

In a general case Λ_{ik} now depends on the momentum, but here, for simplicity, we assume it to be a constant. So, in this case we have

$$U_{a'a} = \Lambda_{ik}u_{ik}^0\,\delta_{nn'}\delta_{p'_y,p_y+\hbar k_y}\delta_{p'_z,p_z+\hbar k_z}.$$

The other quantities depend only on ε and are independent of p_y. Here, $\delta_{p'_y,p_y+\hbar k_y}$ is the Kronecker delta, and therefore $\delta^2 = \delta$ and $\sum_{p''_y}\delta^2 = 1$. The remaining sum over p_y reduces to the multiplication by the number of states with different p_y values. This yields

$$\sum_{p_y} = \frac{L_2\Delta p_y}{2\pi\hbar} = \frac{L_2L_1eH}{2\pi\hbar c}.$$

The summation over p'_z leads to the replacement $p'_z \to p_z + \hbar k_z$ in $\delta(\omega_{a'a} - \omega)$. This being done, we pass from the sum to an integral: $\sum_{p_z} \to L_3(2\pi\hbar)^{-1}\int dp_z$. Thus,

$$Q = -\frac{\pi\omega^2}{2}\left(\frac{\Lambda_{ik}u_{ik}^0}{2\pi\hbar}\right)^2\frac{eH}{c}\sum_n\int dp_z\frac{\partial f_0}{\partial\varepsilon}\delta\left(\frac{k_zp_z}{m} - \omega\right). \tag{12.63}$$

Here use is made of the free-electron model; in that case $\varepsilon = p_z^2/2m + \hbar\Omega(n+\frac{1}{2}) \pm \hbar\Omega \to p_z^2/2m + \hbar\Omega n$ $(n = 0, 1, \ldots)$ provided that all levels with $n > 0$ are doubly degenerate. We assume that $\Omega \gg \omega$ and $\Omega \gg kp_z/m$. Then, the condition $\omega_{a'a} = \omega$ can be satisfied only for $n = n'$. We shall also assume that $T \ll \hbar\Omega \ll \mu$. Substituting into relation (12.63),

$$\frac{\partial f_0}{\partial\varepsilon} = -\frac{1}{4T}\cosh^{-2}\left[\frac{(\varepsilon - \mu)}{2T}\right].$$

we obtain the absorption coefficient in the form

$$\Gamma = \frac{Q}{\frac{1}{2}\rho\omega^2u_0^2s} = \frac{\pi\omega^2}{(2\pi\hbar)^2}\frac{eH}{c}\frac{(\Lambda_{ik}u_{ik}^0)^2}{\frac{1}{2}\rho\omega^2u_0^2s}\frac{1}{4T}$$
$$\times\sum\int dp_z\,\delta\left(\frac{k_zp_z}{m} - \omega\right)\cosh^{-2}\left(\frac{p_z^2/2m + \hbar\Omega n - \mu}{2T}\right).$$

* Moreover, the spin projection must also be specified. We assume that the number n enumerates both the Landau levels and the spin projections, especially because in heavy metals, due to the strong spin orbit coupling, the level numbering is more complicated and the spin is not separated in a pure form.

With $H \to 0$ this will yield the absorption in the absence of a field. Therefore, we may write:

$$\Gamma = \Gamma_0 \frac{k_z}{m} \frac{\hbar \Omega}{4T} \sum_n \int \mathrm{d}p_z \frac{\delta(k_z p_z/m - \omega)}{\cosh^2[(\hbar\Omega n + p_z^2/2m - \mu)2T]}. \tag{12.64}$$

Indeed, with $\Omega \to 0$ the sum over n is replaced by an integral, and $\Gamma \to \Gamma_0$. The integral over p_z gives the value of p_z in accordance with formula (12.60). Assuming $p_z^2/2m \ll \hbar\Omega$, we can simplify the formula for Γ even further:

$$\Gamma = \Gamma_0 \frac{\hbar\Omega}{4T} \sum_n \cosh^{-2}\left(\frac{\hbar\Omega n - \mu}{2T}\right). \tag{12.65}$$

If H is very large (so that $\hbar\Omega \gg T$), then in a case where one of the arguments $\hbar\Omega n$ is close to μ,

$$\Gamma \approx \frac{\Gamma_0 \hbar\Omega}{4T} \gg \Gamma_0. \tag{12.66}$$

But if all $\hbar\Omega n$ are far from μ, then

$$\Gamma = \Gamma_0 \frac{\hbar\Omega}{4T} \mathrm{e}^{-\alpha\hbar\Omega/T} \ll \Gamma_0 \tag{12.67}$$

($\alpha \sim 1$). Thus, Γ as a function of H undergoes very strong oscillations located uniformly along the scale of H^{-1}:

$$\Delta(H^{-1}) = \frac{e\hbar}{mc\mu},$$

The sharp and high maxima are separated by broad minima, where the absorption of sound is negligibly small. However, the exact value of the absorption coefficient at the minimum can be determined only if account is taken of the scattering of electrons. This effect can be estimated using the following procedure. The presence of scattering gives rise to an uncertainty of the order of $\hbar/\tau$ in the electron energy. Therefore, we replace the δ-function in (12.64), which expresses the energy conservation law by a somewhat smeared-out function with a width τ^{-1}. Usually, $\omega \ll \tau^{-1}$ and therefore ω may be neglected. Instead of the δ-function we take

$$\frac{1}{\pi\tau} \frac{1}{\tau^{-2} + \left(\dfrac{k_z p_z}{m}\right)^2}.$$

Now we introduce a new variable, $p_z/(2mT)^{1/2} = y$. The expression for Γ assumes the form

$$\Gamma = \Gamma_0 \frac{\hbar\Omega}{4T} \int \mathrm{d}y \frac{1}{\pi} \frac{B}{1 + B^2 y^2}$$
$$\times \sum_z \cosh^{-2}\left(\frac{y^2 - A_n}{2}\right), \tag{12.68}$$

where

$$B = \left(\frac{2T}{m}\right)^{1/2} k_z \tau, \qquad A_n = \frac{\mu - \hbar\Omega n}{T}.$$

In the integral (12.68) both factors are rapidly varying. The factor $\sum_n \cosh^{-2}$ exhibits many maxima with a width of the order of unity separated by intervals of order $(\hbar\Omega/T)^{1/2}$. The term $B/(1+B^2y^2)$ has a maximum of width B^{-1} at $y=0$.

Consider a case where the width of the curve $B(1+B^2y^2)$ is much smaller that the separation between the two maxima of the other factor (fig. 87). For this purpose it is required that

$$B\left(\frac{\hbar\Omega}{T}\right)^{1/2} \gg 1 \quad \text{or} \quad kl\left(\frac{\hbar\Omega}{\mu}\right)^{1/2} \gg 1. \tag{12.69}$$

For ordinary metals the condition (12.69) is very stringent, since for $H \sim 10^4$ Oe, we have $(\hbar\Omega/\mu)^{1/2} \sim 10^{-2}$, so that a very small wavelength is required. If we take a very pure metal with $l \sim 0.1$ cm, then k must approximately be equal to 10^3 or $\omega = ks \sim 10^8\ \mathrm{s}^{-1}$. This is a rather high frequency for sound, though attainable at present. In semi-metals $m^* \sim 0.01\ m$, $p_0 \sim 0.01\ \hbar/a$, which makes it possible to lower the limiting frequency by two orders of magnitude.

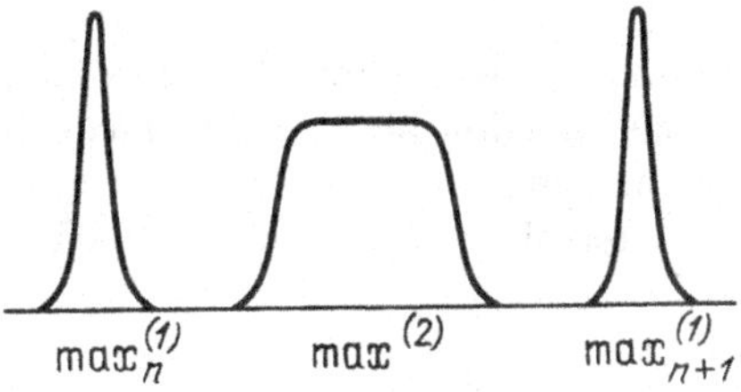

Fig. 87.

Let H be such that one of the maxima of the factor $\sum_n \cosh^{-2}$ appears to be in the position $y=0$. Then, we have one of the maxima of Γ.

If the width B^{-1} were small compared to unity, the factor $(B/\pi)(1+b^2y^2)^{-1}$ could be replaced by the δ-function. This would give formula (12.66). If, however, $B \leqslant 1$, the absorption lineshape will be quite different.

In the case of $B \ll 1$ we obtain

$$\Gamma_{\max} \sim B\frac{\hbar\Omega}{T}\Gamma_0 \sim \hbar\Omega(T\mu)^{-1/2} kl\Gamma_0. \tag{12.70}$$

Let us now consider the minimum. It will evidently occur if the peaks of the first function nearest to $y=0$ are separated maximally from $y=0$. Obviously, the corresponding $A_{n0} \sim \hbar\Omega/T$. Since the width of the functions $\cosh^{-2}$ is of the order of unity, under these conditions they may be replaced by δ-functions. Thus, we find

$$\Gamma_{\min} \sim \Gamma_0 \left(\frac{\hbar\Omega}{T}\right) \frac{1}{BA_{n0}^{3/2}} \sim \Gamma_0 \left(\frac{\mu}{\hbar\Omega}\right)^{1/2} \frac{1}{kl}. \tag{12.71}$$

Thus, for the two cases $B \gg 1$ and $B \ll 1$ we have

$$\begin{aligned} \frac{\Gamma_{\max}}{\Gamma_{\min}} &\sim \frac{\hbar\Omega}{T}\left(\frac{\hbar\Omega}{\mu}\right)^{1/2} kl \quad \text{if} \quad \left(\frac{T}{\mu}\right)^{1/2} kl \gg 1, \\ \frac{\Gamma_{\max}}{\Gamma_{\min}} &\sim \frac{\hbar\Omega}{\mu}\left(\frac{\hbar\Omega}{T}\right)^{1/2} (kl)^2 \quad \text{if} \quad \left(\frac{T}{\mu}\right)^{1/2} kl \ll 1. \end{aligned} \tag{12.72}$$

Up until now we have been concerned with free electrons, which made it possible to estimate the order of magnitude of the effect. Let us now find out when the maxima occur in an anisotropic metal. We obtain

$$\varepsilon(p_z + \hbar k_z, n) - \varepsilon(p_z, n) = \hbar\omega$$

or

$$v_z \hbar k_z = \hbar\omega.$$

Hence,

$$v_z \sim s \ll v.$$

This means that the important cross sections of the Fermi surface are those where the average velocity in the direction of the field practically vanishes. But such a property is exhibited in the general case only by the central cross section of the Fermi surface. Hence, the interval between the absorption maxima will be determined by the relation

$$\Delta(H^{-1}) = 2\pi \frac{e\hbar}{(cS(\mu))_{\text{cent}}}.$$

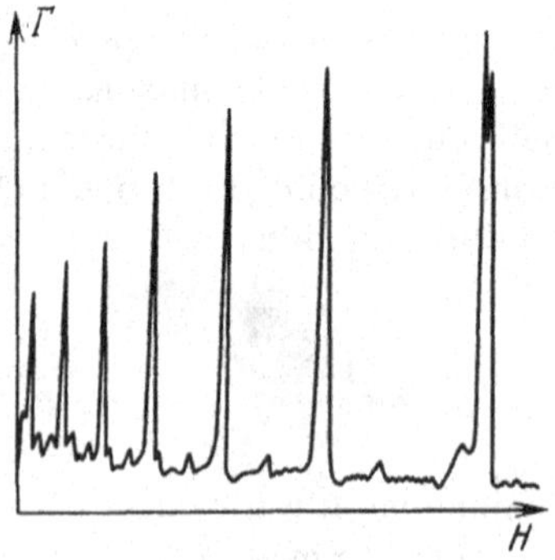

Fig. 88.

Since the maxima are very sharp, this method enables one to deduce the quantum levels themselves. This is especially important for the low-lying levels, where the quasiclassical approximation does not apply, and the levels are not, generally speaking, equidistant. Of course, this is possible only for small portions of the Fermi surface with small masses or in semi-metals.

At present the effect of giant oscillations has been detected in a number of substances (for example, in gallium, see fig. 88) (Shapira and Lax 1965).

13

Fermi-liquid effects

13.1. Interaction of quasiparticles

In ch. 2, while speaking of the correspondence between a liquid consisting of Fermi particles (a Fermi liquid) and an ideal Fermi gas, we deliberately omitted one important point. According to the Landau theory, there is one important distinction between the spectra of the quasiparticles of the Fermi liquid and those of the quasiparticles of the Fermi gas. While in the case of the Fermi gas the shape of the energy spectrum (2.6) is determined by the energy of a free particle alone, in the Fermi liquid a significant role is played by the interaction with other quasiparticles, which is not weak, generally speaking.

In ch. 2 we have found out that that this interaction does not lead to a strong damping if the energy of the quasiparticle is close to the Fermi energy. But this was associated not with the weakness of the interaction but with the form of the Fermi spectrum (the presence of an occupied Fermi sphere). The fact that the interaction is not weak manifests itself, in particular, in that the effective mass of quasiparticles may be appreciably different from the mass of free particles (for example, in liquid ^{3}He at a low temperature $m^* = 3m_{^3\text{He}}$: in metals this difference is smaller: see ch. 14).

According to the idea expressed by Landau (1956), the interaction of quasiparticles may be introduced as a certain self-consistent field generated by surrounding quasiparticles, which acts on a given quasiparticle. But here the energy of the quasiparticle will evidently depend on the state of other quasiparticles, i.e., in other words, it will be a functional of their distribution function*.

In such a case, the following questions arise. First, how can this energy be determined? Second, what is the equilibrium distribution function? Third, how can the nonequilibrium function be found? In other words, strictly speaking, one has to reconsider all the derivations and results obtained before, since they have, in fact, used the gas model without taking account of the dependence of the energy spectrum on the distribution function.

As will be shown in this chapter, this new circumstance, in effect, does not prove to be important, with rare exceptions. In some cases it leads to numerical differences

* In this chapter, the term "quasiparticles" is used to describe particles of the gas model of the Fermi liquid. But, in contrast to the preceding chapters, the interaction of quasiparticles will be taken into account.

but does not alter the order of magnitude of the result. But there are phenomena which occur exclusively due to the dependence of the spectrum on the distribution function. Such phenomena are known as Fermi-liquid effects. For the sake of simplicity, we shall consider the isotropic model in the following.

In this chapter we shall use a different notation for the distribution function, namely $n(\boldsymbol{p}, \boldsymbol{r})$. This is because the letter f in the theory of the Fermi liquid always denotes the Landau function (see below) and the use of another symbol for the latter seems to be unreasonable.

We begin with the energy of quasiparticles. Let the total energy of the system be $E[n]$, where $[n]$ signifies the functional dependence on the distribution function. It is natural to define the energy of quasiparticles as follows.

With a small change in the distribution function δn the variation of the total energy per unit volume is given by

$$\delta E = \sum_{\sigma} \int \varepsilon(\boldsymbol{p}, \boldsymbol{\sigma})\, \delta n(\boldsymbol{p}, \boldsymbol{\sigma}) \frac{\mathrm{d}^3 \boldsymbol{p}}{(2\pi\hbar)^3}, \tag{13.1}$$

where $\sum_{\sigma}$ denotes the summation over spin projections. The quantity $\varepsilon(\boldsymbol{p}, \boldsymbol{\sigma})$ is naturally regarded as the energy of quasiparticles. Indeed, if one quasiparticle of momentum p_1 and spin projection σ_1 appears, then $\delta n(\boldsymbol{p}, \boldsymbol{\sigma}) = (2\pi\hbar)^3 \delta(\boldsymbol{p} - \boldsymbol{p}_1) \delta_{\sigma\sigma_1}$. But in this case $\delta E = \varepsilon(\boldsymbol{p}_1, \boldsymbol{\sigma}_1)$.

Generally speaking, along with the momentum and spin, the coordinate should be introduced here. However, actually, because of the wavelength of electrons being of the order of interatomic distances, the inhomogeneities of the distribution function under consideration, with characteristic dimensions much larger than the interatomic distances, will not affect the energy of quasiparticles.

It is convenient to write formula (13.1) in a different form. If the spin of the electron plays a significant role in the problem, then, instead of this distribution function, one has to employ a so-called statistical operator or a density matrix: $n(p, \sigma)$. Here, the average value of any quantity, to which the operator A dependent on the spin and momentum operators corresponds, is expressed by the formula

$$\bar{A} = \mathrm{Sp}_{\sigma} \int A n(\boldsymbol{p}, \boldsymbol{\sigma}) \frac{\mathrm{d}^3 \boldsymbol{p}}{(2\pi\hbar)^3}. \tag{13.2}$$

In particular, formula (13.1) may also be written in this form:

$$\delta E = \mathrm{Sp}_{\sigma} \int \varepsilon(\boldsymbol{p}, \boldsymbol{\sigma})\, \delta n(\boldsymbol{p}, \boldsymbol{\sigma}) \frac{\mathrm{d}^3 \boldsymbol{p}}{(2\pi\hbar)^3}. \tag{13.3}$$

The determination of the energy of quasiparticles in accordance with (13.3), as we shall prove, gives a Fermi distribution in the case of equilibrium. To this end, we write the formula for the entropy:

$$S = -\mathrm{Sp}_{\sigma} \int \{n \ln n + (1-n) \ln(1-n)\} \frac{\mathrm{d}^3 \boldsymbol{p}}{(2\pi\hbar)^3}. \tag{13.4}$$

The validity of this formula as applied to quasiparticles in a Fermi liquid is proved by the fact that it is of purely combinatorial origin, and for quasiparticles the Pauli principle is valid just as for particles of a Fermi gas. Let us now find the maximum of the entropy, requiring that the number of particles and the total energy be constant. This can be done by seeking the maximum of the expression

$$S' = S + \alpha E + \beta N,$$

where α and β are undetermined Lagrange multipliers. By varying this expression with respect to n, from the condition $\delta S' = 0$ we find the equilibrium distribution function:

$$n_0(\varepsilon) = n_{\mathrm{F}}(\varepsilon) = [\mathrm{e}^{(\varepsilon-\mu)/T} + 1]^{-1}, \tag{13.5}$$

where μ and T are expressed in terms of α and β.

Formulae (13.5) and (13.3) immediately prove the correctness of the definition of heat capacity in section 2.4. Formula (2.18) corresponds only formally to (13.3). In fact, there is a distinction between these formulas, since eq. (2.18) ignores the variation of ε upon variation of the distribution function with temperature. However, this gives only small corrections to the heat capacity, of the order of $(T/\mu)^2$.

13.2. The Landau function

Although the dependence of the spectrum on the distribution function does not manifest itself in the heat capacity, it nevertheless does occur. If E is a functional of n, then this is also valid for ε. With a slight variation of the distribution function the function ε acquires a correction, which may be written in the form

$$\delta\varepsilon(\boldsymbol{p}, \boldsymbol{\sigma}) = \mathrm{Sp}_{\sigma'} \int f(\boldsymbol{p}, \boldsymbol{\sigma}; \boldsymbol{p}', \boldsymbol{\sigma}')\, \delta n(\boldsymbol{p}', \boldsymbol{\sigma}') \frac{\mathrm{d}^3 p}{(2\pi\hbar)^3}. \tag{13.6}$$

Since the function f is a second variational derivative of the total energy, it is symmetrical with respect to the transpose of the arguments, i.e.,

$$f(\boldsymbol{p}, \boldsymbol{\sigma}; \boldsymbol{p}', \boldsymbol{\sigma}') = f(\boldsymbol{p}', \boldsymbol{\sigma}'; \boldsymbol{p}, \boldsymbol{\sigma}).$$

The existence of relation (13.6) manifests itself, first of all, in the kinetic equation. In section 3.2 in deriving the kinetic equation we had

$$\frac{\mathrm{d}n}{\mathrm{d}t} = \frac{\partial n}{\partial t} + \left(\frac{\partial n}{\partial \boldsymbol{r}}\right)\frac{\mathrm{d}\boldsymbol{r}}{\mathrm{d}t} + \left(\frac{\partial n}{\partial \boldsymbol{p}}\right)\frac{\mathrm{d}\boldsymbol{p}}{\mathrm{d}t}.$$

The quantity $\mathrm{d}\boldsymbol{p}/\mathrm{d}t$ was determined by the acting force. It was presumed that the only force that arises comes from the external field. However, in the case under consideration, due to relation (13.6), the energy of quasiparticles begins to depend on the coordinate via the distribution function n. This implies the appearance of a potential energy, which describes the action on a given quasiparticle of the self-consistent field of the other quasiparticles. Hence, we must now write

$$\frac{\mathrm{d}\boldsymbol{p}}{\mathrm{d}t}=\mathrm{e}\boldsymbol{E}-\frac{\partial\varepsilon}{\partial\boldsymbol{r}}$$

$$=e\boldsymbol{E}-\mathrm{Sp}_{\sigma'}\int f(\boldsymbol{p},\boldsymbol{\sigma};\boldsymbol{p}',\boldsymbol{\sigma}')\frac{\partial n}{\partial\boldsymbol{r}}\frac{\mathrm{d}^3\boldsymbol{p}'}{(2\pi\hbar)^3}.$$

Thus, an additional term appears on the left-hand side of the kinetic equation. If we assume that $n=n_0(\varepsilon_0)+n_1$, where ε_0 is the energy at equilibrium at $T=0$, and retain only terms linear in n_1, then the additional term has the form

$$-\frac{\partial n_0}{\partial\varepsilon}\boldsymbol{v}\,\mathrm{Sp}_{\sigma'}\int f(\boldsymbol{p},\boldsymbol{\sigma};\boldsymbol{p}',\boldsymbol{\sigma}')\frac{\partial n_1}{\partial\boldsymbol{r}}\frac{\mathrm{d}^3\boldsymbol{p}'}{(2\pi\hbar)^3}.$$

Finally, if we seek n_1 in the form $n_1=-\psi\partial n_0/\partial\varepsilon$, then by dividing the entire equation by $\partial n_0/\partial\varepsilon$ we obtain the additional term in the equation for ψ in the form

$$-\frac{\boldsymbol{v}}{(2\pi\hbar)^3}\mathrm{Sp}_{\sigma'}\int f(\boldsymbol{p},\boldsymbol{\sigma};\boldsymbol{p}',\boldsymbol{\sigma}')\frac{\partial}{\partial\boldsymbol{r}}\psi(\boldsymbol{r},\boldsymbol{p}',\boldsymbol{\sigma}')\frac{\mathrm{d}S'}{v},$$

where the momenta $\boldsymbol{p}$ and $\boldsymbol{p}'$ are on the Fermi surface.

Thus, the kinetic equation may be written in the form

$$\frac{\partial\psi}{\partial t}+\boldsymbol{v}\frac{\partial}{\partial\boldsymbol{r}}\left(\psi+\mathrm{Sp}_{\sigma'}\int f(\boldsymbol{p},\boldsymbol{\sigma};\boldsymbol{p}',\boldsymbol{\sigma}')\,\psi(\boldsymbol{p}',\boldsymbol{\sigma}')\frac{\mathrm{d}S'}{v(2\pi\hbar)^3}\right)$$
$$-e\boldsymbol{v}\boldsymbol{E}=I(\psi). \tag{13.7}$$

In the presence of not too strong a magnetic field one more term appears on the left-hand side of the kinetic equation:

$$\frac{e}{c}\left[\frac{\partial\varepsilon}{\partial\boldsymbol{p}}\boldsymbol{H}\right]\frac{\partial n}{\partial\boldsymbol{p}}.$$

In the zeroth approximation this gives zero since n_0 depends only on ε. In the first approximation one has to take into account the term with ψ not only in $\partial n/\partial\boldsymbol{p}$ but also in $\partial\varepsilon/\partial\boldsymbol{p}$. A simple calculation shows that this leads to the appearance in eq. (13.7) of the following term:

$$\frac{e}{c}[\boldsymbol{v}\boldsymbol{H}]\frac{\partial}{\partial\boldsymbol{p}}\left(\psi+\mathrm{Sp}_{\sigma'}\int f(\boldsymbol{p},\boldsymbol{\delta};\boldsymbol{p}',\boldsymbol{\sigma}')\psi(\boldsymbol{p}',\boldsymbol{\sigma}')\frac{\mathrm{d}S'}{v(2\pi\hbar)^3}\right).$$

The operator $(e/c)[\boldsymbol{v}\boldsymbol{H}]\partial/\partial\boldsymbol{p}$ is nothing else than $\partial/\partial t_1$, where t_1 is the variable introduced in section 5.1.

Consider the collision integral. It is obvious that it vanishes only if $n_0(\varepsilon)$ with a real energy ε rather than $n_0(\varepsilon_0)$ is substituted into it. We may write

$$n = n_0(\varepsilon_0) + n_1 = n_0(\varepsilon) - \frac{\partial n_0}{\partial \varepsilon} \mathrm{Sp}_{\sigma'} \int f n_1 \frac{\mathrm{d}^3 \boldsymbol{p}'}{(2\pi\hbar)^3} + n_1 .$$

Substituting here $n_1 = -\psi \partial n_0 / \partial \varepsilon$, we obtain

$$n = n_0(\varepsilon) - \frac{\partial n_0}{\partial \varepsilon} \left[\psi + \mathrm{Sp}_{\sigma'} \int f\psi \frac{\mathrm{d}S'}{v(2\pi\hbar)^3} \right].$$

Since the collision integral vanishes upon substitution of $n_0(\varepsilon)$, it evidently follows that it will depend on the function

$$\varphi = \psi + \mathrm{Sp}_{\sigma'} \int \frac{f\psi \, \mathrm{d}S'}{v(2\pi\hbar)^3} \tag{13.8}$$

in the same way as it was dependent on ψ at $f = 0$.

Let us now find the expression for the electric current in terms of ψ:

$$\begin{aligned} \boldsymbol{j} &= e \, \mathrm{Sp}_{\sigma} \int \frac{\partial \varepsilon}{\partial \boldsymbol{p}} n \frac{\mathrm{d}^3 \boldsymbol{p}}{(2\pi\hbar)^3} \\ &= e \, \mathrm{Sp}_{\sigma} \int \boldsymbol{v} n_1 \frac{\mathrm{d}^3 \boldsymbol{p}}{(2\pi\hbar)^3} + e \, \mathrm{Sp}_{\sigma,\sigma'} \int n_0 \frac{\partial}{\partial \boldsymbol{p}} \int f n_1 \frac{\mathrm{d}^3 \boldsymbol{p}}{(2\pi\hbar)^3} \frac{\mathrm{d}^3 \boldsymbol{p}'}{(2\pi\hbar)^3} \end{aligned}$$

($\boldsymbol{v} = \partial \varepsilon_0 / \partial \boldsymbol{p}$). Taking the integral over $\boldsymbol{p}$ by parts, we have

$$\boldsymbol{j} = e \, \mathrm{Sp}_{\sigma} \int \boldsymbol{v} n_1 \frac{\mathrm{d}^3 \boldsymbol{p}}{(2\pi\hbar)^3} - e \, \mathrm{Sp}_{\sigma,\sigma'} \int \frac{\partial n_0}{\partial \varepsilon} \boldsymbol{v} \int f n_1 \frac{\mathrm{d}^3 \boldsymbol{p}}{(2\pi\hbar)^3} \frac{\mathrm{d}^3 \boldsymbol{p}'}{(2\pi\hbar)^3}.$$

Finally, substituting $n_1 = -\psi \partial n_0 / \partial \varepsilon$, we have

$$\boldsymbol{j} = e \, \mathrm{Sp}_{\sigma} \int \left[\psi + \mathrm{Sp}_{\sigma'} \int f\psi \frac{\mathrm{d}S'}{v(2\pi\hbar)^3} \right] \boldsymbol{v} \frac{\mathrm{d}S}{v(2\pi\hbar)^3}. \tag{13.9}$$

Thus, it turns out that with the f-function taken into account the expression for the current and the collision integral depend on the combination of φ in (13.8) in the same way as they were dependent on ψ at $f = 0$. It can easily be shown that this is valid for the heat flux as well. If ψ is independent of the time or ω is much smaller than the characteristic frequency for the problem under consideration, then the term $\partial\psi/\partial t$ is absent on the left-hand side of the kinetic equation (13.7). But in this case too, it contains the same combination of φ (13.8). It thus follows that in all the phenomena considered earlier, for which it may be assumed that $\omega = 0$ in the kinetic equation, the presence of the f-function plays no role whatsoever.

The function f plays a role only in those kinetic phenomena for which the term $\partial\psi/\partial t$ in the kinetic equation is important. However, even in this case it does not always occur. Consider, for example, the anomalous skin effect or cyclotron resonance (see ch. 7). We have found that the function ψ is almost δ-shaped: $\psi \propto \delta(v_x)$, where v_x is the velocity component normal to the surface. Of course, the function ψ has also a smooth part, but in the problems under discussion it is not important. It is not difficult to see that upon substitution of the δ-function into the integral (13.8) we obtain a smooth function. Hence, the δ-shaped parts of the functions ψ and φ are identical. Thus, the difference between ψ and φ may be ignored, which completely justifies the calculations performed with the function f being neglected.

Finally, in some of the problems discussed earlier in this book the term $\partial\psi/\partial t$ is essential and at the same time the ψ-function does not exhibit a δ-shaped character. For example, this refers to the problem of the propagation of magnetoplasmon waves in an even metal. In this case, the term $\boldsymbol{v}\partial\psi/\partial\boldsymbol{r}$ is not essential, but the term $\partial\psi/\partial t$ is present. The difference from the case with $f=0$ consists in the following: in the current and in the collision integral ψ is replaced by $\varphi = L\psi$, where L is a real linear integral operator. By order of magnitude $L \sim 1$.

From the foregoing it follows that the statement that at $\omega\tau \gg 1$ the kinetic coefficients will depend on $-i\omega$ in the same manner as they were dependent on τ^{-1} at $\omega\tau \ll 1$ is still correct in order of magnitude and with respect to the complex nature of the coefficients. But the quantity τ itself was determined only by order of magnitude. Hence, the presence of the function f in this case does not lead to a qualitative change in the results. However, in concrete calculations with a given form of the spectrum the function f must be taken into account.

Thus, practically in all calculations where we can apply the concept of the quasiclassical electron moving along an orbit, the Fermi-liquid effects do not practically manifest themselves. It is for this reason that in order to avoid complications in calculations and in their interpretation we have not introduced the f-function in the preceding chapters.

13.3. The role of the interaction of quasiparticles in paramagnetic susceptibility

In section 10.1 we have considered the simplest quantum phenomenon - spin paramagnetism. According to formula (10.2), the paramagnetic susceptibility χ contains a single characteristic of a given metal: the density of states $\nu(\mu)$, i.e., the same quantity which, according to formula (2.28), determines the electronic heat capacity. As has been noted in section 3.1, the formula for the heat capacity is not changed when account is taken of quasiparticle interaction. However, as we will show below, the function f, while not altering the order of magnitude of χ, leads to a change in its particular value*.

* In metals the exact value of χ is not of great interest since only the total susceptibility can be measured, and the calculation of the diamagnetic part presents serious difficulties. But there exists a neutral Fermi liquid - liquid ^{3}He - which exhibits only spin paramagnetism. Then, by comparing χ with the heat capacity one can extract information about the function f.

The variation of the energy spectrum due to the action of the magnetic field on the spin may be written in the general case as follows:

$$\delta\varepsilon = -\beta_1(\boldsymbol{\sigma H}). \tag{13.10}$$

This change is made up of two components. First, the magnetic field acts on the magnetic moment of the electron, which introduces a contribution $-\beta\boldsymbol{\sigma H}$ (β is the Bohr magneton). Second, the variation of the distribution function has an inverse effect on the energy spectrum. The corresponding variation of the energy is evidently given by

$$\begin{aligned}
&\mathrm{Sp}_{\sigma'}\int f(\boldsymbol{p},\boldsymbol{\sigma};\boldsymbol{p}',\boldsymbol{\sigma}')\,\delta n(\boldsymbol{p}',\boldsymbol{\sigma}')\frac{\mathrm{d}^3\boldsymbol{p}'}{(2\pi\hbar)^3}\\
&=\mathrm{Sp}_{\sigma'}\int f(\boldsymbol{p},\boldsymbol{\sigma};\boldsymbol{p}',\boldsymbol{\sigma}')\frac{\partial n_0}{\partial\varepsilon'}\,\delta\varepsilon(\boldsymbol{p}'\,\boldsymbol{\sigma}')\frac{\mathrm{d}^3\boldsymbol{p}'}{(2\pi\hbar)^3}\\
&=-\mathrm{Sp}_{\sigma'}\int f(\boldsymbol{p},\boldsymbol{\sigma};\boldsymbol{p}',\boldsymbol{\sigma}')\frac{\partial n_0}{\partial\varepsilon'}\,\beta_1(\boldsymbol{p}')\,(\boldsymbol{\sigma}'\boldsymbol{H})\frac{\mathrm{d}^3\boldsymbol{p}'}{(2\pi\hbar)^3}.
\end{aligned}$$

For $\delta\varepsilon$ we have substituted expression (13.10).

Let us now consider the dependence of f on spins. In an isotropic liquid there is no preferred direction for the spin, and therefore there is only one possibility:

$$f=\eta(\boldsymbol{p},\boldsymbol{p}')+(\boldsymbol{\sigma\sigma}')\zeta(\boldsymbol{p},\boldsymbol{p}'), \tag{13.11}$$

where σ_i are Pauli matrices. In fact, even for a real metal one may use the approximately formula (13.11). This is because the term containing the product $(\boldsymbol{\sigma\sigma}')$ is of exchange origin, i.e., is determined by the Coulomb interaction of electrons. At the same time, any anisotropic terms are due either to relativistic spin-orbit interaction, discussed in section 10.7, or to the direct interaction of the magnetic moments. All these effects are approximately $(v/c)^2 \sim 10^{-3}$ to 10^{-4} of the Coulomb interactions. This is also the relative order of the anisotropic terms in the function f. Therefore, expression (13.11) is not a bad approximation in a real metal either.

Substituting expression (13.11) into the preceding formula, we obtain the variation of the energy due to the function f:

$$-2(\boldsymbol{\sigma H})\int\frac{\partial n_0}{\partial\varepsilon}\,\zeta(\boldsymbol{p},\boldsymbol{p}')\,\beta_1(\boldsymbol{p}')\frac{\mathrm{d}^3\boldsymbol{p}'}{(2\pi\hbar)^3}.$$

Hence, both terms in $\delta\varepsilon$ are proportional to $\boldsymbol{\sigma H}$ and we may write the following equation:

$$\beta_1=\beta-2\int\zeta(\boldsymbol{p},\boldsymbol{p}')\beta_1(\boldsymbol{p}')\frac{\mathrm{d}S'}{v(2\pi\hbar)^3}, \tag{13.12}$$

where the momentum $\boldsymbol{p}'$ lies on the Fermi surface.

The magnetic moment is given by

$$M = \beta\, \mathrm{Sp}_\sigma\, \boldsymbol{\sigma} \int n(\boldsymbol{p}, \boldsymbol{\sigma}) \frac{\mathrm{d}^3 p}{(2\pi\hbar)^2}$$

$$= \beta\, \mathrm{Sp}_\sigma\, \boldsymbol{\sigma} \int \delta n(\boldsymbol{p}, \boldsymbol{\sigma}) \frac{\mathrm{d}^3 \boldsymbol{p}}{(2\pi\hbar)^3}$$

$$= \beta\, \mathrm{Sp}_\sigma\, \sigma \int \frac{\partial n_0}{\partial \varepsilon} \delta\varepsilon(\boldsymbol{p}, \boldsymbol{\sigma}) \frac{\mathrm{d}^3 \boldsymbol{p}}{(2\pi\hbar)^3}$$

$$= -\beta\, \mathrm{Sp}_\sigma\, \boldsymbol{\sigma}(\boldsymbol{\sigma H}) \int \frac{\partial n_0}{\partial \varepsilon} \beta_1(\boldsymbol{p}) \frac{\mathrm{d}^3 \boldsymbol{p}}{(2\pi\hbar)^3}$$

$$= 2\beta \boldsymbol{H} \int \beta_1(\boldsymbol{p}) \frac{\mathrm{d}S}{v(2\pi\hbar)^3}, \tag{13.13}$$

where in the last integral $\boldsymbol{p}$ lies on the Fermi surface.

In the isotropic model the function $\zeta(\boldsymbol{p}, \boldsymbol{p}')$ may depend only on the angle between $\boldsymbol{p}$ and $\boldsymbol{p}'$, and $\beta_1(\boldsymbol{p})$ cannot depend on the direction of the vector $\boldsymbol{p}$. Here the result is considerably simplified. Solving eq. (13.12), we obtain

$$\beta_1 = \frac{\beta}{1 + Z_0}, \tag{13.14}$$

where

$$Z_0 = \nu(\mu)\zeta_0, \qquad \zeta_0 = \frac{\int \zeta\, \mathrm{d}S/v}{\int \mathrm{d}S/v} = \frac{\int \zeta\, \mathrm{d}\Omega}{4\pi}.$$

Evidently, $\zeta \sim \mu\hbar^3/p_0^3$ and since

$$\nu = 2 \int \frac{\mathrm{d}S}{v(2\pi\hbar)^2} \sim \frac{p_0^2}{\hbar^3 v} \sim \frac{p_0^3}{\hbar^3 \mu}$$

it follows that $Z_0 \sim 1$.

In what follows, instead of f it will be more convenient to use a dimensionless function of the order of unity:

$$F = f\nu(\mu) \tag{13.15}$$

and its representation in the form of (13.11):

$$F = X(\boldsymbol{p}, \boldsymbol{p}') + (\boldsymbol{\sigma\sigma}')\, Z(\boldsymbol{p}, \boldsymbol{p}').$$

Substituting β_1 into formula (13.13) yields (Landau 1956)

$$\chi = \frac{\beta^2 \nu(\mu)}{1 + Z_0}. \tag{13.16}$$

Thus, the expression found differs from (10.2) and therefore one cannot directly predict the paramagnetic susceptibility from data on heat capacity.

The following possibility is not excluded. The quantity Z_0 must be of the order of unity and may have an arbitrary sign. Evidently, all our reasonings are valid only at $Z_0 > -1$. If Z_0 becomes equal to -1, the quantity χ tends to infinity, i.e., the metal becomes ferromagnetic. In this case, a special approach is needed, and our treatment is invalid.

The physics of this phenomenon is as follows. Usually, the Pauli principle makes electrons exhibit a tendency towards opposite directions of the spins, and if electrons do not interact or their interaction is spin-independent, then in equilibrium there is an equal number of electrons with opposite spins. To produce a total spin different from zero, one has to redistribute the electrons among the states, which is associated with an increase in the energy. But if the electron–electron interaction is spin-dependent with the spins tending to align themselves parallel to each other (a negative sign of Z), then at a sufficient magnitude of Z the decrease in the energy due to the interaction may exceed the increase in the kinetic energy upon redistribution among the states. This gives rise to a new ground state with an uncompensated total spin and magnetic moment. This is what we know as ferromagnetism*. But, as has been noted above, this case requires a special treatment (see Appendix I).

A case is also possible where Z_0 is very close to -1 ($Z_0 + 1 \ll 1$); in such a case χ becomes very large. This occurs in palladium and its alloys. Such metals are called "nearly ferromagnetic" and exhibit a number of anomalous properties. We shall not consider this case here.

13.4. *Landau quantization and quantum oscillations*

We now turn our attention to quantum phenomena in a magnetic field. They all result from the quantization of energy levels, which is associated with both the orbital motion of electrons and the orientation of their spins.

Let us begin with the orbital motion. In section 10.4 we have found the quasi-classical energy levels (with large quantum numbers) without taking the function f into account (formula 10.17). It was pointed out that the rule of quantization of the levels corresponds in coordinate space to quantization of the magnetic flux: the magnetic flux passing inside the spiral electron trajectory may be changed only by an even number of flux quanta:

$$\Delta_n \Phi = 2n\Phi_0,$$

where $2\Phi_0$ is given by formula (10.19). We will show that this rule is quite general and does not change due to the interaction of quasiparticles.

* We are speaking here of ferromagnetic metals. In insulators, a more convenient concept is usually the concept of localized spins, which become aligned under the influence of exchange interaction.

According to the rules of quantum mechanics, the quasiclassical wave function of a particle is

$$\Psi \propto \exp\left(\frac{i}{\hbar}\int \boldsymbol{P}\,\mathrm{d}\boldsymbol{l}\right),$$

where $\boldsymbol{P}$ is a generalized momentum, and the integral is taken along the classical trajectory. In the case of quasiparticles, we deal with a wave packet (section 3.1), but this does not change matters much. In the presence of a magnetic field $\boldsymbol{P}=\boldsymbol{p}-(e/c)\boldsymbol{A}$, where $\boldsymbol{A}$ is the vector potential. We assume the trajectory in momentum space to be closed. This will also refer to the projection of the coordinate trajectory onto the (x, y) plane. The variation of the quasiclassical wave function under the action of the magnetic field is expressed by the factor

$$\exp\left(-\frac{ie}{\hbar c}\int \boldsymbol{A}\,\mathrm{d}\boldsymbol{l}\right). \tag{13.17}$$

The wave function must of necessity be single-valued. If we assume that $\boldsymbol{A}$ lies in the (x, y) plane perpendicular to the magnetic field, in the integral (13.17) the integration runs along the closed projection of the electronic trajectory onto the (x, y) plane. Obviously, while moving along this contour we will periodically return to the same points, and in each of the round trips in the function Ψ there will appear a phase factor

$$\exp\left(-\frac{ie}{\hbar c}\oint \boldsymbol{A}\,\mathrm{d}\boldsymbol{l}\right).$$

The single-valuedness of the wave function requires that this factor be equal to unity, i.e.,

$$\frac{e}{\hbar c}\oint \boldsymbol{A}\,\mathrm{d}\boldsymbol{l}=2\pi n,$$

where n is an arbitrary integer. Transforming the contour integral, we obtain

$$\oint \boldsymbol{A}\,\mathrm{d}\boldsymbol{l}=\int \mathrm{rot}\,\boldsymbol{A}\,\mathrm{d}\boldsymbol{S}=\int \boldsymbol{H}\,\mathrm{d}\boldsymbol{S}=\Phi,$$

where Φ is the magnetic flux across the contour.

Since all this reasoning actually refers only to sufficiently large quantum numbers, we may, strictly speaking, write only

$$\Delta\Phi_n=2\pi\frac{n\hbar c}{e}=2n\Phi_0$$

where $\Delta\Phi_n=\Phi_{m+n}-\Phi_m$ $(m \gg 1)$. Evidently, this leads to the rule (10.17) of quantization in momentum space, established earlier.

However, not only the quantization of the orbital motion is important here, but the spin splitting is so as well. The latter is completely determined by formulae (13.10) and (13.14)*.

Let us take any formula for the oscillatory contribution, for example, eq. (10.33). From this formula it is seen that the period of oscillations is determined by the orbital quantization rule, i.e., it does not change when account is taken of Fermi-liquid effects. As for the amplitude, it includes the factor $\cos(\pi k m^*/m)$ (this is the amplitude of a given harmonic k). The mass of the free electron has arisen from the Bohr magneton $\beta = e\hbar/2mc$. But since, in accordance with (13.14), β is replaced by β_1, it may be concluded that the Fermi-liquid effects will lead to the replacement $m \to m(1+Z_0)$.

13.5. Zero (high-frequency) sound

We have so far proved that the phenomena considered earlier are not changed substantially when account is taken of the function f. We turn now to specific phenomena associated with this function.

In ch. 7 we concluded that in a metal in the absence of a magnetic field no electromagnetic waves can propagate. This was also true in those cases where the wavelength is much smaller than the mean free path (the anomalous skin effect). There arises the question: To what extent is this conclusion valid if the function f is taken into account?

Evidently, the only possibility that may be dealt with is the oscillations of the electron distribution function in the case of $\omega \gg \tau^{-1}$ (Landau 1957) In this case one may neglect the collision term in the kinetic equation. The variation of the distribution function may always be represented in the form

$$\delta n = n_1(\boldsymbol{p}) + \boldsymbol{\sigma n}_2(\boldsymbol{p}). \tag{13.18}$$

Since we are concerned here with small oscillations, the kinetic equation will be linear in δn and therefore the equations for n_1 and $\boldsymbol{n}_2$ are separated.

We begin with the equation for n_1. Such vibrations are called zero sound. Assuming n_1 in the form $n_1 = -\psi \partial n_0/\partial \varepsilon$ and $\psi \propto e^{i(\boldsymbol{kr}-\omega t)}$, we obtain from (13.7)

$$i(-\omega + \boldsymbol{kv})\psi + i\boldsymbol{kv}\, \mathrm{Sp}_\sigma \int f\psi \frac{dS}{v(2\pi\hbar)^3} - e\boldsymbol{vE} = 0. \tag{13.19}$$

The electric field is found from Maxwell's equations. If it is assumed that $\delta n = n_1(\boldsymbol{p})$, then upon substitution of the function f in the form of (13.11) into (13.19) only $\eta(\boldsymbol{p}, \boldsymbol{p}')$ is retained. We consider the isotropic case. The quantity will depend only on the angle between $\boldsymbol{p}$ and $\boldsymbol{p}'$. Finally, we make one more simplifying assumption: $\eta = \text{const} = \eta_0$. In this case, from (13.19) we find

$$i(-\omega + \boldsymbol{kv})\psi + i\boldsymbol{kv} X_0 \psi_0 - e\boldsymbol{vE} = 0, \tag{13.20}$$

* Spin-orbit interaction effects are not considered here; cf. footnote on p. 181 and section 10.7.

where

$$\psi_0 = \frac{\int \psi \, dS/v}{\int dS/v}, \qquad X_0 = \eta_0 \nu(\mu).$$

The electric field is found from the equation

$$\operatorname{div} \boldsymbol{E} = 4\pi e \, \mathrm{Sp}_\sigma \int \delta n \frac{d^3\boldsymbol{p}}{(2\pi\hbar)^3} = 4\pi e \nu(\mu)\psi_0.$$

There are also vortex fields associated with the current, but they are evidently much smaller. Assuming that $\boldsymbol{E} \propto e^{i(\boldsymbol{kr}-\omega t)}$, we obtain

$$\boldsymbol{E} = -4\pi i \boldsymbol{k} e \nu(\mu) \frac{\psi_0}{k^2}.$$

Substituting this into eq. (13.20), we can express ψ through ψ_0:

$$\psi = \frac{\boldsymbol{kv}[X_0 + \pi e^2 \nu(\mu)/k^2]\psi_0}{\omega - \boldsymbol{kv}}.$$

Finally, integrating over the angles, we arrive at a homogeneous equation for ψ_0. The condition of solvability of this equation is

$$1 = \int \frac{\boldsymbol{kv}[X_0 + 4\pi e^2 \nu(\mu)/k^2]}{(\omega - \boldsymbol{kv})} \frac{d\Omega}{4\pi}.$$

As has been said above, $X_0 \sim 1$. The second term in the numerator is of the order of $e^2 p_0^2/k^2\hbar^3 v \sim (e^2/\hbar v)(p_0^2/h^2k^2) \gg 1$, because $e^2/\hbar v \sim 1$, and we are dealing here with vibrations of wavelength much larger than interatomic distances, i.e., $k \ll p_0/\hbar$. Let us also assume (this will be confirmed below) that $kv \ll \omega$. Then, neglecting X_0 as compared with the second term in the numerator and expanding in $\boldsymbol{kv}$, we get

$$1 = \int (\boldsymbol{kv})^2 4\pi e^2 \frac{\nu(\mu)}{k^2\omega^2} \frac{d\Omega}{4\pi} = \tfrac{4}{3}\pi e^2 \nu(\mu) \frac{v^2}{\omega^2}$$

from which it follows that

$$\omega = [\tfrac{4}{3}\pi e^2 \nu(\mu) v^2]^{1/2} = \text{const.} \tag{13.21}$$

The quantity $\hbar\omega$ is of order $(e^2/\hbar v)^{1/2}(p_0^2/m) \sim \mu$. This justifies the above assumption that $kv \ll \omega$, since in order to provide $kv \sim \omega$ the wave vector k must be of order $p_0/\hbar$, and we assume it to be much smaller. Hence, the oscillations considered must have very large frequencies, of the order of $\mu/\hbar \sim 10^{15}\ \mathrm{s}^{-1}$, and in the radio-frequency range such oscillations cannot be observed. As a matter of fact, as has been noted in section 2.2, all inferences concerning the Fermi liquid are valid only in those cases where they refer to a small neighborhood of the Fermi surface. Hence, in a case where $\hbar\omega$ is of order μ, the theory becomes invalid. The only statement we can make here is that, because of the generation of electric fields no low-frequency oscillations of the electron density are present in a metal*.

However, we can raise the question of oscillations of the distribution function that are not accompanied by a variation in electron density, i.e., oscillations with $\psi_0 = 0$. The situation here is as follows. Since in the isotropic model in a case where $\boldsymbol{p}$ and $\boldsymbol{p}'$ lie at the Fermi surface the function $X(\boldsymbol{p}, \boldsymbol{p}')$ may depend only on the angle between $\boldsymbol{p}$ and $\boldsymbol{p}'$, it may be represented in the form

$$X = \sum X_n P_n(\cos\theta), \tag{13.22}$$

where P_n are Legendre polynomials. If this expansion is continued up to $n = N$, it can be shown that, in principle, the kinetic equation allows for solutions proportional to $e^{im\varphi}$ with $m \leqslant N$, where φ is the "longitude" of the momentum $\boldsymbol{p}$ in the polar system of coordinates with the axis along $\boldsymbol{k}$.

Of course, the function ψ, which is proportional to $e^{im\varphi}$, vanishes upon averaging, and therefore no oscillations of the charge and no large electric fields arise in such a wave.

However, such solutions can be obtained only if the coefficients X_n are sufficiently large: $X_n > X_{n,\min}$. For instance, if the function X contains only two harmonics: $X = X_0 + X_1 \cos\theta$, then solutions proportional to $e^{i\varphi}$ are possible; they occur only at $X_1 > 6$. For the subsequent harmonics to appear, even larger values of the subsequent X_n are required.

At the same time, if $X(\theta)$ is a sufficiently smooth function, then the coefficients X_n in (13.22) must decrease with increasing n. This is evident even from the fact that if X is independent of the angle, then only $X_0 \neq 0$, and in the case where all X_n are the same, we have the δ-function. Hence, one may anticipate at best the appearance of only one or two first harmonics.

At this point, however, one should recall the role of vortex electric fields. From Maxwell's equations it is seen that these fields may play a role in the region of long waves (estimates show that this is $\lambda > \lambda_0 \sim 10^{-5}$ cm). Just as with the appearance of a space charge, the electric fields impede the generation of oscillations. In this case the first harmonic (i.e., $\psi \propto e^{i\varphi}$), which gives oscillations of the electric current, will be suppressed. One can thus anticipate only the subsequent harmonics, which, as noted above, is much less probable.

Therefore, it is more probable to find such oscillations in the short-wavelength region, where electric fields have no effect. From the kinetic equation (13.19) it follows that the dispersion law for oscillations, $\omega(k)$, if any, will be linear: $\omega = uk$, with $u \sim v$. Hence, we are speaking here of frequencies $\omega > \omega_0 \sim vk \sim 10^{13}\ \mathrm{s}^{-1}$. This

* This reasoning naturally does not imply that waves of frequency $\omega \sim \mu$ cannot propagate in a metal. Such waves are really observed. Since for many metals we can use the free-electron approximation (ch. 14), the calculation performed above applies for them. Such waves with their frequency slightly dependent on the wavelength are described (to within terms of order $(vk/\omega)^2$) by formula (13.21) are called plasma oscillations or plasmons. For free electrons $\nu(\mu) = p_0 m/(\pi^2 h^3)$ and, hence, the frequency of plasmons is $\omega_0 = (4\pi n_e e^2/m)^{1/2}$, where n_e is the density of valence electrons.

is the infrared frequency region. It should, however, be stressed that even the requirement $X_1 > 6$ is already very stringent since the interaction of quasiparticles in a metal is not really strong (section 14.3). Perhaps it is for this reason that such waves have not yet been detected.

13.6. Spin waves

A more promising case is that where only the spin characteristics of the electron liquid are oscillating, $\delta n = \boldsymbol{n}_2\boldsymbol{\sigma}$. Such oscillations are called spin waves. Here, the picture looks differently for the cases where a magnetic field is applied and in the absence of the field.

We begin with the case without field. Naturally, in spin oscillations no appreciable electric fields are present, and therefore eq. (13.19) does not contain the term with $\boldsymbol{E}$. Only the term with ζ from the function f makes a contribution (see eq. 13.11). Assuming for the sake of simplicity that $\zeta = \text{const.} = \zeta_0$, we obtain

$$\mathrm{i}(-\omega + \boldsymbol{kv})\psi + \mathrm{i}\boldsymbol{kv}Z_0\psi_0 = 0,$$

where $Z_0 = \zeta_0 \nu(\mu)$. This equation is solved for ψ:

$$\psi = \boldsymbol{kv}Z_0 \frac{\psi_0}{(\omega - \boldsymbol{kv})}.$$

Defining ψ_0 from this equation, we obtain the equation for finding $\omega(k)$:

$$1 = Z_0 \tfrac{1}{2} \int_{-1}^{1} \frac{\cos\theta \, \mathrm{d}(\cos\theta)}{s - \cos\theta}, \tag{13.23}$$

where $s = \omega / kv$. This equation may have real roots only if the denominator in the integrand does not vanish at all (the contour around the pole in the opposite case can be found by adding a small imaginary part to the frequency, $\omega \to \omega - \mathrm{i}/\tau$). Thus, $s > 1$. Upon integration we find

$$1 = Z_0 \left(\tfrac{1}{2} s \ln \frac{s+1}{s-1} - 1 \right). \tag{13.24}$$

This equation has a solution only at $Z_0 > 0$. Thus, according to section 13.3, such waves are known in advance to be absent in "nearly ferromagnetic" metals.

So, at $Z_0 > 0$ in a metal in the absence of a magnetic field waves of spin density can propagate with a linear dispersion law $\omega = uk$, with u being of order v (the velocity at the Fermi boundary) but always larger than v. Of course, with a more complicated form of the function Z (see eq. 13.22) spin waves are possible, in principle, with other types of oscillations of the distribution function, but everything that has been said earlier about the difficulty of observation of complicated types of zero sound refers to this case.

Note also the following point. Spin waves can be excited by the action on the electron spin of the magnetic field of the electromagnetic wave incident on the metal. The appearance of spin waves could be recorded from the variation of the impedance (for example, upon setting up a standing wave in the film). The force acting on the electron spin from the side of the magnetic field is $\beta\boldsymbol{\sigma}\nabla\boldsymbol{H} \sim \beta H/\delta$. According to Maxwell's equation $H \sim cE/\delta\omega$. Hence, the ratio of the magnetic field force to the electric field force of the wave is of order $c\beta/\delta^2\omega e \sim \hbar/(m\delta^2\omega)$. For $\omega \sim 10^{10}\,\mathrm{s}^{-1}$ one has $\delta \sim 10^{-4}$ cm and this ratio is of the order of 10^{-2}. It should, however, be noted that any wave is generated at a length of the order of its wavelength λ. In this case, at $\omega \sim 10^{10}\,\mathrm{s}^{-1}$ one has $\lambda \sim u/\omega \sim 10^8/10^{10} \sim 10^{-2}$ cm. Since in the case of the anomalous skin effect the field is damped according to the law $E \approx E_0(\delta/x)^2$ (this follows from formula 7.33), then $E \sim 10^{-4}\,E_0$ for $\lambda \sim 10^{-2}$ cm. Thus, the conditions for observing such waves are unfavorable.

In the infrared region the dielectric constant $\varepsilon \approx -4\pi n_e e^2/(m\omega^2)$. The high-frequency field is damped into the bulk of the metal by the law $\exp[-(\omega/c)|\varepsilon|^{1/2}x]$, i.e., the penetration depth is of order $(\hbar/p_0)(c/v) \sim 10^{-6}$ cm and is independent of ω. If we take a frequency of the order of $10^{14}\,\mathrm{s}^{-1}$, then $\lambda \sim \delta$. But the ratio of the forces will again be equal to 10^{-2}. Hence, although the infrared range is more favorable for observation of the effect, the influence of spin waves on the surface impedance will still be small. Perhaps it is for this reason that they have not been observed so far.

Only one mode of oscillation associated with the function f has been detected so far, namely spin waves in an external magnetic field (Silin 1958).

We must first take into account that in this case the energy of quasiparticles begins to be dependent on their spin and this naturally refers to the equilibrium distribution function. Therefore we have to use the equation for the density matrix, which, in accordance with quantum mechanics, has the form

$$\frac{\partial n}{\partial t} = \frac{i}{\hbar}[\mathscr{H}, n], \tag{13.25}$$

where $\mathscr{H}$ is the Hamiltonian. In this case, the coordinate and momentum operators for quasiparticles are regarded as classical variables, and therefore if n and $\mathscr{H}$ were independent of the spins, the right-hand side of eq. (13.25) would give only the Poisson bracket:

$$-\frac{\partial n}{\partial \boldsymbol{r}}\frac{\partial \varepsilon}{\partial \boldsymbol{p}} + \frac{\partial n}{\partial \boldsymbol{p}}\frac{\partial \varepsilon}{\partial \boldsymbol{r}},$$

which coincides with the corresponding terms of the ordinary kinetic equation.

However, the dependence on the spin alters the situation. We must write explicitly what is given by the right-hand side of eq. (13.25) because of the noncommutativity of the spin operators. Thus, we arrive at the equation

$$\frac{\partial n}{\partial t} - \frac{\mathrm{i}}{\hbar}[\varepsilon, n] + \boldsymbol{v}\frac{\partial n}{\partial \boldsymbol{r}} + \dot{\boldsymbol{p}}\frac{\partial n}{\partial \boldsymbol{p}} = 0. \tag{13.26}$$

Here, because of the presence of a magnetic field, we have

$$\dot{\boldsymbol{p}} = \frac{e}{c}\left(\frac{\partial \varepsilon}{\partial \boldsymbol{p}}\boldsymbol{H}\right) - \frac{\partial \varepsilon}{\partial \boldsymbol{r}}. \tag{13.27}$$

At equilibrium in the presence of not too strong a magnetic field we have in accordance with section 13.8:

$$\varepsilon(\boldsymbol{p}, \boldsymbol{\sigma}) = \varepsilon(\boldsymbol{p}) - \beta_1 \boldsymbol{\sigma} \boldsymbol{H},$$

$$n_0(\boldsymbol{p}, \boldsymbol{\sigma}) = n_0(\boldsymbol{p}) - \frac{\partial n_0}{\partial \varepsilon}\beta_1 \boldsymbol{\sigma} \boldsymbol{H}. \tag{13.28}$$

We substitute into (13.26) $n = n_0 - (\partial n_0/\partial \varepsilon)\boldsymbol{\psi}\boldsymbol{\sigma}$ and retain the terms linear in $\boldsymbol{\psi}$. Here, however, we take into account that the energy ε also acquires a correction:

$$-\mathrm{Sp}_{\sigma'}\int f(\boldsymbol{p}, \boldsymbol{\sigma}; \boldsymbol{p}', \boldsymbol{\sigma}')\frac{\partial n_0}{\partial \varepsilon}\psi(\boldsymbol{p}')\,\boldsymbol{\sigma}'\frac{\mathrm{d}^3 p'}{(2\pi\hbar)^3}$$

$$= \boldsymbol{\sigma}\int Z(\boldsymbol{p}, \boldsymbol{p}')\,\psi(\boldsymbol{p}')\frac{\mathrm{d}\Omega'}{4\pi}.$$

We then multiply the equation obtained by $\boldsymbol{\sigma}$ and take Sp_σ, as a result of which we obtain the equation for $\boldsymbol{\psi}$:

$$-\mathrm{i}\omega\boldsymbol{\psi} - \frac{2\beta_1}{\hbar}\left[\boldsymbol{H}\left(\boldsymbol{\psi}(\boldsymbol{p}) + \int Z(\boldsymbol{p}, \boldsymbol{p}')\,\boldsymbol{\psi}(\boldsymbol{p}')\frac{\mathrm{d}\Omega'}{4\pi}\right)\right]$$

$$+\boldsymbol{v}\frac{\partial}{\partial \boldsymbol{r}}\left(\boldsymbol{\psi}(\boldsymbol{p}) + \int Z(\boldsymbol{p}, \boldsymbol{p}')\,\boldsymbol{\psi}(\boldsymbol{p}')\frac{\mathrm{d}\Omega'}{4\pi}\right)$$

$$+\frac{e}{c}[\boldsymbol{v}\boldsymbol{H}]\frac{\partial}{\partial \boldsymbol{p}}\left(\boldsymbol{\psi}(\boldsymbol{p}) + \int Z(\boldsymbol{p}, \boldsymbol{p}')\,\boldsymbol{\psi}(\boldsymbol{p}')\frac{\mathrm{d}\Omega'}{4\pi}\right)$$

$$= 0. \tag{13.29}$$

This equation contains the following combination everywhere, except in the first term:

$$\boldsymbol{\rho}(\boldsymbol{p}) = \boldsymbol{\psi}(\boldsymbol{p}) + \int Z(\boldsymbol{p}, \boldsymbol{p}')\,\boldsymbol{\psi}(\boldsymbol{p}')\frac{\mathrm{d}\Omega'}{4\pi} \tag{13.30}$$

(recall that $\boldsymbol{p}$ and $\boldsymbol{p}'$ refer here to the Fermi surface).

In the isotropic model we can expand Z in Legendre polynomials $[\theta = (\widehat{\boldsymbol{p}, \boldsymbol{p}'})]$:

$$Z = \sum_n Z_n P_n(\cos\theta). \tag{13.31}$$

The function $\boldsymbol{\psi}$ may be represented in the general case as

$$\boldsymbol{\psi}(\boldsymbol{p}) = \sum_n \boldsymbol{\psi}_n, \qquad \boldsymbol{\psi}_n = \sum_m \boldsymbol{\psi}_{nm} Y_n^m(\theta, \varphi),$$

where ψ_{nm} are coefficients which do not depend on the angles; θ_1 and φ are the polar angles of the vector $\boldsymbol{p}$; $Y_n^m = P_n^m(\cos\theta_1)\,\mathrm{e}^{\mathrm{i}m\varphi}$ are spherical functions. According to a well-known rule,

$$P_n(\boldsymbol{n}_1\boldsymbol{n}_2) = \sum_{m=-n}^{n} \frac{(n+|m|)!}{(n-|m|)!}\, Y_n^m(\boldsymbol{n}_1)\, Y_n^{-m}(\boldsymbol{n}_2) \tag{13.32}$$

($\boldsymbol{n}$ is the unit vector along $\boldsymbol{p}$). The integral of two spherical function is

$$\int Y_n^m(\boldsymbol{n}_1)\, Y_{n'}^{-m'}(\boldsymbol{n}_1) \frac{\mathrm{d}\Omega}{4\pi} = \frac{\delta_{mn'}\delta_{nn'}}{2n+1} \frac{(n-|m|)!}{(n+|m|)!}. \tag{13.33}$$

Using these relations, we find

$$\rho_n(\boldsymbol{p}) = \psi_n(\boldsymbol{p}) \left[1 + \frac{Z_n}{2n+1}\right]. \tag{13.34}$$

This relation may be transformed to

$$\psi(\boldsymbol{p}) = \int L(\boldsymbol{p}, \boldsymbol{p}')\, \rho(\boldsymbol{p}') \frac{\mathrm{d}\Omega'}{4\pi}, \tag{13.35}$$

where

$$L = \sum L_n P_n(\cos\theta), \qquad L_n = \frac{2n+1}{1 + Z_n/(2n+1)}. \tag{13.36}$$

Using these relations, we can simplify eq. (13.23):

$$-\mathrm{i}\omega \int L(\boldsymbol{p}, \boldsymbol{p}')\, \boldsymbol{\rho}(\boldsymbol{p}') \frac{\mathrm{d}\Omega'}{4\pi} - \frac{2\beta_1}{\hbar}[\boldsymbol{H}\boldsymbol{\rho}]$$
$$+ \boldsymbol{v}\frac{\partial}{\partial \boldsymbol{r}}\boldsymbol{\rho} + \frac{e}{c}[\boldsymbol{v}\boldsymbol{H}]\frac{\partial}{\partial \boldsymbol{p}}\boldsymbol{\rho} = 0. \tag{13.37}$$

We shall consider here only transverse oscillations, i.e., $\boldsymbol{\rho} \perp \boldsymbol{H}$. Choosing the axis z along $\boldsymbol{H}$, we may write the equations for ρ_x and ρ_y, from which it is possible to derive equations for $\rho_+ = \rho_x + \mathrm{i}\rho_y$ and for $\rho_- = \rho_x - \mathrm{i}\rho_y$:

$$\omega \int L(\boldsymbol{p}, \boldsymbol{p}')\, \rho_\pm(\boldsymbol{p}') \frac{\mathrm{d}\Omega'}{4\pi} \pm \frac{2\beta_1 H}{\hbar}\rho_\pm$$
$$+ \mathrm{i}\boldsymbol{v}\frac{\partial}{\partial \boldsymbol{r}}\rho_\pm + \mathrm{i}\frac{eH}{m^*c}\frac{\partial}{\partial\varphi}\rho_\pm = 0, \tag{13.38}$$

where θ_1, φ are the polar coordinates of the vector $\boldsymbol{p}$ in a coordinates system with the axis along z; $m^* = p_0/v$ is the effective mass.

Let us first consider the solution which is independent of the coordinate $\boldsymbol{r}$. Assuming that $\rho_{\pm} \propto Y_n^m(\theta_1, \varphi)$, we obtain from (13.38):

$$\omega_{n,m} = (\mp \Omega_0 + m\Omega)\left(1 + \frac{Z_n}{2n+1}\right), \tag{13.39}$$

where

$$\Omega_0 = \frac{2\beta_1 H}{\hbar} = \frac{2\beta H}{(1+Z_0)\hbar}, \qquad \Omega = \frac{2\beta^* H}{\hbar} \quad \left(\beta^* = \frac{e\hbar}{2m^* c}\right). \tag{13.40}$$

Note that for $n = m = 0$ we get

$$\omega_{00} = \Omega_0(1+Z_0) = \frac{2\beta H}{\hbar}, \tag{13.41}$$

that is, the spin precession frequency for a free electron. Hence, in the presence of a magnetic field there are certain resonance frequencies in the metal, which are associated with the precession of the electron spin. This phenomenon is called paramagnetic resonance. Usually, the quantity ω_{00} is involved here since the frequencies with $m \neq 0$ are cyclotron frequencies and waves with $n \neq 0$ are difficult to excite. The detailed theory of paramagnetic resonance in a bulk metal takes account of the skin effect (Dyson 1955, Azbel' et al. 1956, 1957a,b). We shall not expound it here and will only note two circumstances. First, as has been shown, the resonance frequency ω_{00} does not depend on Fermi-liquid effects. Since under the conditions of the anomalous skin effect the motion of electrons is also independent of these effects, it follows that paramagnetic resonance may be treated by using the gas model.

Second, as is always the case, under resonance conditions the energy of the electromagnetic wave is transferred more effectively to electrons, which manifests itself, in particular, in a stronger deviation of the distribution function from equilibrium. This change in the distribution function persists over the mean free path of the electron. In the case of paramagnetic resonance, the spin of the electron rather than its momentum is important. Therefore, the characteristic distance is not the usual mean free path but rather the path on which a collision with a flip of the electron spin takes place. This length l_s by far exceeds the usual mean free path.

For the anomalous skin effect, only those electrons are important which travel their mean free path in the skin layer. So, except for special cases (see ch. 8), the distribution function deviates most strongly from equilibrium only at a depth δ. At the same time there is a small correction, which is attenuated within a depth l - the mean free path. In the case of paramagnetic resonance, the electron first moves a distance l in the skin layer and then leaves it and begins to diffuse into the bulk of the metal. According to the diffusion rules (see formula 5.47) $t \sim x^2/D$, where x is the distance, t is the time, and D is the diffusion coefficient. The total path is $vt = l_s$ and since $D \sim lv$, then it follows that $x \sim (ll_s)^{1/2} \gg l$. Thus, it may be expected that in films of thickness less than $(ll_s)^{1/2}$ the variation of the electron distribution function will reach the opposite surface and will cause emission. In other words, upon paramagnetic resonance such films will display increased transparency. This is used to observe resonance in experiments.

Now consider spin waves. Evidently, each frequency (13.39) is, in effect, the starting point of a function $\omega_{nm}(k)$, i.e., one of the branches of the spectrum of spin waves. Let us consider ρ_- at $n=m=0$ (only this branch has been observed experimentally). We limit ourselves, for simplicity, to the case $Z=Z_0+Z_1\cos\theta$. We seek $\rho_-\propto e^{ikr}$. From eq. (13.38) we obtain

$$\omega\int L(\boldsymbol{p},\boldsymbol{p}')\,\rho_-(\boldsymbol{p}')\frac{d\Omega'}{4\pi}-\Omega_0\rho_--\boldsymbol{kv}\rho_-+i\Omega\frac{\partial}{\partial\varphi}\rho_-=0. \tag{13.42}$$

Consider the long-wavelength limit, i.e., $kv\ll\Omega_0$. We will solve eq. (13.42) by using the method of successive approximations:

$$\rho_-=\rho^{(0)}+\rho^{(1)}+\rho^{(2)}+\cdots,$$

$$\omega=\Omega_0(1+Z_0)+\omega^{(1)}+\omega^{(2)}+\cdots.$$

We write eq. (13.42) in a first approximation

$$\omega^{(1)}\frac{\rho^{(0)}}{1+Z_0}+\Omega_0(1+Z_0)\int L(\boldsymbol{p},\boldsymbol{p}')\,\rho^{(1)}(\boldsymbol{p}')\frac{d\Omega'}{4\pi}$$

$$-\Omega_0\rho^{(1)}-\boldsymbol{kv}\rho^{(0)}+i\Omega\frac{\partial\rho^{(1)}}{\partial\varphi}=0. \tag{13.43}$$

Since the angles θ_1 and φ are defined with respect to the system of coordinates with the axis along $\boldsymbol{H}$, it follows that

$$\boldsymbol{kv}=kv(\cos\alpha\cos\theta_1+\sin\alpha\sin\theta_1\cos\varphi),$$

where $\alpha=(\widehat{\boldsymbol{k},\boldsymbol{H}})$. It follows that $\rho^{(1)}$ will have components proportional to Y_1^1, Y_1^{-1} and Y_1^0. We label them, respectively, by $\rho_1^{(1)}$, $\rho_{-1}^{(1)}$ and $\rho_0^{(1)}$. From (13.42) we obtain the following equations for them:

$$\frac{\omega^{(1)}}{1+Z_0}\rho^{(0)}+\Omega_0\frac{1+Z_0}{1+\frac{1}{3}Z_1}\rho_0^{(1)}-\Omega_0\rho_0^{(1)}-kv\cos\alpha\cos\theta_1\rho^{(0)}=0,$$

$$\frac{\Omega_0(1+Z_0)}{1+\frac{1}{3}Z_1}\rho_1^{(1)}-(\Omega_0+\Omega)\rho_1^{(1)}-\tfrac{1}{2}kv\sin\alpha\sin\theta_1\rho^{(0)}e^{i\varphi}=0,$$

$$\frac{\Omega_0(1+Z_0)}{1+\frac{1}{3}Z_1}\rho_{-1}^{(1)}-(\Omega_0-\Omega)\rho_{-1}^{(1)}-\tfrac{1}{2}kv\sin\alpha\sin\theta_1\rho^{(0)}e^{-i\varphi}=0. \tag{13.44}$$

If the first of these equations is integrated over the angles, we obtain $\omega^{(1)}=0$ since $\rho_0^{(1)}\propto Y_1^0=\cos\theta_1$. Hence, the correction to $\Omega_0(1+Z_0)$ is of order k^2. Solving the equations, we find

$$\rho^{(1)}(\theta_1, \varphi) = kv \cos\alpha \cos\theta_1 \frac{(1+\frac{1}{3}Z_1)\rho^{(0)}}{\Omega_0(Z_0-\frac{1}{3}Z_1)}$$
$$+\tfrac{1}{2}kv \sin\alpha \sin\theta_1(1+\tfrac{1}{3}Z_1)\rho^{(0)}$$
$$\times\{[-\Omega(1+\tfrac{1}{3}Z_1)+\Omega_0(Z_0-\tfrac{1}{3}Z_1)]^{-1}\,\mathrm{e}^{\mathrm{i}\varphi}$$
$$+[\Omega(1+\tfrac{1}{3}Z_1)+\Omega_0(Z_0-\tfrac{1}{3}Z_1]\,\mathrm{e}^{-\mathrm{i}\varphi}\}. \qquad (13.45)$$

In the second approximation, we obtain from eq. (13.43)

$$\omega^{(2)}\frac{\rho^{(0)}}{1+Z_0}+\Omega_0(1+Z_0)\int L(\boldsymbol{p},\boldsymbol{p}')\,\rho^{(2)}(\boldsymbol{p}')\frac{\mathrm{d}\Omega'}{4\pi}$$
$$-\Omega_0\rho^{(2)}-\boldsymbol{kv}\rho^{(1)}+\mathrm{i}\Omega\frac{\partial\rho^{(2)}}{\partial\varphi}=0.$$

If $\rho^{(2)}$ contains a term which does not depend on the angles, it cancels in the second and third terms. As for the terms proportional to spherical functions of higher orders, upon integration with L these spherical functions are reproduced. Therefore, if the last equation is integrated over the angles, we find that

$$\frac{\omega^{(2)}\rho^{(0)}}{1+Z_0}=\int \boldsymbol{kv}\rho^{(1)}\frac{\mathrm{d}\Omega}{4\pi}.$$

Substituting the value of $\rho^{(1)}$ from eq. (13.45), we obtain (Platzman and Wolff 1967)

$$\omega=\frac{2\beta H}{\hbar}\Bigg[1+\tfrac{1}{3}k^2v^2(1+\tfrac{1}{3}Z_1)(Z_0-\tfrac{1}{3}Z_1)$$
$$\times\left(\frac{\cos^2\alpha}{\Omega_0^2(Z_0-\frac{1}{3}Z_1)^2}+\frac{\sin^2\alpha}{\Omega_0^2(Z_0-\frac{1}{3}Z_1)^2-\Omega^2(1+\frac{1}{3}Z_1)^2}\right)\Bigg]. \qquad (13.46)$$

As mentioned above, such spin waves have been detected experimentally (Schultz and Dunifer 1967). This has been done in an investigation of the passage of cyclotron electromagnetic waves through thin films of sodium and potassium. Under the conditions where $\omega=2\beta H/\hbar$, i.e., the frequency of the incident field coincides with the frequency of spin precession, a paramagnetic resonance arises. This manifests itself in the appearance of a maximum of film transparency.

The possibility of propagation of a spin wave is manifested in the appearance of additional transparency peaks, depending on the magnetic field or frequency under the conditions where a standing spin wave is set up in the metal, i.e., $kD=n\pi$ (D is the film thickness). At small n the wave vectors are very small, so that the corresponding frequencies (or fields if the frequency is given) differ little from the paramagnetic resonance frequency, i.e., the peaks of the spin waves are close to the paramagnetic resonance peak.

However, while the paramagnetic peak is independent of the direction of the field (in a cubic metal), the position of the additional peak is changed with a change in the orientation of the field with respect to $\boldsymbol{k}$ (in an experiment with a film $\boldsymbol{k}$ is directed along the normal to the film). From formula (13.46) it follows that at $\Omega|1+\frac{1}{3}Z_1|>\Omega_0|Z_0+\frac{1}{3}Z_1|$ the additional peak is shifted with a change in the field orientation from one side of the main peak to the other. This has been observed experimentally (Schultz and Dunifer 1967).

13.7. The Kondo effect at low temperatures

The ideas of the theory of the Fermi liquid have been applied for the investigation of the behavior of metals containing magnetic impurities (section 4.6) at low temperatures in a case where the electron–impurity interaction has antiferromagnetic sign (Nozières 1974). As has been noted in section 4.6, when the temperature is lowered the effective interaction increases indefinitely, which leads to a complete shielding of the impurity spin by the spin of the conduction electrons.

The study of this shielding in the case of large spins requires taking account of the circumstance that large spins can be created by electrons of the unfilled d- or f-shells of the impurity atoms, which have a nonzero orbital moment. The analysis of this question is complicated (see the review article by Tsvelick and Wiegmann 1983); we will restrict ourselves to the simplest model, in which the impurity spin belongs to the s-shell and is equal to $S=\frac{1}{2}$. Although this never occurs in nature, the true cases of the d- or f-shells are similar in many respects to the model under discussion. For example, for the case of the f-shell, because of a strong spin–orbit coupling, the states are characterized by the total moment $\boldsymbol{J}=\boldsymbol{L}+\boldsymbol{S}$. Due to the action of the crystal field the states with different projections of $\boldsymbol{J}$ onto the crystal axis are separated by their energy to such an extent that in the presence of an inversion center only two-fold degeneracy is preserved, which may be described as a certain spin $\frac{1}{2}$: the two spin projections correspond to two states. The main point is that in all real cases complete shielding of the impurity spin of any magnitude occurs at $T=0$.

We will consider here the case of $T \ll T_K$, when the impurity spin is completely shielded, i.e., the total spin of the electron–impurity complex is practically equal to zero. However, this complex displays polarizability, which provides an indirect interaction between the electrons located in the vicinity of this complex: one of the electrons polarizes the complex; this polarization acts on another electron. Thus, an interaction arises between the electrons, which leads to Fermi-liquid effects. However, it must be kept in mind that these are local effects associated with the impurity atoms and that the corrections introduced by them are of order c_m (atomic concentration of the impurity).

Thus, the problem of the magnetic impurity in the gas model of noninteracting particles is replaced by the interaction between electrons and a nonmagnetic impurity; the electrons are considered to be interacting with one another and this interaction occurs only at sites of the impurity centers.

To begin with, we need certain information from scattering theory (see Landau and Lifshitz 1974). If a flux of free particles is elastically scattered on a centrosymmetric potential with its center at the point $r=0$, at large r values its wave function has the form

$$\Psi \approx \mathrm{e}^{ikz} + f(\theta)\frac{\mathrm{e}^{ikr}}{r} \tag{13.47}$$

(z is the direction of the incident flux). The first term corresponds to freely incident particles ($k = p/\hbar$) and the second to the scattered wave. The differential effective scattering cross section, i.e., the probability that the particle after being scattered will be directed into an element of solid angle $\mathrm{d}\Omega$, is given by

$$\mathrm{d}Q = |f(\theta)|^2\,\mathrm{d}\Omega. \tag{13.48}$$

The integral over $\mathrm{d}\Omega$ gives the total effective cross section.

The wave function (13.47) is axisymmetric with respect to the z axis. Any such function may be represented in the form of an expansion in states with a given orbital moment l:

$$\Psi = \sum_{l=0}^{\infty} a_l P_l(\cos\theta) R_{kl}(r), \tag{13.49}$$

where $P_l(\cos\theta)$ are Legendre polynomials; A_l are constants; and the functions R_{kl} satisfy the equation derived from the Schrödinger equation:

$$r^{-2}\frac{\partial}{\partial r} r^2 \frac{\partial R_{kl}}{\partial r} + \left(k^2 - \frac{l(l+1)}{r^2} - \frac{2m}{\hbar^2} U(r)\right) R_{kl} = 0. \tag{13.50}$$

Consider the solution of eq. (13.50) at large distances. It is not difficult to show that it has the form (see Landau and Lifshitz 1974)

$$R_{kl} \approx \frac{2}{r}\sin(kr - \tfrac{1}{2}l\pi + \delta_l), \tag{13.51}$$

where δ_l is the "scattering phase", which owes its origin to the potential $U(r)$. The coefficients A_l in formula (13.49) must be chosen such that the function $\Psi - \mathrm{e}^{ikz}$ has only terms proportional to e^{ikr}/r, i.e., corresponding to the diverging wave. Using the asymptotics of the expansion of the function e^{ikz} in Legendre polynomials,

$$\begin{aligned} \mathrm{e}^{iks} &= \exp(ikr\cos\theta) \\ &\approx \frac{1}{2ikr}\sum_{l=0}^{\infty}(2l+1)\,P_l(\cos\theta)[(-1)^{l+1}\mathrm{e}^{-ikr} + \mathrm{e}^{ikr}], \end{aligned} \tag{13.52}$$

we obtain a unique definition of the coefficients a_l.

As a result, the wave function at large distances from the center assumes the form (13.47), where

$$f(\theta)=\frac{1}{2ik}\sum_{l=0}^{\infty}(2l+1)[\exp(2i\delta_l)-1]P_l(\cos\theta). \tag{13.53}$$

Substituting into (13.48), integrating over θ and using the property of the Legendre polynomials

$$\int_{-\pi}^{\pi} P_l(\cos\theta)\, P_{l'}(\cos\theta)\sin\theta\, d\theta=\delta_{ll'}\frac{2}{2l+1}, \tag{13.54}$$

we obtain the total effective scattering cross section:

$$Q=\frac{4\pi}{k^2}\sum_{l=0}^{\infty}(2l+1)\sin^2\delta_l. \tag{13.55}$$

Thus, the total cross section is determined by the "scattering phases" of the asymptotics of the wave functions with a specified orbital moment and is the sum of "partial cross sections" corresponding to a certain orbital moment. The maximal value of $\sin x$ is 1 and therefore the maximal value of each partial cross section (called the "unitary limit") is given by

$$Q_{l,\max}=\frac{4\pi}{k^2}(2l+1)=\frac{4\pi\hbar^2}{p^2}(2l+1). \tag{13.56}$$

Suppose now that we are dealing with electrons in a metal containing impurities. Under the action of impurities the electronic wave functions are changed. If, for example, we take a sample in the form of a large sphere of radius L and set up the boundary condition $\Psi=0$ at the boundary, this will lead to a quantization of the wave vector in the asymptotic formula

$$R_{kl}\approx\frac{2}{r}\sin(kr-\tfrac{1}{2}l\pi)$$

namely, $k_{nl}=n\pi/L+\frac{1}{2}l\pi$.

Now let an impurity atom be located in the center of the sphere. The term δ_l will add to the sine argument. Since the boundary condition is not changed, it follows that the values of k must change:

$$k'_{nl}-k_{nl}=-\frac{\delta_l}{L}. \tag{13.57}$$

The number of states in the interval $k'-k$ is $(\delta_l/L)/(\pi/L)=\delta_l/\pi$. If the summation is carried out over all l values, assuming that for each orbital moment l there are $2l+1$ possible values of its projections m and that we have two possible spin projections, we obtain the total change in the number of states:

$$\Delta N=\frac{2}{\pi}\sum_l(2l+1)\,\delta_l.$$

The change in the number of states below the Fermi boundary is equal to the change in the number of electrons. By virtue of electroneutrality this must be equal to the excess charge of the impurity, i.e., to the difference in charge between the impurity ion and the ion of the main lattice. Thus, we have

$$\Delta Z = \frac{2}{\pi}\sum_l (2l+1)\delta_l(p_0) \tag{13.58}$$

(p_0 is the Fermi momentum). This formula is called the Friedel sum rule (Friedel 1952, 1954, 1958).

We thus see that the variation of the electronic characteristics under the action of an impurity may be expressed in terms of the scattering phases. In the case under consideration the electrons interact with the electron–impurity complex. The question of the size of this complex, i.e., the effective radius of the interaction forces, is not quite obvious. The most natural assumption is that at $T \ll T_K$ this interaction may be considered a point interaction. According to quantum mechanics (Landau and Lifshitz 1974), it is sufficient here to take into account only s-scattering. As the electron spin plays a significant role, we rewrite formula (13.58) in the form

$$\Delta Z = \frac{1}{\pi}\sum_\sigma \delta_\sigma(p_0), \tag{13.59}$$

where δ_σ corresponds to a phase with an orbital moment $l=0$ and a spin projection σ ($\sigma = \pm$).

We shall assume, in the spirit of the theory of the Fermi liquid, that due to electron–electron interaction the phases δ_σ depend on the distribution function. According to our assumption, the electron–impurity complex is regarded as a point complex at $T \ll T_K$; therefore, the interaction between electrons achieved by polarization of this complex, occurs at a single point in space. But, according to the Pauli principle, only electrons with opposite spin projection can both be at a single point. Because of this, interaction is possible only between electrons with opposite spins, from which it follows that

$$\delta_\sigma(\xi) = \delta_0(\xi) + \sum_{p'} \varphi(\xi,\xi')\, \delta n_{-\sigma}(\xi'). \tag{13.60}$$

We are interested here in the vicinity of the Fermi boundary; therefore expand all the quantities in $\xi = \varepsilon - \mu$:

$$\delta_0(\xi) \approx \delta_0 + \alpha\xi + O(\xi^2), \tag{13.61}$$

$$\varphi(\xi, \xi') \approx \varphi + O(\xi, \xi'), \tag{13.62}$$

where $\varphi = \text{const}$. We shall see that the subsequent expansion terms are unimportant as $T \to 0$.

Since the impurity acts only on the electron spin, the total electron density $n_{e^+} + n_{e^-}$ (we introduce the notation $n_{e^+} = \int n_+ \, d^3\boldsymbol{p}/(2\pi h)^3$) will be constant, i.e., $\delta n_{e^+} + \delta n_{e^-} = 0$. Hence,

$$\delta n_{e^+} - \delta n_{e^-} = 2\delta n_{e^+} = -2\delta n_{e^-} = n_{e^+} - n_{e^-} \equiv r. \tag{13.63}$$

Taking account of what has been given above, one may rewrite (13.60) as

$$\delta_\sigma(\xi) = \delta_0 + \alpha\xi - \tfrac{1}{2}\sigma\varphi r. \qquad (13.64)$$

Thus, it turns out that the scattering phases may be expressed in terms of two constants: α and φ. These constants are, in fact, interconnected due to the Kondo singularity being tied up to the Fermi level. If we make an equal shift in ξ and μ, then $\delta_\sigma(\xi)$ must not change. Because of this we have

$$\alpha + \tfrac{1}{2}\nu_0\varphi = 0, \qquad (13.65)$$

where $\nu_0 = p_0 m/(\pi^2 h^3)$ is the density of states for a pure metal summed up over the spin projections (the integral (13.60) includes only $\partial n_-/\partial\varepsilon$, which leads to the factor $\frac{1}{2}$).

Let us now examine formula (13.59). On the left-hand side is the change of the number of electrons below the Fermi level. An increase in this number has occurred due to a decrease in the energy level. Hence, for a given spin projection

$$-\tfrac{1}{2}\nu_0 V\,\delta\varepsilon_\sigma = \frac{\delta_\sigma}{\pi} \qquad (13.66)$$

(V is the volume of the system). The variation of the density of states for a given spin projection is

$$\frac{\partial n_{e\sigma}}{\partial(\varepsilon+\delta\varepsilon_\sigma)} - \frac{\partial n_{e\sigma}}{\partial\varepsilon} = \frac{\partial n_{e\sigma}}{\partial\varepsilon}\left(\frac{\partial\varepsilon}{\partial(\varepsilon+\delta\varepsilon_\sigma)} - 1\right)$$

$$= -\tfrac{1}{2}\nu_0\frac{\partial\delta\varepsilon_\sigma}{\partial\varepsilon} = \frac{\alpha}{\pi V}$$

($n_{e\sigma}$ is the density of electrons with spin projection σ). This quantity must be multiplied by N_m, the number of impurities. As a result, V^{-1} is replaced by n_m, which is the number of impurities per unit volume.

Since the heat capacity is proportional to the density of states, it follows that

$$\frac{\delta C}{C} = \frac{\alpha n_m}{\pi}\frac{1}{\frac{1}{2}\nu_0} = \frac{2\alpha n_m}{\nu_0\pi}. \qquad (13.67)$$

Thus, we conclude that in the heat capacity the coefficient in the linear temperature law changes.

Now we shall consider the magnetic susceptibility. In the presence of a magnetic field $n_+ \neq n_-$. According to formulae (13.66) and (13.64), the corrected energy values are equal to

$$\xi'_+ = \xi - \beta H - \frac{2\alpha\xi}{\pi\nu_0 V} + \frac{r\varphi}{\pi\nu_0 V},$$

$$\xi'_- = \xi + \beta H - \frac{2\alpha\xi}{\pi\nu_0 V} - \frac{r\varphi}{\pi\nu_0 V}.$$

At the Fermi boundary these values are identical and equal to zero. The values of ξ obtained here are denoted by $\delta\mu_+$ and $\delta\mu_-$. We have

$$\delta\mu_+ = \frac{\beta H - \dfrac{r\varphi}{\pi\nu_0 V}}{1 - \dfrac{2\alpha}{\pi\nu_0 V}}, \qquad \delta\mu_- = \frac{-\beta H + \dfrac{r\varphi}{\pi\nu_0 V}}{1 - \dfrac{2\alpha}{\pi\nu_0 V}},$$

$$r = n_{e^+} - n_{e^-} = n_{e^+}(\mu + \delta\mu_+) - n_{e^-}(\mu + \delta\mu_-)$$

$$= \frac{\nu_0\left[\beta H - \dfrac{r\varphi}{\pi\nu_0 V}\right]}{1 - \dfrac{2\alpha}{\pi\nu_0 V}}.$$

This is the equation for r. Solving it, we find the magnetic moment $M = \beta r$ and the magnetic susceptibility $\chi = M/H$. As a result, we have

$$\chi = \beta^2\nu_0\left(1 + \frac{2\alpha}{\pi\nu_0 V} - \frac{\varphi}{\pi V}\right) \tag{13.68}$$

(we have taken into account that the corrections corresponding to a single impurity are of order V^{-1}, where V is the volume of the system, i.e., they are small and we can carry out an expansion in them). Again, as before, the correction to χ must be multiplied by N_m, the number of impurities, which leads to the replacement of V^{-1} by n_m. Using relation (13.65), we find

$$\frac{\delta\chi}{\chi} = \frac{4\alpha n_m}{\pi\nu_0}. \tag{13.69}$$

Comparing this formula with eq. (13.67), we obtain

$$\frac{\delta\chi/\chi}{\delta C/C} = 2. \tag{13.70}$$

This relation is derived in the detailed theory (Wilson 1974, 1975; Tsvelick and Wiegmann 1983) as a result of extremely difficult computations*. It contains no constants characterizing a particular metal or impurity. According to formula (13.64), the coefficient α has the dimension of inverse energy. Evidently, by order of magnitude it is equal to T_K^{-1} (T_K is the Kondo temperature; see section 4.6), although its exact relationship with the constant J in the interaction energy of the electron and impurity spins (4.39) can be established only in the detailed theory, which gives (Wilson 1974, 1975):

$$\alpha^{-1} = 3.08\varepsilon_F\left(|J|\frac{\nu_0}{n}\right)^{1/2} \exp\left[-\frac{n}{|J|}\nu_0\right].$$

* The extension of the above calculation to the model with a finite orbital moment (Nozières and Blandin 1980, Mihaly and Zawadowsky 1978) gives $\frac{2}{3}(2l+3)$ on the right-hand side of formula (13.70); at $l = 0$ we have 2.

If we denote $\alpha \equiv T_K^{-1}$, then, considering that $C = \frac{1}{3}\pi^2 \nu_0 T$, we find from (13.67) the impurity part of the heat capacity per unit volume:

$$C_m = \frac{2n_m\alpha}{\nu_0\pi}\frac{\pi^2\nu_0 T}{3} = \frac{2}{3}\pi n_m \frac{T}{T_K}. \tag{13.71}$$

Using formula (13.69) and the expression for the paramagnetic susceptibility of electrons, $\chi = \beta^2\nu_0$, we find the impurity part:

$$\chi_m = \frac{4n_m\alpha}{\nu_0\pi}\cdot\beta^2\nu_0 = \frac{\frac{4}{\pi}n_m\beta^2}{T_K}. \tag{13.72}$$

We now turn to the conductivity. If only s-scattering is possible, the formula for conductivity may be written directly in accordance with section 3.3, with account taken of the fact that the scattering probability depends on ξ and therefore should not be replaced by its value at $\xi = 0$. Thus, we find

$$\sigma = -\frac{1}{6}\nu_0 e^2 v^2 \sum_\sigma \int d\xi \frac{\partial n_0}{\partial \xi} W_\sigma^{-1}(\xi), \tag{13.73}$$

where $W = \tau^{-1}$ is the total probability of all scattering processes.

At $T = 0$ only elastic scattering processes are possible since both the electronic system and the electron-impurity complex are in the ground state. The probability of elastic scattering can be found from the expression for the effective cross section:

$$W_\sigma = n_m v Q_\sigma \tag{13.74}$$

[the validity of this relation can be checked, in particular, in the following way: $W = \tau^{-1} = v/l = v n_m Q$ (see section 4.7)].

Upon substitution of Q_σ in accordance with formulas (13.55), (13.60), and (13.61), we see that one more constant appears, namely δ_0. As regards δ_0, we can make the following assumption, which is, in effect, a consequence of the model adopted. We proceed from the fact that at $T = 0$ complete screening of the impurity spin by electrons takes place. But this implies that the effective interaction between the electrons and the impurity with $T \to 0$ (or $\xi \to 0$) becomes infinitely large. In this case, the effective scattering cross section for electrons must attain a maximum value (a unitary limit), which corresponds to $\delta = \frac{1}{2}\pi$. In view of this, we put $\delta_0 = \frac{1}{2}\pi$. Here

$$\sin^2 \delta_0(\xi) \approx \cos^2 \alpha\xi \approx 1 - \alpha^2\xi^2.$$

Restricting ourselves to the term with $l = 0$ in (13.55), inserting (13.74) into (13.73) and integrating, we find

$$\sigma = \sigma(0)(1 + \frac{1}{3}\pi^2\alpha^2 T^2),$$

where

$$\sigma(0) = \frac{1}{12}\pi\hbar e^2 v^2 \frac{\nu_0^2}{n_m} = \frac{p_0 e^2}{4\pi c_m \hbar^2} \tag{13.75}$$

in which $c_m = n_m/n_e$.

The expression found for the conductivity holds at $T=0$, when only elastic scattering is possible. However, at $T\neq 0$ there are also possible inelastic scattering processes, which introduce corrections of the same order $(\alpha T)^2$ as the correction found above for elastic scattering. Inclusion of inelastic scattering requires the introduction of certain changes in the scattering theory formulas used earlier (see Landau and Lifshitz 1974). For the amplitude of elastic scattering $f(\theta)$ a more general formula is obtained instead of (13.53):

$$f(\theta)=\frac{1}{2ik}\sum_{l=0}^{\infty}(2l+1)[S_l-1]P_l(\cos\theta), \tag{13.76}$$

where S_l are certain complex quantities which are less than unity in modulus. In accordance with this, for the total elastic scattering cross section, instead of eq. (13.55), we have

$$Q_{\mathrm{el}}=\frac{\pi}{k^2}\sum_{l=0}^{\infty}(2l+1)|1-S_l|^2. \tag{13.77}$$

If we represent the asymptotics of the wave function at large distances as a superposition of the converging and diverging waves, then in purely elastic scattering their amplitudes are equal (formula 13.51) and in the presence of inelastic scattering the intensity of the diverging wave with given l is attenuated as compared with the intensity of the converging wave by a factor $|S_l|^2$. Since this weakening is associated with inelastic scattering, the effective cross section of inelastic scattering has the form

$$Q_{\mathrm{in}}=\frac{\pi}{k^2}=\sum_{l=0}^{\infty}(2l+1)(1-|S_l|^2). \tag{13.78}$$

The total cross section is given by

$$Q=Q_{\mathrm{el}}+Q_{\mathrm{in}}=\frac{2\pi}{k^2}\sum_{l=0}^{\infty}(2l+1)(1-\mathrm{Re}\,S_l). \tag{13.79}$$

According to our assumption only scattering with $l=0$ is important.

The temperature is assumed to be low, i.e., inelastic processes are not strong. In this case, $S_0=(1-s)\,\mathrm{e}^{2i\delta_0(\xi)}$, where $s\ll 1$. Substituting this into (13.78) and (13.79), we find that

$$Q_{\mathrm{in}}=\frac{2\pi s}{k^2},$$

$$Q=\frac{2\pi}{k^2}[1-\cos 2\delta_0(\xi)+s\cos 2\delta_0(\xi)]$$

$$=\frac{4\pi}{k^2}\sin^2\delta_0(\xi)+Q_{\mathrm{in}}\cos 2\delta_0(\xi). \tag{13.80}$$

Substituting (13.61) with $\delta_0 = \frac{1}{2}\pi$ and taking into account that, because of the smallness of Q_{in} in the second term, we may limit ourselves to $\delta_0(\xi) \approx \delta_0 = \frac{1}{2}\pi$, we obtain

$$Q = \frac{4\pi}{k^2}(1 - \alpha^2\xi^2) - Q_{\text{in}}. \tag{13.81}$$

At first sight, this formula seems paradoxical in the sense that the inelastic scattering cross section is subtracted, not added. This is associated with the fact that, because of inelastic processes, part of the incident beam is absorbed and, hence, elastic scattering becomes weaker; near the unitary limit ($\delta = \frac{1}{2}\pi$) this decrease appears to be dominant.

Now we are to find Q_{in}. In calculating this correction we proceed from the concept that the binding energy of the singlet electron–impurity complex is of order T_{K}. Therefore, the probability of this complex being broken down is of order $\exp(-T/T_{\text{K}})$, i.e., much less than the correction required. Hence, we may speak of the consequences of the indirect polarization interaction between electrons. As a result, the process shown in fig. 7 becomes possible, i.e., the electron makes a transition into a new state, giving rise to two quasiparticles. It must be borne in mind that the interaction occurs at a given point in space; it means that the total momentum of the electrons is not conserved. If use is made of the gas model, the corresponding contribution to the collision integral is given by (section 4.4)

$$\begin{aligned} I(n) = {} & \frac{2\pi}{\hbar} n_{\text{m}}(\tfrac{1}{2}\nu_0)^3 |g|^2 \\ & \times \int \mathrm{d}\xi_2 \, \mathrm{d}\xi_1' \, \mathrm{d}\xi_2' \delta(\xi - \xi_2 - \xi_1' - \xi_2') \\ & \times [(1-n)(1-n_2)n_1'n_2' - nn_2(1-n_1')(1-n_2')]. \end{aligned} \tag{13.82}$$

Here g is the interaction constant (recall that it is assumed to be a point interaction and to operate only between electrons with opposite spins).

Upon substitution of the equilibrium distribution functions the brackets vanishes. But if we put $n = n_0 + \delta n$, then, taking the integrals, we obtain $I(n) = -\delta n W_{\text{in}}$, where

$$W_{\text{in}} = \frac{\pi}{\hbar}(\tfrac{1}{2}\nu_0)^3 n_{\text{m}}(\pi^2 T^2 + \xi^2)|g|^2. \tag{13.83}$$

Since the scattering probabilities and the effective cross sections are connected by relation (13.74), by multiplying (13.81) by $n_{\text{m}}v$, we find

$$W = \frac{4n_{\text{m}}}{\nu_0 \pi \hbar}(1 - \alpha^2\xi^2) - \frac{\pi}{\hbar}(\tfrac{1}{2}\nu_0)^3 n_{\text{m}}(\pi^2 T^2 + \xi^2)|g|^2.$$

It now remains to express the interaction constant g. It is rather obvious that since g and φ in formula (13.60) originate from a single source – an indirect interaction of electrons through the screened impurities - it follows that to first order in c_{m} they must be proportional to each other. The proportionality factor can be found as follows: $\varphi \int \delta n \, \mathrm{d}^3\boldsymbol{p}(2\pi\hbar)^{-3}$ characterizes the change in phase and $(g/V)\int \delta n \, \mathrm{d}^3\boldsymbol{p}(2\pi\hbar)^{-3}$ characterizes the change in energy (the diagonal matrix element from the interaction). But these quantities are connected by relation (13.66). Hence, it follows that

$$g = \frac{2}{\pi\nu_0}\varphi = -\frac{4}{\pi\nu_0^2}\alpha \tag{13.84}$$

(see formula 13.65). Substituting this into the formula for W, we have

$$W = \frac{4n_{\mathrm{m}}}{\nu_0\pi\hbar}[1 - \tfrac{3}{2}\alpha^2\xi^2 - \tfrac{1}{2}\pi^2\alpha^2T^2]. \tag{13.85}$$

Using this expression in the formula (13.73) for the conductivity, and integrating over ξ, we obtain

$$\sigma = \sigma(0)\,(1 + \pi^2\alpha^2T^2) = \sigma(0)\left[1 + \pi^2\left(\frac{T}{T_{\mathrm{K}}}\right)^2\right], \tag{13.86}$$

where $\sigma(0)$ is expressed by formula (13.75).

The total resistance is measured experimentally. The conductivity (13.85) we have found corresponds to taking account of only the exchange electron–impurity interaction. Therefore, we give here the expression for the total resistance at $T \ll T_{\mathrm{K}}$:

$$\rho = \rho_v + \rho_J(0)\left[1 - \pi^2\left(\frac{T}{T_{\mathrm{K}}}\right)^2\right] \qquad \rho_J(0) = \sigma^{-1}(0).$$

The term ρ_v corresponds to both the scattering from ordinary impurities and the potential scattering from magnetic impurities. Note that $\rho_J(0)$ is of the same order of magnitude as the part ρ_v associated with magnetic impurities. Indeed, the latter is of order

$$\frac{m}{n_{\mathrm{e}}e^2\tau} \sim \frac{p_0}{n_{\mathrm{e}}e^2 l} \sim \frac{p_0 n_{\mathrm{m}} Q}{n_{\mathrm{e}}e^2} \sim \frac{\hbar^2 c_{\mathrm{m}}}{e^2 p_0} \sim \frac{1}{\sigma(0)},$$

where we have substituted $Q \sim a^2 \sim (\hbar/p_0)^2$.

Thus, even in a case where we have no way for finding the relationship between T_{K} and the interaction constant J (see eq. 4.39) and cannot therefore describe in a unified manner the properties of the metal at high and low temperatures*, we have three formulas which characterize the low-temperature dependences $[C_{\mathrm{m}}(T), \chi_{\mathrm{m}}(T)$

* A unified description of the sharply changing behavior of any quantity in various ranges of values of an external parameter (say, temperature) is called a crossover.

and $\sigma(T)/\sigma(0)]$, which contain only one parameter, T_K, thereby making it possible to test the theory in a rigorous way. Of course, as has been said above, we have dealt with the model of an impurity with spin $S=\frac{1}{2}$, and for a quantitative comparison with experiment in particular systems we need an appropriate theory. However, all the temperature dependences and orders of magnitudes found above are in qualitative agreement with experiment for any systems.

14

Methods for calculating electronic spectra of metals

14.1. The orthogonalized plane wave method

From the foregoing it is clear that there exist many methods which enable us to extract detailed characteristics of electronic spectra. We can find practically the entire Fermi surface and the velocity at each point on this surface. The question arises of the possibility to compute these characteristics of spectra on the basis of the picture of interacting electrons and nuclei.

This is also interesting from another point of view. Experimental data always provide us only with information about pieces of the surface or, more exactly, pieces of the contours of certain planar cross sections of the surface or the areas of these cross sections. To compare all these data and to deduce the entire Fermi surface is easy only in the simplest cases, but usually this is a formidable task. Therefore, if there were a theory providing, at least, the general shape of the Fermi surface, the interpretation of experimental data would be substantially simplified.

As has been noted in ch. 1, two limiting cases may be considered: strongly coupled electrons and almost free electrons. Strictly speaking, none of the expansion parameters used in both these methods is really small, and therefore these methods lead to poorly converging or even diverging series. It is for this reason that, beginning from the 1930s, attempts have been made to find a sufficiently reliable and rapidly converging procedure for calculating electronic spectra and possibly also the binding energy of metals and kinetic coefficients. From then, a large and, in fact, rather independent area of the theory of metals has grown that is concerned with these problems. It is intimately tied up with computer calculations, and the mathematical aspect naturally dominates over the physical aspect. A description of this area is beyond the scope of the present book. The reader interested in the corresponding range of problems is referred, for example, to the books by Ziman (1971) and Ziesche and Lehmann (1983). However, for the sake of completeness, we shall briefly describe a number of ideas lying at the basis of the calculation methods. It is convenient to consider first the so-called orthogonalized plane wave (OPW) method (Herring 1940), although this was not the first method. It is a certain combination of the strong- and weak-coupling approximations.

We begin with the Schrödinger equation for an electron in a periodic field:

$$\mathcal{H}\psi = \varepsilon\psi,$$

where

$$\mathscr{H} = -\frac{\hbar^2\nabla^2}{2m} + V(\boldsymbol{r}).$$

This problem is equivalent to finding the minimum of the integral

$$\int \psi^* \mathscr{H}\psi \, \mathrm{d}V - \varepsilon \int \psi^*\psi \, \mathrm{d}V. \tag{14.1}$$

Let us seek this minimum with the aid of a certain trial function ψ_{tr}, which is defined as a linear combination of some well-known functions:

$$\psi_{\mathrm{tr}} = \int_i a_i\chi_i. \tag{14.2}$$

The coefficients a_i are chosen so as to minimize expression (14.1). If we insert eq. (14.2) into eq. (14.1) and differentiate with respect to all a_i, we shall obtain a system of homogeneous linear equations for a_i. The condition of their compatibility is the vanishing of the determinant of the system. The elements of this determinant are the matrix elements of the operator $\mathscr{H} - \varepsilon$ with respect to the functions χ_i. Thus we have

$$\mathrm{Det}\left[\int \chi_i^* \mathscr{H}\chi_k \, \mathrm{d}V - \varepsilon \int \chi_i^*\chi_k \, \mathrm{d}V\right] = 0. \tag{14.3}$$

This just the required equation for the determination of ε.

In the OPW method the functions χ_i are constructed as follows. It is assumed that the strong-coupling approximation applies for inner-shell electrons. While the wave function of an electron in a free atom was

$$u_{nlm} = Y_l^m(\theta, \varphi)R_{nl}(\boldsymbol{r}), \tag{14.4}$$

where Y_l^m is a spherical function, in a metal use may be made of the Wannier functions (section 1.2):

$$\psi_{nlm;\boldsymbol{p}}(\boldsymbol{r}) = N^{-1/2} \sum_\nu \mathrm{e}^{\mathrm{i}\boldsymbol{p}\boldsymbol{a}_\nu/\hbar}\, u_{nlm}(\boldsymbol{r} - \boldsymbol{a}_\nu), \tag{14.5}$$

where Y_l^m is a spherical function, in a metal use may be made of the Wannier functions (section 1.2):

$$\chi_{\boldsymbol{p}} = V^{-1/2}\, \mathrm{e}^{\mathrm{i}\boldsymbol{p}\boldsymbol{r}/\hbar} - \sum_{nlm} B_{nlm;\boldsymbol{p}}\psi_{nlm;\boldsymbol{p}}. \tag{14.6}$$

This is a plane wave, from which a term, which is a linear combination of the wave functions of the inner electrons, has been subtracted.

The coefficients $B_{nlm;\boldsymbol{p}}$ are chosen so as to make the function (14.6) orthogonal to all $\psi_{nlm;\boldsymbol{p}'}$ (with any $\boldsymbol{p}'$). It can be proved that this is provided by subtracting a linear combination with the same $\boldsymbol{p}$ as in the plane wave. The orthogonality condition gives a unique determination of the coefficients $B_{nlm;\boldsymbol{p}}$.

If we wish to determine the energy for a certain value of the quasimomentum $\boldsymbol{p}$, we use the function χ_i in the form $\chi_i = \chi_{p+\hbar K_i}$, where $\boldsymbol{K}_i$ are the reciprocal lattice vectors, and each of the functions $\chi_{p+\hbar K_i}$ has the form (14.6). The larger the number of such functions we use, the higher is the accuracy obtained.

However, in order to calculate the determinant (14.3), one needs not only the function χ_i but also the periodic potential $V(\boldsymbol{r})$, which enters into the Schrödinger equation. There are various ways to accomplish this task. For example, we may take the following expression for this potential:

$$V(\boldsymbol{r}) = \sum_{\nu} V_{\mathrm{AT}}(\boldsymbol{r} - \boldsymbol{a}_\nu), \tag{14.7}$$

where the "atomic" potential V_{AT} is found as the sum of the nuclear potential, the Coulomb potential of electrons and the so-called exchange potential, which takes account of the antisymmetry of the total electronic wave function:

$$V_{\mathrm{AT}}(\boldsymbol{r}) = -\frac{Ze}{|\boldsymbol{r}|} + e^2 \sum_{j=1}^{Z} \frac{u_j^*(\boldsymbol{r}')\, u_j(\boldsymbol{r}')}{|\boldsymbol{r} - \boldsymbol{r}'|}\, \mathrm{d}V' \\ -3e^2 \left[\frac{3}{8\pi} \sum_{j=1}^{Z} u_j^*(\boldsymbol{r})\, u_j(\boldsymbol{r})\right]^{1/3}, \tag{14.8}$$

where u_j are the wave functions of electrons in an isolated atom (i.e., the functions 14.4).

The OPW method gives good agreement with experiment for simple substances, in which the d- and f-states are far enough from the valence band. However, in the case of transition metals with unfilled d- or f-shells and also in the case of noble metals, where the d-states of electrons are close to the valence band and have a marked effect on the spectrum, the OPW method works much worse.

Moreover, the complexity of calculations by this method makes it unsuitable for a rapid interpretation of experimental data.

14.2. The pseudopotential method

A substantial progress with respect to both the refinement of fundamental computations and the elaboration of a semiphenomenological but "high-speed" procedure has been the development of the pseudopotential method (Phillips and Kleinman 1959; Bassani and Celli 1959). It is based on the following idea. Let us seek the wave functions of valence electrons not in the form (14.2) or (14.6) but in the form of some more general and unknown functions, from which the corresponding linear combinations are subtracted for orthogonalization with respect to the atomic functions, as in (14.6):

$$\psi_i = \varphi_i - \sum_n B_n^i \psi_n. \tag{14.9}$$

Here, for simplicity, we have used a single subscript, n, in place of nlm. The functions ψ_n correspond to the electrons of the inner shells. These functions are similar to (14.5). The orthogonality condition gives

$$B_n^i = \int \psi_n^*(\boldsymbol{r})\, \varphi_i(\boldsymbol{r})\, \mathrm{d}V.$$

By substituting the functions (14.9) into the Schrödinger equation, we get

$$\mathcal{H}\varphi_i - \sum_n B_n^i(\varepsilon_n - \varepsilon)\psi_n = \varepsilon\varphi_i.$$

Here we assume that $\mathcal{H}\psi_n = \varepsilon_n\psi_n$. We introduce an integral operator V_i such that

$$V_i\varphi_i = -\sum_n B_n^i(\varepsilon_n - \varepsilon)\psi_n$$

$$= -\sum_n (\varepsilon_n - \varepsilon)\psi_n \int \psi_n^*(\boldsymbol{r}')\, \varphi_i(\boldsymbol{r}')\, \mathrm{d}V'. \tag{14.10}$$

Since $\varepsilon > \varepsilon_n$, then with $\boldsymbol{r}$ close to the positions of the ions this quantity is positive, i.e., near the ions the potential V_i corresponds to repulsion. The Schrödinger equation assumes the form

$$(\mathcal{H} + V_i)\varphi_i = \varepsilon\varphi_i. \tag{14.11}$$

This equation has the form of the Schrödinger equation for the function φ_i with the total potential $W = V(\boldsymbol{r}) + V_i$, which is known as the pseudopotential.

However, the operator V_i and, hence, the pseudopotential W are defined ambiguously because, according to formula (14.9), any linear combination of the functions ψ_n may be incorporated into the function φ_i; this will be compensated for by a corresponding correction to the second term of (14.9). Thus, the function ψ_i does not change in such a case, but the operator V_i will be different. The most general form of the operator V_i is

$$V_i\varphi_i = \sum_n \psi_n(\boldsymbol{r}) \int F_n(\boldsymbol{r}')\, \varphi_i(\boldsymbol{r}')\, \mathrm{d}V' \tag{14.12}$$

where $F_n(\boldsymbol{r})$ are quite arbitrary functions. It can be shown that the energy levels are independent of the choice of $F_n(\boldsymbol{r})$. In view of this, some criterion is needed for the unambiguous choice of the pseudopotential. The most natural requirement is that of the maximum smoothness of the function φ_i, i.e., $\int (\nabla\varphi_i)^2\, \mathrm{d}V = \min$. The result of such a requirement is that the potential V_i almost fully compensates the potential $V(\boldsymbol{r})$, included in $\mathcal{H}$, in the region of the ion core. This is rather obvious. For the orthogonality condition to be fulfilled with respect to the wave functions of the inner shells of ions the functions ψ_i must rapidly oscillate in the vicinity of the ion cores. The latter circumstance leads to the appearance of a large kinetic energy, which is equal to $\hbar^2 \int |\nabla\psi|^2\, \mathrm{d}V/2m$ according to quantum mechanics. The transition to the smooth functions φ_i implies that the greater part of this kinetic energy is incorporated into the potential energy, as a result of which a strong compensation takes place.

A rather simple choice of the pseudopotential, which satisfies the smoothness criterion for the wave function, is the so-called Austin pseudopotential (Austin et al. 1962), with which in (14.12) it is assumed that $F_n(\boldsymbol{r}) = -\psi_n(\boldsymbol{r})\, V(\boldsymbol{r})$. This gives for the total pseudopotential in the equation for φ_i

$$W^{\mathrm{A}}\varphi_i = \int w^{\mathrm{A}}(\boldsymbol{r}, \boldsymbol{r}')\, \varphi_i(\boldsymbol{r}')\, \mathrm{d}V',$$

$$w^{\mathrm{A}}(\boldsymbol{r}, \boldsymbol{r}') = V(\boldsymbol{r}')\left\{\delta(\boldsymbol{r}-\boldsymbol{r}') - \sum_n \psi_n^*(\boldsymbol{r}')\, \psi_n(\boldsymbol{r})\right\}. \tag{14.13}$$

If the system of functions ψ_n were complete, then w^{A} would be equal to zero. In fact, ψ_n corresponds only to deep localized states. The choice of the pseudopotential in the form (14.12), therefore, practically leaves the potential unchanged outside the ion cores and almost entirely elminates the large negative potential inside the core, thereby allowing for the smoothness of the wave function.

Thus, we conclude that the solution of the complete problem may be replaced by the solution of the Schrödinger equation:

$$(-\hbar^2\nabla^2/2m + W)\varphi_p(\boldsymbol{r}) = \varepsilon(\boldsymbol{p})\, \varphi_p(\boldsymbol{r}), \tag{14.14}$$

where W is a certain smooth potential, which is not too large even in the vicinity of the ion core. Although the wave functions obtained in this way will not be correct in the vicinity of the core, the energy levels will nevertheless be determined with sufficient accuracy.

Although in accordance with formula (14.12) the pseudopotential is a nonlocal integral operator of the type

$$W\varphi = \int w(\boldsymbol{r}, \boldsymbol{r}')\, \varphi(\boldsymbol{r}')\, \mathrm{d}V',$$

it turns out that very accurate results can be obtained for $\varepsilon(\boldsymbol{p})$ by using a local operator instead, i.e., simply a certain periodic function $W(\boldsymbol{r})$ and, moreover, a function of the type

$$W(\boldsymbol{r}) = \sum_\nu U(\boldsymbol{r} - \boldsymbol{a}_\nu) \tag{14.15}$$

where the sum runs over the positions of the atoms and the function $U(r)$ is considered to be isotropic (Brust 1964). This function is chosen by way of comparing the results either with experimental findings or with the results of calculations performed according to the OPW method. In a practical realization of this procedure it turned out that the functions $U(r)$ for different metals are very close to one another and the difference in the spectra stems, in fact, from the difference in the number of valence electrons and crystal lattices.

The choice of the function $U(r)$ actually reduces to the determination of several numbers. This is associated with the fact that, due to the smoothness of the functions $\varphi_p(\boldsymbol{r})$, they can be expanded in plane waves (see the Bloch theorem 1.15):

$$\varphi_p(\boldsymbol{r}) = N^{-1/2} \sum_{\boldsymbol{K}} c_{\boldsymbol{p}+\hbar\boldsymbol{K}} \exp\left(\frac{\mathrm{i}(\boldsymbol{p}+\hbar\boldsymbol{K})\boldsymbol{r}}{\hbar}\right).$$

From eq. (14.14) we obtain the equation for $c_{\boldsymbol{p}+\hbar\boldsymbol{K}}$:

$$\left[\frac{(\boldsymbol{p}+\hbar\boldsymbol{K})^2}{2m} - \varepsilon(\boldsymbol{p})\right] c_{\boldsymbol{p}+\hbar\boldsymbol{K}} + \sum_{\boldsymbol{K}_1} W_{\boldsymbol{p}+\hbar\boldsymbol{K},\boldsymbol{p}+\hbar\boldsymbol{K}_1} c_{\boldsymbol{p}+\hbar\boldsymbol{K}_1} = 0. \tag{14.16}$$

According to eq. (14.15), the matrix elements $W_{\boldsymbol{pp}'}$ are given by

$$W_{\boldsymbol{p},\boldsymbol{p}+\hbar\boldsymbol{K}} = U(K)\left(N^{-1} \sum_{\boldsymbol{a}_\nu} \mathrm{e}^{\mathrm{i}\boldsymbol{K}\boldsymbol{a}_\nu}\right) = U(K)\, S(\boldsymbol{K}). \tag{14.17}$$

Here $U(K)$ is the matrix element of the function $U(r)$ and $S(\boldsymbol{K})$ is the so-called structure factor, which depends only on the lattice geometry. If the crystalline lattice coincides with the Bravais lattice (one atom in the unit cell), then $S(\boldsymbol{K}) = 1$. In a more general case $S(\boldsymbol{K}) \neq 1$. Since in practice only a few of the lowest values of K are taken, then, according to eq. (14.17), instead of the entire function it is sufficient to choose several values of $U(K)$.

This is sufficient for a particular substance. But if one proceeds from the supposition that the functions $U(r)$ differ little for different metals, then it is desirable to have the entire function $U(r)$ or, more precisely, its Fourier transform $U(k)$, from which it is possible to find values of $U(K)$ for each particular lattice. The points found originally for germanium (Brust 1964) are nicely interpolated by the function (Falicov and Golin 1965)

$$U(k) = A_1(k^2 - A_2)\frac{1}{\exp[A_3(k^2 - A_4)] + 1}. \tag{14.18}$$

If k and A_i are expressed in atomic units ($m = \hbar = e = 1$, i.e., the length, k^{-1} in units equal to the Bohr radius, $r_B = \hbar^2/me^2 = 0.529$ Å, and the energy in Rydbergs, $\mathrm{Ry} = me^4/\hbar^2 = 27.2$ eV), then the coefficients in this formula will be equal to

$$A_1 = 0.0655, \qquad A_2 = 2.78, \qquad A_3 = 2.38, \qquad A_4 = 3.70.$$

Note that the quantity A_1, which governs the scale of the potential, is actually small since the Fermi energy in atomic units is usually of the order of unity. Moreover, it should be noted that the lattices of metals are usually rather symmetric, so that there are many values of $\boldsymbol{K}$ with the same magnitude. Therefore, even the first few values of $|\boldsymbol{K}|$ provide a sufficiently wide basis for finding the energy spectrum.

This procedure has been used to calculate a number of energy spectra, including the spectra of semimetals, such as antimony and arsenic. Since the number of conduction electrons in semimetals is small ($\sim 10^{-2}$ per atom in As and 10^{-3} per atom in Sb), the accuracy of calculations of the Fermi surface must be very high. It has turned out that the results are in good agreement with experiment.

Perhaps the only example where this method does not apply is bismuth, in which the number of conduction electrons is of the order of 10^{-5} per atom. Here, for the Fermi surface the pseudopotential method gives results which differ from experimental data by a factor of 2–3. Hence, it may be concluded that the accuracy of this method can be brought to a few percent.

It should also be noted that the use of simple model local pseudopotentials chosen according to the electronic spectra makes it possible to reproduce theoretically the measured phonon spectra for nontransition metals (Brovman and Kagan 1974) and the temperature dependences of the kinetic coefficients (Zhernov and Kagan 1978) with an accuracy of 15–20%.

Nevertheless, the use of the simplified pseudopotential (14.15) cannot be regarded as quite satisfactory. First of all, it does not reflect the nonlocal character of the true pseudopotential. Furthermore, in order to find the Fourier component of $U(\boldsymbol{K})$, one has to resort to the OPW method, which calls for unwieldy calculations, or one has to choose the appropriate constants from a comparison with experiment, which strongly reduces the value of this procedure as a microscopic theory. Therefore, methods have been worked out for constructing "model" pseudopotentials.

The most popular method is the Heine-Abarenkov-Shaw method (Heine and Abarenkov 1964, 1965; Shaw 1968), which will be briefly described here. Let us return to formula (14.15), but now the quasiatomic potential U is assumed to be nonlocal. In this case,

$$W\varphi = \int \sum_{\nu} U(\boldsymbol{r}-\boldsymbol{a}_\nu, \boldsymbol{r}'-\boldsymbol{a}_\nu)\, \varphi(\boldsymbol{r}')\, \mathrm{d}V'. \tag{14.19}$$

If we expand φ in plane waves, as was done earlier, we shall obtain eq. (14.16), this time with the coefficients

$$W_{\boldsymbol{p}+\hbar\boldsymbol{K},\boldsymbol{p}+\hbar\boldsymbol{K}_1} = U\left(\frac{\boldsymbol{p}}{\hbar}+\boldsymbol{K}, \frac{\boldsymbol{p}}{\hbar}+\boldsymbol{K}_1\right) S(\boldsymbol{K}_1-\boldsymbol{K}), \tag{14.20}$$

where $U(\boldsymbol{k}, \boldsymbol{k}')$ is the Fourier component of the potential $U(\boldsymbol{r}, \boldsymbol{r}')$ and $S(\boldsymbol{K})$ is the structure factor.

The quasiatomic potential U is constructed as follows. We start with the Schrödinger equation for a particle in the centrosymmetric potential. Representing the wave function in the form (14.4), we obtain the equation for the functions $R_{nl}(r)$:

$$-\frac{\hbar^2}{2m}\left(\frac{1}{r^2}\frac{\partial}{\partial r}r^2\frac{\partial R_{nl}}{\partial r} - \frac{l(l+1)}{r^2}R_{nl}\right) + U(r)R_{nl} = \varepsilon R_{nl}. \tag{14.21}$$

In fact, we are dealing here not with one electron in the field of the given potential but with a many-electron system. It is simulated in the following way. A special potential U_l is chosen for each of eqs. (14.21). At $r > r_l$ it coincides with the atomic potential, i.e., $V_a(r) = -Ze^2/r$ (Ze is the charge on the ion), and at $r < r_l$ it is assumed to be a constant: $-A_l(\varepsilon)$. For smooth matching it is necessary that

$$-A_l(\varepsilon) = V_a(r_l). \tag{14.22}$$

This cut-off of the Coulomb potential at $r < r_l$ is made only for $l \leq l_0$; usually $l_0 = 1$ or 2. This is because the positive centrifugal energy $\hbar^2 l(l+1)/2mr^2$, which is added to $U(r)$ in (14.21), does not allow the electron to approach the ion closely at large l.

The constants A_l or, which is the same, r_l are chosen so that the Schrödinger equation with the cut-off Coulomb potential could give the correct energy levels of the isolated atom, i.e., the electron-ion system. In this way the values of A_l for certain energies are obtained, following which the function $A_l(\varepsilon)$ is found by extrapolation between these values.

On the whole the Heine-Abarenkov-Shaw pseudopotential may be written in the form

$$U(r) = V_a(r) - \sum_{l=0}^{l_0} \theta(r_l - r)[A_l(\varepsilon) + V_a(r)]P_l, \tag{14.23}$$

where

$$\theta(x) = \begin{Bmatrix} 1, & x > 0, \\ 0, & x < 0, \end{Bmatrix}$$

and P_l is the projection operator on states with given l. Since the potential is different for different l values, it cannot be written as a local potential.

In order to obtain the Fourier components, one can use the expansion of the plane wave in spherical functions (see Landau and Lifshitz 1974):

$$e^{i\boldsymbol{k}\boldsymbol{r}} = 4\pi \sum_{l=0}^{\infty} \sum_{m=-l}^{l} i^l j_l(kr)\, Y_l^{m*}(\theta_k, \varphi_k)\, Y_l^m(\theta, \varphi), \tag{14.24}$$

where $j_l(x) = (\pi/2x)^{1/2} \mathrm{J}_{l+1/2}(x)$ and J is the Bessel function. This being done, we obtain

$$e^{-i\boldsymbol{k}'\boldsymbol{r}}\, U e^{i\boldsymbol{k}\boldsymbol{r}} = V_a(r)\, e^{i(\boldsymbol{k}-\boldsymbol{k}')\boldsymbol{r}}$$

$$- \sum_{l=0}^{l_0} \theta(r_l - r)[A_l(\varepsilon) + V_a(r)]$$

$$\times (4\pi)^2 \sum_{m=-l}^{l} \sum_{m_1=-l}^{l} j_l(kr)\, j_l(k'r)$$

$$\times Y_l^{m*}(\theta_k, \varphi_k)\, Y_l^m(\theta, \varphi)\, Y_l^{m_1}(\theta_{k'}, \varphi_{k'})\, Y_l^{m_1*}(\theta, \varphi).$$

Let us consider the second term of this expression. We introduce the variable $x = r/r_l$ and make use of the fact that, according to (14.22),

$$A_l(\varepsilon) + V_a(r) = \frac{Ze^2(r - r_l)}{rr_l} = \frac{Ze^2}{r_l}\frac{x-1}{x}.$$

Integrating over the angles θ and φ and using the orthonormality of the functions $Y_l^m(\theta, \varphi)$, we obtain the sum over m, which we transform in accordance with the rule of convolution of spherical functions (see Landau and Lifshitz 1974):

$$\sum_{m=-l}^{l} Y_l^{m*}(\theta_k, \varphi_k)\, Y_l^m(\theta_{k'}, \varphi_{k'}) = \frac{2l+1}{4\pi} P_l(\cos\theta_{kk'}). \tag{14.25}$$

Then, in order to obtain the Fourier component under consideration, it remains to take the integral over r:

$$\int_0^\infty r^2\, \mathrm{d}r\, \theta(r_l - r) \to r_l^3 \int_0^1 x^2\, \mathrm{d}x.$$

As for the first term, the integral over $\mathrm{d}V$ in it gives $-4\pi Ze^2/|\boldsymbol{k} - \boldsymbol{k}'|^2$.

Thus, on the whole we get

$$U(\boldsymbol{k}, \boldsymbol{k}') = \frac{-4\pi Ze^2}{|\boldsymbol{k} - \boldsymbol{k}'|^2} + 4\pi Ze^2 \sum_{l=0}^{l_0} (2l+1)\, P_l(\cos\theta_{kk'}) \times \int_0^1 x(1-x)\, j_l(kr_lx)\, j_l(k'r_lx)\, \mathrm{d}x. \tag{14.26}$$

The first term depends only on the difference $|\boldsymbol{k} - \boldsymbol{k}'|$ and corresponds to the local potential; the second term depends on $\boldsymbol{k}$ and $\boldsymbol{k}'$ separately and in this lies its nonlocality.

The Fourier component $U(\boldsymbol{k}, \boldsymbol{k}')$ thus found is inserted into formula (14.20) and the result is used in the system of equations (14.16). Terminating the system at some $|\boldsymbol{K}|$ and solving the secular equation (the condition that the determinant of the system of linear homogeneous equations vanish), we obtain the energy spectrum.

There are various other methods of determining the "model pseudopotential", but apparently the above-described method is the most substantiated, and is physically comprehensible and convenient in practice.

Nevertheless, there are cases where the model pseudopotential method does not work. These include metals in which the d- and f-bands either get into the valence band or reside near its edge. The difficulty in such cases consists in the following. The use for a metal of values of A_l and r_l matched to the spectrum of an electron in the field of the ion presupposes the possibility of extrapolation of $A_l(\varepsilon)$ to other energy values, i.e., actually $A_l(\varepsilon)$ are considered to be weakly dependent on the energy within certain energy intervals.

Let us examine a case where the d- or f-band is close to the conduction band. The d- or f-electron in the atom is acted on by a potential consisting of the Coulomb attraction of the ion and of the centrifugal repulsion $\hbar l(l+1)/2mr^2$. This gives rise to a potential well, which has the shape of a spherical layer around the nucleus, where such an electron is localized. In a metal, as a result of the overlap of the valence shells, the potential far from the ion is nonzero and negative. Nonetheless, in this case, the well cannot disappear altogether; it will be partly preserved and separated from the outer external region by a potential barrier. In such a situation, a discrete level with a large orbital momentum is not completely destroyed; instead it is transformed to a "virtual" or "resonance" level.

Although this situation may also be described by a model potential of the type considered above, the corresponding coefficients $A_l(\varepsilon)$ begin to depend strongly on the energy near the resonance and, as a result, the value of the method can be reduced to zero. It is for this reason that for the construction of the energy spectra of transition and rare-earth metals (and of noble metals as well), in which the d- or f-bands are of importance, use is made of much more complicated methods, such as the augmented plane wave (APW) method or the Green's function method proposed by Korringa, Kohn and Rostocker (KKR method). The description of these methods is beyond the scope of this book. The reader is referred to specialized literature (for example, Ziman 1971, Ziesche and Lehmann 1983).

At the same time one should keep in mind that the theory of energy spectra faces two quite different tasks. One is to calculate as strictly as possible the spectra on the basis of elementary interactions of electrons and ions. This was discussed above. The other task is the practical construction of spectra, which facilitates the analysis and systematization of experimental material. The latter task is easily accomplished for simple metals with the aid of the free-electron model described in the next section. In metals, in which the d- or f-bands are important, one can make an attempt to describe them using the formulas derived from the strong-coupling approximation by choosing the coefficients from comparison with experimental data. Such a procedure proves to be very successful in many cases. But one should realize that this is not a proof of the correctness of such an approximation. In fact, the orbits overlap sufficiently strongly, and also a so-called hybridization occurs with the s- and p-bands. The success of the strong-coupling formulas is associated with the fact they take correct account of the symmetry of the crystalline lattice and of the associated possibility of degeneracy of the energy terms $\varepsilon(\boldsymbol{p})$ for symmetric points in the reciprocal lattice space.

14.3. The free-electron model

The general considerations described in the preceding section and also the particular calculations of pseudopotentials for simple metals or their choice on the basis of comparison with experimental data show that in such metals the pseudopotential is substantially lower than the Fermi energy. Hence, to a first approximation the pseudopotential may be considered to be equal to zero and the electrons to be free.

According to the model of almost free electrons (section 1.3), deviations will manifest themselves only in a small neighborhood of the faces of the Brillouin zone and will amount to a "rounding-off" of the corresponding small portions of the Fermi surface. The accuracy of such an approach is not high (~10%), but the procedure is very simple and therefore is ideally suitable for the problem mentioned above: the reproduction of the Fermi surface from fragmentary experimental data.

In order to explain the basic idea of the free-electron model, we start with a one-dimensional case. In the absence of a potential the problem reduces to the division of the parabola $\varepsilon(p) = p^2/2m$ into sections corresponding to (1) $-\frac{1}{2}\hbar K < p < \frac{1}{2}\hbar K$; (2) $-\hbar K < p < -\frac{1}{2}\hbar K$ and $\frac{1}{2}\hbar K < p < \hbar K$; (3), ..., where $K = 2\pi/a$ (see fig. 4) and bringing these sections by way of transfer by $n\hbar K$ to the Brillouin zone $-\frac{1}{2}\hbar K < p < \frac{1}{2}\hbar K$. This can be done by a graphical construction. Let us sketch in the ε, p plane many parabolas with apexes at points 0, $\pm\hbar K$, $\pm 2\hbar K$, ... (fig. 89). We single out the Brillouin zone. Let us assume that a piece of some parabola "belongs" not only to it but also to all the parabolas for which it is "internal", i.e., lies above them. Let us now consider the pieces that enter the Brillouin zone. The lowest piece belongs to a single parabola. It is regarded as the first band. The subsequent overlying pieces belong to two parabolas. On the whole they form the second band. Then comes the third band consisting of pieces belonging to three parabolas, etc. It is not difficult to see that this procedure automatically gives us a picture of the bands in the limit where the potential tends towards zero (cf. figs. 4 and 5).

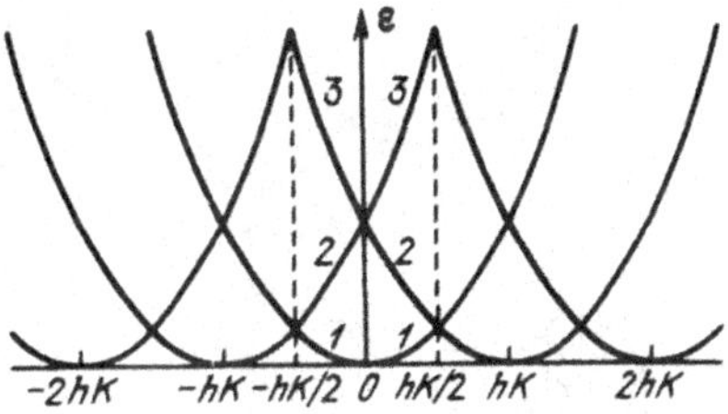

Fig. 89.

Let us now use an analogous procedure for building up a Fermi surface (Harrison 1959, 1960). We sketch the reciprocal lattice. Taking the lattice point $\boldsymbol{K} = 0$, we sketch a Fermi sphere of radius $p_0 = \hbar(3\pi^2 n_e)^{1/3}$, where n_e is the density of valence electrons. Analogous spheres are drawn around the points with other $\boldsymbol{K}$'s. In the general case the spheres intersect. Let us now consider the Brillouin zone. Its various parts enter different numbers of spheres. Let us single out a region which enters a single sphere or a larger number of spheres. The boundary of this region is interpreted as a Fermi surface in the lowest band. If the entire Brillouin zone is obtained, it means that the first band is completely filled and the Fermi surface belongs to the higher-lying bands. Then come portions of the volume of the Brillouin zone that enter two or more spheres. The boundaries of these portions form a Fermi surface in the second band, etc. This is demonstrated for a two-dimensional model in fig. 90.

An example of the three-dimensional case is shown in fig. 91, which gives the Fermi surfaces of fcc metals with different numbers of valence electrons (Harrison 1959, 1960). The geometric construction described above provides good agreement with experimental data for simple polyvalent metals.

Of course, as has been said above, the accuracy of this method becomes worse near the faces of the Brillouin zone, which may lead to incorrect predictions concerning small portions of the Fermi surface or even its topology. This can,

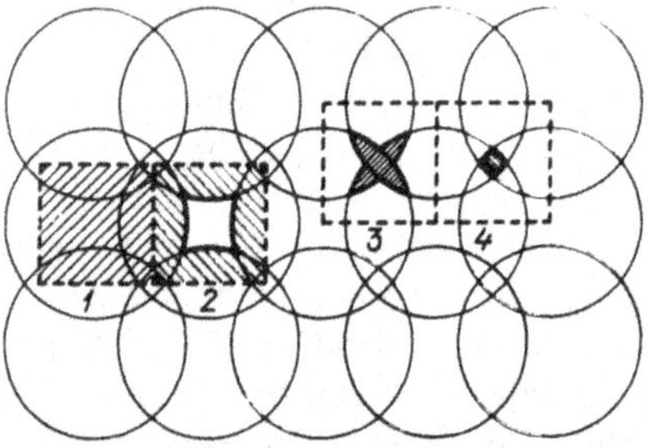

Fig. 90.

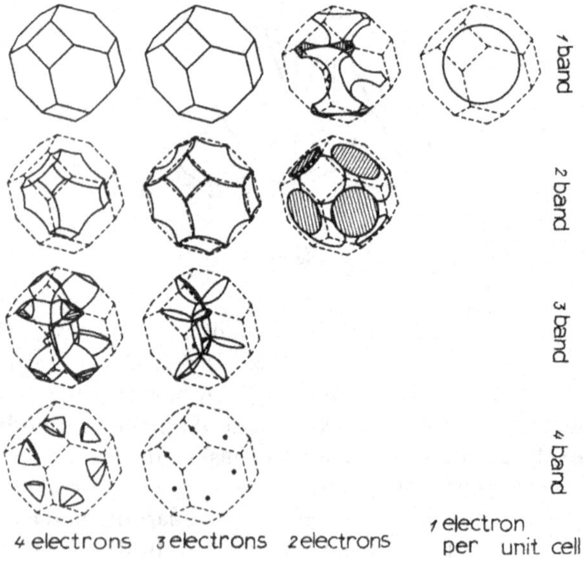

Fig. 91.

however, be corrected. According to the model of weakly bound electrons (section 1.3), the energy near such a face is given by

$$\varepsilon = \frac{1}{2}[\varepsilon(\boldsymbol{p}) + \varepsilon(\boldsymbol{p} - \hbar\boldsymbol{K})] + \{\tfrac{1}{4}[\varepsilon(\boldsymbol{p}) - \varepsilon(\boldsymbol{p} - \hbar\boldsymbol{K})]^2 + V_{\boldsymbol{K}}^2\}^{1/2}. \tag{14.27}$$

The quantity $V_{\boldsymbol{K}}$ in this formula can be chosen either from experimental data or from calculations based on a more accurate method (the OPW or pseuodpotential method). In this way it is possible to overcome the difficulties associated with the vicinity of the Brillouin zone faces.

If we are dealing with the vicinity of the intersection of two faces, i.e., a region close to the edge of the Brillouin zone, then in the model of weakly bound electrons it is necessary to take a linear combination of three rather than two plane waves:

$$\exp\left(\frac{\mathrm{i}\boldsymbol{pr}}{\hbar}\right), \qquad \exp\left(\frac{\mathrm{i}(\boldsymbol{p} - \hbar\boldsymbol{K}_1)\boldsymbol{r}}{\hbar}\right), \qquad \exp\left(\frac{\mathrm{i}(\boldsymbol{p} - \hbar\boldsymbol{K}_2)\boldsymbol{r}}{\hbar}\right).$$

The functions $\varepsilon(\boldsymbol{p})$ in this case are defined as eigenvalues of a third-order matrix and will include three unknown constants: V_{01}, V_{02}, and V_{12}. Near the apex of the Brillouin zone one has to deal with four functions and six unknown constants.

The free-electron model provides an easy interpretation of experimental data. If, for example, we obtain the dependence of the boundary Fermi momentum on the angle along some contours on the Fermi surface, the construction in accordance with the free-electron model makes it possible to classify easily to which portions of the Fermi surface these contours belong. Furthermore, knowing from the model the general shape of a given portion of the surface and having experimental data pertaining to various regions of this surface, one can choose a reasonable interpolation procedure.

But if the construction based on the free-electron model is looked upon as a quantitative method, then, as noted above, its accuracy for the Fermi surface is of the order of 10% (near the Brillouin zone faces this is valid only after introducing the required correction, as pointed out above). The situation with effective masses is much worse. Upon comparison of the data obtained from cyclotron resonance with the free-electron model it is necessary to take into account that the cyclotron period along a certain orbit is only part of the period that a free electron would have if it were moving along the cross section of the entire Fermi surface. It is only this latter period that must be connected with the free mass by the relation

$$T = 2\pi\frac{mc}{eH}.$$

Using the free-electron model, it is not difficult to determine the periods of motion along the cross sections of the true portions of the Fermi surface.

However, such predictions differ from experimental data by 30–40%. It is interesting to note that this difference depends little on the direction for a particular metal. Therefore, by introducing a correction to the free mass, one can then obtain, with good accuracy, all cyclotron masses for various portions of the Fermi surface. The difference between mass and free mass is presumably associated with the interaction of electrons with phonons (section 14.4).

14.4. The strongly compressed matter approximation

According to the foregoing the free-electron approximation provides a rather good description of the Fermi surface of metals. Although this result is quite justified by the reasoning concerning the pseudopotential, it can be looked upon from a somewhat different viewpoint. One may ask: Is not this an evidence that the kinetic energy of valence electrons is higher than their potential energy? As we shall see below, this idea can be justified for many polyvalent metals. This situation occurs in a strongly compressed substance. Indeed, the average potential energy is of order $e^2/\bar{r}$, where $\bar{r}$ is the mean distance between atoms. The kinetic energy is of order $p_0^2/2m$. But, according to the uncertainty principle, $p_0 \sim \hbar/\bar{r}$. Hence, in order of magnitude the kinetic energy is $\hbar^2/m\bar{r}^2$. The ratio of the potential to the kinetic energy is given by

$$\frac{e^2/\bar{r}}{\hbar^2 m\bar{r}^2} \sim \frac{\bar{r}}{r_B}, \tag{14.28}$$

where $r_B = \hbar^2/me^2$ is the so-called Bohr radius. It then follows that at a high density, when $\bar{r} \ll r_B$, the potential energy of electrons becomes much less than their kinetic energy.

It may be presumed that under ordinary conditions too the valence electrons in a metal are in a "strongly compressed" state, as a result of which their potential energy is much lower than their kinetic energy. It should rather be expected in polyvalent metals, because the lattice periods of metals differ little, so that the electron density in polyvalent metals must be higher*.

An advantage of such an approach, if it is correct, consists in the following. In this case, it may be asserted that not only the Fermi surface is well described by the free-electron approximation, but, moreover, all interactions of electrons with

* Such reasoning should not be taken too literally. In fact, without external compression the crystal lattice is balanced by internal forces, i.e., it corresponds to the minimum energy. Since the total energy is the sum of the kinetic energy of electrons and the Coulomb potential energy, it evidently follows that they are both of the same order of magnitude. Here we can only say that the parameter $e^2/(\pi\hbar v)$ obtained with such a balance is not too large. A calculation of the model of strongly compressed matter (Abrikosov 1960, 1961) gives the following energy per atom:

$$E = \frac{3p_0^2 Z}{10m} - \frac{e^2 p_0}{\hbar(3\pi^2 Z)^{1/3}}(1.44Z^2 + 0.74Z^{4/3})$$

where Z is the number of electrons per atom. By minimizing this expression and determining the parameter $e^2/(\pi\hbar v)$, we find the following values: 0.2 for $Z = 2$ and 0.16 for $Z = 3$. This confirms that the suggested approach is reasonable.

lattice ions and with electrons are small compared with their kinetic energy and may be treated with the aid of perturbation theory.

A consideration of the model of strongly compressed matter (Abrikosov 1960, 1961) shows that the expansion parameter is $e^2/(\pi\hbar v)$, where $v = p_0/m$ [of course, by order of magnitude this is the same as (14.20) since $p_0 \sim \hbar/\bar{r}$, but here an exact value is required]. For real metals the momentum can be calculated from the density of valence electrons; in that way we obtain values of the parameter $e^2/(\pi\hbar v)$ (table 14.1).

Table 14.1
Values of the expansion parameter for the model of strongly compressed matter.

Metal	Sn	Pb	Al	Cu	Ag	Au	Na	K
$\dfrac{e^2}{\pi\hbar v}$	0.37	0.38	0.35	0.45	0.51	0.50	0.67	0.83

Thus, in fact, the parameter $e^2(\pi\hbar v)$ is not large; in polyvalent metals it is smaller than in univalent metals.

Let us find the change in the Fermi surface resulting from the finiteness of the expansion parameter. The correction to the energy from electron-ion interaction is equal to

$$\delta\varepsilon(\boldsymbol{p}) = (4\pi n_e e^2)^2 \sum_{\boldsymbol{K}\neq 0} K^{-4}\left[\frac{p^2}{2m} - \frac{(\boldsymbol{p}-h\boldsymbol{K})^2}{2m}\right]^{-1}, \tag{14.29}$$

where n_e is the electron density. The corrections of first order in e^2 from electrons and ions cancel because of the electroneutrality condition. As for the second-order correction due to electron-electron interactions, it is isotropic and, hence, has no effect on the Fermi surface. So the required correction is described by the equation

$$\delta p(\theta, \varphi) = \frac{m}{p_0}[\overline{\delta\varepsilon} - \delta\varepsilon(p_0\boldsymbol{n})], \tag{14.30}$$

where $\overline{\delta\varepsilon}$ is the average over the angles from (14.29) at $|\boldsymbol{p}| = p_0$ and $\boldsymbol{n} = \boldsymbol{p}/p_0$.

From eqs. (14.29) and (14.30) it follows that the correction to the Fermi surface is second order in $e^2/(\pi\hbar v)$. This can fully account for the experimentally observed accuracy of the free-electron approximation in polyvalent metals.

In noble metals the expansion parameter is not so small. Therefore, it is not surprising that, although the Fermi surface must not intersect the faces of the Brillouin zone according to the free-electron model, the surface is, in fact, open (fig. 92; cf. fig. 9). We can try to eliminate this discrepancy by using formula (14.27) with $V_{\boldsymbol{K}} = 4\pi e^2 n_e/K^2$ in the vicinity of the face of the Brillouin zone. In this case, we really obtain an intersection. From a comparison of the radii of the "necks"

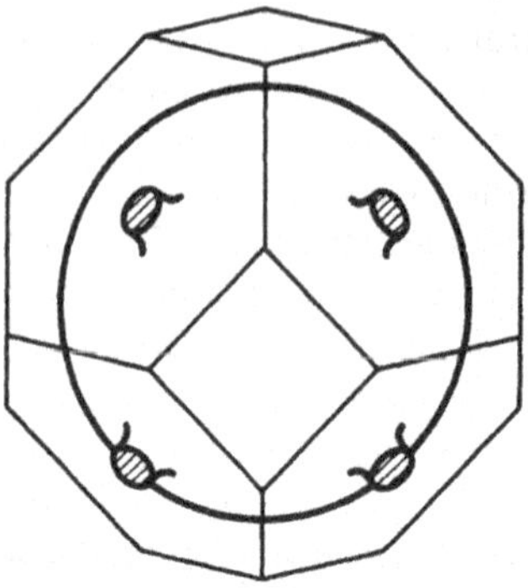

Fig. 92.

with experimental data we find that for such a low approximation the agreement is not bad (table 14.2). However, this similarity should be not be taken too seriously because, as has been noted in section 14.2, the situation in noble metals is strongly complicated because of the admixture of the d-states.

In alkali metals the situation seems somewhat strange from the viewpoint of the approach under discussion. The point is that their spectra are very close to the spectra of free electrons, while their values of $e^2/(\pi\hbar v)$ are relatively high. However, here there is no contradiction whatsoever since at higher values of this parameter the model is simply inapplicable and should not be used for any predictions*.

Finally, let us consider the masses. For strongly compressed matter it has been shown (Abrikosov 1960, 1961) that due to the interaction of electrons with phonons their effective mass must be given by

$$\frac{m^*}{m} = 1 + \alpha_1 \frac{e^2}{\pi\hbar v}, \tag{14.31}$$

Table 14.2

Metal	$\frac{r}{p_0}$ (theory)	$\frac{r}{p_0}$ (experiment)
Cu	0.1	0.19
Ag	0.2	0.14
Au	0.2	0.18

* Good agreement with data for alkali metals can also be obtained by using the model under consideration if we introduce the screening of the Coulomb interaction, i.e., if we replace r^{-1} by $e^{-\kappa r}/r$ (or k^{-2} in the Fourier components by $(k^2+\kappa^2)^{-1}$, where $\kappa^2 = 4\pi e^2(\partial n_e/\partial\mu) = 4e^2 p_0 m/(\pi\hbar^3)$; section 4.1).

where $\alpha_1 \sim 1$, and is approximately isotropic. Hence, the mass varies to a first approximation with respect to the parameter $e^2/\pi\hbar v$, the correction depending little on the direction in the metal. This is in agreement with experimental data.

The data given indicate that the approximation of "strongly compressed" electrons makes it possible to account, at least qualitatively, for various facts from a unified viewpoint. But its main value consists in that it provides a simple model, which allows one to get an insight not only into the electronic spectra but also into the kinetic properties of polyvalent metals.

Part II. Superconducting metals

15

Macroscopic theory of superconductivity

15.1. General properties of superconductors

In Part I we have been concerned with the various mechanisms of electrical resistance. Some of these mechanisms give a gradual decrease in resistance with temperature and other lead to a constant value or even to an increase in resistance at very low temperatures. However, in 1911 H. Kamerlingh Onnes, while studying the variation with temperature of the electrical resistance of mercury, observed that the resistance dropped sharply to zero at a temperature of 4.2 K. The same properties were later detected in some other metals. The new phenomenon was termed "superconductivity" and the corresponding metals were called "superconductors".

The temperature at which the resistance disappears is called the critical temperature, T_c; it is different in different superconductors. The highest critical temperature among pure metals is shown by niobium: $T_c = 9.25$ K; the lowest has been found in tungsten: $T_c = 0.0154$ K. Although these are both low temperatures, in fact the temperature range is very wide, since the two extremes differ by about a factor of a thousand.

In what follows we shall also mention the practical applications of superconductors. Even at this moment they are rather numerous. However the recent discovery of so-called high-temperature superconducting ceramics arouses hopes for a further development of such applications. At the moment this book was written the highest critical temperature was registered for a composition $YBa_2Cu_3O_{7-\delta}$ and is equal to 93–95 K. But the discovery of even higher critical temperatures is not excluded. The importance of these discoveries is based on the fact that the superconductivity in such ceramics can be maintained in cryostats with liquid nitrogen (boiling point 77.4 K).

Subsequent investigations of the properties of superconductors have shown that superconductivity can be destroyed not only by increasing the temperature but also by applying a sufficiently strong magnetic field (Kamerlingh Onnes 1914). The critical field in which superconductivity is destroyed decreases with increasing temperature. It has been established empirically that the dependence $H_c(T)$ is described well by the formula

$$H_c(T) = H_c(0)\left[1 - \left(\frac{T}{T_c}\right)^2\right]. \tag{15.1}$$

Superconductivity is also destroyed by a strong electric current. If the superconductor is not too thin (see below), the critical current, at which resistance appears, satisfies Silsbee's rule (Silsbee 1916): the magnetic field produced by the critical current at the surface of a superconductor must be equal to H_c.

One of the basic properties of superconductors is the so-called Meissner effect (Meissner and Ochsenfeld 1933). If a metal is placed in a magnetic field smaller than H_c, then upon transition into the superconducting state the field is expelled from its interior, i.e., the true field $\boldsymbol{B}=0$ in the superconductor (recall that the magnetic induction $\boldsymbol{B}$ is the average microscopic field)*. This is shown in fig. 93: (a) a superconductor; (b) a normal metal.

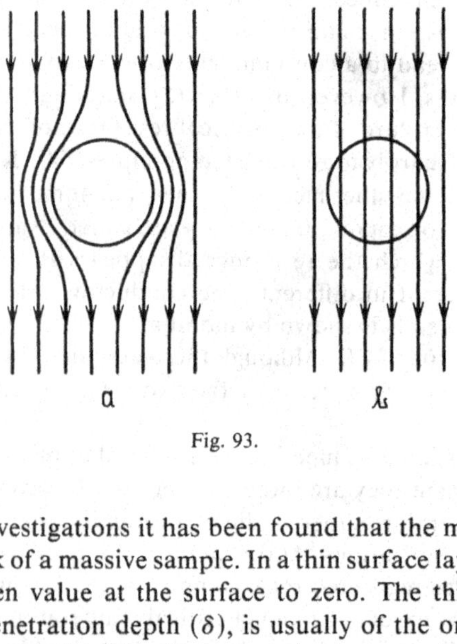

Fig. 93.

In more detailed investigations it has been found that the magnetic field is equal to zero only in the bulk of a massive sample. In a thin surface layer the field gradually decreases from a given value at the surface to zero. The thickness of this layer, which is called the penetration depth (δ), is usually of the order of 10^{-5}–10^{-6} cm.

When a superconductor is placed in an external magnetic field, an undamped current appears in the surface layer, which produces a magnetic field of its own, and the latter compensates completely the external field inside the superconductor.

This can be traced out most simply on a cylinder in a longitudinal magnetic field. Suppose we have a solenoid with current i. Inside the solenoid a magnetic field is created, $H=(4\pi/c)i\cdot n$, where n is the number of turns in a coil per 1 cm of length. Instead of $i\cdot n$ we can write J, which is the current per 1 cm of the solenoid length. If the superconducting cylinder is placed in a field H, a current $J=cH/4\pi$ will

* In superconducting alloys the behavior in a magnetic field is more complicated; in particular, the Meissner effect may be incomplete. We shall defer the discussion of this question until ch. 18.

flow in the surface layer, which is directed so that the field produced by it will compensate the external field H.

The penetration depth δ depends on various factors, in particular on temperature. When the temperature approaches T_c, the penetration depth $\delta \to \infty$. The empirical formula that reflects in general aspects the temperature dependence of δ has the form

$$\delta(T) = \frac{\delta(0)}{[1-(T/T_c)^4]^{1/2}}. \tag{15.2}$$

As for the thermodynamic character of the transition of a metal from the superconducting to the normal state, it has been found that if the transition occurs at $H = 0$, i.e., at $T = T_c$, it is a second-order phase transition (see Appendix II). Upon this transition the latent heat of transition is not absorbed, but a jump in heat capacity takes place instead. But if the transition occurs at $H \neq 0$, i.e., at $T < T_c$, then it is a first-order transition at which the latent heat of transition is absorbed.

This change in the character of the transition may be associated with the behavior of the penetration depth. With $T \to T_c$ the penetration depth $\delta \to \infty$ and a superconductor does not differ at all at the transition point from a normal metal (with respect to the penetration of an infinitely weak field). Hence, the quantity δ that may serve as a characteristic of the extent of difference between a superconductor and a normal metal varies continuously and the transition should be second order. But if the transition occurs in a finite magnetic field, a discontinuous variation takes place in δ from a finite value in the superconducting phase to an infinite value in the normal phase (the field penetrates a normal metal completely, i.e., formally $\delta = \infty$).

The continuous variation of δ at T_c indicates that the state of the electron liquid in a metal varies continuously. This is also evidenced by measurements of thermal conductivity.

In Part I it has been pointed out that usually the major role in the thermal conductivity of a metal is played by electrons. But the vanishing of the resistance might indicate that the electrons cease to interact with the crystal lattice and, hence, are incapable of participating in thermal conductivity. In such a case, the overall coefficient of thermal conductivity at the transition point should decrease abruptly to a value determined by phonons alone. This does not occur. At $T = T_c$ the quantity $\varkappa$ varies continuously.

The last of the fundamental properties that should be mentioned here is the behavior of heat capacity at low temperatures. In Part I we have described a construction (fig. 11), which enables one to separate the electronic part of the heat capacity from the lattice part. In this way we can find the coefficient in the Debye law $C = BT^3$ for the lattice (phonon) heat capacity. Assuming that the lattice heat capacity does not vary upon transition to the superconducting state, it is possible to determine the electronic heat capacity of a superconductor by subtracting BT^3 from the total value. As a result of such measurements it turned out that the electronic heat capacity of a superconductor at temperatures substantially lower than T_c depends on temperature by an exponential law (Corak and Satterthwaite 1954,

Corak et al. 1954, 1956):

$$C_{es} = a \exp\left(\frac{-\Delta}{T}\right). \tag{15.3}$$

This means that the excited states of the electronic system are separated from the ground state by an "energy gap", just as in a semiconductor. On the other hand, a superconductor differs very strongly from a semiconductor in its electrical properties. Moreover, the phase transition at T_c is a second-order phase transition, which implies that the energy gap vanishes at T_c, i.e., the energy spectrum depends strongly on temperature. Such a dependence of the energy spectrum on temperature (in other words, on the state of the electron system) is contained in the Landau theory of the Fermi liquid (ch. 13). We shall use it in the derivation of the microscopic theory.

15.2. *Thermodynamics of the superconducting transition*

Let us consider a cylindrical sample in a longitudinal magnetic field. The condition for the superconducting transition is the equality of the free energies:

$$F_s(H, T) = F_n(H, T). \tag{15.4}$$

Here and henceforth the subscripts s and n designate the superconducting and the normal phase, respectively. Equilibrium may occur only at a field of $H = H_c(T)$.

The magnetic field enters a superconductor only to the penetration depth. In the presence of the field a superconducting current flows in the surface layer and screens the field in the bulk of the sample. This introduces a contribution to the energy of the order of $H_c^2/8\pi$ per unit volume of the surface layer. The total contribution to F_s from the currents and the field is of order $(H_c^2/8\pi)\delta S/V$, where S is the surface area of the sample (the free energy is referred to the unit volume).

The field penetrates a normal metal completely. Since we consider the equilibrium in a given field, we have to use the formula for F_H (Appendix III, formula AIII.10). Usually, superconducting metals are nonmagnetic, i.e., there $\mu \approx 1$. Hence, $F_n(H_c, T) = F_n(T) - H_c^2/8\pi$. From this we see that the magnetic contribution to F_n is much larger than that to F_s, namely D/δ times as larger (D is the diameter of the sample). Therefore, for a massive cylinder the field contribution to F_s may be neglected; as a result, we obtain from eq. (15.4)

$$F_n(T) - F_s(T) = \frac{H_c^2}{8\pi}. \tag{15.5}$$

Differentiating with respect to T, we find the entropy difference ($S = -\partial F/\partial T$):

$$S_n - S_s = \frac{-H_c}{4\pi}\frac{dH_c}{dT}. \tag{15.6}$$

The quantity $q = T(S_n - S_s)$ gives the latent heat of transition. Since $dH_c/dT < 0$ (see eq. 15.1) it turns out that $q > 0$, i.e., the heat is absorbed upon transition from the superconducting to the normal phase.

A further differentiation with respect to temperature gives the difference in heat capacities $(C = T(\partial S/\partial T)_V$

$$C_n - C_s = -\frac{T}{4\pi}\left[H_c \frac{d^2 H_c}{dT^2} + \left(\frac{dH_c}{dT}\right)^2\right]. \tag{15.7}$$

In particular, at $T = T_c$ the field $H_c = 0$ and we have

$$C_s(T_c) - C_n(T_c) = \frac{T_c}{4\pi}\left(\frac{dH_c}{dT}\right)^2. \tag{15.8}$$

It is interesting to note that this simple treatment, in which use is made of only a single experimental fact - the Meissner effect, enables one to derive rigorous formulae relating the critical magnetic field to the thermodynamic characteristics of the superconductor.

15.3. The intermediate state

So far we have considered a cylindrical geometry. Now suppose that the sample has an arbitrary shape or that, even if it is a cylinder, the field is not longitudinal. In this case the superconducting sample distorts the field, which becomes inhomogeneous in space. Let us see what happens if the specimen has the form of an ellipsoid. In such a case, the Maxwell field H_i inside it is homogeneous, although different from the external field at infinity H_0. They are connected by the relation (Appendix III):

$$H_i = H_0 - 4\pi n M, \tag{15.9}$$

where M is the magnetization and n is the demagnetization factor.

Since the metal is in the superconducting state, inside $B = 0$, i.e.,

$$H_i + 4\pi M = H_0 + 4\pi(1-n)M = 0,$$

that is,

$$M = -\frac{H_0}{4\pi(1-n)}, \qquad H_i = \frac{H_0}{1-n}. \tag{15.10}$$

According to the boundary conditions of the Maxwell theory, the normal component of $\boldsymbol{B}$ and the tangential component of $\boldsymbol{H}$ are continuous at the interface between the media. Outside the superconductor $\boldsymbol{B} = \boldsymbol{H}$. From the fact that $\boldsymbol{B} = 0$ in the superconductor one can conclude that $\boldsymbol{H}$ at the superconductor boundary has only a tangential component, i.e., the lines of force "by-pass" the superconductor (fig. 93a). From the continuity of the tangential component $\boldsymbol{H}$ at the interface one may conclude that at those points where the direction of $\boldsymbol{H}$ coincides with the external field $\boldsymbol{H}_0$ the field at the boundary is at a maximum and equals $H_0/(1-n)$, i.e., is larger than H_0. The superconductor moves apart the force lines of the field, which leads to their concentration outside the superconductor. We also see that along the superconductor surface the field varies and in some portions it may attain values of H_c, while $H_0 < H_c$.

Suppose that in a certain portion of the superconductor surface the field exceeds H_c. Then the superconductivity must vanish there. It will be most natural to presume that the picture will be as shown in fig. 94, where the shaded portion of the sample has passed to the normal state. However, this assumption leads to a contradiction. If the metal in the normal state is nonmagnetic, then for the magnetic field there is no difference between the normal metal and vacuum. Hence, departing from the superconducting region the field must decrease. If it is accepted that at the interface between the normal and the superconducting region $H = H_c$, then in the normal region the field $H < H_c$. But in this case the metal must be superconducting.

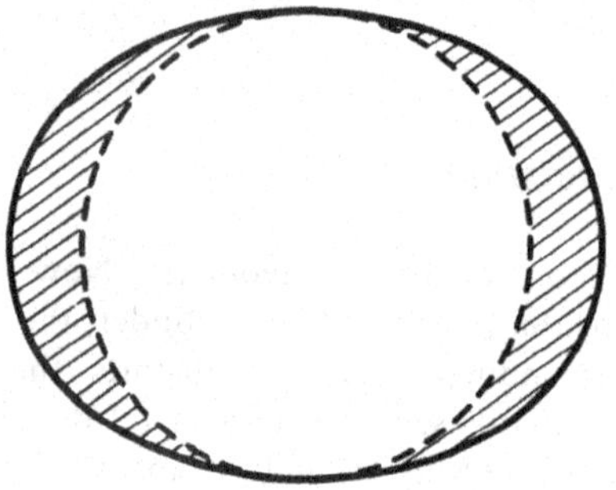

Fig. 94.

The way to remove this contradiction has been suggested by Peierls and London (Peierls 1936, London 1936) who advanced the idea of the "intermediate state". According to this idea, in the region of the external field

$$(1-n)H_c < H_0 < H_c \tag{15.11}$$

the superconductor passes to a special state characterized by the condition

$$H_i = H_c. \tag{15.12}$$

From eq. (15.9) we obtain

$$M = -\frac{H_c - H_0}{4\pi n}. \tag{15.13}$$

Substituting into the formula for the induction, we find

$$B = H_i + 4\pi M = \frac{H_0}{n} - H_c\frac{1-n}{n}. \tag{15.14}$$

From this we see that $B = 0$ at $H_0 = (1-n)H_c$ and $B = H_c$ at $H_0 = H_c$, and in the interval between these values the dependence of B on H_0 is linear. For the cylinder in the field perpendicular to its axis ($n = \frac{1}{2}$) the dependence $B(H_0)$ is shown in fig. 95.

In the Maxwell theory the field H_i does not always have a clear-cut microscopic meaning. Therefore, although the assumption $H_i = H_c$ and formula (15.14), which follows from it, agree well with experiment, they leave a feeling of dissatisfaction. Therefore, the intermediate state has been studied in more detail. It has been found that when a cylindrical specimen is placed in a perpendicular field and undergoes a transition to the intermediate state, its electrical resistance becomes anisotropic. Along the axis of the cylinder there is a finite resistance, and across the axis the resistance is equal to zero. This indicates that the cylinder splits up into thin layers of normal and superconducting phases perpendicular to its axis. When a current is passed along the axis, the layers are connected in series, but when it is passed across the axis they are connected in parallel, and the normal layers are shunted by the superconducting ones.

The picture in an infinite plate perpendicular to the magnetic field must be the simplest. In this case, $n = 1$, i.e., the intermediate state begins with $H_0 = 0$. If it is assumed that in the normal layers the field is H_c and in the superconducting layers it is equal to zero, then the mean field, i.e., the induction in the plate, is $B = xH_c$, where x is the fraction of the normal phase. Outside the superconductor $B = H_0$. In this case, the field is directed normally to the interface. From the continuity condition for the normal component of $\boldsymbol{B}$ (i.e., in this particular case, $\boldsymbol{B}$ itself) we have

$$x = \frac{H_0}{H_c}. \tag{15.15}$$

No further conclusions can be made without studying the structure of the layers.

A detailed investigation of the structure of the intermediate state has been carried out by Landau (1937). We shall present here a simplified version of this theory. Let

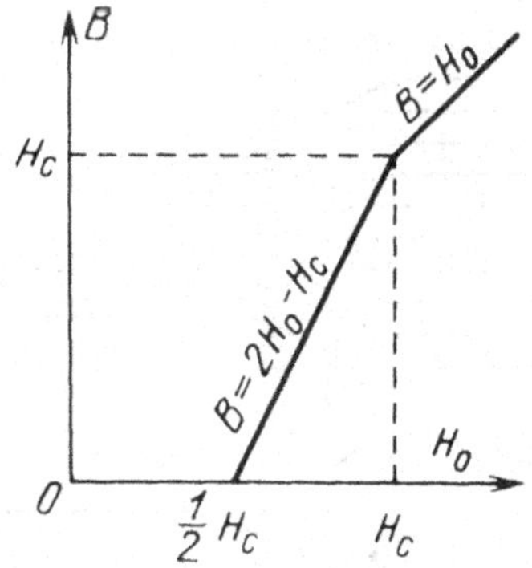

Fig. 95.

us consider a plate in a perpendicular field. If we assume that alternating plane layers of the normal and superconducting phases are formed in the plate, then the interfaces between the layers will coincide with the lines of force of the field. A kink in the lines of force would have led to a large additional energy. Therefore, we have to assume that the normal layers are slightly widened near the surface (fig. 96). If the thickness of the normal layers is a and that of the superconducting layer is b, then the characteristic size of the "corner" is of the order of

$$c \sim \frac{ab}{a+b}. \tag{15.16}$$

As $a \to 0$ or $b \to 0$ the "corners" disappear, as should be expected.

Let us calculate the energy of this structure. We use the formulas of Appendix III (AIII.12) and (AIII.13) for the free energy of a body of an arbitrary shape in a magnetic field. The magnetization M is defined with respect to the field H_0. But since outside the sample $B = H_0$ and B does not vary upon intersection of the boundary, the induction of the body $B = H_0$, i.e., $M = 0$. Hence, we may presume that the normal layers have an energy F_n and the superconducting layers an energy F_s per unit volume. We assume that the thickness of the plate (i.e., the length of the layers) is equal to L. The size in the direction perpendicular to the plane of the drawing is assumed to be equal to unity. If the free energy is counted off from F_s, the superconducting layers give a zero contribution and the total free energy will be equal to $H_c^2/8\pi$ multiplied by the volume of the normal phase. Considering that the number of layers per cm of length is $(a+b)^{-1}$, the energy per unit area of the plate will be

$$\frac{H_c^2}{8\pi}\frac{1}{a+b}\left[La + \gamma\left(\frac{ab}{a+b}\right)^2\right], \tag{15.17}$$

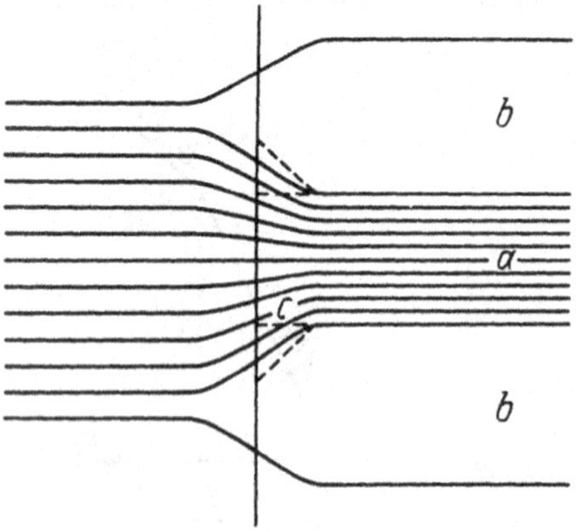

Fig. 96.

where $\gamma \sim 1$. The second term takes account of the "corners", i.e., the expansion of the normal layers near the surface.

However, if only this energy is taken into account, layers of finite size cannot be obtained. Therefore, Landau assumed the existence of an additional surface energy at the interface between the normal and the superconducting phase. The energy per unit area of the interface σ_{ns} has the dimension of erg/cm^2, while $H_c^2/8\pi$ has the dimension of erg/cm^3. This makes it possible to introduce a new convenient parameter ξ having the dimension of length:

$$\sigma_{ns} = \frac{H_c^2}{8\pi}\xi. \tag{15.18}$$

At a later stage we shall see that the parameter ξ plays a fundamental role in the theory of superconductivity.

Considering that each normal layer has two boundaries of length $L + \beta c$, where $\beta \sim 1$ and c is the length of the corner (15.16), we obtain a contribution to the energy per unit area of the plate surface arising from the boundaries:

$$\frac{H_c^2}{8\pi}\xi\frac{1}{a+b}2\left[L + \beta\frac{ab}{a+b}\right]. \tag{15.19}$$

Before finding the minimum of the total free energy, one has to take into account that the widths of the layers a and b are not independent but are determined by the external field H_0. According to eq. (15.19), we find the concentration of the normal phase:

$$x = \frac{a}{a+b} = \frac{H_0}{H_c}. \tag{15.20}$$

In the following all the lengths will be expressed in terms of a. Using eq. (15.20), we have

$$a + b = a\frac{H_c}{H_0}, \qquad b = a\frac{H_c - H_0}{H_0}. \tag{15.21}$$

Substituting this into eqs. (15.17) and (15.19), we obtain the total free energy:

$$F = \frac{H_c^2}{8\pi}\left\{L\frac{H_0}{H_c} + \gamma a\frac{H_0}{H_c}\frac{(H_c - H_0)^2}{H_c^2} + 2\xi\left[\frac{L}{a}\frac{H_0}{H_c} + \beta\frac{H_0(H_c - H_0)}{H_c^2}\right]\right\}. \tag{15.22}$$

In this expression only two terms depend on a. Taking the minimum with respect to a, we find

$$a = \left(\frac{2\xi L}{\gamma}\right)^{1/2}\frac{H_c}{H_c - H_0}, \qquad b = \left(\frac{2\xi L}{\gamma}\right)^{1/2}\frac{H_c}{H_0},$$
$$a + b = \left(\frac{2\xi L}{\gamma}\right)^{1/2}\frac{H_c^2}{H_0(H_c - H_0)}. \tag{15.23}$$

In practice, $\xi \sim 10^{-5}$ cm. For a plate with a thickness of the order of 1 cm the period of the structure is of the order of 10^{-3} cm and it increases with thickness as $L^{1/2}$. Note the similarity between these results and the structure of diamagnetic domains (section 10.6). Chronologically, the theory of the intermediate state had been worked out earlier. According to eq. (15.23), the period of the structure $a+b$ increases at both boundaries of the intermediate state, i.e., with $H_0 \to 0$ and $H_0 \to H_c$. If $H_0 \to 0$, then $b \to \infty$ and $a \to$ const. and, conversely, with $H_0 \to H_c$, $a \to \infty$ and $b \to$ const.

The exact theory (Landau 1937) permits one to determine the shape of the layers from the condition that the interface between the normal and superconducting phase coincides with the line of force. The results of this theory are in qualitative agreement with those given above, the only difference being that at the boundaries of the intermediate state the thickness of the "foreign" layers does not remain constant and decreases slightly. For example, $a \sim [\xi L/\ln(H_c/H_0)]^{1/2}$ as $H_0 \to 0$.

In the experimental determination of the structure of the intermediate state, use is made of the inhomogeneity of the magnetic field which arises due to phase separation. The first experiments made use of a magnetic-field detector in the form of a bismuth wire of a very small size, whose resistance varied, depending on the magnetic field. The use of such a detector made it possible to trace out the emergence of the layers of the intermediate state to the sample surface (Meshkovskii and Shal'nikov 1947). In another method, ferromagnetic powders were used. If the sample is sprinkled with such a powder, the ferromagnetic particles are concentrated at the regions where the normal layers emerge (Sharvin and Balashova 1956, Sharvin 1957). Figure 97 shows a photograph of the normal layers obtained in this way in a superconducting plate. The field $\boldsymbol{H}_0$ is tilted relative to the normal $\boldsymbol{n}$. The layers are parallel to the plane in which $\boldsymbol{n}$ and $\boldsymbol{H}_0$ lie.

A further method of determining the structure of the intermediate state is covering the sample surface with a thin layer of a transparent substance having a large Faraday effect, i.e., capable of rotating the plane of polarization of light in a magnetic field. Illumination with polarized light and the observation of the reflected light with the aid of a polaroid gives a very contrasty picture of the layers of the normal and superconducting phases. It is also possible to follow their dynamics upon variation of the external field (Laeng and Rinderer 1972).

Owing to these experiments it has been possible to establish the existence of layers in the intermediate state and to determine the parameter ξ, i.e., the surface energy at the interface between the phases.

In conclusion, we return to the Maxwell field H_i. Although, as has been pointed out earlier (see Appendix III), the true physical significance is exhibited not by the Maxwell field H_i but by the induction B, which corresponds to the average of the microscopic field in the sample, the condition (15.12) $H_i = H_c$ must have a physical meaning. It turns out that in going from the microscopic description of the intermediate state to the "macroscopic" one, in which the averaging is made over distances larger than the thickness of the layers, it is necessary to introduce an additional

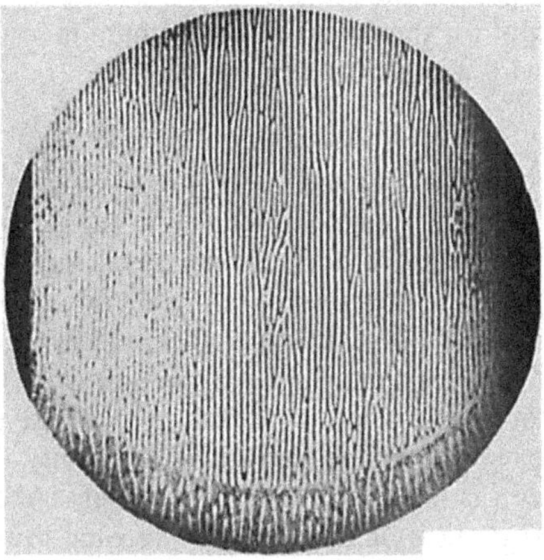

Fig. 97.

variable, namely, the average of the magnetic field taken only over the normal layers (Andreev 1966). It is exactly this average that corresponds to H_i; this clarifies the meaning of the condition (15.12).

15.4. Destruction of superconductivity by a current

Now we shall deal with another problem of the intermediate state. Suppose a current flows in a thick superconducting wire of radius r_0. In accordance with Silsbee's rule (Silsbee 1916), resistance can appear when the field generated by the current reaches the value H_c. Consider the Maxwell equation

$$\mathrm{rot}\, \boldsymbol{H} = \left(\frac{4\pi}{c}\right) j.$$

Integrating over a circle of radius $r < r_0$ and using the rule of transformation of the curl, we obtain

$$\int \mathrm{rot}\, \boldsymbol{H}\, \mathrm{d}\boldsymbol{S} = \oint \boldsymbol{H}\, \mathrm{d}\boldsymbol{l} = H \cdot 2\pi r = \frac{4\pi}{c} \int \boldsymbol{j}\, \mathrm{d}\boldsymbol{S} = \left(\frac{4\pi}{c}\right) J(r).$$

Hence,

$$H(r) = 2\frac{J(r)}{cr}, \tag{15.24}$$

where $H(r)$ is the field at a distance r from the cylinder axis and $J(r)$ is the current flowing through a circle of radius r.

In the superconducting state the current flows only in a thin surface layer. Hence, $H=0$ everywhere, except for this layer. The maximum field exists at the surface. It becomes equal to H_c with a current given by

$$J_c = \tfrac{1}{2} c r_0 H_c.$$

where r_0 is the radius of the conductor.

Let $J > J_c$. In this case, the superconductivity in the surface layer begins to be destroyed. It would appear as if a normal tube and a superconducting core were formed (fig. 98a). As a matter of fact, the entire current flows inside the superconducting cylinder. But it has a smaller radius and from (15.24) it follows that with the same current the field H at the interface will be stronger. Hence, the process of destruction will go through the whole wire. But in this case the entire sample will become normal and the current will be distributed uniformly along the cross section. Then, $J(r) = j\pi r^2$, where j is the constant current density. According to eq. (15.24), the field in the vicinity of the cylinder axis will become lower than H_c; thus, the normal state is also unstable.

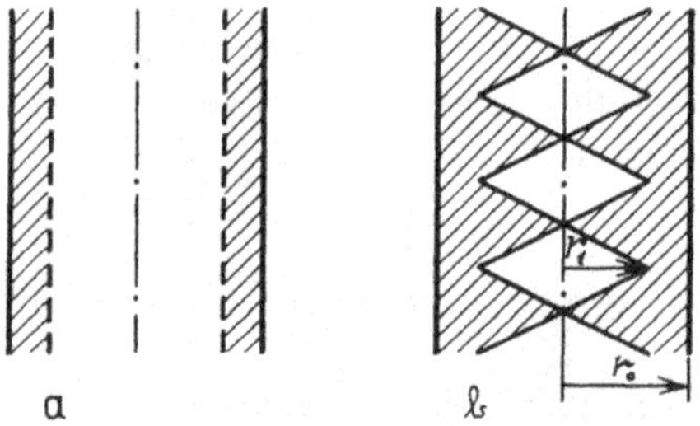

Fig. 98.

The solution lies in the assumption of the existence of the intermediate state (London 1937)*. We shall assume that the interior of the cylinder, with a radius r_1, is in the intermediate state and the external tube is in the normal state (fig. 98b). In the interior $H(r) = H_c$. From eq. (15.24) we obtain

$$J(r) = \tfrac{1}{2} c r H_c. \tag{15.25}$$

The current density is given by

$$j_{r<r_1}(r) = \frac{\mathrm{d}J(r)}{\mathrm{d}S} = \frac{\mathrm{d}J(r)}{2\pi r\,\mathrm{d}r} = cH_c(4\pi r). \tag{15.26}$$

On the other hand, in the outer normal tube the current is distributed uniformly, i.e.,

$$j_{r>r_1}(r) = \frac{J - J(r_1)}{\pi(r_0^2 - r_1^2)} \tag{15.27}$$

* The same idea was put forward at the same time by L.D. Landau (unpublished, see Shoenberg 1955).

From the condition of the continuity of current density at $r = r_1$ we find that

$$\frac{cH_c}{4\pi r_1} = \frac{J - \frac{1}{2}cr_1 H_c}{\pi(r_0^2 - r_1^2)} \tag{15.28}$$

(here we have substituted $J(r_1)$ from eq. 15.25).

We now introduce new variables:

$$\frac{r_1}{r_0} = \rho, \qquad \frac{J}{J_c} = \frac{2J}{cr_0 H_c} = \lambda. \tag{15.29}$$

Equation (15.28) will now be rewritten in the form

$$\rho^2 - 2\lambda\rho + 1 = 0.$$

The roots of this equation are

$$\rho = \lambda \pm (\lambda^2 - 1)^{1/2}.$$

Only one must be retained because the physical solution evidently corresponds to $\rho \to 0$ with $\lambda \to \infty$. Thus,

$$\rho = \lambda - (\lambda^2 - 1)^{1/2}. \tag{15.30}$$

Let us draw two cross sections of the conductor at a certain distance from each other. Evidently, the potential must be constant along each cross section. The potential difference is the product of the current and the resistance. Let us denote the resistance of a normal metal by R_n. In such a case the resistance of the normal tube of $r_1 < r < r_0$ will be $R_n r_0^2/(r_0^2 - r_1^2)$. The potential difference is equal to the product of this resistance and the current flowing through the normal tube, on the one hand, and to the product of the total current and the total resistance R on the other:

$$JR = [J - J(r_1)] R_n \frac{r_0^2}{r_0^2 - r_1^2}. \tag{15.31}$$

Substituting $J(r_1)$ and passing over to the dimensionless variables ρ and λ, we obtain:

$$\frac{R}{R_n} = \tfrac{1}{2}[1 + (1 - \lambda^{-2})^{1/2}]. \tag{15.32}$$

This formula is applicable in the region of the intermediate state, i.e., for $\lambda > 1$. From this formula it follows that at $\lambda = 1$ a jump occurs in the resistance up to $\frac{1}{2}R_n$, and then with increasing λ the resistance approaches asymptotically to R_n (the solid line in fig. 99). This prediction is in qualitative agreement with experiment, but the quantitative results are different. A jump $J = J_c(\lambda = 1)$ occurs up to a value greater than $\frac{1}{2}R_n$, and the subsequent asymptotic approach to R_n occurs more rapidly (the dashed line in fig. 99). This can be accounted for in the following manner. According to formula (15.26), the current density in the core is inversely proportional to r.

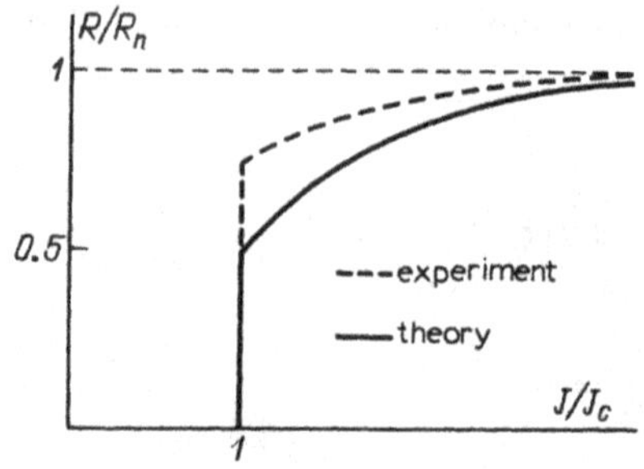

Fig. 99.

Hence, the structure of the intermediate state is such that the thickness of the normal layers increases proportionately to r, as shown in fig. 98b. If we recall the presence of the surface energy σ_{ns}, it will become clear that a boundary with a kink would have led to an infinite energy. Hence, the "discs" of the superconducting phase must have rounded edges, this being responsible for a certain delay in the transition and for the attainment of a larger total resistance. It is interesting to note that with an increase in the wire diameter the initial jump in the resistance decreases and tends towards $\frac{1}{2}R_n$. This is in accord with the conceptions presented above, since the relative role of roundings-off decreases with increasing radius of the layers.

We do not describe here the complete microscopic theory (Andreev 1968). The result of this theory is as follows: the period of the structure at $J=J_c$ is of order $(\xi^2 r_0)^{1/3}$ and falls off with increasing current. The theory also predicts the value of the true critical current and the magnitude of the jump in the resistance at J_c, which are in good agreement with experiment. It is interesting to note that in accordance with the microscopic (BCS) theory the picture of the intermediate state outlined above is observed only in the case of a not-too-pure metal. But if the metal is so pure that the Hall components of the resistivity tensor are comparable with the diagonal components, then the superconducting layers bend along the axis, forming a system of cones (a Christmas tree) instead of a system of planar discs, and move along the axis of the conductor.

In concluding this section, let us consider formula (15.32) from a somewhat different standpoint. Substituting λ into it according to eq. (15.29) and expressing $H_c(T)$ in accordance with formula (15.1), we obtain

$$\frac{R}{R_n}=\tfrac{1}{2}\left\{1+\left[1-\left(\frac{cr_0H_c(0)}{2J}\right)^2\left(1-\frac{T^2}{T_c^2}\right)^2\right]^{1/2}\right\}. \tag{15.33}$$

The formula can be applied as long as the expression in brackets is positive.

Let us specify the current J and consider $R(T)$. If $J\to 0$, then, according to this formula, the resistance appears near T_c and immediately becomes equal to R_n. But if we take a large current, the transition will be extended to a finite temperature range (fig. 100). From this it follows that we may speak of the jump in the resistance

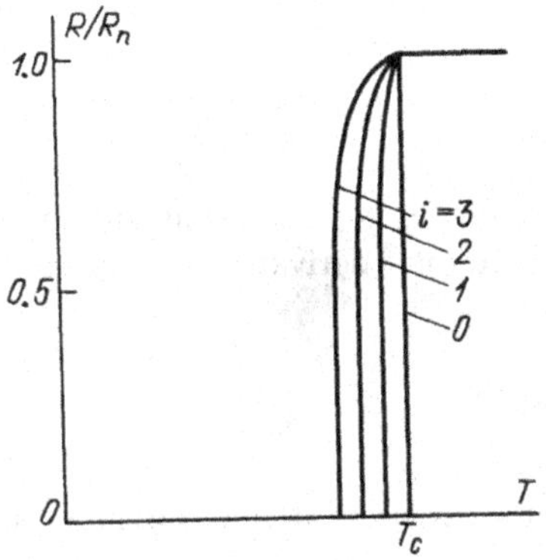

Fig. 100.

from zero to R_n in the superconducting transition only if we mean a measurement in an infinitesimal current.

15.5. The London equations

The first attempt to develop the electrodynamics of superconductors was undertaken by the brothers F. London and H. London (1935). The purpose of this theory was to express in mathematical form the basic experimental facts: the absence of resistance and the Meissner effect, without consideration of the microscopic factors responsible for superconductivity.

If an electron does not undergo scattering, it is accelerated by an external field and, hence,

$$m\frac{d\boldsymbol{v}}{dt} = e\boldsymbol{E}. \tag{15.34}$$

This refers to all electrons. Hence, we can rewrite formula (15.34) for the current density. Introducing $\boldsymbol{j} = n_e e\boldsymbol{v}$, where n_e is the electron density and e is the charge, we get:

$$\frac{d}{dt}(\Lambda \boldsymbol{j}) = \boldsymbol{E},$$

where $\Lambda = m/(n_e e^2)$.

In Part I we have already noted that the full derivative signifies the variation of the quantity in a given volume element moving together with the liquid. It is connected with the partial derivative, which describes the variation at a given point of space by the relation

$$\frac{\mathrm{d}}{\mathrm{d}t}=\frac{\partial}{\partial t}+\boldsymbol{v}\nabla.$$

Since the real current velocities v in a metal are small compared to the Fermi velocity, we can replace the total derivative by a partial derivative, in which case we have

$$\frac{\partial(\Lambda \boldsymbol{j})}{\partial t}=\boldsymbol{E}. \tag{15.35}$$

According to the general equation of electrodynamics:

$$\text{rot}\,\boldsymbol{E}=-c^{-1}\frac{\partial \boldsymbol{H}}{\partial t}. \tag{15.36}$$

Taking the curl of eq. (15.35) and substituting (15.36), we get

$$\frac{\partial}{\partial t}(\text{rot}\,\Lambda \boldsymbol{j}+c^{-1}\boldsymbol{H})=0.$$

This means that the quantity in parentheses does not change with time. But in the bulk of a massive superconductor $\boldsymbol{j}=0$ and $\boldsymbol{H}=0$. Hence, the quantity in parentheses is not simply conserved but is always equal to zero:

$$\text{rot}\,\Lambda \boldsymbol{j}+c^{-1}\boldsymbol{H}=0. \tag{15.37}$$

Equations (15.36) and (15.37) constitute the basis of London electrodynamics.

Let us consider the simplest problem: a bulk superconductor in an external magnetic field. According to the Maxwell equation,

$$\text{rot}\,\boldsymbol{H}=\frac{4\pi}{c}\,\boldsymbol{j}. \tag{15.38}$$

We take the curl of this equation:

$$\text{rot rot}\,\boldsymbol{H}=\text{grad div}\,\boldsymbol{H}-\Delta\boldsymbol{H}=\frac{4\pi}{c}\,\text{rot}\,\boldsymbol{j}.$$

Using the fact that div $\boldsymbol{H}=0$ and substituting rot $\boldsymbol{j}$ from (15.37), we obtain

$$\Delta\boldsymbol{H}-\delta^{-2}\boldsymbol{H}=0, \tag{15.39}$$

where

$$\delta=\left(\frac{\Lambda c^2}{4\pi}\right)^{1/2}=\left(\frac{mc^2}{4\pi n_e e^2}\right)^{1/2} \tag{15.40}$$

(we have used here the definition of Λ). Suppose the superconductor occupies the half-space $x>0$ and the field is applied parallel to its surface in the direction z. From eq. (15.39) we find

$$H = H_0 \exp\left(\frac{-x}{\delta}\right) \tag{15.41}$$

where H_0 is the field at the surface. From eq. (15.37) we get

$$j_y = \frac{cH_0}{4\pi\delta} \exp\left(\frac{-x}{\delta}\right). \tag{15.42}$$

Thus, we conclude that London electrodynamics provides a correct qualitative description of the fact that the field and current are present only in a thin surface layer of the superconductor. Moreover, according to eq. (15.40), the quantity δ is of the order of 10^{-5}–10^{-6} cm, consistent with experimental data.

Of course, in the derivation given above we proceeded from the assumption that all the electrons participate in the superconducting current. This contradicts the fact that δ depends on temperature and also that a second-order phase transition occurs at T_c, i.e., the state of electrons varies continuously. The theory may be "corrected" to a certain extent by introducing, instead of the total number of electrons, a certain number of "superconducting electrons" $n_s(T)$, which decreases with temperature and turns to zero at T_c.

In conclusion, we note that eqs. (15.35) and (15.37) may be written in a different form, which will prove useful at a later stage. We introduce the vector potential $\boldsymbol{A}$; $\boldsymbol{H}$ and $\boldsymbol{E}$ are expressed by the relations

$$\boldsymbol{H} = \operatorname{rot} \boldsymbol{A}, \qquad \boldsymbol{E} = -c^{-1}\frac{\partial \boldsymbol{A}}{\partial t}.$$

Using $\boldsymbol{A}$, we can write both London equations in the form of a single equation:

$$\boldsymbol{j} = -\frac{1}{\Lambda c}\boldsymbol{A} = -\frac{n_e e^2}{mc}\boldsymbol{A}. \tag{15.43}$$

At first glance this means that only one London equation is important and the second one is its corollary. But this is not the case: taking div of the Maxwell equation (15.38), we obtain the law of charge conservation:

$$\operatorname{div} \boldsymbol{j} = 0. \tag{15.44}$$

But this is compatible with eq. (15.43) only if

$$\operatorname{div} \boldsymbol{A} = 0. \tag{15.45}$$

This means that eq. (15.43) is valid not for any vector potential $\boldsymbol{A}$ but only for one that satisfies the condition (15.45). Thus, instead of the two London equations (15.35) and (15.37), we again have two equations, (15.43) and (15.45).

16

Basic ideas of the microscopic theory

16.1. The superfluidity condition

The microscopic theory of superconductivity was developed by J. Bardeen, L.N. Cooper and J.R. Schrieffer in 1957 and independently by N.N. Bogoliubov in 1957 (Bardeen et al. 1957, Bogoliubov 1958, Bogoliubov et al. 1958). It is quite natural to ask why superconductivity remained a puzzling phenomenon for nearly half a century after its discovery. Science was advancing; a great deal of experimental information was obtained and several useful phenomenological theories were developed (for example, the theory described in ch. 1).

The main cause is as follows. The phenomenon of superconductivity is very similar to the superfluidity of liquid helium discovered by P.L. Kapitza (Kapitza 1938). The theory of this phenomenon was developed by L.D. Landau in 1941 (Landau 1941a, b, 1944, 1947). One of the manifestations of superfluidity is the frictionless flow of helium in capillaries. It was natural to interpret superconductivity as the superfluidity of the electron liquid.

The Landau theory introduced a criterion for superfluidity. Imagine helium flowing in a capillary with a velocity $\boldsymbol{v}$. If we pass to the coordinate system associated with helium, it will be at rest while the walls of the tube will move at a velocity $-\boldsymbol{v}$. If viscosity arises, the moving tube will begin to entrain the resting helium. This means that the momentum $\boldsymbol{P}$ and the energy E of the helium will be changed. But, as we know from ch. 2, in a homogeneous quantum system the variation of energy and momentum occurs by way of the appearance of quasiparticles. Suppose a quasiparticle with momentum $\boldsymbol{p}$ and energy $\varepsilon(p)$ has appeared. Let us now pass to the laboratory coordinate system associated with the tube. In this system the energy and momentum are given by

$$\boldsymbol{P}' = \boldsymbol{P} + M\boldsymbol{v}, \qquad E' = E + \boldsymbol{P}\boldsymbol{v} + \tfrac{1}{2}\mathrm{M}v^2.$$

(M is the mass of the liquid). The variation of the energy resulting from the creation of a quasiparticle is $E + \boldsymbol{P}\boldsymbol{v} = \varepsilon(p) + \boldsymbol{p}\boldsymbol{v}$. Hence, in order for the creation of a quasiparticle to be energetically favorable it is necessary that

$$\varepsilon(p) + \boldsymbol{p}\boldsymbol{v} < 0.$$

The lowest value of $\varepsilon + \boldsymbol{p}\boldsymbol{v}$ is attained with the opposite orientation of $\boldsymbol{p}$ and $\boldsymbol{v}$; hence, in any case, $\varepsilon - pv < 0$ or

$$v > \frac{\varepsilon(p)}{p}.$$

This is the lowest velocity at which quasiparticles of momentum $\boldsymbol{p}$ and energy $\varepsilon(p)$ can appear. But we are interested in the velocity at which any quasiparticles can be created. Evidently, for this to occur it is necessary that

$$v > v_c = \min\left(\frac{\varepsilon(p)}{p}\right). \tag{16.1}$$

Thus, viscosity arises if the velocity exceeds v_c. In liquid helium the energy spectrum is such that v_c is a finite quantity. This means that at low flow velocities there is no viscosity.

We shall consider the Fermi liquid from this standpoint. The excitation of such a system consists of the creation of a particle–antiparticle pair. If such a pair is generated directly at the surface of the Fermi sphere, the energy may be as small as possible. At the same time, the total change in the momentum may reach $2p_0$ if the particle and the antiparticle are located on the opposite sides of the Fermi sphere. From this it follows that $v_c = 0$, i.e., in the Fermi system there is viscosity at any flow velocity.

Attempts have been made to find the solution on the assumption that electrons combine into pairs which are Bose particles and are analogous to helium atoms. However, no causes for such a combination could be found, since electrons have the same charge and according to the Coulomb law they repel each other. It is this circumstance that held back the development of the theory of superconductivity.

16.2. Phonon attraction

In 1950 a phenomenon was discovered which is known as the isotope effect (Maxwell 1950; Reynolds et al. 1950). In determining the T_c and H_c of superconductors of different isotopic composition it was found that T_c and H_c depend on the mass of ions forming the crystal lattice:

$$T_c \propto M^{-1/2}, \qquad H_c \propto M^{-1/2}. \tag{16.2}$$

The mass of the ions manifests itself only when lattice vibrations are taken into account. Hence, superconductivity is not a purely electronic phenomenon; phonons are also involved in it. On the basis of this fact Frölich (1950) and Bardeen (1950, 1951a, b) demonstrated independently that the electrons, while residing in the crystal lattice, are also capable of attracting one another*.

Let us consider the process shown schematically in fig. 101a. One electron with momentum $\boldsymbol{p}_1$ emits a phonon of momentum $\hbar\boldsymbol{k}$. The momentum of the electron

* We shall describe here the phonon mechanism of electron attraction which takes place in all superconductors known to date. The possibility is not excluded that the excitonic mechanism is operating in the newly found high-temperature ceramics (section 16.11). However, its principal scheme is very similar to phonon attraction.

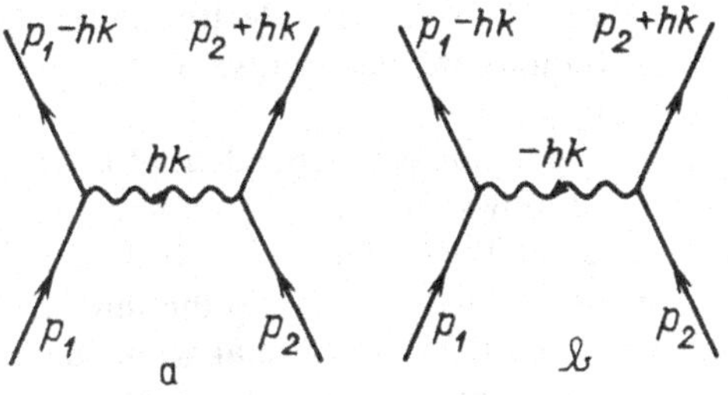

Fig. 101.

then becomes equal to $p'_1 = p_1 - \hbar k$. The phonon is absorbed by a second electron, which has had momentum p_2 before the absorption; after the phonon is absorbed the second electron acquires a momentum $p'_2 = p_2 + \hbar k$. The amplitude of this process, which is of second order according to perturbation theory, is given by

$$\frac{|V_k|^2}{\varepsilon(p_1) - \varepsilon(p_1 - \hbar k) - \hbar\omega(k)},$$

where $V_k = V_{p-\hbar k, p}$ (section 4.3, formula 4.13).

On the other hand, the same finite momenta result if the electron of p_2 emits a phonon with momentum $-\hbar k$, which is then absorbed by the electron of p_1 (fig. 101b). The amplitude of this process is

$$\frac{|V_k|^2}{\varepsilon(p_2) - \varepsilon(p_2 + \hbar k) - \hbar\omega(k)}.$$

Here it is taken into account that $\omega(k)$ and V_k are invariant under the replacement of k by $-k$. The amplitudes obtained above must add up. Here one has to take account of the energy conservation law:

$$\varepsilon(p_1) + \varepsilon(p_2) = \varepsilon(p'_1) + \varepsilon(p'_2).$$

As a result, for the total amplitude we find

$$-\frac{2|V_k|^2 \hbar\omega(k)}{[\hbar\omega(k)]^2 - [\varepsilon(p_1) - \varepsilon(p_1 - \hbar k)]^2}.$$

Substituting expression (4.13) for V_k ($n_k = 0$ in this case), we obtain the order of magnitude of the total amplitude:

$$-\frac{\hbar^3}{p_0 m V} \frac{[\hbar\omega(k)]^2}{[\hbar\omega(k)]^2 - [\varepsilon(p_1) - \varepsilon(p_1 - \hbar k)]^2}. \tag{16.3}$$

Let us analyze this expression. At $\varepsilon(p_1) - \varepsilon(p'_1) \ll \hbar\omega(k)$ we obtain a negative constant, which is independent of k. If the electrons were involved in a direct point interaction with one another, i.e., $U = U_0\delta(r_1 - r_2)$, the matrix element corresponding to the first order perturbation theory for the scattering amplitude would be equal to U_0. Hence, the interaction between the electrons transmitted by phonons is equivalent, with slight variations in the energy, to a point interaction, its sign corresponding to attraction.

Note also that since the interaction is independent of $\boldsymbol{k}$, i.e., of the rotation angle of the electron momenta, it follows that this is an interaction in the s-state, i.e., in a state with an orbital moment $l=0$. Thus, the coordinate wave function of the interacting electrons is symmetric with respect to the interchange of particles. But since the electrons are Fermi particles, their total wave function is antisymmetric under interchange. Hence, their spin wave function must be antisymmetric, i.e., the spins of the interacting electrons must be directed in opposite directions. The same may be put in a different way. The interaction found is equivalent to the attraction of particles if their coordinates coincide. The Pauli principle forbids two identical Fermi particles to be at a single point. Hence, they must differ in the spin direction.

If we take into account that the density of states of phonons is proportional to $k^2(\mathrm{d}k/\mathrm{d}\omega)$, i.e., falls off rapidly with decreasing k, and that the resultant amplitude is independent of k, then it becomes clear that the key role will be played by phonons with the largest k values, i.e., with $k \sim k_\mathrm{D} \sim \pi/a$ and $\hbar\omega \sim \hbar\omega_\mathrm{D}$. At $\varepsilon(\boldsymbol{p})-\varepsilon(\boldsymbol{p}') \gg \hbar\omega_\mathrm{D}$ the interaction rapidly decreases. Since the major role is played by electrons in the neighborhood of the Fermi boundary, it may be said that the phonon attraction occurs in a narrow layer near the Fermi boundary, whose thickness is of order $\hbar\omega_\mathrm{D}$ on the energy scale and of order $\Delta_p \sim \hbar\omega_\mathrm{D}/v$ on the momentum scale.

The fact that the interaction spreads over an energy interval $\Delta\varepsilon \sim \hbar\omega_\mathrm{D}$ implies, according to quantum mechanics, that it is "retarded" or, in other words, it operates during a finite time interval $\Delta t \sim \hbar/(\hbar\omega_\mathrm{D}) \sim \omega_\mathrm{D}^{-1}$. Its magnitude is of order $(p_0 m/\hbar^3)^{-1} \sim \nu^{-1}\mu)$.

Simultaneously, Coulomb repulsion is operating. Taking account of the Debye screening (section 14.1), one can show that it operates at interatomic distances and may be written in the form $e^2a^2\delta(\boldsymbol{r}_1-\boldsymbol{r}_2)$. for the same reason it is retarded with a characteristic time of order $\hbar/\varepsilon_\mathrm{F}$. The ratio of the interaction constants is of order

$$\frac{e^2a^2}{\hbar^3/p_0 m} \sim \frac{e^2\hbar^2}{p_0^2}\frac{p_0 m}{\hbar^3} \sim \frac{e^2}{\hbar v} \sim 1$$

that is, both interactions are of the same order of magnitude. Hence, in some metals phonon attraction may be dominant, while in others Coulomb repulsion predominates. This explains why some metals become superconductive and others remain normal at all temperatures.

To be more precise, it should be said that quantitative calculations always give some preference to the Coulomb repulsion. But its time of action is much shorter than that of the phonon attraction. This leads to an effective damping of the Coulomb repulsion: it is divided by the log of the time ratio, i.e., by $\ln(\varepsilon_\mathrm{F}/\hbar\omega_\mathrm{D}) = \ln(M/m)$.

In most cases, the two interactions differ quantitatively rather strongly to one or the other side, so that in studying superconductivity one may limit oneself to taking account of phonon attraction alone. Instead of the exact formula (16.3), we shall simulate its amplitude by a constant $-g$ $(g \sim \hbar^3/p_0 m \sim (\nu/\mu)]^{-1})$ in the energy interval $\mu - \hbar\omega_D < \varepsilon < \mu + \hbar\omega_D$ and by zero at other energies.

16.3. Cooper pairs

It would appear that after the discovery of the attraction it became immediately possible to work out the theory of superconductivity. Nevertheless, several years were needed for this to be accomplished. The difficulty that had to be overcome was as follows. As has been explained earlier, the superfluidity of the electron liquid could have arisen upon binding of electrons into pairs. But from quantum mechanics it is known that if the interaction is not strong enough or has a small action radius, a bound state cannot be produced. Exceptions are the purely one- and two-dimensional cases when the motion of particles is limited by a straight line or a plane; in these cases, any attraction leads to the formation of a bound state. But at first sight such models have nothing to do with the case under consideration.

A measure of the binding energy of a pair may be the critical temperature. If one considers that it is of the order of degrees, i.e., 10^4 times lower than the kinetic energy of electrons (the Fermi energy), the binding of electrons into pairs seems to be impossible.

The resolution of this paradox was found by Cooper (Cooper 1956), who noticed that pairs are formed not by free isolated electrons but by quasiparticles of the Fermi liquid. We shall present here a slightly modified version of Cooper's derivation which provides an answer correct in order of magnitude, in contrast to the original paper in which actually only quasiparticles of particle type were taken into account and this led to an incorrect result.

For the sake of simplicity we will consider the isotropic case. The energy spectrum of quasiparticles is shown in fig. 102 by a solid line (section 2.2). Here the right branch refers to quasiparticles of particle type and the left to quasiparticles of antiparticle type. On the whole, the energy spectrum may be written in the form $|\xi(p)|$, where $\xi = v(p - p_0)$. One should, however, keep in mind that antiparticles have a different sign of the charge. As will be seen below, we are interested in the

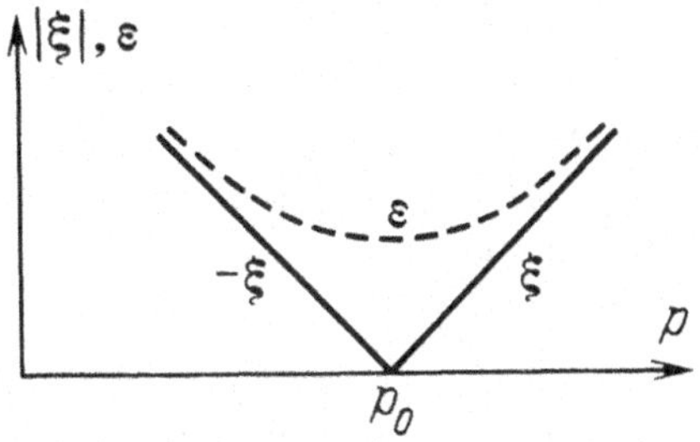

Fig. 102.

interaction of quasiparticles with the same $|\boldsymbol{p}|$, i.e., the interaction of either two particles or two antiparticles. In either case, phonons produce attraction.

Let us write the Schrödinger equation for two quasiparticles:

$$[\mathscr{H}_0(\boldsymbol{r}_1)+\mathscr{H}_0(\boldsymbol{r}_2)+U(\boldsymbol{r}_1,\boldsymbol{r}_2)]\Psi(\boldsymbol{r}_1,\boldsymbol{r}_2)=E\Psi(\boldsymbol{r}_1,\boldsymbol{r}_2). \tag{16.4}$$

Here $\mathscr{H}_0(\boldsymbol{r}_1)$ is the Hamiltonian of a free quasiparticle, i.e.,

$$\mathscr{H}_0(\boldsymbol{r}_1)\,\psi_p(\boldsymbol{r}_1)=|\xi(\boldsymbol{p})|\,\psi_p(\boldsymbol{r}_1) \tag{16.5}$$

(in the free-quasiparticle model $\psi_p=V^{-1/2}\,\mathrm{e}^{\mathrm{i}\boldsymbol{pr}/\hbar}$). In the ground state the total momentum and spin of a bound pair must be equal to zero. Hence, the wave function of the pair will be a superposition of the states of two free quasiparticles having opposite momenta and spins, i.e.,

$$\Psi(\boldsymbol{r}_1,\boldsymbol{r}_2)=\sum_p c_p\psi_{p,+}(\boldsymbol{r}_1)\,\psi_{p,-}(\boldsymbol{r}_2), \tag{16.6}$$

where the subscripts $\pm$ signify the projection of spin $\pm\frac{1}{2}$. Substituting this into eq. (16.4), we obtain

$$2|\xi(\boldsymbol{p})|c_p+\sum U_{pp'}c_{p'}=Ec_p. \tag{16.7}$$

We put

$$U_{pp'}=\begin{cases}-g, & p_0-\dfrac{\hbar\omega}{v}<|\boldsymbol{p}|,|\boldsymbol{p}'|<p_0+\dfrac{\hbar\omega_\mathrm{D}}{v},\\[2mm] 0, & \text{outside this interval,}\end{cases} \tag{16.8}$$

and solve eq. (16.7) for c_p:

$$c_p=\frac{gI}{2|\xi(\boldsymbol{p})|-E}, \tag{16.9}$$

where

$$I=\sum_{|\boldsymbol{p}'|=p_0-\hbar\omega_\mathrm{D}/v}^{|\boldsymbol{p}'|=p_0+\hbar\omega_\mathrm{D}/v} c_{p'}. \tag{16.10}$$

Taking into account that we are looking for a bound state (a negative energy value), denoting $E=-2\Delta$ and substituting c_p from eq. (16.9) into eq. (16.10) we find

$$I=\tfrac{1}{2}\sum_{|\boldsymbol{p}|=p_0-\hbar\omega_\mathrm{D}/v}^{|\boldsymbol{p}|=p_0+\hbar\omega_\mathrm{D}/v}\frac{1}{|\xi(p)|+\Delta}.$$

Passing to integration over ξ and taking into account that the integral is taken over neighborhood of the Fermi surface, we get

$$1=\tfrac{1}{2}g\nu(\mu)\ln\left(\frac{\hbar\omega_\mathrm{D}}{\Delta}\right)$$

(the factor $\frac{1}{2}$ has appeared because the summation is carried out over states of one of the quasiparticles, for which the spin projection is specified, whereas the density of states defined earlier, $\nu=p_0m/(\pi^2\hbar^3)$, included both projections of the spin).

Solving for Δ, we find

$$\Delta = \hbar\omega_D \exp\left(\frac{-2}{g\nu(\mu)}\right). \tag{16.11}$$

Thus, a pair of quasiparticles has a finite binding energy 2Δ. Such bound pairs are called Cooper pairs. At $T = 0$ they form a Bose condensate.

Below we shall obtain Δ from the exact theory. It will be twice as large as (16.11). The difference is associated with the fact that in the derivation given above we were concerned with the creation of a pair against the background of the unreconstructed ground state with a "normal" energy spectrum*.

In spite of a certain inaccuracy, the result (16.11) makes it possible to draw certain conclusions. First of all, note that since $g \sim \nu^{-1}(\mu)$, the expression in the exponent is formally of the order of unity. In fact, $g\nu$ is always somewhat smaller than unity. A certain numerical smallness in the exponent turns to a substantial smallness of the ratio $\Delta/\hbar\omega_D$, which, judging by the known values of T_c (it will be shown below that $\Delta \sim T_c$), is usually 10^{-2} or less**.

Further, from the result obtained it is seen that the binding energy of a Cooper pair 2Δ is finite at any g, which is not consistent with the above-given reasoning concerning the impossibility of the formation of a bound complex upon week interaction. In fact, we are speaking not of isolated particles but of quasiparticles with the Fermi sphere being occupied. This actually leads to the replacement of the three-dimensional problem by a two-dimensional one: the integrals over the momenta are transformed in accordance with the rule $\int d^3\boldsymbol{p}/(2\pi\hbar)^3 \to \frac{1}{2}\nu \int d\xi$, and for the two-dimensional problem $\int d^2\boldsymbol{p}/(2\pi\hbar)^2 \to 2\pi \int p\, dp/(2\pi\hbar)^2 \to 2\pi m \int d\varepsilon/(2\pi\hbar)^2$, since $\varepsilon = p^2/2m$. In a two-dimensional model, as has been said above, any attraction will be sufficient for the particles to be bound.

We thus see that the presence of a filled Fermi sphere plays a most important role in the formation of Cooper pairs.

It is clear that any change in temperature will have an effect on the binding energy. It is because of this that $\Delta(T)$ decreases with increasing temperature and vanishes at $T_c \sim \Delta(0)$. Such is the physical picture. We will now turn our attention to a more detailed description of the Bardeen–Cooper–Schrieffer (BCS) theory***.

16.4. *The energy spectrum*

While considering the quantization of lattice vibrations in section 4.3, we introduced operators u_k and u_k^+. Let us write these operators in the following

* The derivation performed in the original work (Cooper 1956) leads to a substantial error, giving an exponent twice as large in absolute magnitude.

** As for the high-temperature ceramics, if the idea of the excitonic mechanism of their superconductivity is confirmed, $\hbar\omega_D$ in formula (16.11) should be replaced by an essentially larger quantity (section 16.11).

*** As noted above, the same theory was developed practically at the same time by Bogoliubov. We shall refer to it as the BCS theory because this name has been generally adopted.

form:

$$u_k = \left(\frac{\hbar}{2nM\omega(\boldsymbol{k})}\right)^{1/2} b_k, \qquad (b_k)_{n_k}^{n_k-1} = n_k^{1/2}.$$

The operator b_k thus introduced is called the phonon annihilation operator and b_k^+, its Hermitian conjugate, is the phonon creation operator. The operator $b_k^+ b_k$ is diagonal with the matrix elements being equal to n_k. Thus,

$$b_k^+ b_k = n_k$$

is the operator of the number of particles in state $\boldsymbol{k}$. Analogously,

$$b_k b_k^+ = n_k + 1.$$

Hence, the commutator of the operators b_k and b_k^+ is

$$b_k b_k^+ - b_k^+ b_k = [b_k b_k^+] = 1.$$

Since operators with different $\boldsymbol{k}$ commute, we may write:

$$[b_k, b_k^+] = \delta_{kk'}.$$

All this refers to phonons, which obey Bose statistics. Let us now consider electrons, which obey Fermi statistics. For them we can also introduce particle creation and annihilation operators. Consider a system of noninteracting Fermi particles. According to the Pauli principle, no more than one particle may exist in each state. We shall conventionally write the wave function of this system, noting in which states there are particles and in which there are no particles at all. For example,

$$\Phi(1_{p_1}, 0_{p_2}, 1_{p_3}, 1_{p_4}, 0_{p_5})$$

denotes a system in which the states $\boldsymbol{p}_1$, $\boldsymbol{p}_3$ and $\boldsymbol{p}_4$ are occupied and the states $\boldsymbol{p}_2$ and $\boldsymbol{p}_5$ are free.

Let the system be described by the wave function $\Phi(\ldots 0_p \ldots)$. Then the annihilation operator a_p acting on this system gives zero, because there are no particles in the state $\boldsymbol{p}$:

$$a_p \Phi(\ldots 0_p \ldots) = 0.$$

But if there is a wave function $\Phi(\ldots 1_p \ldots)$, then under the action of the operator a_p it is transformed to $\Phi(\ldots 0_p \ldots)$, i.e.,

$$a_p \Phi(\ldots 1_p \ldots) = \Phi(\ldots 0_p \ldots).$$

Let us now act on these functions by means of the operator $a_p^+ a_p$. Since a_p^+ is the creation operator, we obtain

$$a_p^+ a_p \Phi(\ldots 0_p \ldots) = a_p^+ \cdot 0 = 0 = 0 \cdot \Phi(\ldots 0_p \ldots),$$

$$a_p^+ a_p \Phi(\ldots 1_p \ldots) = a_p^+ \Phi(\ldots 0_p \ldots) = \Phi(\ldots 1_p \ldots)$$
$$= 1 \cdot \Phi(\ldots 1_p \ldots).$$

From these formulae it follows that the operator $a_p^+ a_p$ is diagonal and equal to the number of particles:

$$a_p^+ a_p = n_p.$$

Analogously, we find

$$a_p a_p^+ \Phi(\ldots 1_p \ldots) = 0, \qquad a_p a_p^+ \Phi(\ldots 0_p \ldots) = \Phi(\ldots 0_p \ldots),$$

and, hence,

$$a_p a_p^+ = 1 - n_p, \qquad \{a_p, a_p^+\} = a_p a_p^+ + a_p^+ a_p = 1.$$

This rule is known as the anticommutation rule of Fermi operators.

Consider now the action of operators $a_p^+ a_{p'}$ and $a_{p'} a_p^+$ with different $\boldsymbol{p}$ and $\boldsymbol{p}'$. In the first case we have

$$\begin{aligned} a_p^+ a_{p'} \Phi(\ldots 0_p \ldots 1_{p'} \ldots) &= a_p^+ \Phi(\ldots 0_p \ldots 0_{p'} \ldots) \\ &= \Phi(\ldots 1_p \ldots 0_{p'} \ldots) \end{aligned}$$

and in the second

$$\begin{aligned} a_{p'} a_p^+ \Phi(\ldots 0_p \ldots 1_{p'} \ldots) &= a_{p'} \Phi(\ldots 1_p \ldots 1_{p'} \ldots) \\ &= -a_{p'} \Phi(\ldots 1_{p'} \ldots 1_p \ldots) = -\Phi(\ldots 0_{p'} \ldots 1_p) \\ &= -\Phi(\ldots 1_p \ldots 0_{p'} \ldots). \end{aligned}$$

The change in sign is associated with the fact that when two particles are interchanged the wave function changes sign. This must be taken into account when there are several occupied states. From this it follows that the anticommutator $\{a_p^+ a_{p'}\}$ with different $\boldsymbol{p}$ and $\boldsymbol{p}'$ is equal to zero and, hence,

$$\{a_{p'}, a_p^+\} = a_{p'} a_p^+ + a_p^+ a_{p'} = \delta_{pp'}. \tag{16.12}$$

The use of the occupation numbers of the states as variables and the introduction of the operators a_p, a_p^+ is called second quantization.

Using the operators of second quantization we can write the Hamiltonian of an interacting system of electrons. For a normal metal we have used two descriptions: quasiparticles with the energy spectrum $\varepsilon_{\rm qp} = |\xi|$ and the gas model. The difference between these two models consists, in particular, in that the former specifies the chemical potential while the latter specifies the total number of particles. As has been shown in section 2.4, the variation of the chemical potential in cases of physical interest is insignificant. Therefore, both descriptions are practically equivalent.

For superconductors it is more convenient to use the picture of quasiparticles, and this also corresponds to the specification of the chemical potential. With the chemical potential being specified the role of the Hamiltonian is played by $\mathcal{H} - \mu N$, where N is the operator of the total number of particles. As a matter of fact, this reduces to the replacement of the electron energy by $\xi = \varepsilon - \mu$, i.e., to a shift in the origin of the energy. Exactly such a Hamiltonian will be used here for the determination of the energy spectrum of quasiparticles.

Below we shall always consider the isotropic model of the metal because in this case it is easier to trace out the main properties of superconductors. For those cases where anisotropy is of fundamental importance a special analysis will be given.

After what has been said above we can write the Hamiltonian*:

$$\mathcal{H}-\mu N=\sum_{p,\sigma}\xi_p a^{+}_{p,\sigma}a_{p,\sigma}-g\sum_{p_1+p_2=p_1'+p_2'}a^{+}_{p_1',+}a^{+}_{p_2',-}a_{p_2,-}a_{p_1,+}. \tag{16.13}$$

The first term is simply $\sum_{p,\sigma}\xi_p n_{p,\sigma}$. The summation runs over momenta and spins ($\sigma=\pm$). The second term corresponds to the interaction of electrons. It describes how two electrons with initial momenta $\boldsymbol{p}_1$ and $\boldsymbol{p}_2$ acquire new momenta, $\boldsymbol{p}_1'$ and $\boldsymbol{p}_2'$, as a result of the interaction; in other words, electrons with initial momenta disappear and instead of them there are created electrons with new momenta. Here, the law of conservation of the total momentum is taken into account**. The sums over the momenta run over the intervals $p_0\pm\hbar\omega_D/v$; it is taken into account that at a single point only electrons with opposite spins are interacting. The coefficient in this term should have been written in the form g/V, but for simplicity we assume the normalizing volume $V=1$ in this part of the book.

For a normal metal we may pass from electrons to quasiparticles. To do this, we have to assume that $a_{p,\sigma}=\alpha_{p,\sigma}$ at $p>p_0$ and that $a_{p,\sigma}=\alpha^{+}_{-p,-\sigma}$ at $p<p_0$, where $\alpha_{p,\sigma}$ are the operators of second quantization for quasiparticles. As a matter of fact, the destruction of a particle with $p<p_0$ means the creation of a quasiparticle of antiparticle type. Thus, even in an ordinary Fermi system the ground state is such that the transition to quasiparticles requires the redefinition of the creation and annihilation operators.

In a superconductor an even more complicated change in the ground state takes place. For this change to be found and for the energy of quasiparticles to be calculated, we employ a variational method (Abrikosov and Khalatnikov 1958)***. We introduce the creation and annihilation operators for quasiparticles, $\alpha_{p,+}$ and $\alpha_{p,-}$ by using the Bogoliubov relations (Bogoliubov 1958):

$$a_{p,+}=u_p\alpha_{p+}+v_p\alpha^{+}_{-p,-},\qquad a_{p,-}=u_p\alpha_{p,-}-v_p\alpha^{+}_{-p,+}. \tag{16.14}$$

Note that with these transformations the momentum and spin are conserved.

Let us write the anticommutator $\{a_{p,+}a^{+}_{p,+}\}$ and substitute into it formulae (16.14). As a result, we have

$$\{a_{p,+}a^{+}_{p,+}\}=u_p^2\{\alpha_{p,+}\alpha^{+}_{p,+}\}+v_p^2\{\alpha^{+}_{-p,-}\alpha_{-p,-}\}$$
$$+u_pv_p\{\alpha_{p,+}\alpha_{-p,-}\}+u_pv_p\{\alpha^{+}_{-p,-}\alpha^{+}_{p,+}\}=1.$$

* Here and below $\xi_p\equiv\xi(\boldsymbol{p})$.

** In the second term of the Hamiltonian we omitted the term with $\boldsymbol{p}_1=\boldsymbol{p}_1'$. This term is responsible for a slight variation in the chemical potential and can be compensated if we write $\xi-\Delta\mu$ in the first term. For simplicity we dispense with this procedure as being unimportant for the derivation.

*** The method described here is simpler than the one used in the original work of Bardeen, Cooper and Schrieffer and is a generalization of the Bogoliubov method (Bogoliubov 1958; Bogoliubov et al. 1958) to finite temperatures.

For the operators $\alpha_{p,\sigma}$ to satisfy the ordinary anticommutation rules, it is necessary that

$$u_p^2 + v_p^2 = 1. \tag{16.15}$$

Thus, there is only one independent function, say v_p. We choose it so that the average energy be a minimum*. Here for the moment we leave the occupation numbers for quasiparticles undetermined. The average energy corresponds to the diagonal matrix element of the Hamiltonian (16.13). Substituting into it formulas (16.14) and retaining only

$$\langle \alpha^+_{p,-}\alpha_{p,-}\rangle = n_{p,-}, \qquad \langle \alpha^+_{p,+}\alpha_{p,+}\rangle = n_{p,+}$$

(here $\langle\cdots\rangle$ signifies the average over states with given occupation numbers), we obtain

$$\begin{aligned} E - \mu N &= \langle \mathscr{H} - \mu N\rangle \\ &= \sum_p \xi_p [u_p^2 (n_{p,+} + n_{p,-}) + v_p^2 (2 - n_{p,-} - n_{p,+})] \\ &\quad - g \sum_p u_p v_p (1 - n_{p,+} - n_{p,-}) \sum_{p'} u_{p'} v_{p'} (1 - n_{p',+} - n_{p',-}). \end{aligned} \tag{16.16}$$

Expressing u_p in accordance with (16.15) and varying with respect to v_p leads to the condition

$$2\xi_p = \frac{\Delta(1 - 2v_p^2)}{u_p v_p}, \tag{16.17}$$

where we have introduced the notation

$$\Delta = g \sum_p u_p v_p (1 - n_{p,+} - n_{p,-}). \tag{16.18}$$

The solution of eq. (16.17) is

$$u_p^2 = \tfrac{1}{2}\left(1 + \frac{\xi_p}{\varepsilon_p}\right), \qquad v_p^2 = \tfrac{1}{2}\left(1 - \frac{\xi_p}{\varepsilon_p}\right), \tag{16.19}$$

where

$$\varepsilon_p = (\xi_p^2 + \Delta^2)^{1/2}. \tag{16.20}$$

Substitution into eq. (16.18) leads to the condition

$$\Delta = \tfrac{1}{2} g \sum_p \frac{\Delta}{\varepsilon_p} (1 - n_{p,+} - n_{p,-}). \tag{16.21}$$

For calculating the energy of quasiparticles we use the ideas of Landau's Fermi liquid theory (section 13.1). The energy of quasiparticles is defined as a variational

* As has been said in section 14.1, the variational principle with the condition of the minimum average energy is equivalent to the Schrödinger equation.

derivative of the total energy with respect to the distribution function

$$\delta(E-\mu N)=\sum_p \varepsilon_{p,+}\delta n_{p,+}+\sum_p \varepsilon_{p,-}\delta n_{p,-}. \qquad (16.22)$$

With such a definition the equilibrium distribution function will be a Fermi function with $\varepsilon_{p,\sigma}$ as the energy*. Since $n_{p,+}$ and $n_{p,-}$ are included in the functional (16.16) in the same manner, the corresponding $\varepsilon_{p,\sigma}$ will also be the same. Varying (16.16) with respect to $n_{p,+}$, for example**, we obtain

$$\varepsilon_{p,+}=\frac{\delta(E-\mu N)}{\delta n_{p,+}}=\xi_p(u_p^2-v_p^2)+2u_pv_p\Delta$$
$$=(\xi_p^2+\Delta^2)^{1/2}=\varepsilon_p.$$

Thus, ε_p defined by formula (16.20) corresponds to the energy of quasiparticles (fig. 102, the dashed line). At $|\xi_p|\gg\Delta$, ε_p turns to $|\xi_p|$, which corresponds to quasiparticles of a normal metal. This can also be seen from formulae (16.14) and (16.9). At $|\xi_p|\gg\Delta$ we have $u_p\to1$, $v_p\to0$ at $\xi_p>0$ and $u_p\to0$ and $v_p\to1$ at $\xi_p<0$. In such a case $\alpha_{p,+}$ and $\alpha_{p,-}$ play the role of the annihilation operators for quasiparticles of a normal metal.

From formula (16.20) it follows that there is an energy gap Δ in the quasiparticles spectrum. This not only agrees with experimental data but also accounts for the superconductivity. The critical velocity according to the Landau criterion (16.1) is finite: $v_c\sim\Delta/p_0$, i.e., at lower velocities the electron fluid exhibits superfluidity. Note also that the binding energy of a Cooper pair is equal to 2Δ. When it is broken up, two quasiparticles are formed, i.e., an energy Δ has to be consumed for each.

There arises, however, the following question: Are there excitations that are associated not with the breakdown of Cooper pairs but with their translational motion? The study of this question leads to the following results, which are rather understandable from a physical viewpoint. Cooper pairs form an interacting system. At low temperatures it may be regarded as a liquid and, hence, excitations with a low energy correspond to the propagation of sound waves (phonons). But sound is accompanied by density oscillations, and the liquid of Cooper pairs is charged. The density oscillations give rise to electric fields, which must be taken into account.

On the whole, we arrive at a situation resembling the problem of propagation of zero sound in the electron liquid (section 13.5). It has been found in section 13.5 that with the electric field taken into account the spectrum of vibrations is no longer linear, $\omega=ck$, and starts with a finite energy of order μ. This also occurs in

* In accordance with the generally adopted convention, here we denote the energy of quasiparticles by ε_p. It enters into the Fermi distribution without subtracting μ, i.e., $n_p=(e^{\varepsilon_p/T}+1)^{-1}$. It should not be confused with the electron energy in the gas model, $\varepsilon(p)$, which has been used in Part I.

** Upon variation with respect to $n_{p,\sigma}$ the quantities u_p and v_p must be assumed to be constant, because these functions are themselves determined from the condition $\delta(E-N\mu)/\delta v_p=0$.

superconductors. It is clear that quasiparticles with such an energy cannot be of interest in a study of superconductivity.

Of course, just as in a normal metal, oscillations are possible, in principle, with a more complicated variation of the distribution function, which do not lead to a change in the electron density. However, because of a lack of clear-cut experimental evidence indicating their existence we shall not consider them.

A question may also be raised about pairing with a non-zero orbital moment, $l \neq 0$. The study of this question shows that if the amplitude of electron scattering is strongly dependent on the angle and there are negative coefficients in its expansion in spherical harmonics (each corresponding to a certain l), pairing may take place corresponding to the negative coefficient that is largest in absolute magnitude (see Abrikosov et al. 1962). An example of this kind in nature is a Fermi liquid - superfluid ^{3}He. In superfluid ^{3}He pairing actually occurs of quasiparticles with an orbital moment of $l = 1$ and parallel spins, which leads to superfluidity (see the review articles by Leggett 1975 and Wheatley 1975), but ^{3}He is a neutral Fermi liquid. Among metals a pairing with a nonzero orbital moment can possibly occur in so-called "systems with heavy fermions" (section 16.11), but this question has so far received no definite answer.

The difficulty lies in that scattering with a nonzero orbital moment implies a nonlocal interaction. However, as has been indicated in this book, phonon attraction and Coulomb repulsion are, in fact, local (i.e., they occur when electrons reside at a single point). Hence, we may speak only of some new, nonphonon mechanism of electron attraction, whose origin is not known at present (see the discussion of this question in section 16.11). In view of this, we shall not consider pairing with $l \neq 0$.

16.5. Temperature dependence of the energy gap

We will begin the study of the thermodynamics of superconductors with the calculation of $\Delta(T)$. Substituting the Fermi function into formula (16.21) and passing to an integration over momenta, we arrive at the equation for determination of Δ:

$$1 = \tfrac{1}{2} g \int \frac{d^3 p}{(2\pi\hbar)^3} \frac{1 - 2n_F(\varepsilon)}{\varepsilon}$$

$$= \tfrac{1}{2} g \nu(\mu) \int_0^{\hbar\omega_D} \frac{\tanh[(\xi^2 + \Delta^2)^{1/2}/2T]}{(\xi^2 + \Delta^2)^{1/2}} \, d\xi. \qquad (16.23)$$

Here we singled out the integration over ξ, took into account the symmetry with respect to $\xi \to -\xi$ and introduced the density of states $\nu(\mu) = p_0 m / \pi^2 \hbar^3)$ (remember that we are dealing with the isotropic model).

With $T \to 0$, $\tanh \to 1$. Taking the integral, we obtain:

$$1 = \tfrac{1}{2} g \nu(\mu) \ln\left(\frac{2\hbar\omega_D}{\Delta(0)}\right)$$

(we assume that $g\nu(\mu) \ll 1$; therefore $\Delta \ll \hbar\omega_D$). From this it follows that

$$\Delta(0) = 2\hbar\omega_D \exp\left(\frac{-2}{g\nu}\right). \tag{16.24}$$

This formula differs from eq. (16.11) by a factor 2, which results from taking account of the transformation of the energy spectrum.

As $T \to T_c$, the energy gap tends to zero. In this case we have

$$1 = \tfrac{1}{2}g\nu(\mu) \int_0^{\hbar\omega_D} \frac{\tanh(\xi)/2T_c}{\xi}\,d\xi = \tfrac{1}{2}g\nu(\mu)\ln\frac{2\hbar\omega_D\gamma}{\pi T_c}$$

where $\gamma = e^C = 1.78$. It follows that

$$T_c = \frac{2\hbar\omega_D\gamma}{\pi}\exp\left(\frac{-2}{g\nu}\right). \tag{16.25}$$

Comparing this with eq. (16.23), we get

$$\Delta(0) = \frac{\pi}{\gamma}T_c = 1.76T_c. \tag{16.26}$$

Let us consider the behavior of Δ near $T=0$ and $T=T_c$. Near $T=0$ formula (16.23) is rewritten as follows: we divide it by $\frac{1}{2}g\nu(\mu)$ and then subtract and add $(\xi^2+\Delta^2)^{-1/2}$ under the integral. As a result we obtain

$$\ln\frac{\Delta(0)}{\Delta(T)} = \int_0^\infty \frac{1-\tanh[(\xi^2+\Delta^2)^{1/2}/2T]}{(\xi^2+\Delta^2)^{1/2}}\,d\xi = 2f\left(\frac{\Delta}{T}\right), \tag{16.27}$$

where

$$f(x) = \int_1^\infty \frac{dy\,(y^2-1)^{-1/2}}{1+e^{yx}} \tag{16.28}$$

(the variable y corresponds to $(\xi^2+\Delta^2)^{1/2}/\Delta$). Since the integral in (16.27) converges at $\xi \sim \Delta$, we extended its upper limit to ∞. Since we are interested in large values of $x = \Delta/T$, we expand the integrand in (16.28) in e^{-yx}:

$$f(x) = \int_1^\infty \sum_{n=1}^\infty (-1)^{n+1} e^{-nyx}(y^2-1)^{-1/2}\,dy.$$

Using the integral representation of the MacDonald functions (the Hankel function of imaginary argument)

$$K_\nu(z) = \frac{\Gamma(\frac{1}{2})}{\Gamma(\nu+\frac{1}{2})}(\tfrac{1}{2}z)^\nu \int_1^\infty e^{-zy}(y^2-1)^{\nu-1/2}$$

($\Gamma(x)$ is the gamma function), we obtain

$$f(x) = \sum_{n=1}^\infty (-1)^{n+1}K_0(nx). \tag{16.29}$$

With large values of the argument the functions K_0 fall off exponentially. Retaining only the first term and using the asymptotic

$$K_0(x) = \left(\frac{\pi}{2x}\right)^{1/2} e^{-x}$$

we obtain, at $T \ll T_c$,

$$\Delta \approx \Delta(0) - [2\pi\Delta(0)T]^{1/2} e^{-\Delta(0)/T}. \tag{16.30}$$

This dependence is a consequence of the fact that the variation of Δ is associated with the appearance of quasiparticles and the number of quasiparticles is proportional to $\exp[-\Delta(0)/T]$.

If conversely $T \to T_c$, then $\Delta \to 0$. In this case, formula (16.23) can be conveniently expanded in Δ. We will proceed just as in the previous case. We divide (16.23) by $\frac{1}{2}g\nu(\mu)$ and then subtract and add $\xi^{-1}\tanh(\xi/2T)$ in the integrand. As a result, we get

$$\ln\frac{T_c}{T} = \int_0^\infty \left(\frac{\tanh(\xi/2T)}{\xi} - \frac{\tanh[(\xi^2+\Delta^2)^{1/2}/2T]}{(\xi^2+\Delta^2)^{1/2}}\right) d\xi \tag{16.31}$$

(because of the convergence of the integral at $\xi \sim T$, we again extended it up to ∞).

Let us use the following formula:

$$\tanh\tfrac{1}{2}\pi x = \frac{4x}{\pi}\sum_{k=0}^{\infty}\frac{1}{(2k+1)^2+x^2}.$$

Substituting this into (16.31), expanding in Δ and integrating over ξ, we find that

$$\ln\frac{T_c}{T} = 2\sum_{n=1}^{\infty}(-1)^{n+1}\frac{(2n-1)!!}{2n!!}\left(\frac{\Delta}{\pi T}\right)^{2n}\sum_{k=0}^{\infty}\frac{1}{(2k+1)^{2n+1}}, \tag{16.32}$$

where $n!! = n(n-2)(n-4)\ldots$. If we retain the first term, we will have

$$\Delta \approx \pi\left(\frac{8}{7\zeta(3)}\right)^{1/2}[T_c(T_c-T)]^{1/2} = 3.06[T_c(T_c-T)]^{1/2} \tag{16.33}$$

($\zeta(x)$ is the Riemann zeta function: $\zeta(x) = \sum_{n=1}^{\infty} n^{-x}$).

This behavior is rather natural. In fact, Δ, while being finite at $T < T_c$ and equal to zero at $T > T_c$, may serve as the order parameter of the superconducting transition. The temperature dependence (16.33) is consistent with Landau's theory of second-order phase transitions (see Appendix II). The question of the validity of this theory for superconductors will be discussed at a later time.

The graph of the function $\Delta(T)$ is shown in fig. 103.

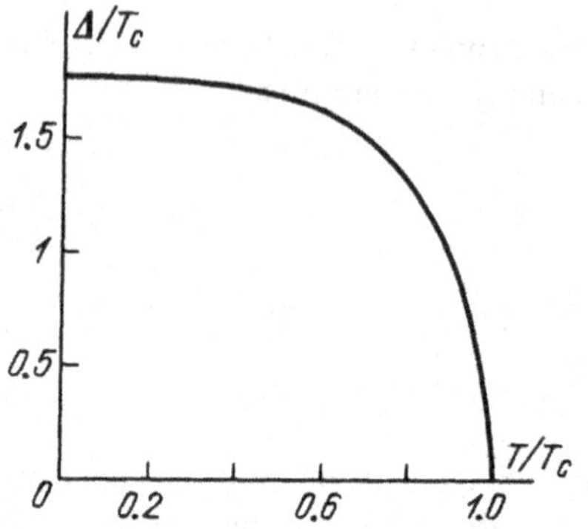

Fig. 103.

16.6. Thermodynamics of superconductors

Statistical physics gives the following expression for the thermodynamic potential*:

$$\Omega = -T \ln Z, \qquad Z = \sum_{n,N} \exp\left(\frac{-E_{nN} + N\mu}{T}\right) \tag{16.34}$$

where $E_{n,N}$ are the energy levels, i.e., the eigenvalues of the Hamiltonian operator of a system of N particles. The expression for the statistical sum may be written in the form of a trace of the matrix $Z_{nm} = \int \Phi_n^* Z \Phi_m \, dV$; $Z = e^{-(\mathcal{H} - N\mu)/T}$ and Φ_n are wave functions of the system (the operation Sp is invariant with respect to the choice of the orthonormalized system of the functions Φ_n):

$$Z = \mathrm{Sp}\, e^{-(\mathcal{H} - N\mu)/T}. \tag{16.35}$$

We substitute $\mathcal{H} - N\mu$ in the form of (16.13) and differentiate Ω (16.34) with respect to the interaction constant g. As a result, we have

$$\frac{\partial \Omega}{\partial g} = g^{-1} \frac{\mathrm{Sp}[\mathcal{H}_{\mathrm{int}}\, e^{-(\mathcal{H} - N\mu)/T}]}{\mathrm{Sp}\, e^{-(\mathcal{H} - N\mu)/T}} = g^{-1} \langle \mathcal{H}_{\mathrm{int}} \rangle, \tag{16.36}$$

where we have denoted the second term in the Hamiltonian (16.23) as $\mathcal{H}_{\mathrm{int}}$ (it describes the interaction of electrons). The angular brackets $\langle \cdots \rangle$ denote the average over the Gibbs distribution.

We express in (16.36) the operators $a_{p,\sigma}$ in terms of $\alpha_{p,\sigma}$ (16.15) and take the average. To within a factor we obtain the second term in formula (16.16). Using the definition of Δ (16.18), we can write:

$$\frac{\partial \Omega}{\partial g} = -\frac{\Delta^2}{g^2}. \tag{16.37}$$

Passing from the variable g to the variable Δ, we obtain

$$\Omega_s - \Omega_n = \int_0^{\Delta} \frac{d g^{-1}(\Delta_1)}{d\Delta_1} \Delta_1^2 \, d\Delta_1. \tag{16.38}$$

* The potential Ω depends on the variables T, V and μ, while the free energy F is a function of T, V and N.

Here we have used the fact that at $\Delta = 0$ the superconductor is converted to a normal metal. The expression for g^{-1} as a function of Δ is obtained from formulae (16.24) and (16.27):

$$g^{-1} = \tfrac{1}{2}\nu(\mu)\left[\ln\left(\frac{2\hbar\omega_D}{\Delta}\right) - 2f\left(\frac{\Delta}{T}\right)\right]. \tag{16.39}$$

The function f is expressed by formula (16.28).

Let us find the asymptotic values of $\Omega_s - \Omega_n$ for $T \to 0$ and $T \to T_c$. In the first case we use formula (16.29) for f. Substituting into eq. (16.38), we find

$$\Omega_s - \Omega_n = -\tfrac{1}{2}\nu(\mu)\left[\tfrac{1}{2}\Delta^2 - 2\sum_{n=1}^{\infty}(-1)^{n+1}\frac{T^2}{n^2}\int_0^{n\Delta/T} K_1(x)x^2\,dx\right],$$

where we have employed the relation $K_0'(x) = -K_1(x)$. The integral in brackets is taken in explicit form, $\int x^2 K_1(x) = -x^2 K_2(x)$. While inserting the lower limit we shall take into account that $[x^2 K_2(x)]_{x\to 0} = 2$. As a result, we obtain the sum

$$2T^2\nu(\mu)\sum_{n=1}^{\infty}(-1)^{n+1}n^2 = \tfrac{1}{6}\pi^2 T^2 \nu(\mu).$$

The free energy of a normal metal on the left-hand side is equal to $-\frac{1}{6}\pi^2 T^2 \nu(\mu)$. This follows from formula (2.27) for the heat capacity and from the relation $C = -T(\partial^2\Omega/\partial T^2)$. Cancelling $-\Omega_n$ on the left- and right-hand sides, we obtain

$$\Omega_s = -\tfrac{1}{2}\nu(\mu)\left[\tfrac{1}{2}\Delta^2 + 2\sum_{n=1}^{\infty}(-1)^{n+1}K_2(n\Delta/T)\Delta^2\right]. \tag{16.40}$$

At $T = 0$ the second term vanishes and we obtain

$$\Omega_s(0) = -\tfrac{1}{4}\nu(\mu)\,\Delta^2(0). \tag{16.41}$$

This formula is quite natural. It corresponds to the fact that the electrons in an energy band with a width of order Δ [their number is of order $\nu(\mu)\Delta$] are bound into pairs with a binding energy of order Δ.

At low temperatures we take into account the correction to $\Delta(0)$ and the term with $n = 1$ in the second term of (16.40). As a result, we find

$$\begin{aligned}\Omega_s(T) - \Omega_s(0) &= \nu(\mu)\Delta^2(0)\left[K_0\left(\frac{\Delta(0)}{T}\right) - K_2\left(\frac{\Delta(0)}{T}\right)\right] \\ &= -2\nu(\mu)\,\Delta(0)\,TK_1\left(\frac{\Delta(0)}{T}\right) \\ &\approx -\nu(\mu)\,[2\pi T^3\Delta(0)]^{1/2}\,e^{-\Delta(0)/T},\end{aligned} \tag{16.42}$$

from which we obtain the entropy per unit volume,

$$S_s = -\frac{\partial\Omega_s}{\partial T} \approx \nu(\mu)\left(\frac{2\pi\Delta^3(0)}{T}\right)^{1/2} e^{-\Delta(0)/T}, \tag{16.43}$$

and the heat capacity,

$$C_s = T\frac{\partial S_s}{\partial T} = \nu(\mu)\left(\frac{2\pi\Delta^5(0)}{T^3}\right)^{1/2} e^{-\Delta(0)/T}. \tag{16.44}$$

Now we pass to the case $T \to T_c$. We use formula (16.32). Expressing T_c via g with the aid of formula (16.25), we find g^{-1} as a function of Δ:

$$g^{-1} = \tfrac{1}{2}\nu(\mu)\left[\ln\left(\frac{2\hbar\omega_D\gamma}{\pi T}\right) - 2\sum_{n=1}^{\infty}(-1)^{n+1}\frac{(2n-1)!!}{(2n)!!}\left(\frac{\Delta}{\pi T}\right)^{2n}\sum_{k=0}^{\infty}(2k+1)^{-(2n+1)}\right].$$

Substitution into eq. (16.38) yields

$$\Omega_s - \Omega_n = -\nu(\mu)\sum_{n=1}^{\infty}(-1)^{n+1}\frac{(2n-1)!!2n}{(2n+2)!!} \times \frac{\Delta^{2n+2}}{(\pi T)^{2n}}\sum_{k=0}^{\infty}(2k+1)^{-(2n+1)}. \tag{16.45}$$

Retaining only the first term, we find from eq. (16.45)

$$\Omega_s - \Omega_n \approx -\nu(\mu)\tfrac{7}{32}\zeta(3)\frac{\Delta^4}{(\pi T)^2}. \tag{16.46}$$

Inserting the value of Δ for $T \to T_c$ (16.33), we obtain

$$\Omega_s - \Omega_n \approx -\frac{2}{7\zeta(3)}\frac{p_0 m}{\hbar^3}(T_c - T)^2, \tag{16.47}$$

from which we find the entropy:

$$S_s(T) \approx S_n(T) + \frac{4}{7\zeta(3)}\frac{p_0 m}{\hbar^3}(T - T_c). \tag{16.48}$$

Differentiating once more, we obtain the heat capacity at the transition point:

$$C_s(T_c) = C_n(T_c) + \frac{4}{7\zeta(3)}\frac{p_0 m}{\hbar^3}T_c. \tag{16.49}$$

Recall (see eq. 2.29) that

$$C_n(T) = S_n(T) = \frac{p_0 m}{3\hbar^3}T. \tag{16.50}$$

From (16.49) it follows that the heat capacity undergoes a jump at the transition point equal to $[4/7\zeta(3)]\,(p_0 m/\hbar^3)T_c$. In order to trace out the temperature course of the heat capacity in the vicinity of T_c, it is necessary to take into account the following terms of the expansion in Δ in formulae (16.45) and (16.32). This being done, we substitute the numbers and obtain

$$\frac{C_s(T)}{C_n(T_c)} = 2.42 + 4.76\,\frac{T - T_c}{T_c}. \tag{16.51}$$

The dependence $C_s(T)$ is shown in fig. 104.

A knowledge of the thermodynamic potentials makes it possible to determine the critical field H_c. For this purpose we use relation (15.5). We take into account that, according to thermodynamics, the corrections to all thermodynamic potentials are identical if they are expressed in terms of the corresponding variables. In particular,

$$(\delta F)_{N,T,V} = (\delta\Omega)_{\mu,T,V}.$$

This enables one to determine H_c from the relation

$$\Omega_n - \Omega_s = \frac{H_c^2}{8\pi}. \tag{16.52}$$

First of all, at $T=0$ we obtain from eq. (16.41) (recall that we assume $\Omega_n(0)=0$)

$$H_c(0) = [2\pi\nu(\mu)]^{1/2}\Delta(0). \tag{16.53}$$

At low temperatures $\Omega_n(T) = -\frac{1}{6}\pi^2 T^2 \nu(\mu)$, and the temperature correction to $\Omega_s(0)$ is exponentially small in accordance with (16.42). Hence,

$$\frac{H_c^2(T)}{8\pi} = \Omega_n(T) - \Omega_s(T) \approx -\tfrac{1}{6}\pi^2 T^2 \nu(\mu) + \tfrac{1}{4}\nu(\mu)\,\Delta^2(0).$$

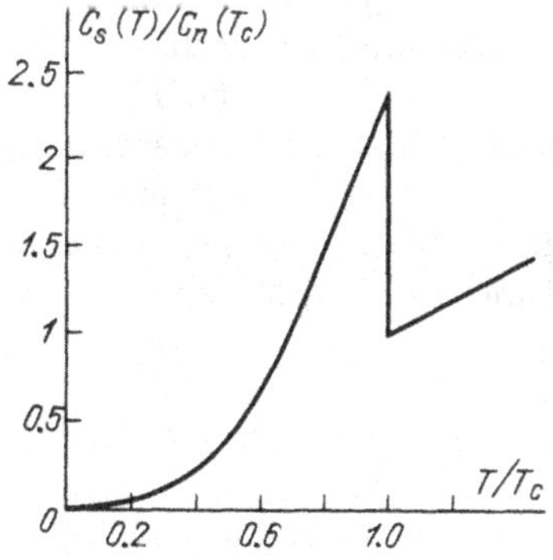

Fig. 104.

With the aid of eq. (16.53) we find from the last relation

$$H_c(T) \approx H_c(0)\left(1 - \tfrac{1}{3}\pi^2 \frac{T^2}{\Delta^2(0)}\right)$$
$$= H_c(0)\left(1 - \tfrac{1}{3}\gamma^2 \frac{T^2}{T_c^2}\right) = H_c(0)\left(1 - 1.05\frac{T^2}{T_c^2}\right). \tag{16.54}$$

Here we have used the relationship between $\Delta(0)$ and T_c (16.26).

Near T_c obtain in accordance with eq. (16.47):

$$H_c \approx \left(16\pi \frac{p_0 m}{7\zeta(3)\hbar^3}\right)^{1/2} (T_c - T)$$
$$= H_c(0)\,\gamma\left(\frac{8}{7\zeta(3)}\right)^{1/2}\left(1 - \frac{T}{T_c}\right)$$
$$= 1.735\, H_c(0)\left(1 - \frac{T}{T_c}\right). \tag{16.55}$$

Formulae (16.55) and (16.49) are consistent with formula (15.8) for the jump in heat capacity, as was to be expected.

Let us consider certain consequences of the results obtained. According to formulas (16.26) and (16.53), the critical temperature T_c and the critical field $H_c(0)$ are proportional to $\Delta(0)$. In accordance with formula (16.24), $\Delta(0)$ is proportional to the Debye frequency ω_D, which depends on the mass of lattice ions by the law $\omega_D \propto M^{-1/2}$. This is in accord with the formulae for the isotope effect (16.2). Further, at low temperatures the heat capacity is described by the exponential law (16.44), this being in agreement with the experimental result (15.3). Finally, although the results for the critical field do not coincide completely with the empirical formula (15.1), the asymptotic formulas for $T \ll T_c$ and $T \to T_c$ (16.54) and (16.55) are rather close to it.

All the results given above refer to the isotropic model, whereas true metals are anisotropic. Therefore, it is surprising that these formulas agree well with experiment not only qualitatively but quantitatively as well. But there are certain differences, which call for consideration. First of all it should be noted that, according to the theory described here, any thermodynamic quantity expressed in dimensionless units, say $C_s(T)/C_n(T_c)$ or $H_c(T)/H_c(0)$, depends only on the ratio T/T_c, where T_c is connected with the gap in the energy spectrum at $T=0$ by relation (16.26).

In an anisotropic metal the energy spectrum of quasiparticles looks somewhat different:

$$\varepsilon(\boldsymbol{p}) = [\Delta^2(\boldsymbol{n}) + \xi^2(\boldsymbol{n})]^{1/2} \tag{16.56}$$

where $\xi(\boldsymbol{n}) = v(\boldsymbol{n})[p - p_0(\boldsymbol{n})]$, and $\boldsymbol{n}$ is the unit vector in the direction $\boldsymbol{p}$. Thus, Δ depends on the direction $\boldsymbol{p}$. At low temperatures the role of the energy gap is played by the minimum value of $\Delta(\boldsymbol{n}, 0)$, which we denote by $\Delta_{\min}$. It is exactly this value that governs the temperature course of the heat capacity with $T \to 0$: $C_s \propto \exp(-\Delta_{min}/T)$. On the other hand, $\Delta(\boldsymbol{n}, T)$ depends on temperature: at $T \to T_c$ $\Delta(\boldsymbol{n}, T) \approx \Delta_0(\boldsymbol{n}) f(T/T_c)$, where $f \to 0$ when $T \to T_c$ (Pokrovskii 1961).

For an anisotropic metal the statement that $C_s(T)/C_n(T_c)$ depends only on T/T_c over the entire temperature range from 0 to T_c does not conform to reality. As a matter of fact, if $\Delta(0)$ is determined from formula (16.26), we obtain a certain averaged value of $\Delta(0)$ and, hence, the low-temperature formula $C_s \propto \exp[-\Delta(0)/T]$ will differ from the true law, which includes Δ_{min}. Therefore, if we compare the experimental data for C_s with the theoretical curve for the isotropic model, then with $T \to 0$ the experimental points should be expected to lie above the theoretical curve, which is what occurs in reality. However, the anisotropy of Δ is not large in reality: $(\overline{\Delta(0)} - \Delta_{min})/\overline{\Delta(0)} < 10\%$. This may be accounted for as follows: the phonon forces of attraction are mainly determined by short-wave length phonons with a frequency of order ω_D. Upon exchange of such phonons with $\hbar k \sim p_0$ the momenta of electrons vary significantly. This leads to a certain effective averaging along the entire Fermi surface.

Another assumption of the theory described above is that the interaction between the electrons is weak:

$$g\nu(\mu) \ll 1.$$

This assumption is not always justifiable. In fact, for lead, for example, $T_c = 7.2$ K and $\hbar\omega_D = 94.5$ K; hence, $g\nu(\mu) = 0.76$. It is not surprising that for such metals the above-described theory differs somewhat from experiment (this refers mainly to the dependence $H_c(T)$). However, after an appropriate correction of the theory the disagreement is markedly reduced.

16.7. *London and Pippard superconductors (qualitative theory)*

Before considering the microscopic derivation of the basic relations that characterize the behavior of superconductors in an electromagentic field, we shall give a simple qualitative analysis. In section 15.5 we have derived the equations of London electrodynamics. In particular, it has been noted that these equations may be written in the form (15.43), i.e.,

$$\boldsymbol{j} = -Q\boldsymbol{A}, \quad (\operatorname{div} \boldsymbol{A} = 0),$$

where

$$Q = \frac{n_e e^2}{mc}$$

In deriving the London equations it was assumed that the superconducting current is transported by electrons. At present, however, we know that the carriers of superconducting current are Cooper pairs. The main difference is that Cooper pairs have a finite size. This size can be determined on the basis of the following considerations. The energy spectrum of quasiparticles $\varepsilon = [v^2(p-p_0)^2 + \Delta^2]^{1/2}$ indicates that binding into pairs involves the momenta in the region of $\delta p \sim \Delta/v$ near the Fermi boundary. According to the uncertainty principle, $\delta p \delta r \sim \hbar$. From this we may infer that the size of the pair is $\hbar v/\Delta$. If we substitute here $v \sim 10^8$ cm/s, $\Delta \sim 10$ K $\sim 10^{-3}$ eV $\sim 10^{-15}$ erg, we obtain

$$\frac{\hbar v}{\Delta} \sim 10^{-4}\ \text{cm}$$

that is, thousands of interatomic distances.

Hence, the pairs are very loose formations. A natural question arises: How could they be accommodated in a metal without interfering with one another? Of course, the Cooper pairs cannot, in fact, be visualized as a classical gas composed of particles, the distance between which is much larger than their size. The pairs penetrate each other and at the same time, just as gas particles, they are free and there is practically no interaction between them. This is, of course, possible only in quantum mechanics, in the same way as the flow of this gas through the lattice without scattering. In order to emphasize this nonclassical behavior, in the rigorous theory one speaks not of pairs having a certain size but of a pair correlation between electrons extending to a certain distance - a "correlation length", which is of order $\xi \sim \hbar v/\Delta$*.

In view of the presence of a correlation the current is determined by the electromagnetic field not only directly at the point of observation but also in the entire region around this point, with a radius of order ξ. Therefore, the relationship between the current and the field becomes nonlocal and is expressed by the relation

$$j_i(\mathbf{r}) = -\sum_k \int Q_{ik}(\mathbf{r}-\mathbf{r}')\, A_k(\mathbf{r}')\, \mathrm{d}V' \tag{16.57}$$

where the kernel Q_{ik} falls off at a distance of order ξ.

In the limiting case, when the field varies very slowly in space, in the integral of (16.57) $A_K(\mathbf{r}')$ may be replaced by $A_k(\mathbf{r})$, which is taken outside the integral sign. Denoting $\int Q_{ik}(\mathbf{r}-\mathbf{r}')\, \mathrm{d}V' = \delta_{ik} Q$, we obtain the London relationship between the current and the field. The characteristic length that determines the variation of the field in a superconductor is obviously the penetration depth δ. As a matter of fact, at this distance the field varies from a given value at the surface of the superconductor to zero. Hence, in a case where $\delta \gg \xi$, we return to London electrodynamics. This limit is called the London limit.

In the opposite limiting case $\delta \ll \xi$ we may apply the inefficiency concept, just as in the anomalous skin effect (section 7.2). Indeed, the electrons (or, more exactly, pairs) that move at large angles to the surface of the superconductor emerge rapidly

* The symbol ξ is generally adopted for the correlation length. We use it because the correlation length and the electron energy $\xi = v(p-p_0)$ will not be encountered in the same formulas.

from the penetration depth and practically do not interact with the field. Only those electrons which move at an angle not exceeding $\theta \sim \delta/\xi$ are effective. But the number of such electrons is $n_e\theta \sim n_e\delta/\xi$. Substituting this number into expression (16.40) for the penetration depth, we obtain

$$\delta \sim \left(\frac{mc^2}{4\pi e^2 n_e \delta\xi^{-1}}\right)^{1/2}.$$

This is an equation for δ. Solving it, we get

$$\delta_P \sim \left(\frac{mc^2}{4\pi n_e e^2}\,\xi\right)^{1/3} \sim \delta_L^{2/3}\xi^{1/3} \tag{16.58}$$

(the London penetration depth defined by eq. (15.40) was denoted as δ_L). The case $\delta \ll \xi$ was first considered by Pippard (1953). Therefore, such superconductors are called Pippard superconductors.

In experiment all pure superconductors are Pippard superconductors or intermediate superconductors, i.e., in them $\xi \gtrsim \delta$. However, as we shall see below, the correlation length can be reduced and δ increased by introducing impurities or structural defects, from which the electrons are scattered. As a result, we can arrive at the London limit. Moreover, near T_c all superconductors become London superconductors.

In concluding this section we note that the expression for the London coefficient Q and the corresponding penetration depth does not depend on whether electrons or pairs are considered to be carriers. Indeed, making the replacement $e \to 2e$, $n_e \to \frac{1}{2}n_e$ and $m \to 2m$, we obtain the same formulas for Q and δ, (15.43) and (15.40), respectively.

16.8. The Meissner effect at $T=0$

Now we shall deal with a quantitative microscopic derivation. Since the equations derived in the general case are cumbersome, we will consider here the simplest case: the behavior of superconductors at $T=0$ in a weak magnetic field, which is independent of time.

The Hamiltonian is known from quantum mechanics to vary with the variations of the vector potential (Landau and Lifshitz 1974):

$$\delta\mathscr{H} = -\frac{1}{c}\int j\delta A\,\mathrm{d}V. \tag{16.59}$$

Here on the left under the integral is the current operator. Averaging this formula over a given state, say the equilibrium state, at temperature T, we obtain

$$\delta E = -\frac{1}{c}\int \langle j\rangle\,\delta\boldsymbol{A}\,\mathrm{d}V, \tag{16.59'}$$

where $\langle \boldsymbol{j} \rangle$ is the average current. Thus, in order to find $\langle \boldsymbol{j} \rangle$ *, we are to find the average energy as a function of $\boldsymbol{A}$, and to vary it with respect to $\boldsymbol{A}$.

Since in the absence of the field $\boldsymbol{j}=0$, it follows that at not too large fields the current is proportional to the field. From this we see that in order to find the current, we are to find the energy in the presence of the field to within terms that are second order in $\boldsymbol{A}$ and then to vary it with respect to $\boldsymbol{A}$.

In the presence of a field the Hamiltonian (16.13) is transformed in such a manner that instead of the momentum operators $-\mathrm{i}\hbar\nabla$ the combination $-\mathrm{i}\hbar\nabla-(e/c)\boldsymbol{A}$ is included. Hence, in the Hamiltonian we obtain

$$\frac{1}{2m}\left[-\hbar^2\nabla^2+\mathrm{i}\hbar\frac{e}{c}(\boldsymbol{A}\nabla+\nabla\boldsymbol{A})+\left(\frac{e}{c}\right)^2\boldsymbol{A}^2\right]. \tag{16.60}$$

While the first term retains the momentum of the particle, the other terms do not. If we expand $\boldsymbol{A}$ into a Fourier integral,

$$\boldsymbol{A}(\boldsymbol{r})=\sum_{\boldsymbol{q}}\boldsymbol{A}_{\boldsymbol{q}}\,\mathrm{e}^{\mathrm{i}\boldsymbol{q}\boldsymbol{r}}, \tag{16.61}$$

we may interpret $\boldsymbol{A}_{\boldsymbol{q}}$ as the operator of the absorption of an electromagnetic quantum with momentum $\hbar\boldsymbol{q}$. In view of this, the operator $\boldsymbol{A}_{\boldsymbol{q}}$ transfers a particle of momentum $\boldsymbol{p}-\hbar\boldsymbol{q}$ to a state of momentum $\boldsymbol{p}$. The terms of the Hamiltonian containing $\boldsymbol{A}$, which are of interest to us, may be written in the following form with the aid of the second-quantization operators:

$$\mathcal{H}_1=-\frac{e}{2mc}\sum_{p,q,\sigma}[(\boldsymbol{p}-\hbar\boldsymbol{q})\overset{+}{a}_{p,\sigma}a_{p-\hbar q,\sigma}+\boldsymbol{p}\overset{+}{a}_{p,\sigma}a_{p-\hbar q,\sigma}]\boldsymbol{A}_{\boldsymbol{q}}, \tag{16.62}$$

$$\mathcal{H}_2=\frac{e^2}{2mc^2}\sum_{p,q,q',\sigma}\overset{+}{a}_{p,\sigma}a_{p-\hbar q-\hbar q',\sigma}\boldsymbol{A}_{\boldsymbol{q}}\boldsymbol{A}_{\boldsymbol{q}'}. \tag{16.63}$$

Here it is taken into account, for example, that

$$-\mathrm{i}\hbar\nabla\boldsymbol{A}_{\boldsymbol{q}}\,\mathrm{e}^{\mathrm{i}\boldsymbol{q}\boldsymbol{r}}=\boldsymbol{A}_{\boldsymbol{q}}\,\mathrm{e}^{\mathrm{i}\boldsymbol{q}\boldsymbol{r}}(-\mathrm{i}\hbar\nabla+\hbar\boldsymbol{q}).$$

Let us express the combinations $\sum_\sigma \overset{+}{a}_{p,\sigma}a_{p',\sigma}$ contained in $\mathcal{H}_1$ and $\mathcal{H}_2$ in terms of the operators α_+ and α_-. Substituting formulas (16.14) and collecting the terms, we find

$$\begin{aligned}\sum_\sigma \overset{+}{a}_{p,\sigma}a_{p',\sigma}={}&u_pu_{p'}(\overset{+}{\alpha}_{p,+}\alpha_{p',+}+\overset{+}{\alpha}_{p,-}\alpha_{p',-})\\&+v_pv_{p'}(\alpha_{-p,-}\overset{+}{\alpha}_{-p',-}+\alpha_{-p,+}\overset{+}{\alpha}_{-p',+})\\&+u_pv_{p'}(\overset{+}{\alpha}_{p,+}\overset{+}{\alpha}_{-p',-}-\overset{+}{\alpha}_{p,-}\overset{+}{\alpha}_{-p',+})\\&+v_pu_{p'}(\alpha_{-p,-}\alpha_{p',+}-\alpha_{-p,+}\alpha_{p',-}).\end{aligned} \tag{16.64}$$

* In the following we will omit the angular brackets for the average current.

Since the part of Hamiltonian $\mathscr{H}_2$ (16.63) is second order in $\boldsymbol{A}$, it is sufficient to take it into account only in the first order of perturbation theory, i.e., it will suffice to average it over the ground state. As a result, only the term with $v_p v_{p'}$ from expression (16.64) makes a contribution, and it appears that $\boldsymbol{p}=\boldsymbol{p}'$, i.e., $\boldsymbol{q}=-\boldsymbol{q}'$. Considering that $n_{p,+}=n_{p,-}=0$ at $T=0$, we obtain from this term

$$E_2=\frac{e^2}{mc^2}\sum_{p,q} v_p^2 \boldsymbol{A}_q \boldsymbol{A}_{-q}. \tag{16.65}$$

The first order of perturbation theory in $\mathscr{H}_1$ turns to zero. Indeed, upon averaging over the ground state we obtain $\sum_p \boldsymbol{p} v_p^2 \delta_{q,0}=0$, since v_p depends only on $|\boldsymbol{p}|$. Therefore, we are to find the second order of perturbation theory:

$$E_1=\sum_{m\neq 0}\frac{|(\mathscr{H}_1)_{m0}|^2}{E_{(0)}-E_{(m)}}.$$

Since there are no quasiparticles in the ground state, the contribution to E_1 comes only from transitions involving the creation of two quasiparticles and their annihilation. Thus, we obtain

$$(\mathscr{H}_1)_{m0}=-\frac{e}{2mc}(u_p v_{p-\hbar q}-v_p u_{p-\hbar q})(2\boldsymbol{p}-\hbar\boldsymbol{q}),$$

$$E_{(0)}-E_{(m)}=-(\varepsilon_p+\varepsilon_{p-\hbar q}),$$

$$E_1=-\left(\frac{e}{2mc}\right)^2\sum_p \frac{[(2\boldsymbol{p}-\hbar\boldsymbol{q})\boldsymbol{A}_q][(2\boldsymbol{p}-\hbar\boldsymbol{q})\boldsymbol{A}_{-q}]}{\varepsilon_p+\varepsilon_{p-\hbar q}} \times(u_p v_{p-\hbar q}-v_p u_{p-\hbar q})^2. \tag{16.66}$$

Here we have used the reality of $\boldsymbol{A}(\boldsymbol{r})$. According to (16.61), it follows that $\boldsymbol{A}_q^*=\boldsymbol{A}_{-q}$. Moreover, it has been taken into account that the operators $\alpha_{p,-}$ and $\alpha_{p',+}$ anticommute, i.e., their permutation leads to a change in sign.

Before performing further calculations one remark should be made. As is known from electrodynamics, a correct theory must have the property of gauge invariance. In the static case this reduces to the invariance of the equations under the transformation $\boldsymbol{A}\to\boldsymbol{A}+\nabla\varphi$, where φ is any scalar function. This is because only the magnetic field $\boldsymbol{H}=\operatorname{rot}\boldsymbol{A}$ has a physical meaning. But if the Fourier component of the vector potential enters explicitly into the equations, then only the following combination is permissible:

$$\boldsymbol{A}_{q\perp}=\boldsymbol{A}_q-\frac{\boldsymbol{q}(\boldsymbol{q}\boldsymbol{A}_q)}{q^2}.$$

Indeed, the gauge transformation in the Fourier components has the form

$$\boldsymbol{A}_q\to\boldsymbol{A}_q+\mathrm{i}\boldsymbol{q}\varphi_q.$$

With such a replacement $\boldsymbol{A}_{q\perp}$ does not change.

The invariance under a gauge transformation is provided in quantum mechanics as a result of the vector potential entering into the Hamiltonian in combination with the momentum operator $\boldsymbol{p}-(e/c)\boldsymbol{A}$. The addition of the gradient to $\boldsymbol{A}$ can be compensated for by a change of the phase of the wave function. In view of this, there is no need to test the gauge invariance of the resultant equations and one can use the gauge of $\boldsymbol{A}$ that is most convenient. We will use a vector potential that satisfies the condition div $\boldsymbol{A}=0$ or, in Fourier components, $\boldsymbol{q}\boldsymbol{A}_q=0$ (section 15.5), because in such a case the derivation is simplified.

In order to obtain explicitly gauge-invariant equations that are valid for any gauge, we should have taken into account the change of the Bogoliubov transformation formulas (16.14) under the action of the field. Since u_p and v_p are scalars, they may have only corrections proportional to $\boldsymbol{q}\boldsymbol{A}_q$, i.e., the ones which vanish in our gauging $\boldsymbol{q}\boldsymbol{A}_q=0$. But if this condition is not imposed, we obtain the same equations, but with $\boldsymbol{A}_q$ being replaced by $\boldsymbol{A}_{q\perp}$. We shall not perform this complicated derivation here, since in choosing the correct initial Hamiltonian the result is known in advance.

We make the replacement $\boldsymbol{p}\to\boldsymbol{p}+\frac{1}{2}\hbar\boldsymbol{q}$ and introduce the notations

$$\xi_{p+\hbar q/2}\equiv\xi_+,\quad \xi_{p-\hbar q/2}\equiv\xi_-,\qquad \xi=v(p-p_0)$$

and analogous ones for the other quantities. Substituting into eq. (16.66) the values of u_p and v_p in accordance with eq. (16.19), we have

$$E_1=-\tfrac{1}{2}\left(\frac{e}{mc}\right)^2\sum_p\frac{(\boldsymbol{p}\boldsymbol{A}_q)(\boldsymbol{p}\boldsymbol{A}_{-q})(\varepsilon_+\varepsilon_--\xi_+\xi_--\Delta^2)}{\varepsilon_+\varepsilon_-(\varepsilon_++\varepsilon_-)}. \tag{16.67}$$

Relation (16.59′) in Fourier components has the form

$$\delta E=-c^{-1}\sum_q\boldsymbol{j}_q\delta\boldsymbol{A}_q. \tag{16.68}$$

From eqs. (16.65) and (16.66) we obtain

$$\boldsymbol{j}_q=-\frac{2e^2}{mc}\sum_p v_p^2\boldsymbol{A}_q+\frac{e}{m^2c}\sum_p\frac{\boldsymbol{p}(\boldsymbol{p}\boldsymbol{A}_q)(\varepsilon_+\varepsilon_--\xi_+\xi_--\Delta^2)}{\varepsilon_+\varepsilon_-(\varepsilon_++\varepsilon_-)}. \tag{16.69}$$

In the first term $2\sum_p v_p^2=n_e$ is the electron density (recall that the normalization volume V has been assumed equal to unity). This equality can be obtained if in the expression for the operator of the number of particles (section 16.4)

$$n_e=\sum_{p,\sigma}a_{p,\sigma}a_{p,\sigma} \tag{16.70}$$

we perform the Bogoliubov transformation and then average it over the ground state.

In the second term we assume that $|\boldsymbol{q}|\ll p_0$. In such a case,

$$\xi_\pm=\frac{1}{2m}[(\boldsymbol{p}+\tfrac{1}{2}\hbar\boldsymbol{q})^2-p_0^2]\approx\xi\pm\tfrac{1}{2}v\hbar q\cos\theta,$$

where $v=p_0/m$; for the vector $\boldsymbol{p}$ we have introduced polar coordinates with the axis along $\boldsymbol{q}$ (the z axis). In accordance with our choice of $\boldsymbol{A}_q$, the current component along $\boldsymbol{q}$ is missing, and we shall find j_x (evidently, $j_x=j_y$). Passing from $\sum_p$ to

$(2\pi\hbar)^{-3}\int d^3\boldsymbol{p}$, we have

$$\boldsymbol{j}_{q\perp} = -Q(q)\,\boldsymbol{A}_{q\perp}, \tag{16.71}$$

where

$$Q(q) = \frac{n_e e^2}{mc} - \frac{1}{(2\pi\hbar)^3}\frac{e^2}{m^2 c} \times \frac{p_0^4}{v}\int_{-\infty}^{\infty} d\xi\, d\Omega \cos^2\varphi \sin^2\theta \frac{\varepsilon_+\varepsilon_- - \xi_+\xi_- - \Delta^2}{\varepsilon_+\varepsilon_-(\varepsilon_+ + \varepsilon_-)}. \tag{16.72}$$

Since ξ of order Δ are important in the integral, we have put $p \approx p_0$ and passed from integration over p to integration over ξ, which may be extended to the interval $(-\infty, \infty)$. Without making further assumptions, we can take the integrals over φ and ξ. This leaves only the integral over $d(\cos\theta)$. We introduce the notation $\cos\theta = t$. As a result, we have

$$Q(q) = \frac{n_e e^2}{mc}\left\{1 - \tfrac{3}{4}\int_{-1}^{1} dt\,(1-t^2) \left[1 - \frac{4\Delta^2(0)}{\hbar q t v}\frac{\operatorname{arcsinh}[\hbar q t v/2\Delta(0)]}{[4\Delta^2(0) + (\hbar q t v)^2]^{1/2}}\right]\right\}$$

$$= \frac{3e^2 n_e \Delta(0)}{mc\hbar q v}\int_0^1 \frac{dt}{t}(1-t^2)\frac{\operatorname{arcsinh}[\hbar q t v/2\Delta(0)]}{\{1+[\hbar q t v/2\Delta(0)]^2\}^{1/2}}. \tag{16.73}$$

The integral over t cannot be taken in explicit form in the general case. Therefore, we confine ourselves to the limiting formulas for large and small q.

The expression $Q(q)$ depends on the combination $q\hbar v/\Delta(0)$. The wave vector $\boldsymbol{q}$ characterizes the rapidity of the spatial variation of the current and field. The corresponding characteristic length is q^{-1}. In physical problems evidently $q^{-1} \sim \delta$ (the penetration depth). The quantity $\xi = \hbar v/\Delta(0)$ is the correlation length, as has been pointed out in the preceding section. Thus, the kernel Q depends on the ratio δ/ξ. A large value of this ratio corresponds to $q \to 0$. In this case we find from eq. (16.73)

$$Q = \frac{n_e e^2}{mc}. \tag{16.74}$$

This formula corresponds to the London limit (see the preceding section and formula 15.43).

Conversely, large values of q corresponds to the Pippard limit $\delta/\xi \ll 1$. In the integral of (16.73) small values of t are important. Neglecting t^2 compared to unity and evaluating the integral, we obtain

$$Q(q) \approx \tfrac{3}{4}\pi^2 \frac{n_e e^2}{mc}\frac{\Delta(0)}{\hbar q} q^{-1}. \tag{16.75}$$

This formula corresponds to the inefficiency concept. Indeed, assuming that $q^{-1} \sim \delta$, we see that formula (16.75) differs from (16.74) by the replacement of n_e by $n_e\delta/\xi$.

In coordinate space the relationship $\boldsymbol{j}(\boldsymbol{q}) = -Q(\boldsymbol{q})\boldsymbol{A}(\boldsymbol{q})$ is expressed by formula (16.57). At $\xi \gg \delta$ the coupling is basically nonlocal because the field $\boldsymbol{A}(\boldsymbol{r}')$ varies rapidly at distances characteristic for the kernel $Q(\boldsymbol{r}-\boldsymbol{r}')$. Here, strictly speaking, the results we have obtained cannot be used because we have found Q for an infinite space, whereas in the problem of the penetration of the field the metal has a surface, i.e., let us say, it occupies a half-space $x>0$. In such problems it is necessary to impose boundary conditions for electrons.

The condition of specular reflection from the boundary proves to be the simplest. In this case, just as for the anomalous skin effect, one may use the kernel Q for an infinite space, but then it is necessary to extend the vector potential in an even manner through the boundary (section 7.3). If the magnetic field is directed parallel to the surface of the metal along the z axis, then we may choose $A_y(x)$ along the y axis. On the strength of what has been said, $A_y(x) = A_y(-x)$.

According to the Maxwell equation

$$\text{rot}\, \boldsymbol{H} = \text{rot rot}\, \boldsymbol{A} = \text{grad div}\, \boldsymbol{A} - \nabla^2 \boldsymbol{A} = \frac{4\pi}{c}\boldsymbol{j}.$$

The first term on the left-hand side is equal to zero, since div $\boldsymbol{A} = 0$. The extension of the vector potential in an even manner through the boundary implies a kink in the $A_y(x)$ curve. Hence, if the field H_0 is given at the metal surface, i.e., $(\mathrm{d}A_y/\mathrm{d}x)_{x=+0} = H_0$, then $(\mathrm{d}A_y/\mathrm{d}x)_{x=-0} = -H_0$. But then the second derivative $\mathrm{d}^2A_y/\mathrm{d}x^2$ acquires a correction $2H_0\delta(x)$. Thus, we have

$$\frac{\mathrm{d}^2 A_y}{\mathrm{d}x^2} = 2H_0\delta(x) - \frac{4\pi}{c} j. \tag{16.76}$$

Passing to Fourier components, substituting eq. (16.71) and solving the resultant equation, we find

$$A_q = -2H_0 \frac{1}{q^2 + \dfrac{4\pi}{c} Q(q)} \tag{16.77}$$

In the coordinate representation we have

$$A(x) = -\frac{2H_0}{\pi} \int_0^\infty \frac{\cos qx\, \mathrm{d}q}{q^2 + \dfrac{4\pi}{c} Q(q)} \tag{16.78}$$

(here it has been taken into account that Q depends on $|\boldsymbol{q}|$ alone). Differentiating with respect to H, we find the magnetic field*

$$H(x) = \frac{2H_0}{\pi} \int_0^\infty \frac{q \sin qx\, \mathrm{d}q}{q^2 + \dfrac{4\pi}{c} Q(q)}. \tag{16.79}$$

* In order to obtain H at the surface one cannot put $x=0$ under the integral; instead, it is necessary to calculate it for small $x>0$, after which x is made to tend towards zero.

We shall not analyze the behavior of the field in the general case; we only find the penetration depth, which we define as follows:

$$\delta = \frac{1}{H_0}\int_0^\infty H\,\mathrm{d}x = -\frac{A(0)}{H_0} = \frac{2}{\pi}\int_0^\infty \frac{\mathrm{d}q}{q^2 + \frac{4\pi}{c}Q(q)}. \tag{16.80}$$

Let us assume that small q are important in the integral. In such a case, $Q(q)$ has the form (16.74) and we obtain from the integral the London penetration depth:

$$\delta_\mathrm{L} = \left(\frac{mc^2}{4\pi n_\mathrm{e} e^2}\right)^{1/2}. \tag{16.81a}$$

Note that in the integral $q \sim \delta^{-1}$ are important and, hence, our calculation is correct at $\delta \gg \hbar v/\Delta \sim \xi$.

In the opposite limiting case we substitute Q from eq. (16.75). Evaluating the integral, we find

$$\begin{aligned}\delta_\mathrm{P} &= 4\times 3^{-11/6}\pi^{-1}\left(\frac{mc^2}{n_\mathrm{e}e^2}\right)^{1/3}\left(\frac{\hbar v}{\Delta(0)}\right)^{1/3}\\ &= 2^{8/3}3^{-11/6}\pi^{-2/3}\delta_\mathrm{L}^{2/3}\left(\frac{\hbar v}{\Delta(0)}\right)^{1/3}.\end{aligned} \tag{16.81b}$$

This formula corresponds to the estimate for the Pippard case (16.58) described in the previous section.

Upon passing to coordinate representation (16.57) in the general case it is essential that the kernel $Q(\boldsymbol{q})$ couples only the transverse components $\boldsymbol{j}$ and $\boldsymbol{A}$ (see eq. 16.71), i.e., strictly speaking, it is necessary to replace $Q(q)$ by $Q(q)\ (\delta_{ik} - q_i q_k/q^2)$. For the Pippard limit this leads to the following form for $Q_{ik}(\boldsymbol{r}-\boldsymbol{r}')$ in formula (16.57):

$$Q_{ik}(\boldsymbol{r}-\boldsymbol{r}') = \tfrac{3}{4}\frac{n_\mathrm{e}e^2}{mc}\frac{\Delta(0)}{\hbar v}\, n_i n_k |\boldsymbol{r}-\boldsymbol{r}'|^{-2}, \tag{16.71'}$$

where

$$\boldsymbol{n} = \frac{(\boldsymbol{r}-\boldsymbol{r}')}{|\boldsymbol{r}-\boldsymbol{r}'|}.$$

16.9. *The relationship between current and field at finite temperatures. The London limit*

We shall not examine the general case at finite temperatures because of its complexity. We only note the following points. When the temperature increases, the role of the correlation length in the kernel $Q(\boldsymbol{r}-\boldsymbol{r}')$ which couples the current to the field (16.57) gradually goes over from $\hbar v/\Delta(0)$ to $\hbar v/T$. Since $\Delta(0) \sim T_\mathrm{c}$, at all temperatures the correlation length is of order $\hbar v/T_c$. Use is commonly made of the standard correlation length equal to

$$\xi_0 = \frac{\hbar v}{\pi \Delta(0)} = \frac{\gamma}{\pi^2} \frac{\hbar v}{T_c} = 0.18 \frac{\hbar v}{T_c}. \tag{16.82}$$

In section 15.1 it has been pointed out that the penetration depth δ increases indefinitely as $T \to T_c$. This will also be obtained below. But it follows that even if $\delta(0) \ll \xi_0$ at $T=0$, then $\delta(T)$ will sooner or later exceed ξ_0 with $T \to T_c$. Here the London situation occurs, i.e., the Fourier component of the kernel $Q(q)$ can be substituted by $Q(0)$. Hence the calculation of $Q(0)$ at a finite temperature enables one to establish the relationship between the current and the field at all temperatures for London superconductors, for which $\delta(0) \gg \xi_0$, and also at a temperature near T_c for Pippard superconductors, for which $\delta(0) \ll \xi_0$ (and, of course, for intermediate superconductors, for which $\delta(0) \sim \xi_0$).

We could have followed the previous path, i.e., we could have found $E\{\boldsymbol{A}\}$ according to perturbation theory and calculated the current by varying with respect to $\boldsymbol{A}$. However, the result can be obtained more simply. At finite temperatures some of the Cooper pairs break down and quasiparticles appear. The latter interact with impurities and phonons and therefore make no contribution to the superconducting current. Hence, the effective number of electrons entering into formula (15.43) or formula (16.74), which coincides with (15.43), must begin to decrease, i.e., is replaced by a certain function $n_s(T)$, which had been known as the "number of superconducting electrons" before the appearance of the microscopic theory. The replacement of n_e by $n_s(T)$ in the London theory made it possible to account qualitatively for the temperature dependence of the penetration depth. We will define this quantity quantitatively.

Suppose that all the quasiparticles move with velocity $\boldsymbol{u}$. In this case, the argument of the distribution function will be $\varepsilon - \boldsymbol{pu}$ rather than $\varepsilon(\boldsymbol{p})$. The motion of the quasiparticles will lead to the appearance of a momentum $\boldsymbol{P}$ given by

$$\boldsymbol{P} = \int \boldsymbol{p}(n_{\boldsymbol{p},+} + n_{\boldsymbol{p},-}) \frac{d^3p}{(2\pi\hbar)^3}$$
$$= 2\int \boldsymbol{p} n_F(\varepsilon - \boldsymbol{pu}) \frac{d^3p}{(2\pi\hbar)^3}$$

(n_F is the Fermi function). Assuming $\boldsymbol{u}$ to be small and expanding in $\boldsymbol{u}$, we obtain

$$\boldsymbol{P} = -2\int \boldsymbol{p}(\boldsymbol{pu}) \frac{\partial n_F}{\partial \varepsilon} \frac{d^3p}{(2\pi\hbar)^3}. \tag{16.83}$$

We transform the integral over $d^3\boldsymbol{p}$ to an integral over $d\xi$ and over the solid angle, as has been done in Part I. Here, however, one has to take into account that ε does not coincide with ξ, the energy of a normal metal. We have

$$2\int \frac{d^3p}{(2\pi\hbar)^3} \to \nu(\mu) \int d\xi \frac{d\Omega}{4\pi}.$$

Here we have taken account of the fact that the integral goes over the neighborhood of the Fermi surface. This also makes it possible to assume that $|\boldsymbol{p}| = p_0$ in formula (16.83). Integrating over the solid angle, we obtain

$$\int \boldsymbol{p}(\boldsymbol{p}\boldsymbol{u})\frac{\mathrm{d}\Omega}{4\pi} \approx \tfrac{1}{3}p_0^2\boldsymbol{u}.$$

Substituting $\varepsilon = (\xi^2 + \Delta^2)^{1/2}$ and introducing a new variable $\xi = \Delta \sinh \varphi$, we find (we have taken into account the rapid fall of the integral with increasing $|\xi|$ and replaced the limits of integration by $-\infty$ and ∞):

$$\boldsymbol{P} = \tfrac{2}{3}p_0^2\boldsymbol{u}\nu(\mu)\frac{\Delta}{T}\int_0^\infty \frac{\cosh \varphi \, \mathrm{d}\varphi}{(\mathrm{e}^{\Delta \cosh \varphi/T}+1)(\mathrm{e}^{-\Delta \cosh \varphi/T}+1)}.$$

The proportionality factor between $\boldsymbol{P}$ and $m\boldsymbol{u}$ is called here the number of "normal electrons" n_n. Substituting $\nu(\mu) = p_0 m/\pi^2\hbar^3$ and taking into account that the total number of electrons is $n_\mathrm{e} = p_0^3/3\pi^2\hbar^3$, we obtain

$$\frac{n_\mathrm{n}}{n_\mathrm{e}} = 2\frac{\Delta}{T}\int_0^\infty \frac{\cosh \varphi \, \mathrm{d}\varphi}{(\mathrm{e}^{\Delta \cosh \varphi/T}+1)(\mathrm{e}^{-\Delta \cosh \varphi/T}+1)}. \tag{16.84}$$

If we make the same replacement of variables in formula (16.27), we get

$$\ln\frac{\Delta(0)}{\Delta(T)} = 2\int_0^\infty \frac{\mathrm{d}\varphi}{\mathrm{e}^{\Delta \cosh \varphi/T}+1}. \tag{16.85}$$

Differentiating with respect to temperature, we have

$$\Delta^{-1}\frac{\mathrm{d}\Delta}{\mathrm{d}T} = 2\int_0^\infty \frac{\cosh \varphi \, \mathrm{d}\varphi}{(\mathrm{e}^{\Delta \cosh \varphi/T}+1)(\mathrm{e}^{-\Delta \cosh \varphi/T}+1}\frac{\mathrm{d}}{\mathrm{d}T}\left(\frac{\Delta}{T}\right).$$

comparing with (16.84), we find

$$\frac{n_\mathrm{n}}{n_\mathrm{e}} = \frac{1}{T}\frac{\mathrm{d}\Delta}{\mathrm{d}T}\frac{1}{\dfrac{d}{\mathrm{d}T}\dfrac{\Delta}{T}}. \tag{16.86}$$

At $T = T_\mathrm{c}$ this quantity is equal to unity (which is seen from the fact that $\Delta \propto (T_\mathrm{c} - T)^{1/2}$), but with $T < T_\mathrm{c}$ it is less than unity. Hence, in contrast to a normal metal, the movement of quasiparticles alone does not correspond to the movement of the entire electron liquid. The remaining number of electrons is just the sought-for "number of superconducting electrons". Thus,

$$\frac{n_\mathrm{s}}{n_\mathrm{e}} = 1 - \frac{n_\mathrm{n}}{n_\mathrm{e}} = -\left[\frac{\mathrm{d}\ln(\Delta/T)}{\mathrm{d}\ln T}\right]^{-1}. \tag{16.87}$$

It is exactly this number that appears in the kernel $Q(T)$ instead of n_e.

Thus, in the London situation at $T \neq 0$ we have

$$Q = \frac{n_\mathrm{s}e^2}{mc}. \tag{16.88}$$

The same result is obtained by way of more lengthy calculations analogous to those in the preceding section. From eq. (16.88) it follows that the penetration depth is expressed, instead of (15.40), by the formula

$$\delta=\left(\frac{mc^2}{4\pi n_s e^2}\right)^{1/2}. \tag{16.89}$$

Let us consider the limiting cases. At $T \ll T_c$ we obtain with the aid of formula (16.30)

$$\frac{n_s}{n_e}\approx 1-\left(\frac{2\pi\Delta(0)}{T}\right)^{1/2} e^{-\Delta(0)/T}, \tag{16.90}$$

$$\delta_L(T)\approx\delta_L(0)\left[1+\left(\frac{\pi\Delta(0)}{2T}\right)^{1/2} e^{-\Delta(0)/T}\right]. \tag{16.91}$$

At $T_c - T \ll T_c$ we find from eq. (16.33) that

$$\frac{n_s}{n_e}\approx\frac{T_c-T}{T_c}, \tag{16.92}$$

$$\delta_L(T)\approx\delta_L(0)\left(\frac{T_c}{2(T_c-T)}\right)^{1/2}. \tag{16.93}$$

In concluding this section we note the following. In the limit $T \to T_c$ all superconductors become London superconductors and their penetration depth is described by the formulas derived above; but in a case where $\delta_L(0) \ll \xi_0$ at $T=0$, the transition to the London limit occurs only in the immediate vicinity of T_c. For such superconductors, the Pippard situation exists practically over the entire temperature range; in particular, it also persists at $T_c - T \ll T_c$; for example, for aluminum the London Limit occurs only at $1-T/T_c \sim 10^{-3}$. The complete derivation for the Pippard case, which is not given here, yields a result qualitatively similar to formula (16.81) for $T=0$ with $\delta_L(0)$ replaced by $\delta_L(T)$. Hence, $\delta_P(T) \sim \delta_P(0)\ (1-T/T_0)^{1/3}$ as $T \to T_c$. The complete formula has the form

$$\delta_P(T)=\delta_P(0)\left[\frac{\Delta}{\Delta(0)}\tanh\left(\frac{\Delta}{2T}\right)\right]^{-1/3}. \tag{16.94}$$

16.10. *The superconducting correlation and surface energy. Two types of superconductors. The role of impurities*

The concept of correlation length introduced in the preceding sections explains the surface energy at the interface between the normal and the superconducting phase, which was proposed by L.D. Landau in his theory of the intermediate state (section 15.3). Imagine a planar interface between the normal and superconducting phases, which is formed in the intermediate state. Let us show how Δ and the magnetic field vary in the vicinity of the interface (fig. 105)*. The superconducting correlation does not allow Δ to vary discontinuously, because the states of electrons are correlated at a distance of order $\xi(T) = \hbar v/\Delta(T)$. It should be noted that the detailed theory yields somewhat different correlation lengths for different quantities. In section 16.8 it has already been said that in electrodynamics this is ξ_0 (16.82), i.e., a temperature-independent quantity. The characteristic length over which Δ varies is $\xi(T) = \hbar v/\Delta(T)$. The difference manifests itself near T_c, where $\xi(T) \to \infty$.

Thus, in the region of contact between the normal and superconducting phases Δ varies as shown in fig. 105. On the other hand, the equilibrium between the normal and superconducting phases can occur only if there is a magnetic field equal to H_c in the normal phase. The field is damped in the superconducting phase at a distance δ. Figure 105 shows the Pippard situation corresponding to $\xi \gg \delta$.

For the convenience of reasoning we replace the true picture by a rougher one with sharp boundaries along H (the line A) and along Δ (the line B) drawn so that the total free energy and the average field are the same as before. In such a case, there arises a region AB of order ξ, in which, on the one hand, $\Delta = 0$, corresponding to the normal phase, and, on the other hand, there is no magnetic field. Taking into account that $F_n - F_s = H_c^2/8\pi$, we obtain an excess energy equal to $H_c^2/8\pi$ per cm^3 of the region AB. But we are interested in the excess energy per cm^2 of the interface area. This is equal to $(H_c^2/8\pi) \cdot$ AB, i.e.,

$$\sigma_{ns} \sim \xi \frac{H_c^2}{8\pi}$$

(it is this parameter, ξ, that we introduced in section 15.3 instead of σ_{ns}).

As has been pointed out, the derivation described, just as fig. 105, refers to the Pippard case $\xi \gg \delta$. But if, conversely, $\delta \gg \xi$, then we find in the same way that at the interface there appears a negative surface energy of order $-\delta H_c^2/8\pi$. Naturally, this also occurs at $\xi = 0$, i.e., in the London theory. The impossibility of explaining the origin of the positive surface energy within the framework of the London theory has long been thought to be one of its main shortcomings.

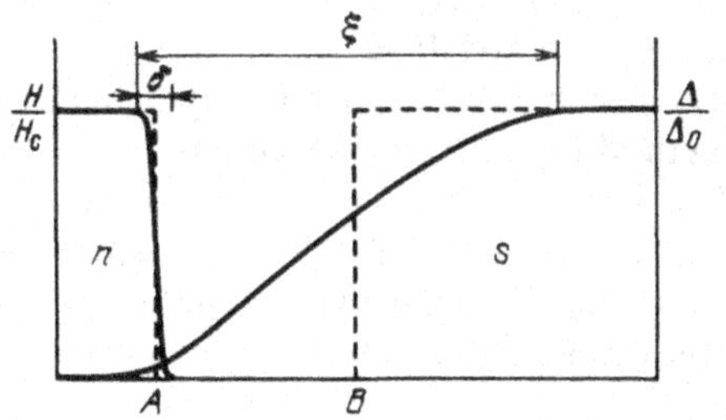

Fig. 105.

* The reasoning given below stems from Pippard (1955).

It has turned out that there really exist superconductors with a negative surface energy. Such superconductors exhibit an extremely unusual behavior in a magnetic field, which will be described in detail in ch. 18. The existence of such superconductors was first predicted by Abrikosov and Zavaritsky (Abrikosov 1952; Zavaritsky 1952) on the basis of the analysis of experimental data on the critical fields of thin superconducting films prepared by different methods. They were called superconductors of the second group. At present they are known as type II superconductors. Accordingly, ordinary superconductors with $\sigma_{ns}>0$ are called type I superconductors.

Since the sign of the surface energy depends on the ratio of the penetration depth δ and the correlation length ξ, it is natural to classify superconductors according to the value of the ratio δ/ξ. Especially attractive is the circumstance that with $T\to T_c$ this ratio tends to a constant, because both $\delta(T)$ and $\xi(T)$ depend on temperature by the law $(T_c-T)^{-1/2}$. The precise boundary between type I and type II superconductors is determined by the condition $\varkappa=\frac{1}{2}\sqrt{2}$, where, with $T\to T_c$, according to Gor'kov (1959) (see the derivation in ch. 17)

$$\varkappa(T_c)=\left(\frac{24}{7\zeta(3)}\right)^{1/2}\frac{\gamma}{\pi}\frac{\delta_L(0)}{\xi_0}=0.96\frac{\delta_L(0)}{\xi_0}. \tag{16.95}$$

The quantity $\varkappa$ was first introduced in the Ginzburg–Landau theory (ch. 17) and has come to be known as the Ginzburg–Landau parameter. Thus, in the vicinity of T_c we have

$$\begin{aligned}&\varkappa<\frac{1}{\sqrt{2}}, \quad \sigma_{ns}>0, \quad \text{type I},\\ &\varkappa>\frac{1}{\sqrt{2}}, \quad \sigma_{ns}<0, \quad \text{type II}.\end{aligned} \tag{16.96}$$

This condition may also be extended to temperature far from T_c. To do this, the quantity $\varkappa$ must be considered to be temperature-dependent, i.e.,

$$\varkappa(T)=\varkappa(T_c)\,A(T). \tag{16.97}$$

Calculations show that the function $A(T)$ varies little with temperature. This dependence is shown in fig. 106 (Gor'kov 1959).

Among pure metals presumably no type II superconductors are encountered. However, there is a method by means of which any type I superconductor can be converted to a type II superconductor. To this end, one has to introduce impurities.

The scattering from impurities first of all reduces the superconducting correlation*. Indeed, imagine an electron moving in the field of impurities. It performs a diffusive motion with a mean free path l and a velocity v; its mean displacement will be of order (section 3.5):

$$x\sim(Dt)^{1/2},$$

* The estimate given below refers to temperatures far from T_c.

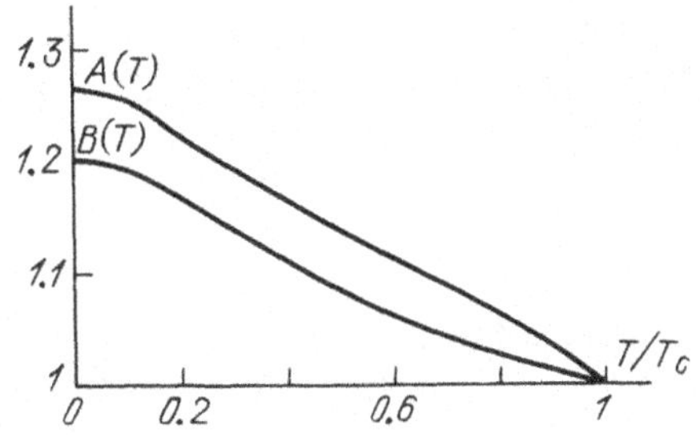

Fig. 106.

where D is the diffusion coefficient, which is of order $D \sim lv$. But its total path length is $t \cdot v$. If we take $tv \sim \xi$, then

$$x \sim \xi' \sim \left(lv \cdot \frac{\xi}{v} \right)^{1/2} \sim (\xi l)^{1/2}. \tag{16.98}$$

In other words, the effective correlation length decreases and becomes equal to ξ'. Of course, all these reasonings are valid as long as $\xi \gg l$. In case $l \gg \xi$, scattering on impurities practically has no effect on the superconducting correlation.

The incorporation of impurities alters the penetration depth too. In a case where $l \ll \xi$, one can estimate the modified quantity δ' rather simply. The presence of a finite mean free path leads to the appearance in the kernel $Q(\boldsymbol{r}-\boldsymbol{r}')$ of a damping at distance l, which can be taken into account by introducing a factor $f(\boldsymbol{r}-\boldsymbol{r}') = \exp(-|\boldsymbol{r}-\boldsymbol{r}'|/l)$. The Fourier component of the product of the functions $Q_0(\boldsymbol{r})f(\boldsymbol{r})$ is equal, as can be easily seen, to $\int Q_0(\boldsymbol{q}-\boldsymbol{q}_1) f(\boldsymbol{q}_1)\, \mathrm{d}^3\boldsymbol{q}_1 (2\pi)^3$ (Q_0 refers to a pure metal). In our case, the characteristic length for $f(\boldsymbol{r})$ is l and, hence, for its Fourier component the characteristic values of $q_1 \sim l^{-1}$. The scale of the kernel $Q_0(\boldsymbol{q}-\boldsymbol{q}_1)$ is ξ^{-1}. If $l > \xi$, then $q_1 \xi \sim \xi/l \gg 1$, i.e., the argument for Q_0 is large and one has to make use of the Pippard limit (16.75).

In subsequent calculations $q \sim \delta^{-1}$ are important. If $\delta/l \ll 1$, then $q \gg q_1$. In such a case

$$\int Q_0(\boldsymbol{q}-\boldsymbol{q}_1) f(\boldsymbol{q}_1) \frac{\mathrm{d}^3\boldsymbol{q}_1}{(2\pi)^3} = Q_0(q) \int f(\boldsymbol{q}_1) \frac{\mathrm{d}^3\boldsymbol{q}_1}{(2\pi)^3}.$$

The last integral is $f(\boldsymbol{r})$ at $\boldsymbol{r} = 0$, i.e., unity. Thus, formula (16.75) is reproduced. But if $\delta/l \gg 1$, then $q \ll q_1$. In this case,

$$\int Q_0(\boldsymbol{q}-\boldsymbol{q}_1) f(\boldsymbol{q}_1) \frac{\mathrm{d}^3\boldsymbol{q}_1}{(2\pi)^3}$$

$$\approx \int Q_0(\boldsymbol{q}_1) f(\boldsymbol{q}_1) \frac{\mathrm{d}^3\boldsymbol{q}_1}{(2\pi)^3} \sim \frac{n_e e^2}{mc} \frac{l}{\xi}$$

(here we have substituted expression (16.75) for Q_0). Since the resultant kernel is independent of q, we have the London situation. The penetration depth is $\delta = (4\pi Q/c)^{-1/2}$. Thus, we find

$$\delta' \sim \delta_L \left(\frac{\xi}{l}\right)^{1/2}. \tag{16.99}$$

Hence, in the presence of impurities

$$\varkappa \sim \frac{\delta'}{\xi'} \sim \frac{\delta_L}{l}.$$

The values given by the exact theory at $l \ll \xi$ are* (Abrikosov and Gor'kov 1958, Gor'kov 1959, Helfand and Werthamer 1966):

$$\delta(T) = \left[\frac{mc^2\hbar}{4\pi n_e e^2 \tau_{tr} \Delta \tanh(\Delta/2T)}\right]^{1/2}, \tag{16.100}$$

$$\tau_{tr}^{-1} = \int W(\theta)(1 - \cos\theta)\frac{d\Omega}{4\pi},$$

$$\varkappa(T) = \varkappa(T_c)\,B(T), \qquad l = v\tau_{tr},$$

$$\varkappa(T_c) = [42\zeta(3)]^{1/2}\pi^{-2}\frac{\delta_L(0)}{l} = 0.72\frac{\delta_L(0)}{l}. \tag{16.101}$$

The plot of the function $B(T)$ is given in fig. 106. This value of $\varkappa$ must be substituted into the criterion (16.97). In the intermediate cases of $\xi \sim l$ one may use the interpolation

$$\varkappa(T) \approx \delta_L(0)\left(0.96\frac{A(T)}{\xi_0} + 0.72\frac{B(T)}{l}\right). \tag{16.102}$$

At $\xi_0^{-1}l \gg 1$ this expression becomes equivalent to eqs. (16.97) and (16.95) and at $\xi_0^{-1}l \ll 1$ it transforms to eq. (16.101).

From what has been said above one may conclude that by reducing the mean free path, i.e., by increasing the impurity concentration (since $l \propto n_i^{-1}$, section 4.1), one can increase $\varkappa$ and convert any type I superconductor to a type II superconductor.

The following question may, however, arise: How do the impurities affect Δ, T_c and other thermodynamic characteristics of the superconductor? Investigations of this topic (Abrikosov and Gor'kov 1958, 1959, Anderson 1963) show that the variation of all these quantities is of order a/l, where a is the atomic dimension. Taking into account that $l \sim (n_i a^2)^{-1} \sim (c_i n a^2)^{-1} \sim a/c_i$, we find (section 4.1) that

* Here $W(\theta)$ is the probability of scattering of an electron from an impurity introduced in section 3.2 and τ_{tr} is the corresponding collision time (in section 3.2 it is denoted as τ; we use here the symbol adopted in the literature on superconductivity, τ_{tr}).

the corrections to the thermodynamic quantities are of a relative order c_i, the atomic concentration of impurities. Hence, as long as $c_i \ll 1$ these corrections are small, in contrast to the correction to δ, which has a relative order of $\delta/l \sim 10^2\text{-}10^3 c_i$. The smallness of the corrections to thermodynamic quantities is a consequence of the fact that they are governed by the local properties of the superconductor, i.e., the properties at a given point, and are not associated with the motion of electrons from one point to another. Evidently, any damping of the type of $\exp(-|\boldsymbol{r}-\boldsymbol{r}'|/l)$ with $\boldsymbol{r} \to \boldsymbol{r}'$ becomes inefficient. As for the local properties, they are determined by the density of states, which will vary only by an amount of order c_i *.

Of course, if $c_i \sim 1$, all the quantities may vary strongly, but in this case we are dealing with a different substance.

The superconducting correlation helps us to understand one more result. In section 16.6 we have calculated the heat capacity, and it has been found to exhibit a discontinuity at T_c but with no singularities. This behavior of the heat capacity is associated with the neglect of fluctuations (Appendix II). However, in this particular case there is every ground to expect that the fluctuational region corresponds to an extrememly small neighborhood of T_c. In Appendix II it is shown that the size of this region is determined by the ratio of the mean separation between the particles and the correlation length far from T_c. In the case under discussion this ratio is of order

$$\frac{\hbar/p_0}{\hbar v/T_c} \sim \frac{T_c}{p_0 v} \sim \frac{T_c}{\mu} \sim 10^{-4},$$

that is, it is very small. The exact size of the fluctuational region will be found in section 17.1.

16.11. High-temperature superconductivity

As has been said in section 15.1, the highest critical temperature attained to date is 93–95 K for the compound $YBa_2Cu_3O_{7-\delta}$. However, this composition is at present not satisfactory in its mechanical properties. Technological superconducting alloys are Nb_3Sn ($T_c = 17\text{–}18$ K) and Nb-Zr-Ti ($T_c = 9\text{–}11$ K). In order to maintain such low temperatures, one has to employ special cryostats with liquid helium (boiling point at atmospheric pressure 4.2 K). Nevertheless, as will be seen in what follows, superconductivity has already found practical applications in spite of the difficulties

* The reasoning given is valid for the isotropic model. For an anisotropic superconductor the situation is different. In this case, the scattering of electrons from an impurity mixes the components of the wave function of a Cooper pair corresponding to different values of the momentum, i.e., different Δ. This leads to a decrease in the binding energy of the pair and to a change of all thermodynamic quantities. The exact theory (Hohenberg 1963) shows that the relative corrections are of order $[(\Delta(\boldsymbol{n})-\bar{\Delta})^2/\bar{\Delta}^2]\xi_0/l$. Since the anisotropy in real superconductors is small, $(\Delta(\boldsymbol{n})-\bar{\Delta})^2/\bar{\Delta}^2 \leq 10^{-2}$, these corrections, in fact, differ little from the corrections for an isotropic superconductor, which are of order c_i.

associated with cooling, since some properties that are important in practice are exhibited only by superconductors.

It is not difficult to visualize the technical revolution that could be brought about by substances exhibiting superconductivity at room temperatures. In the first place, this would include the possibility of producing lines transmitting electrical power over any distance without loss. It would be possible to solve the problem of energy storage in small volumes and, hence, to use electric motors for all transport facilities that run on petroleum products. Many other perspectives could also be visualized.

This is why at present many laboratories in several countries are concerned with the study of high-temperature ceramics. On the one hand, it is necessary to define the limiting critical temperature and to develop standard technology for preparing the substances required. On the other hand, a construction method for superconducting equipment must be designed, which could be of practical use. The possibility is not excluded that the latter will be a formidable task for ceramics and the search for high-temperature superconductors will be continued. Moreover, the mechanism of superconductivity in ceramics has not been definitely established. Therefore, we shall present here some ideas concerning high-temperature superconductivity [detailed information on this topic can be found in the book edited by Ginzburg and Kirzhnits (1977)].

We begin with a more detailed discussion of the familiar interaction between electrons: Coulomb repulsion and phonon attraction. Then we shall consider "quasiphonon" mechanisms for the interaction of electrons and, finally, turn to nonphonon or so-called "exciton" mechanisms for the interaction. At the end of the section we shall discuss the possibility of producing high critical magnetic fields.

The formula for the critical temperature was derived in the BCS theory (16.25). It took into account only electron–phonon interaction in its simplified form. A more detailed formula was found by McMillan (1968), who carried out a more accurate examination of the electron–phonon interaction and took account of Coulomb repulsion; the formula has the form

$$T_c = \frac{\hbar\omega_D}{1.45}\exp\left[-\frac{1.04(1+\lambda)}{\lambda - \mu^*(1+0.62\lambda)}\right], \tag{16.103}$$

where λ characterizes the electron–phonon interaction (λ corresponds to $\frac{1}{2}g\nu$ in the simplified theory) and μ^* characterizes the electron–electron repulsion. Since the Coulomb interaction is screened at interatomic distances, it may be treated as a point interaction and described by a constant μ. The quantity μ^* is connected with this constant by the relation*

$$\mu^* = \frac{\mu}{1+\mu\ln(\varepsilon_F/\hbar\omega_D)}. \tag{16.104}$$

* We use the symbol μ for the dimensionless Coulomb interaction constant ($\mu \sim e^2/\hbar v$) since it is generally accepted in the literature. It should not be confused with the chemical potential, which is denoted here as ε_F.

The large logarithm in the denominator results from the different action times of the Coulomb repulsion ($t_c \sim \hbar/\varepsilon_F$) and phonon attraction ($t_{ph} \sim \omega_D^{-1}$); $t_C/t_{ph} \sim \hbar\omega_D/\varepsilon_F \ll 1$. For $\lambda \ll 1$ and $\mu^* = 0$ formula (16.103) takes a form similar to eq. (16.25).

Since λ and μ^* appear in the exponent, it is clear that these constants are of prime interest. The Coulomb interaction constant μ is formally of the order of unity. However, μ^*, which enters into T_c according to (16.104), is weakened by the large logarithm. Estimates give $\mu^* \sim 0.1$–0.2. As for the constant λ, it is also formally of the order of unity. But it must be taken into account that electron-phonon interaction eventually occurs as a result of the interaction of electrons with ions, which is described by a small pseudopotential. Therefore, λ usually does not reach unity. On the whole, the difference $\lambda - \mu^*(1+0.62\lambda)$ appearing in (16.103) may have any sign, but substances for which this difference is close to zero are rare. At relatively small λ the substance is not a superconductor* and at relatively large λ the term with μ^* may be neglected.

It is therefore reasonable to concentrate our attention on the constant λ. This constant determines not only the critical temperature but also the magnitude of the electrical resistance that appears as a result of the scattering of electrons as phonons. Since at room temperature this mechanism dominates over others, a paradoxical consequence arises: poor conductors (at room temperature) are good superconductors (i.e., they have a high T_c). Examples illustrating this are Pb: $\rho(18°\text{C}) = 20.8 \times 10^{-6}$ ohm · cm, $T_c = 7.2$ K, and Al: $\rho(18°\text{C}) = 3.2 \times 10^{-6}$ ohm · cm, $T_c = 1.2$ K. The superconducting ceramics that have lately been found are extremely bad conductors ($\rho \sim 10^{-3}$ ohm · cm at room temperature). If in formula (16.103) we assume that $\mu^* = 0$, then T_c will increase monotonically with increasing λ. Actually this formula does not apply when $\lambda \gg 1$. It has been shown (Allen and Mitrovic 1982) that

$$T_c \approx 0.18\hbar(\overline{\omega^2})^{1/2}\lambda^{1/2},$$

i.e. grows monotonically with λ. However in reality the value of λ never exceeds unity by order of magnitude. The idea has been proposed that λ can increase strongly if the lattice is close to a structural transition. Such a transition is usually preceded by the formation of a "soft mode", i.e., by a pronounced decrease of $\omega(\boldsymbol{k})$ at a certain $\boldsymbol{k}$. Such a decrease could imply an increase in the action time of phonon attraction and this was the basis for speculations on the subject. However, it should be taken into account that a soft mode at a particular $\boldsymbol{k}$ has a very small statistical weight (the integral includes $k^2\,dk$) and therefore it cannot play an essential role.

* Actually, the situation is in all probability more complicated. The interaction forces between electrons at large distances presumably always correspond to attraction. This may give rise to Cooper pairs with a nonzero orbital moment (if we deal with the isotropic model of course; in the anisotropic case there is no conserved orbital momentum, but there is another analogous classification of states, so that the qualitative conclusions do not change). The binding energy of such pairs is small, so that T_c will also be low. Nevertheless, this reasoning leads one to the conclusion that at sufficiently low temperatures any pure metal will become superconducting.

A special situation could take place for the "nesting" type of transitions in which, due to the change of the lattice period and the formation of a new Brillouin zone, some parts of the Fermi surface would be almost entirely superposed on each other. In the vicinity of such a transition an increase of the state density $\nu(\mu)$ would take place, i.e., an increase of λ in eq. (16.103) [more details can be found in the book edited by Ginzburg and Kirzhnits (1977)]. Such transitions are known to exist in nature, but their relation to superconductivity has not yet been established.

Apart from the increase of λ, one may think also of the increase of $\hbar\omega_D$. Considering that $\hbar\omega_D \sim sp_0 \sim \varepsilon_F(m/M)^{1/2}$, where M is the mass of the ion, an attempt must be made to reduce this mass. A unique possibility in this respect could be provided by solid hydrogen if it were a metal. Unfortunately, under ordinary conditions solid hydrogen is a dielectric crystal consisting of H_2 molecules. Theoretical calculations show that with increased pressure a phase transition must occur from molecular hydrogen to the atomic phase, which is analogous to alkali metals. The pressure of the transition is very high, of the order of 2.5 Mbar.

Of course, such a pressure is difficult to reach and to maintain and it is of no practical value. There exists, however, a possibility that, because of the large difference in specific volume between the molecular and the atomic phase, the latter can be preserved as a metastable phase even when the pressure is removed (an example is diamond, which is the metastable phase of carbon under ordinary conditions). Also, various methods have been proposed for producing metallic hydrogen directly from the gas phase without applying high pressures (adsorption on carbon, condensation in a strong magnetic field, etc.). However, the question of the existence of a metastable metallic phase and the possibility of its prepration in the absence of pressure has not yet been clarified.

The difficulty in calculations for solid hydrogen lies in the fact that, in contrast to all other elements, its ions have no electron shells and are simply protons. Because of this, for hydrogen the pseudopotential coincides with the real coulomb potential and is not small (section 14.2). Usually, calculations for metallic hydrogen are based on the strongly compressed matter approximation (section 14.4). However, the resultant series do not converge rapidly enough, so that it is difficult to choose the energetically most favorable structure and to solve the problem of the possibility of the metastable state. At the same time the strong electron-ion interaction implies also a strong electron-phonon interaction, which may bring about a certain increase in λ in formula (16.103). Preliminary calculations show that $T_c \sim 100$–200 K (Schneider 1969).

It has also been suggested that the difficulties associated with the preparation of metallic hydrogen can be overcome by producing the hydrides, i.e., compounds of metals with hydrogen. In the phonon energy spectrum of such compounds branches may appear that correspond mainly to the vibrations of protons and exhibit high frequencies. However, the hydrides are dielectrics. Therefore, attempts have been made to take advantage of the good solubility of hydrogen in some metals, especially in platinum. Since the natural solubility is still insufficient, special methods have been applied to produce metastable solid solutions with an increased hydrogen concentration. Although these compounds were superconductors, however their T_c was below 10 K. It is possible that the interaction of electrons with the high-frequency vibrations of protons is weak for some reasons.

There is one more theoretical possibility, which in fact, makes use of the same advantages exhibited by metallic hydrogen (Abrikosov 1978). It concerns the production of a substance that might be called "metallic excitonium". Imagine an even metal containing equal numbers of electrons and holes. If the mass of the holes is much larger than the mass of electrons due to the specific nature of the energy spectrum*, their potential energy will then exceed their kinetic energy. As a result, the holes will become classical objects and will form a periodic superstructure, i.e., a "lattice in the lattice" (Herring 1968, see Halperin and Rice 1968). The periods of such a superstructure are determined by the density of electrons and holes and must not necessarily coincide with the periods of the main lattice. For the hole lattice to be stable, it is necessary that the amplitude of the zero-point vibrations of the holes be much less than (not more than $\frac{1}{5}$) the lattice period, which leads to the requirement $m_h^*/m_e^* > 50$. If, besides, m_e is less than the mass of a free electron m, then the ratio m_h^*/m must not be very large.

As a result, a "substance" is produced, which resembles metallic hydrogen, but with the protons being replaced by heavy holes [Abrikosov (1978) proposed the term "metallic excitonium" for this substance]. In such a substance "phonons" can propagate. Their velocity will be determined by the mass of holes, which may be of the order of 10–20m at best, i.e., 100 times less than the mass of the proton. With a not-too-small density of electrons and holes this may lead to $\hbar\omega_D$ in formula (16.103) being 10 times larger than in the case of hydrogen.

At present no such substances are known to exist and, of course, we still do not know whether they can be produced. They should most likely be sought for among compounds containing transition metals (and also metals similar to them, such as Cu, Ag, Au) or rare-earth elements for the valence band to be formed from the weakly overlapping d- or f-shells of the atoms. In this case, the formation of heavy holes is possible. The search for such compounds is not a simple matter. However, on the other hand, there is no need for high pressures and the problem of the stability of the metastable metallic phase does not arise, as is the case with metallic hydrogen.

We have so far been concerned with the mechanism of electron attraction accomplished by phonons (or by similar hole lattice vibrations in the hypothetical "metallic excitonium"). There naturally arises the question of the possibility of a different attraction mechanism. Since the direct interaction of electrons is a repulsion, we are to seek in metals a "transmitting system" different from the phonon system.

The general scheme of the interaction of electrons via the transmitting system A may be outlined as follows:

$$e_1 + A \to e_1' + A^*, \qquad e_2 + A^* \to e_2' + A, \tag{16.105}$$

* This may occur if holes appear in a narrow band generated from the d- or f-levels of the atoms, whereas the electrons come from the s- or p-states of the atoms.

where e_i corresponds to an electron of momentum $\boldsymbol{p}_i$; A is the ground state; A^* is the excited state of the transmitting system. As a result of this double "reaction" the system A reverts to the initial state and the momenta of electrons e_1 and e_2 undergo a change, i.e., they are scattered from each other. It can easily be shown that such an interaction is necessarily an attraction and if it is not too strong, we have

$$T_c \sim \Delta E \exp\left(-\frac{1}{\lambda}\right) \tag{16.106}$$

where ΔE is the energy difference between the states A and A^* and λ depends on the interaction of electrons with the system A. As a matter of fact, phonon attraction occurs according to an analogous scheme.

Let us examine some variants. The role of the transmitting system A could be played by a system of spins in a magnetic metal. The excitation of this system is described by a gas of quasiparticles called magnons or spin waves. Since ferromagnetism is antagonistic to superconductivity (see section 21.2 for more detail), we may speak here of antiferromagnetic metals. However, estimates of the quantities ΔE and λ, which determine T_c, provide no advantages in this case compared to phonons.

A mechanism of a rather similar type is an exchange of "virtual paramagnons". If the substance is close to the magneto-ordered state (ferro- or antiferromagnetic), there can exist strong fluctuations corresponding to the magnetic phase and they produce a strong interaction between quasiparticles. Such virtual paramagnons are involved in explaining the unusual properties of almost ferromagnetic metals (which have been mentioned in ch. 13) and the formation of Cooper pairs in superfluid He^3; they are also proposed as a possible explanation of the high-temperature superconductivity in ceramics (Anderson 1987, Fukuyama and Yoshida 1987) and of the unusual superconductivity in metals with "heavy fermions" (see pp. 378–379).

Another transmitting system could be electrons that are separated in some way from conduction electrons (i.e., their states are not mixed). The excitations of such electron systems are called "excitons" and therefore this fundamentally possible mechanism of superconductivity is called the "exciton" mechanism (Little 1964; Ginzburg 1964, 1968, 1970, 1976).

In particular, the electrons of the inner shells of atoms that are retained in the lattice ions of metals could be such an isolated system. The excitation energy ΔE is here of the order of 10 eV, i.e., 10^5 K. If this mechanism were efficient, all metals would be high-temperature superconductors. This is not the case in reality because of the weak interaction of the conduction electron with the inner-shell electrons and, accordingly, the smallness of λ in (16.105).

Let us consider the matrix element corresponding to the first of the processes (16.105). It is given by

$$M_{12} = \int \psi^*_{A^*}(\boldsymbol{r}_1)\, \psi_A(\boldsymbol{r}_1)\, V(|\boldsymbol{r}_1 - \boldsymbol{r}_2|)\, e^{-i(\boldsymbol{p}'-\boldsymbol{p})\boldsymbol{r}_2/\hbar}$$

$$\times\, u^*_{\boldsymbol{p}'}(\boldsymbol{r}_2)\, u_{\boldsymbol{p}}(\boldsymbol{r}_2)\, dV_1\, dV_2,$$

where $V(r)$ is the interaction; the functions ψ_A and ψ_A^* are the wave functions of one of the ionic electrons; the conduction electron is described here by the Bloch function $e^{i\boldsymbol{pr}/\hbar}u_{\boldsymbol{p}}(\boldsymbol{r})$. The essential values of $\boldsymbol{r}_2$ in the integral over $\boldsymbol{r}_2$ are $r_2 \sim a$, where a is the interatomic separation. The order of magnitude of the quantity r_1 is determined by the ionic radius r_i, which is appreciably smaller than a. Hence, we can expand $V(|\boldsymbol{r}_1 - \boldsymbol{r}_2|)$ in r_1/r_2, i.e.,

$$V(|\boldsymbol{r}_1 - \boldsymbol{r}_2|) \approx V(r_2) - \frac{\boldsymbol{r}_1\boldsymbol{r}_2}{r_2} V'(r_2).$$

Since the functions ψ_A and ψ_A^* are orthogonal, the zeroth expansion term $V(r_2)$ gives no contribution and the matrix element M_{12} may be expected not to be larger than r_i/a in order of magnitude. Such smallness also results from the second process in (16.105). Hence, the effective electron interaction constant g will contain $(r_i/a)^2$, and this smallness will be included in λ (because $\lambda = \frac{1}{2}g\nu(\mu)$). It may be said that the ionic shells exhibit low polarizability.

Thus, an attempt must be made to find a transmitting system of electrons with high polarizability. Little was the first (Little 1964) to suggest the use of long polymeric organic molecules with so-called alternating bonds of carbon atoms, which behave as if they were one-dimensional metals. If the structure of the molecule includes strongly polarizable side atomic groups (fig. 107), these groups could serve as the transmitting system A. In fact, apart from the technical infeasibility of manufacturing such conductors from unruptured polymeric molecules, a fundamental objection arises against this idea. The point is that a purely one-dimensional superconductor cannot exist because fluctuations of the order-parameter phase destroy the long-range order at any finite temperature (Rice 1965)*.

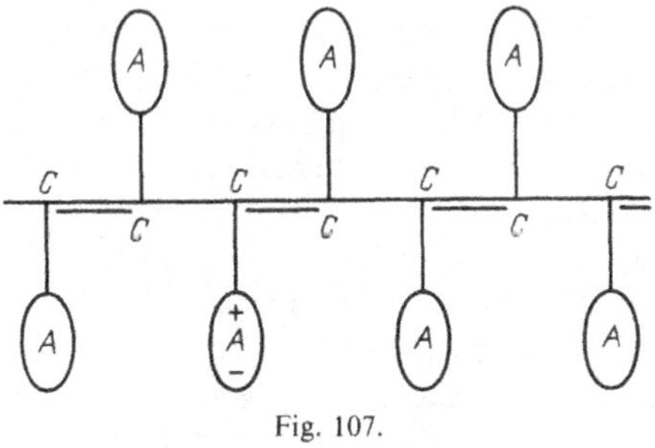

Fig. 107.

* What is really calculated is the correlation function $Z(x_1 - x_2) = \langle \Psi(x_1)\, \Psi(x_2) \rangle$, where Ψ is the Ginzburg-Landau order parameter which is proportional to Δ (ch. 17), and the average value is used with thermodynamic fluctuations taken into account. As a result, it is found that $Z \propto \exp(-|x_1 - x_2| mT/\Psi_0^2 h^2)$ with $|x_1 - x_2| \to \infty$. Here Ψ_0 is the equilibrium value of Ψ (in the one-dimensional problem Ψ_0^2 is normalized to the number of electrons per cm of length). Hence, $Z \to 0$ with $|x_1 - x_2| \to \infty$; this is an indication of the absence of long-range order at a finite temperature. In the presence of long-range order Ψ is coherent over the entire space and, hence, Z has a finite limit with $|x_1 - x_2| \to \infty$. The destruction of long-range order occurs because of the fluctuations in the phase of the complex function $\Psi = |\Psi|\, e^{i\varphi}$.

In order to overcome this difficulty, numerous "quasi-one-dimensional" compounds have been synthesized. These are mostly molecular crystals, which may be identified in the sense of conducting properties with bundles of weakly bound conducting filaments. The conductivity along the filaments is high, but the probability of an electron to be transferred from filament to filament is low, and therefore the conductivity anisotropy $\sigma_{\parallel}/\sigma_{\perp}$ may reach a few thousands. Apart from molecular crystals, polymers (for example, $(SN)_x$ and $(CH)_x$, where x stands for the multiple repetition of the group) have also been studied. Some of these compounds have been found to exhibit superconductivity, but with $T_c \leq 8$ K; the corresponding crystals were presumably more "three-dimensional" and the superconductivity mechanism was the ordinary phonon mechanism.

In this connection, it should be noted that the basic idea of Little was not one-dimensionality, which, as has already been pointed out, interferes with superconductivity, but rather the presence of strongly polarizable side groups that surround the main filament by a thick "coat". In currently available molecular quasi-one-dimensional crystals the conductivity anisotropy occurs not because of the large distances between the filaments but because of the high anisotropy of the electronic wave functions, which overlap strongly along the filaments and weakly in the transverse direction. Hence, in compounds of this type there is no room for a strongly polarizable "coat".

The idea of Ginzburg (Ginzburg 1964) consists in producing heterogeneous structures, in which metal and insulator films would alternate [a semiconductor has also been proposed in place of an insulator (Allender et al. 1973)]. Due to the quantum tunnelling effect some of the electrons of the metal could get into the insulator and exchange the excitons of the insulator. A theoretical estimation of T_c is very difficult because it depends on the detailed assumptions of the properties of the interface. High T_c cannot probably be attained by this method because of the smallness of the layer in which the pairing occurs and because of the weakness of the effective interaction. No substantial increase in T_c has been achieved in experiments with this type of structure.

The following remark may be made concerning the possibility of superconductivity in thin boundary layers of a metal. We have already discussed superconducting correlation (sections 16.7 and 16.10). One of its manifestations is the so-called "proximity effect – the mutual influence of a superconductor and a normal metal which are in contact with each other (see section 20.1 for more detail). In particular, superconductivity in a film applied onto a normal metal will be destroyed if its thickness is smaller than ξ. Hence, even if superconductivity could have arisen in a thin layer near the point of contact between a metal and any polarizable dielectric medium, it would have been suppressed if the thickness of this layer is less than ξ. Superconductivity could be preserved provided the value of T_c attained is so high that $\xi_0 \sim \hbar v/T_c$ is of the order of the thickness of the superconducting layer; note that to room temperature there corresponds $\xi \sim 10^{-5}$–10^{-6}, i.e., several hundreds of interatomic distances. The way out of this situation could be the use of thin superconducting films between layers of an insulator, which is what has been done in experiments, though with no success yet.

The limiting case of heterogeneous structures are quasi-two-dimensional superconductors, in which the conducting layers have a thickness of one molecule and are separated by relatively thick nonconducting layers. Such compounds have readily been synthesized on the basic of lamellar chalcogenides, such as TaS_2 and $NbSe_2$. Due to the weakness of the interaction, large organic molecules (say pyridine, C_6H_5N) can be incorporated into the spaces between the layers of a chalcogenide. Such crystals are called intercalation compounds. Here the separation between the layers may strongly increase (up to 50 Å), which, in particular, affects the conductivity anisotropy (it may increase from 10 up to 10^5).

Although purely two-dimensional conductivity is impossible for the same reason as purely one-dimensional superconductivity [the fluctuational destruction of the long-range order (Rice 1965)], it turns out that even an insignificant interaction between the layers suppresses the fluctuations and restores the superconductivity (Dzyaloshinskii and Katz 1968). In experiments in some cases upon intercalation there has been observed an increase in T_c. For example, upon intercalation of TaS_2 by pyridine T_c increased from 0.3 K up to 3.5 K (Gamble et al. 1976). Although high critical temperatures have not yet been reached in such compounds, this path should nonetheless be regarded as promising*.

In connection with superconductivity of two-dimensional systems mention should be made of the increase in T_c at the twinning plane of crystals (Khaikin and Khlyustikov 1981, 1983). This phenomenon is considered in detail in section 20.2.

The effect of twinning planes may be of prime importance in the high-temperature ceramic $YBa_2Cu_3O_{7-\delta}$, in which, due to a phase transition at 750 K, twinning planes appear spontaneously upon lowering the temperature at a distance of 500–1500 Å. It appears that their critical temperature is 93 K, while the "bulk" T_c is ~89 K; this could explain many peculiarities in the interval 89–93 K. Traces of higher-T_c superconductivity sometimes observed in this ceramic could possibly be due to the occasional formation of dense arrays of twinning planes (Abrikosov and Buzdin 1988).

Recently one more variant has been proposed of the system transmitting the interaction between conduction electrons – the so-called two-level systems in metallic "glasses" with a strongly disordered crystalline structure. In such compounds, entire groups of atoms can be transferred from some positions to others, these transfers being associated with relatively small changes in energy. Such groups are called two-level systems. Usually, at each temperature T those systems are important, which have a level difference of order T.

Theoretical investigations (Harris et al. 1983) show that the interaction of electrons through two-level systems adds to the phonon attraction and brings about an increase

* Taking into account that high-temperature ceramics have a layered structure and that their conductivity is due to layers containing Cu and O, which are separated by relatively large distances, one may suppose that these ceramics are superconductors of the excitonic type.

in T_c. An estimation of this effect in real metallic glasses shows that T_c may increase by 2–7%. This is in accord with experimental facts: the critical temperature of a disordered metal always decreases upon annealing (heating which leads to recrystallization); for example, it drops from 2.34 to 2 K in the alloy $Cu_{33}Zr_{67}$ and from 3.15 K to 3 K in $Zr_{76}Ni_{24}$. Although this effect is not strong, it is interesting from the standpoint of the demonstration of a nonphonon superconductivity mechanism.

Another idea has been proposed associated with the special centers in inhomogeneous metals (Schütter et al. 1987). Let us imagine that there exist centers for which it is energetically favorable to bind two electrons with opposite spins since attraction occurs between electrons of such centers. Although the existence of such centers can be justified, strictly speaking, only in semiconducting glasses, nevertheless their appearance in metals is not ruled out. In the latter case the hybridization of conduction electrons with these states, i.e., the possibility of virtual transitions from the band states to the centers, leads to an attraction and to the formation of Cooper pairs. The effectiveness of such a mechanism depends on the concentration of these special centers. In particular, such special centers may be regular structural units. This idea has also been proposed for the ceramics. For example, as an explanation of high-T_c superconductivity in ceramics it has been proposed that the holes of the conduction band (which transmit the current in those materials) can be localized at the double-negative oxygen ion (Zvezdin and Khomskii 1987).

Unusual superconductivity takes place in a number of compounds containing rare-earth metals and actinides, say YBe_{13} and $CeCu_2Bi_2$ (see Brandt and Moshchalkov 1984). Although the critical temperature of these compounds is not high ($T_c < 1$ K), their other properties are very interesting. First of all, they exhibit relatively strong critical fields. This may be interpreted as a consequence of an increase of the effective mass of carriers; estimates give values of m^* of the order of 200–400 m, where m is the mass of the free electron. It is for this reason that the term "heavy fermions" is used.

The study of the various characteristics in the normal state indicates that in such compounds the density of states has a narrow and large peak at the Fermi level (recall that the density of states $\nu = p_0 m^*/(\pi^2 \hbar^3)$. The origin of this peak is presumably associated with the electron screening of the spin of magnetic atoms of rare-earth elements and actinides. If the magnetic atoms are minor impurities, this leads to the Kondo effect in conductivity (section 4.6), but has no noticeable effect on the energy spectrum and thermodynamic properties. However, if the magnetic atoms become regular structural units and the exchange interaction constant J is so high that the Kondo temperature is higher than the freezing temperature of the spins, the Abrikosov–Suhl resonance in conductivity (see formula 4.49) may manifest itself as a peak in the density of states.

A further interesting property consists in that the low-temperature electronic heat capacity in UBe_{13} has been found to be proportional to T^3 or T^2. This is evidence that at certain points or contours of the Fermi surface the energy gap Δ vanishes. In all probability, the latter is associated with the nonphonon mechanism of electron attraction (probably the virtual paramagnons mentioned before).

Apart from increasing T_c, high critical fields are also important. In principle, in order to attain high critical fields two things are required: (a) the localization of electrons must be as large as possible; and (b) there must be attraction in the localized state. In this case, Cooper pairs could be "prepared in advance" from localized electrons and a certain probability of jumps could provide the supercurrent (Abrahams and Kulik 1977, Kulik and Pedan 1980, 1982, 1983, Bulaevskii et al. 1984, Alexandrov et al. 1986). In section 18.1 we shall see that the strongest critical fields can be attained in type II superconductors, and these are limited by two mechanisms: (1) Larmor twisting of pairs, and (2) the tendency of the magnetic field to flip the electron spins parallel to each other. These two mechanisms of pair collapse could be challenged successfully by superconductors of the "localized" type: the first mechanism could be challenged due to the large effective mass of carriers and the second due to the strong attraction of carriers with opposite spins in a single "well". However, it is still unknown whether such superconductors can be produced*. Metallic glasses in which a random potential could lead to localization of electrons could be of interest from this stand-point (section 11.6) (Kulik 1984).

* The possibility is not excluded that Chevrel phases and superconductors with "heavy fermions" could be such superconductors.

17

The Ginzburg–Landau theory

17.1. Derivation of the Ginzburg–Landau equations

In ch. 16 we have been concerned with the application of the microscopic (BCS) theory for describing the thermodynamic properties of superconductors and their behavior in a weak magnetic field. A generalization of this theory (Gor'kov 1958) allows one to consider the behavior of superconductors in strong fields, including alternating fields. However, the corresponding equations are extremely complicated, which is why they are seldom used in practice for solution of physical problems. Instead, use is made of a simplified theory, which we shall describe in the present chapter.

In 1950, i.e., even before the microscopic (BCS) theory was developed, Ginzburg and Landau had proposed a theory currently known as the Ginzburg–Landau theory (Ginzburg and Landau 1950), which describes the properties of superconductors near T_c; this theory has successfully surmounted the difficulties of London electrodynamics. For example, it explained the origin of the positive surface energy σ_{ns}. The equations of this theory were derived on the basis of the ideas of Landau's theory of second-order phase transitions (Appendix II).

Later, Gor'kov (Gor'kov 1959a,b) demonstrated that the equations of the Ginzburg–Landau theory are the limit of the BCS equations provided that the following two conditions are satisfied: (a) $T_c - T \ll T_c$, and (b) $\delta \gg \xi_0$. For London superconductors the second condition is fulfilled at all temperatures and only the first condition (a) remains. For Pippard superconductors, conversely, the requirement (b) is more stringent. We shall deal here with the simple derivation given in the fundamental work of Ginzburg and Landau (1950).

We start with the definition of the order parameter. We take it as the wave function of Cooper pairs present in the Bose condensate. For an ideal Bose condensate under homogeneous conditions, the ground state is a state with $\boldsymbol{p} = 0$. Below the point of Bose condensation in this state there is a finite number of particles with a wave function $\Psi = \text{const.}$ (i.e., $e^{i\boldsymbol{pr}/\hbar + i\alpha}$ at $\boldsymbol{p} = 0$) identical for all particles; this is called coherence. It is believed that with a slight (long-wavelength) inhomogeneity due to the application of an external field, the coherence is preserved and the function $\Psi(\boldsymbol{r})$ characterizes all particles of the condensate.

Near the critical temperature Ψ is small and the free energy Ω may be expanded into a series in Ψ. In the absence of an external field in a bulk superconductor Ψ

is independent of the coordinates. Since Ψ is a complex quantity and Ω is real, the expansion is carried out in powers of $|\Psi|^2$. Thus, we have (for unit volume)

$$\Omega_s = \Omega_n + a|\Psi|^2 + \tfrac{1}{2}b|\Psi|^4 + \cdots . \tag{17.1}$$

Since the theory is built up for the vicinity of T_c, the coefficients a and b can be expanded in $\tau = (T - T_c)/T_c$ and only the first nonvanishing terms need be retained. Since the minimum of Ω_s corresponds to $\Psi = 0$ above T_c and $\Psi \neq 0$ below T_c, the coefficient a changes sign at the transition point; therefore, $a = \alpha\tau$, where $\alpha > 0$. From the condition that $\Psi = 0$ must correspond to the minimum of Ω_s at the transition point too, we have

$$b \approx b(T_c) > 0.$$

Differentiating with respect to Ψ^*, we obtain the condition of the minimum of Ω_s:

$$\Psi[\alpha\tau + |\Psi|^2] = 0.$$

This yields the equilibrium values:

$$\begin{aligned} \Psi &= 0, \qquad T > T_c, \\ |\Psi|^2 &= -\frac{\alpha\tau}{b} \equiv \Psi_0^2, \qquad T < T_c. \end{aligned} \tag{17.2}$$

Substituting the equilibrium value of $|\Psi|^2 = \Psi_0^2$ into (17.2) and comparing with (15.5), we have

$$\Omega_n - \Omega_s = \frac{(\alpha\tau)^2}{2b} = \frac{H_{cm}^2}{8\pi}. \tag{17.3}$$

In the present chapter we use the notation H_{cm} for the thermodynamic critical field of a bulk superconductor in order to distinguish it from the critical field of a thin film and also from other critical fields that will be introduced later. Comparing eq. (17.3) with formula (16.55), we see that the Ginzburg-Landau theory gives a correct temperature dependence of H_{cm}. From this comparison we can also derive a microscopic formula for a certain combination of the coefficients α and b:

$$\frac{\alpha^2}{b} = \frac{4}{7\zeta(3)} T_c^2 \frac{mp_0}{\hbar^3} = \frac{4\pi^2}{7\zeta(3)} \nu(\mu) T_c^2. \tag{17.4}$$

Suppose now that an external magnetic field is applied to the superconductor. In this case, both the field in the superconductor and Ψ depend on the coordinates. Here one has to examine the total free energy of the entire superconductor. The energy of the field per unit volume is $H^2/8\pi$. Moreover, one has to add the energy associated with the inhomogeneity of Ψ. We will assume that the variation of Ψ in space occurs slowly. This permits one to consider only the correction $|\nabla\Psi|^2$ to the free energy. However, the order parameter Ψ has the meaning of the wave function of Cooper pairs. In view of this, the momentum operator $-i\hbar\nabla$ must necessarily be included in the combination $-i\hbar\nabla-(2e/c)\boldsymbol{A}$, where $\boldsymbol{A}$ is the vector potential*; we took into account that the pair has a charge $2e$.

In order to impart the ordinary form to the new term in the free energy, we write it as the kinetic energy of a particle of mass $2m$ **:

$$\frac{1}{4m}|(-i\hbar\nabla-(2e/c))\boldsymbol{A}\Psi|^2.$$

Thus, we have

$$\int \Omega_s\,dV=\int \Omega_n^{(0)}\,dV$$
$$+\int\left\{\alpha\tau|\Psi|^2+\tfrac{1}{2}b|\Psi|^4+\frac{1}{4m}\left|\left(-i\hbar\nabla-\frac{2e}{c}\boldsymbol{A}\right)\Psi\right|^2+\frac{H^2}{8\pi}\right\}dV \qquad (17.5)$$

(here we have introduced the notation $\Omega_n^{(0)}$ for the free energy of the normal phase without the field).

In order to obtain the minimum of the total free energy, we vary expression (17.5) with respect to Ψ^*. In such a case, we obtain

$$\int\left\{\alpha\tau\Psi\delta\Psi+b|\Psi|^2\delta\Psi^*\right.$$
$$\left.+\frac{1}{4m}\left(-i\hbar\nabla-\frac{2e}{c}\boldsymbol{A}\right)\Psi\left(i\hbar\nabla-\frac{2e}{c}\boldsymbol{A}\right)\delta\Psi^*\right\}dV=0.$$

In order to get rid of $\nabla\delta\Psi^*$, we integrate the corresponding term by parts and use Gauss' theorem. We find

$$\frac{i\hbar}{4m}\int\delta\Psi^*\boldsymbol{n}\left(-i\hbar\nabla-\frac{2e}{c}\boldsymbol{A}\right)\Psi\,dS$$
$$+\frac{1}{4m}\int\delta\Psi^*\left(-i\hbar\nabla-\frac{2e}{c}\boldsymbol{A}\right)^2\Psi\,dV,$$

where the integration in the first term runs over the surface of the superconductor ($\boldsymbol{n}$ is the unit vector of the normal).

The variation of $\delta\Psi^*$ is arbitrary. Assuming that $\delta\Psi^*=0$ at the surface and equating to zero the factor at $\delta\Psi^*$ in the integral over the volume, we obtain the equation for Ψ:

$$\frac{1}{4m}\left(-i\hbar\nabla-\frac{2e}{c}\boldsymbol{A}\right)^2\Psi+\alpha\tau\Psi+b|\Psi|^2\Psi=0. \qquad (17.6)$$

* Recall that the theory must be invariant under a gauge transformation of the vector potential $\boldsymbol{A}\to\boldsymbol{A}+\nabla\varphi$. In the expression $[-i\hbar\nabla-(2e/c)\boldsymbol{A}]\Psi$ this is provided by the fact that $\nabla\varphi$ is compensated for by the change in the phase of the function Ψ.

** The mass m in this term is quite arbitrary, but it is convenient to take it to be equal to the mass of the electron (see below).

If now we assume $\delta\Psi^*$ to be arbitrary at the surface of the superconductor, the surface integral will give the following boundary condition:

$$\boldsymbol{n}\left(-\mathrm{i}\hbar\nabla-\frac{2e}{c}\boldsymbol{A}\right)\Psi|_{\mathrm{b}}=0. \tag{17.7}$$

Variation of (17.5) with respect to $\delta\Psi$ gives an equation and a boundary condition which are complex-conjugate to (17.6) and (17.7).

Now we vary (17.5) with respect to the vector potential $\boldsymbol{A}$, assuming that $\boldsymbol{H}=\operatorname{rot}\boldsymbol{A}$. The variation of H^2 gives $2\operatorname{rot}\boldsymbol{A}\operatorname{rot}\delta\boldsymbol{A}$. We may use the formula $\operatorname{div}[\boldsymbol{ab}]=\boldsymbol{b}\operatorname{rot}\boldsymbol{a}-\boldsymbol{a}\operatorname{rot}\boldsymbol{b}$. This yields $2\delta\boldsymbol{A}\operatorname{rot}\operatorname{rot}\boldsymbol{A}+2\operatorname{div}[\delta\boldsymbol{A}\operatorname{rot}\boldsymbol{A}]$. The volume integral of div is transformed to a surface integral and vanishes. Equating the variation to zero, we find

$$\operatorname{rot}\operatorname{rot}\boldsymbol{A}=\operatorname{rot}\boldsymbol{H}=\frac{4\pi}{c}\boldsymbol{j}, \tag{17.8}$$

$$\boldsymbol{j}=-\frac{\mathrm{i}e\hbar}{2m}(\Psi^*\nabla\Psi-\Psi\nabla\Psi^*)-\frac{2e^2}{mc}|\Psi|^2\boldsymbol{A}. \tag{17.9}$$

Equation (17.8) is a Maxwell equation. The boundary condition is the specification of the field at the superconductor surface. Expression (17.9) corresponds to the quantum-mechanical current in magnetic field if the wave function is equal to Ψ, the charge is $2e$ and the mass $2m$.

We now pass over to new units which will allow us to drop most of the constants in (17.6)–(17.9). We introduce the following notations:

$$\Psi'=\frac{\Psi}{\Psi_0},\qquad \boldsymbol{H}'=\frac{\boldsymbol{H}}{H_{\mathrm{cm}}\sqrt{2}},$$
$$\boldsymbol{r}'=\frac{\boldsymbol{r}}{\delta},\qquad \delta=\sqrt{\frac{2mc^2}{4\pi\Psi_0^2(2e)^2}},$$
$$\boldsymbol{A}'=\frac{\boldsymbol{A}}{H_{\mathrm{cm}}\sqrt{2}\delta},\qquad H_{\mathrm{cm}}=\frac{2\sqrt{\pi}\alpha\tau}{b^{1/2}},$$
$$\Psi_0^2=\alpha\frac{|\tau|}{b}. \tag{17.10}$$

As a result, the equations become (we omit the primes in the new quantities):

$$\left(-\frac{\mathrm{i}\nabla}{\varkappa}-\boldsymbol{A}\right)^2\Psi-\Psi+|\Psi|^2\Psi=0, \tag{17.6$'$}$$

$$\boldsymbol{n}\left(-\frac{\mathrm{i}\nabla}{\varkappa}-\boldsymbol{A}\right)\Psi|_{\mathrm{b}}=0, \tag{17.7$'$}$$

$$\operatorname{rot}\operatorname{rot}\boldsymbol{A}=-\frac{\mathrm{i}}{2\varkappa}(\Psi^*\nabla\Psi-\Psi\nabla\Psi^*)-|\Psi|^2\boldsymbol{A}. \tag{17.8$'$}$$

Equations (17.6′), (17.7′) and (17.8′) contain only one constant, $\varkappa$, which is called the Ginzburg–Landau parameter; it is defined as

$$\varkappa = 2^{3/2} e H_{\mathrm{cm}} \frac{\delta^2}{\hbar c}. \tag{17.11}$$

This is the same constant introduced in section 16.9 (formula 16.95). This point will be discussed later.

Let us consider the simplest problem: the penetration of a weak magnetic field into the bulk of a superconductor with a planar boundary. Let the superconductor occupy the half-space $x = 0$. The field is applied to it along the z axis. The field penetrates the superconductor and decreases rapidly in the bulk, i.e., it depends on x. Therefore, we choose the vector potential $\boldsymbol{A}$ along the direction y. In this case,

$$H(x) = \frac{\mathrm{d}A_y}{\mathrm{d}x}.$$

It is natural to assume that Ψ is also dependent only on x. From eqs. (17.6) and (17.7) we obtain

$$-\varkappa^2 \frac{\mathrm{d}^2\Psi}{\mathrm{d}x^2} - \Psi + A^2\Psi + |\Psi|^2\Psi = 0,$$

$$\left.\frac{\mathrm{d}\Psi}{\mathrm{d}x}\right|_{\mathrm{b}} = 0.$$

We assume that $\varkappa \ll 1$. Then we obtain $\Psi = \text{const}$. From the condition in the bulk of the superconductor it follows that $|\Psi|^2 = 1$. The function Ψ may be chosen real. Equation (17.8′) yields

$$\frac{\mathrm{d}^2 A}{\mathrm{d}x^2} - A = 0.$$

Differentiating with respect to x, we derive a similar equation for H. The solution that satisfies the boundary condition $H = H_0$ is $H = H_0\,\mathrm{e}^{-x}$ or, in conventional units,

$$H = H_0\,\mathrm{e}^{-x/\delta}, \tag{17.12}$$

where δ is defined in (17.10). But we know that this must give the London penetration depth near T_{c}. Choosing m in (17.10) to be equal to the mass of the electron, we find that $|\Psi_0|^2 = \frac{1}{2}n_{\mathrm{s}}$, i.e., the number of Cooper pairs. Comparing (17.2) with the BCS formula (16.92), we find that

$$\frac{\alpha}{b} = n_{\mathrm{e}}. \tag{17.13}$$

Using formulae (17.13) and (17.14), we can define α and b ($n_e = p_0^3/3\pi^2\hbar^3$):

$$\alpha = \frac{12\pi^2}{7\zeta(3)} \frac{mT_c^2}{p_0^2}, \qquad b = \frac{\alpha}{n_e}. \tag{17.14}$$

Substituting formulae (16.55) for $H_{cm}(T)$ and (16.93) for $\delta(T)$ into the expression for $\varkappa$ (17.11), we obtain

$$\varkappa = 3\left(\frac{2}{7\zeta(3)}\right)^{1/2}\left(\frac{\pi\hbar}{v}\right)^{3/2}\frac{cT_c}{\hbar p_0}. \tag{17.15}$$

The constant $\varkappa$ may be expressed through the ratio of $\delta_L(0) = (mc^2/4\pi n_e e^2)^{1/2}$ and $\xi_0 = (\gamma/\pi^2)\hbar v/T_c$ (see formula 16.82). As a result, we obtain formula (16.95).

Thus, all the constants that appear in the theory have a microscopic interpretation. Only the microscopic meaning of the order parameter Ψ remains unclear. However, this can be found by comparison with the microscopic theory. It is rather evident that Ψ is proportional to Δ. These two quantities are complex scalars in the general case, both are equal to zero at $T > T_c$ and proportional to $|\tau|^{1/2}$ at $T_c - T \ll T_c$. Therefore it is sufficient to compare the coefficients at $|\tau|^{1/2}$ for Ψ (17.2) and for Δ when $T \to T_c$ (16.33). From this we get

$$\Psi(\boldsymbol{r}) = \frac{1}{\pi}\left(\tfrac{7}{8}\zeta(3)\right)^{1/2} n_e^{1/2}\frac{\Delta(\boldsymbol{r})}{T_c}. \tag{17.16}$$

The Ginzburg–Landau theory refers also to superconductors that contain impurities; such superconductors are usually called superconducting alloys or dirty superconductors in the literature. As has already been remarked in section 16.9, impurities in a small concentration exert but a slight effect on the thermodynamic properties, i.e., on T_c, Δ, and H_{cm}, but at $l \ll \xi$ they appreciably change the penetration depth. Inserting expressions (16.55) for H_{cm} and (16.100) for δ with $T \to T_c$ into the general formula (17.11) for $\varkappa$, we obtain formula (16.101), where $l = v\tau_{tr}$.

For convenience of comparison with experiment $\varkappa$ may be expressed via the conductivity of a normal metal $\sigma = n_e e^2 l/p_0$ and the coefficient in the linear law of heat capacity $C_n = \gamma_c T$, $\gamma_c = p_0 m/3\hbar^3$ (Gor'kov 1959a,b):

$$\varkappa = \left(\frac{21\zeta(3)}{2\pi}\right)^{1/2}\frac{ce}{\pi^3}\frac{\gamma_c^{1/2}}{\sigma}. \tag{17.17}$$

In conventional units, when the temperature is measured in Kelvin and the heat capacity in erg/cm$^3 \cdot$ K, our γ_c is measured in $(\text{erg}\cdot\text{cm}^{-3}\cdot\text{K}^{-2})/k_B^2$, where k_B is Boltzmann's constant. Substituting the numerical values of c and e and expressing the resistivity $\sigma^{-1} = \rho$ in ohm $\cdot$ cm, we obtain

$$\varkappa = 7.5 \times 10^3[\gamma_c(\text{erg}\cdot\text{cm}^{-3}\cdot\text{K}^{-2})]^{1/2}\rho(\text{ohm}\cdot\text{cm}). \tag{17.17$'$}$$

Using relation (17.4) derived from H_{cm}, which we assume to be constant, and expressions (16.100) and (17.10) for δ together with formula (16.33) for Δ (which remains unchanged too), we find the coefficients α and b and the relationship between Δ and Ψ:

$$\alpha = \frac{6}{\pi}\frac{\hbar T_c}{\tau_{tr}}\frac{m}{p_0^2}, \qquad \frac{\alpha}{b} = \frac{2\pi^3}{7\zeta(3)}\frac{T_c\tau_{tr}}{\hbar}n_e,$$
$$\Psi(\boldsymbol{r}) = (\tfrac{1}{4}\pi)^{1/2}\left(\frac{n_e\tau_{tr}}{hT_c}\right)^{1/2}\Delta(\boldsymbol{r}). \tag{17.18}$$

Inserting these expressions into eq. (17.6), we obtain the form of this equation that is frequently used for alloys:

$$\frac{\pi D\hbar}{8T_c}\left(\nabla - \frac{2ie}{\hbar c}\boldsymbol{A}\right)^2\Delta + \left(\frac{T_c - T}{T_c}\right)\Delta - \left(\frac{7\zeta(3)}{8\pi^2T_c^2}\right)|\Delta|^2\Delta = 0, \tag{17.6''}$$

where $D = \frac{1}{3}v^2\tau_{tr}$ is the diffusion coefficient (section 3.5). The corresponding expression for the current is

$$\boldsymbol{j} = -i\frac{\pi e\nu D}{4T_c}\left(\tfrac{1}{2}(\Delta^*\nabla\Delta - \Delta\nabla\Delta^*) - \frac{2ie}{\hbar c}\boldsymbol{A}|\Delta|^2\right), \tag{17.9'}$$

where $\nu = p_0 m/\pi^2\hbar^3$ is the density of states.

For subsequent applications we transform the expression for the free energy (17.5). Passing to the reduced quantities and using formulas (17.10), we have

$$\int \Omega_s \, dV = \int \Omega_n^{(0)} \, dV + \frac{H_{cm}^2}{4\pi}\int \left\{-|\Psi|^2 + \tfrac{1}{2}|\Psi|^4 + \left|\left(-i\frac{\nabla}{\varkappa} - \boldsymbol{A}\right)\Psi\right|^2 + H^2\right\} dV. \tag{17.19}$$

As before, in the derivation of the equations, we integrate the term with $\nabla\Psi^*$ by parts. Here the surface integral vanishes due to the boundary condition (17.7) and in the volume integral left we use the fact that Ψ satisfies eq. (17.6'). As a result, we have

$$\int \Omega_s \, dV = \int \Omega_n^{(0)} \, dV + \frac{H_{cm}^2}{4\pi}\int (H^2 - \tfrac{1}{2}|\Psi|^4) \, dV. \tag{17.19'}$$

In what follows we will be interested in the transition from the normal to the superconducting state under the influence of an external field. For a cylindrical geometry the free energy in a given external field is obtained by subtracting $\boldsymbol{H}_0\boldsymbol{B}/4\pi$, where $\boldsymbol{H}_0$ is the external field and $\boldsymbol{B}$ is the induction equal to the average field in the sample, $\boldsymbol{B} = V^{-1}\int \boldsymbol{H} \, dV$ (see Appendix III). For the normal phase $\Omega_{nH} = \Omega_n^{(0)} - H_0^2/8\pi$.

Thus, we obtain (in reduced units):

$$\int (\Omega_{sH} - \Omega_{nH}) \, dV = \frac{H_{cm}^2}{4\pi}\int [(H - H_0)^2 - \tfrac{1}{2}|\Psi|^4] \, dV. \tag{17.20}$$

In concluding this section we evaluate the size of the region where the Ginzburg-Landau theory based on the Landau approximation for second-order phase transitions becomes inapplicable because of the fluctuational effects. According to Appendix II, the range of validity of the Landau approximation is determined by condition (AII.20):

$$1 \gg |\tau| \gg \frac{b^2 T_c^2}{\alpha c^3},$$

where $\tau = (T - T_c)/T_c$. In the Ginzburg-Landau theory the coefficients α and b have the same meaning as in the model considered in Appendix II, and the role of the parameter c (the coefficient in front of $\nabla\varphi$, where φ is the real scalar order parameter) is played by $\hbar^2/m$. Hence,

$$1 \gg |\tau| \gg \frac{m^3 b^2 T_c^2}{\alpha \hbar^6}.$$

Substituting α and b for a pure superconductor, we obtain in accordance with (17.14)

$$1 \gg |\tau| \gg \left(\frac{T_c}{\varepsilon_F}\right)^4, \tag{17.21}$$

where $\varepsilon_F \sim p_0^2/m$ is the Fermi energy. Usually, T_c/ε_F does not exceed 10^{-3}, i.e., the quantity that stands on the right is less than 10^{-12}, which is many orders of magnitude smaller than the fluctuations in temperature that can be controlled in experiments.

For superconductors containing impurities at $l \ll \xi$ in accordance with (17.18) we have

$$1 \gg |\tau| \gg \left(\frac{\hbar}{\tau_{tr}}\right)^3 \frac{T_c}{\varepsilon_F^4}. \tag{17.22}$$

This requirement is less stringent than (17.21), but it also practically rules out the fluctuational region.

From this it follows that the fluctuations are unimportant in the thermodynamics of bulk superconductors. The physical cause for this is the large correlation length, as has already been pointed out at the end of section 16.9*. Nevertheless, situations are possible where the role of fluctuations increases noticeably, and these are responsible for the observed effects. This occurs for the kinetic effects in small objects, say for the conductivity of thin films and filaments. This point will be discussed in section 19.6.

Thus, the Ginzburg-Landau theory can be applied practically up to the very critical temperature. What is the range of its validity on the low-temperature side? One of these conditions is known: this is $|\tau| \ll 1$. The other is $\delta \ll \xi_0$, i.e., the London

* There are indications that in high-T_c ceramics the fluctuation region is relatively large. Specific-heat data show that its width is ~2 K, i.e. $|\tau| \sim 0.02$. This could be due to the small correlation length (~30 Å) or to the relatively large value of T_c/ε_F. Nevertheless this region is not so large as to make G-L theory inapplicable.

region. The latter can be seen, for example, from the fact that, according to section 16.7, the relation between the current and the field is nonlocal in the general case and changes to the local theory only in the London region. In the Ginzburg–Landau theory the relation between the current and the field is local.

As has been said above, the Ginzburg–Landau equations are the exact limit of the BCS theory at $|\tau| \ll 1$. We will not give the corresponding derivation here because it is cumbersome (see Gor'kov 1959a, b). We note only that the relation that generalizes formula (16.18) for Δ to the inhomogeneous case $\Delta(\boldsymbol{r})$ (here we are speaking of $T \to T_c$, i.e., small Δ) is a complex integral relation: the right-hand side contains terms of the type

$$\int K(\boldsymbol{r}, \boldsymbol{r}')\Delta(\boldsymbol{r}')\,\mathrm{d}V',$$

$$\int K(\boldsymbol{r}, \boldsymbol{r}', \boldsymbol{r}'', \boldsymbol{r}''')\,\Delta(\boldsymbol{r}')\,\Delta(\boldsymbol{r}'')\,\Delta(\boldsymbol{r}''')\,\mathrm{d}V'\,\mathrm{d}V''\,\mathrm{d}V'''.$$

The distances that are important in the kernels of the integrals are of order ξ_0. On the other hand, the distances over which Δ varies are of order $\xi(T) \sim \hbar v/\Delta(T) \gg \xi_0$ (at $|\tau| \ll 1$). This makes it possible to consider the variations of $\Delta(\boldsymbol{r})$ to be slow and to pass from the integral to differential relations. An analogous situation occurs with the vector potential $\boldsymbol{A}$ if the condition $\delta(T) \gg \xi_0$ is satisfied.

Both requirements can be expressed in the form of a single relation. For this purpose we use the fact that $\xi_0 \sim \xi(0) \sim \delta_L(0)/\varkappa$ and also that $\delta(T) \sim \delta_L(0)|\tau|^{-1/2}$ at $|\tau| \ll 1$. The condition $\delta(T) \gg \xi_0$ may be written in the form $|\tau|^{1/2} \ll \varkappa$. Thus, the general condition for the validity of the Ginzburg–Landau theory reads

$$1 - \frac{T}{T_c} \ll \min(\varkappa^2, 1). \tag{17.23}$$

Although we have been speaking all the time of pure superconductors, formula (17.23) is also valid for dirty superconductors.

The Ginzburg–Landau theory is much simpler than the complete BCS theory for superconductors in an external field. Nevertheless, its content has proved to be very rich, since it has made it possible to account for all the basic equilibrium properties of superconductors. In spite of the limitation (17.23), it has turned out that Ginzburg–Landau theory provides a satisfactory quantitative agreement with experiment even at temperatures well below T_c if all the parameters are expressed in reduced units. This is associated with the slowness of the variation of $\varkappa$ with temperature [the function $A(T)$ in eq. (16.97) and the function $B(T)$ in eq. (16.101)].

17.2. Surface energy at the interface between the normal and superconducting phases

Below we will demonstrate the solution of several problems using the Ginzburg–Landau theory. Consider a case where Ψ and A depend only on a single coordinate, say x, with A being directed perpendicular to x. Here the general equations (17.6′), (17.7′) and (17.8′) simplify and take the form

$$\frac{1}{\varkappa^2}\frac{\mathrm{d}^2\Psi}{\mathrm{d}x^2}+\Psi(1-A^2)-\Psi^3=0, \tag{17.24}$$

$$\left.\frac{\mathrm{d}\Psi}{\mathrm{d}x}\right|_{\mathrm{b}}=0, \tag{17.25}$$

$$\frac{\mathrm{d}^2 A}{\mathrm{d}x^2}-\Psi^2 A=0 \tag{17.26}$$

(it is easy to see that Ψ may be taken to be real).

The integrals of eqs. (17.24) and (17.26) are easily evaluated. To this end, we multiply eq. (17.24) by $\mathrm{d}\Psi/\mathrm{d}x$ and eq. (17.26) by $\mathrm{d}A/\mathrm{d}x$, add up and integrate over x. With the boundary condition in the bulk of the superconductor being taken into account, we get

$$\frac{1}{\varkappa^2}\left(\frac{\mathrm{d}\Psi}{\mathrm{d}x}\right)^2+\left(\frac{\mathrm{d}A}{\mathrm{d}x}\right)^2+\Psi^2(1-A^2)-\tfrac{1}{2}\Psi^4=\text{const.}=\tfrac{1}{2} \tag{17.27}$$

(the value of the constant is taken from the boundary conditions 17.28).

Let us find the surface energy between the normal and superconducting phases (Ginzburg and Landau 1950). Suppose that as $x\to\infty$ there is a superconducting phase and for $x\to-\infty$ there is a normal phase. We assume that $\boldsymbol{H}\parallel z$, $\boldsymbol{A}\parallel y$ and $H=\mathrm{d}A/\mathrm{d}x$. Then we have the following boundary conditions:

$$\begin{aligned} x\to\infty:&\quad \Psi=1,\quad H=A=0,\quad \frac{\mathrm{d}\Psi}{\mathrm{d}x}=0,\\ x\to-\infty:&\quad \Psi=0,\quad H=H_0=\frac{1}{\sqrt{2}},\quad \frac{\mathrm{d}\Psi}{\mathrm{d}x}=0 \end{aligned} \tag{17.28}$$

(in conventional units $\Psi=1$ corresponds to Ψ_0 and $H_0=1/\sqrt{2}$ corresponds to H_{cm}). Since eqs. (17.24), (17.25) and (17.26) are not integrable in the general case, we will consider the limiting case $\varkappa\ll 1$.

This case corresponds to $\xi\gg\delta$ and is shown in fig. 105. Obviously, the major part of the surface energy arises from the region in which H and A are small. Assuming that $A=H=0$ in eq. (17.27), we obtain

$$\frac{\mathrm{d}\Psi}{\mathrm{d}x}=\frac{\varkappa}{\sqrt{2}}(1-\Psi^2). \tag{17.29}$$

The solution of this equation that satisfies the boundary condition $\Psi=1$ for $x\to\infty$ and decreases towards diminishing x is

$$\Psi=\tanh\left(\frac{\varkappa x}{\sqrt{2}}\right). \tag{17.30}$$

Of course, we have to take into account that such a solution becomes incorrect in the region where the field penetrates, but this region makes a small contribution to surface energy. For convenience, we have chosen the origin at point $x=0$, where $\Psi \approx 0$ (the boundary A in fig. 105).

Consider now the formula for the free energy difference (17.30). The integrand vanishes as $x \to \infty$ and also with $x \to -\infty$. A contribution comes only from the transition region. It is this region that corresponds to the excess energy associated with the boundary, i.e., σ_{ns}. Substituting into eq. (17.20) $H_0 = 1/\sqrt{2}$, $H=0$ and Ψ according to (17.30), we obtain

$$\begin{aligned}\sigma_{ns} &= \frac{H_{cm}^2}{8\pi}\int_0^\infty (1-\Psi^4)\,dx \\ &= \frac{H_{cm}^2}{8\pi}\int_0^\infty \left(1-\tanh^4\left(\frac{\varkappa x}{\sqrt{2}}\right)\right) dx \\ &= \frac{H_{cm}^2}{8\pi}\int_0^\infty \cosh^{-2}\left(\frac{\varkappa x}{\sqrt{2}}\right)\left(1+\tanh^2\left(\frac{\varkappa x}{\sqrt{2}}\right)\right) dx \\ &= \frac{H_{cm}^2}{8\pi}\frac{4\sqrt{2}}{3\varkappa}. \end{aligned} \tag{17.31}$$

The lower limit of the integral over x is chosen where Ψ becomes small and the field begins to penetrate, i.e., $x=0$. In conventional units we have

$$\delta_{ns} = \frac{H_{cm}^2}{8\pi}\frac{4\sqrt{2}\delta}{3\varkappa}. \tag{17.32}$$

For $\varkappa \gtrsim 1$ the calculation of σ_{ns} is possible only by a numerical method. It turns out that σ_{ns} vanishes at $\varkappa = 1/\sqrt{2}$, and becomes negative at larger values of $\varkappa$. Since this contradicted the superconductivity picture known by the time the Ginzburg–Landau theory was developed (1950), this case was not considered. The experimental data on the size of the layers in the intermediate state (section 15.3) make it possible to determine $\varkappa$. For pure superconductors the values of $\varkappa$ obtained are small: 0.16 for mercury, 0.15 for tin, and 0.026 for aluminum*.

Since the vanishing of σ_{ns} at $\varkappa = 1/\sqrt{2}$ is important, we shall give here a rigorous proof (Lifshitz and Pitaevskii 1978). Consider the complete free-energy expression (17.20) at $H_0 = H_{cm}$, or in reduced units $H_0 = 1/\sqrt{2}$. Evidently, the integrand and, hence, the whole integral turns to zero at

$$H = \frac{dA}{dx} = \frac{1-\Psi^2}{\sqrt{2}}.$$

Substituting this expression into eq. (17.26), we find

$$\frac{d\Psi}{dx} = -\frac{A\Psi}{\sqrt{2}}.$$

It is easy to see that these formulae satisfy eq. (17.27) at $\varkappa = 1/\sqrt{2}$.

* It is easier to determine $\varkappa$ experimentally from the supercooling field (section 18.4), though, strictly speaking, there is a risk of obtaining overestimated values of $\varkappa$.

17.3. *The critical field and magnetization of a thin film. Supercooling and superheating*

Since $\varkappa \ll 1$ for many pure superconductors, it will be reasonable to solve the problems in this approximation, which considerably facilitates calculations. We shall consider the question of the properties of a thin superconducting film placed in a parallel magnetic field H_0 (Ginzburg and Landau 1950, Ginzburg 1952). Suppose $\boldsymbol{H} \parallel z$ and the film occupies a volume $-\frac{1}{2}d < x < \frac{1}{2}d$. Equation (17.24) is rewritten in the form

$$\frac{\mathrm{d}^2\Psi}{\mathrm{d}x^2} = \varkappa^2\Psi(A^2 - 1 + \Psi^2). \tag{17.33}$$

At $\varkappa \ll 1$ in the zeroth approximation we neglect the right-hand side. As a result, we have (with eq. 17.25 taken into account):

$$\frac{\mathrm{d}^2\Psi}{\mathrm{d}x^2} = 0 \quad \text{and} \quad \left.\frac{\mathrm{d}\Psi}{\mathrm{d}x}\right|_{x=\pm\frac{1}{2}d} = 0.$$

This gives

$$\Psi = \text{const.} \tag{17.34}$$

Substituting this, as yet unknown, constant into eq. (17.26) and using $H = \mathrm{d}A/\mathrm{d}x$ and the boundary condition $H(\pm\frac{1}{2}d) = H_0$, we get

$$H = H_0 \frac{\cosh(\Psi x)}{\cosh(\frac{1}{2}\Psi d)}, \tag{17.35}$$

$$A = \frac{H_0}{\Psi}\,\frac{\sinh(\Psi x)}{\cosh(\frac{1}{2}\Psi d)}. \tag{17.36}$$

Let us consider eq. (17.33) in the next approximation. Substituting $\Psi + \Psi_1$ into it, we have

$$\frac{\mathrm{d}^2\Psi_1}{\mathrm{d}x^2} = \varkappa^2\Psi(A^2 - 1 + \Psi^2).$$

For this inhomogeneous equation to have a solution, it is necessary that the right-hand side be orthogonal to the solution of a homogeneous equation with correct boundary conditions; such a solution is $\Psi_1 = \text{const}$. Hence,

$$\int_{-\frac{1}{2}d}^{\frac{1}{2}d} (A^2 - 1 + \Psi^2)\,\mathrm{d}x = 0, \tag{17.37}$$

whence we find the relation between Ψ and H_0:

$$H_0^2 = \frac{2\Psi^2(1-\Psi^2)\cosh^2(\frac{1}{2}\Psi d)}{(\sinh(\Psi d)/\Psi d) - 1}. \tag{17.38}$$

This relation is valid at any H_0 and gives an implicit expression for Ψ in terms of H_0. From this we can find the magnetization of the film (in units of $H_{cm}\sqrt{2}$):

$$4\pi M = B - H_0 = \frac{1}{d}\int_{-\frac{1}{2}d}^{\frac{1}{2}d} H\,dx - H_0$$

$$= -H_0\left[1 - \frac{\tanh(\frac{1}{2}(\Psi d))}{\frac{1}{2}\Psi d}\right]. \tag{17.39}$$

At $\Psi d \gg 1$, $4\pi M = -H_0$, which corresponds to the Meissner effect – the nonpenetration of the field into the superconductor. If $\Psi d \ll 1$, then from (17.39) and (17.38) we find

$$4\pi M \approx -\tfrac{1}{12}H_0\, d^2\Psi^2, \qquad H_0^2 \approx \frac{12(1-\Psi^2)}{d^2}, \tag{17.40}$$

from which we obtain

$$4\pi M \approx -\tfrac{1}{12}H_0 d^2\left(1 - \frac{H_0^2 d^2}{12}\right). \tag{17.41}$$

For further applications we will find the free-energy difference for the superconducting and normal states in a given external field H_0 and a given $\Psi = \text{const.}$ without using the equation for Ψ and relation (17.38) which follows from it. This means that we are dealing with a nonequilibrium state characterized by some constant Ψ. Adding $-\boldsymbol{H}_0\boldsymbol{B}/4\pi + H_0^2/8\pi$ to (17.19) and passing to the reduced quantities, we obtain

$$\int_{-\frac{1}{2}d}^{\frac{1}{2}d} (\Omega_{sH} - \Omega_{nH})\,dx = \frac{H_{cm}^2}{4\pi}\int_{-\frac{1}{2}d}^{\frac{1}{2}d} [-\Psi^2 + \tfrac{1}{2}\Psi^4 + (H - H_0)^2 + A^2\Psi^2]\,dx. \tag{17.42}$$

Note that the minimization of this expression with respect to Ψ^2 leads to the condition (17.38) and substitution of this condition into (17.42) gives the equilibrium formula (17.20).

Let us substitute the formulae for H and A, (17.35) and (17.36), into (17.42). This yields

$$\int_{-\frac{1}{2}d}^{\frac{1}{2}d} (\Omega_{sH} - \Omega_{nH})\frac{dx}{(H_{cm}^2/4\pi)d}$$

$$= H_0^2\left(1 - \frac{\tanh(\frac{1}{2}\Psi d)}{\frac{1}{2}\Psi d}\right) - \Psi^2 + \tfrac{1}{2}\Psi^4. \tag{17.43}$$

In a field equal to the critical field H_c this expression must be equal to zero. Hence,

$$H_c^2 = \frac{\Psi^2(2-\Psi^2)}{2\left(1 - \dfrac{\tanh\frac{1}{2}\Psi d}{\frac{1}{2}\Psi d}\right)}. \tag{17.44}$$

Formulae (17.38) and (17.39) give the dependence of H_c on the film thickness.

Consider the limiting cases. If the film thickness is large, then at the transition point Ψ is close to unity. We put $\Psi = 1-\varphi$, where $\varphi \ll 1$. Substituting this into (17.38) yields

$$\varphi \approx \frac{H_0^2}{2d},$$

that is, really $\varphi \ll 1$. Inserting $\Psi = 1-\varphi$ into (17.44), we find that

$$H_c^2 \approx \frac{(1-2\varphi)(1+2\varphi)}{2(1-2/d)} \approx \tfrac{1}{2}\left(1+\frac{2}{d}\right).$$

Hence,

$$H_c \approx \frac{1}{\sqrt{2}}\left(1+\frac{1}{d}\right) \tag{17.45}$$

or, in conventional units,

$$H_c \approx H_{cm}\left(1+\frac{\delta}{d}\right). \tag{17.46}$$

For a thin film we have already the expansion of formulae (17.38) in d - this is expression (17.40). From (17.44) at $d \ll 1$ we have

$$H_c^2 \approx \frac{6(2-\Psi^2)}{d^2}. \tag{17.47}$$

Comparing with (17.40), we obtain at the transition point:

$$\Psi = 0, \qquad H_c = \frac{2\sqrt{3}}{d} \tag{17.48}$$

or, in conventional units,

$$H_c = 2\sqrt{6}H_{cm}\frac{\delta}{d}. \tag{17.48'}$$

Thus, while for $d \gg 1$ the transition to the normal state with $\Psi = 0$ occurs from the state with $\Psi \approx 1$ which means a first-order transition for $d \ll 1$ upon approach to the transition point $\Psi \to 0$, i.e., this is a second-order transition. Here H_c increases with decreasing thickness according to the law $H_c \propto d^{-1}$. The complete plot of H_c/H_{cm} against δ/d is given in fig. 108.

Thus, at a certain film thickness the first-order phase transition changes to a second-order phase transition. In order to find out how this occurs, we return to the nonequilibrium formula (17.43). The graph of the function on the right-hand side (we denote it by Ω) versus Ψ^2 at various H_0 is shown in fig. 109, where (a) corresponds to $d/\delta = 4$ and (b) to $d/\delta = 1.6$. From these graphs it is seen that with small thickness Ω has one minimum, which corresponds to $\Psi \neq 0$ for low fields

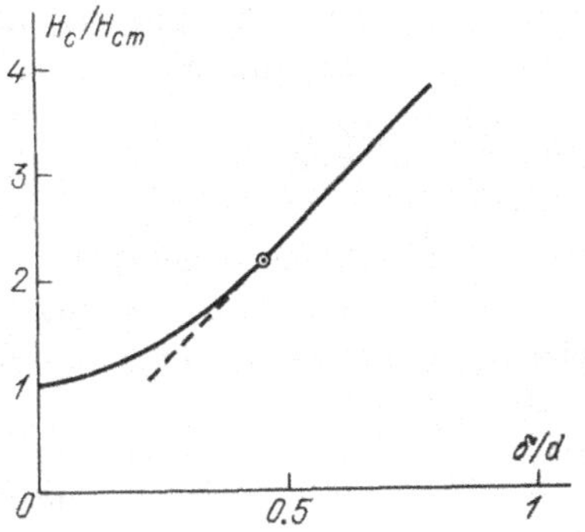

Fig. 108.

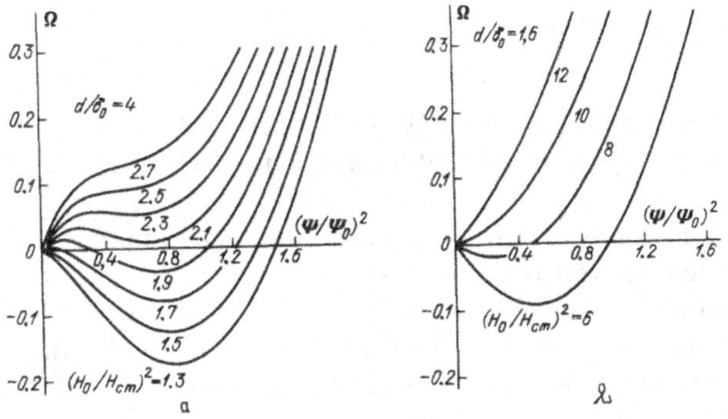

Fig. 109.

and to $\Psi = 0$ for high fields. This is the case of a second-order phase transition. For large film thicknesses in a general case we have two minima, one at $\Psi = 0$ and the other at finite Ψ. The field at which the depths of the minima become equal is just the transition field. This is obviously a first-order phase transition.

The boundary situation between (a) and (b) is the case where both minima merge at the point $\Psi = 0$. Evidently, at this point we have simultaneously $\Psi = 0$, $\mathrm{d}\Omega/\mathrm{d}(\Psi^2) = 0$ and $\mathrm{d}^2\Omega/(\mathrm{d}(\Psi^2))^2 = 0$. From these conditions we obtain

$$d_c = \sqrt{5} \tag{17.49}$$

or, in conventional units,

$$d_c = \sqrt{5}\,\delta \tag{17.49'}$$

(the circle in fig. 108).

In the case of a first-order phase transition, the lower of the two minima corresponds to the equilibrium state, and the higher-lying minimum corresponds to a metastable state. Indeed, a slight deviation of Ψ^2 from the value in the minimum brings about an increase in free energy. When the magnetic field is changed, the metastable phase persists until the high minimum disappears. The result is that a hysteresis may be observed in an investigation of the transition in increasing and decreasing fields. The occurrence of hysteresis is the simplest way for identifying a first-order transition.

The limiting fields of the retardation of metastable phases are called the supercooling field (retardation of the normal phase) and the superheating field (retardation of the superconducting phase). The superheating field is difficult to observe because at the edges of the specimen there always occurs a concentration of the field lines of force, i.e., an increase in the field strength. Hence, the edges may serve as nucleation sites for the normal phase. It is much easier to observe the supercooling field. The disappearance of the minimum at $\Psi = 0$ occurs at $\mathrm{d}\Omega/\mathrm{d}(\Psi^2) = 0$. As has been pointed out, this condition gives formula (17.38). Substituting $\Psi = 0$ into it leads to

$$H_0 = H_{c2} = \frac{2\sqrt{3}}{d}, \tag{17.50}$$

which coincides with (17.48). Hence, the critical field formula for a thickness $d < d_c$ also determines the supercooling field for a thickness $d > d_c$ (the dashed line in fig. 108).

In fact, this conclusion is valid only for not-too-large film thicknesses. The point is that it has been derived from the formula for the nonequilibrium free energy (17.43) obtained under the assumption $\Psi = \text{const}$. At the same time it is clear that in a bulk specimen the minimal nucleus of the superconducting phase will be localized in a certain region of space, i.e., we shall have $\Psi(\boldsymbol{r}) \neq \text{const}$. (this will be shown in sections 18.2 and 18.4). The assumption of $\Psi = \text{const}$. is evidently valid at $d \ll \xi$ or, in our units, $d \ll \varkappa^{-1}$, i.e., at $\varkappa \ll 1$ up to thicknesses much larger than d_c. At thicknesses $d \gg \xi$ the supercooling field coincides with the field H_{c3} (section 18.4.).

The experimental determination of magnetic properties of thin superconducting plates is carried out on films prepared by evaporation of drops of a molten metal on a dielectric substrate. Of course, in such a case it is rather difficult to control the film thickness. In view of this, use may be made of the fact that all the theoretical formulas contain the ratio d/δ, where δ is expressed by formula (18.93), i.e., it increases when the temperature approaches T_c. This makes it possible to use a single film instead of many films, and to carry out measurements at various temperatures. The reduced length that appears in the formulas will then be expressed, according to (16.93), by the following formula:

$$d_{\text{red}} = \frac{\mathrm{d}}{\delta_L(0)} \left[\frac{2(T_c - T)}{T_c}\right]^{1/2}.$$

The results of measurements on films deposited by sputtering onto the substrate at room temperature (Zavaritskii 1951) have confirmed the Ginzburg–Landau theory.

17.4. *The critical current of a thin wire with $\varkappa \ll 1$*

We will now consider another problem of the Ginzburg–Landau theory - the critical current of a thin cylindrical wire of radius R *. According to the Maxwell equation, we have (see eq. 15.23):

$$H_J = \frac{2J}{cR}. \tag{17.51}$$

Hence, the specification of the total current is equivalent to specifying the field at the interface. We introduce cylindrical coordinates: ρ, φ, and z. Since the current flows along z, the field $\boldsymbol{H}$ has only the component H_φ. We choose the vector potential $\boldsymbol{A}$ along the z axis. Then, from the general equations (17.6′), (17.7′) and (17.8′) we obtain

$$\frac{\mathrm{d}^2\Psi}{\mathrm{d}\rho^2} = \varkappa^2(A^2 - 1 + \Psi^2)\Psi, \tag{17.52}$$

$$\left.\frac{\mathrm{d}\Psi}{\mathrm{d}\rho}\right|_{\rho=R} = 0, \tag{17.53}$$

$$\frac{\mathrm{d}^2 A}{\mathrm{d}\rho^2} + \frac{1}{\rho}\frac{\mathrm{d}A}{\mathrm{d}\rho} - \Psi^2 A = 0, \tag{17.54}$$

$$H = -\frac{\mathrm{d}A}{\mathrm{d}\rho}. \tag{17.55}$$

From eqs. (17.52) and (17.53) it follows that in a first approximation with respect to $\varkappa \ll 1$ we have $\Psi = \text{const}$. Substituting this into eq. (17.54), we find that A is proportional to $I_0(\Psi\rho)$, where I_0 is the Bessel function of imaginary argument. According to (17.55), $H(\rho)$ is proportional to $I_1(\Psi\rho)$; due to the boundary condition we have

$$H = \frac{H_J I_1(\Psi\rho)}{I_1(\Psi R)}, \tag{17.56}$$

$$A = -\frac{H_J}{\Psi}\frac{I_0(\Psi\rho)}{I_1(\Psi R)}. \tag{17.57}$$

The solvability condition of the equation for the correction to Ψ is analogous to (17.37):

$$\int_0^R (A^2 - 1 + \Psi^2)\rho \,\mathrm{d}\rho = 0.$$

* In their original work, Ginzburg and Landau (1950) considered a thin film and assumed that the current density varies only in the transverse direction and that along the film width the current is uniform. This does not correspond to reality because the current is redistributed over the film and becomes nonuniform in two directions. It is for this reason that we consider a cylinder with a current.

From this we get

$$H_J^2 = \frac{\Psi^2(1-\Psi^2)I_1(\Psi R)}{I_0^2(\Psi R) - I_1^2(\Psi R)}. \tag{17.58}$$

We begin with the case $R \ll 1$. Assuming that $\Psi \leqslant 1$, we expand the right-hand side of (17.58) in R:

$$H_J^2 \approx \tfrac{1}{4}\Psi^4(1-\Psi^2)R^2. \tag{17.59}$$

Evidently, the value of H_J is small at $R \ll 1$, i.e., superconductivity cannot be destroyed due to H_J becoming equal to H_{cm} (in conventional units), as in the case of a bulk specimen according to Silsbee's rule (section 15.4). The mechanism of destruction of superconductivity is different here. If we consider H_J^2 (17.59) as a function of Ψ^2, then we shall see that it vanishes at $\Psi = 0$ and $\Psi = 1$, and has a maximum in between. Hence, there is a certain maximum value of H_J which is compatible with superconductivity; at larger values of H_J superconductivity cannot exist. What happens at such large currents will be discussed in section 22.9. Here we will find the maximum of H_J.

Differentiating with respect to Ψ^2, we obtain for the maximum point:

$$\Psi^2 = \tfrac{2}{3}, \qquad H_{J,\max} = \frac{R}{3^{3/2}}. \tag{17.60}$$

Passing to conventional units and using formula (17.51), we obtain the value of the critical current:

$$J_c = \frac{cR^2}{3\sqrt{6}} \frac{H_{\text{cm}}}{\delta}. \tag{17.61}$$

Since $J_c \propto R^2$, it is reasonable to introduce the critical current density:

$$j_c = \frac{J_c}{\pi R^2} = \frac{1}{3\pi\sqrt{6}} \frac{cH_{\text{cm}}}{\delta}. \tag{17.62}$$

The quantity j_c is independent of size and is a characteristic of the superconducting metal.

The physical significance of this quantity can be understood if we substitute the BCS expressions for δ and H_{cm}. Then we obtain

$$j_c = \frac{1}{3\sqrt{2}} \frac{n_s e\Delta(T)}{p_0} \tag{17.63}$$

[here we have introduced $\Delta(T)$ in accordance with eq. (17.33) and n_s according to eq. (16.92)]. From this formula it follows that j_c corresponds to the motion of electrons with a velocity of the order of $\Delta(T)/p_0$. Thus, we conclude that Cooper pairing and, hence, superconductivity exists until the pairs begin to move at a velocity larger than Δ/p_0. This occurs not only near T_c but also at any temperature and is consistent with the Landau criterion (16.1).

It should be noted that the critical current density (17.62) is very large. If we substitute typical values for pure superconductors, $H_{cm} \sim 10^2$–10^3 Oe, $\delta \sim 10^{-5}$–10^{-6} cm, we obtain:

$$j_c \sim 10^7\text{–}10^9 \text{ A/cm}^2. \tag{17.64}$$

For large thicknesses, $R \gg 1$, we obtain from formula (17.58):

$$H_J^2 \approx \Psi^3(1-\Psi^2)R.$$

At the maximum value of H_J^2 we have $\Psi^2 = \frac{3}{5}$ and

$$H_{J,\max} \approx 3^{3/4} 2^{1/2} 5^{-5/4} R^{1/2}.$$

Passing to conventional units and expressing J_c, we find

$$J_c \approx 3^{3/4} 5^{-5/4} c H_{cm} R^{3/2} \delta^{-1/2}. \tag{17.65}$$

The complete curve of H_J^2 versus R in reduced units is shown in fig. 110 (curve 1).

The above-described mechanism of superconductivity destruction by the current occurs only as long as the radius of the cylinder is not too large. According to formula (17.59), the magnetic field of the current is of order H_{cm} at $R \sim \delta$. Here the destruction of superconductivity by the magnetic field is possible according to Silsbee's rule. If the sample were massive, an intermediate state would arise (see section 15.4). However, in the intermediate state the sample must consist of layers of the normal and superconducting phases, i.e., the order parameter must vary in space. This is possible only at distances of order $\xi \sim \delta/\varkappa \gg \delta$. Thus, there exists a whole region of thicknesses from δ to ξ (or in reduced units from 1 to $\varkappa^{-1}$) at which neither the superconducting state nor the normal state (the field in the middle of the film is weaker than H_{cm}) nor the intermediate state can be realized at first sight.

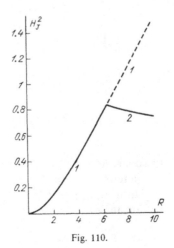

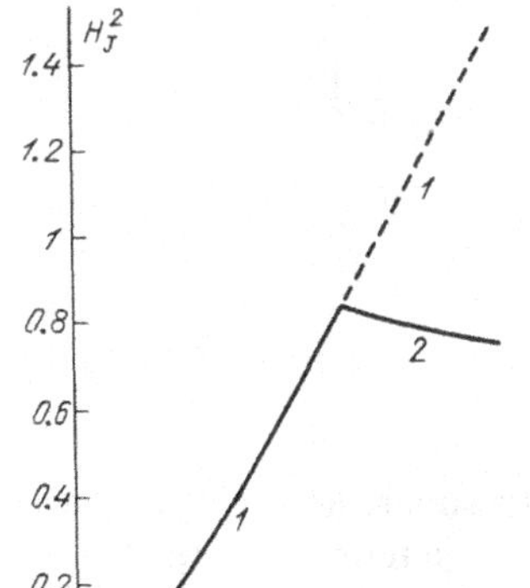

Fig. 110.

This means, in fact, that at a certain thickness a discontinuous transition occurs from the superconducting to the normal state. Strictly speaking, such a transition cannot be treated as a phase equilibrium because in the normal state energy is dissipated in the presence of a current, and the state is not an equilibrium state. However, because of the high conductivity of the metal the departure from equilibrium is very small. For the same reason the electric field is always small compared to the magnetic field. This enables one to consider the transition of the wire with a current as a first-order phase transition and to take into account the magnetic field alone.

In this particular case, the condition for the transition is the equality of the energies in a given field since the specification of the total current is equivalent to the specification of the field at the interface. Since the field is inhomogeneous in the normal phase too, we cannot use formula (17.20) and we have to find Ω_{sH} and Ω_{nH} with an appropriate distribution of H. From eq. (17.19) we have

$$\int \Omega_{sH}\,\mathrm{d}V = \int \Omega_{n}^{(0)}\,\mathrm{d}V + \tfrac{1}{2}H_{cm}^{2}\int_0^R (H^2 - \tfrac{1}{2}\Psi^4 - 2HH_J)\rho\,\mathrm{d}\rho, \tag{17.66}$$

$$\int \Omega_{nH}\,\mathrm{d}V = \int \Omega_{n}^{(0)}\,\mathrm{d}V + \tfrac{1}{2}H_{cm}^{2}\int_0^R (H^2 - 2HH_J)\rho\,\mathrm{d}\rho. \tag{17.67}$$

For a superconductor formula (17.56) must be substituted as $H(\rho)$. For a normal metal we have $j = \text{const.} = J/(\pi R^2)$;

$$H = \frac{2J(\rho)}{c\rho} = \frac{2j\pi\rho^2}{c\rho} = \frac{2J\rho}{cR^2} = \frac{H_J\rho}{R} \tag{17.68}$$

(the last expression is also valid in reduced units).

Let us consider in more detail the case $R \gg 1$. Using the asymptotics of the functions I_ν at $z \gg 1$,

$$I_\nu(z) \approx \frac{1}{(2\pi z)^{1/2}}\,\mathrm{e}^{z}\left(1 - \frac{1}{2z}(\nu^2 - \tfrac{1}{4})\right), \tag{17.69}$$

we find from eqs. (17.58) and (17.56)

$$\Psi \approx 1 - O(R^{-1}), \tag{17.70}$$

$$H \approx H_J\left(\frac{R}{\rho}\right)^{1/2}\mathrm{e}^{\rho - R}. \tag{17.71}$$

From this it follows that in the superconducting phase H is different from zero only in a layer of order δ, i.e., to within terms of order R^{-1} we may replace the integrand in (17.66) by $-\frac{1}{2}\Psi^4 \approx -\frac{1}{2}$. Calculating Ω_{nH} with H (17.68) and using the condition $\Omega_{nH} = \Omega_{sH}$, we obtain $H_{Jc} = \sqrt{\frac{3}{5}}$ or, in conventional units,

$$H_{Jc} = \sqrt{\tfrac{6}{5}}\mathrm{H}_{cm}. \tag{17.72}$$

As should be expected, H_{Jc} is somewhat larger than H_{cm} since the field in the normal state varies from 0 to H_{Jc} at the surface. However, this difference is only 10%. Since H_J is proportional to J, then from eq. (17.68) we obtain

$$J_c = J_{\text{Silsbee}}\sqrt{\tfrac{6}{5}}. \tag{17.73}$$

In the case $R \sim 1$, it is impossible to express H_{Jc} in analytical form. The corresponding curve (2) is given in fig. 110. The intersection of curves 1 and 2 occurs at a rather large value of R ($R = 6.24$). In this region J_c is well described by the asymptotics (17.65) on the left of the point of intersection. Note that continuation of curve 2 beyond the point of intersection to the left is meaningless since $H_{J,\max}$ corresponds to the maximum field compatible with superconductivity.

Just as in any first-order phase transition, hysteresis is likely to occur. As the current increases the transition may be retarded up to the stability limit of the metastable superconducting state corresponding to the maximum of the field H_J (17.58) (curve 1 in fig. 110). When the current decreases, the transition may probably be retarded until $J = 0$.

Note that all the reasonings given in the present section are valid as long as $R \ll \xi$ ($R \ll \varkappa^{-1}$ in reduced units). At larger thicknesses an intermediate state is formed.

17.5. *Quantization of the magnetic flux*

Let us now consider a hollow cylinder or, to put it differently, a tube with wall thickness larger than δ. Suppose this cylinder is placed in an external longitudinal magnetic field weaker than the critical field ($H < H_{c1}$ in the case of a type II superconductor, ch. 18). Imagine that the external field varies with time. We have not considered the extension of the Ginzburg–Landau equations to nonstationary problems, but for our purposes it will suffice to use the London equation (17.36). We integrate this equation along a closed path, lying entirely within the superconductor and surrounding the cavity in the cylinder. Here we obtain

$$\frac{\partial}{\partial t}\oint \Lambda \boldsymbol{j}\,\mathrm{d}\boldsymbol{l} = \oint \boldsymbol{E}\,\mathrm{d}\boldsymbol{l} = \int \mathrm{d}\boldsymbol{S}\,\mathrm{rot}\,\boldsymbol{E} = -\frac{1}{c}\int \mathrm{d}\boldsymbol{S}\frac{\partial \boldsymbol{H}}{\partial t}$$

(we have used the Maxwell equation here). Hence,

$$\oint \Lambda \boldsymbol{j}\,\mathrm{d}\boldsymbol{l} + \frac{1}{c}\int \boldsymbol{H}\,\mathrm{d}\boldsymbol{S} = \text{const.}$$

Since the current can flow only in a surface layer, then along the closed path lying in the bulk of the superconductor $\boldsymbol{j} = 0$. Consequently,

$$\Phi = \int \boldsymbol{H}\,\mathrm{d}\boldsymbol{S} = \text{const.} \tag{17.74}$$

Thus, we have come to the conclusion that the magnetic flux through a loop lying within the superconductor cannot vary with time.

Note that this conclusion is valid not only for the problem of a hollow cylinder under consideration but also for an inhomogeneous superconductor, whose different portions have different critical parameters. In such a specimen multiply connected superconducting regions may form enclosing normal regions, as a result of which a phenomenon known as "trapped magnetic flux" arises with decreasing magnetic field.

Let us return, however, to the problem of the tube. We will now show that the magnitude of the magnetic flux trapped by the superconducting tube cannot be arbitrary. For this purpose we consider the expression for the superconducting current (17.9). We write the function Ψ in the form $\Psi = |\Psi|\,e^{i\chi}$, i.e., we introduce the modulus and the phase. The current is then written in the form

$$j = \frac{\hbar e}{m}|\Psi|^2\left(\nabla\chi - \frac{2e}{\hbar c}\boldsymbol{A}\right). \tag{17.75}$$

We divide $\boldsymbol{j}$ by $|\Psi|^2$ and integrate along the same loop as before, i.e., along a closed path, which encloses the hole and passes in the bulk of the tube wall. Here we obtain

$$\oint \frac{\boldsymbol{j}}{|\Psi|^2}\,\mathrm{d}\boldsymbol{l} = \frac{\hbar e}{m}\left(\oint \nabla\chi\,\mathrm{d}\boldsymbol{l} - \frac{2e}{\hbar c}\oint \boldsymbol{A}\,\mathrm{d}\boldsymbol{l}\right). \tag{17.75'}$$

If the loop passes inside the wall, then $\boldsymbol{j} = 0$ on it. The second integral on the right-hand side is equal to

$$\oint \boldsymbol{A}\,\mathrm{d}\boldsymbol{l} = \int \operatorname{rot}\boldsymbol{A}\,\mathrm{d}\boldsymbol{S} = \int \boldsymbol{H}\,\mathrm{d}\boldsymbol{S} = \Phi.$$

As for the first integral, at first glance it must be equal to zero, just as any integral of the gradient of a function along a closed loop. However, in fact, this is not necessarily the case. For the function $\Psi = |\Psi|\,e^{i\chi}$ to be single-valued it is not required that the phase of χ return to the same value when we make a complete turn around the loop; it is sufficient only that it varies by $2\pi n$, where n is an integer. Thus, in a general case $\oint \nabla\chi\,\mathrm{d}\boldsymbol{l} = 2\pi n$. We obtain

$$\Phi = n\Phi_0, \tag{17.76}$$

$$\Phi_0 = \pi\frac{\hbar c}{e} = 2.07\times10^{-7}\ \mathrm{Oe\cdot cm^2}. \tag{17.77}$$

Thus, the magnetic field passing through the loop may take only a discrete series of values. This phenomenon is called magnetic flux quantization, and the quantity Φ_0 is known as the flux quantum.

The phenomenon of flux quantization in a superconductor was first predicted by F. London (1950). However, being unaware of Cooper pairing, London assumed the charge of the carriers to be equal to e instead of $2e$, and obtained a flux quantum equal to $2\Phi_0$. Recall that a flux quantum also appears in the theory for a normal metal. In a metal placed in a magnetic field, the electrons move along spiral trajectories (in the case of a closed Fermi surface); these trajectories enclose a magnetic flux equal to $n \cdot 2\Phi_0$ (section 10.4). The magnetic flux quantum Φ_0 determines the period of interference oscillations of the resistance of a hollow normal cylinder (section 11.4).

In the case under consideration, the quantized magnetic flux Φ is the sum of the flux passing through the hole and the flux in the surface layer of the superconductor, of thickness δ, which adjoins the cavity. If the radius of the tube is much larger than δ, then the latter may be neglected.

For the experimental determination of the magnitude of the flux through the cavity it is not sufficient to simply vary the external field. According to the condition (17.74) the internal flux will not change in such a case. Therefore, one has to proceed as follows (Doll and Näbauer 1961, Deaver and Fairbank 1961). The superconductor is cooled in a given external field from a temperature above T_c to a certain temperature below T_c. Upon transition to the superconducting state a certain magnetic flux is frozen in it. The magnitude of this flux can be determined from the minimum condition for the free energy difference (17.20):

$$\int (\Omega_{sH} - \Omega_{nH})\,\mathrm{d}V$$

$$= \frac{H_{\mathrm{cm}}^2}{4\pi}\left\{(H_{\mathrm{int}} - H_0)^2 \pi R_1^2 + \int_{R_1}^{R_2} [(H - H_0)^2 - \tfrac{1}{2}|\Psi|^4] 2\pi\rho\,\mathrm{d}\rho\right\}, \tag{17.78}$$

where R_1 and R_2 are the inner and outer radii of the tube; H_{int} is the field inside the tube. The first term in brackets refers to the cavity, and the second to the superconducting tube. We are to find H_{int} satisfying the quantization condition (17.76) and corresponding to the minimum of expression (17.78). Since in the second term the dependence on H_{int} extends only to the layer of thickness δ, we may neglect it.

The minimization of the first term leads to the following result: a magnetic flux through the core equal to $\Phi_{\mathrm{int}} = n\Phi_0$, is realized in that interval of the field H_0 in which $\Phi_{\mathrm{ext}} = H_0 \pi R_1^2$ lies within the interval $(n - \frac{1}{2})\Phi_0 < \Phi_{\mathrm{ext}} < (n + \frac{1}{2})\Phi_0$. When the boundary of this interval is crossed, the transition to the next n occurs. This situation is shown in fig. 111.

According to the derivation given above, the quantization of magnetic flux is a consequence of the variation of the phase of the order parameter Ψ by $2\pi n$ upon making a complete turn around the closed loop. This behavior of the phase has been demonstrated in a very ingenious experiment (Little and Parks 1962, 1964). Let us take a thin cylindrical superconducting film whose thickness is much less than the penetration depth and apply the magnetic field along the axis of the cylinder. Such a thin film does not practically screen out the magnetic field (its magnetization in the field is equal to zero according to formula 17.40) and, hence, the field inside and outside the film is the same. The radius of the cylinder is assumed to be much

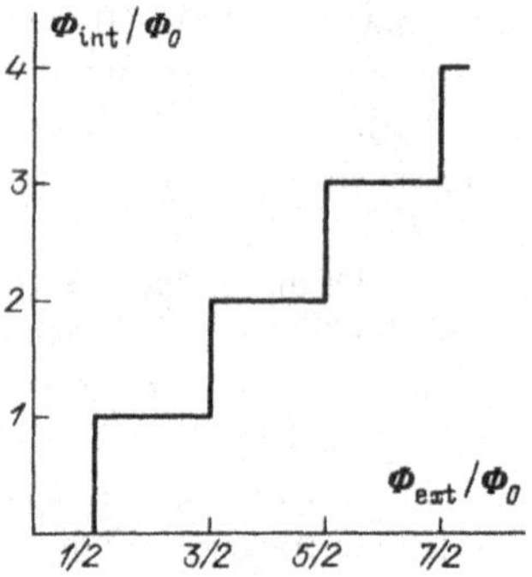

Fig. 111.

larger than the penetration depth. In this case all the quantities, including the vector potential $\boldsymbol{A}$, do not practically vary along the film thickness.

Let us choose $\boldsymbol{A}$ such that it has only the component A_φ in cylindrical coordinates. We substitute $\Psi = |\Psi|\,e^{i\chi}$ into the free energy (17.5). The integrand has the form

$$\Omega_s - \Omega_n^{(0)} = \alpha\tau|\Psi|^2 + \tfrac{1}{2}b|\Psi|^4 + \frac{\hbar^2}{4m}\left(\nabla\chi - \frac{2e}{c\hbar}\boldsymbol{A}\right)^2|\Psi|^2 + \frac{H^2}{8\pi} \tag{17.79}$$

(in accordance with the result of section 17.4 it may be assumed that $|\Psi| = \text{const.}$). The combination $\boldsymbol{A} - (c\hbar/2e)\nabla\chi$ is gauge invariant since any gradient added to $\boldsymbol{A}$ may be compensated by the variation of the phase.

Using the fact that $\nabla\chi$ and $\boldsymbol{A}$ are constant along the superconducting film, we write the combination that appears in parentheses in eq. (17.79) in the form of the average value:

$$\nabla\chi - \frac{2e}{c\hbar}\boldsymbol{A} = \frac{1}{2\pi R}\oint\left(\nabla\chi - \frac{2e}{c\hbar}\boldsymbol{A}\right)\mathrm{d}\boldsymbol{l}$$

$$= \frac{1}{2\pi R}\left(2\pi n - \frac{2e}{c\hbar}\Phi\right) = \frac{1}{R}\left(n - \frac{\Phi}{\Phi_0}\right),$$

where the integral is taken over the contour of the cylinder cross section perpendicular to the axis; $\Phi = H\pi R^2$ is the magnetic flux through the cylinder; Φ_0 is the flux quantum. Substituting into eq. (17.79) gives

$$\Omega_s - \Omega_n^{(0)} = \alpha\tau|\Psi|^2 + \frac{\hbar^2}{2mR^2}\left(n - \frac{\Phi}{\Phi_0}\right)^2|\Psi|^2$$

$$+ \tfrac{1}{2}b|\Psi|^4 + \frac{H^2}{8\pi}. \tag{17.80}$$

The number n has to be chosen so as to correspond to the minimum of the free energy, i.e., a given n is realized at

$$n-\tfrac{1}{2}<\frac{\Phi}{\Phi_0}<n+\tfrac{1}{2}. \tag{17.81}$$

In the formula (17.80) the second term can be added to the first term with τ being replaced by τ':

$$\tau'=\tau+\frac{\hbar^2}{4mR^2\alpha}\left(n-\frac{\Phi}{\Phi_0}\right)^2.$$

From the condition $\tau'=0$ we obtain the variation of the critical temperature:

$$\frac{\Delta T_c}{T_c}=-\frac{\hbar^2}{4mR^2\alpha}\left(\frac{\Phi}{\Phi_0}-n\right)^2. \tag{17.82}$$

Substituting the BCS expression for α from formulae (17.14) and (17.18), we find

$$-\frac{\Delta T_c}{T_c}=\begin{cases} C_1\dfrac{\xi_0^2}{R^2}\left(\dfrac{\Phi}{\Phi_0}-n\right)^2, & \xi_0\ll l, \\ C_2\dfrac{\xi_0 l}{R^2}\left(\dfrac{\Phi}{\Phi_0}-n\right)^2, & \xi_0\gg l, \end{cases} \tag{17.83}$$

where

$$C_1=\frac{7\pi^2\zeta(3)}{48\gamma^2}=0.55, \qquad C_2=\frac{\pi^3}{24\gamma}=0.72, \qquad l=v\tau_{tr}.$$

From (17.83) it follows that the dependence of $-\Delta T_c/T_c$ on Φ/Φ_0 consists of periodically recurring parabolic pieces, each of which corresponds to the interval (17.81) with the corresponding n (fig. 112). This prediction is consistent with experiment (Little and Parks 1962, 1964). Note that in the experiment carried out by Little and Parks the superconductor cannot trap the magnetic flux because of its small thickness, which is why no flux quantization occurs; however, the nonuniqueness of the phase is observed and it is the phase that is measured in the experiment.

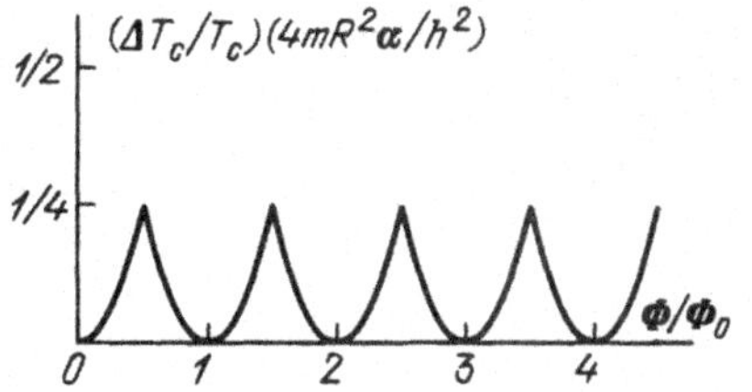

Fig. 112.

18 Type II superconductivity

18.1. Magnetic properties of type II superconductors. The qualitative picture

In this chapter we shall be concerned with the properties of type II superconductors, for which, according to eq. (16.96), $\varkappa > 1/\sqrt{2}$ and $\sigma_{ns} < 0$. First of all, we note that with $\sigma_{ns} < 0$ a first-order phase transition to the normal state is impossible, even if we are dealing with a superconducting cylinder in a longitudinal field. Indeed, suppose that the superconductor has been divided up into layers of the normal and superconducting phases parallel to the magnetic field. In section 17.4 we have defined the critical field of a thin layer, which was found to be proportional to $H_{cm}\delta/d$; this may considerably exceed H_{cm}. Hence, it is energetically favorable for such layers to retain superconductivity in fields stronger than H_{cm}. In ordinary type I superconductors the splitting up into layers does not occur because of the surface energy σ_{ns} at the layer boundaries. But if $\sigma_{ns} < 0$, then nothing will hinder the splitting up of the superconductor into layers of the n- and s-phases.

Thus, the phase transition of a type II superconductor will proceed by way of a gradual expulsion of the superconducting phase, and the metal will become normal at a field in which even an infinitesimal superconducting region can no longer exist. Thus, it may be concluded that the transition will extend over an interval of magnetic fields, and in this interval the external magnetic field will partially penetrate the superconductor, i.e., the Meissner effect will be incomplete. The superconductor will be in a special state, called the mixed state, which adjoins the superconducting phase at a certain field H_{c1} (the lower critical field) and the normal phase at another field H_{c2} (the upper critical field) (Abrikosov 1957).

The fields H_{c1} and H_{c2} can be determined in order of magnitude from qualitative considerations. We begin with H_{c2}, the field at which the superconductivity disappears altogether. The breakup of Cooper pairs occurs as a result of curving of their trajectory in the magnetic field. Evidently, a pair can be preserved as long as the Larmor radius is larger than its size, i.e.,

$$r_L \sim \frac{cp_\perp}{eH} > \xi.$$

In this formula $p_\perp$ is the perpendicular component of the momentum corresponding to the motion of the pair as a whole. Therefore, $p_\perp \leq p \sim mv_s$, where v_s is the velocity of the pair. From the Landau criterion (see also formula 17.63) it follows that $v_s < \Delta/p_0$, otherwise the pair will break up. The same condition may be written in the form $p_\perp < \hbar/\xi$. Thus, we have

$$\xi < \frac{cp_\perp}{eH} < \frac{ch}{\xi eH}$$

or

$$H < \frac{c\hbar}{e\xi^2} \sim H_{c2}. \tag{18.1}$$

The field H_{c2} may be expressed in terms of H_{cm}:

$$H_{c2} \sim \frac{c\hbar}{e\xi}\frac{\Delta}{hv} \sim \frac{c}{e\xi v}\frac{H_{cm}\hbar^{3/2}}{(mp_0)^{1/2}}$$

$$\sim H_{cm}\frac{\delta_L}{\xi} \sim H_{cm}\varkappa. \tag{18.2}$$

We have so far considered a pure superconductor. But if it contains impurities and $l \ll \xi$, then in formula (18.1) ξ must be replaced by $\xi' \sim (\xi l)^{1/2}$. In this case we obtain

$$H_{c2} \sim \frac{c\hbar}{e\xi l} \sim \frac{c\hbar}{el}\frac{\Delta}{\hbar v} \sim H_{cm}\frac{\delta_L}{l} \sim H_{cm}\varkappa. \tag{18.3}$$

Thus, in both cases the field H_{c2} is of order $H_{cm}\varkappa$. However, the second case is more important. It follows from eq. (18.3) that by increasing the impurity concentration and, hence, by decreasing the mean free path (i.e. $l \propto n_i^{-1}$), one can strongly increase H_{c2}. Up to H_{c2} the metal retains superconducting regions, i.e., a zero-resistance supercurrent can flow in it. The mean free path can, in principle, be decreased down to interatomic distances. Considering that the usual values of $H_{cm} \sim 10^2$–10^3 Oe and that $\delta_L \sim 10^{-5}$–10^{-6} cm, we obtain a limiting estimate for the fields H_{c2} (Gor'kov 1963):

$$H_{c2} < 10^3 \frac{10^{-5}}{10^{-8}}\,\text{Oe} \sim 10^6\,\text{Oe}. \tag{18.4}$$

These fields are so high that a question arises: Would not the pairs be broken up earlier due to orientation of the spins of both electrons parallel to the magnetic field? (the Clogston paramagnetic limit; Clogston 1962). The order of magnitude of this limit is evidently determined by the condition

$$\beta H_{cp} \approx 2\Delta, \tag{18.5}$$

where β is the Bohr magneton. Considering that the highest T_c value is of the order of 20 K, we obtain

$$H_{cp} < 10^6\,\text{Oe} \tag{18.6}$$

hence, the upper limits for H_{cp} and H_{c2} coincide and the value 10^6 Oe may be regarded as limiting.

What happens when the field is decreased below H_{c2}? If the field penetrates the superconductor partially, then persistent eddy currents must exist in it. Near the limiting field H_{c1} the metal must be almost completely superconducting, i.e., regions with eddy currents must be rare. The field H_{c1} can be determined if it is taken into account that an elementary vortex is a quantum entity. This was first detected in the study of the superfluidity of liquid helium (Onsager 1949, Feynman 1955).

Suppose that Cooper pairs execute a vortex motion around a certain axis. Let us draw a circular loop around the axis. The velocity of the pairs along this loop is evidently constant. According to the Bohr quantization rule,

$$\oint p\, dq = 2m \oint \boldsymbol{v}_s\, d\boldsymbol{l} = 2mv_s \cdot 2\pi\rho = 2\pi n\hbar.$$

Hence,

$$v_s = n\hbar/(2m\rho),$$

that is, the velocity of the pairs falls off proportionately to the distance from the axis.

This inference becomes invalid when $\rho \lesssim \xi$, the size of the pairs. On the other hand, we ignored the fact that the pairs are charged and their motion implies the generation of electric current and a magnetic field. The latter must vanish at a distance of the order of the penetration depth δ. Hence, the formula derived is valid only at $\rho < \delta$. Below we shall assume that $\delta \gg \xi$, i.e., $\varkappa \gg 1$.

As regards the number n, it may, in principle, be arbitrary. However, it is rather clear that we must take $n = 1$. First of all, it is evident that a small vortex is created more readily than a large one and therefore the lower critical field H_{c1} corresponds to the creation of the smallest vortex. With higher fields the problem is solved from the condition of the minimum free energy. From this condition one has to find the positions of the vortex axes. Suppose, for example, that $n = 2$. This corresponds to two fused vortices with $n = 1$. In order to find the position of the vortex axis we have to find the minimum free energy varying the coordinates of the axes, i.e., the points x and y of its intersection with the plane perpendicular to the magnetic field (the z-axis). But if the vortices did not stick together, we would be able to minimize Ω with respect to x_1y_1 and x_2y_2, i.e., a larger number of variables. In a general case this leads to a lower minimum of Ω. Thus, we will assume $n = 1$ and, hence*,

$$v_s = \frac{\hbar}{2m\rho}, \quad \xi \ll \rho \ll \delta. \tag{18.7}$$

* Quantum vortices of supercurrent which are attenuated at a distance δ and which carry one magnetic flux quantum Φ_0 (see sections 18.2 and 18.3) are usually called fluxoids or Abrikosov vortices.

The energy of such a vortex is basically the kinetic energy of Cooper pairs. The energy per unit length of the vortex is given by (the density of the pairs is $\frac{1}{2}n_s$):

$$\begin{aligned}\xi_0 &= \tfrac{1}{2}n_s \int_\xi^\delta \frac{2mv_s^2}{2} 2\pi\rho\, d\rho \\ &= \frac{\pi n_s h^2}{4m}\int_\xi^\delta \frac{d\rho}{\rho} = \frac{\pi n_s \hbar^2}{4m}\ln\left(\frac{\delta}{\xi}\right) \\ &= \frac{\pi n_s \hbar^2}{4m}\ln \varkappa\end{aligned}$$

(here we have substituted formula 18.7). The magnetic moment of the vortex per unit length is equal to

$$\begin{aligned}M &= \tfrac{1}{2}n_s \frac{2e}{c}\int_\xi^\delta \rho v_s 2\pi\rho\, d\rho = \pi\frac{e\hbar n_s}{mc}\int_\xi^\delta \rho\, d\rho \\ &\approx \frac{\pi e\hbar n_s}{2mc}\delta^2\end{aligned}$$

(we have neglected ξ^2 as compared to δ^2).

In the presence of a magnetic field H the vortex acquires a magnetic energy $-MH$. The smallest field at which an elementary vortex can appear corresponds to $\varepsilon_0 - MH = 0$ or

$$H_{c1} = \frac{\varepsilon_0}{M} \sim \frac{c\hbar}{e\delta^2}\ln \varkappa.$$

Using the BCS formulae, we obtain

$$H_{c1} \sim \frac{H_{cm}}{\varkappa}\ln \varkappa. \tag{18.8}$$

Just as in the case of H_{c2}, this formula also refers both to a pure superconductor and to the case $l \ll \xi_0$. From formula (18.8) it follows that the higher H_{c2} (i.e., the greater $\varkappa$) the lower ΔH_{c1}, i.e., the earlier the magnetic field begins to penetrate the superconductor.

In the following two sections we shall present the quantitative theory of the magnetic properties of type II superconductors based on the Ginzburg–Landau equations (Abrikosov 1957). Although this theory is valid, strictly speaking, only near T_c, yet, as has already been noted, all the basic conclusions are valid at any temperatures.

18.2. Magnetic properties of type II superconductors. Quantitative theory for the vicinity of H_{c2}

We begin with H_{c2}. Since this is the limiting field for the existence of superconductivity, it follows that an infinitesimal nucleus of the superconducting phase must decrease with time at $H > H_{c2}$ and increase with time at $H < H_{c2}$. This means that at $H = H_{c2}$ there may exist a stationary infinitesimal superconducting nucleus.

Let us consider the homogeneous Ginzburg–Landau equations (17.24), (17.25) and (17.26)*. Assuming Ψ to be infinitesimal, we retain only terms of lower order in Ψ. In this case, we get

$$\frac{1}{\varkappa^2}\frac{d^2\Psi}{dx^2} + \Psi(1 - A^2) = 0, \tag{18.9}$$

$$\frac{d^2 A}{dx^2} = 0. \tag{18.10}$$

From eq. (18.10) we obtain $A = H_0 x$ (we assume that $\boldsymbol{H} \| z$, $\boldsymbol{A} \| y$); the origin $x = 0$ may be chosen at an arbitrary point. Substituting into (18.9), we have

$$-\frac{d^2\Psi}{dx^2} + \varkappa^2 H_0^2 x^2 \Psi = \varkappa^2 \Psi. \tag{18.11}$$

We must find the solution that vanishes as $x \to \pm\infty$.

We have deliberately placed the terms of eq. (18.22) in such a manner that it becomes similar to the Schrödinger equation for a harmonic oscillator:

$$-\frac{\hbar^2}{2m}\frac{d^2\Psi}{dx^2} + \tfrac{1}{2}kx^2\Psi = \varepsilon\Psi. \tag{18.12}$$

As is well known from quantum mechanics, the last equation has solutions which vanish with $x \to \pm\infty$ at

$$\varepsilon = \hbar\omega(n + \tfrac{1}{2})$$

where $\omega = (k/m)^{1/2}$. If we denote the coefficient in front of the first term of (18.12) as a and that the second term as b, then $\hbar\omega = 2\sqrt{ab}$. Comparing eq. (18.11) with eq. (18.12), we obtain $2\sqrt{ab} = 2\varkappa H_0$ and, hence, the required solutions exist at $\varkappa^2 = 2\varkappa H_0(n + \frac{1}{2})$ or**

$$H_0 = \frac{\varkappa}{2n+1}.$$

* At first sight, it may seem that the search for a nucleus which is dependent only on a single coordinate is a limitation on generality. In fact, for homogeneous magnetic field the vector potential depends linearly on the coordinates. Choosing $A_y = H_0 x$, we see that the equation for Ψ contains explicitly only the coordinate x.

** The possibility of superconducting regions with small Ψ at $H_0 = \varkappa(2n+1)$ was pointed out by Ginzburg and Landau (1950). But they considered only type I superconductors with $\varkappa < 1/\sqrt{2}$, for which these values of H_0 were lower than H_{cm}, and the statement of the problem of small Ψ was meaningless (the supercooling field was not studied by Ginzburg and Landau).

We are interested in the strongest field at which the solution $\Psi \neq 0$ is possible. It corresponds to $n = 0$ and is equal to

$$H_{c2} = \varkappa \tag{18.13}$$

or, in conventional units,

$$H_{c2} = \varkappa\sqrt{2}H_{cm}. \tag{18.14}$$

This value is consistent with the qualitative estimate (18.2) and (18.3). Note that $H_{c2} > H_{cm}$ at $\varkappa > 1/\sqrt{2}$.

The solution corresponding to the lowest level of the oscillator has the form $\Psi \propto \exp[-(\sqrt{km}/2\hbar x^2]$. Again, denoting the coefficients in (18.12) as a and b, we have $\Psi \propto \exp[-\frac{1}{2}(b/a)^{1/2}x^2]$. In our case we obtain $\Psi \propto \exp(-\frac{1}{2}\varkappa H_0 x^2)$ or, in view of $H_0 = \varkappa$,

$$\Psi \propto \exp(-\tfrac{1}{2}\varkappa^2 x^2). \tag{18.15}$$

As has been pointed out above, we may choose the coordinate origin $x = 0$ at any point, i.e., the nucleus of the superconducting phase may appear at any point in the superconductor. The change in the coordinate origin corresponds to the change in the gauge of the vector potential if we wish to use real Ψ. But if we wish to use a unified gauge, then we have to take a complex Ψ. Indeed, writing $\Psi = |\Psi| e^{i\chi}$ and substituting into (17.6′), we obtain

$$(-i\varkappa^{-1}\nabla + \varkappa^{-1}\nabla\chi - \boldsymbol{A})^2|\Psi| - |\Psi| + |\Psi|^3 = 0. \tag{18.16}$$

It is sufficient to take $\chi = ky$ and we obtain the same equation with the replacement $A_y \to A_y - k/\varkappa = H_0(x - k/\varkappa H_0)$. Hence, the small nucleus may not only have the form (18.15) but also

$$\Psi \propto \exp\left[iky - \tfrac{1}{2}\varkappa^2\left(x - \frac{k}{\varkappa^2}\right)^2\right]. \tag{18.17}$$

Evidently, the solution may also be any linear combination of functions of the type (18.17). Since the conditions along the entire superconductor are uniform, it is most natural to suppose a linear combination of solutions centered through equal intervals, i.e.,

$$\Psi = \sum_{n=-\infty}^{\infty} C_n \exp\left[ikny - \tfrac{1}{2}\varkappa^2\left(x - \frac{kn}{\varkappa^2}\right)^2\right]. \tag{18.18}$$

Let us now consider the solution of eq. (17.6′) at H_0 slightly smaller than $\varkappa$. We apply the method of successive approximations. First of all, we take into account the term with Ψ in equation (17.8′) for A. Substituting eq. (18.18), we obtain two equations:

$$\frac{\partial^2 A}{\partial x \partial y} = \frac{\partial H}{\partial y}$$

$$= \frac{i}{2\varkappa} \sum_{n,m} C_n C_m^* \exp\left[ik(n-m)y - \tfrac{1}{2}\varkappa^2 \left(x - \frac{kn}{\varkappa^2} \right)^2 \right.$$

$$\left. - \tfrac{1}{2}\varkappa^2 \left(x - \frac{km}{\varkappa^2} \right) \right] \cdot k(m-n),$$

$$-\frac{\partial^2 A}{\partial x^2} = -\frac{\partial H}{\partial x}$$

$$= \frac{1}{2\varkappa} \sum_{n,m} C_n C_m^* \exp\left[ik(n-m)y - \tfrac{1}{2}\varkappa^2 \left(x - \frac{kn}{\varkappa^2} \right)^2 \right.$$

$$\left. - \tfrac{1}{2}\varkappa^2 \left(x - \frac{km}{\varkappa^2} \right)^2 \right] \cdot [k(n+m) + 2\varkappa^2 x]$$

(we have taken $A = H_0 x$ on the right-hand side of eq. (17.8′)). From this it is easy to see that

$$H = \frac{\partial A}{\partial x} = H_0 - \frac{|\Psi|^2}{2\varkappa}, \qquad A = H_0 x - \frac{1}{2\varkappa} \int |\Psi|^2 \, \mathrm{d}x. \tag{18.19}$$

Consider now the next approximation to eq. (17.6′). We write

$$\Psi = \Psi^{(0)} + \sum_n \mathrm{e}^{ikny} \psi_n^{(1)}(x),$$

where $\Psi^{(0)}$ is the function (18.18). We shall assume the second term to be a small correction and write eq. (17.6′) to first order in this correction. Here we obtain a linear inhomogeneous equation, which may be split into equations for individual $\psi_n^{(1)}$. The vector potential is inserted in the form (18.19), the first term being written in the form $\varkappa x + (H_0 - \varkappa)x$ and the second term is assumed to be a first-order correction, just as the second term in (18.19). As a result, we find

$$\left(\frac{kn}{\varkappa} - \varkappa x \right)^2 \psi_n^{(1)} - \frac{1}{\varkappa^2} \frac{d^2 \psi_n^{(1)}}{\mathrm{d}x^2} - \psi_n^{(1)}$$

$$= 2(\varkappa - H_0)\varkappa x \left(x - \frac{kn}{\varkappa^2} \right) C_n \psi_n(x)$$

$$+ \sum_{m,p} C_p C_m^* C_{n-p+m} \left[\left(x - k\frac{n - \frac{1}{2}(p-m)}{\varkappa^2} \right) \psi_{n-p+m}(x) \int^x \psi_p(x')\, \psi_m(x')\, \mathrm{d}x' \right.$$

$$\left. - \psi_p(x)\, \psi_m(x)\, \psi_{n-p+m} \right], \tag{18.20}$$

where

$$\psi_n(x) = \exp\left[-\tfrac{1}{2}\varkappa^2\left(x - \frac{kn}{\varkappa^2}\right)^2\right]. \tag{18.21}$$

In order for this inhomogeneous equation to have a solution, it is necessary that its right-hand side be orthogonal to the solution of a homogeneous equation with the same boundary conditions. Such is $\psi_n(x)$. From the orthogonality condition we obtain

$$2^{-1/2}\left(\frac{1}{2\varkappa^2} - 1\right) \sum_{p,m} C_p C_m^* C_{n-p+m} \exp\left(-k^2 \frac{(p-n)^2 + (p-m)^2}{2\varkappa^2}\right) + C_n \frac{\varkappa - H_0}{\varkappa} = 0. \tag{18.22}$$

We multiply this equation by C_n^* and sum over n. The resultant relation may be written in the form (see eq. 18.18)

$$\left(\frac{\varkappa - H_0}{\varkappa}\right)\overline{|\Psi|^2} + \left(\frac{1}{2\varkappa^2} - 1\right)\overline{|\Psi|^4} = 0. \tag{18.23}$$

The induction, i.e., the average field, is obtained from formulae (18.19) and (18.23):

$$B = \bar{H} = H_0 - \frac{\overline{|\Psi|^2}}{2\varkappa} = H_0 - \frac{(\overline{|\Psi|^2})^2}{\overline{|\Psi|^4}} \frac{\overline{|\Psi|^4}}{2\varkappa\overline{|\Psi|^2}} = H_0 - \frac{1}{\beta_A} \frac{\varkappa - H_0}{2\varkappa^2 - 1}, \tag{18.24}$$

where

$$\beta_A = \frac{\overline{|\Psi|^4}}{(\overline{|\Psi|^2})^2} \tag{18.25}$$

is a quantity which does not depend on H_0.

Let us find the free energy using formula (17.19). Substituting eqs. (18.18) and (18.19) into it and expressing H_0 in terms of B, according to (18.24), we obtain (with all $\overline{|\Psi|^2}$ and $\overline{|\Psi|^4}$ calculated using formulae 18.25 and 18.23):

$$\frac{\Omega_s - \Omega_n^{(0)}}{H_{cm}^2/4\pi} = B^2 - \frac{(\varkappa - B)^2}{1 + (2\varkappa^2 - 1)\beta_A}. \tag{18.26}$$

This is the thermodynamic potential with respect to B. In order to obtain the external field from it, we have to calculate $\frac{1}{2}\partial\Omega_s/\partial B$. So, we have

$$H_{ext} = B + \frac{\varkappa - B}{1 + (2\varkappa^2 - 1)\beta_A}. \tag{18.27}$$

Solving formula (18.24) for H_0, we obtain the same expression, i.e., $H_0 = H_{ext}$.

The minimum of free energy (18.26) is reached at the minimum value of β_A. This allows one to choose the coefficients C_n and k in formula (18.18) for Ψ. Note that, according to eq. (18.25), $\beta_A > 1$. Indeed, if $|\Psi|^2 = \overline{|\Psi|^2} + q$, then $\bar{q} = 0$. At the same time

$$\overline{|\Psi|^4} = \overline{(\overline{|\Psi|^2} + q)^2} = (\overline{|\Psi|^2})^2 + \overline{q^2} > (\overline{|\Psi|^2})^2.$$

It is not difficult to see that eq. (18.22), which we derived from the solvability condition for the inhomogeneous equation (18.22), may be written in the following form:

$$\frac{\partial \beta_A}{\partial C_n^*} = 0, \tag{18.28}$$

that is, it follows from the requirement of the minimum of β_A.

A detailed analysis of the problem of choosing the correct solution is very time-consuming, so we do not give it here (see Saint-James et al. 1969). If we consider the simplest variant, namely if all the coefficients C_n in (18.18) are assumed to be equal, it turns out that

$$\beta_A = \frac{k}{\varkappa\sqrt{2\pi}}\left[\sum_{n=-\infty}^{\infty} \exp\left(\frac{-k^2 n^2}{2\varkappa^2}\right)\right]^2.$$

The minimum value is reached at $k = \varkappa(2\pi)^{1/2}$ and is equal to

$$\beta_A^{\square} = \left[\sum_{n=-\infty}^{\infty} \exp(-\pi n^2)\right]^2 = \vartheta_3^2(1; 0) = 1.18, \tag{18.29}$$

where ϑ_3 is the theta function defined in the general case by

$$\vartheta_3(X;\, Y) = \sum_{n=-\infty}^{\infty} \exp(-\pi X n^2 + 2\pi i n Y). \tag{18.30}$$

The function Ψ itself may be written in the form

$$\Psi^{\square} = C \exp(-\tfrac{1}{2}\varkappa^2 x^2)\,\vartheta_3(1;\, (2\pi)^{1/2}\varkappa i(x+iy)). \tag{18.31}$$

Using the properties of the ϑ_3-functions, one can show that upon rotation of the coordinate system by $\frac{1}{2}\pi$ the function Ψ is multiplied by a phase factor $\exp(i\varkappa^2 xy)$, remaining unaltered in other respects. Thus, $|\Psi|^2$ has the symmetry of a square lattice. At points $x = (\sqrt{2\pi}/\varkappa)(m+\frac{1}{2})$, $y = (\sqrt{2\pi}/\varkappa)(n+\frac{1}{2})$ (m and n are integers) the function Ψ vanishes. According to formula (18.19), the magnetic field at these points reaches the maximum value H_0.

As a result of a detailed analysis (Kleiner et al. 1964), it has been found that the lowest value of β_A is exhibited by a solution that has the symmetry of a triangular lattice; it may be written in the form

$$\Psi^{\triangle} = C \sum_{n=-\infty}^{\infty} \exp\left[\tfrac{1}{2}\pi i n(n-1) - \tfrac{1}{2}\varkappa^2\left(x - \frac{kn}{\varkappa^2}\right)^2 + iky\right], \tag{18.32}$$

where $k = \varkappa(\pi\sqrt{3})^{1/2}$. Other notations may also be used, but they differ only in the choice of the coordinate system. Comparison with formula (18.18) shows that in this case the coefficients C_n are chosen such that $C_{n+4} = C_n$ and $C_0 = C_1 = C$, $C_2 = C_3 = -C$. The solution (18.32) may be written in the following form:

$$\Psi^{\triangle} = \exp(-\tfrac{1}{2}\varkappa^2 x^2)\,\vartheta_3\left[\exp(-\tfrac{1}{6}i\pi);\, \tfrac{1}{2}\varkappa\left(\frac{\sqrt{3}}{\pi}\right)^{1/2}(y - ix) - \tfrac{1}{4}\right]. \tag{18.33}$$

The triangular lattice has a period $a=(2/\varkappa)(\pi/\sqrt{3})^{1/2}$ (the side of the triangle). The function $\Psi^{\triangle}$ vanishes at points $x=-(\frac{1}{2}a\sqrt{3})(m+\frac{1}{2})$, $y=a(1+\frac{1}{2}m+n)$, where m and n are integers. For this lattice

$$\beta_A^{\triangle}=1.16. \tag{18.34}$$

We see that the two solutions considered are periodic and both have points at which $\Psi=0$. In the neighborhood of each of these points Ψ is proportional to $x+iy$, where x and y are counted off from the point where $\Psi=0$. This may also be written in the form $\Psi \propto \rho\, e^{i\theta}$, where $\rho=(x^2+y^2)^{1/2}$, $\theta=\arctan(y/x)$. The phase χ coincides with the angle of the cylindrical coordinate system. We see that the phase varies by 2π upon each trip around such a point. Let us now examine formula (17.75) for the current written in conventional units and expressed in terms of the modulus and the phase Ψ. In cylindrical coordinates $\nabla\chi=\rho^{-1}\partial\chi/\partial\theta=\rho^{-1}$. At sufficiently small ρ, $\nabla\chi \gg (2e/\hbar c)A$. Hence,

$$j=\frac{\hbar e}{m}\frac{|\Psi|^2}{\rho}.$$

If we write $j=en_s v_s$, we obtain ($|\Psi|^2=\frac{1}{2}n_s$):

$$v_s=\frac{\hbar}{2m\rho},$$

which coincides exactly with formula (18.7) for a quantum vortex.

At the point $\rho=0$, v_s tends to infinity. For the physical quantity – electric current – to be finite, it is necessary that $\Psi\to 0$ with $\rho\to 0$, and this does really occur. Another argument may be also be given: the vanishing of Ψ at $\rho=0$ is necessary for the uniqueness of Ψ at this point. Such behavior of the wave function of the Bose condensate in the presence of a quantum vortex was first predicted by Feynman (1955) for superfluid liquid helium.

A natural question arises: we constructed the solutions of the Ginzburg–Landau equations without any preliminary assumptions and, in spite of this, in all the solutions there appear points where $\Psi=0$ and the phase varies by 2π in each trip around this point. What is this associated with? The answer consists in the following. The equations contain the vector potential $\boldsymbol{A}$. If the magnetic field inside the superconductor is directed parallel to the external field, then $A_y=\int^x H\,dx$ must increase with increasing x. But the physical quantity is not the vector potential but the magnetic field, which does not increase on the average in the specimen. Hence, the increase of the vector potential must be compensated. This is what really occurs. The point is that the equations always include the combination $\varkappa^{-1}\nabla\chi - A$ (in reduced units).

Consider the solution $\Psi^{\square}$ (18.31) in the xy-plane (fig. 113). The solid circles designate the points where $\Psi=0$. For the phase to be single-valued, we draw through these points branch cuts parallel to the y-axis. If one moves $\parallel y$ staying on the right

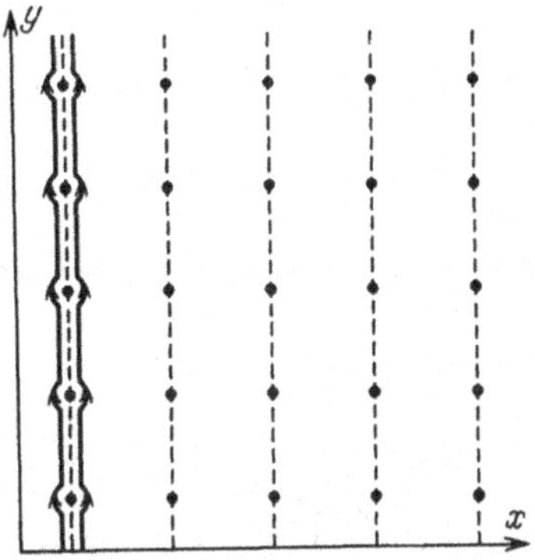

Fig. 113.

of such a branch cut, χ will vary according to the law

$$\chi(y) = \chi_{\text{reg}}(y) + \pi\frac{y}{a},$$

where a is the period of the structure; $\chi_{\text{reg}}(y)$ is the regular part not associated with the vortices. But if one moves $\| y$ staying on the left of the branch cut, then

$$\chi(y) = \chi_{\text{reg}}(y) - \pi\frac{y}{a}.$$

From this it follows that at each of the branch cuts the phase gradient undergoes a discontinuity:

$$\Delta\left(\frac{\partial\chi}{\partial y}\right) = \frac{2\pi}{a}.$$

At the same time the vector potential increases by $\varkappa a$ per period with increasing x ($A_y \approx H_0 x$, $H_0 \approx \varkappa$). In order to compensate this increase, it is necessary that

$$\varkappa a = \frac{1}{\varkappa}\Delta\left(\frac{\partial\chi}{\partial y}\right) = \frac{2\pi}{\varkappa a}.$$

It follows then that

$$a^{\square} = \frac{\sqrt{2\pi}}{\varkappa}.$$

This is what was exactly obtained for the square lattice.

As for the triangular lattice, the separation between the branch cuts is $\frac{1}{2}a\sqrt{3}$ and the compensation condition is

$$\tfrac{1}{2}\varkappa a\sqrt{3} = \frac{2\pi}{\varkappa a}.$$

Hence,

$$a^{\triangle}=2\left(\frac{\pi}{\sqrt{3}}\right)^{1/2}\frac{1}{\varkappa},$$

which corresponds to the solution of (18.33).

Let us find the current lines. Substituting eq (18.18) into the expression for the current (the right-hand side of eq. 18.8′), we obtain

$$j_x=\sum_{n,m} C_n C_n^*\left(k\frac{n+m}{2\varkappa}-\varkappa x\right)$$
$$\times\exp\left[\mathrm{i}(n-m)y-\tfrac{1}{2}\varkappa^2\left(x-\frac{kn}{\varkappa^2}\right)^2-\tfrac{1}{2}\varkappa^2\left(x-\frac{km}{\varkappa^2}\right)^2\right]$$
$$=-\frac{1}{2\varkappa}\frac{\partial|\Psi|^2}{\partial y},$$
$$j_y=\frac{1}{2\varkappa}\frac{\partial|\Psi|^2}{\partial x}.$$

The equation for the current lines is

$$\frac{dx}{j_x}=\frac{dy}{j_y}.$$

Substituting j_x and j_y, we obtain

$$\frac{\partial|\Psi|^2}{\partial x}\mathrm{d}x+\frac{\partial|\Psi|^2}{\partial y}\mathrm{d}y=0$$

or

$$|\Psi|^2=\mathrm{const.}$$

According to eq. (18.19), these are simultaneously the lines of constant field. These lines are shown in fig. 114 for the quadratic solution (the numbers are the values of $|\Psi|^2/|\Psi|^2_{\max}$).

Consider the square contour in fig. 114 with vertices at the points marked by 1. It is not difficult to see that the integral $\oint \boldsymbol{j}\,\mathrm{d}\boldsymbol{l}$ taken along this contour is equal to zero. Hence, according to section 17.6, an integer number of magnetic flux quanta passes through this contour. We will show that this is one flux quantum. Indeed, we have

$$\Phi=BS^{\square}\approx\varkappa\frac{2\pi}{\varkappa^2}=\frac{2\pi}{\varkappa}. \tag{18.35}$$

Using conventional units and substituting the value of $\varkappa$ (17.11), we get

$$\Phi=\frac{2\pi}{\varkappa}H_{\mathrm{cm}}\sqrt{2}\delta^2=\frac{\pi\hbar c}{e}=\Phi_0. \tag{18.35′}$$

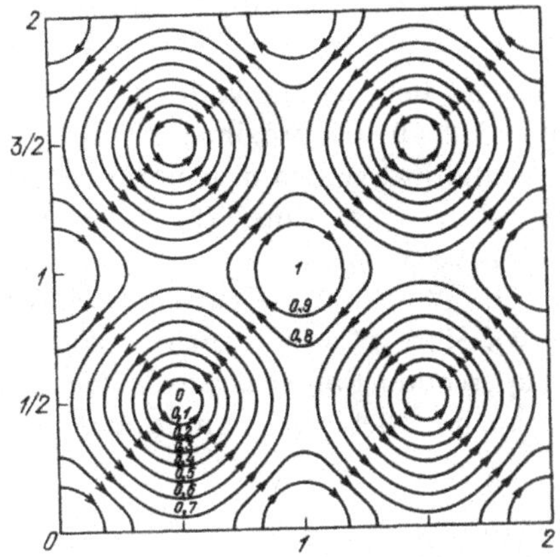

Fig. 114.

The situation is analogous for the triangular lattice (the contour is a hexagon which surrounds the point at which $\Psi = 0$).

In conclusion, we write formula (18.24) in a form more convenient for comparison with experiment:

$$-4\pi M = \frac{H_{c2} - H_0}{(2\varkappa^2 - 1)\beta_A}, \tag{18.36}$$

that is, the dependence $M(H_0)$ near $H_0 = H_{c2}$ is linear.

18.3. *Magnetic properties of type II superconductors. Metals with $\varkappa \gg 1$*

In the present section we shall use an approximation that can be applied with fields smaller than H_{c2}. In order for this approximation to correspond to a real mixed state, it is necessary that $H_{c1} \ll H_{c2}$ or, according to the results of section 18.1, $\varkappa \gg 1$.

In the preceding section we have found the solution at fields close to H_{c2}. It is rather obvious that the main qualitative property of the function Ψ - the presence of points where $\Psi = 0$ and the fact that in going around these points the phase varies by 2π - must be preserved at lower fields too. This is associated with the necessity of the compensation of the increase of the vector potential by discontinuities in the phase gradient. But the lower the field the slower is the increase of A_y and, hence, the more seldom are the discontinuities. Indeed, the compensation condition (for the square lattice) with an arbitrary field has the form

$$\frac{2\pi}{\varkappa a} = A_y(a) - A_y(0) = \int_0^a H(x)\,\mathrm{d}x = \bar{H}a = Ba.$$

From this it seen that $a=(2\pi/\varkappa B)^{1/2}$, i.e., the period of the structure increases with decreasing B.

At small fields the distance between the vortex lines becomes so large that they are almost independent and to a first approximation we may consider a single vortex line. We introduce the cylindrical system of coordinates (ρ, θ, z) and the z-axis is directed along the axis of the vortex line. The vector potential is chosen to be perpendicular to the radius-vector ρ, i.e., $A=A_\theta$.

If we write $\Psi=f\,e^{i\chi}$ $(f=|\Psi|)$, then the current in Ginzburg–Landau units has the form $\boldsymbol{j}=f^2(\nabla\chi/\varkappa-\boldsymbol{A})$. Since $f^2=\frac{1}{2}n_s$ is the density of pairs in the condensate (n_s is the density of "superconducting electrons"), then it follows that $\nabla\chi/\varkappa-\boldsymbol{A}$ plays the role of their velocity. Therefore we introduce the following notation:

$$\boldsymbol{v}_s=\frac{\nabla\chi}{\varkappa}-\boldsymbol{A}.$$

We emphasize that the vector $\boldsymbol{v}_s$, which is convenient for the solution of equations, is not a true physical quantity, in contrast to the current $\boldsymbol{j}=f^2\boldsymbol{v}_s$, which can be measured. Equations (17.6) and (17.8) assume the following form in terms of the new notation:

$$-\frac{1}{\varkappa^2\rho}\frac{d}{d\rho}\left(\rho\frac{df}{d\rho}\right)+v_s^2 f=f-f^3, \tag{18.37}$$

$$-\frac{dH}{d\rho}=v_s f^2, \tag{18.38}$$

$$H=(\text{rot}\,\boldsymbol{A})_z=-\frac{1}{\rho}\frac{d}{d\rho}(\rho v_s)+\frac{2\pi}{\varkappa}\delta(\rho). \tag{18.39}$$

In the last formula it is taken into account that in the vicinity of an individual filament $\chi=\theta$, i.e., it varies by 2π on encircling the line $\rho=0$. Because of this, $\nabla\chi$ has an unusual property. If we take the integral of $\nabla\chi$ over an infinitesimal loop around the point $\rho=0$, this will give, on the one hand

$$\oint\nabla\chi\,d\boldsymbol{l}=2\pi$$

and on the other hand

$$\oint\nabla\chi\,d\boldsymbol{l}=\int\text{rot}\,\nabla\chi\,d\boldsymbol{S}.$$

Hence,

$$(\text{rot}\,\nabla\chi)_z=2\pi\delta(\boldsymbol{\rho}). \tag{18.40}$$

The magnetic field $\boldsymbol{H}=\text{rot}\,\boldsymbol{A}=-(\text{rot}\,\boldsymbol{v}_s-\varkappa^{-1}\,\text{rot}\,\nabla\chi)$, from which we obtain (18.39).

The boundary conditions for eqs. (18.37), (18.38) and (18.39) are as follows. With $\rho\to\infty$ we have $f\to1$ and $H\to0$ (and, consequently, $v_s\to0$). With $\rho\to0$ we have

$v_s \to (\varkappa\rho)^{-1}$; this is associated with the fact that

$$\nabla_\theta \chi = \frac{1}{\rho}\frac{\partial \chi}{\partial \theta} \to \frac{1}{\rho}.$$

As for f, it must not increase indefinitely at any ρ.

The solution of the equations in a general case can be obtained only by numerical integration. However, at $\varkappa \gg 1$ the situation is substantially simplified. To demonstrate this, we note that the distances over which the function v_s varies significantly is $\rho \sim 1$. As regards the function f, it varies appreciably at distances of $\rho \sim \varkappa^{-1} \ll 1$. In view of this, in eq. (18.38) the function f may be supposed to have reached the value $f=1$ already. Substituting (18.39) for $\rho \neq 0$, we obtain the equation for v_s; its solution is

$$v_s = \frac{K_1(\rho)}{\varkappa} \tag{18.41}$$

where K_1 is the MacDonald function, which has been introduced in section 16.5.

In the equation for f we may use the asymptotic form of v_s for small distances, i.e., $v_s \approx (\varkappa\rho)^{-1}$. Inserting this into eq. (18.37) gives

$$\frac{1}{\varkappa^2}\left[\frac{1}{\rho}\frac{\mathrm{d}}{\mathrm{d}\rho}\left(\rho\frac{\mathrm{d}f}{d}\right) - \frac{f}{\rho^2}\right] = f^3 - f. \tag{18.42}$$

The solution of this equation at distances $\rho \gg \varkappa^{-1}$ (but, of course, $\rho \ll 1$) is

$$f^2 = 1 - \frac{1}{(\varkappa\rho)^2}. \tag{18.43}$$

At $\rho \ll \varkappa^{-1}$ we assume that $f \ll 1$ and find that

$$f = C\varkappa\rho, \tag{18.44}$$

where C is a numerical constant, which can be found by numerical solution of eq. (18.42). The result $\Psi \propto \rho$ corresponds to what was observed in the vicinity of H_{c2}.

Let us determine the free energy per unit length of the vortex line. It must be measured from the energy of the superconductor in the absence of magnetic field. Using formula (17.19′), we obtain (in units of $H_{\mathrm{cm}}^2\delta^2/4\pi$):

$$\varepsilon_0 = 2\pi \int [\Omega_s - \Omega_s^{(0)}]\rho\,\mathrm{d}\rho = 2\pi \int [H^2 + \tfrac{1}{2}(1-f^4)]\rho\,\mathrm{d}\rho. \tag{18.45}$$

The principal part of this integral arises from the second term and is associated with distances much larger than $\varkappa^{-1}$. Substituting eq. (18.43) yields

$$\varepsilon_0 \approx \frac{2\pi}{\varkappa^2}\int \frac{1}{\rho}\,\mathrm{d}\rho \approx \frac{2\pi}{\varkappa^2}\ln \varkappa.$$

As a result of numerical integration, we find a more accurate result:

$$\varepsilon_0 = \frac{2\pi}{\varkappa^2}(\ln \varkappa + 0.081). \tag{18.46}$$

For the formation of vortex lines to be energetically favorable, we must have a field strength at which the free energy $\Omega_s - \Omega_{ns}^{(0)} - 2H_0B$ becomes negative. If the density of the vortex lines, i.e., the number of lines that intersect the unit area, is labeled as n, then

$$\Omega_s - \Omega_s^{(0)} = n\varepsilon_0, \qquad B = \bar{H} = n\int \boldsymbol{H}\,\mathrm{d}\boldsymbol{S} = n\oint \boldsymbol{A}\,\mathrm{d}\boldsymbol{l}. \tag{18.47}$$

Because the contour passes at a distance $\rho = \infty$ and at such distances $v_s = 0$ (see 18.41), we conclude that $\boldsymbol{A} = \nabla\chi/\varkappa$ and is equal in magnitude to $(\varkappa\rho)^{-1}$. Hence,

$$B = \frac{2\pi n}{\varkappa}. \tag{18.48}$$

But since B is also the magnetic flux per unit area, it follows that the magnetic flux per vortex line is $2\pi/\varkappa$ or, according to eq. (18.35), one flux quantum Φ_0.

From the condition $\Omega_s - \Omega_s^{(0)} - 2H_0B = 0$ we obtain the value of the first critical field:

$$H_{c1} = \frac{\varepsilon_0\varkappa}{4\pi}. \tag{18.49}$$

This formula applies for any $\varkappa$. In the case $\varkappa \gg 1$ we can substitute formula (18.46) for ε_0. In conventional units we have

$$H_{c1} = \frac{H_{cm}}{\varkappa\sqrt{2}}(\ln\varkappa + 0.08). \tag{18.50}$$

It is interesting to note that the field in the center of the vortex line is twice as large as H_{c1}. Indeed, according to eq. (18.38), $H(0) = \int_0^\infty v_s f^2\,\mathrm{d}\rho$. The principal part of the integral is taken over the region $\varkappa^{-1} \ll \rho \ll 1$. Inserting $f \approx 1$ and $v_s \approx (\varkappa\rho)^{-1}$, we find $H(0) \approx \varkappa^{-1}\ln\varkappa$. Numerical integration gives a more accurate formula:

$$H(0) = \frac{1}{\varkappa}(\ln\varkappa - 0.18). \tag{18.51}$$

The distribution of the field and $f(\rho)$ in the vortex is shown in fig. 115.

When $H_0 > H_{c1}$, there appear many vortices which begin to interact. In order to find the distribution of the fields and currents in this case and also the dependence

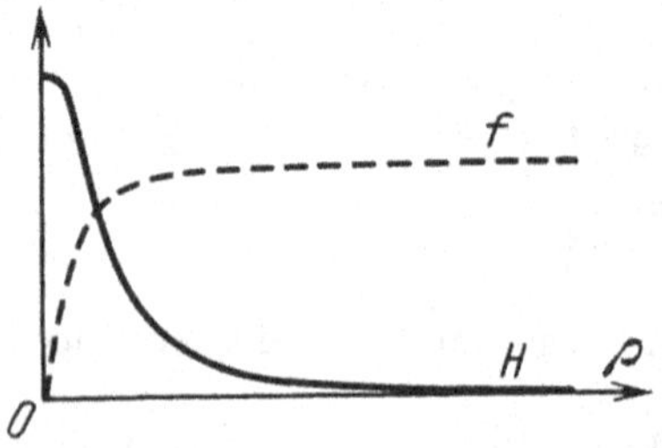

Fig. 115.

$B(H_0)$, we use the fact that the vortex axes come closer together to a distance of order $\varkappa^{-1}$ only with $H_0 \to H_{c2}$, and at smaller fields the intervortex distance is substantially larger. In view of this, the vortex cores (of order $\varkappa^{-1}$) will simply play the role of singularities in the solution, and their detailed structure will be unimportant. We write the Maxwell equation in the form

$$\operatorname{rot} \boldsymbol{H} = f^2 \boldsymbol{v}_s,$$

substitute $f = 1$, and take the curl. Taking into account that each vortex line gives a δ-function according to eq. (18.40), we obtain

$$\Delta H - H = -\frac{2\pi}{\varkappa} \sum_m \delta(\boldsymbol{\rho} - \boldsymbol{\rho}_m), \tag{18.52}$$

where $\boldsymbol{\rho}_m$ corresponds to the centers of the vortices ($\boldsymbol{\rho} = (x, y)$). The solution of this equation is

$$H = \frac{1}{\varkappa} \sum_m K_0(|\boldsymbol{\rho} - \boldsymbol{\rho}_m|), \tag{18.53}$$

where K_0 is the MacDonald function.

The free energy will be measured from the value for the superconducting phase in the absence of the field. We use formula (17.19), in which we assume that $\Psi = f\,\mathrm{e}^{\mathrm{i}\chi}$ and $f = 1$. So, we have

$$\int (\Omega_s - \Omega_s^{(0)})\,\mathrm{d}V = \int (v_s^2 + H^2)\,\mathrm{d}V. \tag{18.54}$$

Considering that in our approximation $\boldsymbol{v}_s = \operatorname{rot} \boldsymbol{H}$, integrating the first term by parts and substituting (18.52), we find that

$$\bar{\Omega}_s - \Omega_s^{(0)} = \overline{H(H - \Delta H)} = \frac{2\pi n}{\varkappa^2} \sum_m K_0(|\boldsymbol{\rho}_m|) \tag{18.55}$$

(n is the number of vortices intersecting 1 cm^2). The term that corresponds to $\boldsymbol{\rho}_m = 0$ must be replaced by $n\varepsilon_0$.

Just as in the previous section, we are to choose the vortex lattice using the criterion of the minimal free energy. The triangle lattice turns out to be the most favorable. The period of the lattice is chosen in accordance with formula (18.48), i.e., on the basis of the fact that $B = n\Phi_0 = n(2\pi/\varkappa)$. So, we obtain $n^{-1} = a \cdot a\sqrt{3}/2$, from which it follows that

$$a = 2\left(\frac{\pi}{\sqrt{3}}\right)^{1/2} \frac{1}{(B\varkappa)^{1/2}} \tag{18.56}$$

(a is the side of the triangle).

The vector $\boldsymbol{\rho}_m$ is given by

$$\boldsymbol{\rho}_m = \boldsymbol{a}_1 q + \boldsymbol{a}_2 l,$$

where l and q are integers, and the basis vectors $\boldsymbol{a}_1$ and $\boldsymbol{a}_2$ have a length a and make an angle of 60°. In accordance with this,

$$|\boldsymbol{\rho}_m| = a(q^2+l^2+ql)^{1/2}.$$

Thus, we obtain

$$\Omega_s - \Omega_s^{(0)} = \frac{\varkappa B\varepsilon_0}{2\pi} + \left(\frac{B}{\varkappa}\right) \sum_{x_{q,l}\neq 0} K(x_{q,l}), \tag{18.57}$$

where

$$x_{q,l} = \left(4\pi \frac{q^2+l^2+ql}{\sqrt{3}\, B\varkappa}\right)^{1/2}. \tag{18.58}$$

The external field is expressed with the aid of the relation

$$H_0 = \tfrac{1}{2}\frac{\partial(\Omega_s - \Omega_s^{(0)})}{\partial B}. \tag{18.59}$$

Substituting eq. (18.56) and using $K_0'(x) = -K_1(x)$, we obtain

$$\varkappa(H_0 - H_{c1}) = \tfrac{1}{4} \sum_{x_{q,l}\neq 0} [2K_0(x_{q,l}) + x_{q,l}K_1(x_{q,l})]. \tag{18.60}$$

We see that the dependence of $\varkappa B$ on $\varkappa(H_0 - H_{c1})$ is a universal function. Let us examine the asymptotic forms of this dependence. With $H_0 - H_{c1} \to 0$, B tends to zero, i.e., $x_{q,l} \to \infty$. Because of the exponential decrease of the MacDonald function it will be sufficient to retain only terms with the lowest x corresponding to $q^2+l^2+q^l=1$. Since in the triangular lattice there are 6 neighbors at each point, a factor 6 appears in eq. (18.60). Using the asymptotics of the MacDonald functions $K_0(x) \approx K_1(x) \approx (\pi/2x)^{1/2}\, e^{-x}$, we get

$$\varkappa(H_0 - H_{c1}) \approx \tfrac{3}{2}(\tfrac{1}{2}\pi)^{1/2}\left(\frac{4\pi}{\sqrt{3}\, B\varkappa}\right)^{1/4} \exp\left[-\left(\frac{4\pi}{\sqrt{3}\, B\varkappa}\right)^{1/2}\right]. \tag{18.61}$$

It is not difficult to see that with $H_0 \to H_{c1}$, $B \to 0$ and $\mathrm{d}B/\mathrm{d}H_0 \to \infty$, i.e., the $B(H_0)$ curve has a vertical tangent at point H_{c1}.

Another limit that can be obtained is the relation between H_0 and B for fields at which the lattice period is much larger than ξ but less than δ, i.e., in reduced units $\varkappa^{-1} \ll a \ll 1$ or

$$\varkappa \gg B \gg \frac{1}{\varkappa}. \tag{18.62}$$

At $\varkappa \gg 1$ this occurs in most of the region of the mixed state, except the immediate neighborhood of H_{c1} and H_{c2}. We pass in formula (18.60) to the Fourier transforms of K_0 and K_1:

$$\varkappa(H_0 - H_{c1}) = \pi \int \frac{2+k^2}{(1+k^2)^2} \sum_{\boldsymbol{\rho}_{q,l}\neq 0} \mathrm{e}^{i\boldsymbol{k}\boldsymbol{\rho}_{q,l}} \frac{\mathrm{d}^2\boldsymbol{k}}{(2\pi)^2}. \tag{18.63}$$

If we add the term with $\boldsymbol{\rho}_{q,l} = 0$ to the sum over $\boldsymbol{\rho}_{q,l}$, we obtain

$$\sum_{\boldsymbol{\rho}_{q,l}} \mathrm{e}^{i\boldsymbol{k}\boldsymbol{\rho}_{q,l}} = (2\pi)^2 n \sum_{\boldsymbol{K}} \delta(\boldsymbol{k} - \boldsymbol{K}), \tag{18.64}$$

where n is the number of vortices per cm^2; $\boldsymbol{K}$ are the reciprocal lattice vectors with respect to the planar lattice formed by the vectors $\boldsymbol{\rho}_{q,l}$. These vectors are given by

$$\boldsymbol{K} = p_1\boldsymbol{K}_1 + p_2\boldsymbol{K}_2,$$
$$\boldsymbol{K}_1 = 2\pi[\boldsymbol{a}_1 \times z_0]n, \qquad K_2 = 2\pi[z_0 \times \boldsymbol{a}_2]n, \tag{18.65}$$

where n^{-1} is the area of the unit cell; $\boldsymbol{a}_1$ and $\boldsymbol{a}_2$ are basis vectors; z_0 is the unit vector with respect to z. Since $n = \varkappa B/2\pi$, it follows that $\boldsymbol{K}_1$ and $\boldsymbol{K}_2$ are equal, in magnitude, to $2(\pi/\sqrt{3})^{1/2}(\varkappa B)^{1/2}$ and the angle between them is equal to $\pi/3$, just as in the main lattice. Substituting (18.64) into (18.63), we get

$$\varkappa(H_0 - H_{c1}) = \pi\left(n\sum_{\boldsymbol{K}}\frac{2+K^2}{(1+K^2)^2} - \int_0^\infty \frac{2+k^2}{(1+k^2)^2}\frac{d^2\boldsymbol{k}}{(2\pi)^2}\right) \tag{18.66}$$

(the subtraction of the last integral corresponds to absence of the term with $\boldsymbol{\rho}_{q,l} = 0$ in (18.63).

Each of the expressions in brackets diverges logarithmically, but taken together they give a finite expression. We single out out from the first sum the term with $\boldsymbol{K} = 0$. It is equal to $2\pi n = \varkappa B$. In the remaining terms and in the integral we shall first confine ourselves to the values $|\boldsymbol{K}|$, $|\boldsymbol{k}| < K_1 P = 2(\pi/\sqrt{3})^{1/2}(\varkappa B)^{1/2}P$; we shall then find the difference and pass to the limit $P \to \infty$. Since in the remaining sum over $\boldsymbol{K}$ the lowest K^2 is $K_1^2 \gg 1$ (according to eq 18.62), it may be replaced by

$$n\sum_{\boldsymbol{K}}\frac{1}{K^2} = \frac{n}{K_1^2}\sum_{p_1,p_2}\frac{1}{p_1^2+p_2^2+p_1p_2}$$
$$= \frac{\sqrt{3}}{8\pi^2}\sum_{p_1,p_2}\frac{1}{p_1^2+p_2^2+p_1p_2}.$$

Evaluating the integral over k with the limit $K_1 P \gg 1$, we get

$$\int_0^{K_1 P}\frac{2+k^2}{(1+k^2)^2}\frac{d^2\boldsymbol{k}}{(2\pi)^2} = \frac{1}{2\pi}(\ln K_1 + \ln P + \tfrac{1}{2}).$$

Collecting together all the terms, we find

$$\varkappa(H_0 - H_{c1}) \approx \varkappa B - \tfrac{1}{4}\ln(\varkappa B) - \tfrac{1}{4}\ln\left(\frac{4\pi}{\sqrt{3}}\right) - \tfrac{1}{4}$$
$$+ \lim_{P\to\infty}\left(\frac{\sqrt{3}}{8\pi}\sum_{0<(p_1^2+p_2^2+p_1p_2)<P^2}\frac{1}{p_1^2+p_2^2+p_1p_2} - \tfrac{1}{2}\ln P\right)$$
$$= \varkappa B - \tfrac{1}{4}\ln(\varkappa B) - 0.49. \tag{18.67}$$

If we substitute here formula (18.49) for H_{c1}, we obtain

$$H_0 \approx B + \frac{1}{4\varkappa}\ln\left(\frac{\varkappa}{B}\right) - 0.45\frac{1}{\varkappa}. \tag{18.68}$$

From this formula we see that at $B \sim \varkappa$, $H_0 \approx B$ to within terms of order $\varkappa^{-1}$, as should be expected. On the basis of the limiting expressions (18.68) and (18.27) one can write an interpolation formula which reflects the logarithmic dependence of $H_0 - B$ on B in the region where the distance between the vortex cores is large and which changes to (18.27) with $B \to \varkappa$ $(\varkappa \gg 1)$:

$$H_0 \approx B + \frac{1}{2\beta_A \varkappa} \ln\left(\frac{\varkappa}{B}\right). \tag{18.69}$$

Of course, first this is only an interpolation and second it does not apply near H_{c1}.

In concluding this section we will show that at $\varkappa = 1/\sqrt{2}$ the critical fields H_{c1}, H_{c2} and H_{cm} coincide with each other. We put $\varkappa = 1/\sqrt{2}$ in eq. (18.37). It can be shown that

$$H = 2^{-1/2}(1 - f^2). \tag{18.70}$$

Indeed, substituting into eq. (18.38), we get

$$f v_s = 2^{1/2} \frac{\partial f}{\partial \rho}. \tag{18.71}$$

We multiply eq. (18.37) (with $\varkappa = 2^{-1/2}$) by f and substitute eq. (18.71):

$$-\frac{1}{\rho} f \frac{\mathrm{d}}{\mathrm{d}\rho}\left(\rho \frac{\mathrm{d}f}{\mathrm{d}\rho}\right) + \left(\frac{\mathrm{d}f}{\mathrm{d}\rho}\right)^2 = \tfrac{1}{2}(1 - f^2) f^2.$$

On the other hand, if we multiply eq. (18.39) by f^2 and substitute eqs. (18.70) and (18.71), the same relation is obtained; this confirms the assumption (18.70). Now let us find ε_0 by using eq. (18.45). Substitution of eq. (18.70) yields

$$\varepsilon_0 = 2\pi \int (1 - f^2)\rho \, \mathrm{d}\rho = 2\pi\sqrt{2} \int H\rho \, \mathrm{d}\rho$$

$$= \sqrt{2}\,\Phi = \sqrt{2}\frac{2\pi}{\varkappa}. \tag{18.72}$$

Here we have taken into account that there is one flux quantum per vortex, i.e., $2\pi/\varkappa$. Substituting this into (18.49), we obtain $H_{c1} = 2^{-1/2}$ or, in conventional units, $H_{c1} = H_{cm}$. For H_{c2} this follows from (18.14). Thus, when $\varkappa$ decreases to $1/\sqrt{2}$, the region of the mixed state is narrowed until it eventually disappears.

Thus, on the basis of qualitative arguments and quantitative calculations we have come to the conclusion that in type II superconductors in a magnetic field a mixed state arises with an incomplete Meissner effect, when the external field penetrates the superconductor in the form of filaments of the magnetic flux. Each filament is a tiny vortex current and carries one quantum of the magnetic flux. The order parameter is equal to zero on the vortex axis and is restored to the equilibrium value in the absence of the field at a distance $\xi = \delta/\varkappa$. This region is called the vortex core. The region of the mixed state disappears at $\varkappa \leq 1/\sqrt{2}$ and becomes large at $\varkappa \gg 1$. Typical magnetization curves ($-4\pi M$ as a function of H_0) are shown in fig. 116.

Direct observation of the vortex structure was first made with the aid of neutron diffraction (Cribier et al. 1964). Since neutrons have a magnetic moment, they can be used for the analysis of magnetic structures in the same way as X-rays are used for the investigation of the electron density distribution. It is in this way that the existence of the vortex lattice was proved directly for the first time.

At a later time ferromagnetic powders were used for the observation of the vortex structure (Essmann and Träuble 1967), similar to what had been done for the intermediate state of type I superconductors. However, since the period of the structure is much smaller in this case, powders consisting of very small particles (diameter ~40 Å) had to be used and the resultant picture was studied with the aid of an electron microscope using the replica technique (fig. 117). Usually, a triangular vortex lattice is obtained. However, when the magnetic field is directed along a fourth-order axis, a square lattice results. This is due to the small difference in energy between the two structures, which leads to the possibility of reconstitution under the influence of crystal anisotropy.

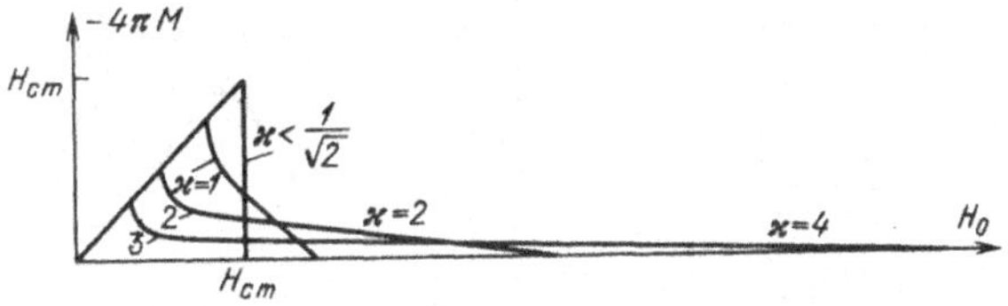

Fig. 116.

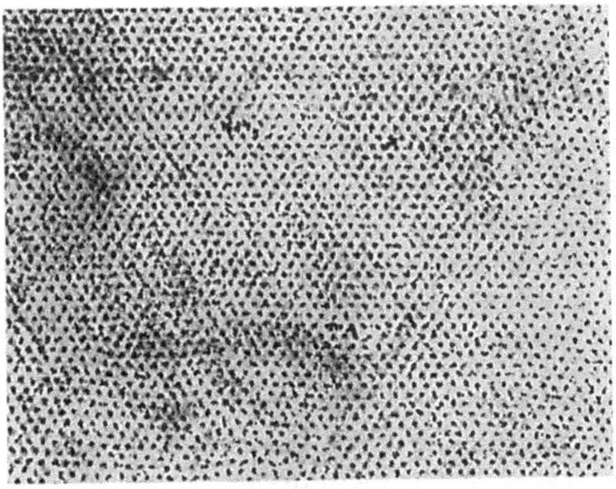

Fig. 117.

18.4. Surface superconductivity

Up until now we have dealt with an infinite superconducting space or, in other words, with phenomena that take place in the interior of a bulk superconductor. Investigations show that in the surface layer superconductivity may be retarded up to fields larger than H_{c2} (Saint-James and De Gennes 1963). The physical cause of this phenomenon consists in the following. While considering the nucleus of the superconducting phase in infinite space in section 18.2, we used the boundary condition $\Psi \to 0$ for $x \to \pm\infty$. In the presence of a surface the boundary condition is $\mathrm{d}\Psi/\mathrm{d}x|_{\mathrm{b}} = 0$.

Figure 118 shows the various possible cases: (a) a nucleus in the bulk; (b) a nucleus with its center at the surface; (c) a nucleus near the surface, but with a shifted center and, finally, (d) a nucleus in a thin film. In the last two cases the boundary condition on the surface requires an increase in Ψ near the surface as compared to the case of infinite space. This leads to an increase in the eigenvalue in the equation for Ψ (18.11) (the role of the energy is played by $\varkappa^2$), as a result, the value of the critical field increases. An example is the critical field of thin films (section 17.4), which is the same for type I and II superconductors with thicknesses of $d \ll \min(1, \varkappa^{-1})$ (section 18.6) and which increases with decreasing thickness.

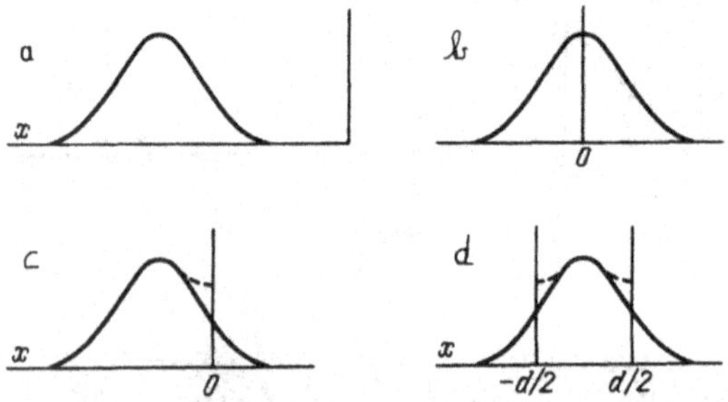

Fig. 118.

It is easy to see that the solutions with the center far from the surface and with the center directly at the surface differ only by the shift of the coordinate origin (in both cases $\mathrm{d}\Psi/\mathrm{d}x|_{\mathrm{b}} = 0$, see fig. 118) and, hence, they correspond to the same value of the field $H_0 = \varkappa$. The solution with the maximum slightly shifted relative to the surface corresponds to a larger value of the field. From this it follows that there exists a certain single solution, for which the field will have a maximum. Let us find this solution.

We begin with the case of the external field parallel to the surface. Here we may take the vector potential in the form $A = H_0(x - x_0)$ and use the boundary conditions $\Psi = 0$ for $x \to \infty$ and $\mathrm{d}\Psi/\mathrm{d}x = 0$ at $x = 0$. In this case, Ψ will depend only on x. Instead of this, we could have taken $A = H_0 x$, looking for Ψ in the form $\mathrm{e}^{iky}\,\Psi(x)$, but, as has been shown in sections 18.2, these are equivalent statements of the problem. Thus, substituting $A = H_0(x - x_0)$ into eq. (18.9), we obtain

$$-\frac{d^2\Psi}{dx^2} + \varkappa^2 H_0^2 (x - x_0)^2 \Psi = \varkappa^2 \Psi. \tag{18.73}$$

The solution of this equation in the general case is

$$\Psi = \exp(-\tfrac{1}{2}\zeta^2)\{C_1 F[\tfrac{1}{4}(1-\beta), \tfrac{1}{2}, (\zeta - \zeta_0)^2] + \zeta C_2 F[\tfrac{1}{4}(3-\beta), \tfrac{3}{2}, (\zeta-\zeta_0)^2]\}, \tag{18.74}$$

where

$$\zeta = (\varkappa H_0)^{1/2} x, \qquad \zeta_0 = (\varkappa H_0)^{1/2} x_0, \qquad \beta = \frac{\varkappa}{H_0},$$

$F(\alpha, \gamma, z)$ is a degenerate hypergeometric function satisfying the equation

$$zF'' + (\gamma - z)F' - \alpha F = 0, \tag{18.75}$$

and C_1 and C_2 are arbitrary constants (see Landau and Lifshitz 1974).

The boundary conditions enable one to determine the ratio C_2/C_1 and β as a function of x_0. After this, minimizing β with respect to x_0, we obtain $\beta_{\min}$ or $H_{0,\max}$. This turns out to be equal to

$$H_{c3} = 1.695\varkappa = 1.695 H_{c2}. \tag{18.76}$$

This is the critical field of surface superconductivity. It is larger than H_{c2}.

The procedure described allows one to obtain the exact value of H_{c3}, but requires cumbersome numerical calculations. We can, instead, use a simpler procedure, which will prove useful later on (Abrikosov 1964b). We write eq. (18.73) using the variables introduced earlier, ζ, ζ_0 and β:

$$-\frac{d^2\Psi}{d\zeta^2} + (\zeta - \zeta_0)^2 \Psi = \beta \Psi. \tag{18.77}$$

The problem of solving this equation may be presented as a variational problem of finding the minimum of a quantity:

$$\beta = \frac{\displaystyle\int_0^\infty \left[\left(\frac{d\Psi}{d\zeta}\right)^2 + (\zeta - \zeta_0)^2 \Psi^2\right] d\zeta}{\displaystyle\int_0^\infty \Psi^2\, d\zeta}.$$

Indeed, varying with respect to Ψ, we arrive at eq. (18.77).

We substitute into the functional (18.77) the trial function $\Psi = \exp(-\frac{1}{2}b\zeta^2)$, which satisfies the correct boundary conditions: $\Psi(\infty) = 0$, $d\Psi/d\zeta = 0$. The coefficient b

must be determined from the minimum condition for β. Evaluating the integrals, we get

$$\beta = \tfrac{1}{2}b + \frac{1}{2b} - \frac{2\zeta_0}{(\pi b)^{1/2}} + \zeta_0^2.$$

Minimizing with respect to b, we obtain $\beta(\zeta_0)$. Then, we are to find the minimum with respect to ζ_0. Instead, we proceed in the reverse order, i.e., first we find the minimum with respect to ζ_0 and then with respect to b. By simple calculations we obtain

$$\beta_{\min} = b = \left(1 - \frac{2}{\pi}\right)^{1/2}, \qquad \zeta_0 = \frac{1}{(\pi b)^{1/2}}. \tag{18.78}$$

This yields

$$H_0 = 1.66\varkappa,$$

which differs from the true value (18.76) by 2%.

Thus, we conclude that the superconductivity is retained in the surface layer up to fields larger than H_{c2} and, hence, in this layer persistent currents may be circulating. However, these currents cannot screen out the field inside the specimen. In fact, if the field inside the specimen were different from the external field, this would give rise to an additional bulk energy:

$$\int (H - H_0)^2 \, \mathrm{d}V,$$

which could not be compensated by a decrease in energy in the surface sheath. Hence, the current density in the surface sheath is distributed so that $\int_0^\infty j_y \, \mathrm{d}x = 0$, i.e., j_y changes sign with depth. This is seen from the fact that j_y is equal to $-A\Psi^2 = -H_0(x - x_0)\Psi^2$. If the result of the variational calculation is substituted, we obtain

$$\int_0^\infty (\zeta - \zeta_0) \exp(-b\zeta)^2 \, \mathrm{d}\zeta = \tfrac{1}{2}\left[\frac{1}{b} - \zeta_0\left(\frac{\pi}{b}\right)^{1/2}\right].$$

According to eq. (18.79) this integral is equal to zero. This is also valid for the exact solution.

It is interesting to note that at $1/\sqrt{2} > \varkappa > (1.69\sqrt{2})^{-1}$ the superconductor will belong to type I, i.e., it passes to the normal state discontinuously at H_{cm}. However, the superconductivity in the surface sheath will persist up to $H_{c3} = 1.69\sqrt{2}\,\varkappa H_{\mathrm{cm}}$. From all this it follows that the determinations of the transition to the normal state in a magnetic field from the disappearance of the diamagnetic moment at the one hand and the appearance of resistance between contacts applied to the superconductor surface on the other may give different results.

The field H_{c3} plays another important role. In section 17.3 we have noted that the normal phase in type I superconductors may be retained when the field is decreased below the critical value. This retardation may occur up to the limiting supercooling field, when the free energy minimum corresponding to the normal phase vanishes. In the case of a bulk type I superconductor the supercooling field is determined from the condition of the possibility of existence of a small stationary nucleus of the superconducting phase in a normal conductor. As a matter of fact, at larger fields it must decrease with time and increase at lower fields. But this is exactly the condition from which the fields H_{c2} and H_{c3} were determined. Since $H_{c3} > H_{c2}$, the supercooling field is defined exactly as the field H_{c3}. Of course, this makes sense only for those type I superconductors for which $H_{c3} < H_{cm}$, i.e., $\varkappa < (1.69\sqrt{2})^{-1}$. But if $1/\sqrt{2} > \varkappa > (1.69\sqrt{2})^{-1}$, the retardation of the normal phase does not occur because a superconducting nucleus appears at the surface even before the transition.

However, the surface can be "spoiled" deliberately, say by way of absorption of magnetic atoms (section 21.1) or by sputtering a film of a normal metal (section 20.1). In such a case, surface superconductivity is impossible, and the role of the supercooling field will be played by H_{c2}.

Let us now examine a case where the external field is directed at an angle to the sample surface. Let the superconductor occupy the half-space $x > 0$ and let the magnetic field have the following components:

$$H_z = H_0 \cos\theta, \qquad H_x = H_0 \sin\theta. \tag{18.79}$$

The vector potential is chosen along the y axis:

$$A_y = H_0(x\cos\theta - z\sin\theta). \tag{18.80}$$

The equation for Ψ in the linear approximation is written as

$$-\frac{\partial^2\Psi}{\partial x^2} + \left[\mathrm{i}\frac{\partial}{\partial y} + \varkappa H_0(x\cos\theta - z\sin\theta)\right]^2\Psi - \frac{\partial^2\Psi}{\partial z^2} = \varkappa^2\Psi. \tag{18.81}$$

The boundary conditions are the same as before.

Along the boundary plane, in particular along the z direction, the conditions must be uniform. It is not difficult to see that the variation of z in eq. (18.81) can be compensated by the phase factor $\exp(\mathrm{i}ky)$ in the function Ψ, i.e., we are dealing here with a situation very similar to what occurs close to H_{c2} (section 18.2). But in order to obtain the critical field, it is sufficient to solve the linear equation, and the solution may be a superposition of the solutions localized near any z, and in particular, one solution of such a type. Therefore, it is sufficient to find Ψ which vanishes as $z \to \pm\infty$. Substituting $\Psi(x, y, z) = \exp(\mathrm{i}\varkappa H_0 x_0 y)\Psi(x, z)$ and introducing the variables

$$\zeta = (\varkappa H_0)^{1/2}x, \qquad \zeta_0 = (\varkappa H_0)^{1/2}x_0, \qquad \eta = (\varkappa H_0)^{1/2}z, \qquad \beta = \varkappa / H_0,$$

we obtain the generalization of eq. (18.77):

$$-\frac{\partial^2\Psi}{\partial\zeta^2}-\frac{\partial^2\Psi}{\partial\eta^2}+[(\zeta-\zeta_0)\cos\theta-\eta\sin\theta]^2\Psi=\beta\Psi.$$

This equation is equivalent to the variational principle for determination of the minimum of β:

$$\beta=\frac{\int_0^\infty d\zeta\int_{-\infty}^\infty d\eta\left\{\left(\frac{\partial\Psi}{\partial\zeta}\right)^2+\left(\frac{\partial\Psi}{\partial\eta}\right)^2+[(\zeta-\zeta_0)\cos\theta-\eta\sin\theta]^2\Psi^2\right\}}{\int_0^\infty d\zeta\int_{-\infty}^\infty d\eta\,\Psi^2}. \tag{18.82}$$

We shall assume the trial function $\Psi=\exp(-\frac{1}{2}a\eta^2-\frac{1}{2}b\zeta^2)$, which would satisfy the boundary conditions. We get

$$\beta=\tfrac{1}{2}\left(b+\frac{1}{b}\cos^2\theta\right)+\tfrac{1}{2}\left(a+\frac{1}{a}\sin^2\theta\right)+\cos^2\theta\left(\zeta_0^2-\frac{2\zeta_0}{(\pi b)^{1/2}}\right).$$

Minimizing this expression with respect to a, b and ζ_0, we find

$$\beta=\cos\theta\left(1-\frac{2}{\pi}\right)^{1/2}+\sin\theta,\qquad b=\cos\theta\left(1-\frac{2}{\pi}\right)^{1/2},$$

$$a=\sin\theta,\qquad \zeta_0=\frac{1}{(\pi b)^{1/2}}. \tag{18.83}$$

Substituting

$$\beta=\frac{\varkappa}{H_{c3}(\theta)},\qquad H_{c3}(0)=\varkappa\left(1-\frac{2}{\pi}\right)^{-1/2},\qquad H_{c2}=\varkappa,$$

we have

$$\frac{1}{H_{c3}(\theta)}=\frac{1}{H_{c3}(0)}\cos\theta+\frac{1}{H_{c2}}\sin\theta, \tag{18.84}$$

that is, a smooth variation of H_{c3} from H_{c3} (0) at $\theta=0$ to H_{c2} at $\theta=\frac{1}{2}\pi$. This gives, in particular,

$$-\left(\frac{1}{H_{c3}}\frac{dH_{c3}}{d\theta}\right)_{\theta=0}=\frac{H_{c3}(0)}{H_{c2}}=1.66.$$

An exact calculation gives the value 1.35 (Saint-James 1965).

In order to determine the current distribution at fields somewhat lower than $H_{c3}(\theta)$, we have to solve the nonlinear Ginzburg–Landau equation. The solution is again taken in the form of a superposition of solutions of the type we found earlier, centered around various z:

$$\Psi = \exp(i\varkappa H_0 x_0 y) \sum_{n=-\infty}^{\infty} C_n \, e^{ikny} \, \Psi_0\left(x, z - \frac{nk}{\varkappa H_0 \sin\theta}\right). \tag{18.85}$$

A superposition of the same type, (18.18), was obtained close to H_{c2}. The subsequent line of reasoning is quite analogous to section 18.2, and we shall not give it here. The result is a solution containing vortices, i.e., points which make the phase Ψ change by 2π when they are encircled (Kulik 1968a).

The periods of the vortex structure can be found using the idea that the rise of the vector potential is compensated by jumps of the phase gradient at branch cuts drawn through the vortex points parallel to the y axis. If the period along the y axis is a, then the jump in $\partial\varphi/\partial y$ at each line is $2\pi/a$. Compensation of the increase of the term $H_0 z \sin\theta$ in (18.81) occurs with the condition

$$-\frac{2\pi}{a} + \varkappa H_0 b \sin\theta = 0,$$

where b is the period in the z direction, or

$$ab = \frac{2\pi}{\varkappa H_0 \sin\theta}. \tag{18.86}$$

In the case under consideration, as opposed to the neighborhood of H_{c2}, anisotropy arises: the directions y and z are not equivalent because the field $\boldsymbol{H}$ lies in the xz plane. Nevertheless, if we assume a rectangular vortex lattice, the minimization of the free energy will lead to the result $a = b$, i.e., it will give a square lattice with a period given by

$$a = b = \left(\frac{2\pi}{\varkappa H_0 \sin\theta}\right)^{1/2}. \tag{18.87}$$

Other variants have not been studied.

From formula (18.87) it follows that the period of the vortex structure increases with decreasing θ according to the law $a \sim \theta^{-1/2}$ (at small θ). Moreover, one of the basis vectors of the vortex lattice is directed along the component H in the plane of the surface (this is the z axis in the case under consideration).

The field H_{c3} has been detected in numerous experiments. There are also data in favor of the vortex structure of the surface currents with a tilted field. They refer to the critical current for surface superconductivity. A theoretical calculation of the critical current for the field parallel to the surface (Abrikosov 1964b), which is not given here, yields

$$J_c \sim j_c \xi, \tag{18.88}$$

where J_c is the critical current flowing through a cross-sectional area of length 1 cm (along the surface) and extending to infinity into the bulk of the metal; j_c is the absolute critical value of the current density corresponding to the Landau criterion (see eq. 17.63). This result is obvious since the thickness of the superconducting sheath in the case of surface superconductivity is of order ξ. Formula (18.88) is an estimate of the value of the critical current within the interval $H_{c2} < H_0 < H_{c3}$. Actually the current depends on H_0 and vanishes at H_{c3}.

Much lower values of the critical current are always obtained in experiments. This may be accounted for by the fact that a field orientation exactly parallel to the surface cannot be realized in practice, and with a tilted field a vortex structure is formed. When current is flowing, the vortices are acted on by a force, which makes them move. As a result, electrical resistance arises (this point will be discussed in more detail in section 18.8). Strictly speaking, we must obtain zero critical current. However, in fact, the vortices cling to surface irregularities (pinning, see section 18.8) and therefore with a sufficiently small current they remain immobile. This effect can be enhanced by deliberately making the surface rougher. Indeed, in this case the critical current increases (Akhmedov et al. 1969).

18.5. Type II superconductors at low temperatures

In this section we shall draw certain conclusions concerning the behavior of type II superconductors at temperatures far from T_c, where the Ginzburg–Landau theory does not apply. We use conventional units.

It is rather obvious that at low temperatures the qualitative picture of the magnetic properties of superconductors is the same as in the vicinity of T_c. There is a surface energy σ_{ns}, which may be positive or negative. In type I superconductors it is positive; a first-order transition occurs at the field H_{cm}, and there is a supercooling field H_{c3} or H_{c2}. For type II superconductors σ_{ns} is negative, and the transition to the normal phase takes place by way of a second-order transition in the field H_{c2}. Therefore, the ratio

$$\frac{H_{c2}(T)}{H_{cm}(T)\sqrt{2}} = \varkappa(T) \tag{18.89}$$

is a quantity that distinguishes the two types of superconductors (with $\varkappa < 1/\sqrt{2}$ and with $\varkappa > 1/\sqrt{2}$), and exactly this quantity was given in formulae (16.97) and (16.101).

In type II superconductors at all temperatures there arises a mixed state in a magnetic field. By order of magnitude the field H_{c1} has been defined in section 18.1. However, in case $\varkappa \gg 1$ (recall that, according to formulae 16.97 and 16.101 $\varkappa$ varies little with temperature) one can determine the value of H_{c1} rather precisely. To do this, we take into account that the major part of the vortex energy builds up in the region $\xi \ll \rho \ll \delta$, and in this region the detailed structure of its core, which has a radius of order ξ, is unimportant. The only important point is that the order parameter and, hence, the current, begins to decrease at such distances and vanishes in the center.

Just as near T_c, each vortex carries one quantum of magnetic flux. This follows from the fact that at $\delta \gg \xi$ far from the vortex axis the London equation (section 16.7) is valid. According to this equation, the current is connected with the vector potential by the relation $\boldsymbol{j} = -Q\boldsymbol{A}$, but in such a form the current is not gauge-invariant, i.e., it depends on the choice of $\boldsymbol{A}$. To provide gauge invariance, it is necessary that $\boldsymbol{A}$ be included in the combination with the gradient of the phase of the complex order parameter $\Delta = |\Delta| \, e^{i\chi}$:

$$\boldsymbol{j} = -Q\left(\boldsymbol{A} - \frac{\hbar c}{2e}\nabla\chi\right). \tag{18.90}$$

But this expression is quite analogous to eq. (17.75).

Considering that far from the vortex axis $|\Delta|$ and, hence, Q are equal to their values in a superconductor in the absence of the field, i.e., are independent of the coordinates, it may be asserted that the flux is equal to an integer number of flux quanta inside the loop, for which $\oint \boldsymbol{j}\, d\boldsymbol{l} = 0$. If we are dealing with a single vortex, then it is sufficient to choose the loop at $\rho \gg \delta$, where $\boldsymbol{j} = 0$. In a case where there are many vortices and they form a lattice, one can find such a loop around each vortex, on which $\oint \boldsymbol{j}\, d\boldsymbol{l} = 0$ due to the lattice symmetry. Evidently, each vortex carries only one flux quantum.

As has been noted, at $\rho \gg \xi$ we may consider Δ to be equal to its value in the absence of the field, which is why for H the London equation (15.39) is valid. Thus, we have

$$\Delta H - \delta^{-2} H = -\delta^{-2} \Phi_0 \delta(\boldsymbol{\rho}). \tag{18.91}$$

The right-hand side takes account of the presence of a vortex with a magnetic flux Φ_0. Integrating over the cross-sectional area perpendicular to the vortex axis, we can see that the choice of the right-hand side is correct. The solution of eq. (18.91) is

$$H = \frac{1}{2\pi\delta^2} \Phi_0 K_0\left(\frac{\rho}{\delta}\right). \tag{18.92}$$

Now we must write the expression for the free energy corresponding to the London equation. It is made up of the field energy and the kinetic energy of the supercurrent. For unit volume we have

$$\Omega_s - \Omega_s^{(0)} = \frac{H^2}{8\pi} + \tfrac{1}{2} n_s m v_s^2 = \frac{H^2}{8\pi} + \tfrac{1}{2} n_s m \left(\frac{j}{n_s e}\right)^2 = \frac{H^2}{8\pi} + \tfrac{1}{2} \Lambda j^2, \tag{18.93}$$

where use is made of the expression $\Lambda = m/n_s e^2$ (see eq. 15.35). Note that with this derivation it is immaterial whether we use the concept of superconducting electrons with density n_s, charge e and mass m or the concept of pairs with density $\tfrac{1}{2} n_s$, charge $2e$ and mass $2m$.

The integral over the volume of the last term can be transformed using the Maxwell equation rot $\boldsymbol{H}=(4\pi/c)\,\boldsymbol{j}$:

$$\tfrac{1}{2}\Lambda\int j^2\,\mathrm{d}V=\tfrac{1}{2}\Lambda\left(\frac{c}{4\pi}\right)^2\int(\mathrm{rot}\,\boldsymbol{H})^2\,\mathrm{d}V$$
$$=\frac{\delta^2}{8\pi}\int\{\mathrm{div}[\boldsymbol{H}\,\mathrm{rot}\,\boldsymbol{H}]+\boldsymbol{H}\,\mathrm{rot}\,\mathrm{rot}\,\boldsymbol{H}\}\,\mathrm{d}V$$
$$=\frac{\delta^2}{8\pi}\left\{\int_S[\boldsymbol{H}\,\mathrm{rot}\,\boldsymbol{H}]\,\mathrm{d}\boldsymbol{S}-\int\boldsymbol{H}\Delta\boldsymbol{H}\,\mathrm{d}V\right\}.$$

Here we have used the formula div $[\boldsymbol{ab}]=\boldsymbol{b}$ rot $\boldsymbol{a}-\boldsymbol{a}$ rot $\boldsymbol{b}$, transformed the integral of div to an integral over the sample surface and took advantage of the fact that rot rot $\boldsymbol{H}=$ grad div $\boldsymbol{H}-\Delta\boldsymbol{H}$ and div $\boldsymbol{H}=0$. Thus, we have

$$\int\Omega_s\,\mathrm{d}V=\int\Omega_s^{(0)}\,\mathrm{d}V+\frac{\delta^2}{8\pi}\int_S[\boldsymbol{H}\,\mathrm{rot}\,\boldsymbol{H}]\,\mathrm{d}\boldsymbol{S}$$
$$+\frac{1}{8\pi}\int\boldsymbol{H}(\boldsymbol{H}-\delta^2\Delta\boldsymbol{H})\,\mathrm{d}V. \tag{18.94}$$

We give also another form of this expression. Since

$$[\boldsymbol{H}\,\mathrm{rot}\,\boldsymbol{H}]=\tfrac{1}{2}\nabla(H^2)-(\boldsymbol{H}\nabla)\boldsymbol{H},\qquad \boldsymbol{H}\Delta\boldsymbol{H}=\mathrm{div}[\tfrac{1}{2}\nabla(H)^2]-(\nabla\boldsymbol{H})^2,$$

we can obtain from eq. (18.94), for a case where $(\boldsymbol{H}\nabla)\boldsymbol{H}=0$, i.e., $\boldsymbol{H}$ is directed along z and depends only on ρ:

$$\int\Omega_s\,\mathrm{d}V=\int\Omega_s^{(0)}+\frac{1}{8\pi}\int[H^2+\delta^2(\nabla\boldsymbol{H})^2]\,\mathrm{d}V. \tag{18.94$'$}$$

In the case of a single vortex under consideration the surface integral in (18.94) is equal to zero. We must substitute formulae (18.91) and (18.92) into the second integral of (18.94). The logarithmic infinity $K_0(\rho)$ at $\rho=0$ must be replaced by $\ln(\delta/\xi)$ because at a distance of order ξ from the axis eq. (18.91) is no longer valid and H ceases to depend on the coordinate (with a logarithmic accuracy). As a result, we obtain an extra energy per unit length of the vortex:

$$\varepsilon_0\approx\frac{1}{(4\pi\delta)^2}\Phi_0^2\ln\left(\frac{\delta}{\xi}\right). \tag{18.95}$$

We have limited ourselves to the calculation of the dominant term with $\ln(\delta/\xi)$ because higher accuracy requires a knowledge of the core structure.

The condition of the nucleation of a vortex is

$$\varepsilon_0-\frac{1}{4\pi}H_0\int H\,\mathrm{d}S=\varepsilon_0-\frac{1}{4\pi}H_0\Phi_0=0.$$

From this relation we obtain

$$H_{c1} = 4\frac{\pi\varepsilon_0}{\Phi_0} \approx \frac{1}{4\pi}\frac{\Phi_0}{\delta^2}\ln\left(\frac{\delta}{\xi}\right). \tag{18.96}$$

This formula is consistent in order of magnitude with eqs. (18.8) and (18.50). Note that, in accordance with formula (18.91), the field in the vortex core is equal with logarithmic accuracy to

$$H(0) \approx \frac{1}{2\pi}\frac{\Phi_0}{\delta^2}\ln\left(\frac{\delta}{j}\right),$$

that is, just as near T_c, $H(0)$ is about twice as large as H_{c1}.

The expression for the current is derived from the Maxwell equation $\boldsymbol{j} = (c/4\pi)$ rot $\boldsymbol{H}$. In the case under consideration,

$$j_\varphi = \frac{c}{2\delta}\frac{1}{(2\pi\delta)^2}\Phi_0 K_1\left(\frac{\rho}{\delta}\right). \tag{18.97}$$

At $\rho \ll \delta$, $K_1 \approx \delta/\rho$. Substituting $\Phi_0 = \pi\hbar c/e$, $\delta = (mc^2/4\pi n_s e^2)^{1/2}$ gives

$$j_\varphi \approx n_s e\frac{\hbar}{2m\rho}.$$

From this it follows that the velocity of Cooper pairs in the vortex at a distance of $\xi \ll \rho \ll \delta$ is equal to $\hbar/2m\rho$ and this is in agreement with formula (18.7) deduced from the Bohr quantization rule. At distances of $\rho \gtrsim \delta$ it becomes important that the liquid consists of charged particles and its motion leads to generation of a magnetic field. As a result, at $\rho \gg \delta$ the current is damped according to the law $\exp(-\rho/\delta)$.

Taking into account the agreement between eq. (18.52) near T_c and eq. (18.91) and also between expressions (18.55) and (18.94) for the free energy, and considering that each vortex carries one flux quantum, we conclude that formulae (18.60) and (18.58) are also valid, provided that correct units are used, namely $\varkappa$ has to be replaced by $2\pi\delta^2/\Phi_0$. This refers also to the limiting expressions (18.61) and (18.67) and also to formula (18.68) without the last term, which cannot be determined accurately because of the logarithmic accuracy in the calculation of H_{c1}. All these results characterize the dependence $B(H_0)$ at $H_{c1} \leqslant H_0 \ll H_{c2}$.

In what follows the estimates for critical fields expressed in terms of Φ_0 and characteristic lengths will be useful. Formula (18.96) gives an estimate for H_{c1}. With the field H_{c2} the vortices are moved together to distances of order ξ and since there is one flux quantum per vortex, it follows that $H_{c2} \sim \Phi_0/\xi^2$. Finally, because of $H_{c2} \sim H_{cm}\varkappa \sim H_{cm}\delta/\xi$, we have $H_{cm} \sim \Phi_0/\delta\xi$. Thus,

$$H_{c1} \sim \frac{\Phi_0}{\delta^2}, \qquad H_{cm} \sim \frac{\Phi_0}{\delta\xi}, \qquad H_{c2} \sim \frac{\Phi_0}{\xi^2}. \tag{18.98}$$

While comparing the theory with experiment, it is necessary to take into account a number of circumstances. First of all, large $\varkappa$ values are attained only in superconductors with a large impurity concentration. We have already noted in section 16.9 that impurities present in a small concentration have a very slight effect on the thermodynamic properties of the superconductor, in particular on H_{cm}. However, at a large impurity concentration of the order of tens of percent, we are actually dealing with a new substance.

The boundaries of the mixed state are the fields H_{c1} and H_{c2}, which are sufficiently clearly seen in the $B(H_0)$ curve. It turns out that from this curve one can also easily determine H_{cm}. Indeed, at H_{c2}

$$\Omega_s(H_{c2}) = \Omega_n(H_{c2}).$$

From this it follows (see Appendix III) that

$$\Omega_s^{(0)} - \frac{1}{4\pi}\int_0^{H_{c2}} B(H_0)\,\mathrm{d}H_0 = \Omega_n^{(0)} - \frac{H_{c2}^2}{8\pi}.$$

But since $\Omega_n^{(0)} - \Omega_s^{(0)} = H_{cm}^2/8\pi$, we find that

$$\tfrac{1}{2}H_{cm}^2 = \int_0^{H_{c2}} [H_0 - B(H_0)]\,\mathrm{d}H_0 = \int_0^{H_{c2}} [-4\pi M(H_0)]\,\mathrm{d}H_0. \tag{18.99}$$

In other words, H_{cm} is determined by the area under the curve of $-4\pi M$ versus H_0.

In experiments one often has to work with specimens of noncylindrical shape. If the specimen has the form of an ellipsoid, the role of the external field H_0 is played by the "internal" field H_i (Appendix III):

$$H_i = H_0 - 4\pi n M,$$

where M is the magnetization and n is the demagnetization factor. The magnetization M is connected with B by the relation

$$B = H_i + 4\pi M.$$

Expressing M by this relation and substituting into the preceding formula, we obtain

$$(1-n)H_i + nB = H_0.$$

If the dependence $B(H_0)$ for cylindrical geometry obtained earlier is written in the form $H_0 = f(B)$, the same function will now relate H_i to B. As a result, we have

$$(1-n)f(B) + nB = H_0. \tag{18.100}$$

This is just the relation between B and H_0 for the ellipsoid. In particular, the onset of the mixed state corresponds to $B=0$, $f(B)=H_{c1}$. Hence, for the ellipsoid it begins at

$$H_0 = (1-n)H_{c1}. \tag{18.101}$$

In particular, for an infinite plate in a perpendicular field ($n=1$) it begins at $H_0=0$. The termination of the mixed state corresponds to $B=f(B)=H_0=H_{c2}$.

While speaking of comparison with experiment, one more circumstance must be borne in mind. The variation of the induction with H_0 in the mixed state is the result of the variation of the vortex density. Hence, upon gradual variation of H_0 there is a displacement of the vortices. This displacement can be hindered by the microscopic irregularities of the specimen (pinning). We defer a more detailed discussion of this point until section 18.7; for the moment we only note that the dependence $B(H_0)$ for real specimens may prove to be rather far from the theoretical equilibrium curve, so that comparison of theory with experiment may be made only for specially prepared homogenized specimens. Homogenization can be achieved, in particular, by prolonged annealing at a temperature close to the melting point.

However, even in this case there is a certain inevitable potential barrier which impedes the free motion of the vortices. It is associated with the sample surface (Bean and Livingston 1964). Let us consider a vortex which is approaching the surface (fig. 119). We assume that $\varkappa \gg 1$, that the vortex axis is parallel to the sample surface (along the z axis), and that the distance x_1 from the axis to the surface by far exceeds the radius of the core. In the absence of the vortex the field is equal to $H_0\,\mathrm{e}^{-x/\delta}$. The vortex adds its own magnetic field, which is distorted by the surface so as, first, not to produce an extra field at the surface (because the field is specified and equals H_0) and, second, to make the current normal to the surface vanish.

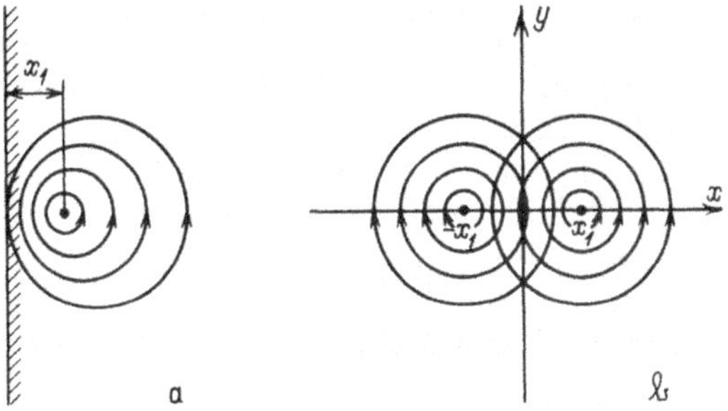

Fig. 119.

This can be realized if we add to the vortex its mirror image relative to the surface with the opposite direction of the field and current (fig. 119b). The vortex field will satisfy the London equation with two sources:

$$\Delta H - \frac{1}{\delta^2} H = -\frac{1}{\delta^2} \Phi_0[\delta(\boldsymbol{\rho} - \boldsymbol{\rho}_1) - \delta(\boldsymbol{\rho} - \boldsymbol{\rho}_2)], \tag{18.102}$$

where $\boldsymbol{\rho}_1 = (x, 0)$ and $\boldsymbol{\rho}_2 = (-x_1, 0)$; the total magnetic field will be given by

$$H = H_0\,\mathrm{e}^{-x/\delta} + \frac{\Phi_0}{2\pi\delta^2}\left[K_0\left(\frac{|\boldsymbol{\rho} - \boldsymbol{\rho}_1|}{\delta}\right) - K_0\left(\frac{|\boldsymbol{\rho} - \boldsymbol{\rho}_2|}{\delta}\right)\right]. \tag{18.103}$$

Let us find the energy in a given field using these formulas:

$$\int \Omega_{sH}\,dV = \int \Omega_s\,dV - \frac{1}{4\pi}H_0 \int H\,dV.$$

We shall calculate the energy per unit length along z and integrate over the half-plane $x>0$. In this case, as opposed to a single vortex in infinite space, the integral over the surface does not vanish, so we must be careful.

Let us, first, integrate eq. (18.102) over the half-plane $x>0$. Then we obtain (the normal to the surface is directed along $-x$):

$$\int H\,dV = \Phi_0 + \delta^2 \int \Delta H\,dx\,dy = \Phi_0 - \delta^2 \int_{x=0} \frac{\partial H}{\partial x}\,dy.$$

Hence,

$$-\frac{H_0}{4\pi}\int H\,dV = -\frac{H_0\Phi_0}{4\pi} + \frac{\delta^2 H_0}{4\pi}\int_{x=0} \frac{\partial H}{\partial x}\,dy.$$

The surface integral in (18.94) is equal to

$$\frac{\delta^2}{8\pi}\int [\boldsymbol{H}\,\mathrm{rot}\,\boldsymbol{H}]\,d\boldsymbol{S} = -\frac{\delta^2}{8\pi}H_0\int_{x=0}\frac{\partial H}{\partial x}\,dy.$$

The volume integral in eq. (18.94) is evaluated with the aid of formulae (18.102) and (18.103). Here we replace $K_0(0)$ by $\ln(\delta/\xi)$ because at distances smaller than ξ the theory presented here is no longer valid and we have to extrapolate H from $x \sim \xi$. We also make use of formula (18.96). Substituting formulas (18.103) for H into the integral $(\delta^2/8\pi)H_0\int_{x=0}(\partial H/\partial x)\,dy$, we see that the first term gives $-(\delta H_0^2/8\pi)\int dy$, which corresponds to the field energy in the layer δ without a vortex. This energy is neglected here.

As a result, the energy associated with the vortex is given by

$$U(x_1) = \frac{\Phi_0}{4\pi}\left\{H_0 \exp\left(-\frac{x_1}{\delta}\right) - \frac{\Phi_0}{4\pi\delta^2}K_0\left(\frac{2x_1}{\delta}\right) - H_0 + H_{c1}\right\}. \tag{18.104}$$

At distances $x_1 \leqslant \xi$, $K_0(2x_1/\delta)$ must be replaced by $\ln(\delta/\xi)$. Comparing with (18.96), we see that over such distances $U(x_1)$ vanishes. With $x_1 \to \infty$ the first two terms of (18.104) vanish; $U(\infty)<0$ at $H_0 > H_{c1}$ and $U(\infty)>0$ at $H_0 < H_{c1}$, as should be expected. The graph of the function $U(x_1)$ is given in fig. 120. We see that a potential barrier appears near the surface, which hinders both the entry of the vortex into the superconductor and its escape from it. An increase in the field leads to the disappearance of the barrier only at a certain field $H_0 = H'_{c1} > H_{c1}$.

This field can be estimated as follows. The condition $U'(x_1)=0$ yields

$$\frac{\Phi_0}{4\pi\delta^2}\frac{2}{\delta}K_1\left(\frac{2x_1}{\delta}\right) - \frac{H_0}{\delta}\exp\left(-\frac{x_1}{\delta}\right) = 0.$$

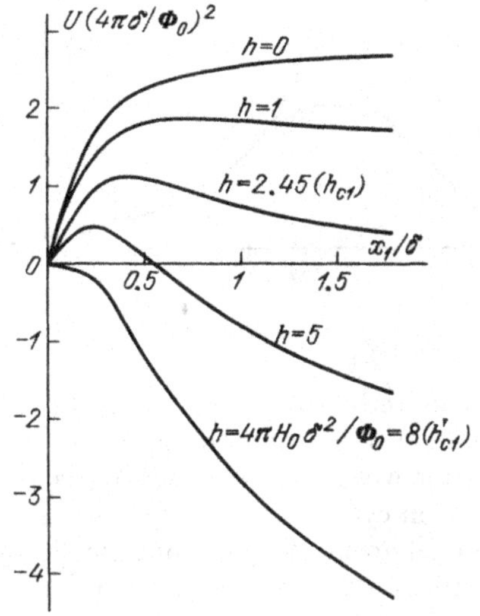

Fig. 120.

We are interested in the case where the top of the barrier reaches the surface, i.e., $x_1=0$. But as $x_1\to 0$, $K_1(2x_1/\delta)\approx\delta/2x_1$. As has already been pointed out, at small x_1 we have to replace x_1 by ξ, i.e., $K_1(2x_1/\delta)\sim\delta/\xi$. As a result, we have (see eq. 18.97):

$$H'_{c1}\sim\frac{\Phi_0}{\delta\xi}\sim H_{cm}.$$

As the field decreases the barrier disappears only at $H_0=0$.

In real experiments, the field lines of force are always concentrated at the edges of the sample, and thus at these sites the external field may exceed H'_{c1}. It is these edges that will be "weak places" through which the vortices will enter the sample. Therefore, the $B(H_0)$ curve taken in an increasing field may be regarded as an equilibrium curve*. However, when the field falls off, the sample surface has no portions where the field is zero and, hence, the barrier is preserved everywhere. Therefore, the magnetization curve taken in a decreasing field even in the most homogenized samples will be nonequilibrium. This is supported by experimental data (fig. 121).

We will now consider another property of the mixed state. As has been found earlier, at the vortex axis $\Delta=0$. But the spectrum of quasiparticles in a superconductor begins with the energy Δ. In view of this, the possibility is not excluded that there exist special quasiparticles localized near the vortex axis, with zero excitation

* In specially prepared samples with a very smooth surface it is possible to succeed in retarding the transition in an increasing field. This provides good agreement with the result of the exact theory, $H'_{c1}=H_{cm}$.

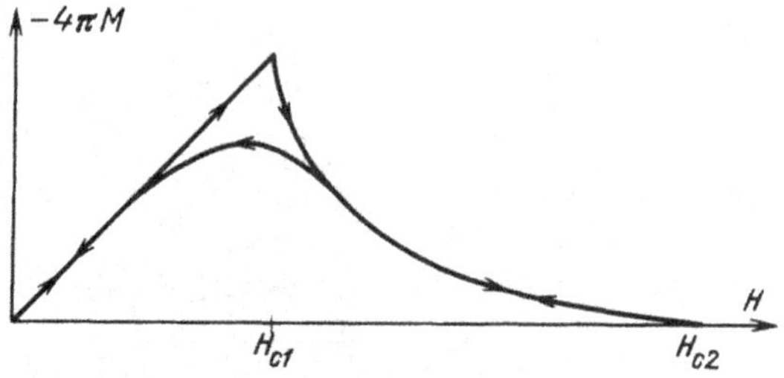

Fig. 121.

energy. This idea is confirmed by rigorous theoretical calculations (Caroli and Matricon 1965, Caroli et al. 1965). Indeed, in the mixed state the superconductor has quasiparticles which move along the vortex axis and which have finite density of states as $\varepsilon \to 0$, just as in cylinders of a normal metal with a radius of order ξ.

Since the cross-sectional area of the core of a single vortex is of the order of ξ^2 and the number of vortices that pass through 1 cm^2 is $n = B/\Phi_0$, the fraction of "normal metal" is $\xi^2 B/\Phi_0$ and, accordingly, the density of states is given by

$$\nu \sim \nu_n \xi^2 \frac{B}{\Phi_0},$$

where ν_n is the density of states for a normal metal. At the field H_{c2} the vortices approach each other to a distance of the order of ξ; in this case $B \approx H_{c2} \sim \Phi_0/\xi^2$ (see eq. 18.97). Hence, the density of states may be written as

$$\nu = \nu_n \frac{B}{H_{c2}} \tag{18.105}$$

(the equality sign is used because of the condition of matching with a normal metal at $B = H_{c2}$). But the finite density of states leads to a linear dependence of heat capacity on temperature. Thus, we arrive at the conclusion that at $T \ll T_c$ in type II superconductors in the mixed state the heat capacity is linear in temperature:

$$\Delta C(B) = \tfrac{1}{3}\pi^2 \nu_n \frac{TB}{H_{c2}} = C_n(T)\frac{B}{H_{c2}}. \tag{18.106}$$

In conclusion we note that, just as in the vicinity of T_c, the generation of a superconducting nucleus near the surface is possible at fields larger than H_{c2}. The ratio H_{c3}/H_{c2} depends on temperature and on electron scattering conditions at the sample surface. A calculation performed by means of a variational method similar to that employed in section 18.4 gives, at $T = 0$, $H_{c3}/H_{c2} \geqslant 1.99$ for specular reflection of electrons from the surface and $H_{c3}/H_{c2} \geqslant 2.09$ for diffuse reflection (Kulik 1968). We see that all the magnetic properties of type II superconductors are not only qualitatively similar to those observed in the vicinity of T_c but differ little quantitatively from them (in the corresponding units).

18.6. A thin film of a type II superconductor in a magnetic field

The critical field of thin films of type I superconductors with $\varkappa \ll 1$ has been calculated in section 17.3. The condition $\varkappa \ll 1$ enabled us to assume $\Psi \approx$ const. throughout the superconductor. In all subsequent computations $\varkappa$ did not appear. Some of the results of section 17.3 may also be extended to type II superconductors. Indeed, with thicknesses of $d \ll \xi$ (in reduced units $d \ll \varkappa^{-1}$) Ψ must not vary throughout the film thickness. But for type II superconductors $\xi \ll \delta$ and, hence, one must employ the formulas for $d \ll \delta$ (in reduced units $d \ll 1$). In particular, the critical field of the transition to the normal state is expressed by formulae (17.48) or (17.48′). Note that for such films the ratio δ/d plays the role of the "effective" parameter $\varkappa$, therefore formula (17.48) is completely consistent with formulae (18.2) and (18.3) for the upper critical field of type II superconductors.

As far as real experimental objects are concerned, films of type II superconductors are prepared by vaporization of a small piece of metal and condensation of the vapor onto a non-metallic substrate kept at helium temperature. It turns out that the atoms of the metal upon contact with the substrate immediately stick to it and are not displaced. Therefore, the film formed is almost entirely amorphous, i.e., has a small mean free electron path and therefore belongs to type II, even if the parent pure metal was a type I superconductor and had $\varkappa \ll 1$. Thus, we see that a film of a type II superconductor is a rather common object. When such a film is annealed, i.e., its temperature is raised to several tens of degrees, recrystallization occurs, and it becomes a type I superconductor with H_{cm} and T_c close to the characteristics of the bulk material. The same occurs if the metal is vaporized onto a substrate kept at room temperature.

A question arises as to the possibility of realizing a mixed state in a film with $\varkappa > 1/\sqrt{2}$. The qualitative answer to this question is rather obvious: the mixed state is possible with thicknesses of $d \gtrsim \xi$. From this it follows that with decreasing film thickness the field H_{c1} must increase more rapidly than the field H_{c2} and they must become equal at $d \sim \xi$.

Let us now consider the problem quantitatively. In order to find H_{c1}, we are to find the conditions for the creation of one vortex. From symmetry considerations it is clear that the lowest energy will be exhibited by a vortex in the center of the film. In the preceding section it has been shown that the boundary conditions on a planar boundary may be fulfilled by introduction of a specularly reflected vortex with an opposite direction of the field and current. In the case under consideration, there are two boundaries and, as a result, there arises an infinite system of reflections which are alternating in field and current, with a distance d between the nearest vortices. Instead of eq. (18.102), we have

$$\Delta H - \frac{1}{\delta^2} H = \frac{1}{\delta^2} \Phi_0 \sum_{n=-\infty}^{\infty} (-1)^n \delta(\boldsymbol{\rho} - \boldsymbol{\rho}_n), \tag{18.107}$$

where $\boldsymbol{\rho}_n = (nd, 0)$ and n assumes all integer values, positive and negative.

The total magnetic field, which satisfies the conditions $H = H_0$ at $x = \pm\frac{1}{2}d$ will be equal to (see eq. 18.102 and 18.103)

$$H = H_0 \frac{\cosh(x/\delta)}{\cosh(d/2\delta)} + \frac{\Phi_0}{2\pi\delta^2} \sum_{n=-\infty}^{\infty} (-1)^n K_0\left(\frac{|\boldsymbol{\rho} - \boldsymbol{\rho}_n|}{\delta}\right). \tag{18.108}$$

Let us now find the free energy. Using the same procedure as in the preceding section, we find the energy of one vortex in the field (per unit length):

$$\int (\Omega_{sH} - \Omega_{sH}^{(0)})\, \mathrm{d}V = \int (\Omega_s - \Omega_s^{(0)})\, \mathrm{d}V - \frac{1}{4\pi} H_0 \int [H(\boldsymbol{\rho}) - H^{(0)}(\boldsymbol{\rho})]\, \mathrm{d}V,$$

where $\Omega_{sH}^{(0)}$ and $\Omega_s^{(0)}$ correspond to the absence of a vortex and $H^{(0)}(\rho)$ is the first term in (18.108). Calculation gives

$$\int_{-d/2}^{d/2} \mathrm{d}x \int_{-\infty}^{\infty} \mathrm{d}y\, (\Omega_{sH} - \Omega_{sH}^{(0)}) = \frac{1}{4\pi} \frac{\Phi_0 H_0}{\cosh(d/2\delta)} + 2\left(\frac{\Phi_0}{4\pi\delta}\right)^2 \sum_{n=1}^{\infty} (-1)^n K_0\left(\frac{nd}{\delta}\right) + \frac{1}{4\pi}[H_{c1}(\infty) - H_0]\Phi,$$

where $H_{c1}(\infty)$ corresponds to an infinite sample. Equating this expression to zero, we obtain the first critical field for a sample of thickness d (Abrikosov 1964a):

$$H_{c1}(d) = \frac{H_{c1}(\infty) - \dfrac{\Phi_0}{2\pi\delta^2} \displaystyle\sum_{n=1}^{\infty} (-1)^{n+1} K_0\left(\frac{nd}{\delta}\right)}{1 - \dfrac{1}{\cosh(d/2\delta)}}. \tag{18.109}$$

The limiting formulas are as follows:

$$H_{c1}(d) \approx H_{c1}(\infty)\left[1 + 2\exp\left(\frac{-d}{2\delta}\right)\right], \quad d \gg \delta,$$

$$H_{c1}(d) \approx \frac{2\Phi_0}{\pi d^2} \ln\left(\frac{d}{\xi}\right), \quad d \ll \delta. \tag{18.110}$$

In the last formula we used the integral representation of the MacDonald function (preceding eq. 16.29) and expression (18.95) for H_{c1} at $\varkappa \gg 1$ (the expression is written with logarithmic accuracy). The asymptotic of (18.110) for $d \ll \delta$ is suitable only at $d \gg \xi$. From this asymptotic it follows that H_{c1} increases with decreasing thickness proportionately to d^{-2}, whereas H_{c2} increases proportionately to d^{-1} (see eq. 17.48). From the BCS theory it follows that $H_{cm} \sim \Phi_0/(\delta\xi)$ (see eq. 18.98). Comparing $H_{c2} \sim H_{cm}\delta/d \sim \Phi_0/(\xi d)$ with $H_{c1} \sim \Phi_0/d^2$, we see that the two fields become equal at $d \sim \xi$.

We shall now consider a thin plate in a perpendicular magnetic field. We assume the plate thickness to be much smaller than the penetration depth. Let us show that in this case vortices are also generated, and not only in a type II superconductor but in a type I superconductor as well (Pearl 1964).

Suppose that such a vortex has been formed and that the field and current are damped at a distance much larger than ξ. In this case, outside the core with a radius of order ξ we may use the London equation with the vortex (18.91), which is more conveniently written in the form

$$\boldsymbol{H} + \frac{4\pi\delta^2}{c} \operatorname{rot} \boldsymbol{j} = \Phi_0 z^0 \delta(\boldsymbol{\rho}) \tag{18.111}$$

(we assume that the plate is perpendicular to the z axis and the vortex is at the point $\rho = 0$). We introduce a vector potential: $\boldsymbol{H} = \operatorname{rot} \boldsymbol{A}$. If use is made of a cylindrical system of coordinates, the current will apparently have only the component j_θ. In view of this, $\boldsymbol{A}$ is also chosen in the form $\boldsymbol{A} = (0, A_\theta, 0)$.

Equation (18.111) may be integrated and the current density expressed through it:

$$\boldsymbol{j} = \frac{c}{4\pi\delta^2}(\boldsymbol{S} - \boldsymbol{A}), \tag{18.112}$$

where $\boldsymbol{S} = (0, S_\theta, 0)$,

$$S_\theta = \frac{\Phi_0}{2\pi\rho} \tag{18.113}$$

(in fact, $\operatorname{rot} \boldsymbol{S} = 0$ everywhere, except for $\boldsymbol{\rho} = 0$, and $\int \operatorname{rot} \boldsymbol{S}\, \mathrm{d}\boldsymbol{s} = \oint \boldsymbol{S}\, \mathrm{d}\boldsymbol{l} = \Phi_0$ for any contour around $\boldsymbol{\rho} = 0$).

If the plate thickness d is smaller than the penetration depth, the current and field vary slightly along the plate thickness. As will be shown later, the characteristic distance over which the field varies significantly is much larger than d and δ, and therefore we may assume the plate to be infinitesimally thin and use the following as the current density:

$$\boldsymbol{j}(\boldsymbol{r}) = \boldsymbol{J}\delta(z) = \boldsymbol{j}\, d\delta(z) = \frac{cd}{4\pi\delta^2}(\boldsymbol{S} - \boldsymbol{A})\, \delta(z). \tag{18.114}$$

This current produces a magnetic field throughout the entire surrounding space, which is determined from the Maxwell equation

$$\operatorname{rot} \boldsymbol{H} = \frac{4\pi}{d}\boldsymbol{j} = \frac{d}{\delta^2}(\boldsymbol{S} - \boldsymbol{A})\, \delta(z).$$

Substituting $\boldsymbol{H}=\operatorname{rot}\boldsymbol{A}$ and using the condition $\operatorname{div}\boldsymbol{A}=0$, we obtain the equation for $\boldsymbol{A}$:

$$-\Delta\boldsymbol{A}+\delta_{\text{eff}}^{-1}\boldsymbol{A}\,\delta(z)=\delta_{\text{eff}}^{-1}\boldsymbol{S}\,\delta(z), \tag{18.115}$$

where the following definition is introduced:

$$\delta_{\text{eff}}=\frac{\delta^2}{d}. \tag{18.116}$$

This parameter, which has the dimension of length, will be the characteristic dimension of the problem. In order for our treatment to make sense, it is necessary that this characteristic dimension be much larger than ξ. For a type II superconductor, this follows from the fact that $\delta>\xi$ and $d\ll\delta$, and for a type I superconductor it is necessary that

$$d\ll\frac{\delta^2}{\xi}\sim\delta\varkappa\sim\xi\varkappa^2. \tag{18.117}$$

In order to solve eq. (18.115), we introduce the Fourier transformation

$$\boldsymbol{A}(\boldsymbol{\rho},z)=\int\boldsymbol{A}_{qk}\exp(\mathrm{i}\boldsymbol{q}\boldsymbol{\rho}+\mathrm{i}kz)\,\mathrm{d}^2\boldsymbol{q}\frac{\mathrm{d}k}{(2\pi)^3}.$$

Here

$$\boldsymbol{A}(\boldsymbol{\rho},z)\,\delta(z)=\boldsymbol{A}(\boldsymbol{\rho},0)\,\delta(z)=\int\boldsymbol{A}_{qk}\exp(\mathrm{i}\boldsymbol{q}\boldsymbol{\rho})\,\mathrm{d}^2\boldsymbol{q}\frac{\mathrm{d}k}{(2\pi)^3}\int_{-\infty}^{\infty}\exp(\mathrm{i}k_1z)\frac{\mathrm{d}k_1}{2\pi}.$$

From eq. (18.115) we get

$$(q^2+k^2)\boldsymbol{A}_{qk}+\frac{1}{\delta_{\text{eff}}}\boldsymbol{A}_q=\frac{1}{\delta_{\text{eff}}}\boldsymbol{S}_q, \tag{18.118}$$

where

$$\boldsymbol{A}_q=\int_{-\infty}^{\infty}\boldsymbol{A}_{qk}\frac{\mathrm{d}k}{2\pi} \tag{18.119}$$

and $\boldsymbol{S}_q$ is the Fourier component of $\boldsymbol{S}(\boldsymbol{\rho})$.

Solving eq. (18.118), we find

$$\boldsymbol{A}_{qk}=\frac{(\boldsymbol{S}_q-\boldsymbol{A}_q)\delta_{\text{eff}}^{-1}}{q^2+k^2}. \tag{18.120}$$

Substitution of eq. (18.120) into eq. (18.119) gives

$$\boldsymbol{A}_q=\frac{\boldsymbol{S}_q}{1+2q\delta_{\text{eff}}}. \tag{18.121}$$

The Fourier component S_q can easily be obtained from the condition

$$\operatorname{rot} \boldsymbol{S} = \Phi_0 z^0 \delta(\boldsymbol{\rho}).$$

Inserting the Fourier expansion, we have

$$\boldsymbol{S}_q = \mathrm{i}\Phi_0 \frac{[\boldsymbol{q}z^0]}{q^2}. \tag{18.122}$$

According to formulae (18.112), (18.119) and (18.122), we obtain

$$\boldsymbol{J}_q = \frac{c}{4\pi\delta_{\text{eff}}}(\boldsymbol{S}_q - \boldsymbol{A}_q) = \frac{c}{2\pi}\frac{q}{1+2q\delta_{\text{eff}}}\boldsymbol{S}_q$$
$$= \frac{c}{2\pi}\frac{\mathrm{i}\Phi_0[\boldsymbol{q}z^0]}{q(1+2q\delta_{\text{eff}})}. \tag{18.123}$$

Considering that $\boldsymbol{H} = \operatorname{rot} \boldsymbol{A}$, we have for the field at $z = 0$

$$\boldsymbol{H}_q = \frac{\mathrm{i}[\boldsymbol{q}\boldsymbol{S}_q]}{1+2q\delta_{\text{eff}}} = \frac{z^0\Phi_0}{1+2q\delta_{\text{eff}}}. \tag{18.124}$$

From these formulas it is seen that there are two asymptotic regions: $q \gg \delta_{\text{eff}}^{-1}$, which corresponds to $\rho \ll \delta_{\text{eff}}$, and $q \ll \delta_{\text{eff}}^{-1}$, which corresponds to $\rho \gg \delta_{\text{eff}}$. In the first of these regions

$$\boldsymbol{J}_q \approx \frac{c}{4\pi\delta_{\text{eff}}}\boldsymbol{S}_q.$$

Hence, $\boldsymbol{J}(\boldsymbol{\rho})$ is simply proportional to $\boldsymbol{S}(\boldsymbol{\rho})$ and, according to eq. (18.113),

$$J(\boldsymbol{\rho}) = \frac{\Phi_0 c}{8\pi^2\delta_{\text{eff}}\rho}, \quad \rho \ll \delta_{\text{eff}}. \tag{18.125}$$

At large distances, from (18.123) we obtain

$$\boldsymbol{J}(\boldsymbol{\rho}) = \int \boldsymbol{J}_q \,\mathrm{e}^{\mathrm{i}\boldsymbol{q}\boldsymbol{\rho}} \frac{\mathrm{d}^2\boldsymbol{q}}{(2\pi)^2} = \operatorname{rot}\left[\frac{c\Phi_0 z^0}{2\pi}\int_0^\infty \mathrm{d}q \int_0^{2\pi} \mathrm{d}\theta \frac{\mathrm{e}^{\mathrm{i}q\rho\cos\theta}}{(2\pi)^2}\right]$$
$$= \operatorname{rot}\left[\frac{c\Phi_0 z^0}{4\pi^2}\int_0^\infty J_0(q\rho)\,\mathrm{d}q\right] = \frac{\Phi_0 c}{4\pi^2}\operatorname{rot}\left(\frac{z^0}{\rho}\right)$$

($J_0(q\rho)$ is the Bessel function). From this we find that the current $\boldsymbol{J}$ has only a θ-component, equal to

$$J(\rho) \approx \frac{\Phi_0 c}{4\pi^2\rho^2} \quad \rho \gg \delta_{\text{eff}}. \tag{18.126}$$

The limiting expression for the magnetic field at $z = 0$ and $\rho \ll \delta_{\text{eff}}$ is derived with the aid of formula (18.124):

$$H(\rho, 0) \approx \frac{\Phi_0}{4\pi\delta_{\text{eff}}}\int_0^\infty \mathrm{d}q\, J_0(q\rho) = \frac{\Phi_0}{4\pi\delta_{\text{eff}}\rho}, \quad \rho \ll \delta_{\text{eff}}. \tag{18.127}$$

For the case $\rho \gg \delta_{\text{eff}}$ it is better to use the London equation (18.111) and formula (18.126):

$$H(\rho, 0) = -\frac{4\pi\delta_{\text{eff}}}{c}\,\text{rot}_z\,\boldsymbol{J} = \frac{\Phi_0\delta_{\text{eff}}}{\pi\rho^3}, \quad \rho \gg \delta_{\text{eff}}. \tag{18.128}$$

The distribution of the field outside the plate can be deduced using formula (18.120). Substituting eqs. (18.121) and (18.122), we obtain

$$\boldsymbol{A}_{qk} = \frac{2q\boldsymbol{S}_q}{(1+2q\delta_{\text{eff}})(q^2+k^2)} = \frac{2\mathrm{i}\Phi_0[\boldsymbol{q}\boldsymbol{z}^0]}{q(1+2q\delta_{\text{eff}})(q^2+k^2)}.$$

Taking the curl, we find

$$H_{zqk} = \mathrm{i}[\boldsymbol{q}\boldsymbol{A}_{qk}]_z = \frac{2\Phi_0 q}{(1+2q\delta_{\text{eff}})(q^2+k^2)},$$

$$\boldsymbol{H}_{\perp qk} = -\frac{2\Phi_0 \boldsymbol{q}k}{q(1+2q\delta_{\text{eff}})(q^2+k^2)}. \tag{18.129}$$

In the coordinate representation we have

$$H = -\nabla\varphi \,\text{sign}\, z$$

$$\varphi = \frac{\Phi_0}{2\pi}\int_0^\infty \frac{\mathrm{d}q}{1+2q\delta_{\text{eff}}} J_0(q\rho)\,\mathrm{e}^{-q|z|}. \tag{18.130}$$

At large distances we have $q \ll \delta_{\text{eff}}^{-1}$. Here

$$\varphi = \frac{\Phi_0}{2\pi r}, \qquad \boldsymbol{H} = \frac{\Phi_0}{2\pi}\frac{\boldsymbol{r}}{r^3}\,\text{sign}\, z. \tag{18.131}$$

It should be noted that, according to eq. (18.130), $\boldsymbol{H}_\perp \neq 0$ at $z=0$. But if from the start we set $z=0$, i.e., if we find $\int_{-\infty}^{\infty} \boldsymbol{H}_{\perp qk}\,\mathrm{d}k/2\pi$, we shall obtain zero. This is associated with the replacement of the plate by an infinitesimally thin plane with a current. For such a film we have from the Maxwell equation rot $\boldsymbol{H} = (4\pi/c)\boldsymbol{j}$:

$$\boldsymbol{H}_\perp(\boldsymbol{\rho}, +0) - \boldsymbol{H}_\perp(\boldsymbol{\rho}, -0) = \frac{4\pi}{c}\boldsymbol{J}(\boldsymbol{\rho}).$$

This condition is satisfied by formulae (18.129) and (18.123). The changeover from $\boldsymbol{H}_\perp(\rho, +0)$ to $\boldsymbol{H}_\perp(\rho, -0)$ occurs at a distance $\Delta z \sim d$. Here, in the center of the plate $\boldsymbol{H}_\perp = 0$.

We have considered one isolated vortex. In fact, of course, vortices arise when the plate is placed in a perpendicular magnetic field. This gives rise to a regular vortex structure. From the symmetry of the problem it is seen that, as before, around each vortex a contour can be drawn on which $\oint \boldsymbol{j}\,\mathrm{d}\boldsymbol{l} = 0$ and, hence, each vortex carries one flux quantum Φ_0. If the core centers are located at the points $\boldsymbol{\rho}_m$, then in place of eq. (18.111) we obtain an equation with $\sum \delta(\boldsymbol{\rho} - \boldsymbol{\rho}_m)$ instead of $\delta(\boldsymbol{\rho})$.

All the subsequent expressions in which $\boldsymbol{S}_q$ is not specified are retained. A new $\boldsymbol{S}_q$ results from the condition

$$\operatorname{rot} \boldsymbol{S} = \Phi_0 \boldsymbol{z}^0 \sum_m \delta(\boldsymbol{\rho} - \boldsymbol{\rho}_m)$$

and is given by (compare with eq. 18.122)

$$\boldsymbol{S}_q = \mathrm{i}\Phi_0[\boldsymbol{q}\boldsymbol{z}^0] \sum_m \frac{\exp(-\mathrm{i}\boldsymbol{q}\boldsymbol{\rho}_m)}{q^2}. \tag{18.132}$$

Using formulae (18.121) and (18.120), we obtain $\boldsymbol{A}_{qk}$ and $\boldsymbol{H}_{qk}$. In particular, at $z > 0$

$$H_z(\boldsymbol{\rho}, z) = -\frac{\partial}{\partial z} \frac{\Phi_0}{2\pi} \int_0^\infty \frac{\sum_m J_0(q|\boldsymbol{\rho} - \boldsymbol{\rho}_m|)\, e^{-qz}}{1 + 2q\delta_{\text{eff}}} \mathrm{d}q.$$

At $z \gg \delta_{\text{eff}}$ we put $q\delta_{\text{eff}} \ll 1$. Taking the integral, we get

$$H_z(\boldsymbol{\rho}, z) \approx \frac{\Phi_0}{2\pi} z \sum_m [z^2 + (\boldsymbol{\rho} - \boldsymbol{\rho}_m)^2]^{-3/2}. \tag{18.133}$$

At a distance z much larger than the period of the structure we can make the replacement

$$\sum_m \to n \int \mathrm{d}^2\boldsymbol{\rho}_m,$$

where n is the number of vortices per cm^2. Taking the integral, we obtain the obvious result

$$H_z(\boldsymbol{\rho}, z) \approx n\Phi_0 = H_0 \tag{18.134}$$

(H_0 is the external field at infinity). The same result could have been obtained from the continuity condition for the normal component of the induction.

The formulae obtained are valid as long as the intervortex spacings are much larger than ξ, i.e., at

$$H_0 \ll \frac{\Phi_0}{\xi^2} \sim H_{c2}$$

(recall that for type I superconductors H_{c2} is the supercooling field).

18.7 *Anisotropic type II superconductor in a magnetic field*

Up until now our analysis was confined to isotropic superconductors. Since however the new high-temperature superconductors are highly anisotropic we shall consider here some properties of such substances. We shall limit ourselves to the case of uniaxial symmetry, because the new substances have layered structures of either tetragonal or orthorhombic symmetry. However even in the latter case the anisotropy in the basal plane is small.

In the GL free energy (17.5) one has instead of the electron mass m the mass tensor

$$m_{ik} = m_t \delta_{ik} + (m_l - m_t) n_i n_k, \tag{18.135}$$

where $\boldsymbol{n}$ is the unit vector along the main axis (it is normal to the basal plane or to the conducting layers). The principal values of m_{ik} are m_t and m_l. The corresponding term in (17.5) takes the form

$$\tfrac{1}{4}(\hat{m}^{-1})_{ik}\left(\mathrm{i}\hbar\frac{\partial}{\partial x_i} - \frac{2e}{c}A_i\right)\Psi^*\left(-\mathrm{i}\hbar\frac{\partial}{\partial x_k} - \frac{2e}{c}A_k\right)\Psi.$$

In accordance with this the first term in the GL equation (17.6) is replaced by

$$\tfrac{1}{4}(\hat{m}^{-1})_{ik}\left(-\mathrm{i}\hbar\frac{\partial}{\partial x_i} - \frac{2e}{c}A_i\right)\left(-\mathrm{i}\hbar\frac{\partial}{\partial x_k} - \frac{2e}{c}A_k\right)\Psi$$

and the current instead of (17.9) becomes

$$j_i = e(\hat{m}^{-1})_{ik}\left[-\frac{\mathrm{i}\hbar}{2}\left(\Psi^*\frac{\partial}{\partial x_k}\Psi - \Psi\frac{\partial}{\partial x_k}\Psi^*\right) - \frac{2e}{c}A_k|\Psi|^2\right]. \tag{18.136}$$

We start with the field H_{c2}. In this case Ψ is small and we must write eq. (17.6) for Ψ in the linear approximation. We can take the field to be uniform. We choose the z-axis along the magnetic field and we place the main crystal axis in the xz-plane. As before, we take the vector potential $A_y = Hx$. Assuming that Ψ depends only on x we obtain

$$-\tfrac{1}{4}\hbar^2\left(\frac{n_z^2}{m_t} + \frac{n_x^2}{m_l}\right)\frac{\mathrm{d}^2\Psi}{\mathrm{d}x^2} + \frac{e^2H^2}{c^2m_t}x^2\Psi = \alpha|\tau|\Psi.$$

Similarly to section 18.2 we find the critical field H_{c2} (Morris et al. 1972):

$$H_{c2} = \frac{\Phi_0}{2\pi\xi_t^2}(\cos^2\theta + \varepsilon^2\sin^2\theta)^{-1/2}, \tag{18.137}$$

where

$$\xi_t = \frac{\hbar}{\sqrt{4m_t\alpha|\tau|}}, \qquad \xi_l = \frac{\hbar}{\sqrt{4m_l\alpha|\tau|}}$$

are the correlation lengths,

$$\varepsilon = \sqrt{\frac{m_t}{m_l}} = \frac{\xi_l}{\xi_t},$$

θ is the angle between the main axis and the magnetic field. The limiting values of H_{c2} are $\Phi_0/(2\pi\xi_t^2)$ for $\boldsymbol{n}\|\boldsymbol{H}$ and $\Phi_0/(2\pi\xi_t\xi_l)$ for $\boldsymbol{n}\perp\boldsymbol{H}$. Although these formulae were derived for the vicinity of T_c they reflect also the situation at low temperatures.

Now we pass to the first critical field. Assuming $\varkappa \gg 1$ and limiting ourselves to logarithmic accuracy we can make calculations for arbitrary temperatures, like in section 18.5. We can put $\Psi = \sqrt{(\frac{1}{2} n_s)}\, e^{i\chi}$ where $n_s = \text{const}$. We introduce also the penetration depths

$$\delta_t = \sqrt{\frac{m_t c^2}{4\pi n_s e^2}}, \qquad \delta_l = \sqrt{\frac{m_l c^2}{4\pi n_s e^2}}.$$

Inserting into eq. (18.136) expression (18.135) for m_{ik}, taking rot from the Maxwell equation $\text{rot}\, \boldsymbol{H} = (4\pi/c)\boldsymbol{j}$ and assuming one vortex line along $\boldsymbol{l}$ we obtain the generalisation of eq. (18.91):

$$\boldsymbol{H} + \delta_t^2 \,\text{rot rot}\, \boldsymbol{H} + (\delta_l^2 - \delta_t^2)\, \text{rot}(\boldsymbol{n}(\boldsymbol{n}\, \text{rot}\, \boldsymbol{H})) = \boldsymbol{l}\Phi_0 \delta(\boldsymbol{\rho}), \tag{18.138}$$

where ρ is the radius vector in the plane perpendicular to $\boldsymbol{l}$. From eq. (18.138) it follows that $\text{div}\, \boldsymbol{H} = 0$.

The important part of the free energy has the form

$$\varepsilon_0 = \int \left[\frac{H^2}{8\pi} + \tfrac{1}{8}(\hat{m}^{-1})_{ik} n_s \left(\hbar \frac{\partial \chi}{\partial x_i} - \frac{2e}{c} A_i \right) \left(\hbar \frac{\partial \chi}{\partial x_k} - \frac{2e}{c} A_k \right) \right] \mathrm{d}V$$

$$= \frac{1}{8\pi} \int [H^2 + \delta_t^2 (\text{rot}\, \boldsymbol{H})^2 + (\delta_l^2 - \delta_t^2)(\boldsymbol{n}\, \text{rot}\, \boldsymbol{H})^2]\, \mathrm{d}V.$$

Integrating by parts and using eq. (18.138) we get

$$\varepsilon_0 = \frac{1}{8\pi} \Phi_0 \boldsymbol{H}(0)\, \boldsymbol{l}. \tag{18.139}$$

The first critical field is determined by the condition

$$n\varepsilon_0 - \frac{\boldsymbol{B}\boldsymbol{H}_{c1}}{4\pi} = n\varepsilon_0 - \frac{n\Phi_0 \boldsymbol{l}\boldsymbol{H}_{c1}}{4\pi} = 0, \tag{18.140}$$

where n is the number of vortices per cm^2 of the cross-sectional plane (actually we consider only one vortex; n is introduced only for convenience of presentation).

Due to axial symmetry the vortex $\boldsymbol{l}$ will be oriented in the plane of the vector $\boldsymbol{n}$ and the external field $\boldsymbol{H}_{c1}$ but in general will coincide with neither of them. We denote this plane by xz and we direct the z-axis along $\boldsymbol{l}$. Performing the Fourier transformation

$$\boldsymbol{H} = \int \boldsymbol{H}_k\, e^{i\boldsymbol{k}\boldsymbol{\rho}} \frac{\mathrm{d}^2 \boldsymbol{k}}{(2\pi)^2}$$

with $\boldsymbol{k} = (k_x, k_y)$ we get from eq. (18.138)

$$(1 + k^2 \delta_t^2)\boldsymbol{H}_k + (\delta_l^2 - \delta_t^2)[\boldsymbol{n} \times \boldsymbol{k}](\boldsymbol{H}_k[\boldsymbol{n} \times \boldsymbol{k}]) = \boldsymbol{l}\Phi_0.$$

Multiplying by $[\boldsymbol{n}\times\boldsymbol{k}]$ and solving the resulting equation we find

$$\boldsymbol{H}_k[\boldsymbol{n}\times\boldsymbol{k}]=\frac{\boldsymbol{l}[\boldsymbol{n}\times\boldsymbol{k}]\Phi_0}{1+\delta_t^2k^2+[\boldsymbol{n}\times\boldsymbol{k}]^2(\delta_l^2-\delta_t^2)}$$

$$=\frac{n_xk_y\varphi_0}{1+\delta_l^2k^2+(\delta_t^2-\delta_l^2)n_x^2k_x^2}.$$

Multiplying eq. (18.141) by $\boldsymbol{l}$ and substituting $\boldsymbol{H}_k[\boldsymbol{n}\times\boldsymbol{k}]$ we get

$$\boldsymbol{l}\boldsymbol{H}_k=\frac{\Phi_0[1+(\delta_t^2n_x^2+\delta_l^2n_z^2)k^2]}{(1+\delta_t^2k^2)[1+\delta_l^2k^2+(\delta_t^2-\delta_l^2)n_x^2k_x^2]}. \tag{18.142}$$

Expression (18.139) for ε_0 contains $\boldsymbol{H}(0)\,\boldsymbol{l}=(2\pi)^{-2}\int \boldsymbol{H}_k\boldsymbol{l}\,\mathrm{d}^2\boldsymbol{k}$. According to eq. (18.142) this is a logarithmic integral. It must be limited from above by $k\sim\xi^{-1}$. Within logarithmic accuracy we obtain

$$\varepsilon_0=\left(\frac{\Phi_0}{4\pi\delta_t}\right)^2\ln\varkappa\,\frac{\sqrt{\delta_t^2\sin^2\theta+\delta_l^2\cos^2\theta}}{\delta_l}, \tag{18.143}$$

where θ is the angle between $\boldsymbol{n}$ and $\boldsymbol{l}$ ($n_z=\cos\theta$, $n_x=\sin\theta$). Denoting the angle between $\boldsymbol{H}_{c1}$ and $\boldsymbol{n}$ by θ_1 we get from eq. (18.140)

$$H_{c1}(\theta_1,\theta)=\frac{\Phi_0\ln\varkappa}{4\pi\delta_t^2}\,\frac{(\varepsilon^2\sin^2\theta+\cos^2\theta)^{1/2}}{\cos(\theta_1-\theta)}, \tag{18.144}$$

where as before $\varepsilon=(m_t/m_l)^{1/2}=\delta_t/\delta_l$. The true value of $\boldsymbol{H}_{c1}$ is obtained by minimization over θ. We find (Balatskii et al. 1986)

$$\tan\theta_0=\frac{1}{\varepsilon^2}\tan\theta_1, \tag{18.145}$$

$$H_{c1}=\frac{\Phi_0\ln\varkappa}{4\pi\delta_t^2}\,\frac{1}{\sqrt{\cos^2\theta_1+\varepsilon^{-2}\sin^2\theta_1}}. \tag{18.146}$$

Consider the limiting cases. The case $\varepsilon\ll1$ corresponds to a layered structure with $m_l\gg m_t$. The electrons in this case rarely jump from one layer to another. Therefore the field normal to the layers is easily screened while a field having a component within the layer plane is screened with much greater difficulty. Hence the vortices corresponding to field penetration exhibit a trend towards lying in the layer plane and at $\boldsymbol{H}_{c1}\perp\boldsymbol{n}$ they are formed at relatively small fields (at $\theta_1=\frac{1}{2}\pi$ one has $H_{c1}=\Phi_0\ln\varkappa/4\pi\delta_l\delta_t$). The opposite case $\varepsilon\gg1$ corresponds to a quasi-one-dimensional structure. Here the field directed along the main axis is poorly screened. Therefore the vortices appear preferably in this direction ($\theta_0\to0$) even if the external field makes a large angle with the axis.

We shall not analyse here the whole structure of the mixed state at $H_{c1}<H<H_{c2}$. Physically it is clear that with increasing vortex density their orientation will approach that of the external field.

In conclusion we consider the anisotropy of the penetration depth of a weak magnetic field. To this purpose we must use the reduced eq. (18.138) or eq. (18.141) without the right-hand side. This time we direct **k** along the z-axis (this is the normal to the surface of the superconducting half-space) and we choose **n** in the xz-plane. One can easily prove that the components H_y and H_x have different penetration depths, namely

$$\delta(H_x)=\delta_t, \qquad \delta(H_y)=\delta_t\sqrt{\cos^2\theta+\varepsilon^{-2}\sin^2\theta}. \tag{18.147}$$

This behaviour is again easily understood from the viewpoint of the anisotropy of electron hopping. For $\varepsilon \to 0$, i.e. for the layered structure, the field is poorly screened by currents demanding electron hopping between the layers, and the corresponding penetration depth increases.

18.8. Superconducting magnets. Pinning

The idea of the application of superconductors instead of ordinary metals with the purpose of eliminating ohmic losses has been put forward since the discovery of superconductivity. The obstacle is the necessity of attaining low temperatures. The high cost of the equipment and of the operation of superconducting cables with helium cooling makes the use of superconductivity for the transmission of electric power economically inefficient*.

The situation is different as far as the cooling of a limited volume is concerned. It is for this reason that at present the use of superconductors for the production of powerful magnets is rapidly expanding. A conventional electromagnet, which is a coil made of copper or aluminum wire, consumes a power which increases proportionately to H^2, where H is the magnetic field it produces. Maintaining a field of 100 kOe in an aperture of diameter 2.5 cm thus require a power of the order of a megawatt. This energy is almost completely released in the form of Joule heat, and in order to remove this heat, a huge and expensive water cooling system is needed**.

If the coil (solenoid) is made of a superconductor, energy is required only for maintaining the helium temperature. This may be several orders of magnitude smaller than the energy consumed by the corresponding electromagnet; besides, the entire installation is much more compact. It should be noted that the superconducting coil can be short-circuited and in this case it is converted to a permanent magnet. As we know, the field flux through a hollow superconductor does not vary with time because any change in the external conditions is compensated for by supercurrents. Therefore, a superconducting permanent magnet is characterized by unique constancy of the magnetic field produced.

* This conclusion will probably change if ways are found for the technology of application of ceramics which are superconducting at liquid nitrogen temperatures.

** It must be noted that water pumps generate vibrations which are very detrimental in most cases.

The idea of superconducting magnets was first advanced by H. Kamerlingh Onnes. But the superconductors known at that time had low critical fields, of the order of hundreds of oersteds. The situation changed in the 1960s, when superconductors appeared with critical fields above 100 kOe; also, cryostats, i.e., devices for maintaining helium temperatures, were highly improved and became less expensive. The subsequent investigations showed that the new materials exhibiting high critical fields are type II superconductors, which are well described by the theory presented in the preceding sections.

The highest critical fields known have been produced in the so-called Chevrel phases: compounds of the type $Me_x(Mo_6S_8)_y$ (Me = metals: Pb, Li, and others). They reach 600 kOe. There are predictions that superconducting ceramics may have even higher critical fields, up to 2 MOe. However, good mechanical properties of alloys are required for practical needs and therefore at present use is made of Nb-Zr-Ti alloys with critical fields of up to 110–120 kOe and Nb_3Sn with a critical field up to 230 kOe.

Nevertheless, the possibility of using type II superconductors to produce high fields is not quite evident. In fact, as has been established in the preceding sections, at $\varkappa \gg 1$ there is a large region of the mixed state, starting with the field $H_{c1} \sim H_{cm}/\varkappa$ and ending with $H_{c2} \sim H_{cm}\varkappa$. Hence, if there is a high field, the superconductor is certainly in the mixed state and is threaded by vortices. The current that creates a field in the solenoid flows perpendicular to the vortices. But in this case, the vortices are acted on by a Lorentz force in a direction perpendicular to $\boldsymbol{H}$ and $\boldsymbol{j}$.

This force can be found as follows. Suppose there is a single vortex with current $\boldsymbol{j}$ in the superconductor and that an external current $\boldsymbol{j}^{ext}$ is passed through it. The total current will then be $\boldsymbol{j}+\boldsymbol{j}^{ext}$. Substituting into eq. (18.93), we obtain the current interaction energy:

$$\Lambda \int \boldsymbol{j}^{ext}\boldsymbol{j}\, dV$$

(the interaction energy for magnetic fields is relatively small). In this particular case, $\boldsymbol{j}$ depends on $\boldsymbol{\rho}-\boldsymbol{\rho}_0$, where $\boldsymbol{\rho}_0$ is the coordinate of the vortex axis. The component of the force that acts on the vortex axis is a derivative of the potential energy with respect to its coordinate with a minus sign:

$$-\frac{\partial}{\partial x_{0k}}\Lambda \int \boldsymbol{j}^{ext}\boldsymbol{j}\, dV = -\Lambda \sum_{i=1,2} \int j_i^{ext} \frac{\partial}{\partial x_{0k}} j_i\, dV$$

$$= \Lambda \sum_{i=1,2} \int j_i^{ext} \frac{\partial}{\partial x_k} j_i\, dV$$

$$= \Lambda \sum_{i=1,2} \int j_i^{ext}\left(\frac{\partial}{\partial x_k} j_i - \frac{\partial}{\partial x_i} j_k\right) dV + \Lambda \sum_{i=1,2} \int j_i^{ext} \frac{\partial}{\partial x_i} j_k\, dV$$

(we have used the fact that j depends on $\boldsymbol{\rho}-\boldsymbol{\rho}_0$).

The last term is equal to zero, because integrating by parts we obtain $-\Lambda \int j_k \operatorname{div} \boldsymbol{j}^{\text{ext}} \, dV = 0$ ($\operatorname{div} \boldsymbol{j}^{\text{ext}} = 0$). The factor in parentheses in the first term is equal to zero at $i = k$, and at $i \neq k$ it is the component of rot $\boldsymbol{j}$ along the z axis (because i, $k = x$, y). It is not difficult to see that on the whole the integral is equal to

$$\Lambda \int [\boldsymbol{j}^{\text{ext}} \operatorname{rot} \boldsymbol{j}] \, dV.$$

But for the vortex

$$\operatorname{rot} \boldsymbol{j} = -\frac{1}{\Lambda c} \operatorname{rot}\left(\boldsymbol{A} - \frac{\hbar c}{2e} \nabla \chi \right)$$

$$= \frac{1}{\Lambda c} \frac{\hbar c}{2e} 2\pi \delta(\boldsymbol{\rho} - \boldsymbol{\rho}_0)$$

(see formulae 18.90 and 18.40).

Hence, the force that acts on a unit length of the vortex is

$$f_{\text{L}} = \frac{\Phi_0}{c} [\boldsymbol{j}^{\text{ext}} \cdot \boldsymbol{b}], \tag{18.148}$$

where $\Phi_0 = \pi \hbar c / e$ is the flux quantum and $\boldsymbol{b}$ is the unit vector along the vortex axis. Multiplying by the number of vortices per unit cross-sectional area, $n = B / \Phi_0$, we obtain the Lorentz force per unit volume ($n\boldsymbol{b} = \boldsymbol{B} / \Phi_0$):

$$\boldsymbol{F}_{\text{L}} = \frac{1}{c} [\boldsymbol{j}^{\text{ext}} \boldsymbol{B}]. \tag{18.149}$$

Under the action of this force the vortex lattice must begin to move. Suppose that this motion is opposed by a viscous force, which is proportional to the velocity. Let us assume that for one vortex

$$f_{\text{v}} = -\eta \boldsymbol{v}_{\text{L}}, \tag{18.150}$$

where $\boldsymbol{v}_{\text{L}}$ is the velocity of vortex motion and η is a certain "viscosity coefficient". At equilibrium $f_{\text{L}} + f_{\text{v}} = 0$, i.e.,

$$\boldsymbol{v}_{\text{L}} = \frac{\Phi_0}{\eta c} [\boldsymbol{j}^{\text{ext}} \boldsymbol{b}]. \tag{18.151}$$

Hence, the vortices move perpendicular to $\boldsymbol{B}$ and $\boldsymbol{j}^{\text{ext}}$. This motion of the vortex lattice leads to the generation of an electric field. Indeed, if in the coordinate system moving together with the lattice with a velocity $\boldsymbol{v}_{\text{L}}$ there is an average magnetic field (induction) $\boldsymbol{B}$, then in accordance with the rules of electrodynamics an extra electric field arises upon transformation to the laboratory coordinate system:

$$\boldsymbol{E} = \frac{1}{c} [\boldsymbol{B} \boldsymbol{v}_{\text{L}}]. \tag{18.152}$$

Substituting eq. (18.138) and taking into account that $\boldsymbol{b} = \boldsymbol{B}/B$ and $\boldsymbol{B} \perp \boldsymbol{j}^{\text{ext}}$, we obtain

$$\boldsymbol{E} = \frac{\Phi_0 B}{\eta c^2} \boldsymbol{j}^{\text{ext}}. \tag{18.153}$$

But the generation of an electric field along $\boldsymbol{j}^{\text{ext}}$ implies the appearance of electrical resistance:

$$\rho = \frac{E}{j^{\text{ext}}} = \frac{\Phi_0 B}{\eta c^2}. \tag{18.154}$$

If the superconductor is completely homogeneous, then upon attainment of the field H_{c2} the resistance ρ must be equal to ρ_n, the resistance of a normal metal. From this it follows that

$$\rho_n = \frac{\Phi_0 H_{c2}}{\eta c^2}$$

and therefore we find that

$$\eta = \frac{\Phi_0 H_{c2}}{\rho_n c^2} \tag{18.155}$$

and

$$\rho = \rho_n \frac{B}{H_{c2}}. \tag{18.156}$$

The last formula is very simple and resembles the formula for the heat capacity (18.106). It could also be interpreted as arising from the fact that the vortex cores with a diameter of the order of ξ are in the normal state. However, if the vortex lattice were at rest, the supercurrent would easily go around the normal cores and no resistance would arise. The appearance of resistance is the result of the motion of the vortices under the action of the Lorentz force.

In fact, however, formulae (18.155) and (18.156) are valid only by order of magnitude, mostly far from T_c and H_{c2}. The true viscosity η depends on many mechanisms and not only on the energy dissipation in the normal cores, and is a complex function of field and temperature. Besides, the simple linear relation (18.150) does not always hold.

Thus, we come to the conclusion that superconductivity is destroyed as soon as vortices appear, i.e., at a field equal to $H_{c1} \sim H_{cm}/\varkappa$. This means that not only we gain nothing with increasing $\varkappa$ but, on the contrary, we strongly lose in the critical field. However, there is a way for coping with this shortcoming. To do this, the motion of the vortex lattice must be stopped, in which case, in accordance with formula (18.152), $\boldsymbol{E} = 0$ and, hence, $\rho = 0$ too. To this end, we have to fasten the vortices at certain points in the sample. This phenomenon is known as pinning.

We have already dealt in section 18.5 with the surface barrier that hinders the entry of vortices into the sample and their escape from it. From this example it is seen that irregularities may be responsible for the formation of barriers impeding the motion of vortices. It is clear that irregularities with a size larger than ξ will be very effective for pinning*.

Consider, as an example, a pore in a superconductor with a size of $d \gg \xi$. We know that the core energy per unit length of the vortex is of order $(\Phi_0/\delta)^2$ (see eq. 18.95). This may be put in a different way. As soon as the core is in the normal state there is an excess energy $H_{\text{cm}}^2/8\pi$ per cm^3 and an excess energy $H_{\text{cm}}^2 \cdot \xi^2$ per unit length. The identity of these expressions follows from (18.97).

If the vortex passes through the pore, the corresponding part of the core energy is absent. This means that the total energy is lower, i.e., the vortex is attracted to the pore. The force of this attraction can be found if one takes into account that the vortex disappears, merging with its image at a distance of order ξ from the edge of the pore (section 18.5). Hence, the force equal to the energy gradient will be of the order of the energy itself divided by ξ. For a pore with a size d,

$$f_{\text{p}} \sim H_{\text{cm}}^2\, d\xi. \tag{18.157}$$

When the current is passing, the vortex is acted on by the Lorentz force, which, in accordance with (18.148), after being multiplied by the length d is of the order of $(H_{\text{cm}} \sim \Phi_0/\delta\xi)$:

$$f_{\text{L}} d \sim \Phi_0 j^{\text{ext}} \frac{d}{c} \sim H_{\text{cm}} j^{\text{ext}}\, \delta\xi \frac{d}{c}.$$

Equating these forces, we find the critical current density capable of unpinning the vortex from the pore:

$$j^{\text{ext}} \sim H_{\text{cm}} \frac{c}{\delta}. \tag{18.158}$$

Comparing with formula (17.62), we see that this value is comparable with the current density required to break up Cooper pairs.

In fact, of course, we must take into account that the pinning force acts on a limited portion of the vortex line, whereas the Lorentz force pulls the entire vortex. Moreover, in real cases there are many interacting vortices that form a lattice, and therefore in calculations of the critical current density one has to take account of the lattice distortion and of the elastic forces that arise in this case. We shall not discuss particular models and the results obtained, especially as the nature of the pinning centers in superconductors employed in technology has not been studied in detail.

Moreover, it is not always clear how the current is actually distributed throughout the superconductors in the mixed state with pinning. Although we know, for example, that in the absence of a magnetic field the current flows only in a surface layer of

* A detailed investigation shows that smaller-scale irregularities can exert a collective action on the entire vortex lattice and also lead to pinning.

thickness δ, experiment shows that in the case under consideration the situation is different, and a considerable part of the current flows through the inner portion of the superconductor. This is associated with the fact that the current is a function of the field: $j_c(B)$. The passage of the current leads to the formation of a "critical profile" of the current, such that the induction and current at each point of the cross section of the conductor corresponds to $j_c(B)$. At $\varkappa \gg 1$ and a high defect density the current is distributed almost uniformly along the cross section. In a general case, we may speak only of the "current-carrying capacity", which is conventionally expressed in terms of the average current density. So far, critical values have been attained of the order of 10^6 A/cm^2 (for ceramics much less, $<10^3$ A/cm^2).

Thus, according to what has been said above, we get the impression that before the critical current is attained the resistance is equal to zero, and then resistance arises in a jumpwise manner. It is easy, however, to see that this occurs only at $T=0$. In fact, the pinning centers create for the vortex lattice a potential relief consisting of "valleys" separated by "mountain chains" - potential barriers. At finite temperature a transition from one valley to another is possible due to thermal fluctuations. The result is that even before the true critical current density is achieved there arises a "creep": the fluctuational jumpwise displacement of the vortex lattice. This gives rise to resistance. It should be emphasized that this occurs at any finite temperature, and therefore in the strict sense of the word the critical current in the mixed state at $T>0$ is equal to zero.

However, in reality the situation is not that hopeless. Suppose we have a superconducting permanent magnet - a short-circuited solenoid. It is used under operating conditions for a limited period of time, because during all this time the energy has to be consumed on cooling. The resistance associated with the creep of vortices is so small that the current flowing through the solenoid does not change appreciably during, say 100 years; from a practical viewpoint the resistance is equal to zero.

We shall not describe here the creep theory (see Anderson 1962; Anderson and Kim 1964). We note only that according to this theory the field in the hole of the solenoid falls off by a very slow law:

$$B(t) = B(0)(1-\alpha \ln t)$$

with the coefficient $\alpha \sim 10^{-3}$. Hence, the decrease of the current in the solenoid and of the magnetic field created by it takes enormous periods of time.

We may conclude that the creep is not dangerous and that it may be neglected, i.e., we may presume that from the practical standpoint the situation is the same as at $T=0$, and unless the critical current is not attained the sample is superconducting. As a matter of fact, the creep may cause another hazard. If a bundle of vortices slips at some point, heat is released. This leads to a sharp decrease in the values of the critical field and current and may drive a portion of the magnet normal. In such a case, the heat release increases drastically and, as a result, an instability arises, which may destroy the magnet.

In order to avoid these processes, superconducting cables or ribbons for magnets are made multicore, and a normal metal (usually copper) is used as an insulator. [In comparison with a superconductor, any substance with finite resistance is an insulator.] However, while the superconductor exhibits low thermal conductivity, because the heat can be transferred only by quasiparticles, whose number at low temperatures is exponentially small (section 19.1), a normal metal is a good heat conductor. It is the normal metal that removes heat in the case of sudden heating. Moreover, if current ceases to flow through the superconductor, the thick normal conductor connected in parallel will play the role of a shunt.

19

Kinetics of superconductors

Up until now we have been concerned only with equilibrium phenomena taking place in superconductors, i.e., with their behavior upon variation of temperature and of the external static magnetic field. The only exception was section 18.8, where we considered qualitatively the kinetics of the vortex structure upon passage of current.

A rigorous treatment of the kinetics of superconductors calls for the application of detailed equations of the BCS microscopic theory, which requires very complicated calculations. In most cases the answer cannot be found in analytical form. A certain simplification can be achieved for fields that vary gradually in space and in time: $\omega \ll \Delta/h$, $k \ll \xi_0^{-1}$, where ω and k are the frequency and the wave vector of the field.

Nevertheless, even in this case the resultant "kinetic equation" (see Aronov et al. 1981) is rather complicated and does not cover all the interesting phenomena. For this reason, we will confine ourselves in some cases to a qualitative description and try, where possible, to give estimates on the basis of simple physical considerations. We begin with phenomena that occur in weak fields or with a low temperature gradient, when the state of the superconductor deviates little from equilibrium; then we will discuss certain effects associated with strong nonequilibrium.

19.1. Electronic thermal conductivity

The transport of heat in superconductors is accomplished, just as in normal metals, by quasiparticles of the electron liquid and by phonons. At not too low temperatures the major contribution comes from electrons. As for the lowest temperatures, in this case too the electron contribution can be separated from the phonon contribution. For example, in a magnetic field above H_c the electrons become normal and thus we can find the difference in thermal conductivity between the superconducting and the normal state. As regards the electronic thermal conductivity of a normal metal, this can be either extrapolated from the region of $T > T_c$ or separated with the aid of a strong magnetic field, as described in Part I. Thus, the electronic thermal conductivity of a superconductor can be deduced from experimental data.

In calculating the thermal conductivity of a normal metal we were faced with the fact that the temperature gradient leads to the generation of an electric field, which is determined from the condition of the absence of current. However, this effect introduces only a small correction, of the order of $(T/\mu)^2$ (see section 6.1). In superconductors there is no electric field, but the current of quasiparticles is balanced by a countercurrent of Cooper pairs (this point will be discussed in detail in section 19.2). In calculating thermal conductivity one may neglect the motion of Cooper pairs since it makes only a correction of order $(T/\mu)^2$.

At temperatures below T_c the dominant mechanism of electron scattering is the scattering from impurities. Therefore we will be concerned only with this mechanism here. The kinetic equation for quasiparticles has the usual form

$$\frac{\partial \varepsilon}{\partial \boldsymbol{p}} \frac{\partial n}{\partial \boldsymbol{r}} - \frac{\partial \varepsilon}{\partial \boldsymbol{r}} \frac{\partial n}{\partial \boldsymbol{p}} = I(n), \tag{19.1}$$

where ε is the energy of quasiparticles and $-\partial\varepsilon/\partial \boldsymbol{r}$ is the force that acts on quasiparticles.

The form of the collision integral is determined as follows. Suppose that the impurities have no spin. In this case, the electrons conserve their spin on scattering. Moreover, the collisions with impurities are elastic, i.e., the energy of quasi-particles is conserved. In view of this, the scattering of electrons from impurities may be described by the term in the Hamiltonian:

$$\mathcal{H}_{\text{imp}} = \sum_{\sigma,\boldsymbol{p},\boldsymbol{p}'} a^{+}_{\boldsymbol{p}',\sigma} a_{\boldsymbol{p},\sigma} V_{\boldsymbol{p}'\boldsymbol{p}}, \tag{19.2}$$

where $V_{\boldsymbol{p}'\boldsymbol{p}}$ is the interaction matrix element, which is a function only of $\theta = (\widehat{\boldsymbol{p}, \boldsymbol{p}'})$. Expression (19.2) corresponds to the disappearance of an electron of momentum $\boldsymbol{p}$ and to the creation, in place of it, of another electron of momentum $\boldsymbol{p}'$.

In order to pass to quasiparticles, we substitute the Bogoliubov transformation (16.14). Here only those terms are retained which are not associated with a change in energy, i.e., the terms describing the creation or annihilation of two quasiparticles are omitted. As a result, we find

$$\mathcal{H}_{\text{imp}} = \sum_{\sigma,\boldsymbol{p}\cdot\boldsymbol{p}'} \alpha^{+}_{\boldsymbol{p}'\sigma} \alpha_{\boldsymbol{p}\sigma} V_{\boldsymbol{p}'\boldsymbol{p}} (u_p u_{p'} - v_p v_{p'}). \tag{19.3}$$

The energy of quasiparticles is conserved upon scattering from impurities. Hence, $\varepsilon_p = \varepsilon_{p'}$. From this it follows that $\xi_p = \pm\xi_{p'}$. If $\xi_{p'} = \xi_p$, then $u_{p'} = u_p$, $v_{p'} = v_p$, $u_p^2 - v_p^2 = \xi_p/\varepsilon_p$. But if $\xi_{p'} = -\xi_p$, then $v_{p'} = u_p$ and $u_{p'} = v_p$ and expression (19.3) vanishes.

Let us now find the scattering probability per unit time, in other words, the function that replaces τ^{-1} in section 3.2*. So, we have

$$\tau_{\text{str}}^{-1} = \frac{2\pi}{\hbar} \int |V_{\boldsymbol{p}'\boldsymbol{p}}|^2 (1 - \cos\theta)(u_p u_{p'} - v_p v_{p'})^2$$
$$\times \delta(\varepsilon_p - \varepsilon_{p'}) \frac{d^3 \boldsymbol{p}'}{(2\pi\hbar)^3}$$

* Recall that in Part II we use the notation τ_{tr} instead of τ.

$$= \frac{\pi}{\hbar} \nu(\mu) \left(\frac{\xi_p}{\varepsilon_p}\right)^2$$

$$\times \left|\frac{d\xi_p}{d\varepsilon_p}\right| \int |V(\theta)|^2 (1-\cos\theta) \frac{d\Omega}{4\pi}$$

$$= \tau_{\mathrm{tr}}^{-1} \frac{|\xi_p|}{\varepsilon_p}; \tag{19.4}$$

we have made the replacements

$$\mathrm{d}^3\boldsymbol{p}' \to p_0^2\, \mathrm{d}p'\, \mathrm{d}\Omega \to p_0^2 \left(\frac{\mathrm{d}\xi'}{v}\right) \mathrm{d}\Omega,$$

$$\delta(\varepsilon-\varepsilon') = \left|\frac{\mathrm{d}\xi}{\mathrm{d}\varepsilon}\right| \delta(|\xi|-|\xi'|) \to \left|\frac{\mathrm{d}\xi}{\mathrm{d}\varepsilon}\right| \delta(\xi-\xi'),$$

since for $\xi' = -\xi$ we have $\tau_{\mathrm{str}}^{-1} = 0$. From eq. (19.4) it follows that the collision integral is given by

$$I(n) = -\frac{|\xi_p|}{\varepsilon_p} \frac{n-n_{\mathrm{F}}}{\tau_{\mathrm{tr}}}. \tag{19.5}$$

Substituting into the left-hand side of the kinetic equation the equilibrium function

$$n_{\mathrm{F}} = \frac{1}{\exp(\varepsilon/T)+1} \tag{19.6}$$

and using the fact that

$$\frac{\partial n_{\mathrm{F}}}{\partial \boldsymbol{r}} = -\frac{\varepsilon}{T} \frac{\partial n_{\mathrm{F}}}{\partial \varepsilon} \nabla T, \tag{19.7}$$

$$\frac{\partial \varepsilon}{\partial \boldsymbol{p}} = \frac{\partial \varepsilon}{\partial \xi} \frac{\mathrm{d}\xi}{\mathrm{d}\boldsymbol{p}} = \boldsymbol{v}\left(\frac{\xi}{\varepsilon}\right), \tag{19.8}$$

where $\boldsymbol{v} = \boldsymbol{p}/m$, we find*:

$$-\frac{\xi}{T} \frac{\partial n_{\mathrm{F}}}{\partial \varepsilon} \boldsymbol{v} \nabla T = -(n-n_{\mathrm{F}}) \left(\frac{|\xi|}{\varepsilon}\right) \frac{1}{\tau_{\mathrm{tr}}}, \tag{19.9}$$

from which it follows that the nonequilibrium correction is

$$n^{(1)} = n - n_{\mathrm{F}} = \frac{\varepsilon}{T} \frac{\partial n_{\mathrm{F}}}{\partial \varepsilon} \tau_{\mathrm{tr}} \operatorname{sign} \xi \boldsymbol{v} \nabla T, \tag{19.10}$$

$$\operatorname{sign} \xi = \frac{\xi}{|\xi|}.$$

* While substituting the equilibrium function n_{F} into the left-hand side of eq. (19.1), we must, strictly speaking, take into account that $\varepsilon = (\xi^2 + \Delta^2)^{1/2}$, where Δ depends on T and, hence, on $\boldsymbol{r}$. However, the corresponding contributions from the two terms on the left-hand side of eq. (19.1) cancel.

The heat flux is expressed by

$$q = 2\int \varepsilon \frac{\partial \varepsilon}{\partial p} n \frac{d^3 p}{(2\pi\hbar)^3}. \tag{19.11}$$

The factor 2 arises as a result of summation over the spin projections. Since n_F depends only on ε, the only contribution here comes from $n^{(1)}$. Substituting (19.10) and (19.8) and passing to the integration over ε, we obtain

$$\begin{aligned} q &= \frac{2\nu}{T}\int_{\Delta}^{\infty} d\varepsilon \int \frac{d\Omega}{4\pi} \varepsilon^2 \tau_{tr} v(v\nabla T)\frac{\partial n_F}{\partial \varepsilon} \\ &= \frac{2\nu\nabla T}{3T} v^2 \tau_{tr} \int_{\Delta}^{\infty} d\varepsilon\, \varepsilon^2 \frac{\partial n_F}{\partial \varepsilon} \\ &= -\left(n_e \frac{\tau_{tr}}{2mT^2}\right)\int_{\Delta}^{\infty} d\varepsilon \frac{\varepsilon^2}{\cosh^2(\varepsilon/2T)} \nabla T \end{aligned} \tag{19.12}$$

(here the 2 arises from the integration over the regions $\xi > 0$ and $\xi < 0$; $n_e = p_0^3/3\pi^2\hbar^3$). Hence, we have (Geilikman 1958)*:

$$\varkappa = \frac{n_e \tau_{tr} T}{2m}\int_{\Delta/T}^{\infty} \frac{x^2}{\cosh^2(\frac{1}{2}x)} dx.$$

With $T \to T_c$, from the above equation we obtain expression (3.28) for a normal metal. But if $T \to 0$, then $\cosh(\frac{1}{2}x) \approx \frac{1}{2}\exp(\frac{1}{2}x)$ and

$$\varkappa \approx \frac{2 n_e \tau_{tr} T}{m}\left(\frac{\Delta(0)}{T}\right)^2 \exp\left(\frac{-\Delta(0)}{T}\right). \tag{19.13}$$

The graph of $\varkappa/\varkappa_n$ against T/T_c is given in fig. 122.

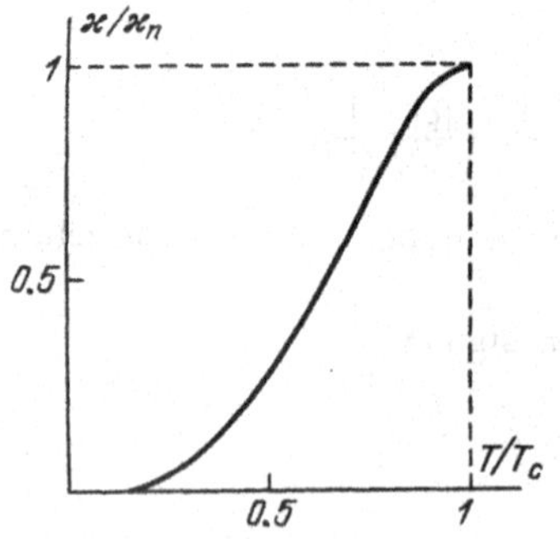

Fig. 122.

* In this and subsequent sections $\varkappa$ is the coefficient of thermal conductivity.

19.2. Thermoelectric phenomena

As has been pointed out in the preceding section, a temperature gradient leads to the appearance not only of a heat flux but also of a flux of quasiparticles, i.e., a "normal current". However, it is balanced by a superconducting current which flows in the opposite direction, because in the bulk of the superconductor the current must be equal to zero (otherwise the Meissner effect would not have occurred). Therefore, it has long been thought that there are no thermoelectric phenomena in a super-conductor. It has turned out, in fact, that the appearance of a supercurrent is amenable to detection.

The presence of a supercurrent implies that the momentum of Cooper pairs is no longer equal to zero, i.e., the Bose condensate of Cooper pairs comes into motion. In this case, the Bogoliubov transformation formulae (16.14) must be replaced by the following:

$$\begin{aligned} a_{p,+} &= u_p \alpha_{p,+} + v_p \alpha^{+}_{-p+q,-} \\ a_{p,-} &= u_{-p+q} \alpha_{p,-} - v_{-p+q} \alpha^{+}_{-p+q,+}. \end{aligned} \tag{19.14}$$

The meaning of these transformations is as follows: the term with v in each of the formulae corresponds to the binding of electrons into a Cooper pair with momentum $\boldsymbol{q}$ and to its escape into the condensate (evidently $\boldsymbol{q} = 2m\boldsymbol{v}_s$). Instead, an electron appears with momentum $-\boldsymbol{p}+\boldsymbol{q}$ and opposite spin projection.

The subscripts for u and v in the second formula are chosen such that the following condition is satisfied:

$$\{a_{p,+}, a_{-p+q,-}\} = -u_p v_p + u_p v_p = 0.$$

The coefficients u and v are connected by the same relation as before:

$$|u_p|^2 + |v_p|^2 = 1. \tag{19.15}$$

Substituting formulae (19.14) into the Hamiltonian (16.13) and averaging, we obtain

$$\begin{aligned} \langle \mathscr{H} - \mu N \rangle = \sum_p \{ &\xi[n_{p,+} + |v_p|^2(1 - n_{p,+} - n_{-p+q,-})] \\ &+ \xi'[n_{-p+q,-} + |v_p|^2(1 - n_{p,+} - n_{-p+,q-})]\} - \frac{|\Delta|^2}{g}, \end{aligned} \tag{19.16}$$

where

$$\Delta = g \sum_p u_p v_p (1 - n_{p,+} - n_{-p+q,-}), \tag{19.17}$$

$$\xi' = \xi(-\boldsymbol{p}+\boldsymbol{q}). \tag{19.18}$$

We will find the real solution for u and v (this does not alter the result but simplifies the computations). In this case, Δ is also real. Minimizing with respect to v_p and using (19.15), we get

$$u^2 = u^2_{-p+q} = \tfrac{1}{2}\left(1 + \frac{\xi + \xi'}{2\varepsilon}\right), \tag{19.19}$$

$$v_p^2 = v^2_{-p+q} = \tfrac{1}{2}\left(1 - \frac{\xi + \xi'}{2\varepsilon}\right), \tag{19.20}$$

$$\varepsilon = [\tfrac{1}{4}(\xi + \xi')^2 + \Delta^2]^{1/2}. \tag{19.21}$$

Varying eq. (19.16) with respect to $n_{p,+}$ and $n_{p,-}$, we find the value of energy in the Fermi distribution functions (19.17):

$$\varepsilon_{p,+} = \varepsilon + \tfrac{1}{2}(\xi - \xi'), \qquad \varepsilon_{-p+q,-} = \varepsilon - \tfrac{1}{2}(\xi - \xi'). \tag{19.22}$$

Let us calculate the electric current. Here we must take into account the difference between charge transfer and energy transfer. Whereas the energy is transferred by quasi-particles, which have an energy $\varepsilon = (\xi^2 + \Delta^2)^{1/2}$ and velocity $\partial\varepsilon/\partial \boldsymbol{p}$, the charge is transferred by initial electrons with velocity $\boldsymbol{p}/m$ (the corresponding operators are a and a^+ and not α and α^+). Thus, the current operator has the form

$$\boldsymbol{j} = \frac{e}{m}\sum_{p,\sigma} \boldsymbol{p} \overset{+}{a}_{p,\sigma} a_{p,\sigma}. \tag{19.23}$$

The same result has been obtained earlier (see formulae 16.59′ and 16.62 at $\boldsymbol{A} = 0$ and $\boldsymbol{q} = 0$).

Substituting here the transforms (19.14) and averaging gives

$$\boldsymbol{j} = \sum_p \boldsymbol{p}[u_p^2 n_{p,+} + v_p^2(1 - n_{-p+q,-}) + u_n^2 n_{p,-} + v_p^2(1 - n_{-p+q,+})]$$

(here we have used qualities 19.20, 19.21). We pass now to a new variable: $\boldsymbol{p} \to \boldsymbol{p} + \tfrac{1}{2}\boldsymbol{q}$. Assuming that $|\boldsymbol{q}| \ll p_0$, we obtain

$$\begin{aligned} &\xi \to \xi + \tfrac{1}{2}\boldsymbol{vq}, && \xi' \to \xi - \tfrac{1}{2}\boldsymbol{vq}, && \tfrac{1}{2}(\xi + \xi') \to \xi, \\ &\varepsilon \to (\xi^2 + \Delta^2)^{1/2}, && \varepsilon_p \to \varepsilon + \tfrac{1}{2}\boldsymbol{vq}/2, && \varepsilon_{-p+q} \to \varepsilon - \tfrac{1}{2}\boldsymbol{vq}. \end{aligned} \tag{19.24}$$

The expression for the current takes the form

$$\boldsymbol{j} = 2\sum_p (\boldsymbol{p} + \tfrac{1}{2}\boldsymbol{q})[u_p^2 n_p + v_p^2(1 - n_{-p+q})]. \tag{19.25}$$

The momentum of the pairs is connected with the velocity of the condensate by the relation $\boldsymbol{q} = 2m\boldsymbol{v}_s$. Since $\boldsymbol{v} = \boldsymbol{p}/m$, we have

$$\tfrac{1}{2}\boldsymbol{vq} = \boldsymbol{pv}_s. \tag{19.26}$$

The result obtained may be interpreted as follows. In the coordinate system associated with the moving condensate the energy of quasiparticles is ε and the momentum of electrons is $\boldsymbol{p}$. But if we pass to the laboratory coordinate system, the energy of quasiparticles becomes equal to $\varepsilon + \boldsymbol{pv}_s$ and the electron momentum to $\boldsymbol{p} + m\boldsymbol{v}_s$.

In the preceding section, while calculating the thermal conductivity we should have replaced, strictly speaking, ε in expression (19.11) by $\varepsilon + \boldsymbol{p}\boldsymbol{v}_s$ and $\partial\varepsilon/\partial\boldsymbol{p}$ by $\partial\varepsilon/\partial\boldsymbol{p} + \boldsymbol{v}_s$. But here one must take into account that the superconducting current arises only as a result of the appearance of a temperature gradient, i.e., $\boldsymbol{v}_s \propto \nabla T$. Substitution into eq. (19.11) of the equilibrium distribution function n_F instead of n under the integral yields the gradient in momentum space of a certain function $f(\varepsilon + \boldsymbol{p}\boldsymbol{v}_s)$. The integral over $d^3\boldsymbol{p}$ is transformed to a surface integral and is equal to zero because $n_F(\infty) = 0$. Therefore, only the integral of the nonequilibrium contribution $n^{(1)}$ is left, which itself is proportional to ∇T. Hence, in this integral we may put $\boldsymbol{v}_s = 0$.

Let us return to formula (19.25). We substitute into it $n = n_F + n^{(1)}$, where $n^{(1)} \ll n_F$, and limit ourselves to the terms of first order in $n^{(1)}$ and $\boldsymbol{v}_s$ (of order ∇T). This gives

$$\boldsymbol{j} = 2e\boldsymbol{v}_s \sum_p [u_p^2 n_F + v_p^2(1 - n_F)] + 2\frac{e}{m}\sum_p \boldsymbol{p}(\boldsymbol{p}\boldsymbol{v}_s)\frac{\partial n_F}{\partial\varepsilon} + 2\frac{e}{m}\sum_p \boldsymbol{p}n^{(1)}. \tag{19.27}$$

In the last term it is taken into account that $n_{-p}^{(1)} = -n_p^{(1)}$ (see 19.10). We have also used the condition (19.15).

If we perform the Bogoliubov transformation (16.14) in the electron density operator

$$n_e = \sum_p \overset{+}{a}_{p,\sigma} a_{p,\sigma}$$

and average, we obtain

$$n_e = 2\sum_p [u_p^2 n_F + v_p^2(1 - n_F)].$$

Thus, the first term in (19.27) is equal to $n_e \boldsymbol{v}_s e$. We compare the second term in (19.27) with formula (16.83) for the momentum of quasiparticles moving with velocity $\boldsymbol{u}$ (in our case the role of $\boldsymbol{u}$ is played by $-\boldsymbol{v}_s$). The proportionality factor between $\boldsymbol{P}$ and $m\boldsymbol{u}$ is the "number of normal electrons" n_n. Hence, the first two terms in (19.27) give

$$(n_e - n_n)e\boldsymbol{v}_s = n_s e\boldsymbol{v}_s = \boldsymbol{j}_s.$$

This is just the superconducting current.

The last term in eq. (19.27) is the current of quasiparticles, which arises under the action of the temperature gradient. Substituting (19.10) and passing to integration over ξ, we get

$$\boldsymbol{j}_n = e\int \frac{\varepsilon}{T}\frac{\partial n_F}{\partial\varepsilon}\tau_{tr}\,\mathrm{sign}\,\xi\,\boldsymbol{v}(\boldsymbol{v}\nabla T)\nu\,d\xi\frac{d\Omega}{4\pi}.$$

This integral vanishes in the first approximation in ξ because of the presence of the factor sign ξ. Therefore, just as in a normal metal, we must expand the factor $\tau_{tr}v^2\nu$ in the integrand in ξ:

$$\tau_{tr}v^2\nu=(\tau_{tr}v^2\nu)_\mu+\xi\frac{\partial}{\partial\mu}(\tau_{tr}v^2\nu)_\mu .$$

The second term gives a nonzero result. Here

$$\int_{-\infty}^{\infty}\xi\ \mathrm{sign}\ \xi\ \mathrm{d}\xi\to\int_{-\infty}^{\infty}|\xi|\ \mathrm{d}\xi\to 2\int_0^{\infty}|\xi|\ \mathrm{d}|\xi|\to 2\int_\Delta^{\infty}\varepsilon\ \mathrm{d}\varepsilon .$$

The result is given by (Gal'perin et al. 1973, 1974)

$$\begin{aligned} j_n &= \tfrac{2}{3}e\left[\frac{\partial}{\partial\mu}(\tau_{tr}v^2\nu)_\mu\right]\frac{\nabla T}{T}\int_\Delta^{\infty}\varepsilon^2\frac{\partial n}{\partial\varepsilon}\mathrm{d}\varepsilon \\ &= -\tfrac{2}{3}e\left(\frac{\partial}{\partial\mu}(\tau_{tr}v^2\nu)_\mu\right)T\nabla T\int_{\Delta/T}^{\infty}\frac{x^2}{\cosh^2(\frac{1}{2}x)}\mathrm{d}x \\ &= -e\left(\frac{\partial}{\partial\mu}\ln(\tau_{tr}v^2\nu)_\mu\right)\varkappa\nabla T, \end{aligned} \tag{19.28}$$

where $\varkappa$ is the thermal conductivity coefficient. Taking into account that in a normal metal $\boldsymbol{j}_n=-\beta\nabla T$ and comparing formulas (3.24) and (6.19), we see that the expression for $\boldsymbol{j}_n$ written in terms of $\varkappa$ does not differ from a normal metal. Since $T\to T_c$ and $\varkappa\to\varkappa_n$, the same is valid for $\boldsymbol{j}_n$. Taking into account that for impurity scattering $\tau_{tr}=l/v\propto\mu^{-1/2}$, $v^2\propto\mu$ and $\nu\propto p_0\propto\mu^{1/2}$, we finally obtain

$$\boldsymbol{j}_s=-\boldsymbol{j}_n=\frac{e}{\mu}\varkappa\nabla T. \tag{19.29}$$

The asymptotics are found with the aid of the formulas for $\varkappa$: (3.24) with $T\to T_c$ and (19.13) with $T\to 0$.

A question now arises as to whether we can determine experimentally the presence of a supercurrent. To this end, we use the expression for the supercurrent in terms of the phase of the order parameter (17.75):

$$\boldsymbol{j}_s=\frac{\hbar e n_s}{2m}\left(\nabla\chi-\frac{2e}{\hbar c}\boldsymbol{A}\right) \tag{19.30}$$

(here we have substituted $|\Psi|^2=n_s/2$, the number of Cooper pairs). Comparing expressions (19.30) and (19.29), we find

$$\nabla\chi=\frac{2m}{\hbar e n_s}\frac{e\varkappa}{\mu}\nabla T+\frac{2e}{\hbar c}\boldsymbol{A}. \tag{19.31}$$

Suppose now that we have a ring consisting of two different superconductors (fig. 123). We assume that the junctions are at different temperatures, T_1 and T_2. We integrate expression (19.31) over a closed loop C, which passes through the bulk

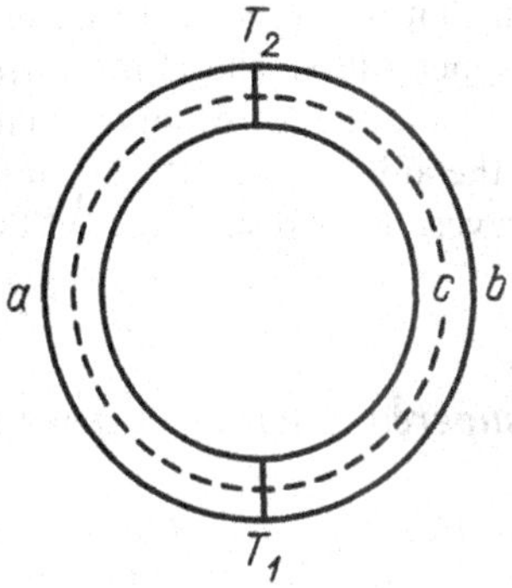

Fig. 123.

of superconductors. The change of the phase along the closed loop must be equal to $2\pi n$, where n is an integer. Hence, we have

$$2\pi n=\frac{2m}{\hbar}\oint_C \frac{\varkappa}{\mu n_s}\nabla T\,\mathrm{d}\boldsymbol{l}+\frac{2e}{\hbar c}\oint_C \boldsymbol{A}\,\mathrm{d}\boldsymbol{l}$$

$$=\frac{2m}{\hbar}\int_{T_1}^{T_2}\left[\left(\frac{\varkappa}{\mu n_s}\right)_b-\left(\frac{\varkappa}{\mu n_s}\right)_a\right]\mathrm{d}T+2\pi\frac{\Phi}{\Phi_0}$$

or

$$\frac{\Phi}{\Phi_0}=n+\frac{m}{\pi\hbar}\int_{T_1}^{T_2}\left[\left(\frac{\varkappa}{\mu n_s}\right)_a-\left(\frac{\varkappa}{\mu n_s}\right)_b\right]\mathrm{d}T. \tag{19.32}$$

From this formula it follows that the magnetic flux passing through the loop is not equal to an integer number of flux quanta and has a correction which depends on temperature. This correction is a manifestation of thermoelectricity in superconductors.

At first glance it seems paradoxical that a magnetic flux can pass through the ring in the absence of current. In fact, the statement that $\boldsymbol{j}=0$ refers only to the bulk of the superconductor. A superconducting current is circulating along the inner surface of the ring and it is this supercurrent that produces the magnetic flux.

Let us estimate the magnitude of the correction to the magnetic flux. We assume that $T\sim T_c$ and, hence, $n_s\sim n_e$ and $\varkappa\sim\varkappa(T_c)$. The expression for $\varkappa(T)$ is derived from formula (3.28). We have

$$\frac{\Delta\Phi}{\Phi_0\Delta T}\sim\frac{m}{\hbar}\frac{\varkappa}{\mu n_e}\sim\frac{m}{\hbar}\frac{n_e\tau_{tr}}{m\mu n_e}T_c$$

$$\sim\frac{T_c}{\mu}\frac{\tau_{tr}}{\hbar}\sim 10^{-4}\frac{l}{v\hbar}\sim 10^{12}\,\mathrm{erg}^{-1}\sim 10^{-4}\,\mathrm{K}^{-1}$$

(we assume here that the mean free path $l\sim 10^{-3}$ cm and $T_c/\mu\sim 10^{-4}$). Thus, the temperature contribution is very small (10^{-4} flux quantum per 1 K). Its measurement is hindered also by the fact that it is difficult to protect the superconducting ring against a small trapped magnetic flux. Therefore, in practice one has to measure the difference effect with $T_1 \rightleftharpoons T_2$. The corresponding measurement has been carried out (Zavaritskii 1974) with the aid of a SQUID (section 22.8) and has qualitatively confirmed the theory. However, other results exceed the theoretical predictions*.

19.3. The behavior of superconductors in a weak high-frequency field

The behavior of superconductors in a weak, time-independent field has been considered in sections 16.7 and 16.8. In principle, the same method can be used to find the relationship between the current and a field of finite frequency. We do not give the relevant calculations here because they are very cumbersome (see, for example, Abrikosov and Khalatnikov 1958). Instead, we present some of the fundamental results (Abrikosov et al. 1958, 1959, Mattis and Bardeen 1958).

First of all let us consider the case $T=0$. In this case the superconductor does not contain quasiparticles that could have absorbed quanta of any energy. The absorption of electromagnetic waves begins only when

$$\hbar\omega = 2\Delta(0). \tag{19.33}$$

This is the energy required for the breakdown of Cooper pairs. In the limit, when $\hbar\omega \gg 2\Delta$ the difference between a super-conductor and a normal metal disappears, i.e., the relationship between the current and the field tends to that existing in a normal metal.

The absorption of electromagnetic energy is proportional to the real part of the impedance. For a pure superconductor in the limiting Pippard case the following relation holds:

$$\frac{Z_s}{R_n} = -2i\left[\frac{\pi\hbar\omega}{\Delta Q_1(\omega)}\right]^{1/3}, \tag{19.34}$$

where $Q_1(\omega)$ enters into the proportionality factor between the current and the field:

$$j = -Q(q, \omega)\, A(q, \omega), \qquad Q(q, \omega) = \tfrac{3}{4}\frac{n_e e^2}{mc}\frac{\Delta}{\hbar v q}\, Q_1(\omega), \tag{19.35}$$

and R_n is the value of R for a normal metal.

The real part of the impedance manifests itself only if the imaginary part Q_1 is nonzero. A plot of Im $Q_1(\omega)$ at $T=0$ is shown in fig. 124. The $R_s(\omega)/R_n$ plot has

* V. I. Kozub (1985) proposed the following explanation. In order to enhance the effect, the superconductors making up the ring are usually chosen such that they differ in their parameters, particularly in T_c. For one of the superconductors the temperature is very low and the energy transport is mainly realized by phonons. These phonons are irradiated into the other superconductor and produce a phonon wind, which drags the quasiparticles. An estimate shows that this effect can increase the observed magnetic flux by two orders of magnitude.

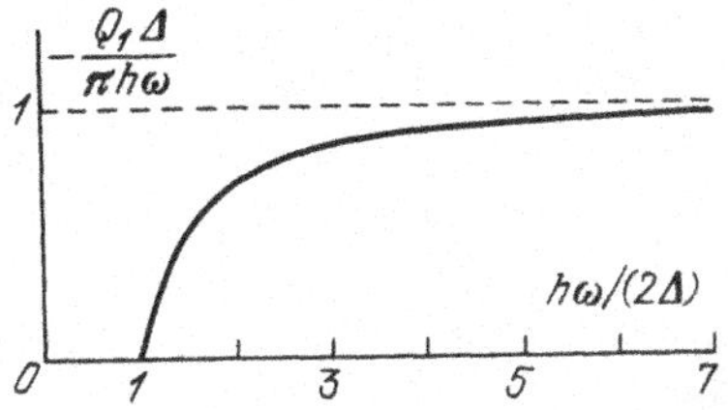

Fig. 124.

qualitatively the same form. Near the absorption threshold

$$\operatorname{Im} Q_1 \approx -\pi^2 \left(\frac{\hbar\omega}{2\Delta} - 1\right). \qquad (19.35')$$

The ratio $R_s(\omega)/R_n$ is also proportional to this quantity.

At $T > 0$ there are quasiparticles in the superconductor which can absorb electromagnetic quanta of any frequency. If $T \ll T_c$, the number of quasiparticles is exponentially small ($\sim \exp[-\Delta(0)/T]$). In the case $\omega \ll \Delta$ the same exponential is contained in the imaginary part of $Q_1(\omega)$ and in the real part of the impedance. The expression for $\operatorname{Im} Q_1$ for ω, $T \ll \Delta$ reads

$$\operatorname{Im} Q_1 \approx -4\pi \sinh\left(\frac{\omega}{2T}\right) K_0\left(\frac{\omega}{2T}\right) \exp\left(\frac{-\Delta(0)}{T}\right), \qquad (19.35'')$$

where K_0 is the MacDonald function.

Starting with $\hbar\omega = 2\Delta(0) = 3.52\,T_c$ a finite absorption also occurs at $T = 0$. This is clearly seen in the experimental plots obtained for aluminum at various frequencies and temperatures (fig. 125) (Biondi and Garfunkel 1959). Note that with $\omega \to 0$ the dependence $R_s(T)$ assumes the same form as the resistivity curve at constant current: $R = 0$ at $T < T_c$ and $R_s = R_n$ at $T > T_c$.

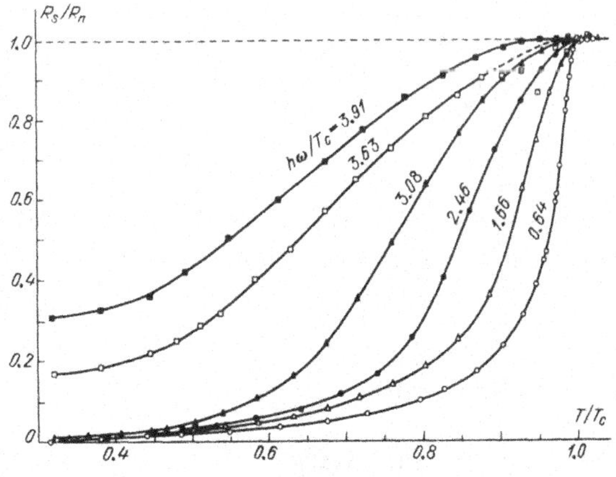

Fig. 125.

The characteristic wavelengths corresponding to the condition (19.33) lie within a range of the order of 1–0.1 mm and the corresponding frequencies are of the order of $\nu = \omega/2\pi \sim 10^{11}$–$10^{12}$ Hz. Note that in the optical range $\nu \sim 10^{15}$ Hz. For these frequencies the impedance of superconductors does not differ from that of normal metals.

However, in principle, apart from elastic reflection at an unshifted frequency, one can also observe Raman light scattering when part of the quantum energy is transferred to the electronic system. In this case (Abrikosov and Fal'kovsii 1961, 1987, Abrikosov and Genkin 1973) the spectral dispersion of the reflected light contains "satellites", i.e., quanta with an altered frequency. In this case, we are speaking of the breakup of the pairs and therefore at low temperatures a "Stokes satellite" appears, with a continuous frequency distribution $\hbar\omega \leqslant \hbar\omega_0 - 2\Delta$ (ω_0 is the frequency of the incident light). The intensity of the satellite is rather low because the interaction of light with the electronic system occurs only in a thin skin layer. Here, special care must be taken in order for the line of the incident light to be narrow (in the sense of the spectral width) and the intensity not to be very high, otherwise the specimen will be heated. The value of Δ is also important. For substances with large Δ the line needs not be so narrow. Thus Raman scattering was first definitely observed in the high-T_c ceramic $YBa_2Cu_3O_{7-\delta}$ (Bazhenov et al. 1987, Lyons et al. 1987). In an anisotropic superconductor Δ_{min} is established in this way (Abrikosov and Fal'kovskii 1987).

In conclusion, we give one more relation between the current and the field for a dirty superconductor in the London limit ($q = 0$) in the vicinity of the critical temperature and also for a low frequency: $\omega \ll \Delta \ll T_c$ (this case is of interest for what follows). We have

$$\boldsymbol{j} = \left(\frac{-c}{4\pi\delta^2} + \frac{i\omega\sigma}{c} \right) \boldsymbol{A},$$

where $\sigma = n_e e^2 \tau_{tr}/m$ is the conductivity of a normal metal; δ depends on the ratio of ξ_0 to l, but in all cases it corresponds to the static penetration depth (sections 16.9 and 16.10). In other words, for the case under consideration

$$\boldsymbol{j} = \boldsymbol{j}_s + \boldsymbol{j}_n$$

where $\boldsymbol{j}_s$ is expressed by the formula for constant field and $\boldsymbol{j}_n = \sigma\boldsymbol{E}(\boldsymbol{E} = -c^{-1}\partial\boldsymbol{A}/\partial t = (i\omega/c)\boldsymbol{A}$. This is consistent with the concepts of London electrodynamics. However, with other relationships between ω, Δ and T there is no such agreement.

19.4. Absorption of ultrasound

In accordance with section 12.1, the variation of the electron energy in a field of sound waves may be considered to be equal to $\Lambda_{ik}(\boldsymbol{p})u_{ik}(\boldsymbol{r}, t)$, where u_{ik} is the strain tensor and Λ_{ik} is the deformation potential. This quantity plays the role of the potential energy. The relevant operator may be written similarly to eq. (19.2). The frequency of the sound that can actually be generated does not exceed 10^9 Hz, whereas $\Delta/(2\pi\hbar) \sim 10^{11}$ Hz. Hence, $\hbar\omega \ll \Delta$. In view of this, in transformating to operators of quasiparticles we have again to drop the terms corresponding to creation and annihilation of two quasiparticles. As regards the terms corresponding to scattering, one has to take account of the fact that the electron energy upon absorption of a phonon increases by an amount $\hbar\omega$. Therefore, $\varepsilon \neq \varepsilon'$ in formula (19.3).

We assume that $\omega\tau_{tr} \gg 1$ and apply the quantum approach described in section 12.2. The only difference from formula (12.20) lies in the appearance of the factor $(uu' - vv')^2$. Inserting the expressions for u and v, we obtain (assuming that the normalization volume $V = 1$):

$$Q = -\pi\omega^2 \sum_{p,p'} |U_{p'p}|^2 \delta(\omega_{p'p} - \omega) \frac{\partial n_F}{\partial \varepsilon} \tfrac{1}{2} \left(1 + \frac{\xi_p \xi_{p'} - \Delta^2}{\varepsilon_p \varepsilon_{p'}}\right). \tag{19.36}$$

The matrix element $U_{p'p}$ is given by (at $k \ll p_0$)

$$U_{p'p} = \Lambda_{ik}(\boldsymbol{p})\, u^0_{ik} \delta_{p',p+\hbar k}, \tag{19.37}$$

where $u^0_{ik} = \frac{1}{2}\mathrm{i}(k_i u^0_k - k_k u^0_i)$ and $\boldsymbol{u}^0$ is the amplitude of the sound wave. Passing from the sums over momenta to integrals $\sum_p = (2\pi\hbar)^{-3} \int \mathrm{d}^3\boldsymbol{p}$ and considering that

$$\delta^2_{p'p} = \delta_{p'p} \to (2\pi\hbar)^3(\delta(\boldsymbol{p}' - \boldsymbol{p}),$$

we get

$$Q = -\pi\omega^2\hbar \int \frac{\mathrm{d}^3\boldsymbol{p}}{(2\pi\hbar)^3} |\Lambda_{ik}(\boldsymbol{p}) u^0_{ik}|^2 \delta(\varepsilon_{p'} - \varepsilon_p - \hbar\omega) \frac{\partial n_F}{\partial \varepsilon} \tfrac{1}{2} \left(1 + \frac{\xi_p \xi_{p'} - \Delta^2}{\varepsilon_p \varepsilon_{p'}}\right), \tag{19.38}$$

where $\boldsymbol{p}' = \boldsymbol{p} + \hbar\boldsymbol{k}$.

In formula (19.38) $\int \mathrm{d}^3\boldsymbol{p}(2\pi\hbar)^{-3}$ corresponds to $\frac{1}{4}\nu \int \mathrm{d}\xi\, \mathrm{d}\cos\frac{1}{2}\theta$, where $\theta = (\widehat{\boldsymbol{v}, \boldsymbol{k}})$. Taking into account that $\xi' = \xi + vk\cos\theta$ and $vk \gg \omega$ (because $v \gg s$; s is the sound velocity), we can pass to the integration over ξ and ξ' and write

$$\int \frac{\mathrm{d}^3\boldsymbol{p}}{(2\pi\hbar)^3} \to \frac{\nu}{4\hbar vk} \int_{-\infty}^{\infty} \mathrm{d}\xi \int_{-\infty}^{\infty} \mathrm{d}\xi'$$

(instead of $|\Lambda_{ik}(\boldsymbol{p})|^2$ we must take $\overline{|\Lambda_{ik}(\boldsymbol{p})|^2}$, which signifies the average over the line on the Fermi surface, where $\cos\theta = 0$ or $\boldsymbol{v} \perp \boldsymbol{k}$). The integration over ξ and ξ' leads to the cancellation of the odd terms with respect to ξ and ξ' in eq. (19.38). After this we may pass over to the integration over ε, taking account of the two possible signs of ξ: $\int_{-\infty}^{\infty} \mathrm{d}\xi \to 2\int_{\Delta}^{\infty} \varepsilon(\varepsilon^2 - \Delta^2)^{-1/2}\, \mathrm{d}\varepsilon$. As a result, we obtain

$$Q = -\frac{\pi\omega^2\nu}{vk} \int_{\Delta}^{\infty} |\Lambda_{ik} u^0_{ik}|^2 [\varepsilon(\varepsilon + \hbar\omega) - \Delta^2]$$
$$\times (\varepsilon^2 - \Delta^2)^{-1/2} [(\varepsilon + \hbar\omega)^2 - \Delta^2]^{-1/2} \frac{\partial n_F}{\partial \varepsilon}\, \mathrm{d}\varepsilon. \tag{19.39}$$

Since $\hbar\omega \ll \Delta$, all the brackets under the integral sign give unity. Comparing with the expression for a normal metal (at $\Delta = 0$) and taking into account that for the absorption coefficient to be obtained the quantity Q is divided in both cases by the same sound energy flux I, we obtain (Bardeen et al. 1957):

$$\frac{\Gamma_s}{\Gamma_n} = -2\int_{\Delta}^{\infty} \frac{\partial n_F}{\partial \varepsilon}\, d\varepsilon = 2n_F(\Delta) = \frac{2}{e^{\Delta/T}+1}. \tag{19.40}$$

As $T \to T_c$ this ratio tends to unity and at low temperatures it falls off as $\exp(-\Delta/T)$, i.e., proportionally to the number of quasiparticles. The curve showing the dependence of Γ_s/Γ_n on T/T_c is given in fig. 126.

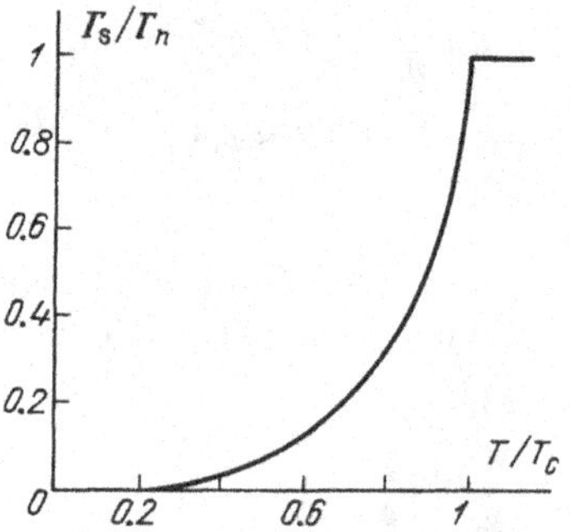

Fig. 126.

19.5. Stimulation of superconductivity by high-frequency, high-intensity field and sound

Up until now, while dealing with the behavior of superconductors under the influence of various external factors, we have limited our treatment to linear problems. Examples are the calculation of the current under the action of a small external field, and heat flux and electric current in the presence of a small temperature gradient. Such calculations are called calculations of the linear response of the system.

However, as the amplitude of external perturbations increases the nonlinear response may also become significant. In particular, under the action of a varying electromagnetic field a current with a frequency twice as high may be generated, which will lead to the emission of an electromagnetic wave at a doubled frequency. Hence, when an electromagnetic wave falls on the metal surface, a reflected wave of doubled frequency may appear. Other nonlinear effects may also arise.

We shall consider a specific nonlinear action of a high-frequency field (electromagnetic wave or sound), on superconductors which leads to a qualitatively new effect – an increase in the critical temperature of the superconducting transition, or, stimulation of superconductivity (Eliashberg 1970, Ivlev and Eliashberg 1971).

In sections 16.4 and 16.5 we have found the energy spectrum of quasiparticles in a superconductor and the equation for Δ. Formula (16.21), which relates Δ to the distribution function, is valid for an arbitrary distribution and not only for the equilibrium distribution expressed by the Fermi function. Thus, we have

$$1=\tfrac{1}{2}g\int\frac{d^3p}{(2\pi\hbar)^3}\,\frac{1-2n}{\varepsilon}, \tag{19.41}$$

where $\varepsilon=(\xi^2+\Delta^2)^{1/2}$.

With an equilibrium distribution we have seen that in the case $T=0$, when no quasiparticles are present, Δ is at a maximum. At finite temperatures the appearance of quasiparticles leads to a decrease in Δ. Here, in the integral (19.41) quasi-particles with an energy of the order of Δ were the most important.

Suppose now that the electronic system absorbs quanta of energy $\hbar\omega$. If $\hbar\omega>2\Delta$, new quasiparticles will be created and, hence, Δ will decrease. However, if $\hbar\omega\ll\Delta$, new quasi-particles cannot be created and the quanta are absorbed by the quasiparticles already present. This leads to their redistribution over the energies. If an appreciable number of quasiparticles receive an extra energy larger than Δ, it means that they move into another energy region. Hence, they leave the region which is relevant in the integral (19.41), and in this sense the situation begins to resemble that observed at $T=0$. As a result, Δ increases.

To simplify the calculation, we assume that the temperature is close to T_c, i.e., $\Delta\ll T_c$ but still $\omega\ll\Delta$. Besides, we presume that the intensity of the external field is not too high, so the relative change of Δ is small.

We substitute $n=n_F(\varepsilon)+n-n_F$ into eq. (19.41). The integral with n_F is expanded in Δ, as was done in section 16.5 (formula 16.32). So, we obtain

$$\ln\left(\frac{T_c}{T}\right)-\tfrac{7}{8}\zeta(3)\frac{\Delta^2}{\pi^2T_c^2}-2\int_\Delta^\infty d\varepsilon\,(\varepsilon^2-\Delta^2)^{-1/2}\overline{(n-n_F)}=0, \tag{19.42}$$

where the overbar in the last integral signifies averaging over the angles. We used here formula (16.25) for T_c; in the integral of $n-n_F$ we passed to the variable ε and took into account two signs of ξ. With our assumptions $\ln(T_c/T)\approx(T_c-T)/T_c$. If the distribution function depends on time, then in the integral we have to take the time-averaged quantity.

The distribution function n can be determined from the kinetic equation.

The calculations based on the detailed microscopic (BCS) theory, which takes account of the dependence of the quantities on time, involve very complicated kinetic equations, which are different, depending on the character of the external force. The difficulty is aggravated by the fact that the scattering from impurities, while being elastic, does not give rise to the relaxation of quasiparticles with respect to energy, and therefore we must take into account the interaction with phonons. The latter circumstance leads to complication of the collision integral. Therefore, instead of detailed calculations, we shall make a simple estimate*.

Suppose we have a thin specimen, which is smaller than the penetration depth, onto which a high-frequency electromagnetic wave falls. As has been noted at the end of section 19.3, a normal current $j_n = \sigma E$ arises in it, which gives rise to ohmic losses. The energy received by the electronic system per unit time is given by

$$\frac{\partial Q}{\partial t} = \sigma \overline{E^2} = \tfrac{1}{2}\sigma E_0^2, \tag{19.43}$$

where $E = E_0 \cos \omega t$. Near T_c the quantity σ is equal to the conductivity of a normal metal, i.e., $\sigma = n_e e^2 \tau_{tr}/m$. This includes the time of scattering from impurities, because this process is the dominant one.

The electronic system transfers the received energy to the lattice in the form of phonons. However, this process is slow, taking a time τ_{ph}. Therefore, the effective temperature of the electronic system increases. It is rather obvious that

$$\tau_{ph} \frac{\partial Q}{\partial t} \sim C_e \delta T, \tag{19.44}$$

where C_e is the electronic heat capacity, which has the same order of magnitude as that in a normal metal, i.e., $C_e \sim p_0 m T/\hbar^3$.

From formulae (1943) and (19.44) we find

$$\delta T \sim \frac{\tau_{ph}\sigma E_0^2}{C_e} \sim \frac{\tau_{ph} n_e e^2 \tau_{tr} E_0^2 \hbar^3}{m \cdot m p_0 T} \sim \frac{(evE_0)^2 \tau_{ph}\tau_{tr}}{T}. \tag{19.45}$$

When the quasiparticles are heated by the field, their number is not changed because $\omega \ll \Delta$. This may be taken into account if we introduce the effective chemical potential. The new distribution will have the form

$$n = \frac{1}{\exp\left(\dfrac{\varepsilon - \delta\mu}{T + \delta T}\right) + 1}$$

From the condition

$$\int_0^\infty n\, d\xi = \int_\Delta^\infty \frac{n\varepsilon}{(\varepsilon^2 - \Delta^2)^{1/2}}\, d\varepsilon = \int_\Delta^\infty \frac{n_F \varepsilon}{(\varepsilon^2 - \Delta^2)^{1/2}}\, d\varepsilon = \text{const.}$$

we find, by expanding in $\delta\mu$ and δT,

$$-\delta\mu \int_\Delta^\infty \frac{\partial n_F}{\partial \varepsilon} \frac{\varepsilon}{(\varepsilon^2 - \Delta^2)^{1/2}}\, d\varepsilon - \frac{\delta T}{T} \int_\Delta^\infty \frac{\partial n_F}{\partial \varepsilon} \frac{\varepsilon^2}{(\varepsilon^2 - \Delta^2)^{1/2}} = 0.$$

Since $\Delta \ll T$, we may put $\Delta = 0$ everywhere. So, we obtain

$$\delta\mu = -2\delta T \ln 2.$$

* The reasoning given here stems from B. I. Ivlev.

Substituting

$$n \approx n_{\mathrm{F}} - \frac{\partial n_{\mathrm{F}}}{\partial \varepsilon}\left(\delta\mu + \frac{\varepsilon}{T}\,\delta T\right)$$

into the last term of eq. (19.42), we find

$$\int_{\Delta}^{\infty} \mathrm{d}\varepsilon \frac{1}{(\varepsilon^2-\Delta^2)^{1/2}} \frac{-\partial n_{\mathrm{F}}}{\partial \varepsilon}\left(\delta\mu + \frac{\varepsilon}{T_{\mathrm{c}}}\,\delta T\right)$$

$$\approx \frac{\delta\mu}{4T_{\mathrm{c}}} \ln\left(\frac{T_{\mathrm{c}}}{\Delta}\right) + \frac{\delta T_{\mathrm{c}}}{2T_{\mathrm{c}}} \approx -\frac{\delta T}{2T_{\mathrm{c}}} \ln 2 \ln\left(\frac{T_{\mathrm{c}}}{\Delta}\right).$$

We have confined ourselves to logarithmic accuracy with respect to $\ln(T_{\mathrm{c}}/\Delta)$. It is important that the integral is negative. Equation (19.42) is now written in the form

$$\frac{T-T_{\mathrm{c}}}{T_{\mathrm{c}}} = a\tau_{\mathrm{ph}}\tau_{\mathrm{tr}} \frac{(evE_0)^2}{T_{\mathrm{c}}^2} - [\tfrac{7}{8}\zeta(3)]\frac{\Delta^2}{\pi^2 T_{\mathrm{c}}^2}, \tag{19.46}$$

where a is a coefficient of the order of unity (we assume that $\ln(T_{\mathrm{c}}/\Delta)$ is not large). A plot of $(T-T_{\mathrm{c}})/T_{\mathrm{c}}$ against Δ is given in fig. 127. The lower curve corresponds to the equilibrium case $E_0 = 0$. In the presence of a wave, i.e. at $E_0 \neq 0$, we obtain the upper curve, which reaches a certain positive value with decreasing Δ. However, the theory outlined here may be applied only for $\Delta > \frac{1}{2}\omega$. At lower Δ the creation of new quasiparticles begins. The coefficient of Δ^2 in eq. (19.46) decreases and may even change sign (the dashed curve).

Thus, we come to the conclusion that the curve in fig. 127 exhibits a maximum at $\Delta \neq 0$ corresponding to $T - T_{\mathrm{c}} > 0$. This means that if the film is kept at a temperature above T_{c} and irradiated, it can pass in a jumpwise manner to the superconducting state with a finite Δ. Of course, this effect may occur only at temperatures not too far from T_{c}.

The results obtained can be extended without any difficulty to the case where the excitation is accomplished by a sound wave. Comparing the two terms on the

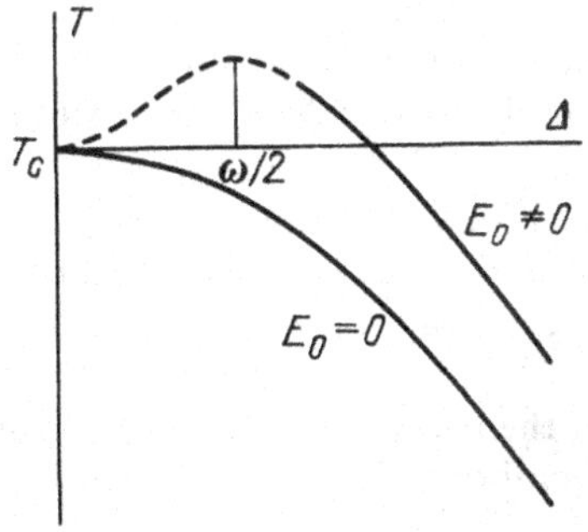

Fig. 127.

right-hand side of the kinetic equation (12.6), we arrive at the conclusion that in this case evE_0 in (19.46) must be replaced by $\Lambda_{ik}\dot{u}_{ik} \sim \omega\varepsilon_F ku^0 = \omega^2\varepsilon_F u^0/s$, where u^0 is the wave amplitude and s is the sound velocity.

The ideas presented here are in qualitative agreement with experiment (Wyatt et al. 1966). The measurement of the critical current of a film subjected to electromagnetic irradiation has shown that the current increases, depending on the irradiation dose. The effect has been observed near T_c and at not too high intensities.

It should, however, be noted that the experiments have been carried out on so-called bridges, i.e., films with a constriction. In this case, apart from the mechanism described, there is another mechanism of stimulation of superconductivity (Aslamazov and Larkin 1978), which will not be described here.

19.6. *Paraconductivity*

In section 17.1 we have given an estimate of the fluctuational region near T_c. According to formulae (17.21) and (17.22), it proved to be extremely small. As has already been said, this is caused by a large correlation length. The estimate made was based on a fluctuational correction to the heat capacity of a bulk metal. However, relative fluctuations are always larger for systems of smaller dimensions, i.e., filaments and films having a thickness smaller than δ and ξ. Actually, in such specimens it is not convenient to determine the heat capacity and it is much simpler to measure the conductivity. As will be shown below, the fluctuational correction to the conductivity of thin films is found to be so high that it can be measured. The physical cause for extra conductivity lies in the possibility of fluctuational creation of Cooper pairs, which produces an additional "channel" for current. This effect has become known as paraconductivity (Aslamazov and Larkin 1968).

In order to find the value of paraconductivity, a time-dependent generalization of the Ginzburg-Landau equations is required. This is associated with the fact that the electric field can be defined as $\boldsymbol{E} = -c^{-1}\partial \boldsymbol{A}/\partial t$, where $\boldsymbol{A}$ is the vector potential; but in this case, $\boldsymbol{A}$ has to be regarded as being dependent on time. It may also be assumed that $\boldsymbol{E} = -\Delta\varphi$ and $\boldsymbol{A} = 0$, but, as will be shown below, the scalar potential φ is contained in the equation for Ψ in combination with $\partial\Psi/\partial t$. In other words, the electric field in superconductors leads necessarily to nonstationary phenomena. The London equation $\partial(\Lambda \boldsymbol{j})/\partial t = \boldsymbol{E}$ also corresponds to this.

The general nonstationary BCS equations are very complicated, even in the limit of slow time and space variations of the field and order parameter. It is for this reason that we shall not give their derivation here; instead, we write the model equation for the vicinity of T_c, which in general correctly reflects the qualitative aspects of the behavior of the order parameter and in some cases is exact.

If a departure from equilibrium is assumed, then $\delta \int \Omega_s \, dV/\delta\Psi^*$ is no longer equal to zero (in ch. 17 we derived the Ginzburg-Landau equation for Ψ on the condition that this variational derivative be zero). At the same time, in the absence of equilibrium Ψ begins to depend on time. At small deviations from equilibrium it is natural to assume that $\partial\Psi/\partial t$ is proportional to $\delta \int \Omega_s \, dV/\delta\Psi^*$.

However, gauge invariance requires that $\partial\Psi/\partial t$ should be included in the equation in the following combination:

$$\frac{\partial\Psi}{\partial t}+2i\frac{e}{\hbar}\varphi\Psi,$$

where φ is the scalar potential of the electric field. The complete condition of gauge invariance requires, in fact, the invariance of the electric and magnetic fields:

$$\boldsymbol{E}=-\frac{1}{c}\frac{\partial \boldsymbol{A}}{\partial t}-\nabla\varphi, \qquad \boldsymbol{H}=\operatorname{rot}\boldsymbol{A}$$

upon variation of the potentials $\boldsymbol{A}\to\boldsymbol{A}+\nabla\eta$, and $\varphi\to\varphi-c^{-1}\partial\eta/\partial t$, where η is a scalar. It is not difficult to see that in a case where $\boldsymbol{A}$ is included in the equation in the combination $[-i\hbar\nabla-(2e/c)\boldsymbol{A}]\Psi$ and φ in the combination $[\partial/\partial t+2i(e/\hbar)\varphi]\Psi$, the guage transformation is compensated by the variation of the phase of the function Ψ. Thus, we have

$$-\gamma\left(\frac{\partial\Psi}{\partial t}+2i\frac{e}{\hbar}\varphi\Psi\right)=\frac{\delta\int\Omega_s\, dV}{\sigma\Psi^*}. \tag{19.47}$$

The microscopic derivation shows that in the paraconductivity problem eq. (19.47) gives an exact answer if the correct choice of the constant γ is made.

The constant γ can be determined from the following considerations. On the right-hand side of eq. (19.47) there is a term $-(\hbar^2/4m)\Delta\Psi$. On the whole this equation resembles the diffusion equation (see section 3.5):

$$\frac{\partial n}{\partial t}=D\Delta n.$$

The role of the diffusion coefficient D is played by $\hbar^2/(4m\gamma)$. On the other hand, in the presence of impurities the Ginzburg-Landau equation may be written in the form (17.6″). It is easy to see that in this equation the coefficient $\pi D\hbar/8T_c$ plays the same role as the coefficient $\hbar^2/4m\alpha$ in a pure superconductor. Writing

$$\frac{\hbar^2}{4m}=\frac{\pi D\hbar\alpha}{8T_c}=D\gamma$$

we have

$$\gamma=\frac{\pi\hbar\alpha}{8T_c}. \tag{19.48}$$

This value coincides with the value obtained in the BCS microscopic theory.

We are interested in the fluctuational correction to Ψ that arises under the action of a constant electric field. It is assumed to be small and therefore to first order the correction we look for will be proportional to $\boldsymbol{E}$. But since $\boldsymbol{E}$ does not depend on time, the same will also refer to the correction to Ψ. In view of this, the derivative $\partial\Psi/\partial t$ may be omitted from (19.47). Assuming Ψ to be small, we retain the terms linear in Ψ. Substituting the Ginzburg–Landau functional into (19.47) yields

$$2\gamma \mathrm{i}\frac{e}{\hbar}\varphi\Psi + \alpha\tau\Psi - \frac{\hbar^2}{4m}\Delta\Psi = 0. \tag{19.49}$$

We have chosen the gauge at which $\boldsymbol{A}=0$ and $\boldsymbol{E}=-\nabla\varphi$. Since $\boldsymbol{E}$ is homogeneous, it follows that $\varphi=-\boldsymbol{Er}$. Recall that $\tau=(T-T_c)T_c$.

In the absence of an electric fleld Ψ undergoes equilibrium fluctuations. The probability of fluctuations is proportional to (see Appendix II)

$$\begin{aligned} W &\propto \exp\left(-\frac{1}{T}\int \Omega_s \,\mathrm{d}V\right) \\ &\propto \exp\left[-\frac{1}{T}\sum_p (\alpha\tau + p^2/4m)|\Psi_p|^2\right], \end{aligned} \tag{19.50}$$

where $\Psi_p = \int \Psi(\boldsymbol{r})\,\mathrm{e}^{-\mathrm{i}\boldsymbol{pr}/\hbar}\,\mathrm{d}V$. Hence, the average equilibrium fluctuation $|\Psi_p^{(0)}|^2$ is given by

$$\begin{aligned} \langle|\Psi_p^{(0)}|^2\rangle &= \frac{\int |\Psi_p|^2 \exp\left[-\frac{1}{T}\left(\alpha\tau+\frac{p^2}{4m}\right)|\Psi_p|^2\right]\mathrm{d}|\Psi_p|^2}{\int \exp\left[-\frac{1}{T}\left(\alpha\tau+\frac{p^2}{4m}\right)|\Psi_p|^2\right]\mathrm{d}|\Psi_p|^2} \\ &= \frac{T}{\alpha\tau+p^2/4m}, \end{aligned} \tag{19.51}$$

where $\langle\cdots\rangle$ signifies the statistical average.

In the presence of a weak electric field we have $\Psi_p = \Psi_p^{(0)} + \Psi_p^{(1)}$. Since the momentum representation of the quantity $\boldsymbol{r}\Psi(\boldsymbol{r})$ is $\mathrm{i}\hbar(\partial/\partial\boldsymbol{p})\Psi_p$, we obtain in the first approximation in $\boldsymbol{E}$:

$$2\gamma e\boldsymbol{E}\frac{\partial\Psi_p^{(0)}}{\partial\boldsymbol{p}} + \left(\alpha\tau+\frac{p^2}{4m}\right)\Psi_p^{(1)} = 0,$$

from which it follows that

$$\Psi_p^{(1)} = -2\frac{\gamma e}{\alpha\tau+p^2/4m}\boldsymbol{E}\frac{\partial\Psi(0)}{\partial\boldsymbol{p}}. \tag{19.52}$$

The average electric current in the Ginzburg–Landau theory is

$$\begin{aligned} \boldsymbol{j} &= -\mathrm{i}\frac{e\hbar}{2m}\langle\Psi^*\nabla\Psi - \Psi\nabla\Psi^*\rangle \\ &= \sum_p \frac{e\boldsymbol{p}}{m}\langle|\Psi_p|^2\rangle. \end{aligned} \tag{19.53}$$

If we substitute $\langle|\Psi_p^{(0)}|^2\rangle$ into this formula in accoradance with (19.51), we obtain zero. In the next approximation we have

$$j=\sum_p \frac{ep}{m}\langle \Psi_p^{(0)}\Psi_p^{(1)*}+\Psi_p^{(1)}\Psi_p^{(0)*}\rangle$$

$$=-2\frac{\gamma e^2}{m}\sum_p \frac{p}{\alpha\tau+p^2/4m}\left\langle E\frac{\partial \Psi_p^{(0)*}}{\partial p}\Psi_p^{(0)}+E\frac{\partial \Psi_p^{(0)}}{\partial p}\Psi_p^{(0)*}\right\rangle$$

$$=-2\frac{\gamma e^2}{m}\sum_p \frac{p}{\alpha\tau+p^2/4m}\left(E\frac{\partial}{\partial p}\right)\langle|\Psi_p^{(0)}|^2\rangle.$$

Substituting eq. (19.51) into the last formula, we find

$$j=\frac{T\gamma e^2}{m^2}\sum_p p\frac{(pE)}{(\alpha\tau+p^2/4m)^3}. \tag{19.54}$$

The change from summation to integration is carried out according to the following rules*:

$$\sum_p \to \begin{cases} \displaystyle\int \frac{\mathrm{d}^3 p}{(2\pi\hbar)^3} & \text{three-dimensional case,} \\ \displaystyle\frac{1}{d}\int \frac{\mathrm{d}^2 p}{(2\pi\hbar)^2} & \text{thin film, thickness } d \ll \dfrac{\hbar}{(4m\alpha\tau)^{1/2}} \sim \xi, \\ \displaystyle\frac{1}{S}\int \frac{\mathrm{d}p}{2\pi\hbar} & \text{thin wire, cross section } S \ll \xi^2. \end{cases}$$

* These rules are established as follows. We have the sum over the momenta, which converges at $p^2 \sim (\hbar/\xi)^2$. In a specimen with dimensions L_1, L_2 and L_3 the components of the momentum take on the values

$$p_x = 2\pi\frac{\hbar n_x}{L_1}, \qquad p_y = 2\pi\frac{\hbar n_y}{L_2}, \qquad p_z = 2\pi\frac{\hbar n_z}{L_3},$$

where n_x, n_y and n_z are integers. If one of the dimensions (say $L_z = d$) is found to be smaller than ξ, then in the sum over the corresponding n a substantial contribution comes only from $n_z = 0$. Therefore, the sum over momenta, divided by the volume, assumes the form

$$\frac{1}{V}\sum_p \to \frac{1}{dL_1L_2}\int \mathrm{d}^2 p\frac{L_1L_2}{(2\pi\hbar)^2}.$$

We have an analogous situation with a wire. This must be taken into account because we do not explicitly write out the normalization volume, assuming it to be unity in the ordinary three-dimensional case.

Taking the integrals in (19.54) for all three cases and substituting γ according to (19.48), we obtain the values of paraconductivity:

$$\sigma' = \begin{cases} \dfrac{1}{8\pi}\left(\dfrac{e}{\hbar}\right)^2\left(\dfrac{\alpha m}{\tau}\right)^{1/2} & \text{three-dimensional case,} \\ \frac{1}{16}\dfrac{e^2}{\hbar\, d\tau} & \text{film, thickness } d \ll \xi, \\ \dfrac{\pi}{32}\dfrac{e^2}{\alpha^{1/2}\tau^{3/2}Sm^{1/2}} & \text{wire, cross section } S \ll \xi^2. \end{cases} \qquad (19.55)$$

We can substitute here the formulae for α in accordance with eq. (17.14) for a pure superconductor ($\xi_0 \ll l$) and eq. (17.18) for an alloy ($\xi_0 \gg l$). To order of magnitude we have: $\sigma' \sim (e^2/\xi_0\hbar)\tau^{-1/2}$ for the three-dimensional case, $\sigma' \sim (e^2/d\hbar)\tau^{-1}$ for the two-dimensional case, and $(e^2\xi_0/S\hbar)\tau^{-3/2}$ for the one-dimensional case.

The two-dimensional case is the most interesting since formula (19.55) for this case contains no characteristics of the film, except its thickness and T_c. Moreover, as has already been pointed out, the fluctuations in a film are stronger than in a three-dimensional specimen; this is also seen from formulae (19.55). Finally, such a film is easy to prepare. Experimental data (Glover 1967) have completely confirmed the formula (19.55) for this case.

Formulae (19.55) turn to infinity as $T \to T_c$, which is evidence that they are not valid in the immediate vicinity of T_c. The range of validity is set up by the requirement that the fluctuational correction be much less than the normal conductivity $\sigma = n_e e^2 \tau_{tr}/m$.

Another source of error is the neglect of the interaction of electrons with fluctuating Cooper pairs. This effect also enhances the conductivity, but it cannot be interpreted so simply as in the above-described paraconductivity of fluctuational Cooper pairs. The corresponding "anomalous Maki-Thompson conductivity" (Maki 1968, Thompson 1970) proves to be small in many cases*, in particular under experimental conditions (Glover 1967), which makes it possible to obtain a confirmation of formulae (19.55).

* The Maki-Thompson conductivity disappears if, for example, the superconductor contains magnetic impurities.

20

The interface between a superconductor and a normal metal

20.1. The proximity effect

The superconducting correlation manifests itself not only in a superconductor but also at the interface between a superconductor and a normal metal. It is rather natural to guess that the normal metal has an effect on the superconductor, bringing about a decrease in Δ at a distance of order ξ. In particular, if the thickness of the superconductor is less than ξ, it ceases to be superconductive altogether. On the other hand, the superconductor must be expected to exert an effect on the normal metal. A thin film of a normal metal deposited onto a bulk superconductor must be superconducting. These phenomena have been detected experimentally (Meissner 1958, 1959) and have come to be known as proximity effects.

We shall consider the proximity effects using the Ginzburg-Landau equations. In order not to go beyond the framework of the Ginzburg-Landau theory, we deal with two models. In the first model we consider a superconductor at a temperature close to T_c, which has an interface with a normal metal that either does not become a superconductor at all or has a critical temperature far below T_c. The second model is the interface between two superconductors with close critical temperatures, T_{c1} and T_{c2}; it is assumed that $T_{c1} < T < T_{c2}$, i.e., one metal is in the superconducting state and the other in the normal state.

Let us consider the first model. We assume first that we have two bulk metals. In section 17.1 we obtained the boundary condition (17.7) on the surface of the superconductor, assuming the variation of $\delta\Psi$ at the interface to be arbitrary. This condition makes sense at the interface with a vacuum or a dielectric, since the influence of such an interface on electrons in the superconductor is extended over distances of the order of interatomic spacings. From the boundary condition (17.7), it follows, in particular, that the normal component of the supercurrent vanishes at the interface:

$$\boldsymbol{jn}|_b = 0. \tag{20.1}$$

Condition (20.1) must also be fulfilled at the interface with a normal metal, because no nondissipative current can flow in a normal metal. However, the mutual effect of the electrons of the two metals can be extended much farther than the interatomic distances. In view of this we apply, instead of eq. (17.7), a different boundary condition:

$$\boldsymbol{n}\left[\nabla - \frac{2ie}{c\hbar}\boldsymbol{A}\right]\Psi|_{\mathrm{b}} = \frac{1}{t}\Psi|_{\mathrm{b}}. \tag{20.2}$$

This condition is also consistent with (20.1). The coefficient t must be of the order of ξ, but its exact value can be determined only with the aid of the BCS theory. The result proves to be (Zaitsev 1965):

$$t = \begin{cases} 0.6\xi_{0\mathrm{n}} & \text{pure superconductor,} \\ 0.3(\xi_{0\mathrm{n}}l)^{1/2} & \text{superconductor with } \xi_0 \gg l, \end{cases} \tag{20.2'}$$

where $\xi_{0\mathrm{n}} = (\gamma/\pi^2)(\hbar v_n/T_c) = 0.18\hbar v_n/T_c$ (formula 16.82) and v_n is the Fermi velocity in a normal metal.

Consider a planar interface. In this case, the boundary condition (20.2) assumes the following form:

$$\left.\frac{\mathrm{d}\Psi}{\mathrm{d}x}\right|_{x=0} = \frac{1}{t}\Psi(0). \tag{20.3}$$

Instead of the Ginzburg–Landau equation for Ψ we may take its integral (17.27), which in the absence of a magnetic field ($A = 0$) is given by

$$\xi^2\left(\frac{\mathrm{d}\Psi}{\mathrm{d}x}\right)^2 + \Psi^2 - \tfrac{1}{2}\Psi^4 = \tfrac{1}{2} \tag{20.4}$$

[here we have expressed Ψ in units of $\Psi_0 = (\alpha|\tau|/b)^{1/2}$ and introduced the notation $\xi^2 = \hbar^2/(4m\alpha|\tau|)$]; the constant on the right-hand side is determined from the condition that in the bulk of the superconductor $\Psi = 1$, $\mathrm{d}\Psi/\mathrm{d}x = 0$.

Equation (20.4) has been solved in section 17.2 in calculations of the surface energy at the interface between a superconductor and a normal metal. The solution has the form

$$\Psi = \tanh\left(\frac{x - x_0}{\sqrt{2}\xi}\right). \tag{20.5}$$

The value of x_0 is determined from the boundary condition (20.3). This gives

$$\sinh\left(\frac{x_0\sqrt{2}}{\xi}\right) = -\frac{2t}{\xi}. \tag{20.6}$$

According to eq. (20.5), the characteristic scale of variation of Ψ is the length $\xi(T)$, which increases upon approach to T_c as $(T_c - T)^{-1/2}$. At the same time, the quantity t does not depend on temperature. According to formulae (17.14) and (17.18) for α and formulae (20.2′) for t, the ratio t/ξ is of order $\tau^{1/2}$ in all cases. In accordance with eq. (20.6), this means that upon approach to T_c, the relation x_0/ξ falls off. This implies that the value of the function Ψ at the interface decreases with increasing ξ. It should be noted that the Ginzburg–Landau theory can actually be applied only to a case where all the quantities involved vary over distances much larger than ξ_0 or $(\xi_0 l)^{1/2}$, i.e., we necessarily have $\xi \gg t$ and $x_0 \approx 0$ (the Fermi velocities are all of the same order of magnitude in all good metals, which is why $\xi_{0n} \sim \xi_{0s}$). This means that, strictly speaking, within the range of validity of the Ginzburg-Landau theory the boundary condition is $\Psi(0)=0$ rather than (20.3).

Let us now consider the problem of a superconducting film deposited onto the surface of a bulk metal. The ordinary boundary condition $\mathrm{d}\Psi/\mathrm{d}x=0$ is fulfilled at the interface with vacuum. It is not difficult to see that the thin-film problem is equivalent to the problem of a sandwich, i.e., a superconducting layer between two bulk layers of the same normal metal. Evidently, the function Ψ must be symmetrical, i.e., $\mathrm{d}\Psi/\mathrm{d}x=0$ in the middle of the film. A deposited film of thickness d is thus equivalent to a sandwiched film of thickness $2d$.

We introduce a new variable, $u=x/(\xi\sqrt{2})$ and write the integral of the Ginzburg-Landau equation for Ψ. In a specimen of finite thickness we can no longer assume that the constant on the right-hand side of (20.4) is $\frac{1}{2}$. We designate it by $\frac{1}{2}(1-s^2)$. So, we obtain

$$\left(\frac{\mathrm{d}\Psi}{\mathrm{d}u}\right)^2=(1-\Psi^2)^2-s^2. \tag{20.7}$$

Let the interface with the metal lie at $x=0$ and that with the vacuum at $x=d$. From the foregoing it is clear that near the interface with the metal Ψ is small and in moving away from it Ψ increases. On this basis we find from eq. (20.7) that

$$\frac{\mathrm{d}\Psi}{\mathrm{d}u}=[(1-\Psi^2)^2-s^2]^{1/2}, \tag{20.8}$$

$$\int_0^{\Psi}\frac{1}{(1-s-\Psi_1^2)^{1/2}}\frac{1}{(1+s-\Psi_1^2)^{1/2}}\,\mathrm{d}\Psi_1=u. \tag{20.9}$$

Here it is taken into account that $\Psi\approx 0$ at $u=0$. At $u=d/(\xi\sqrt{2})$, $\mathrm{d}\Psi/\mathrm{d}u=0$, i.e., according to eq. (20.8),

$$\Psi\frac{d}{\xi\sqrt{2}}=(1-s)^{1/2}. \tag{20.10}$$

Substitution into eq. (20.9) gives

$$\int_0^{(1-s)^{1/2}}\frac{1}{(1-s-\Psi_1^2)^{1/2}}\frac{1}{(1+s-\Psi_1^2)^{1/2}}\,\mathrm{d}\Psi_1=\frac{d}{\xi\sqrt{2}}. \tag{20.11}$$

We make the replacement of the integration variable $\Psi_1=(1-s)^{1/2}\sin\varphi$. The last formula then becomes

$$(1+s)^{-1/2}\int_0^{\frac{1}{2}\pi}\mathrm{d}\varphi\frac{1}{(1-k^2\sin\varphi)^{1/2}}=\frac{1}{(1+s)^{1/2}}K(k)=\frac{d}{\xi\sqrt{2}}, \tag{20.12}$$

where $k^2=(1-s)/(1+s)$ and $K(k)$ is the complete elliptic integral of the first kind. Let d be fixed. Upon approach to T_c the value of ξ increases, i.e., the right-hand side in eq. (20.12) decreases. The lowest value of the elliptic integral is obtained at $k=0$, i.e., at $s=1$, and is equal to $\frac{1}{2}\pi$. Hence, ξ cannot increase indefinitely and the permissible maximum value is

$$\xi_{\max}=\frac{2}{\pi}d. \tag{20.13}$$

But this means that superconductivity cannot be preserved up to T_c. In other words, the critical temperature must decrease. In accord with the definition of ξ we obtain (Werthamer 1963):

$$\frac{T_c-T_c(d)}{T_c}=\tfrac{1}{16}\pi^2\frac{\hbar^2}{m\alpha}\frac{1}{d^2}. \tag{20.14}$$

On the other hand, with $k\to 1$, i.e., $s\to 0$, the elliptic integral increases indefinitely, which corresponds to large values of d. Putting $s=0$ in formula (20.9), we arrive at the dependence (20.5) with $x_0=0$.

Let us examine the behavior of Ψ near T_c. To do this, we assume that $1-s=r\ll 1$. Then, $k^2\approx\frac{1}{2}r\ll 1$ and $K(k)\approx\frac{1}{2}\pi(1+\frac{1}{4}k^2)\approx\frac{1}{2}\pi(1+\frac{1}{8}r)$. Hence, according to eq. (20.12),

$$r\approx\frac{16}{\pi}\left(\frac{d}{\xi}-\frac{d}{\xi_{\max}}\right).$$

From (20.9) we obtain in this case

$$\Psi\approx\sqrt{r}\sin(u\sqrt{2})=\frac{4}{\sqrt{\pi}}\left(\frac{d}{\xi}-\frac{d}{\xi_{\max}}\right)^{1/2}\sin\left(\frac{x}{\xi_{\max}}\right). \tag{20.15}$$

This formula can naturally be applied only for $r\ll 1$. It enables one to assert that the transition to the normal state is still a second-order phase transition. If we express formula (20.15) in the form of a temperature dependence and assume that $T_c(d)-T\ll T_c-T_c(d)$, the result will be

$$\Psi\approx 2\left(\frac{T_c(d)-T}{T_c-T_c(d)}\right)^{1/2}\sin\left(\frac{\pi x}{2d}\right). \tag{20.16}$$

Although the derivation made is valid only for the neighborhood of T_c, we can draw a qualitative conclusion concerning what will occur at lower temperatures. Evidently, as $T\to 0$, $\xi\to\xi_0$ (or $(\xi_0 l)^{1/2}$ for a dirty superconductor). From eq. (20.13) (which may be written in the form $d_{\min}=\frac{1}{2}\pi\xi$) it follows that at thicknesses smaller than ξ_0 the superconductivity of the deposited film will be absent at any temperatures.

In order to trace out the behavior of a thin film of a normal metal on the surface of a bulk superconductor, one has to use the second of the models described at the beginning of this section. Suppose that two superconductors with slightly different critical temperatures ($T_{c2} - T_{c1} \ll T_{c1}$) are in contact with each other and let us assume that $T_{c1} < T < T_{c2}$. In such a case, metal (1) will be normal and metal (2) will be a superconductor. According to Zaitsev (1965), in the case of slightly different parameters of the metals, the boundary conditions reduce to the continuity of Ψ and $\mathrm{d}\Psi/\mathrm{d}x$:

$$\Psi_1(0) = \Psi_2(0), \qquad \left.\frac{\mathrm{d}\Psi_1}{\mathrm{d}x}\right|_{x=0} = \left.\frac{\mathrm{d}\Psi_2}{\mathrm{d}x}\right|_{x=0}, \tag{20.17}$$

where the interface is assumed at $x = 0$.

First, just as in the first model, we assume that the two metals are bulk metals. Let the normal metal occupy the half-plane $x > 0$ and the superconductor the half-plane $x < 0$. Then, at $x < 0$ eq. (20.4) is satisfied. The solution of this equation is

$$\Psi_2 = \tanh\left[\frac{x_0 - x}{\sqrt{2}\,\xi_2}\right], \quad x < 0. \tag{20.18}$$

At $x > 0$ in the Ginzburg–Landau equation $\tau > 0$. Therefore, in passing to dimensionless variables the term with Ψ^2 in (20.4) will have a different sign. Moreover, the constant on the right-hand side will be equal to zero and not to $\frac{1}{2}$, since $\Psi \to 0$ and $\mathrm{d}\Psi/\mathrm{d}x \to 0$ as $x \to \infty$. Thus, we obtain

$$\xi_1^2\left(\frac{\mathrm{d}\Psi_1}{\mathrm{d}x}\right)^2 = \Psi_1^2 + \tfrac{1}{2}\Psi_1^4, \quad x > 0. \tag{20.19}$$

Suppose that $x_0 \ll \xi_2$ in formula (20.18). Then, $\Psi_2(0) \ll 1$ at $x = 0$ and, hence, according to (20.17) $\Psi_1(0)$ is also much less than unity, but since according to (20.19) Ψ_1 decreases monotonically as $x \to \infty$, it follows that everywhere $\Psi_1(x) \ll 1$. Solving eq. (20.19) under this assumption, we obtain

$$\Psi_1 = C\exp\left(-\frac{x}{\xi_1}\right).$$

From the boundary conditions (20.17) we find $C = x_0(\sqrt{2}\,\xi_2)$, $x_0 \approx \xi_1$ and, hence,

$$\begin{aligned} \Psi_1 &= \frac{\xi_1}{\sqrt{2}}\,\xi_2\exp\left(-\frac{x}{\xi_1}\right), \qquad && x > 0, \\ \Psi_2 &= \tanh\left(\frac{\xi_1 + |x|}{\sqrt{2}\,\xi_2}\right), && x < 0. \end{aligned} \tag{20.20}$$

From formulae (20.19) it follows that Ψ falls off rapidly in the bulk of the normal metal but also that it decreases significantly in the superconductor near the interface.

Let us now assume that the normal plate has a finite thickness d, i.e., $\mathrm{d}\Psi/\mathrm{d}x = 0$ at $x = d$. In this case eq. (20.19) will contain an unknown constant and we write it in the form

$$2\xi_1^2\left(\frac{\mathrm{d}\Psi_1}{\mathrm{d}x}\right)^2 = (1 + \Psi_1^2)^2 - s^2,$$

which leads to

$$\frac{d\Psi_1}{dx} = -\frac{1}{\sqrt{2}\,\xi_1}[\Psi_1^2-(s-1)]^{1/2}[\Psi_1^2+s+1]^{1/2} \tag{20.21}$$

and

$$\int_{(s-1)^{1/2}}^{\Psi_1} d\Psi_3 \frac{1}{[\Psi_3^2-(s-1)]^{1/2}} \frac{1}{(\Psi_3^2+s+1)^{1/2}} = \frac{d-x}{\sqrt{2}\,\xi_1} \tag{20.22}$$

(here we have already taken into account that $d\Psi_1/dx = 0$ at $x = d$, i.e., $\Psi_1 = (s-1)^{1/2}$).

Suppose, as before, that the condition (20.20) is satisfied. It is clear that as long as the thickness of the normal film is not too small we still have $x_0 \ll \xi_2$ in (20.18), i.e.,

$$\Psi_2(0) \approx \frac{x_0}{\sqrt{2}\,\xi_2}, \qquad \left.\frac{d\Psi_2}{dx}\right|_{x=0} \approx -\frac{1}{\sqrt{2}\,\xi_2}.$$

From the boundary conditions (20.17) we determine s and x_0. Since $\Psi_1(0) = \Psi_2(0) \ll 1$, it follows that, as before, $\Psi_1(x) \ll 1$. Hence, in the second factor of the integrand in eq. (20.22) we may put $\Psi_3 \approx 0$. The integral is then taken and we obtain

$$\operatorname{arccosh}\left(\frac{\Psi_1}{(s-1)^{1/2}}\right) = \frac{d-x}{\xi_1}. \tag{20.23}$$

It is obvious that s is close to unity. This has been taken into account in eq. (20.23).

Substituting $\Psi_1(0)$ and $x = 0$, we have

$$\operatorname{arccosh}\left(\frac{x_0}{\sqrt{2}\,\xi_2(s-1)^{1/2}}\right) = \frac{d}{\xi_1}. \tag{20.24}$$

Taking into account that Ψ_1^2 is small and that s is close to unity, we obtain from the equality of the derivatives at $x = 0$ and formula (20.21)

$$\left[\left(\frac{x_0}{\sqrt{2}\,\xi_2}\right)^2 - (s-1)\right]^{1/2} = \frac{\xi_1}{\sqrt{2}\,\xi_2}. \tag{20.25}$$

First, let $d \gg \xi_1$. Then, according to eq. (20.24), $x_0/\xi_2 \gg (s-1)^{1/2}$. From eq. (20.25) we obtain $x_0 = \xi_1$. Substituting into eq. (20.24), we find $(s-1)^{1/2} = (\sqrt{2}\,\xi_1/\xi_2)\exp(-d/\xi_1)$. Substitution into eq. (20.23) yields

$$\Psi_1 = \frac{\xi_1}{\sqrt{2}\,\xi_2}\left[\exp\left(-\frac{x}{\xi_1}\right) + \exp\left(\frac{x-2d}{\xi_1}\right)\right]. \tag{20.26}$$

But if $d \ll \xi_1$, then the argument of arccosh in (20.24) must be close to unity. Expanding in d/ξ_1, we obtain

$$(s-1)^{1/2} = \left(\frac{x_0}{\sqrt{2}\,\xi_2}\right)\left(1 - \frac{d^2}{2\xi_1^2}\right).$$

We insert this expression into eq. (20.25) and find

$$x_0 = \frac{\xi_1^2}{d}.$$

Finally, from eq. (20.23) we obtain Ψ_1:

$$\Psi_1 = \frac{\xi_1^2}{\sqrt{2}\,\xi_2 d}\left(1 - \frac{dx}{\xi_1^2} - \frac{x^2}{2\xi_1^2}\right). \tag{20.27}$$

In order that the condition $x_0 \ll \xi_2$ be satisfied, it is necessary that

$$\xi_1 \gg d \gg \frac{\xi_1^2}{\xi_2}. \tag{20.28}$$

Thus, it turns out that even with the thickness of the normal film being smaller than ξ_1 the function Ψ_1 in it will still be small. Only at still smaller thicknesses does Ψ_1 begin to approach unity. Under the assumption $x_0 \gg \xi_2$, which corresponds to $d \ll \xi_1^2/\xi_2$, we obtain, just as in the preceding case,

$$\Psi_1 = 1 - \sqrt{2}\frac{d\xi_2}{\xi_1^2} - \frac{2dx}{\xi_1^2} + \frac{x^2}{\xi_1^2}. \tag{20.29}$$

Within the framework of the Ginzburg–Landau theory we cannot consider the behavior of a normal film with the critical temperature being much lower than in a superconductor or in a nonsuperconducting film. However, from the foregoing it is sufficiently clear that due to the proximity effect it will be superconductive if its thickness is sufficiently small, and that with a large thickness it will remain normal.

20.2. *Superconductivity of twinning planes*

As was already mentioned in section 16.11, in several metals an increase of T_c at twinning planes of crystals has been reported (Khaikin and Khlyustikov 1981; see also Khlyustikov and Buzdin 1987). When a "bicrystal" was prepared, i.e., two single crystals of different orientation were grown approaching each other, a contact plane appeared. In the general case of arbitrary relative orientation such a boundary behaves by no means extraordinarily. However, if the relative orientation is especially chosen so that the boundary is a crystallographic plane (a plane filled by crystalline atoms) for both crystals then they are called "twins". The appearance of such a plane means a new symmetry element which was absent in the crystallographic group (see e.g. fig. 128).

It appeared that, in the vicinity of the twinning boundary thus created in tin, niobium, rhenium and thallium, superconductivity prevails up to a temperature slightly higher than T_{c0} – the critical temperature of the sample without such a boundary. For tin this increase was 0.04 K ($T_{c0} = 3.7$ K), for niobium it was 0.11 K

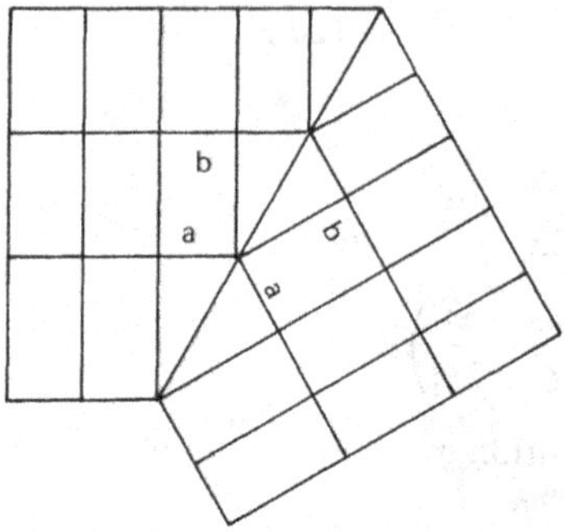

Fig. 128.

($T_{c0} = 9.3$ K). Of course this increase is not large. However, it should be kept in mind that the superconducting region is surrounded by the normal metal, which suppresses its superconductivity due to the proximity effect. To reduce this effect, samples were prepared consisting of small particles with sizes less than ξ or pellets of powders consisting of small particles, which were pressed so that one could expect twin boundaries with a small spacing. In such a way it was possible to detect signs of superconductivity (diamagnetism) in tin up to 12 K, i.e. to a temperature exceeding T_{c0} by more than a factor of 3.

The properties of twinning planes became of special interest recently because they appear spontaneously in the high-temperature superconductor $YBa_2Cu_3O_{7-\delta}$. The reason is that such samples must be prepared at high temperature, over 900°C. This is necessary either for sintering of the ceramic or for growing single crystals. When the temperature is lowered a phase transition takes place (at ~750°C) from a tetragonal to an orthorhombic phase: $a = 3.823$ Å, $b = 3.884$ Å, $c = 11.674$ Å. An abrupt reduction of the volume takes place and the crystal is split into flat domains with different orientation of the a and b axes. The domains are separated by twin boundaries parallel to the c-axis and dividing the basal rectangular face along the diagonal (see fig. 128)*. The spacing between these boundaries depends on the preparation method and varies between 500 and 1500 Å. Since the correlation length in this substance is small (in the basal plane $\xi_0 \approx 16$ Å), one can practically assume these planes to be isolated. As will be shown below the spontaneous appearance of twinning planes leads to several interesting phenomena near T_{c0}.

The increase of T_c in the vicinity of the twin boundary can be explained in various ways. First of all special types of lattice vibrations may exist which are connected particularly with the twinning planes, similar to the Rayleigh waves at the surface. The exchange by these phonons can enhance the electron attraction. Secondly a weakening of Coulomb repulsion close to the boundary is possible. Finally special electronic states are possible in which the electrons move freely along the boundary

* Recently it has become clear that domain boundaries in YBaCuO can be of several kinds, and that some of these suppress superconductivity. Here we consider only true twinning planes.

but cannot leave it due to the rapid damping of the wave function. Such states are called Tamm states. Superconductivity can be associated with such electrons. If the boundary is not a twinning plane then it contains a large concentration of dislocations or other defects which can destroy the singular electronic or phonon states. This can explain why T_c does not increase at an arbitrary boundary.

In what follows we shall assume that the reason of the increase of T_c is the local increase of interaction between electrons due either to new types of phonons or to the reduction of Coulomb repulsion. Let us assume that the dimensionless constant of Cooper pairing entering eq. (16.23) for Δ, $\lambda = \frac{1}{2}g\nu$, has a sharp peak near the twinning plane: $\lambda = \lambda_0 + \delta\lambda(x)$, where x is directed along the normal to the boundary plane. Let us denote the relative increase of the critical temperature by τ_0:

$$\tau_0 = \frac{T_c - T_{c0}}{T_{c0}}. \tag{20.30}$$

This quantity can be estimated from the following argument. In the vicinity of the boundary local superconductivity appears in a layer of the order of $\xi_c \approx \xi_0 \tau_0^{-1/2}$, where ξ_0 is defined by formula (16.82). If the peak of λ has a width d ($d \ll \xi_0 \ll \xi_c$; actually, d should be of the order of interatomic distances) then due to the proximity effect the effective increase of λ will be of the order of $\delta\lambda d/\xi_c$. But since T_c is proportional to $\exp(-1/\lambda)$ we get the estimate

$$\frac{\delta T_c}{T_c} = \tau_0 \approx \frac{\delta\lambda}{\lambda_0^2}\frac{d}{\xi_c}.$$

Substituting ξ_c we get

$$\tau_0 \sim \left(\frac{\delta\lambda}{\lambda_0^2}\frac{d}{\xi_0}\right)^2. \tag{20.31}$$

Now we consider this question in greater detail. The order parameter for a uniform superconductor was obtained from eq. (16.21). Particularly, at $\Delta \to 0$ the expression for T_c was derived. In the inhomogeneous case, assuming g to depend on x, one can obtain a generalisation of (16.21) for the nonuniform case (limit $\Delta \to 0$):

$$\begin{aligned}\Delta(x) &= \int \lambda(x)\, K(\boldsymbol{r}-\boldsymbol{r}')\, \Delta(x')\, \mathrm{d}V' \\ &= \lambda_0 \int K(\boldsymbol{r}-\boldsymbol{r}')\, \Delta(x')\, \mathrm{d}V' + \int \delta\lambda(x)\, K(\boldsymbol{r}-\boldsymbol{r}')\, \Delta(x')\, \mathrm{d}V'.\end{aligned} \tag{20.32}$$

Here we suppose that $\lambda = \lambda_0 + \delta\lambda(x)$ where $\delta\lambda(x)$ is finite in an interval of the order of d. Since the kernel $K(\boldsymbol{r}-\boldsymbol{r}')$ varies on a scale of order of ξ_0 (see section 16.9) and $\Delta(x)$ varies on a scale ξ_c, both of these distances being much larger than d, it is possible to substitute in the second term of (20.32) x' entering $K(\boldsymbol{r}-\boldsymbol{r}')\,\Delta(x')$ by zero. Denoting

$$\Lambda = \int \delta\lambda\,(x)\,\mathrm{d}x \sim \delta\lambda d$$

one arrives at

$$\Delta(x) = \lambda_0 \int K(x-x', \boldsymbol{\rho})\,\Delta(x')\,\mathrm{d}x'\,\mathrm{d}\boldsymbol{\rho} + \Lambda\Delta(0) \int K(x, \boldsymbol{\rho})\,\mathrm{d}\boldsymbol{\rho}. \tag{20.33}$$

We have now separated the dependences on x and on $\boldsymbol{\rho} = y, z$.

To solve eq. (20.33) we pass to the Fourier transform,

$$\Delta(p) = \lambda_0 K(p, \boldsymbol{q}_\perp = 0)\,\Delta(p) + \Lambda\Delta(x=0)\,K(p, \boldsymbol{q}_\perp = 0).$$

The solution is

$$\Delta(p) = \frac{\Lambda\Delta(x=0)\,K(p,0)}{1-\lambda_0 K(p,0)}. \tag{20.34}$$

Here p is the momentum along x and $\boldsymbol{q}_\perp$ is the momentum in the transverse direction. Taking into account that

$$\Delta(x=0) = \int \Delta(p)\frac{\mathrm{d}p}{2\pi}$$

we get the self-consistency relation, defining the critical temperature,

$$1 = \Lambda \int \frac{K(p)}{1-\lambda_0 K(p)}\frac{\mathrm{d}p}{2\pi}. \tag{20.35}$$

It is easy to see that the critical temperature for a bulk superconductor is obtained from the condition

$$1 - \lambda_0 K(T_{\mathrm{c}0}, p=0) = 0. \tag{20.36}$$

We transform the denominator of formula (20.35) accordingly:

$$\begin{aligned} 1-\lambda_0 K(T_\mathrm{c}, p) &= \lambda_0[K(T_{\mathrm{c}0}, 0) - K(T_\mathrm{c}, 0)] + \lambda_0[K(T_\mathrm{c}, 0) - K(T_\mathrm{c}, p)] \\ &= \lambda_0\left[\ln\left(\frac{T_\mathrm{c}}{T_{\mathrm{c}0}}\right) + \frac{7\zeta(3)}{48\pi^2}\left(\frac{vp}{T_\mathrm{c}}\right)^2\right]. \end{aligned} \tag{20.37}$$

In the last formula the first term corresponds to the first bracket and it is obtained just as in the derivation of formula (16.25) for T_c. In order to find the second term, it is necessary to know K for a finite p, which corresponds to a moving Bose-condensate of Cooper pairs. The corresponding derivation is done in section 21.2 and the expression for $K(p)$ corresponds to the right-hand side of formula (21.24), divided by $\lambda_0 = \frac{1}{2}g\nu$, with $I=0$ and $|\boldsymbol{q}| = p$. Expanding in powers of p, we get the expression in (20.37). Since in the integral (20.35) only small p's are essential, we can put in the numerator $K(p) \approx K(0) \approx \lambda_0^{-1}$, after which the integral takes the form

$$1 = \frac{\Lambda}{\lambda_0^2} \int \frac{dp}{2\pi} \left[\tau_0 + \frac{7\zeta(3)}{48\pi^2} \frac{v^2 p^2}{T_c^2} \right]^{-1} \tag{20.38}$$

(here, as before, $\tau_0 = (T_c - T_{c0})/T_{c0}$). Calculating the integral and expressing τ_0 we obtain

$$\tau_0 = 0.45 \left(\frac{\Lambda}{\lambda_0^2 \xi_0} \right)^2, \tag{20.39}$$

which corresponds to the estimate (20.31).

A similar derivation can be performed for a superconductor containing impurities with $l \ll \xi_0$. In this case in (20.31) $\xi_0^2 \to \xi_0 l$. For pure tin $\xi_0 = 3.1 \times 10^{-5}$ cm, $\lambda \approx 0.3$. In order to have $\Delta T_c = 0.04$ K, i.e. $\tau_0 = 0.011$, one should have $\Lambda \approx 4 \times 10^{-7}$. Even if we put $d \sim 40$ Å we get $\delta\lambda \approx 1$; but d is probably even smaller. Thus, a seemingly small displacement of the critical temperature leads to a considerable value of the interaction constant.

Since we consider a narrow vicinity of T_{c0} we may apply Ginzburg–Landau theory. As we shall see below, this will be insufficient for calculating the critical magnetic field of the twinning plane superconductivity at $T \ll T_{c0}$, but we shall not consider this case. The appearance of a twinning plane with a larger critical temperature can be naturally taken into account by adding a term $-\frac{1}{2}\gamma\,\delta(x)|\Psi|^2$ to the GL free energy. After that the GL functional in a given external field H_0 takes the form

$$\int \Omega_{sH}\, dV = \int \Omega_{nH}\, dV + \int \left\{ \frac{(H - H_0)^2}{8\pi} + \frac{1}{4m} \left| \left(-i\hbar\nabla - \left(\frac{2e}{c}\right) \boldsymbol{A} \right) \Psi \right|^2 + \alpha\tau|\Psi|^2 + \tfrac{1}{2} b |\Psi|^4 - \gamma\,\delta(x)|\Psi|^2 \right\} dV. \tag{20.40}$$

Variation over Ψ^* yields

$$\frac{1}{4m} \left[-i\hbar\nabla - \left(\frac{2e}{c}\right) \boldsymbol{A} \right]^2 \Psi + \alpha\tau\Psi + b|\Psi|^2 \Psi = \gamma\,\delta(x)\, \Psi. \tag{20.41}$$

First let us consider the case where there is no magnetic field. It is natural to assume in this case that Ψ depends only on x and that it is real. Then we find

$$-\frac{\hbar^2}{4m} \frac{d^2\Psi}{dx^2} + \alpha\tau\Psi + b\Psi^3 = \gamma\delta(x)\, \Psi. \tag{20.42}$$

The right-hand side of this equation corresponds to a discontinuity of $d\Psi/dx$.

Therefore, instead of the δ-function term we can write a boundary condition

$$\left.\frac{\mathrm{d}\Psi}{\mathrm{d}x}\right|_0 = -\frac{2m\gamma}{\hbar^2}\gamma\Psi(0). \tag{20.43}$$

Actually, instead of adding the δ-function term to the free energy we could write the "surface invariant" (Andreev 1987):

$$\int \mathrm{d}S\,[A|\Psi_+|^2 + B|\Psi_-|^2 + C(\Psi_+^*\Psi_- + \Psi_-^*\Psi_+)],$$

where

$$\Psi_+ = \Psi(x+0), \qquad \Psi_- = \Psi(x-0).$$

After that variation of Ψ on both sides of the surface would yield more general boundary conditions than (20.43). However, it should be remembered that the twinning plane is necessarily connected with the appearance of a new symmetry element, e.g. a mirror reflection plane. In the corresponding transformation $|\Psi_+|^2 \rightleftarrows |\Psi_-|^2$. Thus, for Ψ_+ and Ψ_- themselves a symmetrical transformation is possible, namely $\Psi_+ = \Psi_-$, as well as an antisymmetrical one: $\Psi_+ = -\Psi_-$. In the last case Ψ must turn to zero at interatomic distances from the twinning plane. This would lead to an energy increase, but there is no necessity for such a phase jump (such as an increasing vector potential, ch. 18, or a phase slip center, ch. 22). The symmetrical case coincides completely with the addition of the $\delta(x)$ term, and only this case are we going to consider.

First of all we show that the new parameter γ in (20.41) is uniquely connected with the increase of the critical temperature at the twinning plane. To this purpose consider eq. (20.42) in the linear approximation,

$$-\frac{h^2}{4m}\frac{d^2\Psi}{dx^2} + \alpha\tau\Psi = 0.$$

The solution of this equation, descending at $x \to \pm\infty$, is proportional to $\exp(-q|x|)$, where $q = \sqrt{4m\alpha\tau}/\hbar$. From the boundary condition (20.43) we find $q = 2m\gamma/\hbar^2$ and hence

$$\tau_0 = \frac{T_c - T_{c0}}{T_{c0}} = \frac{m\gamma^2}{\hbar^2\alpha}. \tag{20.44}$$

Now we introduce new dimensionless variables:

$$\varphi = \frac{\Psi}{\Psi_0}, \qquad t = \frac{T - T_{c0}}{T_c - T_{c0}}, \qquad x' = \frac{x}{\xi_c},$$

where

$$\xi_c = \frac{1}{q(T_c)} = \frac{h}{\sqrt{4m\alpha\tau_0}}, \qquad \Psi_0 = \sqrt{\frac{\alpha\tau_0}{b}} = \Psi(-\tau_0).$$

With these variables equation (20.42) does not contain any characteristics of the superconductor:

$$-\frac{d^2\varphi}{dx^2}+t\varphi+\varphi^3=2\delta(x)\,\varphi \tag{20.45}$$

(we omit the prime on x). The right-hand side is equivalent to the boundary condition

$$\varphi'(0)=-\varphi(0). \tag{20.46}$$

Equation (20.45) (without the right-hand side) has a first integral. Multiplying by $d\varphi/dx$ and integrating with respect to x we get

$$-\left(\frac{d\varphi}{dx}\right)^2+t\varphi^2+\tfrac{1}{2}\varphi^4=\text{const.} \tag{20.47}$$

Since at $x=\pm\infty$ we have $\varphi=0$, it follows that const. $=0$. Solving the equation thus obtained we have

$$\varphi=\frac{\sqrt{2t}}{\sinh(|x|\sqrt{t}+p)}. \tag{20.48}$$

The constant p is found from the boundary condition (20.46)

$$p=\tfrac{1}{2}\ln\left(\frac{1+\sqrt{t}}{1-\sqrt{t}}\right). \tag{20.49}$$

So the function φ in the general case decays exponentially while departing from the twinning plane, with a characteristic distance $\xi_c/\sqrt{t}=\xi(T)$, as expected. However, this happens only far from T_{c0}, the critical temperature of the bulk specimen. Close to T_{c0} we have $t\to 0$, $p\approx\sqrt{t}$, $\varphi\approx\sqrt{2}/(|x|+1)$, i.e. the exponential decay is replaced by a power law.

Now we shall consider the thermodynamics of twinning-plane superconductivity. It is evident that a single plane makes a negligible contribution to the thermodynamic functions, of the order of ξ_c/L, where L is the size of the sample. However, as mentioned above, in the high-temperature oxide $YBa_2Cu_3O_{7-\delta}$ a whole system of parallel twinning planes appears spontaneously. Therefore the corresponding contribution will not be negligible, particularly close to T_{c0}.

So we assume that there exists a system of parallel periodic twinning planes perpendicular to the x-axis, with a spacing L. Let us put the origin in the middle between two planes (fig. 129). Evidently φ will have a minimum at this point which we denote as φ_1. At this point $d\varphi/dx=0$. Therefore in the first integral (20.47)

$$\text{const.}=t\varphi_1^2+\tfrac{1}{2}\varphi_1^4. \tag{20.50}$$

At the twinning plane φ reaches a maximal value, which we denote by φ_0. According to the boundary condition (20.46) $(\varphi')^2=\varphi_0^2$. Inserting this into (20.47) we find

$$\text{const.}=(t-1)\varphi_0^2+\tfrac{1}{2}\varphi_0^4. \tag{20.51}$$

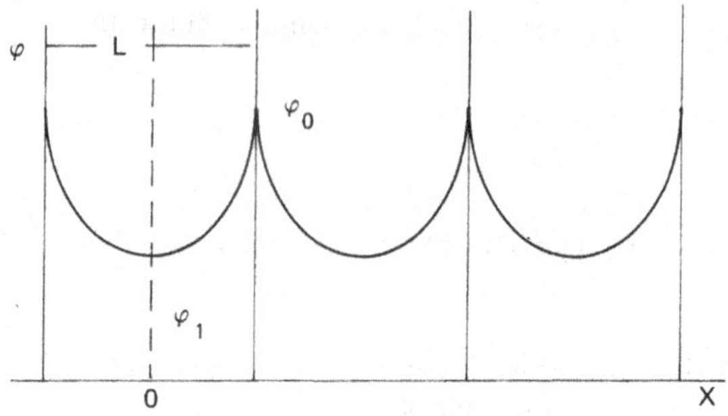

Fig. 129.

Comparing eqs. (20.50) and (20.51) we get the first relation

$$2t\varphi_1^2+\varphi_1^4=2(t-1)\varphi_0^2+\varphi_0^4. \tag{20.52}$$

Substituting (20.50) into (20.47) and expressing $\mathrm{d}x$ in terms of $\mathrm{d}\varphi$ we find

$$\mathrm{d}x=\frac{\sqrt{2}\,\mathrm{d}\varphi}{[(\varphi^2-\varphi_1^2)(2t+\varphi^2+\varphi_1^2)]^{1/2}}.$$

Integrating from $x=0$ to $x=\frac{1}{2}L$ (i.e. $x=L/2\xi_c$ in dimensionless units) we find

$$\frac{L}{2\xi(T_c)}=\int_{\varphi_1}^{\varphi_0}\frac{\sqrt{2}\,\mathrm{d}\varphi}{[(\varphi^2-\varphi_1^2)(2t+\varphi^2+\varphi_1^2)]^{1/2}}. \tag{20.53}$$

Relations (20.52) and (20.53) are implicit formulae defining φ_1 and φ_0 as functions of t for a given spacing between the planes L.

The entropy is equal to

$$\int S_s\,\mathrm{d}V=-\frac{\partial}{\partial T}\int \Omega_s\,\mathrm{d}V$$

$$=-\int\left(\frac{\partial\Omega_s}{\partial T}\right)_\Psi\mathrm{d}V-\int\frac{\partial\Psi}{\partial t}\left(\frac{\partial\Omega_s}{\partial\Psi}\right)_T\mathrm{d}V.$$

But the latter term turns to zero since the condition $\partial\Omega_s/\partial\Psi$ defined the GL equation. So only the first term remains and we get per unit volume

$$S=-\frac{\alpha}{T_{c0}}\int_0^{L/2}|\Psi|^2\frac{\mathrm{d}x}{\frac{1}{2}L}.$$

Transforming to dimensionless variables, we get

$$S=-\frac{2\alpha\xi_c|\Psi_0|^2}{LT_{c0}}\int_0^{L/2\xi_c}\varphi^2\,\mathrm{d}x$$

$$=-\frac{2\xi_c\alpha^2\tau_0}{LT_{c0}b}\int_{\varphi_1}^{\varphi_0}\frac{\sqrt{2}\,\varphi^2\,\mathrm{d}\varphi}{[(\varphi^2-\varphi_1^2)(2t+\varphi^2+\varphi_1^2)]^{1/2}}. \tag{20.54}$$

According to Landau phase transition theory (see Appendix II) the jump of the heat capacity of the bulk specimen at the transition point is

$$\Delta C_0 = \frac{\alpha^2}{bT_{c0}}$$

(this formula can also be obtained from relations 15.8 and 12.3). Differentiating S with respect to temperature and dividing by ΔC_0 we get (Abrikosov and Buzdin 1988)

$$\frac{C}{\Delta C_0} = -\frac{2\xi_c}{L}\frac{T}{T_{c0}}\frac{\partial}{\partial T}\left\{\int_{\varphi_1}^{\varphi_0} \frac{\sqrt{2}\,\varphi^2\,d\varphi}{[(\varphi^2-\varphi_1^2)(2t+\varphi^2+\varphi_1^2)]^{1/2}}\right\}. \tag{20.55}$$

Hence formulae (20.52), (20.53) and (20.55) present the temperature dependence of the heat capacity and depend on the parameter L/ξ_c. Figure 130 presents the experimental data on the temperature dependence of the heat capacity of a single crystal $YBa_2Cu_3O_{7-\delta}$ (Inderhees et al. 1987) together with the theoretical curve for $L/\xi_c = 12$ giving the best fit. Excluding the small peak in the vicinity of T_{c0}, which

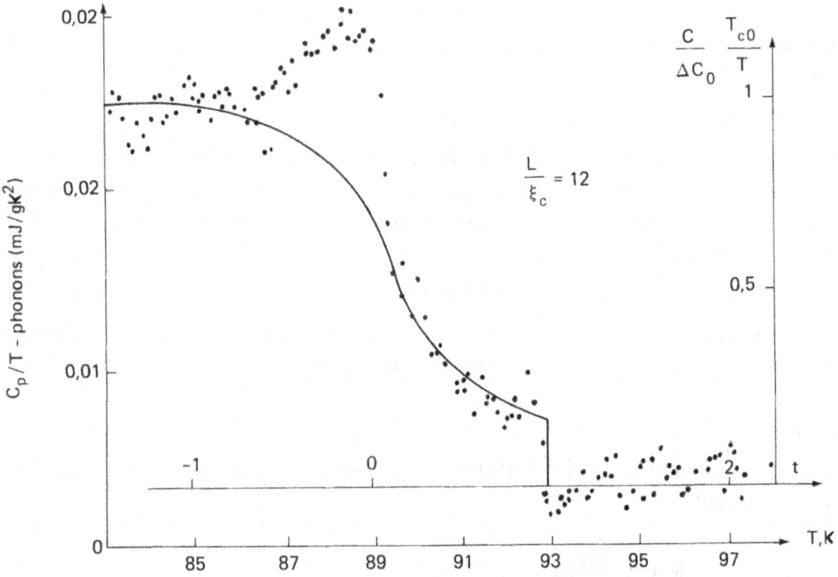

Fig. 130.

can be ascribed to fluctuations not taken into account in GL theory, the fitting is good. So one may conclude that 93 K corresponds to T_c i.e. to the critical temperature of the twinning plane superconductivity, whereas the bulk critical temperature T_{c0} is 89 K. Since $\xi_c \sim \xi_0 \tau_0^{-1/2}$, we have, with $\xi_0 \approx 16$ Å, $\xi_c \approx 75.5$ Å. So $L \approx 900$ Å, which is reasonable.

We now proceed to the calculation of the critical magnetic field. Let us consider only the case when the magnetic field is parallel to the twinning plane. Obviously, if the superconductor is of type II the disappearance of twinning plane superconductivity will proceed as a second-order transition. Let us assume H along the z-axis and choose the vector potential as $A_y(x)$. Then we should add a term e^2A^2/mc^2 to the left-hand side of (2.41). Let us introduce some quantities:

- the penetration depth $$\delta = \sqrt{\frac{2mc^2}{4\pi\psi_0^2(2e)^2}};$$

- the Ginzburg–Landau parameter $$\kappa = \frac{\delta}{\xi_c} = \frac{mc}{e}\sqrt{\frac{b}{2\pi}};$$

- the thermodynamic critical field of the bulk superconductor at a temperature symmetric to T_c with respect to T_{c0} $$H_{cm}(-\tau_0) = 2\alpha\tau_0\sqrt{\frac{\pi}{b}};$$

- the scale for the critical field $$H_{c2}(-\tau_0) = \kappa\sqrt{2}\,H_{cm}(-\tau_0) = 2\alpha\tau_0\frac{mc}{e}.$$

Close to the critical field we can assume $A_y(x) = H_0x$. With dimensionless variables we get

$$-\varphi''(x) + h^2x^2\varphi(x) + t\varphi(x) - 2\delta(x)\,\varphi(x) = 0, \tag{20.56}$$

where $h = H_0/H_{c2}(-\tau_0)$. We first find the solutions of eq. (20.56) in the limiting cases.

First we consider the vicinity of T_c, i.e. $t \to 1$. In this case h will be small and therefore the corresponding term can be treated as a perturbation. Let us rewrite eq. (20.56) in the following form:

$$-\varphi''(x) + \varphi(x) + h^2x^2\varphi(x) = (1-t)\,\varphi(x)$$

with the boundary condition $\varphi'(0) = -\varphi(0)$. In the case under consideration $1-t$ plays the role of energy. In the zeroth approximation $1-t=0$ and the eigenfunction is

$$\varphi \propto \mathrm{e}^{-|x|}.$$

In the first approximation the perturbation has to be averaged over the unperturbed state. So we get

$$1 - t = \frac{\displaystyle\int_{-\infty}^{\infty} h^2x^2\,\mathrm{e}^{-2|x|}\,\mathrm{d}x}{\displaystyle\int_{-\infty}^{\infty} \mathrm{e}^{-2|x|}\,\mathrm{d}x}.$$

We find therefrom

$$1 - t = \tfrac{1}{2}h^2,$$

i.e.

$$h = [2(1-t)]^{1/2}. \tag{20.57}$$

Now we pass to conventional units. Since H_{c2} varies linearly with temperature close to T_{c0} we can write

$$H_{c2}(-\tau_0)=\left.\frac{\mathrm{d}H_{c2}}{\mathrm{d}T}\right|_{T_{c0}}(T_c-T_{c0}).$$

Keeping in mind that $t=(T-T_{c0})/(T_c-T_{c0})$ we find:

$$1-t=\frac{T_c-T}{T_c-T_{c0}}.$$

Hence the critical field of the twinning plane is

$$H^{t}_{c2}(T)=\left|\frac{\mathrm{d}H_{c2}}{\mathrm{d}T}\right|[2(T_c-T)(T_c-T_{c0})]^{1/2}. \tag{20.58}$$

One sees that $H^{t}_{c2}\sim(T_c-T)^{1/2}$. This behaviour corresponds to a thin film having a thickness ξ_c (see 17.48′).

Another limiting case is the field at temperature $t<0$, $|t|\gg 1$ i.e. on the scale T_c-T_{c0} deep in the superconducting region, not so far however from T_{c0} as to violate the applicability condition of GL theory: $T_{c0}-T\ll T_{c0}$. Under these conditions the twinning plane itself is a perturbation. Let us rewrite eq. (20.56) in the form

$$-\varphi''(x)+h^2x^2\varphi(x)-2\delta(x)\,\varphi(x)=-t\varphi(x). \tag{20.56′}$$

Neglecting the term with $\delta(x)$ we have solutions of (20.56′) descending with $x\to\pm\infty$ if $-t=h(2n+1)$. The smallest eigenvalue is $-t=h$. The corresponding eigenfunction is $\varphi_0=(h/\pi)^{1/4}\exp(-\frac{1}{2}hx^2)$. Considering the term with $\delta(x)$ as a perturbation we obtain the first approximation,

$$-t=h+\int\varphi^2[-2\delta(x)]\,\mathrm{d}x=h-2\sqrt{\frac{h}{\pi}}.$$

Since h is large in this case the second term is in fact a correction. Solving with respect to h we find

$$h=-t+\sqrt{\frac{2|t|}{\pi}}. \tag{20.59}$$

Multiplying by $H_{c2}(-\tau_0)=|\mathrm{d}H_{c2}/\mathrm{d}T|_{T_{c0}}(T_c-T_{c0})$ we see that $h=-t$ corresponds to $H^{t}_{c2}(T)=|\mathrm{d}H_{c2}/\mathrm{d}T|(T_{c0}-T)=H_{c2}$. Hence

$$\frac{H^{t}_{c2}(T)-H_{c2}(T)}{H_{c2}(T)}=\frac{2}{\sqrt{\pi|t|}}. \tag{20.60}$$

It follows that $H^{t}_{c2}(T)>H_{c2}(T)$ at any temperature (fig. 131) and hence in type-II superconductors twinning plane superconductivity can exist in the absence of bulk superconductivity in the proper interval of magnetic fields up to $T=0$. Of course the critical field H^{t}_{c2} at low temperature cannot be calculated from the GL equations and demands a more complete microscopic theory.

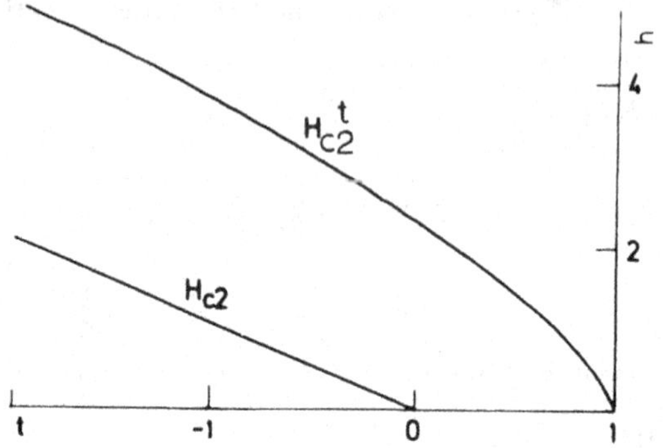

Fig. 131.

We have considered until now the limiting cases. An expression for h which is valid for any t (within the applicability region of GL-theory) can be obtained in the following way. We assume for the general solution of eq. (20.56′) the following form:

$$\varphi(x)=\sum_n c_n\varphi_n(x), \tag{20.61}$$

where φ_n are eigenfunctions, corresponding to the eigenvalues of the parameter t: $\varepsilon_n=h(2n+1)$; substituting into (20.56′), multiplying by $\varphi_n(x)$ and integrating with respect to x we find

$$c_n=\frac{2\varphi(0)\,\varphi_n(0)}{t+\varepsilon_n}. \tag{20.62}$$

Taking into account the self-consistency condition

$$\varphi(0)=\sum_n c_n\varphi_n(0)$$

we find

$$1=\sum_n\frac{2\varphi_n^2(0)}{t+\varepsilon_n}. \tag{20.63}$$

The normalized eigenfunctions of eq. (20.56) (without the δ-term) are expressed by (see Landau and Lifshitz 1974)

$$\varphi_n=\left(\frac{h}{\pi}\right)^{1/4}(2^n n!)^{-1/2}\exp(-\tfrac{1}{2}hx^2)\,\mathrm{H}_n(x\sqrt{h}), \tag{20.64}$$

where H_n are the Hermite polynomials

$$\mathrm{H}_n(x)=(-1)^n\,\mathrm{e}^{x^2}\frac{\mathrm{d}^n}{\mathrm{d}x^n}\mathrm{e}^{-x^2}. \tag{20.65}$$

Putting $x=0$ we have

$$\mathrm{H}_{2n}(0)=(-1)^n 2^n (2n-1)!!, \qquad \mathrm{H}_{2n+1}(0)=0.$$

Substituting into eq. (20.63) we get an equation for h:

$$h^{1/2}=\frac{2}{\pi^{1/2}}\sum_{k=0}^{\infty}\frac{(2k)!}{(k!)^2 2^{2k}(t/h+4k+1)} = \frac{1}{2\sqrt{\pi}}B\left(\tfrac{1}{2},\tfrac{1}{4}\left(1+\frac{t}{h}\right)\right) \tag{20.66}$$

where $\mathrm{B}(x,y)=\Gamma(x)\,\Gamma(y)/\Gamma(x+y)$ is the beta-function. From the general expression the limiting cases (20.52) and (20.59) can also be obtained.

With respect to the high-temperature oxide Y-Ba-Cu-O, the formula (20.58) is of the main interest. Experiments with samples having parallel c-axes of the crystallites and the magnetic field also in the same direction (Fang et al. 1987) showed that around the critical temperature the critical field is proportional to $(T_c-T)^{1/2}$ rather than to T_c-T. Hence it can be concluded that this is due to twinning plane superconductivity. If we use the value $|\mathrm{d}H_{c2}/\mathrm{d}T|\sim 10\,\mathrm{kOe/K}$ which is known from other sources then we obtain $T_c-T_{c0}\approx 5$-6 K, which fits the result for the specific heat.

Now let us consider the case when the twinning boundary is formed in a type-I superconductor. First of all it should be mentioned that there exists an effective GL parameter for twinning plane superconductivity, namely

$$\kappa_{\mathrm{eff}}\sim\frac{\lambda_{\mathrm{L}}[\psi]}{\xi_c}\sim\frac{\lambda_{\mathrm{L0}}}{\psi\xi_c}\sim\frac{\lambda_{\mathrm{L0}}}{\xi_c[(T_c-T)/T_{c0}]^{1/2}}. \tag{20.67}$$

Hence close to T_c we have $\kappa_{\mathrm{eff}}\gg 1$ and there is a second-order phase transition which is described by the preceding formulae. At some temperature where $\kappa_{\mathrm{eff}}\approx 1$ a "tricritical point" occurs and the transition becomes first-order.

In order to calculate the first-order transition critical field we apply a method similar to the one we used for the calculation of the surface energy (section 17.2). Let us assume $\kappa\ll 1$, i.e., $\xi\gg\delta$. Then the field practically does not penetrate into the superconducting region. We shall suppose this region to be $-L<x<L$ (its size is still to be determined). So we arrive at eq. (20.42) or (20.45) but with another boundary condition: $\varphi(\pm L/\xi_c)=0$. We have to calculate the free energy of this region which is equal to (see 20.40; we have passed to reduced variables)

$$\Omega\equiv\int\Omega_{\mathrm{sH}}\,\mathrm{d}x-\int\Omega_{\mathrm{nH}}\,\mathrm{d}x = \frac{H_c^2}{4\pi}\xi_c\left\{\int_{-L/\xi_c}^{L/\xi_0}[\varphi'^2+t\varphi^2+\tfrac{1}{2}\varphi^4+\tfrac{1}{2}h_0^2]\,\mathrm{d}x-2\varphi^2(0)\right\}, \tag{20.68}$$

where $h_0=H_0/H_{\mathrm{cm}}(-\tau_0)$.

This time we shall use the full equations containing the field. Choosing the vector potential as $A_y(x)$ we get

$$-\frac{\hbar^2}{4m}\frac{d^2\Psi}{dx^2}+\frac{e^2}{mc^2}A^2\Psi+\alpha\tau\Psi+\tfrac{1}{2}b\Psi^3=\gamma\delta(x)\ \Psi, \tag{20.69}$$

$$\frac{d^2A}{dx^2}=\frac{8\pi e^2}{mc^2}\Psi^2 A. \tag{20.70}$$

We introduce a new scale for A:

$$A_0=H_c(-\tau_0)\ \delta(-\tau_0)=\left(\alpha\tau_0\frac{mc^2}{2e^2}\right)^{1/2}.$$

Multiplying eq. (20.69) without the right-hand side by $d\Psi/dx$ and eq. (20.70) by $(8\pi)^{-1}\,dA/dx$, adding them up and integrating with respect to x we obtain in dimensionless variables the first integral

$$-2\varphi'^2+(2t+A)^2\varphi^2+\varphi^4-(A')^2x^2=-h_0^2. \tag{20.71}$$

The constant is chosen from the condition at the boundary of the superconducting region $\varphi=0$, $H=H_0$.

From this integral we have inisde the region under consideration at $x<0$

$$\frac{d\varphi}{dx}=-(t\varphi^2+\tfrac{1}{2}\varphi^4+\tfrac{1}{2}h_0^2)^{1/2} \tag{20.72}$$

(we have put $A=A'=0$). On the other hand, using the boundary condition at the twinning plane we get from eq. (20.72):

$$h_0^2=2(1-t)\ \varphi^2(0)-\varphi^4(0). \tag{20.73}$$

Using the first integral (20.72) we get from eq. (20.68)

$$\Omega=\frac{H_{cm}^2(-\tau_0)}{4\pi}\xi_c\left\{\int_{-L/\xi_c}^{L/\xi_c}2(\varphi')^2\,dx-2\varphi^2(0)\right\}.$$

With the aid of eq. (20.72) we pass over to the integration over φ and introduce the new variable $y=\varphi/\varphi(0)$; we apply also eq. (20.73):

$$\Omega=\frac{1}{\pi}H_{cm}^2(-\tau_0)\ \varphi^2(0)\xi_c\left\{\int_0^1[1-t(1-y^2)+\tfrac{1}{2}(y^4-1)\ \varphi^2(0)]^{1/2}\,dy-\tfrac{1}{2}\right\}.$$

The first-order transition is defined by the condition $\Omega=0$. This gives an implicit definition of $\varphi(0)$:

$$\int_0^1[1-t(1-y^2)+\tfrac{1}{2}(y^4-1)\ \varphi^2(0)]^{1/2}\,dy=\tfrac{1}{2}. \tag{20.74}$$

Having obtained $\varphi(0)$ we can find the critical field from eq. (20.73). Actually this can be done only numerically. It appears that the transition curve crosses the curve

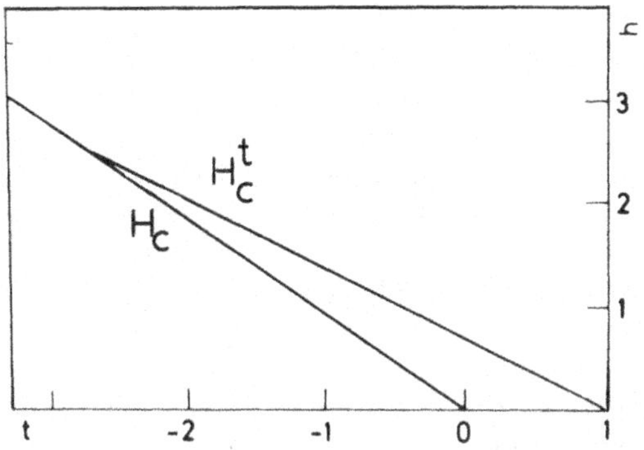

Fig. 132.

for the transition of the bulk material (fig. 132), i.e., the region of twinning plane superconductivity is thus limited to the vicinity of T_{c0}.

The width of the superconducting region is of the order of ξ_c; it can be found easily from the integral (20.72)

$$\frac{2L}{\xi_c}=\int_0^{\varphi(0)}\frac{d\varphi}{(t\varphi^2+\frac{1}{2}\varphi^4+\frac{1}{2}h_0^2)^{1/2}}. \tag{20.75}$$

Since the field is completely expelled from a region of width $2L$, the magnetic moment of such a plane per unit area is equal to

$$\mu_d=-\frac{1}{2\pi}HL(H). \tag{20.76}$$

The calculation performed here is valid for $\kappa=0$. However it can be generalized to finite κ's. Since nothing is used here apart from the constants of the bulk superconductor the results are universal. A numerical calculation for $\kappa=0.13$ (tin) gives excellent agreement with experiment (fig. 133).

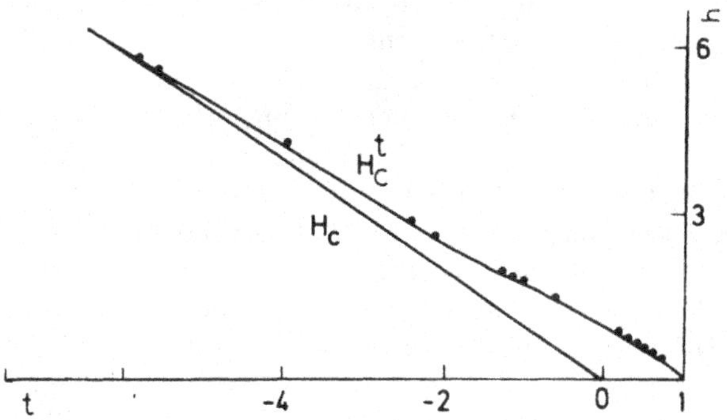

Fig. 133.

The existence of twinning planes can explain some other properties of the superconducting oxide Y-Ba-Cu-O. For example, films which are prepared by another technology contain no twinning planes and their critical temperature does not exceed 89 K. Furthermore, as already mentioned the superconductivity of an isolated twinning plane is suppressed by its normal surroundings due to the proximity effect. Hence the critical temperature could be raised if the twinning planes could be placed at distances less than ξ_c or if they were formed in granules of the proper size. As was mentioned already, Khaikin and Khlyustikov (1981) in pressing samples from a powder of pure tin managed to discover traces of superconductivity up to 12 K, i.e. at temperatures higher than $3T_{c0}$. It is not excluded that communications about signs of superconductivity in Y-Ba-Cu-O up to 500 K appearing from time to time, are connected with the existance of regions with dense arrays of twinning planes.

For more details about twinning-plane superconductivity see the review by Khlyustikov and Buzdin (1987).

20.3. *Andreev reflection*

In section 20.1 we have been concerned with the behavior of the superconducting order parameter in the vicinity of the contact between a superconductor and a normal metal. Let us now consider another aspect of such a contact.

As has been said in Part I, the chemical potentials of electrons in both metals must be equal. But the electrons in the ground state of a superconductor are bound into Cooper pairs, and in a normal metal they exist one by one. Hence, the chemical potentials of two separate electrons in a normal metal and of a Cooper pair in a superconductor must be equal. Suppose an electron has been transferred from the normal metal to the superconductor. In this case, it will have no partner, with which it could have been bound into a Cooper pair, which means that its energy will be higher by half the binding energy, i.e., by an amount Δ. Hence, for an individual electron the energy levels in the superconductor begin by Δ higher than in the normal metal. Thus, if its excitation energy above the Fermi level is less than Δ, it must be reflected from the interface.

Such reflection occurs in a rather peculiar manner. In order to ascertain this, it will be better to use the language of the quasiparticle concept. Suppose, for the sake of simplicity, that we deal with the interface between the normal and superconducting layers in the intermediate state of a superconductor. In this case the metals are identical, in principle, and Δ varies from zero to an equilibrium value in the bulk superconductor at a distance $\xi \sim \hbar v/\Delta$. The energy spectrum of quasiparticles expressed by the formula $\varepsilon = (\xi^2 + \Delta^2)^{1/2}$ varies in the manner shown in fig. 134. Quasiparticles with an energy less than Δ in the bulk superconductor must be reflected back into the normal metal. Upon such reflection the energy must be conserved.

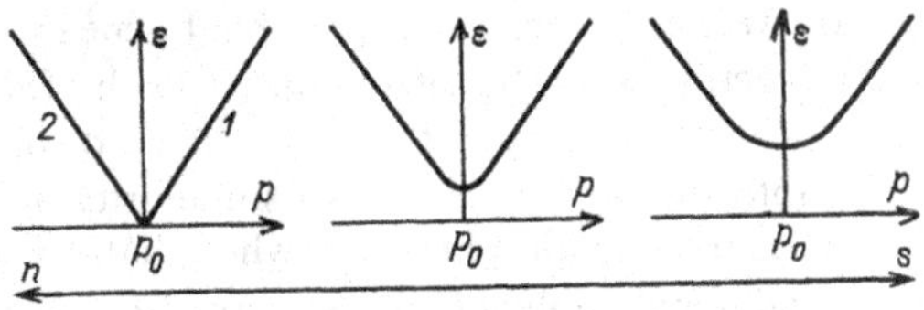

Fig. 134.

As for the variation of the quasimomentum δp, it can be evaluated in the following way. The quantity δp is of order $\mathrm{d}p/\mathrm{d}t$ multiplied by the time passed in the interface region, i.e., ξ/v. As regards $\mathrm{d}p/\mathrm{d}t$, it is equal to the acting force: $-\mathrm{d}V/\mathrm{d}x \sim \Delta/\xi$. This gives

$$\delta p \sim \frac{\xi}{v}\frac{\Delta}{\xi} \sim \frac{\Delta}{v} \sim p_0 \frac{\Delta}{p_0 v} \sim p_0 \frac{\Delta}{\varepsilon_F} \ll p_0 .$$

This means that the variation of the quasimomentum is smaller than the quasimomentum itself. But the motion of the electron after reflection must be directed away from the interface.

All these conditions can be fulfilled only in a single way, namely, if the quasiparticle is transferred, say, from branch 1 of the energy spectrum to branch 2 (fig. 134), i.e., is converted from a "particle" to an "antiparticle" with a close quasimomentum. The energy and the quasimomentum do not change. As regards the velocity, it will be equal after reflection to

$$\boldsymbol{v}_{\text{ref}} = \frac{\partial[v(p_0 - p)]}{\partial \boldsymbol{p}} = -\frac{v\boldsymbol{p}}{p} = -\boldsymbol{v}_{\text{inc}} .$$

Thus, upon such reflection the "particle" is converted to an "antiparticle". The quasimomentum and energy are conserved and the velocity changes sign. Such reflection is radically different from the specular reflection of electrons, where only the normal component of the electron velocity changes sign with respect to the interface. The specificity of this reflection, called Andreev reflection (Andreev 1964), is associated with the slowness of the variation of the potential, the role of which is played by Δ.

A question may arise as to the conservation of charge. The answer is: an electron residing at the Fermi boundary slightly above it picks up another electron with an opposite quasimomentum, which lies slightly below the Fermi boundary, and forms with it a Cooper pair, which goes into the superconductor condensate. An "antiparticle" is left, with a momentum opposite to the momentum of the picked-up electron, i.e., a momentum coinciding with the momentum of the original electron.

Evidently, upon Andreev reflection the transfer from branch 2 to branch 1 of the energy spectrum is also possible, i.e., the antiparticle may be converted to a particle with the opposite velocity. Moreover, it is clear that due to the proximity effect the slowness of variation of the potential will also be conserved for quasiparticles upon contact of the normal metal with a completely foreign superconductor, which means that the reflection law will remain valid.

Direct observation of Andreev reflection is possible by means of the nonresonant size effect described in section 8.3. The effect consist in the following: as soon as the extremal diameter of the trajectory in the magnetic field (or an integer number of such diameters) fits into the plate thickness, a singularity of surface impedance arises. Figure 135 sketches the situation in a case where half the extremal trajectory fits into a layer of a normal metal applied onto a superconductor surface. An electron approaching the interface is converted to an antiparticle with an opposite velocity. This antiparticle travels the same electron trajectory in the opposite direction and eventually approaches the interface. The antiparticle is then converted to an electron and the process repeats itself.

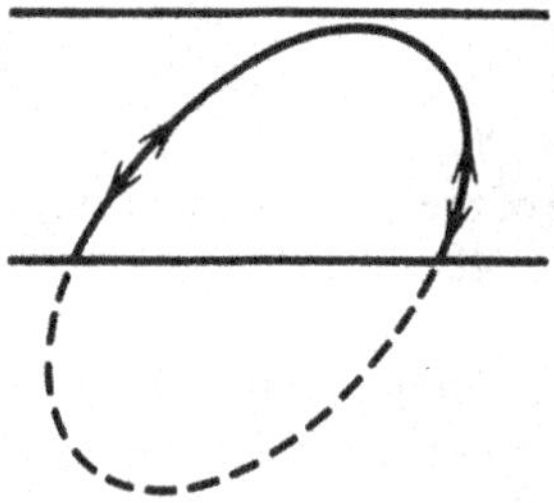

Fig. 135.

Thus, the singularities of the impedance will arise not only in those fields where an integer number of diameters fit into the thickness of the normal plate but also when their number is half-integral. In other words, extra singularities of $Z(H)$ will arise in the middle between the original ones. This is actually observed in experiments (Krylov and Sharvin 1970, 1973).

Note that if the extremal cross section is not the central one, then in an ordinary experiment on the size effect a spiral trajectory corresponds to it in coordinate space, i.e., the electron is displaced along the field. But if the trajectory corresponds to an extra resonance, the velocity component along the field will also change sign upon reflection and the antiparticle will pass its half of the turn of the coil in the opposite direction. Hence, each quasiparticle stays within one half of the turn of the helix, and on the average, no displacement occurs along the field.

20.4. *Low-temperature heat capacity of the intermediate state*

In the previous section we have already considered the case where the reflection from the n–s (normal-superconducting) interface leads to the replacement of translational motion by periodic motion. Let us now consider another case of the same kind: the intermediate state. We assume that the layers are plane-parallel. For microscopic samples the layer thickness is usually of the order of 10^{-3} cm, i.e., much larger than ξ, and the n–s interfaces may be taken into account by using the boundary condition. We obtain a periodic picture of the motion of quasiparticles in a normal layer: a quasiparticle of particle-type reaches the interface and is converted to an antiparticle; It then flies back to another interface and is converted to a particle, and the entire process is repeated.

The presence of a magnetic field in the normal layers does not alter the situation because the curved trajectory of the antiparticle exactly repeats the trajectory of the particle that has travelled in the opposite direction. As a matter of fact, in real type I superconductors the Larmor radius in the critical field (~100–1000 Oe) is not less than 10^{-2} cm and this implies that the trajectories may be assumed to be straight (fig. 136).

The periodic motion is quantized according to the Bohr rule:

$$\oint p \, \mathrm{d}l = 2\pi\hbar[n + \gamma(n)], \tag{20.77}$$

where $\gamma(n) < 1$. Since for the particle $\varepsilon = \xi = v(p_p - p_0)$, it follows that its quasimomentum is given by (here we use the symbol ε in place of ε_{qp}):

$$p_p = p_0 + \frac{\varepsilon}{v}.$$

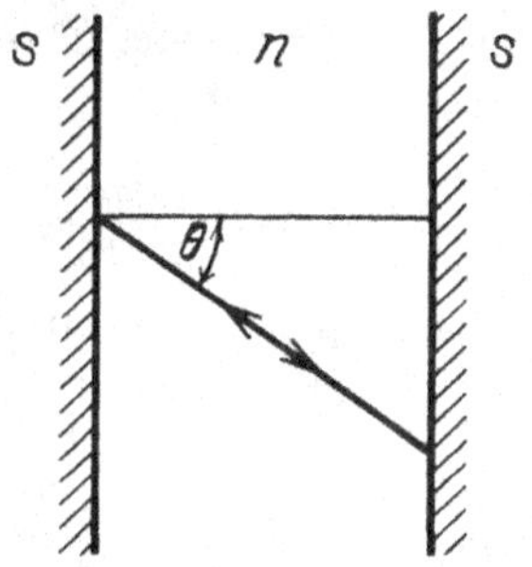

Fig. 136.

The reflected antiparticle has an energy $\varepsilon = -\xi = v(p_0 - p_a)$ and, hence, its quasimomentum is

$$p_a = p_0 - \frac{\varepsilon}{v}.$$

In the integral (20.77) p is the same on both halves of the trajectory, but $\mathrm{d}l$ changes sign. If the width of the layer is a, then the pathlength in one direction is $a/|\cos\theta|$, where θ is the angle between the trajectory and the normal to the interface (fig. 136). As a result, we obtain from eq. (20.77)

$$(p_p - p_a) \cdot \frac{a}{|\cos\theta|} = 2\pi(n + \gamma).$$

Substituting the expressions for p_p and p_a, we obtain (Andreev 1965):

$$\varepsilon(n, \theta) = \pi\hbar(n+\gamma)v\frac{|\cos\theta|}{a}. \tag{20.78}$$

This is the formula for the energy spectrum of quasiparticles in the intermediate state. The energy depends on the discrete variable n and the continuous angle variable θ.

Let us now find the density of states. The vector $\boldsymbol{n} = \boldsymbol{p}/p$ is uniquely determined by its projections onto the plane of the layer boundary. The number of states is expressed by the formula

$$2S\,\mathrm{d}p_x\frac{\mathrm{d}p_y}{(2\pi\hbar)^2},$$

where S is the area of the interface and the factor 2 stems from two spin projections. We introduce the angle φ in the p_x, p_y plane. In such a case, for the number of states we have

$$2Sp_\perp\,\mathrm{d}p_\perp\frac{\mathrm{d}\varphi}{(2\pi\hbar)^2},$$

where $p_\perp = (p_x^2+p_y^2)^{1/2}$. But since $p_\perp^2+p_z^2 = p_0^2$, it follows that $p_\perp\,\mathrm{d}p_\perp \to |p_z|\,\mathrm{d}p_z = p_0^2|\cos\theta|\,\mathrm{d}(\cos\theta)$. Thus, we find that the number of states per quantum number n in eq. (20.78) is

$$2Sp_0^2|\cos\theta|\,\mathrm{d}(\cos\theta)\frac{\mathrm{d}\varphi}{(2\pi\hbar)^2} = 2Sp_0^2|\cos\theta|\frac{\mathrm{d}\Omega}{(2\pi\hbar)^2}, \tag{20.79}$$

where $\mathrm{d}\Omega$ is an element of the solid angle for the directions of the vector $\boldsymbol{p}$.

For the free energy per unit volume of the normal layer we obtain (see eq. 10.11):

$$\Omega(T) = -\frac{T}{a}\sum_n\int\mathrm{d}\Omega\,2p_0^2\frac{|\cos\theta|}{(2\pi\hbar)^2}\ln\left[1+\exp\left(-\frac{\varepsilon(n,\theta)}{T}\right)\right].$$

Since ε depends only on $|\cos\theta|$, the integral $\int\mathrm{d}\Omega = \int_{-1}^{1}\mathrm{d}\cos\theta\int\mathrm{d}\varphi$ may be replaced by $4\pi\int_0^1\mathrm{d}\cos\theta$. Here we find

$$\Omega(T) = -\frac{2p_0^2T}{a\pi\hbar^2}\sum_n\int_0^1 x\,\mathrm{d}x\ln\left[1+\exp\left(-\frac{\varepsilon(n,x)}{T}\right)\right]. \tag{20.80}$$

Further we consider two temperature regions. If $T \gg \hbar v/a$, the summation over n may be replaced by an integration. In this case, introducing the variable ε in place n, we have in accordance with (20.78):

$$\Omega(T) = -\frac{2p_0^2T}{v\pi^2\hbar^3}\int_0^\infty\mathrm{d}\varepsilon\ln(1+\mathrm{e}^{-\varepsilon/T})$$

$$= -\frac{p_0^2T}{v\pi^2\hbar^3}\int_{-\infty}^{\infty}\mathrm{d}\varepsilon\ln(1+\mathrm{e}^{-|\varepsilon|/T}).$$

This is the usual expression for a normal metal. Evaluation of the integral gives

$$\Omega(T) = -\tfrac{1}{6}\frac{p_0^2}{v\pi^2\hbar^3}\pi^2 T^2.$$

This yields the usual formula for the electronic heat capacity of a normal metal (see eq. 2.29):

$$C = -T\frac{\partial^2\Omega}{\partial T^2} = \frac{p_0 m}{3\hbar^3}T.$$

But if $T \ll \hbar v/a$, the procedure is different. In the integral (20.79), instead of the variable x we introduce, using formula (20.78), the variable $y = \varepsilon/T$. So, we obtain

$$\Omega(T) = -\frac{2p_0^2T^3a}{\pi^3\hbar^4v^2}\sum_n \frac{1}{(n+\gamma)^2}\int_0^{\pi\hbar(n+\gamma)v/aT} y\,dy\,\ln(1+e^{-y}).$$

The limit of integration over y may be put equal to infinity because of the rapid convergence of the integral. The sum over n converges and gives a number of the order of unity, which we designate by α. On the whole, after taking the integral over y we obtain

$$\Omega(T) = -3\zeta(3)\frac{\alpha p_0^2T^3a}{2\pi^3\hbar^4v^2}$$

from which we have

$$C = -T\frac{\partial^2\Omega}{\partial T^2} = \beta\frac{p_0^2T^2a}{\hbar^4v^2},$$

where

$$\beta = 9\frac{\zeta(3)}{\pi^3}\sum_n (n+\gamma)^{-2}.$$

Taking into account that the fraction of normal layers is always equal to $x_n = B/H_{cm}$ (the dependence of B on the external field being determined by the geometry of the experiment), we find for the heat capacity of the entire sample (Andreev 1965):

$$C = \begin{cases} \dfrac{p_0 mB}{3\hbar^3H_{cm}}T, & T \gg \dfrac{\hbar v}{a}, \\[2ex] \beta\dfrac{m^2aB}{\hbar^4H_{cm}}T^2, & T \ll \dfrac{\hbar v}{a}. \end{cases} \tag{20.81}$$

Experimental measurements of the heat capacity of superconductors in the intermediate state (Zavaritskii 1965) have confirmed formulae (20.81).

21

Superconductivity and magnetism

21.1. Superconductors containing magnetic impurities. Gapless superconductivity

As has been pointed out in section 16.9, impurities in a small concentration exert only a slight effect on the thermodynamic properties of superconductors. The variation of these properties is of the relative order of $a/l \sim c_i$, where a is the atomic dimension, l is the mean free path, and c_i the atomic concentration. The behavior of the magnetic impurities described in sections 4.6 and 13.7 is completely different. The action of these impurities is measured by the quantity ξ/l_s, where l_s is the mean free path relative to the spin flip associated with these impurities. The same may be written in the form $\hbar/(\tau_s T_c)$, where $\tau_s = l_s/v$ is the corresponding mean free time. This is associated with the completely different action of ordinary nonmagnetic (potential) impurities and magnetic atoms on Cooper pairs.

Usually, impurities act only on the electric charge and, hence, scatter both electrons of a Cooper pair identically, and this does not lead to its collapse. As for magnetic impurities, they are capable of flipping the electron spin. Hence, upon scattering from such an impurity the pair can pass to a state with parallel spins (triplet state), where the Pauli principle requires that the electrons comprising the pair be not at a single point and be described by an antisymmetric coordinate wave function. The pair is thus broken up.

It has been shown in section 4.6 that to first order the probability of scattering of an electron from a magnetic impurity is proportional to $(J/n)^2 S(S+1)$. This quantity replaces $|V_{p'p}|^2$ in the probability of scattering from an ordinary impurity. In fact, the interaction of an electron with each impurity contains both the potential part (V) and the exchange part ($-(J/n)(\boldsymbol{\sigma}\mathbf{S})$); the former is, as a rule, larger than the latter; however, in the effect of interest to us the potential part does not play a role and will be neglected here. The scattering time associated only with the exchange interaction will be denoted as τ_s.

The theory that provides an accurate description of the action of magnetic impurities (Abrikosov and Gor'kov 1960a) is based on the application of the methods of quantum field theory and it is too complicated to be described here. Therefore, we shall make an attempt to present the basic features of such objects with the aid of simple reasonings.

We begin with the critical temperature. The general formula for the dependence $\Delta(T)$ for pure superconductors is given in section 16.5 (formula 16.23). It may be rewritten in a different form by using the property of the hyperbolic tangent

$$\tanh(\tfrac{1}{2}\pi x)=\frac{4x}{\pi}\sum_{k=1}^{\infty}\frac{1}{(2k-1)^2+x^2}. \tag{21.1}$$

On the basis of this formula we obtain from eq. (16.23):

$$1=2g\nu(\mu)T\sum_{\omega_k=\pi T}^{\hbar\omega_D}\int_0^{\infty}\frac{d\xi}{\omega_k^2+\xi^2+\Delta^2}, \tag{21.2}$$

where $\omega_k=\pi T(2k-1)$. Here we have transferred the summation in finite limits to ω_k, which will be convenient in what follows and, strictly speaking, is more correct than the limitation on ξ.

If the integrand is treated formally, it has imaginary poles at $\omega_k=\pm i(\xi^2+\Delta^2)^{1/2}=\pm i\varepsilon$. The possibility of the breakup of pairs under the action of magnetic impurities leads to a finite lifetime. This point has already been discussed in section 2.2. The wave function of the state $\exp(-i\xi t/\hbar)$ takes on a damping factor $\exp(-\gamma t/\hbar)$. This may be interpreted as if the energy of quasiparticles acquires an imaginary correction $-i\gamma$. In view of this, it becomes clear why in the exact theory ω_k is replaced by $\omega_k+\hbar\tau_s^{-1}$ in expression (21.2). In fact, the new integrand has poles at points $\omega_k=\pm i(\varepsilon\pm i\hbar\tau_s^{-1})$.

In order to obtain T_c, we put $\Delta=0$. So, we have

$$1=2g\nu T\sum_{\omega_k=\pi T}^{\hbar\omega_D}\int_0^{\infty}\frac{d\xi}{(\omega_k+\hbar\tau_s^{-1})^2+\xi^2}.$$

The integral over ξ gives

$$1=g\nu\pi T\sum_{\omega_k=\pi T}^{\hbar\omega_D}\frac{1}{\omega_k+\hbar\tau_s^{-1}}=g\nu\sum_{k=1}^{\hbar\omega_D/2\pi T}\frac{1}{2k-1+\rho},$$

where we have substituted $\omega_k=\pi T(2k-1)$ and introduced the dimensionless parameter $\rho=\hbar/(\tau_s\pi T)$.

We subtract and add to the last sum a similar sum with $\rho=0$. The difference rapidly converges, and therefore the limit of the sum over k may be extended to infinity. Substituting the value of the sum of $(2k-1)^{-1}$, we have

$$1=\tfrac{1}{2}g\nu\ln\left(\frac{2\gamma\omega_D\hbar}{\pi T_c}\right)+g\nu\sum_{k=1}^{\infty}\left(\frac{1}{2k-1+\rho}-\frac{1}{2k-1}\right),\quad \gamma=1.78.$$

Let us use the following property of the Γ-functions:

$$\psi(x)-\psi(y)=\sum_{k=0}^{\infty}\left(\frac{1}{y+k}-\frac{1}{x+k}\right), \tag{21.3}$$

where $\psi(x) = \Gamma'(x)/\Gamma(x)$ is the logarithmic derivative of the Γ-function; the latter is determined as follows: $\Gamma(n) = (n-1)!$ for integer values of $x = n$, and for noninteger values we have

$$\Gamma(x) = \int_0^\infty e^{-t} t^{x-1}\, dt. \tag{21.4}$$

Dividing by $\frac{1}{2}g\nu$, rearranging the terms and using formula (16.25), we obtain

$$\ln\left(\frac{T_{c0}}{T_c}\right) = \psi(\tfrac{1}{2}+\tfrac{1}{2}\rho) - \psi(\tfrac{1}{2}), \tag{21.5}$$

where T_{c0} is the critical temperature of a pure superconductor. This formula is exact and holds for any $\rho = \hbar/(\pi T_c \tau_s)$. As for τ_s itself, from the foregoing it is clear that τ_s^{-1} is proportional to $\pi\nu(\mu)(J/n)^2 S(S+1)n_m$ and the exact value of the quantity appearing in eq. (21.5) is

$$\frac{\hbar}{\tau_s^{-1}} = \tfrac{7}{24}\pi\nu(\mu)\frac{J}{n}\,S(S+1)n_m, \tag{21.6}$$

where n_m is the concentration of magnetic atoms.

Let us now consider the asymptotic values of formula (21.5). At $\rho \ll 1$ we use the following expansion:

$$\psi(\tfrac{1}{2}+\tfrac{1}{2}\rho) \approx \psi(\tfrac{1}{2}) + \tfrac{1}{2}\psi'(\tfrac{1}{2})\rho = \psi(\tfrac{1}{2}) + \tfrac{1}{4}\pi^2\rho.$$

Expanding $\ln(T_{c0}/T_c) \approx (T_{c0} - T_c)/T_{c0}$, we obtain

$$T_c \approx T_{c0} - \frac{\pi\hbar}{4\tau_s}. \tag{21.7}$$

From this we see that, in contrast to ordinary impurities, magnetic impurities lower T_c significantly.

Obviously, at $\hbar\tau_s^{-1} \sim T_{c0}$ the superconductivity must disappear altogether, i.e., T_c must vanish. In the vicinity of this point in formula (21.5) $\rho \gg 1$. Consider the corresponding asymptotics. At large x

$$\psi(x) \approx \ln x - \frac{1}{2x} - \frac{1}{12x^2}$$

and $\psi(\frac{1}{2}) = -\ln(4\gamma)$. Expanding in ρ^{-1}, we have

$$\ln\left(\frac{T_{c0}\pi\tau_s}{2\gamma\hbar}\right) \approx \tfrac{1}{6}\left(\frac{\pi T_c \tau_s}{\hbar}\right)^2. \tag{21.8}$$

From this it follows that the critical concentration at which $T_c = 0$ corresponds to

$$\frac{\hbar}{\tau_{sc}^{-1}} = \frac{\pi T_{c0}}{2\gamma}. \tag{21.9}$$

In the vicinity of the critical concentration the critical temperature is given by

$$T_c \approx \sqrt{6}\,\frac{\hbar}{\pi}\,\frac{(\tau_s - \tau_{sc})^{1/2}}{\tau_{sc}^{3/2}}. \tag{21.10}$$

A further fundamental consequence of the presence of magnetic impurities is gapless superconductivity. The point is that magnetic impurities not only reduce the binding energy of Cooper pairs but, as exact calculations show (Abrikosov and Gor'kov 1960a), the result is that not all pairs have the same binding energy. In this sense, the situation is similar to a weakly nonideal Bose gas (see, for example, Abrikosov et al. 1962). Whereas in an ideal Bose gas at $T=0$ all the particles are present in the condensate, i.e., have $\boldsymbol{p}$ and ε equal to zero and are described by a coherent wave function $\Psi = \text{const.}$, in a nonideal gas at $T=0$ only a certain part of the particles are present in the condensate, and the remaining particles have an energy and momentum different from zero. However, in this case too the order parameter is the coherent wave function of the condensate particle.

In superconductors containing magnetic impurities the situation is the same. The Bose condensate of Cooper pairs does not contain all the pairs: some of these have smaller binding energies. Evidently, for such pairs to be broken up it is necessary to consume a smaller energy than that required for the collapse of the pairs present in the condensate. Up until now the quantity Δ alone has described both the number of pairs in the condensate and the binding energy. However, in application to superconductors with magnetic impurities this statement becomes invalid. Whereas one quantity described the order parameter (we still denote it as Δ), i.e., the wave function of the condensate, a completely different quantity characterizes the minimum binding energy of the pairs or the energy gap, which manifests itself in the low-temperature heat capacity, thermal conductivity, absorption of electromagnetic radiation and ultrasound, etc.

It is exactly the order parameter that determines the basic specific features of superconductors: the Meissner effect and the absence of electrical resistance. As long as Δ is finite all these properties are operative. At the same time, the energy gap may vanish; this is actually observed in superconductors containing magnetic impurities in a given concentration range (n_{mc} corresponds to τ_{sc} according to formulas 21.9 and 21.6):

$$0.91\,n_{mc} < n_m < n_{mc}.$$

A natural question may arise: How does this agree with the Landau criterion? According to this criterion superconductivity is impossible in the absence of a gap in the energy spectrum. However, the Landau criterion has been deduced for pure systems, in which spatial homogeneity exists and the momentum is conserved. For systems containing impurities this criterion cannot be applied.

We give here, without derivation, the formula for the energy gap,

$$\hbar\omega_0 = \left[\Delta_0^{2/3} - \left(\frac{\hbar}{\tau_s}\right)^{2/3}\right]^{3/2}, \tag{21.11}$$

and the formula for the electronic heat capacity in the gapless region:

$$C=\tfrac{1}{3}\pi^2\nu(\mu)\left[1-\left(\frac{\tau_s\Delta_0}{\hbar}\right)^2\right]^{1/2}T, \tag{21.12}$$

where Δ_0 is the order parameter at $T=0$, which depends in a complicated way on τ_s. At the boundary of the gapless region $\Delta_0(\tau_s)=\hbar/\tau_s$.

The disappearance of the gap can conveniently be traced out from the density of states. For a normal metal $\nu_n=p_0m/(\pi^2\hbar^3)=\text{const}$. For a pure superconductor

$$\nu(\varepsilon)=\frac{1}{\pi^2\hbar^3}p_0m\frac{d\xi}{d\varepsilon}=\nu_n\varepsilon(\varepsilon^2-\Delta^2)^{-1/2}$$

at $\varepsilon>\Delta$ and 0 at $\varepsilon<\Delta$. Note that $\nu(\varepsilon)$ has a square root singularity for $\varepsilon\to\Delta$. The density of states $\nu(\varepsilon)$ is given in fig. 137 at various values of τ_s. The ordinate represents $\nu(\varepsilon)/\nu_n$ and the abscissa represents $\varepsilon/\Delta_0(\tau_s)$. The curves are labeled with the corresponding values of

$\hbar/\tau_s\Delta_0$.

It should be noted that magnetic impurities do not provide the only pair-breaking mechanism. It was later found that the same effect is exhibited by any perturbations giving rise to terms in the Hamiltonian that are noninvariant under change of the sign of time (see Maki 1969).

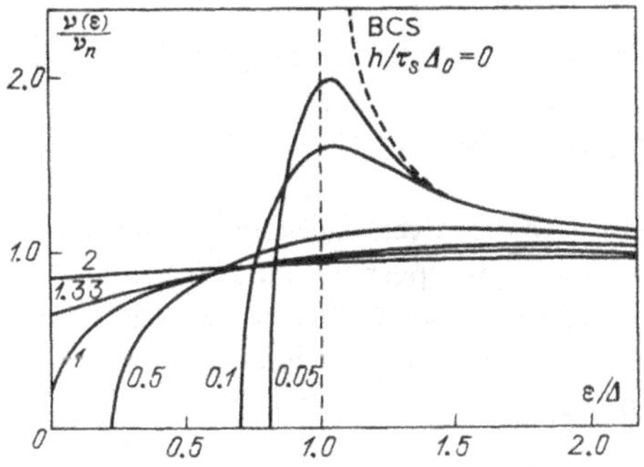

Fig. 137.

Cooper pairs constitute a superposition of the states of electrons with opposite momenta and spins. These states result from one another by inversion of the sign of time (the momentum and spin change sign with $t\to -t$). A perturbation which is noninvariant under the change of the sign of time acts differently on the electrons contained in a Cooper pair and, as it were, pulls it apart.

An example is an external magnetic field. Of course, the field does not penetrate a bulk superconductor, and therefore we assume that we are dealing here with a thin film with a thickness smaller than δ and ξ. The main perturbation from the magnetic field is $-(e/mc)\boldsymbol{pA}$. The wave function of the electron is proportional to $\exp(-i\varepsilon t/\hbar)$. From this we see that in the presence of a magnetic field between the electrons that form a pair, i.e., have momenta $\boldsymbol{p}$ and $-\boldsymbol{p}$, there arises a phase difference, which varies with time according to the law

$$\frac{d\theta}{dt}=\frac{2e}{\hbar c}\boldsymbol{v}\boldsymbol{A}, \quad \text{with} \quad v=p_0/m. \tag{21.13}$$

However, the electrons are scattered by impurities or at the boundaries of the sample. Therefore, they execute a Brownian diffusive motion with, according to sections 3.5 and 5.5,

$$\overline{x^2}\sim Dt\sim lvt\sim v^2t\tau_{tr},$$

where $D\sim lv\sim v^2\tau_{tr}$ is the diffusion coefficient. This relation may be extended to any quantity: the mean square of the sought-for quantity is of the order of the square of the instantaneous rate of its change multiplied by $t\tau_{tr}$. In our case,

$$\overline{\theta^2}\sim\left(\frac{d\theta}{dt}\right)^2 t\tau_{tr}.$$

Hence, for θ to be of the order of 2π, a time τ_H is needed, with

$$\frac{1}{\tau_H}\sim\left(\frac{d\theta}{dt}\right)^2\tau_{tr},$$

where $(d\theta/dt)^2$ is the mean square value of eq. (21.13). Thus, we have $(A\sim Hd)$

$$\frac{1}{\tau_H}\sim v^2\tau_{tr}\left(\frac{e}{\hbar c}\right)^2 H^2d^2\sim D\left(\frac{e}{\hbar c}\right)^2 H^2d^2. \tag{21.14}$$

Other examples could also be given, namely the mixed state in a bulk type II superconductor and in a film in a perpendicular magnetic field, small particles, etc. It can be shown (see Maki 1969) that all the formulas derived for magnetic impurities remain to be valid for all pair-breaking mechanisms. Only the physical meaning of the constant ρ is changed.

21.2. The inhomogeneous superconducting state

In the preceding section we have said that magnetic impurities destroy superconductivity. The situation is even worse in a case where there is ferromagnetic ordering. In such a case, the number of electrons with different orientations of the spin is different, which means that the corresponding Fermi surfaces are also different. If ferromagnetism is strong, i.e., the Curie temperature is of the order of hundreds of degrees, as is the case with ordinary ferromagnets based on transition metals, superconductivity is impossible.

However, with rare-earth metals or actinides the situation may be different. The magnetism of such atoms is associated with the unfilled 4f and 5f shells. These are inner shells with a small radius. In neighboring atoms they practically do not overlap. In view of this, the exchange interaction of the spins is realized by way of an indirect exchange through conduction electrons. In other words, one may proceed from the picture of localized spins which interact with conduction electrons by the mechanism discussed in section 4.1.

In the next section we shall consider in more detail the interaction between two localized spins, which is transmitted by conduction electrons (RKKY). Of importance at this point is only the circumstance that the resultant magnetism is "weak" with a Curie temperature of the order of a few degrees. It is for this reason that the question arises as to the mutual effect of ferromagnetism and superconductivity.

Two extreme situations may be encountered. If the Curie temperature of the ferromagnetic transition θ is somewhat higher than the critical temperature of the superconducting transition T_c (not much higher), superconductivity may arise against the background of existing ferromagnetism. But if the temperature of the superconducting transition is higher than the Curie temperature, ferromagnetism arises against the background of superconductivity. It is possible that the phases that will be considered in this and subsequent sections do not cover the entire variety of existing possibilities, but they give an idea of the specific features of the mutual effect of such antagonistic phenomena, as ferromagnetism and superconductivity.

We begin with the case where superconductivity appears against the background of ferromagnetism. Instead of eq. (16.13), we write the Hamiltonian

$$\mathscr{H} - \mu N = \sum_{p,\sigma} (\xi_p - \sigma_z I) a^+_{p,\sigma} a_{p,\sigma} - g \sum a^+_{p'_1,+} a^+_{p'_2,-} a_{p_2,-} a_{p_1,+}, \tag{21.15}$$

where I is the self-consistent "molecular field", under the action of which the electron energy begins to depend on the projection spins ($\sigma_z = \pm 1$). Further, we introduce the operators $\alpha_{p,+}$ and $\alpha_{p,-}$ in accordance with formulas (16.14) and repeat the entire derivation given in section 16.4 taking account of the altered Hamiltonian. In such a case we obtain the same formulas for the coefficients u and v and formula (16.21) for Δ, but the energies included in the Fermi distributions $n_{p,+}$ and $n_{p,-}$ have the form

$$\varepsilon_{p,+} = \varepsilon_p - I, \qquad \varepsilon_{p,-} = \varepsilon_p + I,$$

as before,

$$\varepsilon_p = (\xi_p^2 + \Delta^2)^{1/2}.$$

Let us consider only the case $T = 0$. Since $\varepsilon_p > 0$, it follows that $n_{p,-} = 0$. Therefore, the sum over $\boldsymbol{p}$ in (16.21) is bounded from below by the condition $\varepsilon_{p,+} > 0$ or $\varepsilon_p > I$. Two cases are possible. If $\Delta > I$, the condition $\varepsilon_{p,+} > I$ does not impose restrictions on the domain of integration over $\boldsymbol{p}$ and we obtain $\Delta = \Delta_0$. But if $\Delta < I$, then, passing to the integral, we have

$$\frac{2}{g\nu} = \ln\frac{2\hbar\omega_D}{\Delta_0} = \int_{(I^2-\Delta^2)^{1/2}}^{\hbar\omega_D} \frac{d\xi}{\varepsilon} = \ln\frac{2\hbar\omega_D}{I+(I^2-\Delta^2)^{1/2}}. \tag{21.16}$$

From this we see that $I+(I^2-\Delta^2)^{1/2} = \Delta_0$ or

$$\Delta = [\Delta_0(2I-\Delta_0)]^{1/2}. \tag{21.17}$$

Thus, we obtain two branches of the solution of eq. (21.16), which are shown in fig. 138. However, the nonzero solution for Δ implies only the local minimum of the free energy; this may correspond to a metastable rather than to the equilibrium state. In order to find out whether the transition to the superconducting state really occurs, we must find the free-energy difference $\Omega_s - \Omega_n$. For this purpose we use formula (16.38).

In the region where $\Delta_1 > I$ we have, as before,

$$\frac{\partial g^{-1}(\Delta_1)}{\partial \Delta_1} = -\frac{\nu}{2\Delta_1}. \tag{21.18}$$

In the region where $\Delta_1 < I$ we obtain from (21.16)

$$\begin{aligned}\frac{\partial g^{-1}(\Delta_1)}{\partial \Delta_1} &= \tfrac{1}{2}\nu \frac{\Delta_1}{I+(I^2-\Delta_1^2)^{1/2}} \frac{1}{(I^2-\Delta_1^2)^{1/2}} \\ &= \frac{\nu}{2\Delta_1} \frac{[I-(I^2-\Delta_1^2)^{1/2}]}{(I^2-\Delta_1^2)^{1/2}}.\end{aligned} \tag{21.19}$$

On the lower branch of fig. 138, formula (21.19) holds for all Δ_1; substituting it into (16.38), we find

$$\begin{aligned}\Omega_s - \Omega_n &= \tfrac{1}{4}\nu[I-(I^2-\Delta^2)^{1/2}]^2 \\ &= \tfrac{1}{4}\nu(2I-\Delta_0)^2,\end{aligned} \tag{21.20}$$

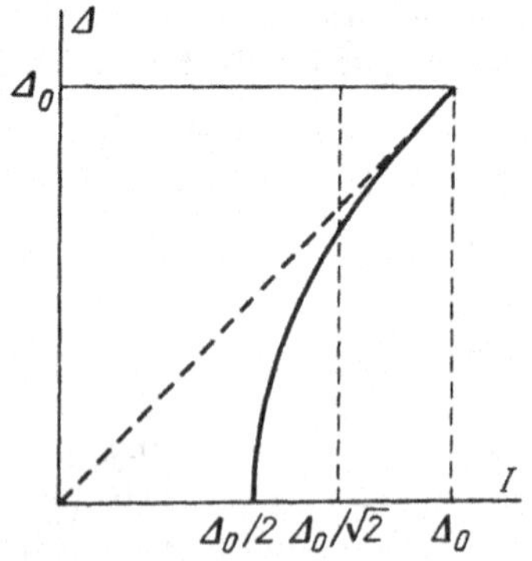

Fig. 138.

where we have substituted (21.17) and have taken into account that $I < \Delta_0$. So, in this case $\Omega_s - \Omega_n > 0$, i.e., the superconducting transition does not occur.

On the upper branch of fig. 132, $\Delta = \Delta_0 > I$. Therefore, there are two domains of integration: $\Delta_1 < I$ and $\Delta_1 > I$. Inserting formulas (21.18) and (21.19) into (16.38), we obtain

$$\Omega_s - \Omega_n = \tfrac{1}{4}\nu(2I^2 - \Delta_0^2), \tag{21.21}$$

from which follows that the upper branch in fig. 138, i.e. $\Delta = \Delta_0$, corresponds to equilibrium superconductivity at

$$I < \frac{\Delta_0}{\sqrt{2}} = 0.707\Delta_0. \tag{21.22}$$

Thus, in the model under discussion, as I increases at the point $I = \Delta_0/\sqrt{2}$ a first-order phase transition occurs from the superconducting phase with $\Delta = \Delta_0$ to the normal phase with $\Delta = 0$. The remaining portions of the solid curves in fig. 138 correspond to the metastable phases.

However, a more detailed investigation shows that the real situation is more complicated. In the preceding treatment we were concerned only with the possibility of the appearance of superconductivity with $\Delta = \text{const}$. It turns out that at values of I larger than the critical value (21.22) there may appear a superconducting phase with an inhomogeneous order parameter $\Delta(r)$ by way of a second-order phase transition (Larkin and Ovchinnikov 1964, Fulde and Ferrell 1964). This new phase is sometimes called the LOFF phase.

In order to demonstrate this transition, we make use of the fact that near the transition point, Δ is small and the equation for Δ will necessarily be linear. This will permit us to use a Fourier expansion and raise the question of the generation of a single Fourier harmonic, i.e.

$$\Delta(\boldsymbol{r}) = \Delta \exp(i\boldsymbol{q}\boldsymbol{r}/\hbar). \tag{21.23}$$

If we recall that $\Delta(\boldsymbol{r})$ plays the role of the coherent wave function of the condensate of Cooper pairs, then formula (21.23) corresponds to the moving condensate, in which the pairs have a momentum $\boldsymbol{q}$, i.e., the velocity $\boldsymbol{v}_s = \boldsymbol{q}/2m$.

The problem of the moving condensate has been considered in section 19.2. We substitute formulae (19.19)-(19.21) into the equation for Δ derived in section 19.2, following which we make the transformation of the momentum variable $\boldsymbol{p} \to \boldsymbol{p} + \frac{1}{2}\boldsymbol{q}$ and use formulas (19.24). Taking into account that for quasiparticles with spin $+\frac{1}{2}$ the energy contains a term with $-I$, and in the opposite case a term with $+I$, we finally obtain

$$1 = \tfrac{1}{2}g \sum_{p} \frac{1 - n_+ - n_-}{\varepsilon}, \tag{21.24}$$

where n_+ and n_- are the equilibrium Fermi functions with energies given by

$$\varepsilon_+ = \varepsilon + \tfrac{1}{2}vqx - I, \qquad \varepsilon_- = \varepsilon - \tfrac{1}{2}vqx + I, \tag{21.25}$$

where $\varepsilon=(\xi^2+\Delta^2)^{1/2}$ and $x=\cos(\boldsymbol{p},\boldsymbol{q})$. We assumed that $vq>2I$ (this is confirmed by subsequent calculations). Since we are interested in the transition region, we assume that $\Delta<I$. Just as before, we set $T=0$.

The integral on the right-hand side of eq. (21.24) is different from zero if $n_+=n_-=1$, i.e. $\varepsilon_+, \varepsilon_-<0$, or $n_+=n_-=0$, i.e., $\varepsilon_+, \varepsilon_->0$; the latter condition, which gives $2(I+\varepsilon)/vq>x>2(I-\varepsilon)/vq$, is realized. Taking into account that x varies in the region $-1\leqslant x\leqslant 1$, and $0<|\xi|\leqslant\hbar\omega_{\mathrm{D}}$, we obtain from (21.24):

$$\begin{aligned}\frac{2}{g\nu}=\ln\frac{2\hbar\omega_{\mathrm{D}}}{\Delta_0}&+\tfrac{1}{2}\ln\frac{2\hbar\omega_{\mathrm{D}}}{\frac{1}{2}vq-I+[(\frac{1}{2}vq-I)^2-\Delta^2]^{1/2}}\\&+\tfrac{1}{2}\ln\frac{2\hbar\omega_{\mathrm{D}}}{\frac{1}{2}vq+I+[(\frac{1}{2}vq+I)^2-\Delta^2]^{1/2}}\\&-\frac{I}{vq}\ln\frac{\frac{1}{2}vq+I+[(\frac{1}{2}vq+I)^2-\Delta^2]^{1/2}}{\frac{1}{2}vq-I+[(\frac{1}{2}vq-I)^2-\Delta^2]^{1/2}}\\&+\frac{1}{vq}\{[(\tfrac{1}{2}vq-I)^2-\Delta^2]^{1/2}+[(\tfrac{1}{2}vq+I)^2-\Delta^2]^{1/2}\}.\end{aligned}\tag{21.26}$$

Transposing terms and introducing new variables $z=I/\Delta_0$, $y=vq/2I$ and $s=\Delta/I$, we see that

$$\begin{aligned}\ln(1/z)=\{&(y-1)\ln[y-1+((y-1)^2-s^2)^{1/2}]\\&+(y+1)\ln[y+1+((y+1)^2-s^2)^{1/2}]\}(2y)^{-1}\\&+\{((y-1)^2-s^2)^{1/2}+((y+1)^2-s^2)^{1/2}\}(2y)^{-1}.\end{aligned}\tag{21.27}$$

At the transition point $s=0$, so we have

$$\ln(1/z)+1-\ln 2=[(y-1)\ln(y-1)+(y+1)\ln(y+1)](2y)^{-1}.\tag{21.28}$$

The right-hand side has the asymptotics of $\ln y$ with $y\to\infty$ and $-(y-1)\ln[1/(y-1)]$ with $y\to 1$. From this it can be seen that at a certain y it has a minimum. Evidently, it has a corresponding maximum value of $z=I/\Delta_0$ at which $\Delta\neq 0$ appears for the first time. If the minimum of the right-hand side is found numerically, we obtain $y_{\mathrm{c}}=1.19$ and $z_{\mathrm{c}}=0.754$. Thus, it turns out that $\Delta\neq 0$ appears for the first time at

$$I_{\mathrm{c}}=0.754\Delta_0,\tag{21.29}$$

$$(vq)_{\mathrm{c}}=2.38I_{\mathrm{c}}.\tag{21.30}$$

The quantity I_{c} is larger than the values obtained earlier, (eq. 21.22), for an immobile condensate at rest (i.e. $\Delta(\boldsymbol{r})=\text{const.}$).

Now we have to check whether the appearance of a finite Δ corresponds to a decrease in free energy. Here, however, we should make a remark. When we were concerned with the problem of solving the equation for Δ in the limit $\Delta\to 0$, we could use the linearity of this equation and consider a single Fourier harmonic. We found out that a harmonic with a definite $q_{\mathrm{c}}=2.38I/v$ appears for the first time.

But the isotropy and linearity of the problem implies that any linear combination of harmonics with different $\boldsymbol{q}$ equal in magnitude to q_c may also appear equally well.

The question of the choice of a particular combination is solved by comparing their free energies, where Δ is expressed as a function of $I_c - I$. However, this requires the solution of a nonlinear problem, for which the method used here is inapplicable. Of course, we can generalize the Bogoliubov transformation, just as has been done in the literature (De Gennes 1966). However, we shall not give the cumbersome calculations here; we shall only show that even in the simplest case of a single Fourier harmonic, which has been considered above, the free energy decreases as a result of superconducting pairing. We also note that a detailed analysis of the problem of the most favorable structure of $\Delta(\boldsymbol{r})$ has not yet been carried out, which to a certain extent is accounted for by the lack of experimental observations of the LOFF phase.

Thus, we are to find the free energy. Using formula (21.26), we obtain

$$\frac{\partial g^{-1}}{\partial \Delta} = \frac{\nu}{2\Delta}\left\{-1 + \frac{1}{vq}[(\tfrac{1}{2}vq - I)^2 - \Delta^2]^{1/2} + \frac{1}{vq}[(\tfrac{1}{2}vq + I)^2 - \Delta^2]^{1/2}\right\}. \tag{21.31}$$

Expanding in Δ and substituting into (16.38) lead to

$$\Omega_s - \Omega_n = -\frac{\omega\Delta^4}{16I_c^2(y_c^2 - 1)}. \tag{21.32}$$

In order to express Δ in terms of $I_c - I$, we expand eq. (21.27) in $s = \Delta/I$. The result is

$$\frac{I_c - I}{I_c} = \frac{1}{4(y_c^2 - 1)}\left(\frac{\Delta}{I_c}\right)^2.$$

Substituting $y_c = 1.19$, we find

$$\Delta \approx 1.3[(I_c - I)I_c]^{1/2}. \tag{21.33}$$

Using this value and $I_c = 0.754\Delta_0$, we obtain

$$\Omega_s - \Omega_n = -0.42\nu(I_c - I)^2. \tag{21.34}$$

Thus, as a result of the formation of an inhomogeneous superconducting condensate, the free energy is really lowered.

As has been said above, the structure considered is only the simplest example. As a matter of fact, one should expect the generation of a three-dimensional structure (for example, with a set of $\boldsymbol{q}$ directed along the principal axes of the tetrahedron) with periods of the order of $\hbar v/\Delta_0 \sim \xi(0)$. The possibility is not excluded that in such a case $\Delta(\boldsymbol{r}) = 0$ at certain points, i.e., the superconductivity will appear to be gapless. This will lead to a non-exponential temperature dependence of the heat capacity. For instance, $C \propto \ln^{-3}(\Delta/T)$ for $\Delta(\boldsymbol{r}) \propto \cos(\boldsymbol{qr}/\hbar)$ (Larkin and Ovchinnikov 1964).

Since under actual conditions it is difficult to avoid the generation of structural defects, the question arises as to the stability of the inhomogeneous superconducting phase towards defects. Analysis shows (Aslamazov and Larkin 1968) that, in principle, the LOFF phase persists in the presence of any defect concentration, i.e., the free energy decreases with the appearance of $\Delta(\boldsymbol{r})$. However, the corresponding value of I_c decreases with defect concentration, and at a concentration corresponding to the mean free path $l < \xi(0)$, I_c becomes less than the value determined by formula (21.22). Here the equilibrium phase will be the homogeneous superconducting phase with $\Delta = \Delta_0$; it is formed by way of a transition of first order in I. The quantity I_c in this case plays the role of the supercooling limit for the normal phase. When I_c further decreases, the inhomogeneous superconductivity of the LOFF phase may appear as a metastable phase.

The LOFF phase has so far not been observed experimentally and it would be rather difficult to create the necessary conditions (see section 21.3).

21.3. Ferromagnetic superconductors

Let us now consider another possibility, namely the case $\theta < T_c$ (θ is the Curie temperature), where ferromagnetism appears against the background of superconductivity (see the review article by Bulaevskii et al. 1985). This situation occurs, for example, in the compounds $HoMo_6S_8$ and $ErRh_4B_4$. In compounds of this type, as the temperature decreases a superconducting transition occurs first (at temperature T_{c1}).

As the temperature decreases further, a magnetic transition arises against the superconductivity background, which leads to a very peculiar lamellar or domain magnetic structure, in which the magnetization in neighboring layers has an opposite direction (temperature T_M). Since the superconducting current screens off the magnetic field, this magnetic structure can be recognized only with the aid of neutron diffraction. Neutrons, which have a magnetic moment, "feel" the magnetic structure and are diffracted on it, just as light on a diffraction grid or X-radiation in a crystal.

Finally, when the temperature continues to decrease, a first-order phase transition to a nonsuperconducting ferromagnetic state (temperature T_{c2}) occurs. The values of these temperatures are $T_{c1} = 1.8$ K, $T_M = 0.7$ K, $T_{c2} = 0.65$ K for $HoMo_6S_8$; $T_{c1} = 8.7$ K, $T_M = 0.8$ K, $T_{c2} = 0.75$ K for $ErRh_4B_4$. In order to understand the sequence of these transitions, let us examine, first of all, the mechanism of an indirect exchange interaction of spins via conduction electrons.

We begin with the Hamiltonian of the interaction of spins with conduction electrons (eq. 4.39), which is written in terms of the operators a and a^+:

$$\mathscr{H}_{eS} = -\frac{J}{n} \sum_{\substack{p,p' \\ \sigma,\sigma'}} a^+_{p',\sigma'}, \sigma_{\sigma',\sigma} a_{p,\sigma} \exp\left[\frac{i}{\hbar}(\boldsymbol{p}-\boldsymbol{p}')\boldsymbol{r}_i\right] \boldsymbol{S}_i, \tag{21.35}$$

where $\boldsymbol{\sigma}$ are the Pauli matrices; $\boldsymbol{S}_i$ are the spin operators of magnetic atoms*; σ, σ' are the projections of the electron spin; $\boldsymbol{r}_i$ are the positions of the atoms.

Expression (21.35) describes the interaction of an electron with spins, in which the electron's momentum is $\boldsymbol{p}$ in the initial state and $\boldsymbol{p}'$ in the final state, σ and σ' being the initial and final spin projections, respectively.

As for the factor that depends on $\boldsymbol{r}_i$, the replacement of the sum over $\boldsymbol{r}_i$ by an integral over the volume would have led to the appearance of the Fourier component of the function $\boldsymbol{S}(\boldsymbol{r})$. In this case, the resultant expression obtained would be analogous to the Hamiltonian (16.60), which describes the interaction of electrons with an electromagnetic field. It may be treated as the change of the momentum and spin of the electron upon absorption of a quantum of the external field. The presence of the sum over $\boldsymbol{r}_i$ in (21.35) takes into account the discreteness in the position of the spins.

Let us first consider a nonsuperconducting metal. Taking the second order of perturbation theory and separating in it the terms with $\boldsymbol{r}_i \neq \boldsymbol{r}_k$, which describe the interaction of the spins, we obtain

$$E_{\mathrm{SS}} = \sum_{m \neq 0} \frac{|(\mathcal{H}_{\mathrm{eS}})_{m0}|^2}{E_0 - E_m}$$

$$= \left(\frac{J}{n}\right)^2 \sum_{\substack{p,p';\sigma,\sigma' \\ i \neq k}} \frac{n_p(1-n_{p'})}{\xi_p - \xi_{p'}}$$

$$\times \exp\left(\frac{\mathrm{i}}{\hbar}(\boldsymbol{p}-\boldsymbol{p}')(\boldsymbol{r}_i - \boldsymbol{r}_k)\right) \sigma^l_{\sigma',\sigma} \sigma^m_{\sigma,\sigma'} S^l_i S^m_k$$

$$= 2\left(\frac{J}{n}\right)^2 \sum_{\substack{p,p' \\ i \neq k}} \frac{n_p(1-n_{p'})}{\xi_p - \xi_{p'}}$$

$$\times \exp\left(\frac{\mathrm{i}}{\hbar}(\boldsymbol{p}-\boldsymbol{p}')\boldsymbol{r}_{ik}\right) \boldsymbol{S}_i \boldsymbol{S}_k,$$

where $\boldsymbol{r}_{ik} = \boldsymbol{r}_i - \boldsymbol{r}_k$ and the factor $n_p(1-n_{p'})$ takes account of the presence of electrons in the initial state and their absence in the final state. We introduce new variables $\boldsymbol{p} \to \boldsymbol{p} + \frac{1}{2}\boldsymbol{q}$, $\boldsymbol{p}' = \boldsymbol{p} - \frac{1}{2}\boldsymbol{q}$.

We take into account the fact that finite temperature leads only to corrections of order $(T/\mu)^2$ and we therefore set $T = 0$. Then $\xi_p < 0$ and $\xi_{p'} > 0$. The expression obtained may be written in the form

$$E_{\mathrm{SS}} = -2\left(\frac{J}{n}\right)^2 \sum_{i \neq k} \int \frac{\mathrm{d}^3\boldsymbol{q}}{(2\pi\hbar)^3} F(q) \exp\left(\frac{\mathrm{i}\boldsymbol{q}\boldsymbol{r}_{ik}}{\hbar}\right) \boldsymbol{S}_i \boldsymbol{S}_k, \tag{21.36}$$

* Here we do not consider the Kondo effect and therefore we assume that $\boldsymbol{S}_i$ are classical vectors.

where

$$F(q)=\int \frac{d^3p}{(2\pi\hbar)^3}\,\frac{1}{\xi'-\xi}\,\theta(\xi')\theta(-\xi), \tag{21.37}$$

and $\theta(x)=1$ at $x>0$ and $\theta(x)=0$ at $x<0$, $\xi=\xi(\boldsymbol{p}+\frac{1}{2}\boldsymbol{q})=(2m)^{-1}[(\boldsymbol{p}+\frac{1}{2}\boldsymbol{q})^2-p_0^2]$, $\xi'=\xi(\boldsymbol{p}-\frac{1}{2}\boldsymbol{q})$. Integration gives

$$F(q)=\tfrac{1}{4}\nu f(q/2p_0), \tag{21.38}$$

where $\nu=p_0m/(\pi^2\hbar^3)$ is the density of states and

$$f(x)=\tfrac{1}{2}\left(1+\frac{1-x^2}{2x}\ln\left|\frac{1+x}{1-x}\right|\right) \tag{21.39}$$

is the so-called Lindhard function.

Substitution of eqs. (21.38) and (21.39) into (21.36) and integration over $\boldsymbol{q}$ leads to

$$\begin{aligned} E_{SS}=&\left(\frac{J}{n}\right)^2\nu\frac{p_0^3}{\pi\hbar^3}\\ &\times\sum_{i\neq k}\left[\cos\left(\frac{2p_0r_{ik}}{\hbar}\right)\left(\frac{2p_0r_{ik}}{\hbar}\right)^{-3}\right.\\ &\left.-\sin\left(\frac{2p_0r_{ik}}{\hbar}\right)\left(\frac{2p_0r_{ik}}{\hbar}\right)^{-4}\right]\boldsymbol{S}_i\boldsymbol{S}_k, \end{aligned} \tag{21.40}$$

which is called the RKKY interaction after the authors (Ruderman and Kittel 1954, Kasuya 1956 and Yosida 1957).

Suppose that the interaction of the spins (eq. 21.40), leads to the formation of a helicoidal magnetic structure, for which $\boldsymbol{S}=\boldsymbol{S}(\boldsymbol{r})$, with

$$\begin{aligned} S_x&=S\cos(qz/\hbar),\\ S_y&=S\sin(qz/\hbar),\\ S_z&=0. \end{aligned} \tag{21.41}$$

This means that the end of the vector $\boldsymbol{S}$ describes in space a spiral line around the z axis. Upon substitution into (21.36) we pass from the summation over i, k to integration according to the formula $\sum_i \to n\int dV$, where n is the number of spins per unit volume. As a result, we have per unit volume

$$E_{SS}=-2J^2S^2F(q)=-\tfrac{1}{2}J^2\nu S^2 f(q/2p_0). \tag{21.42}$$

A maximum of $f(q)$ corresponds to the lowest energy. The Lindhard function (21.39) is equal to unity at $q=0$ and decreases monotonically with increasing x [the asymptotic with $x\to\infty$ is $f(x)\approx 2/(3x^2)$]. Hence, for the energy minimum we have $q=0$, i.e. ferromagnetic ordering.

This conclusion is not general, as a matter of fact. If we take into account that the spins occupy certain positions in space, we cannot pass from the summation over $\boldsymbol{r}_i$ to the integration over the volume. For certain lattices, other types of spin ordering are more favorable, namely antiferromagnetic or helicoidal ordering with $\boldsymbol{q} \neq 0$. But here we confine ourselves to the simplest case.

How does the situation change in a superconductor? In order to find the answer to this question, we must express in the Hamiltonian (21.35) the operators a and a^+ in terms of α and α^+ in accordance with the Bogoliubov formulas (16.14). We assume from the start that spin ordering is helocoidal and is described by formulas (21.41). In this case, we obtain

$$\mathscr{H}_{eS} = -\frac{J}{n} \sum_{\boldsymbol{p},\boldsymbol{p}',i} (S_i^x - \mathrm{i}S_i^y)(u_{p'}\alpha^+_{\boldsymbol{p}',+} + v_{p'}\alpha_{-\boldsymbol{p}',-})(u_p\alpha_{\boldsymbol{p},-} - v_p\alpha^+_{-\boldsymbol{p},+})$$
$$+ (S_i^x + \mathrm{i}S_i^y)(u_{p'}\alpha^+_{\boldsymbol{p}',-} - v_{p'}\alpha_{-\boldsymbol{p}',+})(u_p\alpha_{\boldsymbol{p},+} + v_p\alpha^+_{-\boldsymbol{p},-}).$$

Let us find, as before, the second order of perturbation theory and retain the terms with $i \neq k$. We assume that $T = 0$. This means that all the numbers $n_{p,+}$ and $n_{p,-}$ for quasiparticles are taken to be equal to zero. Of course, strictly speaking, this is justifiable only for the limiting case, when the temperature of magnetic ordering T_M is much lower than T_{c1}. As a result, we obtain

$$E_{SS} = -\sum_{\substack{\boldsymbol{p},\boldsymbol{p}' \\ i \neq k}} \frac{(u_p v_{p'} - u_{p'} v_p)^2}{\varepsilon + \varepsilon'} \exp\left[\frac{\mathrm{i}}{\hbar}(\boldsymbol{p} - \boldsymbol{p}')\boldsymbol{r}_{ik}\right] \boldsymbol{S}_i \boldsymbol{S}_k .$$

Substituting formulas (16.19) for u and v yields

$$E_{SS} = -\tfrac{1}{2}\left(\frac{J}{n}\right)^2 \sum_{\boldsymbol{p},\boldsymbol{p}';i \neq k} \left(1 - \frac{\xi\xi' + \Delta^2}{\varepsilon\varepsilon'}\right)$$
$$\times \frac{1}{\varepsilon + \varepsilon'} \exp\left[\frac{\mathrm{i}}{\hbar}(\boldsymbol{p} - \boldsymbol{p}')\boldsymbol{r}_{ik}\right] \boldsymbol{S}_i \boldsymbol{S}_k . \tag{21.43}$$

We introduce the same variables as before. Then, instead of formula (21.36), we obtain a formula of the same type, which however, will contain $F_1(q)$:

$$F_1(q) = \tfrac{1}{4} \int \frac{\mathrm{d}^3 p}{(2\pi\hbar)^3} \left(1 - \frac{\xi\xi' + \Delta^2}{\varepsilon\varepsilon'}\right) \frac{1}{\varepsilon + \varepsilon'} . \tag{21.44}$$

First of all, note that $F_1(q) = 0$ at $\boldsymbol{q} = 0$. In fact, here $\xi = \xi'$ and $\xi^2 + \Delta^2 = \varepsilon^2$. The physical explanation of this circumstance consists in the following. An indirect exchange interaction between the spins arises from the fact that the spin $\boldsymbol{S}_i$ at point $\boldsymbol{r}_i$ polarizes the surrounding electrons and they in turn polarize the other spin $\boldsymbol{S}_k$. But if we are dealing with a superconductor, the electrons in the ground state are not polarized, and the excited state is separated from the ground state by an energy gap. The static spin cannot change the energy of the electronic system and, hence, cannot polarize it. From this it follows that while ferromagnetism hinders superconductivity, the latter in turn impedes the generation of ferromagnetism.

In what follows, we need, as before, the maximum of function $F_1(q)$. It will be seen that this maximum corresponds to $\Delta/v \sim \hbar/\xi \ll q \ll 2p_0$. Therefore, instead of a detailed calculation of function F_1, we can find its asymptotics, assuming that $\Delta \ll vq \ll 2p_0 v$. To this end, we add and subtract in eq. (21.44) the corresponding expression with $\Delta = 0$. The difference converges at $p \sim \Delta/v$, $\boldsymbol{vq} \sim \Delta$. In view of this, we may differentiate the difference with respect to Δ and then, introducing the variables ξ and ξ' in place of p and $\theta = (\boldsymbol{p}, \boldsymbol{q})$, we assume the limits of integration over these variables to be infinite. As a result, we find

$$\frac{\partial}{\partial \Delta^2}[F_1(q) - F(q)] = \frac{\nu}{4vq}\int_0^\infty \mathrm{d}\xi \int_0^\infty \mathrm{d}\xi' \times \frac{-3 + \dfrac{1}{\varepsilon^2} + \dfrac{1}{\varepsilon'^2} + \dfrac{1}{\varepsilon\varepsilon'}}{2\varepsilon\varepsilon'(\varepsilon+\varepsilon')}$$

$$= \frac{\nu}{8vq\Delta}\int_0^\infty \mathrm{d}\varphi_1 \int_0^\infty \mathrm{d}\varphi_2 \times \frac{-3 + \dfrac{1}{\cosh^2\varphi_1} + \dfrac{1}{\cosh^2\varphi_2} + \dfrac{1}{\cosh\varphi_1 \cosh\varphi_2}}{\cosh\varphi_1 + \cosh\varphi_2}$$

where we have introduced the variables $\xi = \Delta \sinh\varphi_1$, $\xi' = \Delta \sinh\varphi_2$. The resultant dimensionless integral over φ_1 and φ_2 is found to be equal to $-\frac{1}{2}\pi^2$.

Integrating over Δ, we obtain

$$F_1(q) - F(q) = -\frac{\nu\pi^2\Delta}{8vq} = -\frac{\nu\pi\hbar}{8\xi_0 q},$$

where we have introduced the notation $\xi_0 = \hbar v/\pi\Delta$, which has already been used before. The remaining part equal to $F(q)$ may be expanded in a series in q because of the smallness of q. Using expressions (21.38) and (21.39), we have

$$F_1(q) \approx \frac{\nu}{4}\left(1 - \frac{q^2}{12p_0^2} - \frac{\pi\hbar}{2\xi_0 q}\right). \tag{21.45}$$

In order to avoid confusion, we should remark that this formula cannot be applied for both too small and too large values of q.

Analogously to eq. (21.42), the energy per unit volume is given by

$$E_{\mathrm{SS}} = -2J^2S^2F_1(q). \tag{21.46}$$

The maximum of function $F_1(q)$ is attained at

$$q = \left(\frac{2\pi\hbar p_0^2}{\xi_0}\right)^{1/3}. \tag{21.47}$$

This means that magnetic ordering has a helicoidal structure with a period given by

$$d = \frac{2\pi\hbar}{q} = \frac{2}{3^{1/2}}\left(\frac{\pi\hbar}{p_0}\right)^{2/3} \xi_0^{1/3} \sim a^{2/3}\xi_0^{1/3}, \tag{21.48}$$

where $a \sim \hbar/p_0$ is of the order of atomic dimensions (Anderson and Suhl 1959).

The resultant period is large compared to atomic dimensions but it is small compared to ξ_0. Since the events that occur at distances of order ξ_0 are important in superconductors, it may be said that magnetization changes so rapidly that it averages out at distances of order ξ_0 and therefore may exist together with superconductivity.

In fact, apart from the exchange interaction of the spins, there also occurs a spin–orbit interaction, which is very strong in rare-earth elements because of their large atomic number. In contrast to the exchange interaction, this interaction depends on the spin orientation relative to the crystallographic axes. This gives rise to a so-called anisotropy field. The latter selects, among the spin directions, the so-called easy directions of magnetization. The simplest case is a uniaxial crystal with an easy direction of magnetization along the axis. In such a crystal, the orientations of the spin along the axis in the forward and reverse directions are energetically the most favorable. The presence of such an anisotropy field "remodels" the helicoidal magnetic structure into a domain structure composed of planar layers (domains) with opposite magnetization of the successive layers.

Since the magnetization in each domain is homogeneous, it follows that in formula (21.45), in place of the second term, we must write the energy of the interdomain boundary. In each such wall there occurs a gradual spin flip from one easy direction of magnetization to another. This can happen by gradual rotation of the magnetic moment with conservation of its magnitude as well as by a gradual change of its magnitude with the direction staying parallel to the easy magnetization axis. In the substances under consideration a change of direction of the moment is unfavourable due to the large anisotropy energy; therefore walls of the second type actually occur (they are called "linear walls"). Since the exchange energy is minimal if the moment does not vary in space, the wall has an additional exchange energy $J^2S^2\nu(a/\delta_w)^2\delta_w$, where δ_w is the thickness of the wall and a is the interatomic distance [this energy corresponds to the term $\int c(\nabla\varphi)^2\,dV$ in formula (AII.15)]. On the other hand the value of the order parameter in the wall is less than the bulk value and this contributes an additional energy of the order of $J^2S^2|\tau|\nu\delta_w$, where $\tau = (T - T_m)/T_m$ [this corresponds to $\int \alpha\tau\varphi^2\,dV$ in (AII.15)]. Minimizing the total energy with respect to δ_w,

$$E_w = J^2S^2\nu\left(\frac{a^2}{\delta_w} + \delta_w|\tau|\right)$$

we get

$$\delta_w \sim \frac{a}{|\tau|^{1/2}}, \qquad E_w \sim J^2S^2\nu a|\tau|^{1/2}$$

Strictly speaking this is true for $|\tau| \ll 1$, but in fact it gives a correct estimate even for $|\tau| \sim 1$.

If the period of the domain structure is d, then, on the average, we have $E_w/d \sim J^2S^2\nu a/d$ per unit volume. Moreover, we have to add the exchange energy that arises from the superconductivity, represented by the third term in eq. (21.45). Taking into account that $q \sim \hbar/d$, we obtain the total extra energy dependent on d:

$$\Delta E(d) \sim J^2S^2\nu\left(\frac{a}{d}+\frac{d}{\xi_0}\right).$$

Minimizing with respect to d, we find

$$d \sim (a\xi_0)^{1/2}, \tag{21.49}$$

$$\Delta E(d) \sim J^2S^2\nu\left(\frac{a}{\xi_0}\right)^{1/2}. \tag{21.50}$$

Again, the period of the structure is found to be smaller than ξ_0, and therefore the magnetization has a slight effect on superconductivity.

In order to solve the problem of the transition to a normal ferromagnetic material, let us compare the corresponding energies at $T=0$. From formula (21.42) at $q=0$ we obtain $E_n=\frac{1}{2}J^2S^2\nu$. The excess energy associated with the domain walls is expressed by formula (21.50). However, the onset of superconductivity gives rise to an extra negative energy given by formula (16.41). If

$$J^2S^2\nu\left(\frac{a}{\xi_0}\right)^{1/2} < \nu\Delta^2(0) \tag{21.51}$$

the superconductivity with the domain magnetic structure will persist up to $T=0$. This is what happens in $HoMo_6Se_8$. If the opposite inequality is fulfilled, then at some finite temperature T_{c2} a first-order phase transition takes place from the domain superconducting structure to a normal ferromagnetic substance. This is observed in $HoMo_6S_8$ and $ErRh_4B_4$. The criterion (21.51) may be rewritten by taking advantage of the fact that

$$\Delta(0) \sim T_{c1}, \qquad \xi_0 \sim \frac{\hbar v}{T_{c1}}, \qquad a \sim \frac{\hbar}{p_0}, \qquad p_0 v \sim \varepsilon_F, \qquad \frac{J^2S^2}{\varepsilon_F} \sim \theta$$

(θ is the Curie point) (see below):

$$\theta\frac{\varepsilon_F^{1/2}}{T_{c1}^{3/2}} < 1. \tag{21.51$'$}$$

In superconductors that pass to the normal ferromagnetic phase below T_{c2} the following interesting phenomenon can be observed. If a current is passed through such a superconductor at a temperature below T_{c2}, then due to finite resistance it will heat the sample and the temperature will become higher than T_{c2}. But in this case the heating will stop and the temperature will fall below T_{c2} and the process will repeat itself. This may result in the following: alternating domains of the normal and the superconducting phase will move throughout the sample with the current; such domains are called autowaves (Buzdin and Mikhailov 1986).

Let us now consider the lower boundary of the domain phase. Since the energy of domain walls is only a small correction to the main exchange energy of the spin–spin interaction, the Curie temperature θ will differ little from the value for a normal ferromagnetic material. To find the latter value, we employ the self-consistent field method (Abrikosov and Gor'kov 1962b). Suppose a ferromagnetic state has been formed and, as a result, the electron spin has been polarized, i.e., $\langle\sigma\rangle$, where $\langle\cdots\rangle$ signifies the equilibrium average at a given temperature referred to unit volume. In accordance with the Hamiltonian (4.39), this is equivalent to the action of an effective magnetic field on each spin of a rare-earth ion, in which case the role of $g\beta H$ is played by $(J/n)\langle\sigma\rangle$ ($g\beta S$ is the magnetic moment of the ion, g is the gyromagnetic ratio and β is the Bohr magneton). The free energy of the entire system of ionic spins per unit volume is ($n_{\mathrm{m}}=n$):

$$F_{\mathrm{i}}=-nT\ln\sum_{M=-S}^{S}\exp\left(\frac{J}{n}\langle\sigma\rangle\frac{M}{T}\right)$$

$$=-nT\left[\ln\sinh\left(\frac{J}{n}\langle\sigma\rangle\frac{S+\frac{1}{2}}{T}\right)-\ln\sinh\left(\frac{J}{n}\frac{\langle\sigma\rangle}{2T}\right)\right]. \tag{21.52}$$

On the other hand, the polarization of electron spins implies their redistribution among the energy levels, i.e., the generation of different Fermi surfaces for $\sigma=+1$ and -1, which is also equivalent to the action of a certain magnetic field. According to section 10.1, under the action of the magnetic field H a paramagnetic moment arises in the electronic system ($s=\frac{1}{2}\sigma$):

$$M=\beta\langle\sigma\rangle=\beta^2\nu H=\chi H,$$

where χ is the magnetic susceptibility. On the other hand, the change in free energy for a given moment M is (see eq. 10.2):

$$F_{\mathrm{e}}=\int H\,\mathrm{d}M=\tfrac{1}{2}\chi H^2=\beta^2\frac{\langle\sigma\rangle^2}{2\chi}=\frac{\langle\sigma\rangle^2}{2\nu}. \tag{21.53}$$

Let us find the minimum of $F_{\mathrm{e}}+F_{\mathrm{i}}$. Differentiating with respect to $\langle\sigma\rangle$, we get

$$\frac{\langle\sigma\rangle}{\nu}-JSB_{S}\left(\frac{J}{n}\frac{\langle\sigma\rangle}{T}\right)=0, \tag{21.54}$$

where $\mathrm{B}_S(x)$ is the Brillouin function defined by

$$\mathrm{B}_S(x)=\left(\frac{S+\frac{1}{2}}{S}\right)\coth[(S+\tfrac{1}{2})x]-\frac{1}{2S}\coth(\tfrac{1}{2}x). \tag{21.55}$$

In order to find the Curie temperature θ, we assume that $\langle\sigma\rangle \to 0$. Taking into account that at $x \ll 1$

$$B_S(x) \approx \tfrac{1}{3}(S+1)x$$

we obtain from (21.52)

$$\theta = J^2 \nu S \frac{(S+1)}{3n}. \qquad (21.56)$$

Taking into account that $\nu \sim n_e/\varepsilon_F$ and $n_e \sim n$, we find

$$\theta \sim \frac{J^2 S^2}{\varepsilon_F}. \qquad (21.56')$$

Two situations are possible. If ε_F is of the same order as in ordinary metals, then at $\theta \sim 1$ K we have $JS \sim 10^2$ K, i.e. $JS \gg \Delta(0)$ (cf., with 21.22, $JS = I$). In this case, the coexistence of superconductivity and ferromagnetism is possible only at $\theta < T_{c1}$, and the change of the phases with decreasing temperature will occur in the sequence described in the present section. But if $\theta > T_c$, then superconductivity will not appear at all.

However, there is still another possibility, in principle. As has already been said in section 16.10, a sharp peak may appear in the density of states at the Fermi boundary; this peak, known as the Abrikosov-Suhl resonance, manifests itself in various properties at low temperatures. It may be interpreted as an increase in the effective mass - "heavy fermions". Evidently, in such a case in formula (21.56) ν will be anomalously high and, hence, JS may be found to be of order Δ. In this case, superconductivity may arise at $\theta \gtrsim T_c$ and also the inhomogeneous superconducting LOFF phase described in the preceding section may appear. Actually, for this to occur it is necessary that $m^* > 10^3 m$, θ, $T_c > 10$ K, which has never been observed.

The exchange interaction of electrons with ions (4.39) in the case of a large negative J may give rise to the unusual effect of reentrant superconductivity (Jaccarino and Peter 1962). We are speaking here of antiferromagnetic superconductors or of ferromagnetic superconductors at a temperature above the Curie point. Under the action of an external field an average polarization of the nuclear spin $\langle \boldsymbol{S} \rangle$ appears. Because of this, the electron spins will be acted on by an effective field given by

$$\boldsymbol{I} = \frac{J}{n}\langle \boldsymbol{S} \rangle + \beta \boldsymbol{H}. \qquad (21.57)$$

If J is sufficiently large in absolute magnitude, then the first term in this formula is larger than the second. Under the influence of the field the superconductivity is destroyed by a paramagnetic mechanism. As H increases further the first term tends towards saturation because $|\langle \boldsymbol{S} \rangle|$ cannot exceed nS, where S is the spin of an individual ion. If $J < 0$, this may lead to a decrease in the magnitude of I and to the onset of superconductivity. In still larger fields the superconductivity is completely destroyed, due mainly to the second term in eq. (21.57) with the aid of the paramagnetic mechanism or due to the orbital effect of pair rotation. An example is the compound $Eu_{0.75}Sn_{0.25}Mo_6S_{7.2}Se_{0.8}$.

More detailed information about ferromagnetic and antiferromagnetic superconductors can be found in the literature (Bulaevskii et al. 1985, Buzdin and Bulaevskii 1986).

21.4. The Knight shift

One of the highly sensitive methods of determination of internal magnetic fields in a solid, which is also frequently used to determine structural details is the nuclear magnetic resonance (NMR) method. If the nucleus has a spin, it also has a magnetic moment.

$$\mu_n = \gamma \hbar S_n . \tag{21.58}$$

Nuclear magnetic moments are thousands of times lower than the electronic moments:

$$\gamma \hbar \sim \frac{e\hbar}{M_p c} \sim \frac{\beta m}{M_p},$$

where m is the mass of the electron and M_p is the mass of the proton. If the nuclear moment is placed in a magnetic field, its energy has the usual form

$$\varepsilon = -\boldsymbol{\mu}_n \boldsymbol{H} = -\gamma \hbar M H, \tag{21.59}$$

where M is the magnetic quantum number (the projection of the nuclear spin onto H). This system with equidistant levels is capable of absorbing electromagnetic quanta with an energy $\gamma \hbar H$. Considering that $\beta \sim 10^{-20}$ and $\gamma \hbar \sim 10^{-23}$, we obtain $\omega \sim \gamma H$, i.e., $\omega \sim 10^8\ s^{-1}$ in a field of the order of 10^4 Oe. This makes nuclear magnetic resonance a convenient tool for the various investigations.

The nuclear spin interacts with electrons when the electrons create a magnetic field. In a nonmagnetic metal in the absence of a magnetic field the electronic system has no magnetic moment and the nuclear spin does not interact with the electrons. If an external magnetic field is turned on, it is capable not only of directly acting on the nuclear spin but, through the orientation of the electron spin, it can also produce a nonzero electronic field in the region of the nucleus. However, in practice only the field created by the spin of s-electrons is important because only these electrons have a finite wave function at the point of location of the nucleus (according to quantum mechanics, $\psi(r) \propto r^l$, where l is the orbital moment).

In metals, the electron spins are polarized under the action of a magnetic field. Since among the valence electrons there are always electrons for which the wave function in the vicinity of the nucleus is close to the atomic s-function, their polarization generates an extra field, which acts on the nuclear spin. This leads to a shift of the NMR frequency; this shift is known as the Knight shift (Knight 1949).

The magnitude of the Knight shift can be calculated as follows. The Hamiltonian of the magnetic dipolar interaction of electrons with a nucleus may be written in the form

$$\mathcal{H}_{\text{en}} = -2\beta\left[3(\boldsymbol{s}\boldsymbol{r})\frac{\boldsymbol{\mu}_{\text{n}}\boldsymbol{r}}{r^5} - \frac{\boldsymbol{s}\boldsymbol{\mu}_{\text{n}}}{r^3}\right] = -2\beta\boldsymbol{s}\,\text{rot}\,\boldsymbol{A}, \tag{21.60}$$

where $\boldsymbol{A}$ is the vector potential of the field created by the spin:

$$\boldsymbol{A} = \frac{[\boldsymbol{\mu}_{\text{n}}\boldsymbol{r}]}{r^3} = \text{rot}\left(\frac{\boldsymbol{\mu}_{\text{n}}}{r}\right). \tag{21.61}$$

Substituting into (21.60), we obtain

$$\begin{aligned}\mathcal{H}_{\text{en}} &= -2\beta\boldsymbol{s}\,\text{rot}\,\text{rot}\left(\frac{\boldsymbol{\mu}_{\text{n}}}{r}\right) = -2\beta\boldsymbol{s}(\text{grad div} - \nabla^2)\frac{\boldsymbol{\mu}_{\text{n}}}{r} \\ &= -2\beta[(\boldsymbol{s}\boldsymbol{\nabla})(\boldsymbol{\mu}_{\text{n}}\boldsymbol{\nabla}) - (\boldsymbol{s}\boldsymbol{\mu}_{\text{n}})\nabla^2]\frac{1}{r} \\ &= -2\beta[(\boldsymbol{s}\boldsymbol{\nabla})(\boldsymbol{\mu}_{\text{n}}\boldsymbol{\nabla}) - \tfrac{1}{3}(\boldsymbol{s}\boldsymbol{\mu}_{\text{n}})\boldsymbol{\nabla}^2]\frac{1}{r} + \tfrac{4}{3}\beta(\boldsymbol{s}\boldsymbol{\mu}_{\text{n}})\boldsymbol{\nabla}^2\frac{1}{r}.\end{aligned} \tag{21.62}$$

The Hamiltonian obtained must be averaged over the electronic wave function. Here the first term in (21.62) gives a nonzero result only with wave functions that behave in the vicinity of the nucleus as atomic states with $l \neq 0$; therefore, it makes a small contribution. The second term, by virtue of Coulomb's law

$$\nabla^2\frac{1}{r} = -4\pi\delta(r)$$

is equal to

$$\mathcal{H}_{\text{en}} = -\tfrac{16}{3}\pi\beta\boldsymbol{s}\boldsymbol{\mu}_{\text{n}}\delta(\boldsymbol{r}).$$

On averaging this gives*

$$E_{\text{en}} = -\tfrac{16}{3}\pi\beta\langle\boldsymbol{s}\rangle\boldsymbol{\mu}_{\text{n}}\frac{|\psi_{\text{s}}(0)|^2}{n}, \tag{21.63}$$

where n is the density of atoms.

In a magnetic field the average electron spin of unit volume is equal to

$$\langle\boldsymbol{s}\rangle = \frac{\chi H}{2\beta},$$

* In order to derive formula (21.63), use may be made of the electron wave function written in terms of the Wannier function (1.22) and account may be taken of the fact that $w_n(r) \approx \psi_s(\boldsymbol{r} - \boldsymbol{a}_n)$ falls off rapidly as $\boldsymbol{r}$ moves away from $\boldsymbol{a}_n$.

from which we have

$$E_{en} = -\tfrac{8}{3}\pi\chi \frac{|\psi_s(0)|^2}{n} (\boldsymbol{\mu}_n \boldsymbol{H}).$$

Thus, the nuclear spin is acted on by an extra magnetic field:

$$\Delta H = \tfrac{8}{3}\pi\chi \frac{|\psi_s(0)|^2}{n} H. \tag{21.64}$$

Experimental determination of the NMR frequency and, hence, of the Knight shift, makes it possible to measure the spin magnetic susceptibility χ directly. Of course, for a normal metal this method becomes less valuable because the magnitude of $|\psi_s(0)|^2$, the probability density of finding electrons at the nucleus, cannot be calculated exactly. But if the metal passes to the superconducting state, the ratio $\chi_s(T)/\chi_n$ can be studied.

Of the greatest interest is $\chi_s(0)$. At first sight it must be equal to zero. Indeed, in the ground state the electrons are bound into Cooper pairs with zero spin and the excited states are separated from the ground state by an energy gap. Hence, a small magnetic field is not capable of generating magnetization, i.e., $\chi_s(0) = 0$. However, in practice $\chi_s(0)$ is always found not only to be finite but often even to be close to χ_n (recall that at low temperatures χ_n is practically independent of temperature).

Attempts have been made to explain this paradox by the assumption of triplet pairing, i.e., Cooper pairs with parallel spins and $l = 1, 3 \ldots$ (the oddness of l is the result of the symmetry of the wave function of Fermi particles: if it is symmetric with respect to spin transposition, then it must be antisymmetric with respect to the coordinates), but this idea has not been confirmed.

In order to understand the origin of finite $\chi_s(0)$, we must take into account that NMR experiments with a superconductor must be carried out on small particles (or very thin films) with a size less than the penetration depth. Otherwise, the magnetic field will not penetrate the superconductor or will be strongly inhomogeneous, which will not make it possible to observe the resonance. In small particles, even if they are single-crystalline, the electrons are inevitably scattered from the boundaries, i.e., we certainly have $l \ll \delta$. It can be shown that ordinary potential scattering does not give rise to $\chi_s(0)$.

However, apart from this there necessarily is spin-orbit scattering, which has already been considered in section 11.4. The Hamiltonian of the spin-orbital part of the electron-impurity interaction is proportional to $(mc)^{-1}[-i\hbar\nabla V, \boldsymbol{p}]\boldsymbol{s}$, where V is the potential of the impurity*; this Hamiltonian gives a correction to the scattering amplitude proportional to $[\boldsymbol{p}\boldsymbol{p}']\boldsymbol{s}$, where $\boldsymbol{p}$ and $\boldsymbol{p}'$ are the electron momentum before and after scattering. Since the correction in the Hamiltonian is symmetric with respect to a change in sign of time (with $t \to -t$, $\boldsymbol{p} \rightleftarrows -\boldsymbol{p}'$ and $\boldsymbol{s} \to -\boldsymbol{s}$), it does not

* The explanation of the physical origin of the spin-orbit interaction and its estimate are given at the beginning of section 10.7.

lead to a noticeable change in the thermodynamic quantities of superconductors (section 21.1).

However, in the presence of this interaction the electron spin is not conserved. Hence, the states cannot be classified in accordance with the total spin of the electronic system $\boldsymbol{S}$, and in the ground state of the superconductor there is an admixture of states with $\boldsymbol{S} \neq 0$. In view of this, polarization in a weak field is also possible.

The results of calculation of this effect (Abrikosov and Gor'kov 1960b, 1962a) are as follows. The ratio χ_s/χ_n depends on temperature and on the parameter l_{s0}/ξ_0, where l_{s0} is the spin-orbital mean free path associated with the corresponding contribution to the scattering amplitude:

$$f(\boldsymbol{p}, \boldsymbol{p}') = a + \frac{\mathrm{i}b}{p_0^2}([\boldsymbol{p}\boldsymbol{p}']\sigma), \qquad l_{s0} = v\tau_{s0},$$
$$\frac{1}{\tau_{s0}} = n_i \nu \frac{\pi}{\hbar} \int |b|^2 \sin^2\theta \frac{\mathrm{d}\Omega}{4\pi}, \tag{21.65}$$

where n_i is the impurity concentration.

At $T = 0$ we obtain

$$\frac{\chi_s(0)}{\chi_n} \approx \tfrac{1}{6}\pi^2 \frac{\xi_0}{l_{s0}}, \qquad \xi_0 \ll l_{s0},$$
$$\frac{\chi_s(0)}{\chi_n} \approx 1 - \tfrac{3}{4}\frac{l_{s0}}{\xi_0}, \qquad \xi_0 \gg l_{s0} \tag{21.66}$$

(recall that $\xi_0 = \hbar v/(\pi\Delta(0))$. From formulae (21.63) it follows that with a weak spin-orbital scattering $\chi_s(0)$ is small and with a strong scattering $\chi_s(0)$ approaches unity. This result is in agreement with experiment.

22

Tunnel junctions. The Josephson effect

22.1. Single-particle tunnel current

The tunneling effect is one of the most familiar phenomena of quantum mechanics, since it demonstrates clearly the difference between quantum mechanics and classical mechanics. According to quantum mechanics, a particle can pass through a potential barrier even in those cases where the barrier height exceeds its energy. This phenomenon has already been considered in section 10.7 in connection with the discussion of magnetic breakdown.

The probability of passing through the barrier is primarily determined by an exponential factor:

$$W \sim \exp\left(-\frac{2}{\hbar}\int \mathrm{Im}(p_x)\,\mathrm{d}x\right)$$

$$= \exp\left\{-\frac{2(2m)^{1/2}}{\hbar}\int_{x_1}^{x_2}[U(x)-E]^{1/2}\,\mathrm{d}x\right\}. \qquad (22.1)$$

The integration is carried out over the region where $U(x) > E$ (see fig. 139). From the above formula it follows that the probability of passing through the barrier falls off with increasing barrier height $(U_{\max} - E)$ and width $(x_2 - x_1)$ and also with increasing mass of the particle.

We may actually speak of electrons and if the barrier is a layer of an insulator, then its thickness must not exceed a few interatomic distances. It is simpler to use a natural oxide layer which is formed at the surface of many metals: aluminum,

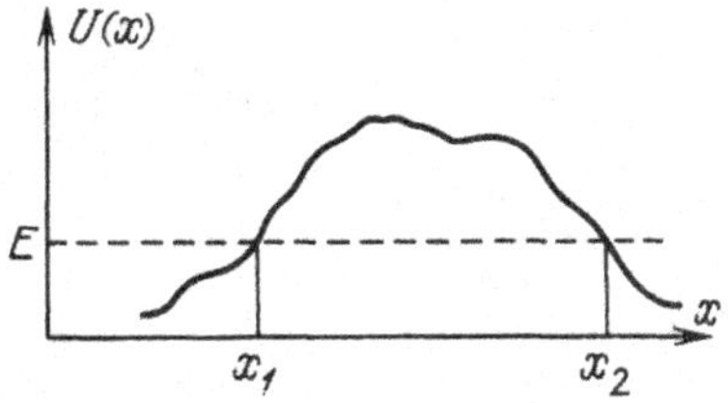

Fig. 139.

tin, lead, etc. Vaporization of a layer of the same or another metal gives rise to a so-called tunnel junction or contact (Giaever 1960a); this simple electronic device is an exceedingly valuable tool for the investigation of superconductivity.

We begin the study of the tunnel junction with two normal metals. At equilibrium the chemical potentials are equal (fig. 140a). If we apply a potential difference, this will fall entirely over the high-resistance dielectric layer. The chemical potentials of the two metals will then be at different heights (the level difference eV, fig. 140b). Assuming the barrier height to be much larger than eV, we find that the number of electrons that can be transferred to the free levels of another metal is proportional to eV and that all of them pass through the barrier with equal probability. From this it follows that the current will be proportional to V, and this is consistent with Ohm's law (fig. 141, curve 1).

Now let one of the metals be normal and the other a superconductor. We assume that $T=0$. At equilibrium in the absence of a potential difference the chemical potentials must be equal. But the electrons in the superconductor are bound into Cooper pairs, and these are in the Bose condensate. This energy level is just the chemical potential of the pair. From the side of the normal metal, this pair corresponds to two free quasiparticles of the gas model at the Fermi level. Suppose that an electron has been transferred from the normal metal into the superconductor. But, as a matter of fact, it does not enter into the composition of the pair. If the binding energy is 2Δ, then the energy of one unpaired electron exceeds by Δ the energy of an electron in a Cooper pair. Hence, in order to transport an electron from a normal metal into a superconductor, it must receive extra energy.

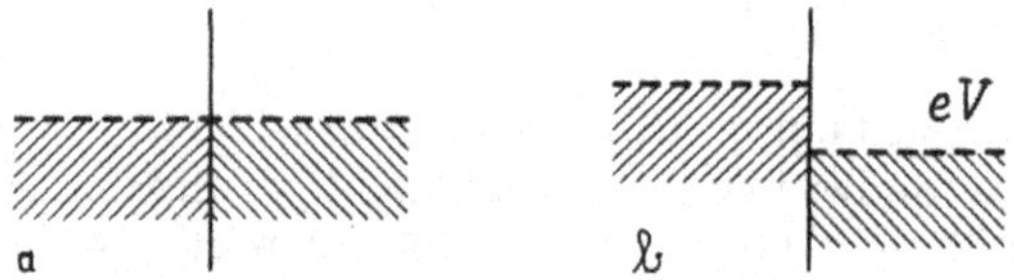

Fig. 140.

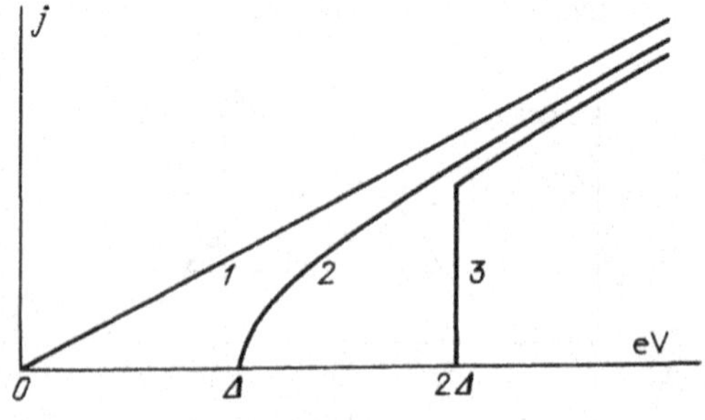

Fig. 141.

The same must be done in order to transport an electron from a superconductor to a normal metal. First we must break up a pair, which requires an energy Δ per electron, following which the electron can pass into the normal metal.

A natural question arises here: Why cannot the pairs be transferred as a whole? Within the framework of the concepts outlined above this is associated with the fact that the pair has a double charge (this increases U in formula 22.1) and a double mass as compared to an individual electron, and therefore the probability of passing through the barrier is vanishingly small for it.

The situation just described may be sketched by an electron level diagram (fig. 142). The superconductor is on the left and the normal metal on the right. The hatched regions correspond to filled levels. The band $\pm\Delta$ on the left around the dashed line is the forbidden band. Above it are the allowed states. The diagram in fig. 142 reflects the fact that for one electron to be transported through the barrier in any direction, so being transferred from a filled state to a free state, it must acquire an energy Δ. With this diagram we need not think of its origin (i.e., of the pairs); we are to deal only with quasiparticles.

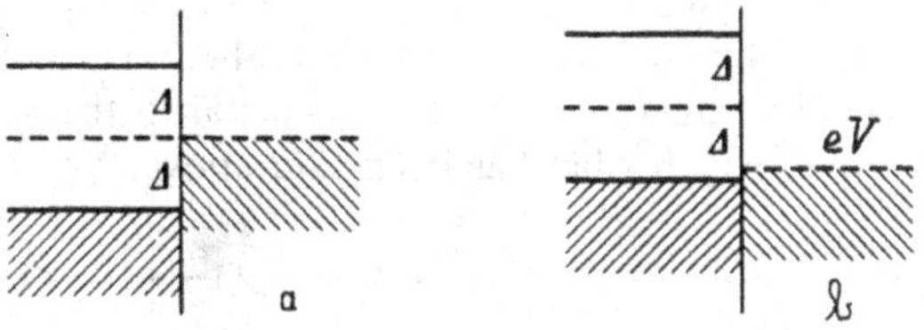

Fig. 142.

When a potential difference is applied, one side of the diagram begins to shift relative to the other and when the filled levels are found to be opposite the empty levels, an electric current will flow. Hence, the condition for the flow of current is

$$eV = \Delta. \tag{22.2}$$

Let us now find the complete dependence $j(V)$. In order to be able to integrate over the energy, we must know the density of states. For the electrons of a normal metal it is given by

$$\nu_n = \frac{p_0 m}{\pi^2 \hbar^3} = \text{const.}$$

For the quasiparticles of a superconductor we have

$$\nu_s = 8\pi p^2 \frac{dp}{(2\pi\hbar)^3\, d\varepsilon}$$

$$= \frac{p_0^2\, d\xi}{v\pi^2\hbar^3\, d\varepsilon} = \frac{p_0 m}{\pi^2\hbar^3}\frac{d\xi}{d\varepsilon}$$

$$= \frac{\nu_n \varepsilon}{(\varepsilon^2 - \Delta^2)^{1/2}}, \quad \varepsilon > \Delta$$

(this formula has already been given in section 21.1).

This is the true density of states. However, if we wish to use the level diagram given in fig. 136, we have to assume that ε may be negative and

$$\nu_s = \begin{cases} \nu_n |\varepsilon| / (\varepsilon^2 - \Delta)^{1/2}, & |\varepsilon| > \Delta, \\ 0, & |\varepsilon| < \Delta. \end{cases} \tag{22.3}$$

Here $\varepsilon > 0$ corresponds to the generation of quasiparticles in a superconductor upon transport of electrons from the normal metal and $\varepsilon < 0$ corresponds to the breakup of the pairs and to the transport of the electrons in the opposite direction*. Of importance is the fact that $\nu_s(|\varepsilon| \to \Delta) \to \infty$, although this infinity is integrable, i.e., the total number of states within an infinitesimal energy interval near $|\varepsilon| = \Delta$ is an infinitesimal quantity.

The current that flows from one metal to another must be proportional to the tunneling probability, the number of occupied states in the first metal, the number of empty states in the second, and also to the densities of states in both metals.

Suppose a potential difference is applied as in fig. 142. Let us find the total current flowing from the superconductor to the normal metal in the forward and reverse directions. In accordance with what has been said above, we have

$$j \propto \int W\nu_1(\varepsilon - eV)\nu_2(\varepsilon)\{n_1(\varepsilon - eV)[1 - n_2(\varepsilon)] - n_2(\varepsilon)[1 - n_1(\varepsilon - eV)]\}\, d\varepsilon$$

$$= \int W\nu_1(\varepsilon - eV)\nu_2(\varepsilon)[n_1(\varepsilon - eV) - n_2(\varepsilon)]\, d\varepsilon$$

$$= W\nu_{1n}\nu_{2n} \int |\varepsilon - eV|[(\varepsilon - eV)^2 - \Delta^2]^{-1/2}[n_1(\varepsilon - eV) - n_2(\varepsilon)]\, d\varepsilon. \tag{22.4}$$

Since $T = 0$, it follows that $n(\varepsilon) = 0$ at $\varepsilon > 0$ and $n(\varepsilon) = 1$ at $\varepsilon < 0$. Hence, the difference $n_1 - n_2$ gives unity in the interval $0 < \varepsilon < eV$ and zero outside. The density of states under the integral is different from zero at $|\varepsilon - eV| > \Delta$, i.e., in this case $\varepsilon < eV - \Delta$. From this it follows that there is no current at $eV < \Delta$. At $eV > \Delta$ we have

$$j \propto W\nu_{1n}\nu_{2n} \int_0^{eV - \Delta} (eV - \varepsilon)[(eV - \varepsilon)^2 - \Delta^2]^{-1/2}\, d\varepsilon$$

$$= W\nu_{1n}\nu_{2n}[(eV)^2 - \Delta^2]^{1/2}. \tag{22.5}$$

At $eV \gg \Delta$ the difference between the normal metal and the superconductor must disappear, i.e., the current must be the same as in the tunnel junction between two

* The introduction of negative ε is analogous to the transition in a normal metal from $|\xi|$ to ξ, i.e., from quasiparticles of particle- and antiparticle-type to the gas model.

normal metals, in which case we have

$$\frac{j}{j_n}=\frac{[(eV)^2-\Delta^2]^{1/2}}{eV}. \tag{22.6}$$

The result is given in fig. 141 (curve 2).

Let us now consider the tunnel junction between two superconductors, which are assumed here, for simplicity, to be identical. Analogously to the foregoing, we find

$$j \propto W\nu_n^2 \int_{\Delta}^{eV-\Delta} (eV-\varepsilon)[(eV-\varepsilon)^2-\Delta^2]^{-1/2}\varepsilon(\varepsilon^2-\Delta^2)^{-1/2}\,d\varepsilon. \tag{22.7}$$

The current is different from zero at $eV > 2\Delta$. Evaluation of the integral yields the following result:

$$\frac{j}{j_n}=\mathrm{E}\left(\frac{[(eV)^2-(2\Delta)^2]^{1/2}}{eV}\right) -2\left(\frac{\Delta}{eV}\right)^2 \mathrm{K}\left(\frac{[(eV)^2-(2\Delta)^2]^{1/2}}{eV}\right), \tag{22.8}$$

where K and E are complete elliptic integrals of the first and second kind:

$$\mathrm{K}(k)=\int_0^{\pi/2} d\varphi\,(1-k^2\sin^2\varphi)^{-1/2}, \qquad \mathrm{E}(k)=\int_0^{\pi/2} d\varphi\,(1-k^2\sin^2\varphi)^{1/2}. \tag{22.9}$$

As must be the case, $j/j_n \to 1$ as $eV\to\infty$. However, in contrast to the junction between a superconductor and a normal metal, the threshold for the generation of current is at $eV=2\Delta$ and at the threshold the current remains finite, i.e., as eV increases gradually, at the point $eV=2\Delta$ a jump occurs in current from zero to a finite value. According to eq. (22.8), at $eV=2\Delta$

$$\frac{j}{j_n}=\tfrac{1}{4}\pi. \tag{22.10}$$

The results are shown in fig. 141 (curve 3).

In principle, one can easily find the expressions for the tunnel current at $T\neq 0$. To this end, one must substitute into the general formula (22.4) the corresponding densities of states (with $\Delta(T)$) and the distribution functions at finite temperatures. We will not perform this calculation, because it involves cumbersome formulae with complex integrals; instead, we will describe the physical picture.

In the case of the s–n junction at finite temperature in a superconductor there is a certain number of broken Cooper pairs. Therefore, individual electrons can pass through the barrier into the normal metal without an extra energy Δ. On the other hand, the electrons of the normal metal can pass into the superconductor, where there are partners with which they can form pairs. From this it is seen that electric current is generated at an arbitrary small potential difference and increases with increasing V. If the temperature is not close to T_c, the current increases especially noticeably at $eV = \Delta(T)$, since the pair-breaking mechanism comes into play. In view of this, $\Delta(T)$ can be found experimentally directly from the maximum of the derivative $\mathrm{d}j/\mathrm{d}V$. It is even better to use for this purpose the junction between two identical superconductors, in which the current increases more drastically in the vicinity of $V = 2\Delta(T)/e$.

Consider now the characteristic of the tunnel junction of two different superconductors. It is easy to see that at $T = 0$ it will have a form similar to curve 3 in fig. 141, with a discontinuity in the current at $V = (\Delta_1 + \Delta_2)/e$. The characteristic at finite temperature is more interesting. We shall assume that $T_{c2} \gg T_{c1}$ and $T \lesssim T_{c1}$. Instead of the level diagram shown in fig. 142, we plot the density of states along the horizontal axis; for the left metal it will be plotted to the left and for the right metal to the right (fig. 137). Thermal excitation is represented conventionally as a certain number of carriers in the left upper band and a certain number of empty states in the left lower band (in the superconductor on the right the excitation is small because of $T \ll T_{c2}$).

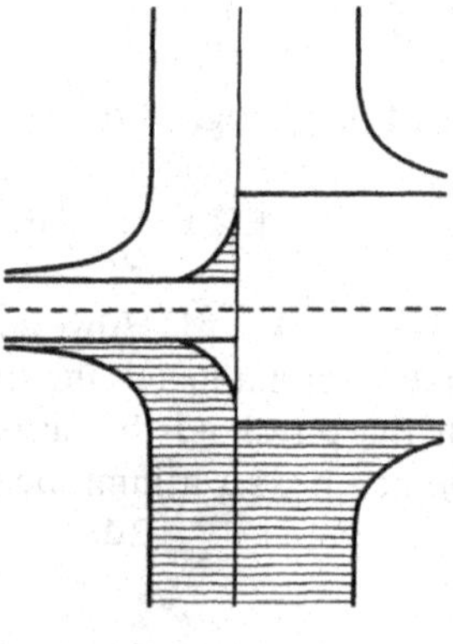

Fig. 143.

As the potential difference increases the left part of fig. 143 is shifted upwards relative to the right part. The current increases due to an increase in the number of thermally excited electrons which reside opposite the empty states in the right upper band. This will continue until the edge of the left upper band coincides with the edge of the right one, i.e., up to $V = (\Delta_2 - \Delta_1)/e$. Following this, the number of electrons capable of passing from left to right will remain constant and the number of empty states, which is proportional to the density of states, will decrease. Hence, the current-voltage characteristic $j(V)$ will bend downwards. This will continue until the upper edge of the lower left band is opposite the lower edge of the upper right band, i.e., up to $V = (\Delta_1 + \Delta_2)/e$, following which $j(V)$ will begin to increase. So, in the case under consideration the current-voltage characteristic will be non-monotonic (fig. 144) (Giaever 1960b, Nicol et al. 1960).

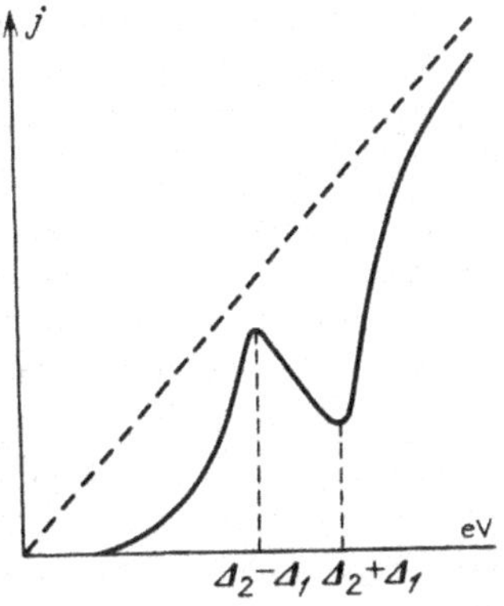

Fig. 144.

Tunnel junctions have played an important role in the direct determination of the energy gap Δ and the density of states and of their temperature dependence. In particular, with the aid of tunneling measurements it has been proved that gapless superconductivity is possible and the theory (Abrikosov and Gor'kov 1960a) has been confirmed quantitatively. Moreover, it has turned out that in the more exact theory, which takes account of the phonon spectrum and the energy dependence of the electron–phonon interaction, the tunneling characteristic makes it possible to determine the parameter $\alpha^2(\omega)\,\rho(\omega)$, where α is the electron–phonon interaction constant and $\rho(\omega)$ is the density of states of phonons. Since $\rho(\omega)$ can be determined independently, one can find $\alpha(\omega)$.

In concluding this section, we note that, apart from the tunnel current considered here, which is associated with the transfer of quasiparticles from one metal to another and which arises in the presence of a finite potential difference (a single-particle current), there also exists a superconducting current flowing through the junction. The latter current will be considered in the sections that follow. There is no superconducting current in experiments if the insulating layer is rather thick (though in any case we may speak only of a few interatomic distances). In the case of a very thin insulating layer the two types of current are both generated, which manifest themselves in a peculiar manner in different circumstances. This point will be discussed below.

22.2. *The Josephson effect*

Up until now we have considered the metals on both sides of the junction separately, assuming that the coherent electronic states are formed on both sides independently. In fact, the passage of electrons through the barrier is the result of penetration of the electron wave function through the junction, and therefore a consistent theory must deal with the formation of a coherent state in the electronic system as a whole. It may be said that at the junction between two superconductors the formation of Cooper pairs of electrons belonging to different metals is possible.

This circumstance leads to the possibility of pair tunneling with a probability comparable with the probability of tunneling of one electron and of the formation of a general condensate. Hence, at $V=0$ a finite supercurrent can flow across the junction. This phenomenon was predicted by Josephson (1962, 1964).

The magnitude of the Josephson superconducting current can be found from the following considerations (Josephson 1962, 1964). Let us find the excess energy associated with the existence of a tunnel junction and the presence of superconductivity. We shall evidently obtain an expression proportional to Δ_1 and Δ_2 and symmetrical with respect to these quantities. The simplest form is

$$E = C\int [|\Delta_1\Delta_2| - \tfrac{1}{2}(\Delta_1\Delta_2^* + \Delta_1^*\Delta_2)]\,\mathrm{d}y\,\mathrm{d}z$$

$$= 2C\int |\Delta_1\Delta_2|[1-\cos(\chi_2-\chi_1)]\,\mathrm{d}y\,\mathrm{d}z$$

(C is a constant and yz is the junction plane). We have assumed that $\Delta_{1,2}$ are complex and that the energy is normalized to zero at $\chi_1-\chi_2=0$ (see below).

The dependence of the energy on the phase difference can be understood as follows. Suppose that a magnetic field parallel to the junction plane is applied; it is described by a vector potential $\boldsymbol{A}$. From the condition of gauge invariance it follows that the phase enters into the following combination with the vector potential:

$$\nabla\chi - \frac{2e}{\hbar c}\boldsymbol{A}.$$

We choose the vector potential to be normal to the junction plane (the x axis) and integrate the above-given combination over x from point 2 in the bulk of the left superconductor to point 1 in the bulk of the right one. As a result, we obtain

$$\chi_1-\chi_2-\frac{2e}{\hbar c}\int_2^1 A_x\,\mathrm{d}x.$$

Thus, in the presence of a magnetic field at the junction the energy is

$$E = C\int |\Delta_1\Delta_2|\left[1-\cos\left(\chi_1-\chi_2-\frac{2e}{c\hbar}\int_2^1 A_x\,\mathrm{d}x\right)\right]\mathrm{d}y\,\mathrm{d}z \qquad (22.11)$$

By varying with respect to A_x we have

$$\delta E = -\frac{2e}{c\hbar}C\int |\Delta_1\Delta_2|\sin\left(\chi_1-\chi_2-\frac{2e}{c\hbar}\int_2^1 A_x\,\mathrm{d}x\right)\delta A_x\,\mathrm{d}V.$$

However, according to the general formula of electrodynamics, the change in energy upon variation of the vector potential is given by

$$\delta E = -\frac{1}{c}\int j\delta A \, \mathrm{d}V,$$

where j is the current density. Comparing the last two formulas, we obtain (we have changed the sign of the argument):

$$j = 2\frac{e}{\hbar}C|\Delta_1\Delta_2| \sin\left(\chi_1 - \chi_2 - \frac{2e}{c\hbar}\int_2^1 A_x \, \mathrm{d}x\right).$$

The maximum magnitude of the sine is unity. Therefore we may write the following expression for the current:

$$j = j_c \sin\left(\chi_1 - \chi_2 - \frac{2e}{c\hbar}\int_2^1 A_x \, \mathrm{d}x\right), \qquad (22.12)$$

where j_c is the critical current across the junction. In particular, in the absence of a field

$$j = j_c \sin(\chi_1 - \chi_2). \qquad (22.13)$$

Thus, a phase difference for the functions Δ_1 and Δ_2 corresponds to the presence of a superconducting current.

According to the above derivation, j_c must be proportional to Δ_1 and Δ_2; it must also be proportional to the probability of electron tunneling through the junction. But the conductivity of the junction in the normal state, or R^{-1}, where R is the junction resistance, is proportional to the same quantity. In contrast to ordinary conductivity, it will be reasonable to write here $j = R^{-1}V$ (V is the potential difference) instead of $j = \sigma E$. The quantity R corresponds to the total resistance of 1 cm^2 of the junction area and has the dimensions of s · cm (or ohm · cm^2).

The complete formula for two different superconductors at $T = 0$ has the form (Ambegoakar and Baratoff 1963)

$$j_c = \frac{2}{eR}\frac{\Delta_1\Delta_2}{\Delta_1 + \Delta_2}\mathrm{K}\left(\frac{|\Delta_1 - \Delta_2|}{\Delta_1 + \Delta_2}\right), \qquad (22.14)$$

where K is a complete elliptic integral of the first kind (see formula 22.9). For two identical superconductors, we substitute $\mathrm{K}(0) = \frac{1}{2}\pi$ and obtain

$$j_c = \frac{\pi\Delta}{2eR}. \qquad (22.15)$$

At a finite temperature, instead of (22.15) we have (Ambegoakar and Baratoff 1963)

$$j_c = \frac{\pi\Delta(T)}{2eR}\tanh\left(\frac{\Delta(T)}{2T}\right). \qquad (22.16)$$

In particular, near T_c,

$$j_c \approx \frac{\pi \Delta^2(T)}{4eRT_c}. \tag{22.17}$$

The microscopic derivation of formulas (22.14), (22.15) and (22.16) will be given in the next section.

For the Josephson effect to be observed in experiments, junctions with a resistivity of 0.1 ohm · mm^2 are required. Actually, it is possible to attain even lower resistivities: 10^{-4} ohm · mm^2 and lower. The corresponding critical current density can reach 10^2–10^3 A/cm^2. If we compare this value with the current density corresponding to the breakup of the pairs in a bulk superconductor (17.63), which is of the order of 10^8 A/cm^2, then the maximum Josephson effect appears to be much lower. Therefore, the Josephson effect and the associated phenomena are sometimes called weak superconductivity.

Thus, apart from the single-particle current found in the preceding section, a superconducting current not exceeding j_c may also flow across the tunnel junction. The Josephson effect is actually observed as follows. The current across the junction is varied and the potential difference is measured. At low magnitudes of the current V remains equal to zero. When the current exceeds j_c, there occurs a switching to the single-particle characteristic and when the current continues to increase the $V(j)$ follows the single-particle curve. However, when the current falls off, a hysteresis occurs rather commonly, namely the single-particle characteristic continues down to the point $j=0$, $V=2\Delta/e$ as V decreases, following which V vanishes in a jumpwise manner (see the oscillogram in fig. 145 for the two directions of the current).

Even in this simple example we see that the theoretical dependence of the current on the potential difference and the experimental determination of $V(j)$ give, in fact, different results, which are accounted for by an abrupt jump of the characteristics measured from one regime to another and by the appearance of "steps", i.e., by a

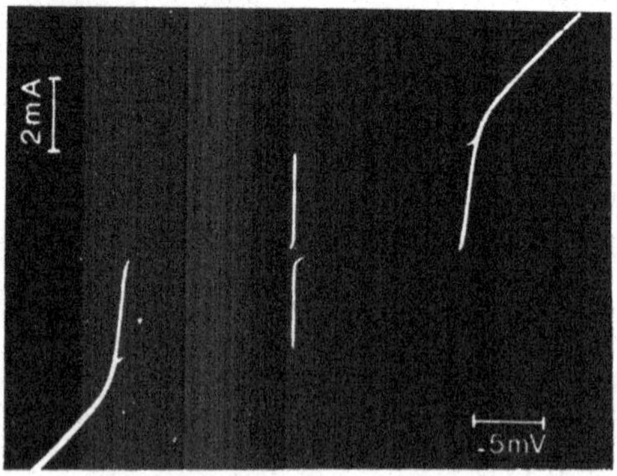

Fig. 145.

discontinuous variation of V at a given current. Since our purpose is to describe the fundamental aspect of the main physical phenomena, we will be concerned, as a rule, with the dependence $j(V)$, making qualitative reservations regarding the events that occur in actual measurements of $V(j)$.

Let us now examine the following problem. At first sight it seems that since the Josephson current and the single-particle current are proportional to the same quantity, namely R^{-1}, they must be observed under identical conditions. However, as has already been pointed out, in practice the Josephson effect disappears much earlier than the single-particle current with increasing R. The cause is the fluctuational potential difference in the junction. This potential difference destroys the Josephson superconducting current but it has little effect on the single-particle current. This is associated with the fact that the single-particle current is generated at a finite potential difference and therefore small fluctuations are not important, whereas the Josephson current occurs at $V = 0$. According to the Nyquist theorem, the fluctuations in V increase proportionately to R (see Lifshitz and Pitaevskii 1978), which just accounts for the difference in behavior of the Josephson and single-particle currents. We shall confine ourselves to this qualitative explanation without describing the theory of fluctuations in the Josephson junctions (see Barone and Paterno 1982).

In concluding this section, we give a simple quantitative derivation of the Josephson current based on the Ginzburg–Landau theory (Aslamazov and Larkin 1969). The Josephson effect can be observed not only in a tunnel junction but also in any superconductor with a "weak link". One example is a film with a constriction (fig. 146) called a bridge. Upon passage of the current its density in the bridge may exceed the critical value (section 17.4). As a result, the bridge begins to play the same role with respect to the wider portions of the film as does the insulating layer in the tunnel junction between two superconductors. In particular, a Josephson current can flow across the bridge.

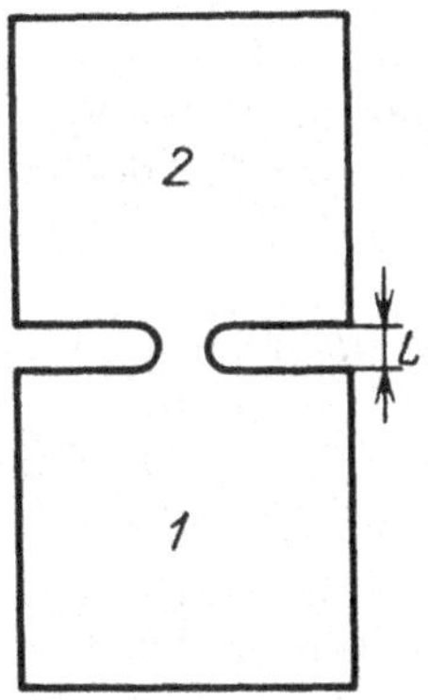

Fig. 146.

Because of the weakness of the Josephson current we may ignore the magnetic field generated by it. Let us write the first Ginzburg–Landau equation (17.6′) in the absence of the field:

$$-\frac{1}{\varkappa^2}\Delta\Psi-\Psi+|\Psi|^2\Psi=0. \tag{22.18}$$

Suppose that the bridge length L is much smaller than ξ (or, in reduced units, $L/\delta \ll \varkappa^{-1}$). In such a case the first term in eq. (22.18), which is of order $(\delta/\varkappa L)^{-2}$, is considerably larger than the remaining ones and we may retain only this term in eq. (22.18). So we obtain

$$\Delta\Psi=0. \tag{22.19}$$

Far from the constriction $\Psi\to$ const. We assume that the constant is equal to $\exp(\mathrm{i}\chi_1)$ in the lower part and to $\exp(\mathrm{i}\chi_2)$ in the upper one (since in the bulk the current density is low and $|\Psi|=1$). The solution of eq. (22.19) will be sought for in the form

$$\Psi=f(\boldsymbol{r})\exp(\mathrm{i}\chi_1)+[1-f(\boldsymbol{r})]\exp(\mathrm{i}\chi_2). \tag{22.20}$$

Evidently, the function $f(\boldsymbol{r})$ tends to unity in the bulk of film 1 and to zero in the bulk of film 2. Substituting into eq. (22.19) gives

$$\Delta f=0.$$

The solution of this equation depends on the particular assumptions concerning the shape of the bridge.

Since we are interested only in the fundamental aspect of the problem, we will not consider the particular models; we substitute Ψ in the form (22.20) into the expression for the current (17.9) (with $\boldsymbol{A}=0$) and use the ordinary units (here eq. 22.20 is multiplied by Ψ_0). As a result, we have

$$j=\frac{e\hbar}{m}\Psi_0^2\nabla f\sin(\chi_1-\chi_2). \tag{22.21}$$

This expression corresponds in form to (22.13) ($\nabla f\sim L^{-1}$), with $j_c\propto\Psi_0^2(T)\propto\Delta^2(T)$, just as in formula (22.17) for the tunnel junction in the neighborhood of T_c.

22.3. *Microscopic derivation of the Josephson current*

We shall employ the method of the so-called tunneling Hamiltonian. We assume that the creation and annihilation of electrons in the left superconductor is described by the operators $a^+_{p,\sigma}$ and $a_{p,\sigma}$ and in the right one, accordingly, by $b^+_{p,\sigma}$ and $b_{p,\sigma}$ and introduce an additional term into the Hamiltonian:

$$\mathscr{H}_\mathrm{T}=\sum_{p,q,\sigma}T_{pq}a^+_{p,\sigma}b_{q,\sigma}+T^*_{pq}b^+_{q,\sigma}a_{p,\sigma}. \tag{22.22}$$

This additional term describes the transport of electrons from one metal to the other. Such a Hamiltonian, in fact, does not completely correspond to the problem under consideration because instead of the real potential barrier (22.22) it actually implies the transfer of free electrons from one phase space to another. However, with correct normalization of the transition matrix T_{pq} it is equivalent to the physical problem under consideration and therefore makes it possible to express the super-current in terms of the junction resistance in the normal state. It is this problem that we shall begin with.

The total current $\langle I \rangle$ from left to right is evidently equal to the rate of decrease of the number of electrons in the left metal multiplied by the electron charge. In accordance with quantum mechanics (see Landau and Lifshitz 1974), the current operator is given by

$$J = -e\dot{N}_{\mathrm{L}} = -\frac{ie}{\hbar}[\mathscr{H}N_{\mathrm{L}}] = -\frac{ie}{\hbar}[\mathscr{H}_{\mathrm{T}}N_{\mathrm{L}}], \tag{22.23}$$

where only the part $\mathscr{H}_{\mathrm{T}}$ of the total Hamiltonian is left because it is only this part of the Hamiltonian that leads to nonconservation of the numbers of electrons in the left and the right metal separately, i.e., it does not commute with N_{L} and N_{R}. Substituting expression (22.22) for $\mathscr{H}_{\mathrm{T}}$ into eq. (22.23) and

$$N_{\mathrm{L}} = \sum_{p,\sigma} a^{+}_{p,\sigma} a_{p,\sigma} \tag{22.24}$$

(section 16.4), we obtain

$$J = \frac{ie}{\hbar}\left(\sum_{p,q,\sigma} T_{pq} a^{+}_{p,\sigma} b_{q,\sigma} - \sum_{p,q,\sigma} T^{*}_{pq} b^{+}_{q,\sigma} a_{p,\sigma} \right). \tag{22.25}$$

The above expression is the operator which is to be averaged over the state of the system. For a normal metal, to within the terms of order $(T/\mu)^2$ we can average over the state at $T = 0$. Thus,

$$\langle J \rangle = \langle \psi_0^* J \psi_0 \rangle. \tag{22.26}$$

Because of the smallness of the electron tunneling probability, we could have taken ψ_0 in a first approximation for the principal part of the Hamiltonian without $\mathscr{H}_{\mathrm{T}}$. However, such an average is equal to zero. In view of this, we find the first approximation to the function ψ_0.

According to perturbation theory (see Landau and Lifshitz 1974),

$$\psi_0 \approx \psi_0^{(0)} + \psi_0^{(1)} = \psi_0^{(0)} + \sum_m \frac{(\mathscr{H}_{\mathrm{T}})_{m0}}{E_0^{(0)} - E_m^{(0)} + \mathrm{i}\delta} \psi_m^{(0)}, \tag{22.27}$$

where the sum runs over the excited states of the system and the superscript (0) signifies that we are dealing with the eigenstates and functions of the Hamiltonian without $\mathscr{H}_{\mathrm{T}}$. The correction $\mathrm{i}\delta$ in the denominator takes account of the circumstance that the spectrum is, in fact, continuous (see Landau and Lifshitz 1974). Substitution into (22.26) yields

$$\langle J\rangle = \langle\psi_0^{(1)*}J\psi_0^{(0)}\rangle + \langle\psi_0^{(0)*}J\psi_0^{(1)}\rangle. \tag{22.28}$$

Using formulae (22.25), (22.27) and (22.22), we easily find that only those terms are nonzero which enter into the products $T_{pq}T_{pq}^* = |T_{pq}|^2$. The resultant expression contains the differences

$$\frac{1}{E_0^{(0)} - E_m^{(0)} + \mathrm{i}\delta} - \frac{1}{E_0^{(0)} - E_m^{(0)} - \mathrm{i}\delta} = -2\pi\mathrm{i}\delta(E_0^{(0)} - E_m^{(0)}),$$

because the principal values of such denominators cancel and only the residue in the pole is retained. The final result is

$$\langle J\rangle = \frac{2\pi e}{\hbar}\sum_{m,p,q,\sigma}|T_{pq}|^2\delta(E_0^{(0)} - E_m^{(0)})$$
$$\times[(a_{p,\sigma}^{+}b_{q,\sigma})_{0m}(b_{q,\sigma}^{+}a_{p,\sigma})_{m0} - (b_{q,\sigma}^{+}a_{p,\sigma})_{0m}(a_{p,\sigma}^{+}b_{q,\sigma})_{m0}]. \tag{22.29}$$

The expression in brackets gives

$$n_{p,\sigma}(1 - n_{q,\sigma}) - n_{q,\sigma}(1 - n_{p,\sigma}) = n_{p,\sigma} - n_{q,\sigma}$$
$$= n_{\mathrm{F}}(\xi_p - eV) - n_{\mathrm{F}}(\xi_q)$$

where n_{F} are the Fermi functions and their arguments correspond to the fact that a potential difference is applied to the junction. Passing from the summation over the momenta to integration and taking into account that only the electrons in the vicinity of the Fermi boundary are involved in the current, we obtain

$$\sum_p \to V_1(\tfrac{1}{2}\nu_1)\int \mathrm{d}\xi_p$$

where V_1 is the volume of the left metal and ν_1 is the density of states; then

$$\langle J\rangle = \frac{4\pi e}{\hbar}V_1V_2(\tfrac{1}{2}\nu_1)(\tfrac{1}{2}\nu_2)\overline{|T_{pq}|^2}$$
$$\times\int \mathrm{d}\xi_p\int \mathrm{d}\xi_q\,\delta(\xi_p - \xi_q)[n_{\mathrm{F}}(\xi_p - eV) - n_{\mathrm{F}}(\xi_q)],$$

where $\overline{|T_{pq}|^2}$ signifies the average over both Fermi surfaces. Taking the integral and dividing by the junction area, we obtain the current density

$$j = \frac{\pi e}{\hbar}eV\nu_1\nu_2\overline{|T_{pq}|^2}\frac{V_1V_2}{S}.$$

Writing this expression in the form $j = V/R$, we have

$$\frac{1}{R} = \frac{\pi e^2}{\hbar}\nu_1\nu_2\overline{|T_{pq}|^2}\frac{V_1V_2}{S}. \tag{22.30}$$

It is not difficult to see that since T_{pq} has the dimensions of the energy R has the correct dimensions cm.s. At the same time it is clear that the correct normalization of T_{pq} must include the factor $(V_1 V_2/S)^{-1/2}$.

Let us now find the Josephson current for two superconductors, expressing it first in terms of $|T_{pq}|^2$ and then with the aid of formula (22.30) in terms of R. To do this, we return to formulas (22.28), (22.27) and (22.25). The expressions thus obtained contain terms of two types: terms that contain a^+a and b^+b and those which include the combination a^+a^+bb or aab^+b^+. We have used terms of the first type in the derivative of the current flowing across the normal junction and they vanish at $V=0$. In a superconductor they give a single-particle current, which we derived with the aid of a more concise method in section 22.1, and they are also equal to zero in the absence of a potential difference.

The remaining terms have the form

$$\mathrm{i}\frac{e}{\hbar}\sum_{p,q,\sigma,m}\Bigg[T_{pq}T_{p_1q_1}(a^+_{p,\sigma}b_{q,\sigma})_{0m}(a^+_{p_1,\sigma_1}b_{q_1,\sigma_1})_{m0}$$
$$\times\left(\frac{1}{E_0^{(0)}-E_m^{(0)}+\mathrm{i}\delta}+\frac{1}{E_0^{(0)}-E_m^{(0)}-\mathrm{i}\delta}\right)$$
$$-T^*_{pq}T^*_{p_1q_1}(b_{q,\sigma}a_{p,\sigma})_{0m}(b^+_{q_1,\sigma_1}a^+_{p_1,\sigma_1})_{m0}$$
$$\times\left(\frac{1}{E_0^{(0)}-E_m^{(0)}+\mathrm{i}\delta}+\frac{1}{(E_0^{(0)}+E_m^{(0)}+\mathrm{i}\delta}\right)\Bigg].$$

We transform the operators $a_{p,\sigma}$ and $b_{q,\sigma}$ using the Bogoliubov formulas (16.14). However, in this particular case, there is no ground for considering the coefficients u_p and v_p to be real since, in contrast to the thermodynamic quantities, the Josephson current depends not only on the moduli but also on the phases of these quantities. In view of this, we write the Bogoliubov transformation in a somewhat modified form:

$$a_{p,+}=u_p\alpha_{p,+}+v_p^*\alpha^+_{-p,-},\qquad a_{p,-}=u_p\alpha_{p,-}-v_p^*\alpha^+_{-p,+}, \tag{22.31}$$

where $|u_p|^2+|v_p|^2-1$. Here, having performed all the calculations of section 16.4, we get

$$|u_p|^2=\tfrac{1}{2}\left(1+\frac{\xi_p}{\varepsilon_p}\right),\qquad |v_p|^2=\tfrac{1}{2}\left(1-\frac{\xi_p}{\varepsilon_p}\right),$$
$$\varepsilon_p=(\xi_p^2+|\Delta|^2)^{1/2},\qquad \Delta=g\sum_p u_pv_p^*(1-n_{p,+}-n_{p,-}). \tag{22.32}$$

Averaging of the operator at $T\neq 0$ must be carried out, strictly speaking, in two stages: first over the given quantum states and then over the Gibbs distribution. However, we can use the fact that the entire system composed of a large number of particles is in a certain quantum state with an accuracy of up to small fluctuations, this state being characterized by the equilibrium occupation numbers $n_F(\varepsilon)$.

One more property of the matrix T_{pq} will be required here. As has been noted in section 1.1, the Schrödinger equation is invariant under the transformation $t \to -t$ and $\psi \to \psi^*$. With this transformation the momenta $\boldsymbol{p}$ and $\boldsymbol{q}$ change sign $[\exp(i\boldsymbol{pr}/\hbar) \to \exp(-i\boldsymbol{pr}/\hbar)]$ and there also occurs the inversion of the tunneling transition, i.e., $b_q \rightleftarrows a_{-p}$. This means that T_{pq} has the following property:

$$T_{pq} = T^*_{-p-q}. \tag{22.33}$$

After performing rather simple but time-consuming calculations we obtain an expression proportional to $u_p^* v_p u_q v_q^* - u_p v_p^* u_q^* v_q$. If we assume that $u_p^* v_p = |u_p||v_p| \exp(i\chi_1)$, $u_q^* v_q = |u_q||v_q| \exp(i\chi_2)$, then this difference is equal to $|u_p v_p u_q v_q| 2i \sin(\chi_1 - \chi_2)$. As regards the energy denominators, it follows that, in contrast to a normal metal, the terms with $+i\delta$ and $-i\delta$ add up, i.e., the doubled principal value of the integral is retained. On the whole, we obtain

$$\langle J \rangle = 8 \frac{e}{\hbar} \sum_{p,q} |u_p v_p u_q v_q| |T_{pq}|^2 \sin(\chi_1 - \chi_2)$$

$$\times \left[\frac{n_p(1-n_q)}{\varepsilon_p - \varepsilon_q} - \frac{n_p n_q}{\varepsilon_p + \varepsilon_q} + \frac{(1-n_p)(1-n_q)}{\varepsilon_p + \varepsilon_q} - \frac{(1-n_p) n_q}{\varepsilon_p - \varepsilon_q} \right].$$

Collecting terms with identical denominators, using the formula

$$|u_p v_p| = \frac{|\Delta|}{2\varepsilon_p} \tag{22.34}$$

and passing over from summation to integration first over ξ_p and ξ_q and then over ε_p and ε_q (with account taken of two possible signs of ξ), and, finally, expressing the Fermi functions via the hyperbolic tangents, we obtain

$$j = -\frac{2e}{\hbar} \overline{|T_{pq}|^2} \frac{V_1 V_2}{S} \sin(\chi_1 - \chi_2) \nu_1 \nu_2 \Delta_1 \Delta_2$$

$$\times \int_{\Delta_1}^{\infty} d\varepsilon_1 \int_{\Delta_2}^{\infty} d\varepsilon_2 \frac{\left[\varepsilon_1 \tanh\left(\frac{\varepsilon_2}{2T}\right) - \varepsilon_2 \tanh\left(\frac{\varepsilon_1}{2T}\right) \right]}{(\varepsilon_1^2 - \varepsilon_2^2)\xi_1 \xi_2},$$

where $\xi_{1,2} = (\varepsilon_{1,2}^2 - \Delta_{1,2}^2)^{1/2}$.

Let us first consider the integral over ε_1. It may be written in the form of a contour integral in a complex plane with branch cuts along the real axis from Δ to ∞ and from $-\infty$ to $-\Delta$ (fig. 147a). Following this, the contour of integration can may be "straightened up" (fig. 147b) and, as a result, we obtain the sum of residues from

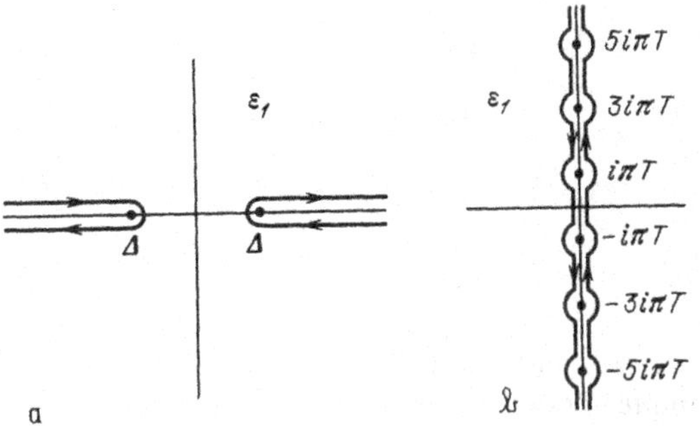

Fig. 147.

$\tanh(\varepsilon_1/2T)$, which are located at the points $\mathrm{i}\pi T(2n+1)$:

$$\int_\Delta^\infty \mathrm{d}\varepsilon_1 \cdots = \tfrac{1}{4}\int_C \mathrm{d}\varepsilon_1 \cdots$$

$$= \tfrac{1}{4}\cdot 2T\cdot 2\pi \sum_{n=-\infty}^{\infty} \varepsilon_2 \frac{1}{[\pi^2T^2(2n+1)^2+\varepsilon_2^2][\pi^2T^2(2n+1)^2+\Delta_1^2]^{1/2}(\varepsilon_2^2-\Delta_2^2)^{1/2}}.$$

Now the same procedure will be used for the integral over ε_2. In each of the terms of the sum over n there are two residues with respect to ε_2 at $\pm\mathrm{i}\pi T(2n+1)$. As a result, we get $j=j_c\sin(\chi_1-\chi_2)$, where

$$j_c = \frac{2}{eR}\pi T\Delta_1\Delta_2 \sum_{n=0}^{\infty} \frac{1}{[\pi^2T^2(2n+1)^2+\Delta_1^2]^{1/2}}\frac{1}{[\pi^2T^2(2n+1)^2+\Delta_2^2]^{1/2}}.$$

Let us consider two cases. Suppose the metals are identical. In this case, using relation (21.1),

$$\sum_{n=1}^{\infty}\frac{1}{(2n-1)^2+x^2} = \frac{\pi}{4x}\tanh(\tfrac{1}{2}\pi x),$$

we have

$$j_c = \frac{\pi}{2eR}\Delta\tanh\left(\frac{\Delta}{2T}\right),$$

which corresponds to formulae (22.13) and (22.16).

Suppose now that the metals are different but $T\to 0$. Here the intervals between the values of $\pi T(2n+1)$ are small as compared with Δ and we may replace the sum by an integral. As a result, we have

$$j_c = \frac{1}{eR}\Delta_1\Delta_2\int_0^\infty \frac{\mathrm{d}\omega}{(\omega^2+\Delta_1^2)^{1/2}(\omega^2+\Delta_2^2)^{1/2}}$$

$$=\frac{2}{eR}\frac{\Delta_1\Delta_2}{\Delta_1+\Delta_2}\mathrm{K}\left(\frac{|\Delta_1-\Delta_2|}{\Delta_1+\Delta_2}\right)$$

that is, formula (22.14).

22.4. The Josephson effect in a magnetic field

Suppose that a magnetic field is applied parallel to the plane of the tunnel junction (yz). Let, for example, H be directed along z. In this case, we may introduce a vector potential:

$$A_x = H_z(x)y. \tag{22.35}$$

The area of the junction is assumed to be so small that the effect of the field created by the current may be neglected. The relevant condition will be derived later. In the presence of the field the current density has the form of (22.12). Taking into account that H_z does not depend on x in the insulating layer (d') and falls off in the superconductor by the law $\exp(-x/\delta)$, we get

$$\int_2^1 A_x\,\mathrm{d}x = H_{z0}yd, \tag{22.36}$$

where

$$d = 2\delta(T)+d' \approx 2\delta(T). \tag{22.37}$$

Here we have taken into account that $d' \sim 10^{-7}$–10^{-8} cm and $\delta(T) \sim 10^{-5}$–10^{-6} cm. Hence, the current density in the presence of a steady field has the form ($H \equiv H_{z0}$)

$$j = j_c \sin\left(\chi_1-\chi_2+\frac{2eHdy}{c\hbar}\right). \tag{22.38}$$

Thus, the current density in the presence of a field varies in a direction perpendicular to the field and even changes sign.

In practice, naturally it is the total current that is measured or, what is the same thing, the average current density. Since in the direction of the field the current is homogeneous, the average current density is given by

$$\bar{j} = \frac{1}{L}\int_0^L j(y)\,\mathrm{d}y$$

$$= j_c\frac{c\hbar}{2eHLd}\left[-\cos\left(\chi_1-\chi_2+\frac{2e}{c\hbar}HLd\right)+\cos(\chi_1-\chi_2)\right], \tag{22.38'}$$

where L is the length of the junction along y. It is not difficult to see that $HLd = \Phi$ is the field flux across the junction. Introducing the flux quantum $\Phi_0 = \pi\hbar c/e$, we may write expression (22.38) in the form

$$\bar{j}=j_c\frac{\Phi_0}{\pi\Phi}\sin\left(\frac{\pi\Phi}{\Phi_0}\right)\sin\left(\theta_0+\frac{\pi\Phi}{\Phi_0}\right), \tag{23.39}$$

where $\theta_0=\chi_1-\chi_2$.

From eq. (22.39) it follows that $\bar{j}$ vanishes at $\Phi=n\Phi_0$, i.e., when the field flux across the junction is equal to an integer number of quanta. The maximum value of current in the intervals between these values is obtained by the choice of the phase difference and evidently corresponds to $\sin[\theta_0-(\pi\Phi/\Phi_0)]=\pm1$. As a result, it turns out that (Rowell 1963)

$$\bar{j}_{\max}=j_c\left|\frac{\sin(\pi\Phi/\Phi_0)}{\pi\Phi/\Phi_0}\right|. \tag{22.40}$$

The graph of this function is given in fig. 148. The importance of the dependence (22.40) lies in the fact that it allows one to measure the value of d, i.e., $\delta(T)$ from the points in which $\bar{j}_{\max}=0$. In principle, one could also use this dependence for measuring the magnetic field, but in practice Josephson interferometers are used for this purpose (section 22.8).

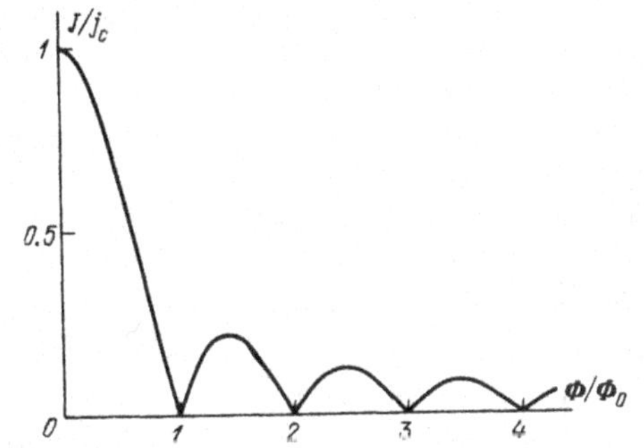

Fig. 148.

Now we shall be concerned with the study of a wide junction, in which case we cannot neglect the field generated by the Josephson current. Here, even in the absence of the external field the distribution of current along the junction becomes nonuniform. Let θ denote the complete argument in formula (22.12). Although the inhomogeneity of the field along y no longer permits us to take the vector potential in the form (22.35), the difference $\theta(y+dy)-\theta(y)$ may still be written as

$$\theta(y+\mathrm{d}y)-\theta(y)=\frac{2e}{\hbar c}H(y)\,d\cdot\mathrm{d}y=\frac{2\pi}{\Phi_0}H(y)\,d\cdot\mathrm{d}y$$

(we assume that H has only one component, H_z). From this we see that

$$H(y)=\frac{\Phi_0}{2\pi d}\frac{\mathrm{d}\theta}{\mathrm{d}y}. \tag{22.41}$$

According to Maxwell's equation

$$j=\frac{c}{4\pi}\frac{\mathrm{d}H}{\mathrm{d}y}. \tag{22.42}$$

At the same time, $j=j_c \sin\theta$. Substituting this into eq. (22.42) and using eq. (22.41), we arrive at the equation for θ (Ferrell and Prange 1963)

$$\frac{\mathrm{d}^2\theta}{\mathrm{d}y^2}=\frac{\sin\theta}{\delta_J^2}, \tag{22.43}$$

where

$$\delta_J=\left(\frac{c\Phi_0}{8\pi^2 j_c d}\right)^{1/2}. \tag{22.44}$$

Let us consider a case where there is a field $H_y=H$ but there is no net current across the junction. If the field is low, the quantity θ may be assumed to be small. Replacing $\sin\theta$ by θ, we obtain

$$\frac{\mathrm{d}^2\theta}{\mathrm{d}y^2}=\frac{\theta}{\delta_J^2}, \tag{22.45}$$

the solution of which is

$$\theta=\theta_0 \exp\left(\frac{-y}{\delta_J}\right).$$

Inserting this into eq. (22.41) gives

$$H(y)=H_0 \exp\left(\frac{-y}{\delta_J}\right). \tag{22.46}$$

The result obtained is easy to interpret. If the super-conducting current can flow across the junction, it can screen out the applied field. The corresponding penetration depth δ_J is given by formula (22.44). If we substitute into it the orders of magnitudes of the parameters, then for $j_c \sim 10^2\ \mathrm{A/cm^2}$ we obtain $\delta_J \sim 10^{-2}$ cm, which is much larger than the ordinary penetration depth. The junction may be regarded as narrow if its width $L \ll \delta_J$. In this case, the field of the current may be neglected and no screening of the external magnetic field takes place. But if $L>\delta_J$, the junction is wide and the screening must be taken into account.

Thus, the Josephson junction is, to a certain extent, a two-dimensional superconductor with the corresponding Meissner effect. As a matter of fact, of course, the current lines are closed in the three-dimensional space: the current flows in both superconductors perpendicular to the field in a layer of thickness δ (fig. 149a).

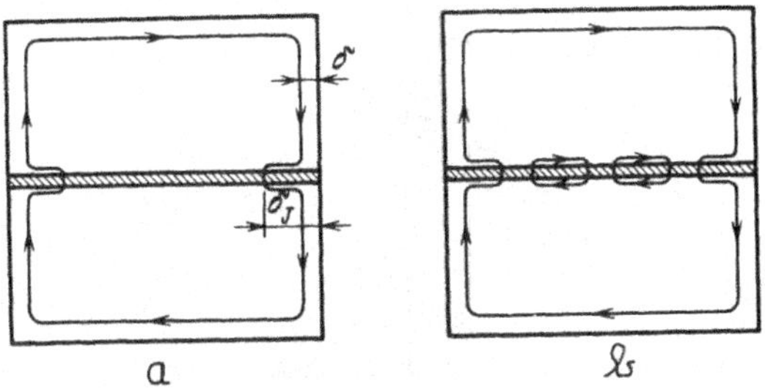

Fig. 149.

So far we have considered the field to be weak and made the replacement $\sin\theta \approx \theta$. In the case of a strong field this cannot be done. Let us return to eq. (22.43); we multiply both sides by $\mathrm{d}\theta/\mathrm{d}y$ and integrate over y. As a result, we get

$$\left(\frac{\mathrm{d}\theta}{\mathrm{d}y}\right)^2 = -2\frac{\cos\theta}{\delta_J^2} + C \tag{22.47}$$

(C is a constant). This gives (it is convenient here to fix the point where $\theta = \pi$)

$$y - y_0 = \int_\pi^\theta \mathrm{d}\theta_1 \left(C - 2\frac{\cos\theta_1}{\delta_J^2}\right)^{-1/2}. \tag{22.48}$$

Let us find a solution of the "soliton" type, in which the current is nonzero in a narrow vicinity of some point and is equal to zero far from it. Then, $\mathrm{d}\theta/\mathrm{d}y \to 0$ as $y \to \pm\infty$. This can be attained by assuming that $\theta \to 0$ with $y \to -\infty$ and $\theta \to 2\pi$ with $y \to \infty$. In this case $C = 2\delta_J^{-2}$ in eq. (22.47). The integral (22.48) is easily evaluated in such a case; as a result, we obtain

$$\theta(y) = 4\arctan\left[\exp\left(\frac{y - y_0}{\delta_J}\right)\right]. \tag{22.49}$$

The plots for $\theta(y)$, the field and current are given in fig. 150. From these plots it is seen that the solution obtained corresponds to a vortex with the center at y_0 *. The picture of current lines (for two vortices) is shown in fig. 149b.

Let us find the field flux in a single vortex. According to (22.41),

$$\Phi = d\int_{-\infty}^{\infty} H\,\mathrm{d}y = \frac{\Phi_0}{2\pi d}\,d[\theta(\infty) - \theta(-\infty)] = \Phi_0.$$

Thus, the vortex carries a magnetic flux quantum. We come to the conclusion that the Josephson junction in the external field is the two-dimensional analog of a type II superconductor.

* The idea of the appearance of such vortices was advanced by Josephson (Josephson 1965).

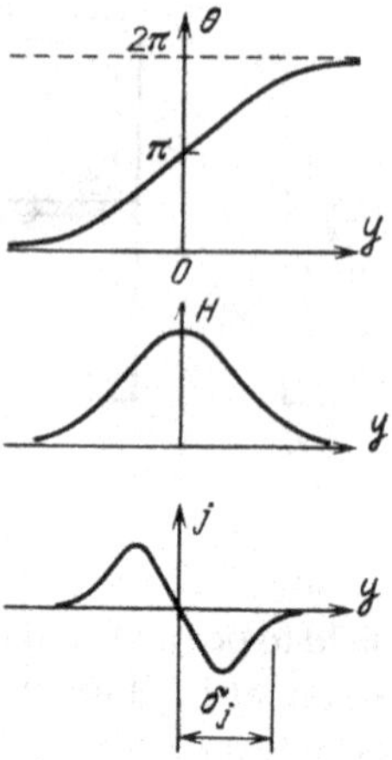

Fig. 150.

Vortices begin to be formed in an external field $H = H_{c1}$, which is of the order of the average field in the vortex, i.e., $H_{c1} \sim \Phi_0/(d\delta_J)$. The exact formula reads (Josephson 1965):

$$H_{c1} = \frac{2\Phi_0}{\pi^2 \delta_J d}. \tag{22.50}$$

The vortex concentration increases with increasing field, but since the Josephson vortex has no "normal core", in which the order parameter decreases strongly, it follows that no transition occurs to the normal state at any field H_{c2}.

In the limiting case at high fields the vortices overlap and the field in the junction is almost homogeneous*. From (22.41) it follows that mainly $\theta \approx H(2\pi d/\Phi_0)(y - y_0)$, i.e., $d\theta/dy = \text{const}$. According to eqs. (22.47) and (22.48), the second term in parentheses of the integrand (22.48) is much smaller than the first and we can carry out an expansion. Finally, we find

$$\theta \approx H\frac{2\pi d}{\Phi_0}(y - y_0) - \left(\frac{H_{c1}\pi}{4H}\right)^2 \sin\left(2\pi d H \frac{y - y_0}{\Phi_0}\right), \tag{22.51}$$

where we have used formula (22.50). Thus, in addition to the regular increase of the phase in the junction plane in a direction perpendicular to the field there occur small oscillations of the phase, which fall off rapidly in amplitude with increasing field.

* In this sense, a strong field leads to the same situation that occurs in a narrow junction. This is natural: in both cases the Josephson current cannot screen out the field.

22.5. The ac Josephson effect

Suppose that the current passing through the Josephson junction becomes larger than the critical current. In this case, superconductivity is destroyed and a potential difference appears. From the condition of gauge invariance it follows that in the equations the potential always appears together with the derivative of the phase with respect to time in the combination (section 19.6)

$$2e\varphi + \hbar\frac{\partial\chi}{\partial t}$$

that is, $(2e/\hbar)\int^t \varphi(t_1)\,\mathrm{d}t_1$ is added to the phase χ. This means that in the presence of the potential difference the phase difference satisfies the equation

$$\frac{\partial\theta}{\partial t} = \frac{2e}{\hbar}V \tag{22.52}$$

where $V = \varphi_1 - \varphi_2$. If $V = \text{const.}$, then the phase difference is given by

$$\theta = \theta_0 + \frac{2eV}{\hbar}t. \tag{22.53}$$

This relation was first derived by Josephson (Josephson 1962, 1964) for the tunnel junction, but, as was found out later, it describes the so-called resistive state of superconductors in various cases (for example, see formula 22.91, section 22.9). Thus, if a constant potential difference is maintained at the junction, an alternating current will flow across the junction:

$$j = j_c \sin\left(\theta_0 + \frac{2eV}{\hbar}t\right). \tag{22.54}$$

The frequency

$$\omega = 2eV/\hbar \tag{22.55}$$

corresponds to $10^{11}\ \mathrm{s}^{-1}$ for $V \sim 10^{-4}$ V.

As has been noted in section 22.2, the potential difference is measured experimentally at a given current. The current is made up of a normal (single-particle) and a superconducting current. The normal current is V/R (we are considering a case where V is not too close to $2\Delta/e$) or, according to eq. (22.52), $(\hbar/2eR)\partial\theta/\partial t$. Thus, we have

$$j = j_c \sin\theta + \frac{\hbar}{2eR}\frac{\partial\theta}{\partial t}. \tag{22.56}$$

Integration of this equation gives

$$\theta = 2\arctan\left\{\left[1-\left(\frac{j_c}{j}\right)^2\right]^{1/2}\tan\left[eRt\frac{(j^2-j_c^2)^{1/2}}{\hbar}\right]+\frac{j_c}{j}\right\}. \tag{22.57}$$

From this, using $V=(\hbar/2e)\,\mathrm{d}\theta/\mathrm{d}t$, we find (Aslamazov and Larkin 1969)

$$V(t)=\frac{Rj(j^2-j_c^2)}{[j^2+j_c^2\cos\omega t+j_c(j^2-j_c^2)^{1/2}\sin\omega t]}, \tag{22.58}$$

where

$$\omega=\frac{2eR}{\hbar}(j^2-j_c^2)^{1/2}. \tag{22.59}$$

This formula may be rewritten in the form

$$V(t)=R\frac{(j^2-j_c^2)}{j+j_c\cos(\omega t-\theta_1)}, \tag{22.58'}$$

where $\theta_1=\arccos(j_c/j)$. As $j\to\infty$, $V\to Rj$, i.e., tends to the value for a normal metal. The plot of V/Rj_c against ωt is given in fig. 151 for several values of j_c/j.

Averaging of formula (22.58) over time gives

$$\bar{V}=\frac{1}{2\pi}\int_0^{2\pi}V(t)\,\mathrm{d}(\omega t)=R(j^2-j_c^2)^{1/2}=\frac{\omega\hbar}{2e}. \tag{22.60}$$

This dependence $\bar{V}(j)$ has been confirmed experimentally. Note also that when a steady current is maintained the average voltage and the frequency ω (22.59) are connected by the Josephson relation (22.55). The average values $\bar{V}/(Rj_c)$ are marked by dashed lines in fig. 151.

Since the tunnel junction is simultaneously a planar capacitor, it has a certain capacitance, i.e., there is a charge on the capacitor plates equal to $Q=CV$, the capacitance being given by $C=\varepsilon/(4\pi d')$, where d' is the thickness of the insulator

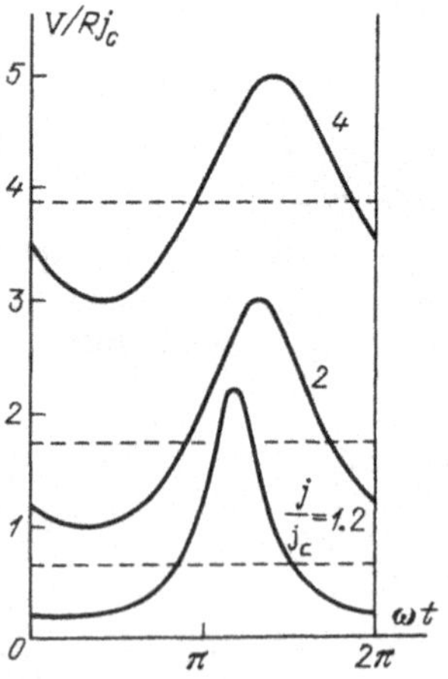

Fig. 151.

layer and ε is its dielectric constant. The change of this charge Q with time also makes a contribution to the current, which is equal to $\partial Q/\partial t = C\partial V/\partial t = (C\hbar/2e)\partial^2\theta/\partial t^2$, so that the complete equation is written, instead of (22.56), in the form (McCumber 1968, Stewart 1968, Johnson 1968)

$$j = j_c \sin\theta + \frac{\hbar}{2eR}\frac{\partial\theta}{\partial t} + \frac{C\hbar}{2e}\frac{\partial^2\theta}{\partial t^2}. \tag{22.61}$$

We shall not consider her all the possible limiting cases. The limit considered above corresponds to a small capacitance or, more exactly, to a case where the third term in (22.61) is small compared to the second term, i.e., $\omega C/R \sim CeR^2 j/\hbar \ll 1$.

Associated with the ac Josephson effect is also another peculiar phenomenon: the appearance of steps in the voltage-current characteristic when a varying external field is applied (Josephson 1962, 1964, Shapiro 1963, Shapiro et al. 1964)*. The experiment is carried out as follows. A steady current j is passed across the Josephson junction and, besides, a varying potential difference $v \cos \Omega t$ is applied. The time-averaged current-voltage characteristic, i.e., $\overline{V(t)}$, is taken as a function of j. Here, for the sake of simplicity, we state the problem in a different manner; we assume that the following potential difference is applied to the junction:

$$V(t) = V_0 + v\cos\Omega t, \tag{22.62}$$

and we find the current.

From (22.52) we have

$$\theta(t) = \frac{2e}{\hbar}\left(V_0 t + \frac{v}{\Omega}\sin\Omega t + \theta_0\right). \tag{22.63}$$

The Josephson current is equal to

$$\begin{aligned} j = j_c \sin\theta(t) &= j_c \sin\left[\frac{2e}{\hbar}(V_0 t + \theta_0)\right]\cos\left(\frac{2ev}{\hbar\Omega}\sin\Omega t\right) \\ &+ j_c \cos\left[\frac{2e}{\hbar}(V_0 t + \theta_0)\right]\sin\left(\frac{2ev}{\hbar\Omega}\sin\Omega t\right). \end{aligned} \tag{22.64}$$

The anharmonic factor appearing in this expression may be expanded in Fourier series:

$$\cos(a\sin\Omega t) = \sum_{n=-\infty}^{\infty} A_n \exp(in\Omega t),$$

$$\sin(a\sin\Omega t) = \sum_{n=-\infty}^{\infty} B_n \exp(in\Omega t),$$

* They are usually called Shapiro steps after the author of the experimental discovery.

where

$$A_n = \frac{1}{2\pi}\int^{2\pi} \cos(a \sin x)\exp(-inx)\,dx = J_{2k}(a), \quad n = 2k,$$

$$B_n = \frac{1}{2\pi}\int_0^{2\pi} \sin(a \sin x)\exp(-inx)\,dx$$
$$= -iJ_{2k+1}(a), \quad n = 2k+1$$

where J_n are Bessel functions. Substituting into formula (22.64) and using the rule $J_n(x) = (-1)^n J_{-n}(x)$, we obtain

$$j = j_c \sin\left[\frac{2e}{\hbar}(V_0 t + \theta_0)\right]\left[J_0\left(\frac{2ev}{\hbar\Omega}\right) + 2\sum_{k=1}^{\infty} J_{2k}\left(\frac{2ev}{\hbar\Omega}\right)\cos(2k\Omega t)\right]$$
$$+ j_c \cos\left[\frac{2e}{\hbar}(V_0 t + \theta_0)\right]\left[2\sum_{k=1}^{\infty} J_{2k-1}\left(\frac{2ev}{\hbar\Omega}\right)\sin[(2k-1)\Omega]\right]. \tag{22.65}$$

Separating some particular term of the first sum, we get

$$j_c \sin\left[\frac{2e}{\hbar}(V_0 t + \theta_0)\cdot 2J_{2k}\left(\frac{2ev}{\hbar\Omega}\right)\cos(2k\Omega t)\right]$$
$$= j_c J_{2k}\left(\frac{2ev}{\hbar\Omega}\right)\left\{\sin\left[\left(\frac{2ev_0}{\hbar} + 2k\Omega\right)t + \theta_0\right]\right.$$
$$\left. + \sin\left[\left(\frac{2eV_0}{\hbar} - 2k\Omega\right)t + \theta_0\right]\right\}.$$

Any term of the second sum transforms to the following form in an analogous way:

$$j_c J_{2k-1}\left(\frac{2ev}{\hbar\Omega}\right)\left\{\sin\left[\left(\frac{2eV_0}{\hbar} + (2k-1)\Omega\right)t + \theta_0\right]\right.$$
$$\left. - \sin\left[\left(\frac{2eV_0}{\hbar} - (2k-1)\Omega\right)t + \theta_0\right]\right\}.$$

Thus, we see that in a case where

$$\frac{2eV_0}{\hbar} = n\Omega, \tag{22.66}$$

where n is an integer, a time-independent term appears in the current, which is proportional to

$$\bar{j}_n = (-1)^n j_c J_n\left(\frac{2ev}{\hbar\Omega}\right)\sin\theta_0. \tag{22.67}$$

This means that in the $\bar{j}_n(V_0)$ curve resonance peaks add up to the single-particle current (section 22.1) at values of V_0 given by formula (22.66).

As has already been said, in experiments actually the potential difference is measured at a given current. In order to obtain an exact answer, we have to solve the equation for the phase, as has been done in the absence of a high-frequency field. However, if resonance corrections are not too large, we may use the following procedure. We plot the $\bar{j}(V_0)$ curve by adding to the single-particle characteristic vertical portions corresponding to the maximum values of resonance corrections, i.e., $\max \bar{j}_n = j_c|J_n(2eV/\hbar\Omega)|$ at $V_0 = V_{0n} = n\hbar\Omega/2e$. An increase in the current leads to a jumpwise change in regime, from purely superconducting ($V_0 = 0$) to the case with a finite V_0; this occurs whenever the current reaches the end of the corresponding vertical portion. This gives rise to a stepped curve (fig. 152)*. Actually, this was the first experimental demonstration of the ac Josephson effect (Shapiro 1963, Shapiro et al. 1964).

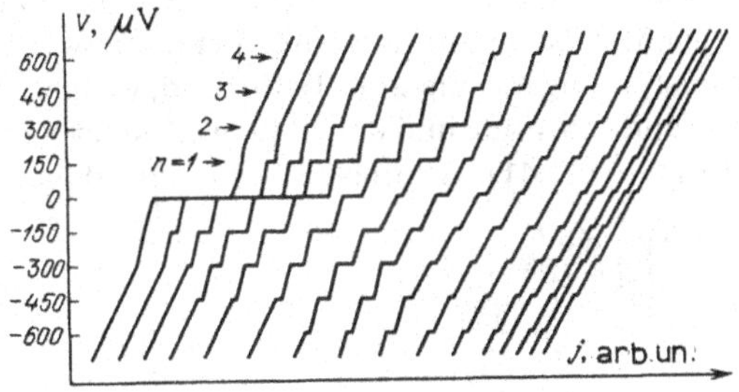

Fig. 152.

In concluding this section, we give a physical interpretation of relations (22.55) and (22.66). While passing the static potential difference, the Cooper pair receives an excess energy and can emit it in the form of a quantum with $\hbar\omega = 2eV$. But if the junction is in an external field, there occurs an exchange of energy with the external field, in which emission and absorption of many quanta of the external field are possible. This is described by formula (22.66).

22.6. Waves in the Josephson junction

Suppose that the Josephson junction is simultaneously subjected to the action of a steady magnetic field and a steady potential difference. In such a case, the phase

* Figure 146 shows the experimental characteristics of a point Nb–Nb junction (Grimes and Shapiro 1968). The abscissa is the current and the ordinate represents the potential difference. Each of the curves corresponds to both directions of the current. The different curves correspond to different amplitudes of the high-frequency field and are shifted relative to one another along the abscissa (the null current is taken as the center of symmetry of a given curve). Actually, in the problem with a given current the steps are not exactly rectangular: instead of a jump at some critical j there is a variation $V(j)$.

difference has simultaneously space and time components. In accordance with formulas (22.38) and (22.54) we find

$$j = j_c \sin\left[\theta_0 + \frac{2eV}{\hbar}t + \frac{2eHd}{\hbar c}y\right]. \tag{22.68}$$

But this is nothing but the moving wave formula. The frequency of this wave is $\omega = 2eV/\hbar$ and the wave vector $k = 2eHd/\hbar c$. Hence, the phase velocity is given by

$$v_0 = \frac{\omega}{k} = \frac{cV}{Hd}. \tag{22.69}$$

The physical meaning of this formula consists in that upon application of the potential difference the sinusoidal current density (22.38) begins to shift along the junction with a constant velocity (22.69) (Eck et al. 1964).

The result (22.68) refers to a case where a steady potential difference and a steady magnetic field are maintained in the junction. Let us now find the waves that propagate spontaneously along the junction. To this end, we make use of the formula for the current (22.61) and substitute into it the Maxwell equation (22.42), expressing H with the aid of formula (22.41). As a result, we obtain the equation

$$\frac{\partial^2\theta}{\partial y^2} - \frac{1}{c_0^2}\left(\frac{\partial^2\theta}{\partial t^2} + \gamma\frac{\partial\theta}{\partial t}\right) = \frac{\sin\theta}{\delta_J^2}, \tag{22.70}$$

where

$$c_0 = \left(\frac{ec}{4\pi^2 dC\hbar\Phi_0}\right)^{1/2} = c\left(\frac{d'}{\varepsilon d}\right)^{1/2}, \tag{22.71}$$

$$\gamma = \tau^{-1} = \frac{1}{RC}. \tag{22.72}$$

The penetration depth δ_J is expressed by formula (22.44) and the capacitance is given by $C = \varepsilon/(4\pi d')$.

If we assume that there is no Josephson current i.e., if we put $j_c = 0$, then $\delta_J = \infty$ and the right-hand side of eq. (22.70) vanishes. If we also neglect damping, i.e., if we set $\gamma = 0$, we will obtain the equation for the propagation of waves with a velocity c_0. From formula (22.71) it follows that this velocity is much less than the velocity of light. Although the equation is written for a somewhat unusual quantity – the phase difference – we may rewrite it in the form of an equation for H by differentiating with respect to z, and in the form of the an equation for E by differentiating with respect to t. Thus, we see that in the absence of a Josephson current along the junction between two superconductors separated by an insulating barrier, waves can propagate with the velocity c_0. Such waves are called Swihart waves (Swihart 1961).

Let us now consider eq. (22.70) in detail, i.e., we assume that a Josephson current is possible. Without damping (i.e., at $\gamma = 0$) this equation is usually called the sine-Gordon equation (upon replacement of sin θ by θ it corresponds to the Klein–Gordon–Fock equation for elementary particles with zero spin). It is rather evident that in the presence of a potential difference and magnetic field a peculiar resonance may arise when the phase velocity of Josephson waves v_0 coincides with the velocity c_0 (22.71) (Eck et al. 1964). In order to find the result, we assume that the right-hand side of (22.70) is small and look for a solution in the form (Kulik 1965)

$$\theta = \theta^{(0)} + \theta^{(1)},$$

where $\theta^{(1)} \ll \theta^{(0)}$ and

$$\theta^{(0)} = \theta_0 + \frac{2eV}{\hbar} t + \frac{2eHd}{\hbar} y = \theta_0 + \omega t + ky.$$

To first order we obtain from eq. (22.70)

$$\frac{\partial^2 \theta^{(1)}}{\partial y^2} - \frac{1}{c_0^2}\left(\frac{\partial^2 \theta^{(1)}}{\partial t^2} + \gamma \frac{\partial \theta^{(1)}}{\partial t}\right) = \frac{\sin(\theta_0 + \omega t + ky)}{\delta_J^2}. \tag{22.73}$$

From this we find

$$\theta^{(1)} = \frac{c_0^2}{\delta_J^2} \operatorname{Im}\left\{\frac{\exp[i(\theta_0 + \omega t + ky)]}{\omega^2 - c_0^2 k^2 - i\omega\gamma}\right\}. \tag{22.74}$$

We substitute expression (22.74) into the complete formula for the Josephson current $j = j_c \sin\theta$ and average it over time and coordinate*:

$$\bar{j} = \lim_{\substack{T\to\infty \\ L\to\infty}} \frac{1}{T}\int_0^T dt \frac{1}{L}\int_0^L dy\, j_c \sin\theta(y, t)$$

$$\approx \lim_{\substack{T\to\infty \\ L\to\infty}} \frac{1}{T}\int_0^T dt \frac{1}{L}\int_0^L dy\, j_c \cos\theta^{(0)}(y, t)\theta^{(1)}(y, t)$$

$$= j_c \frac{\omega c_0^2 \gamma}{2\delta_J^2} \frac{1}{(\omega^2 - c_0^2 k^2)^2 + \omega^2\gamma^2}. \tag{22.75}$$

Thus, we obtain a sharp maximum of the average current at $\omega = c_0 k$ or $v_0 = c_0$, i.e., according to eq. (22.69),

$$v_0 = \left(\frac{c_0}{c}\right) Hd. \tag{22.76}$$

(fig. 153). In this case, there is a resonance with respect to the phase velocities of Josephson and Swihart waves. The condition for the applicability of the result obtained is the smallness of $\bar{j}$ as compared with j_c. Taking $\bar{j}$ in the maximum, we find

$$\frac{c_0^2}{\delta_J^2} \ll \omega\gamma, \tag{22.77}$$

that is, the damping γ must not be too small.

* The remaining current components (the single-particle current and the capacitive current) are proportional to the time derivatives of θ and upon averaging they give zero.

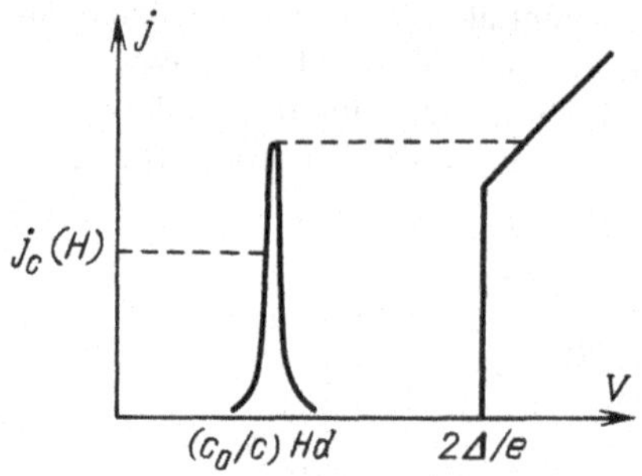

Fig. 153.

As has been repeatedly noted, in practice a steady current across the junction is specified and the steady part of the arising potential difference is recorded. Just as upon application of a varying field, in the case under consideration the current increases up to a limiting value at a given voltage, which is followed by a discontinuous switch to a different regime with a different voltage, i.e., steps appear in the characteristic (see fig. 153).

In order for the above-described resonance effect $v_0 = c_0$ to be observed, it is necessary that the principal Josephson maximum at $V = 0$ be not greater than the resonance maximum with the condition (22.76). Otherwise, a switch will occur of the characteristics with $V = 0$ to the single-particle branch. In view of this, one has to choose the value of the magnetic field close to the value that makes the principal Josephson current vanish (see formula (22.40).

Up until now we have considered the junction to be infinite. A finite junction of width L permits the generation of standing electromagnetic waves if there is an integer number of half-waves within a distance L, i.e., $L = \frac{1}{2} n\lambda$ or $k_n = n\pi/L$. The corresponding frequencies are

$$\omega_n = \frac{c_0 n\pi}{L}. \tag{22.78}$$

Such frequencies correspond to potential differences $V_n = \hbar\omega_n/2e$, i.e.,

$$V_n = \frac{\hbar c_0}{2e}\frac{n\pi}{L}. \tag{22.79}$$

Therefore, instead of a single resonance for an infinite junction, we obtain a set of discrete resonances. We shall not give here a detailed derivation; we confine ourselves to a description of the qualitative picture.

Calculation (Kulik 1965, Dmitrenko et al. 1965) shows that at V_n defined by condition (22.79) sharp maxima of the average current $\bar{j}$ appear at $V = V_n$. The height of these maxima depends on n in such a manner that the highest maximum is that which most closely corresponds to the resonance condition for phase velocities, (22.76), i.e., that for which

$$\left| V_{n_0} - \frac{c_0}{c} Hd \right| = \min. \tag{22.80}$$

Hence, by varying H we can shift the modulation of the "fence" of the maxima (22.79). Since in actual experiments one is concerned with the measurement of $\bar{V}(j)$ instead of $\bar{j}(V)$, a stepped characteristic is obtained (fig. 154), whose shape is essentially governed by the magnetic field H (the term adopted in the literature is "Fiske steps" after the author who discovered them (Fiske 1964).

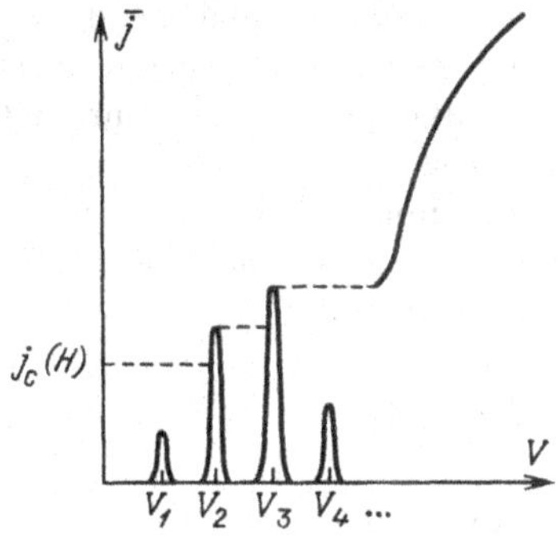

Fig. 154.

As the width of the junction increases the spacing between the adjacent maxima decreases in accordance with formula (22.79). The maxima begin to overlap, and in the limit $L \to \infty$ the curve (22.75) is obtained (cf. figs. 154 and 153).

These properties of the characteristic of finite junctions have been used for the generation of electromagnetic waves with the aid of Josephson junctions (Yanson et al. 1965; Langenberg et al. 1965). One of the resonances (22.79) was used; its amplitude was increased by choosing the magnetic field in accordance with condition (22.80).

22.7. *Josephson junctions with a normal metal or a semiconductor layer*

In the section that follows we will consider various methods used to prepare Josephson junctions. One of the most widespread is still a tunnel junction with an insulating barrier. However, it is rather difficult to prepare such a junction in a controlled manner. As has already been said, the insulator barrier must be very thin and have no short-circuited portions. The usual procedure is the oxidation of a metal followed by sputtering a film of another superconducting metal. But here gaps may appear in the oxide layer and, hence, short-circuited regions. But if the oxide layer is made too thick, no Josephson current will flow.

In view of this, attempts are being made to use a normal metal or semiconductor layer instead of an insulator. The transparency of such a layer, as will be seen below, is much larger than that of an insulator and, hence, it may be made thicker. In such a case it is easier to control the barrier transparency and, hence, to prepare junctions with required properties.

We begin with a normal layer (s–n–s junctions). We use the model described in section 20.1: the junction between two metals with slightly differing critical temperatures. We assume that $T_{c2} - T_{c1} \ll T_{c1}$ and $T_{c1} < T < T_{c2}$. Suppose that the metals form a sandwich, in which the first, normal metal forms a layer of thickness d between two superconducting plates of the second metal. We assume, as before, that $\xi_2 \gg \xi_1$ and put $d \gg \xi_1$. In this case, the critical current flowing across the junction [which will evidently turn out to be proportional to $\exp(-d/\xi_1)$] will be so small that it will not affect the Ψ functions in both superconductors. Therefore, they may be taken in the form (20.18) with some phase factor. Thus, for the metals on the right and left sides we have respectively:

$$\Psi_r = \tanh\left(\frac{x - \frac{1}{2}d + x_{0r}}{\sqrt{2}\xi_2}\right)\exp(i\chi_1), \quad x > \tfrac{1}{2}d,$$

$$\Psi_l = \tanh\left(\frac{\frac{1}{2}d + x_{0l} - x}{\sqrt{2}\xi_2}\right)\exp(i\chi_2), \quad x < -\tfrac{1}{2}d. \tag{22.81}$$

In the normal metal which forms the strip $-\frac{1}{2}d < x < \frac{1}{2}d$, use may be made of the linearized Ginzburg–Landau equation

$$\xi_1^2 \frac{d^2\Psi}{dx^2} = \Psi,$$

of which the solution is

$$\Psi = A\exp\left(\frac{x}{\xi_1}\right) + B\exp\left(\frac{-x}{\xi_1}\right). \tag{22.82}$$

Using the continuity condition $d\Psi/dx$ at the boundaries (the continuity of Ψ is determined by x_{0r} and x_{0l}, which are of no interest in this particular problem) and taking into account that $x_{0r}, x_{0l} \ll \xi_2$ (section 20.1), we find

$$A \approx \frac{\xi_1}{\sqrt{2}\xi_2}\exp\left(i\chi_1 - \frac{d}{2\xi_1}\right),$$

$$B \approx \frac{\xi_1}{\sqrt{2}\xi_2}\exp\left(i\chi_2 - \frac{d}{2\xi_1}\right). \tag{22.83}$$

The current is given by

$$j = -\frac{ie\hbar}{2m}\Psi_0^2(\Psi^*\nabla\Psi - \Psi\nabla\Psi^*),$$

where $\Psi_0^2 = \alpha|\tau|/b$. Substituting eqs. (22.82) and (22.83), we see that the current does not depend on x and is equal to

$$j = -\frac{\mathrm{i}e\hbar\alpha|\tau|}{2mb}(B^*A - A^*B)$$

$$= \frac{e\hbar\alpha|\tau|}{2mb}\frac{\xi_1}{\xi_2^2}\exp\left(-\frac{d}{\xi_1}\right)\sin(\chi_1 - \chi_2). \tag{22.84}$$

Considering that Ψ^2 at the boundary is equal to $(\alpha|\tau|/b)(\xi_1^2/2\xi_2^2)$ (see eq. 20.19), we may write

$$j = \frac{e\hbar}{m\xi_1}|\Psi(\tfrac{1}{2}d)|^2 \exp\left(-\frac{d}{\xi_1}\right)\sin(\chi_1 - \chi_2), \tag{22.85}$$

which is strongly reminiscent of formula (22.21).

In this case, the smallness of the Josephson current is due primarily to the violation of the superconducting correlation in the normal metal, this being responsible for the appearance of the factor $\exp(-d/\xi_1)$. Although the derivation given refers to a very special model, the basic qualitative inference about the critical current being proportional to $\exp(-d/\xi_1)$, where ξ_1 is the correlation length in the normal metal, is valid in all cases. Note that, in contrast to an ordinary tunnel junction with an insulating layer, where the smallness of the Josephson current is due to the small permeability of the potential barrier to electrons, here it is due to the loss of the coherence of the Cooper pairs in the normal metal.

s–n–s junctions have not found wide application in practice. This is due to the low resistance of the metal to normal current, which proves to be inconveient for instruments that make use of Josephson elements. This explains why Josephson elements with a semiconductor barrier s–sm–s have lately been studied intensively (see Alfeev 1979; Aslamazov and Fistul' 1979, 1981, 1982, 1984).

As is known, the electric properties of semiconductors are determined to a considerable extent by the presence of impurities. The impurities in semiconductors create additional electron energy levels in the forbidden region between the last filled band (valence band) and the first empty band (conduction band). The levels may be shallow, i.e., they may lie close to the edge of the valence band (acceptors) or of the conduction band (donors), or deep, i.e., they may lie in the bulk of the forbidden region. If the impurity concentration that creates shallow levels is high, the impurity levels extend into the band that "creeps" on the conduction or valence band. Here, the semiconductor becomes "degenerate", i.e., a finite number of current carriers arises in a band at $T = 0$. Such a semiconductor is, in fact, a semimetal, i.e., in this case the s–sm–s junction becomes a s–n–s junction.

At lower impurity concentrations the electrons or holes in the bands are generated only by way of thermal excitation. However, since the critical temperatures of superconductors are low as compared with the excitation energy, the presence of thermally activated carriers may be neglected. In such a case, the semiconductor is, in fact, an insulator and the height of the potential barrier is equal to the lowest energy required for the generation of electrons in the valence band or of holes in the conduction band or, in other words, to the energy gap. Comparing the exponents of the temperature excitation (V/T) and of the tunneling probabilities ($d\sqrt{mV}/\hbar$) (see section 22.1), where V corresponds to $U(x)-\mu$, we see that for $V>10^{-2}$ eV and $d<10^{-3}$ cm the thermal excitation may really be ignored.

In real determinations of the tunneling probability (Aslamazov and Fistul' 1979, 1981, 1982, 1984) one must take into account the effects that arise at the metal-semiconductor interface. In section 6.1 we have considered the contact between two metals. It has been pointed out that part of the electrons are transferred from one metal to the other, as a result of which an electrical double layer is formed in the vicinity of the interface; at this double layer there is a jump in the potential - a contact potential difference. The thickness of the double layer is of the order of interatomic distance because it is at such distances that the electric field is screened.

The situation is different in semiconductors, where the screening is much weaker. If the semiconductor layer is very thin, it may happen that all the electrons residing at the impurity levels will move into the metal and, as a result, the potential barrier will increase almost up to the spacing between the valence band and the conduction band. On the other hand, with the escape of electrons the semiconductor will acquire an electric charge, which in turn will reduce the effective barrier to a certain extent.

In the thicker layers of the semiconductor the electrons will pass into the metals from the boundary regions. In such a case the potential barrier in the middle will be determined by the energy gap for the impurity levels, and at the edges high but narrow "Schottky barriers" will be generated.

Besides, there exist other effects which exert an influence on the tunneling probability. They include fluctuations in the impurity concentration, leading to the formation of narrow channels with lowered barriers (Aslamazov and Fistul' 1982), the formation of "resonance" trajectories along the periodic arrays of impurity states in a semiconductor (Aslamazov and Fistul' 1984) and, finally, the illumination effects. It has been found experimentally that some s–sm–s junctions, in which no Josephson effect has been observed, exhibit the effect as a result of illumination and retain it when the illumination is turned off (see Barone and Paterno 1982). This is presumably associated with the formation of long-lived metastable electronic states on impurities.

22.8. *Practical applications of the Josephson effect*

It has been shown in section 22.4 that the critical Josephson current depends strongly on the magnetic field. This property can be used for the measurement of small magnetic fields. In fact, for this purpose use is made not of Josephson elements but of so-called Josephson interferometers of SQUIDs (superconducting quantum interference devices).

One of the schemes of such an interferometer is given in fig. 155. It shows two Josephson elements connected in parallel. The total current across the first element is equal to $I_{c1} \sin \theta_1$ and that flowing across the second is $I_{c2} \sin \theta_2$. The overall current across both junctions is given by

$$I = I_{c1} \sin \theta_1 + I_{c2} \sin \theta_2. \tag{22.86}$$

Suppose that a magnetic field is applied normal to the junction plane. Let us carry out a derivation analogous to that made in section 17.6, where magnetic flux quantization was discussed. From the fact that $\boldsymbol{j} = 0$ in the bulk of a superconductor it may be concluded that the order-parameter phase satisfies the condition

$$\nabla \chi = \frac{2e}{c\hbar} \boldsymbol{A}.$$

Integrating along the dashed contour, we see that the variation of the phase along the entire contour is

$$\theta_1 - \theta_2 + \frac{2d}{c\hbar} \Phi,$$

where Φ is the field flux across the contour and θ_1 and θ_2 are the jumps in the phase at the Josephson junctions.

For the order parameter to be single-valued, it is necessary that the total change in the phase be equal to $2\pi n$, where n is an integer, i.e.,

$$\theta_1 - \theta_2 + \frac{2\pi\Phi}{\Phi_0} = 2\pi n$$

(we have substituted $\Phi_0 = \pi c\hbar/e$, the magnetic flux quantum). We may introduce a certain quantity θ and write

$$\theta_1 = \theta - \frac{\pi\Phi}{\Phi_0} + 2\pi n, \qquad \theta_2 = \theta + \frac{\pi\Phi}{\Phi_0}.$$

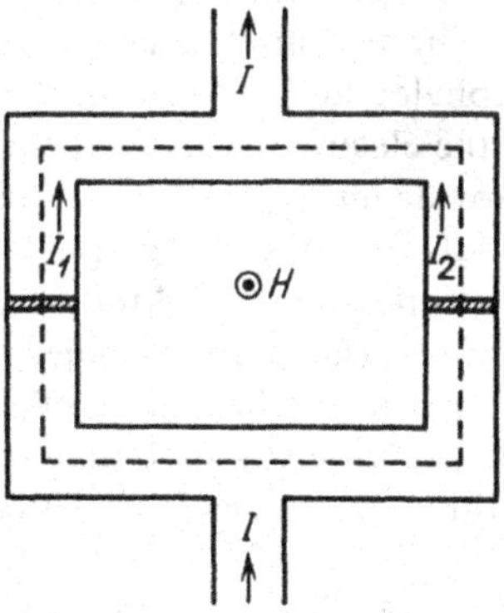

Fig. 155.

Substituting this into formula (22.86) (we assume the magnetic field to be so small that it has no effect on I_{c1} and I_{c2}), we have

$$I = I_{c1} \sin\left(\theta - \frac{\pi\Phi}{\Phi_0}\right) + I_{c2} \sin\left(\theta + \frac{\pi\Phi}{\Phi_0}\right).$$

If both junctions are identical, i.e., $I_{c1} = I_{c2} = I_c$, then

$$I_{max} = \max\left[2I_c \sin\theta \cos\left(\frac{\pi\Phi}{\Phi_0}\right)\right] = 2I_c \left|\cos\left(\frac{\pi\Phi}{\Phi_0}\right)\right|. \tag{22.87}$$

It follows from this that the total critical current vanishes at

$$\Phi = (n + \tfrac{1}{2})\Phi_0. \tag{22.88}$$

But in this case we are dealing not with the magnetic flux across an isolated junction, HLd, but with the flux through a large cavity, $\Phi = HS$ (S is the area of the cavity). Since the magnetic flux quantum is a very small quantity, $\Phi_0 = 2 \times 10^{-7}$ Oe · cm^2, it follows that with the aid of the Josephson interferometer we can measure a field as small as 10^{-10} Oe (recall that the magnetic field of the Earth is 0.5 Oe).

The scheme described above is not the only one possible. There are others, which are more convenient in some cases.

One of the important problems in measuring such small fields is the protection against numerous noises, i.e., random fields that result from various sources, especially in cities. Of course, the creation of such protection would have limited the application of SQUIDs considerably. However, it has been possible to avoid this in practice. The point is that noise sources are usually at a large distance from the object studied and the fields resulting from such sources vary very little in space. Circuits have been developed, which record only the magnetic field gradient or even its second derivative. Although the sensitivity is reduced to a certain extent, the protection may be dispensed with.

Apart from applications for physical investigations, SQUIDs have been employed in biology and medicine. The weak currents that flow in organisms produce, in space, magnetic fields that can be measured. Thus, it is possible to obtain magnetocardiograms and magnetoencephalograms (records of the variation with time of magnetic fields associated with the electrical activity of the heart and the brain). These methods of investigation provide much more details than do the ordinary measurements of the electric potentials (electrocardiography and electroencephalography) and are carried out by a contactless method. Moreover, by recording the magnetic fields at various points of space relative to the object under examination one can calculate the distribution of currents producing these fields with the aid of computers. Although these investigations have been started only recently, it is clear even today that SQUIDs have provided biology and medicine with unique methods of investigation and diagnostics.

The extraordinary sensitivity of SQUIDs to magnetic fields makes it possible to use them for developing supersensitive galvanometers and voltmeters. It has become possible to measure alternating currents of down to 10^{-14} A and potential differences of down to 10^{-15} V.

The capability of closed superconducting circuits to retain undamped current and also the possibility of destroying the superconductivity by magnetic field are used to develop superconducting memory elements for computers - cryotrons. Some of these devices make use of Josephson elements. Two circumstances are important here: miniaturization and high-speed response. For miniaturization use is made of sputtered superconducting films. The smallest contact area produced in this way is $10^{-9}\ \mathrm{cm}^2$ (edge contact, fig. 156a). As regards the high-speed response, the switch of the tunnel junction from the Josephson mode, $j<j_c$, $V=0$, to the single-particle mode, $j>j_c$, $V>2\Delta/e$, is accomplished during a time less than 10^{-10} s.

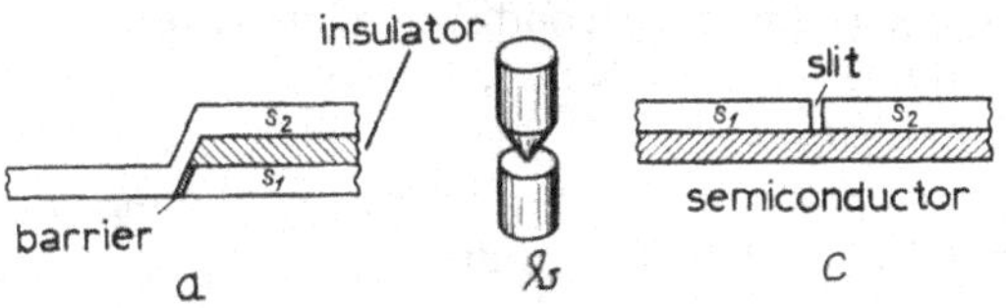

Fig. 156.

One of the key problems in the development of increasingly compact microcircuits, i.e., containing an ever increasing number of elements in a given volume, is the removal of heat released inside such a microcircuit during its operation. The contact with the cooling liquid or gas occurs only along the surface, and heat conduction cannot provide efficient heat removal from the volume. And an increase in temperature can lead to the destruction of the entire structure. In this sense, Josephson elements possess an important advantage because the heat release upon their switching is negligible.

We have so far been concerned with the use of the dc Josephson effect. As regards the ac Josephson effect, in spite of the existence of many ideas, it is still used in rare cases. It would be most natural to use Josephson elements for the generation of electromagnetic oscillations with a retunable frequency. The main difficulty is bringing the waves outside the junction. The use of the resonance of Josephson waves with Swihart waves (section 22.6) has made it possible to obtain a radiation power of up to 10^{-10} W with an efficiency of 2%. Of course, this figure is very low, and at present there exist other devices which enable one to obtain a much higher power at the same frequencies.

In principle, the following two applications of Josephson generation are possible: (a) in low-temperature low-noise devices that do not require a high power; and (b) for measurement of potential differences smaller than 10^{-15} V using the relation $\omega=2eV/\hbar$, since frequency determinations are carried out with a high accuracy and allow one to bring the measured voltage down to 10^{-17} V.

A further application of the ac Josephson effect is the detection of external radiation. The simplest application is to use the decrease of the critical Josephson current in the presence of an external electromagnetic field. According to formula (22.67),

$$j_{0c} = j_c \left| J_0\left(\frac{2eV}{\hbar\Omega}\right)\right|, \tag{22.89}$$

where J_0 is the Bessel function. In principle, one can also use the other steps of the characteristic (at $V_n = n\hbar\Omega/2e$).

In conclusion, let us consider various methods for the preparation of Josephson elements. We have already considered two types: the tunnel junction and the bridge. In the first case, the surface of a superconductor is subjected to oxidation or a layer of a normal metal or a semiconductor is deposited onto it by sputtering. This is followed by sputtering a film of a second superconductor over the first one. One of the versions of the tunnel junction is the edge, in which the junction is made along the edge of a thin film (fig. 156a).

As for the bridge, it can be prepared in the form of a film with a constriction (fig. 146) or a film of varying thickness. A further type of bridge is a point junction (fig. 156b).

Finally, the use of a normal metal or semiconductor as the barrier makes it possible to produce so-called palanar elements (fig. 156c), in which a superconductor film with a very narrow slit is sputtered onto a semiconductor substrate. Here, the substrate serves as the barrier. In planar elements it is easier to control the properties of the barrier mateirial; this may be an ordinary bulk single-crystalline sample with a certain amount of dopant (if we deal with a semiconductor).

22.9. The dynamical resistive state of a thin superconductor at supercritical currents

It has been shown in section 17.5 that for a very thin superconductor there exists a critical current density above which the superconductivity is destroyed. On the basis of this result one could come to the conclusion that when the current increases up to this value there occurs a transition to the normal phase and the characteristic changes to the linear dependence $\boldsymbol{j} = \sigma\boldsymbol{E}$ (fig. 157). In fact, the experimental findings are dependences of the type shown in fig. 158 (Meyer and Minnigerode 1972), i.e., a gradual approach to the normal resistance, and jumps in the electric field. This strongly resembles the picture considered in section 22.5 for the Josephson junction, and shows that superconductivity is retained in some form in the presence of the electric field as well.

At first sight this conclusion seems paradoxical. As a matter of fact, from the London equation $\partial(\Lambda\boldsymbol{j})/\partial t = \boldsymbol{E}$ it follows that in the presence of an electric field the current must increase with time. Hence, it will certainly exceed the critical value,

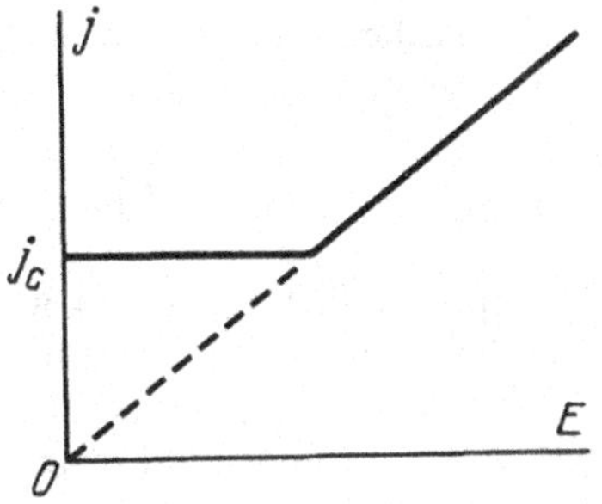

Fig. 157.

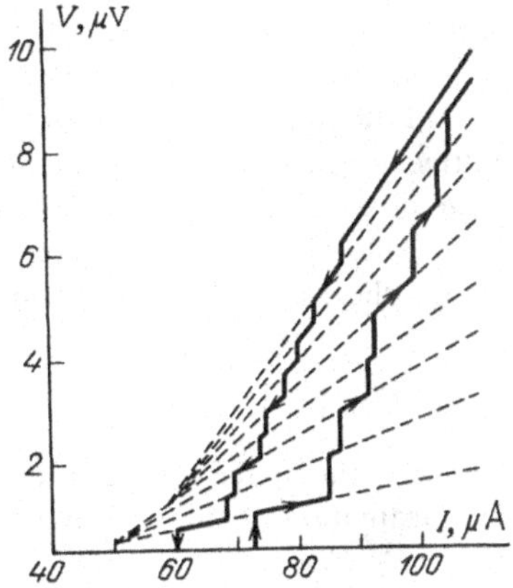

Fig. 158.

which will lead to the destruction of superconductivity. On the other hand, we have already encountered a similar phenomenon: the motion of the vortex lattice in a type II superconductor under the action of a magnetic field and a current perpendicular to it leads to the generation of an electric field with the superconducting order parameter being retained. Another example is the Josephson junction at $j > j_c$.

The appearance of an electric field certainly leads to nonstationarity, i.e., time dependence of the quantities. As has already been noted, the nonstationary equations of the theory of superconductivity are very complicated, even in the limit of slow time and space variations. Therefore, we use the model equation (19.47) for the vicinity of T_c.

However, in this case we have to overcome the following difficulty. Suppose we have an infinite superconducting wire. It is clear that the electric field that arises in it must be, on an average, homogeneous along the wire. But in this case Ψ must also be homogeneous on an average. At the same time, the potential φ appearing in eq. (19.47) must increase in magnitude.

A difficulty of the same kind has been enountered in ch. 18, where we dealt with type II superconductors in a magnetic field. It has been shown that the rise of the vector potential is compensated by the appearance of vortices. They give rise to jumps in the phase gradient at the cut lines that pass through the vortex centers (see fig. 113).

We shall now consider a superconducting wire (a whisker) having a diameter much smaller than both δ and ξ. Here all the quantities will depend only on the coordinate along the wire and on time. Thus, instead of the (x, y) plane in this case we have the (x, t) plane. Since the magnetic field of the current in the whisker is small, we may choose $\boldsymbol{A}=0$ and $\boldsymbol{E}=-\nabla\varphi$. In this case we can establish the similarity with the vortex picture:

$$-\mathrm{i}\frac{\partial}{\partial x}\to-\mathrm{i}\frac{\partial}{\partial x},\qquad -\mathrm{i}\frac{\partial}{\partial y}-\frac{2e}{\hbar c}A_y\to-\mathrm{i}\frac{\partial}{\partial t}+\frac{2e}{\hbar}\varphi. \tag{22.90}$$

Comparing with fig. 113 and the reasonings given in section 18.2, we come to the conclusion that a vortex lattice appears in the (x, t) plane (Ivlev and Kopnin 1978). On encircling each vortex the phase must vary by an amount 2π. The compensation of the increase of the scalar potential will occur due to the jumps in the phase derivative at the cut line along the t axis. The compensation condition reads

$$\frac{2\pi}{t_0}=\frac{2e}{\hbar}\langle E\rangle a, \tag{22.91}$$

where t_0 is the time period, a is the space period and $\langle E\rangle$ is the average electric field (with respect to the coordinate and time). Solving condition (22.91) for the electric field, we obtain

$$\langle E\rangle=\frac{\Phi_0}{ct_0}\frac{n}{L}, \tag{22.92}$$

where Φ_0 is the flux quantum, L is the length of the conductor and n is the number of "vortex" centers ($n=L/a$).

The physical interpretation of the resultant picture is different from that observed with magnetic vortices. At certain points along the conductor there a jump occurs in the time derivative of the phase $\partial\chi/\partial t$, i.e., the rate of variation of the phase with time. This process is periodic. During the larger part of the period the rate of change of the phase on two sides of the points on the conductor is the same. However, at certain moments the phase changes more sharply, and this occurs differently on the two sides from this point, so that a phase difference equal to 2π results. Therefore, such points are called "phase slip centers"*. Since the function $\Psi=|\Psi|\exp(\mathrm{i}\chi)$ must be single-valued, it follows that at the center itself $|\Psi|=0$ at the moment of phase slip.

* The idea of the phase slip centers was first put forward by Anderson (1966) and Little (1967) and described in detail by Langer and Ambegoakar (1967).

Note that fig. 113 shows a rectangular lattice corresponding to the time-independent location of the phase slip centers along the superconductor. However, generally speaking, the lattice may also be oblique, which would correspond to the motion of the entire structure along the superconductor with a constant velocity. This is possible, in principle, for a finite superconductor if there is a mechanism of creation and destruction of the phase slip centers at the edges.

Let us now consider the events that occur in a thin superconductor of finite length. At $\boldsymbol{A}=0$ the expression for the current has the form

$$j=|\Psi|^2\frac{\hbar e}{m}\frac{\partial\chi}{\partial x}. \tag{22.93}$$

If we plot a three-dimensional graph in which x is the coordinate along the superconductor, z corresponds to Re Ψ, and y corresponds to Im Ψ, then the spiral curve will correspond to a state with current (fig. 159a). However, in the presence of an electric field the current increases with time; this means that the spiral is "compressed", i.e., the density of its turns increases. But this process cannot continue for an indefinitely long time because the current reaches a value j_c. There is only one possibility: to dispose of one of the turns of the spiral. But to do this, we have to "straighten out" the spiral at some place. This exactly corresponds to the appearance of a phase slip center and, in particular, $|\Psi|=0$ (fig. 159b).

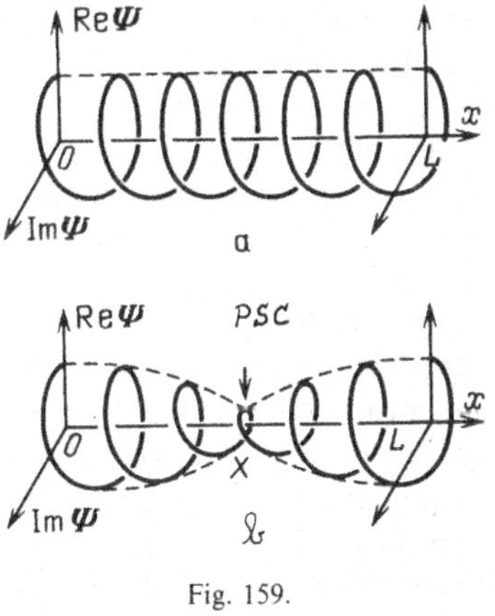

Fig. 159.

Let us consider a simple model (Skocpol et al. 1974), which explains the characteristic curve in fig. 158 with jumps in the potential difference. Suppose that the thin superconductor under study is connected on both sides to bulk specimens of the same metal. Such a construction is called a bridge and we have already described it as a possible realization of the Josephson junction. In this particular case, the constricted portion is assumed to be long. For bulk superconductors the flowing current is weak and they are in a state of equilibrium. If the temperature is close to T_c and eq. (19.47) is applicable, then in the junctions its right-hand side must equal zero and, hence, the left-hand side must also vanish. Substituting $\Psi = |\Psi| e^{i\chi}$, we obtain

$$\frac{\partial |\Psi|}{\partial t} + i\left(\frac{\partial \chi}{\partial t} + \frac{2e}{\hbar}\varphi\right)|\Psi| = 0.$$

Separating the real and imaginary parts, we have

$$\frac{\partial |\Psi|}{\partial t} = 0, \tag{22.94}$$

$$\frac{\partial \chi}{\partial t} + \frac{2e}{\hbar}\varphi = 0. \tag{22.95}$$

These equation are the boundary conditions for the thin superconductor under study.

Condition (22.95) is also realized at temperatures far from T_c. It may be interpreted as the equality of the chemical potentials of "normal electrons" ($\mu_n = e\varphi$) and of electrons constituting the Cooper pairs ($\mu_s = -\frac{1}{2}\hbar\partial\chi/\partial t$). We introduce the notation

$$\eta = \mu_n - \mu_s = e\varphi + \tfrac{1}{2}\hbar\frac{\partial \chi}{\partial t}. \tag{22.96}$$

According to (22.95), $\eta = 0$ in a bulk superconductor. However, just as in the case of a magnetic field, this quantity cannot vanish abruptly and has a finite penetration depth.

Indeed, let us visualize the interface between a superconductor and a normal metal. If an electron is "injected" into the superconductor, then in order to assume an equilibrium state it has to lose an energy Δ because the equilibrium electrons are bound into pairs with a binding energy 2Δ (per two particles).

The electron loses its energy in collisions. However, collisions with impurities are inefficient because they are elastic, and collisions with quasiparticles are rare ($\propto e^{-\Delta/T}$). Therefore, the only process that provides energy relaxation is the emission of phonons. At low temperatures the characteristic time of such a process, τ_{ph}, is large. As a result of this, the penetration depth for the quantity η is large compared to the correlation length ξ (of order $\xi(T_c\tau_{ph}/\hbar)^{1/2}$ for a pure superconductor). We label this quantity by δ_E, since it also determines the spatial variation of the electric field in the superconductor.

By analogy with the London equation we write

$$\frac{\partial^2 \eta}{\partial x^2} - \delta_E^{-2}\eta = 0 \tag{22.97}$$

(the exact theory confirms this equation; see Ivlev and Kopnin 1984). Suppose the junctions of the superconductor are located at points 0 and L. Solving eq. (22.97) with the boundary condition $\eta = 0$ at $x = 0$, we get

$$\eta = C_1 \sinh\left(\frac{x}{\delta_E}\right). \tag{22.98}$$

On the other hand, we can solve eq. (22.97) by using the boundary condtion $\eta = 0$ at $x = L$. This gives

$$\eta = C_2 \sinh\left(\frac{x - L}{\delta_E}\right). \tag{22.99}$$

If there are no phase slip centers in the superconductor, the general solution is $\eta = 0$. This is possible only as long as the current is less than j_c.

When the current exceeds j_c, a phase slip center appears. Let it be located at point X. Here we obtain

$$\eta(X-0) = C_1 \sinh\left(\frac{X}{\delta_E}\right), \qquad \eta(X+0) = C_2 \sinh\left(\frac{X-L}{\delta_E}\right).$$

But at the phase slip center the phase derivative $\partial\chi/\partial t$ undergoes a discontinuity: $\Delta(\partial\chi/\partial t) = 2\pi/t_0$. In accordance with (22.96), it follows that

$$C_2 \sinh\left(\frac{X-L}{\delta_E}\right) - C_1 \sinh\left(\frac{X}{\delta_E}\right) = \frac{\pi\hbar}{t_0}. \tag{22.100}$$

Due to the phase slip all the quantities depend periodically on time with a period t_0. In experiments the time-averaged current and potential difference are measured (such averages are designated by a bar). According to (22.96), $\eta = e\bar{\varphi}$ (since $\overline{\partial\chi/\partial t} = 0$). The average electric field on the left from the point of phase slip ($0 < x < X$) is deduced from formula (22.98):

$$\bar{E} = -\frac{\partial\varphi}{\partial x} = -\frac{1}{e}\frac{\partial\eta}{\partial x} = -\frac{C_1}{e\delta_E}\cosh\left(\frac{x}{\delta_E}\right). \tag{22.101}$$

On the right of X ($X < x < L$) we have

$$\bar{E} = -\frac{C_2}{e\delta_E}\cosh\left(\frac{x-L}{\delta_E}\right). \tag{22.102}$$

The constants C_1 and C_2 are found on the basis of the following considerations.

In experiments the total current j is specified most easily. It consists of the superconducting and normal parts:

$$j = j_s + j_n.$$

This formula is exact near T_c (see the end of section 19.3); j_s was determined in section 16.8 ($\boldsymbol{j}_s = -Q\boldsymbol{A}$, where Q is expressed by formulae 16.88 and 16.92) and $j_n = \sigma E$, where σ is the normal conductivity. However, in principle, an analogous formula may be written at any temperatures. To do this, one must use the relation

$\boldsymbol{j}=-Q\boldsymbol{A}$ in the limit $\boldsymbol{q}=0$, $\omega \ll \Delta$ (here we mean the gauge $\varphi=0$, $\boldsymbol{E}=-c^{-1}\partial \boldsymbol{A}/\partial t$) and expand in ω:

$$\boldsymbol{j}=-Q(0)\boldsymbol{A}+\frac{\mathrm{i}\omega}{c}\sigma \boldsymbol{A},$$

where σ is a certain constant, which depends on Δ. The second term plays the role of $\boldsymbol{j}_\mathrm{n}$ since $(\mathrm{i}\omega/c)\boldsymbol{A}=-c^{-1}\partial \boldsymbol{A}/\partial t=\boldsymbol{E}$.

According to fig. 153, at the phase slip center the supercurrent fluctuates from the maximum value to zero. The maximum value must not exceed j_c. In view of this, in such a center $\bar{j}_\mathrm{s}=\alpha j_\mathrm{c}$, where $\alpha<1$. We assume that even in the presence of several centers in each of them $\bar{j}_\mathrm{s}=\alpha j_\mathrm{c}$ with the same α, which is independent of the number of centers. Of course, this is a model assumption but, as will be seen below, it enables one to account for the stepped characteristic in fig. 158. At this point we will consider a single center. Thus, we have

$$\bar{j}_\mathrm{s}(X)=j-\bar{j}_\mathrm{n}(X)=\alpha j_\mathrm{c}. \tag{22.103}$$

Substituting $j_\mathrm{n}=\sigma E$ and E into eq. (22.103) in accordance with eqs. (22.101) and (22.102), we find

$$C_1=-\frac{(j-\alpha j_\mathrm{c})e\delta_E}{\sigma \cosh(X/\delta_E)}, \qquad C_2=-\frac{(j-\alpha j_\mathrm{c})e\delta_E}{\sigma \cosh[(L-X)/\delta_E]}. \tag{22.104}$$

The period t_0 is associated with the potential difference. Inserting the length of the conductor L as a space period into formula (22.91) gives

$$\frac{2\pi}{t_0}=\frac{2e}{\hbar}\langle E\rangle L=\frac{2e}{\hbar}\int_0^L \bar{E}\,\mathrm{d}x=\frac{2e}{\hbar}[\varphi(0)-\bar{\varphi}(L)]=\frac{2eV}{\hbar}. \tag{22.105}$$

This is none other than the Josephson relation (22.55).

Formula (22.105) can also be derived directly from the problem under examination. Since at the ends of the conductor $\eta=0$, it follows that

$$\varphi(0)-\varphi(L)=\frac{\hbar}{2e}\left[\left(\frac{\partial \chi}{\partial t}\right)_L-\left(\frac{\partial \chi}{\partial t}\right)_0\right].$$

If we take the time average, then the regular part of the derivative $\partial\chi/\partial t$ gives zero and there will remain only a jump equal to $2\pi/t_0$, i.e., $\bar{\varphi}(0)-\bar{\varphi}(L)=\hbar\pi/et_0$, which is consistent with formula (22.105).

Inserting the expressions for C_1 and C_2 into eq. (22.100) and expressing the right-hand side in terms of V, we obtain

$$V=(j-\alpha j_\mathrm{c})\frac{\delta_E}{\sigma}\left[\tanh\left(\frac{X}{\delta_E}\right)+\tanh\left(\frac{L-X}{\delta_E}\right)\right].$$

From symmetry considerations it may be presumed that the phase slip center is located in the middle of the conductor, i.e., $X=\frac{1}{2}L$. Here we have

$$V=2(j-\alpha j_\mathrm{c})\frac{\delta_E}{\sigma}\tanh\left(\frac{L}{2\delta_E}\right). \tag{22.106}$$

In accord with eqs. (22.101) and (22.102) $\bar{E}$ and also the normal current increase from the edges to the middle portion of the conductor. Hence, the supercurrent is at a maximum at the edges. Here it cannot exceed j_c. This leads to the following condition:

$$j_s(0) = j - \frac{(j - \alpha j_c)}{\cosh(L/2\delta_E)} \leqslant j_c.$$

At the same time, we must have $j \geqslant j_c$. From this we obtain the interval of values of the current j within which formula (22.106) applies:

$$j_c \leqslant j \leqslant j_c \frac{1 - \alpha \operatorname{sech}(L/2\delta_E)}{1 - \operatorname{sech}(L/2\delta_E)}. \tag{22.107}$$

At $j = j_c$ there is a jump in the potential from zero to

$$V_1 = 2(1 - \alpha) j_c \frac{\delta_E}{\sigma} \tanh\left(\frac{L}{2\delta_E}\right). \tag{22.108}$$

This is followed by a linear portion of the characteristic up to the upper limit of the current given by formula (22.107). Here we have

$$V_1' = \frac{2(1 - \alpha) j_c (\delta_E/\sigma) \tanh(L/2\delta_E)}{1 - \operatorname{sech}(L/2\delta_E)}. \tag{22.109}$$

At a current j exceeding the upper limit in (22.107) two phase slip centers appear. It is rather natural to assume that they will be located so that the current j_n is symmetric with respect to either of them. This is possible if they are located at $X_1 = \frac{1}{4}L$, $X_2 = \frac{3}{4}L$. In such a case,

$$\eta = C_1 \sinh\left(\frac{x}{\delta_E}\right), \qquad 0 < x < X_1,$$

$$\eta = C_2 \sinh\left(\frac{x - \frac{1}{2}L}{\delta_E}\right), \qquad X_1 < x < X_2,$$

$$\eta = C_3 \sinh\left(\frac{x - L}{\delta_E}\right), \qquad X_2 < x < L.$$

Let us define the coefficients C_1, C_2 and C_3, from the condition requiring that in the phase slip centers $\bar{j}_s = j - \bar{j}_n$ be equal to αj_c. This gives

$$C_1 = C_2 = C_3 = -(j - \alpha j_c) \frac{(e\delta_E/\sigma)}{\cosh(L/4\delta_E)}.$$

From the condition of the jump in η in the phase slip center we obtain the relation between the coefficients C_i and the period t_0, and from a formula analogous to

(22.105) (but this time we have $4\pi/t_0$ on the left-hand side) we deduce the relation between t_0 and V. This gives the following portion of the characteristic:

$$V = 4(j - \alpha j_c)\frac{\delta_E}{\sigma}\tanh\left(\frac{L}{4\delta_E}\right). \tag{22.110}$$

Formula (22.110) holds at currents starting from the upper limit (22.107) up to the point where the maximum of the supercurrent reaches the value j_c. Such a maximum is attained at $x = 0, \frac{1}{2}L$ and L. This yields

$$j_c\frac{1-\alpha\,\mathrm{sech}(L/2\delta_E)}{1-\mathrm{sech}(L/2\delta_E)} \leqslant j \leqslant \frac{1-\alpha\,\mathrm{sech}(L/4\delta_E)}{1-\mathrm{sech}(L/4\delta_E)}. \tag{22.111}$$

At the lower edge of this interval there occurs a jump in the voltage from the value V_1' (22.109) to

$$V_2 = \frac{4(1-\alpha)j_c(\delta_E/\sigma)\tanh(L/4\delta_E)}{1-\mathrm{sech}(L/2\delta_E)} \tag{22.112}$$

(it is not difficult to see that $V_2 > V_1'$). This is followed by a linear portion of the characteristic in accordance with formula (22.110), which continues up to the upper limit of the interval (22.111), on which the voltage becomes equal to

$$V_2' = \frac{4(1-\alpha)j_c(\delta_E/\sigma)\tanh(L/4\delta_E)}{1-\mathrm{sech}(L/4\delta_E)}. \tag{22.113}$$

A further increase in the current leads to the appearance of a large number of phase slip centers and to the analogous dependences $V(j)$. A plot of $\sigma E/j_c$ against j/j_c $(E = V/L)$ for $L/\delta_E = 16$ is given in fig. 160. Here we take $\alpha = 0.09$ (see below).

Thus, it turns out that the characteristic consists of jumps in voltage and portions with a linear dependence $V(j)$. This is in agreement with experiment. As L increases

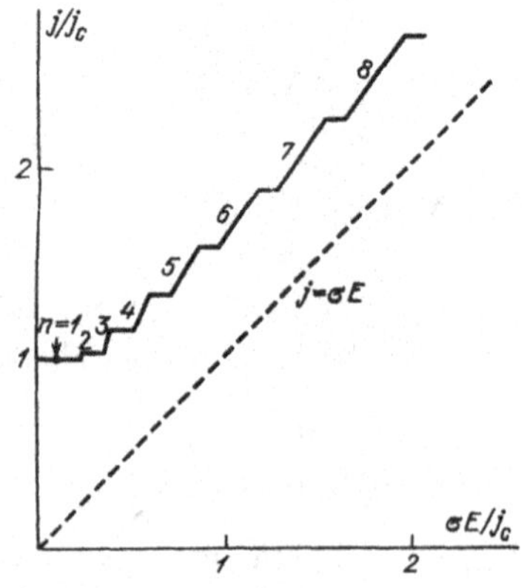

Fig. 160.

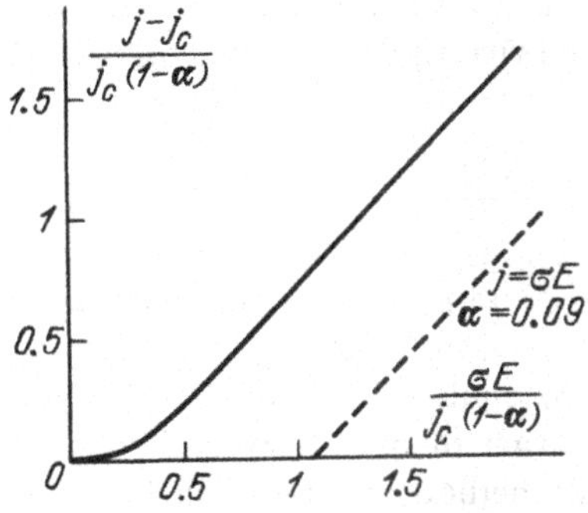

Fig. 161.

the intervals (22.107) and (22.111) and all the subsequent ones decrease and the characteristic approaches a continuous $V(j)$ curve. This dependence is not difficult to find by generalizing the preceding derivation to n phase slip centers. The linear portion of the characteristic of the number n has the form

$$V_n(j) = 2n \tanh\left(\frac{L}{2n\delta_E}\right)(j - \alpha j_c)\frac{\delta_E}{\sigma}. \tag{22.114}$$

The magnitude of the total current at which the maximum supercurrent reaches the critical value is given by

$$j_n = j_c \frac{1 - \alpha\ \mathrm{sech}(L/2n\delta_E)}{1 - \mathrm{sech}(L/2n\delta_E)}. \tag{22.115}$$

Inserting (22.115) into (22.114) yields

$$V_n(j_n) = \frac{2nj_c(1-\alpha)(\delta_E/\sigma)\sinh(L/2n\delta_E)}{\cosh(L/2n\delta_E) - 1}. \tag{22.116}$$

With $n \to \infty$ the number n may be treated as a continuous variable. It is even more convenient to introduce the variable $z = L/(2n\delta_E)$. From eqs. (22.115) and (22.116) we obtain

$$\frac{(j - j_c)}{j_c(1-\alpha)} = \frac{1}{\cosh z - 1}, \tag{22.117}$$

$$\frac{\sigma E}{j_c(1-\alpha)} = \frac{\sinh z}{z}\frac{1}{\cosh z - 1}, \tag{22.118}$$

where $E = V/L$. This is a parametric representation of the dependence $E(j)$ in the limit $L \gg \delta_E$.

With a large current $j \gg j_c$ we have $z \ll 1$. In such a case we find from eqs. (22.117) and (22.118)

$$j = \sigma E + \tfrac{1}{3}(2+\alpha)j_c, \tag{22.119}$$

that is, the characteristic is a straight line parallel to Ohm's law. Conversely, large z values correspond to the vicinity of j_c. From eqs. (22.117) and (22.118) we obtain in this case

$$\frac{j-j_c}{j_c} \approx 2(1-\alpha) \exp\left(-\frac{j_c(1-\alpha)}{\sigma E}\right). \qquad (22.120)$$

The $j(V)$ curve with $V \to 0$ and $j \to j_c$ has a horizontal tangent. The plot of the curve (22.117) and (22.118) is given in fig. 161.

The results of the phenomenological model described above are in good agreement with the results of the exact theory (Ivlev and Kopnin 1984)*. Comparing the formulae derived by both methods, we may conclude that $\alpha \leqslant 0.09$.

* The exact theory is not free from some arbitrariness either, but many of the data provide evidence in favor of the principle of "minimum entropy production". If this principle is used, in the exponential law (22.119) we obtain the coefficient 0.91 (as $1-\alpha$) and in the formula for $j \gg j_c$ we obtain an extra current $\Delta j = 0.68 j_c$. If $\alpha = 0.09$ is inserted into formula (22.119), we obtain $\Delta j = 0.696 j_c$.

Appendix

I

The ferromagnetic metal model

Real ferromagnetic metals have complicated (i.e. consisting of several bands) and strongly anisotropic electronic spectra. However, some of the properties of such metals can be demonstrated by the example of an isotropic Fermi liquid (see ch. 13) with a sufficiently large exchange part of the function f (Abrikosov and Dzyaloshinskii 1958).

A ferromagnetic substance is characterized by the fact that in a state of equilibrium it has a nonzero net spin and magnetic moment or, as is usually said, it possesses a spontaneous magnetization. Let the unit vector $\boldsymbol{m}$ characterize the direction of the spontaneous moment. The energy of quasiparticles depends on the orientation of the spin with respect to $\boldsymbol{m}$ and, in view of this, it may be written in the form

$$\varepsilon(\boldsymbol{p}, \boldsymbol{\sigma}) = \varepsilon_0(\boldsymbol{p}) - b(\boldsymbol{p})\boldsymbol{\sigma m}. \tag{AI.1}$$

According to this formula, the energy of the electron with a spin parallel to $\boldsymbol{m}$ is $\varepsilon_0 - b$ and, accordingly, the equilibrium distribution function is $n_{\rm F}(\varepsilon_0 - b) \equiv n^+$. For an electron with an opposite spin the energy is equal to $\varepsilon_0 + b$ and the equilibrium distribution function is $n_{\rm F}(\varepsilon_0 + b) \equiv n^-$. The operator having eigenvalues n^+ and n^- corresponding to appropriate spin orientation can be written as

$$n_0(\boldsymbol{p}, \boldsymbol{\sigma}) = \tfrac{1}{2}(n^+ + n^-) + \tfrac{1}{2}(n^+ - n^-)\boldsymbol{\sigma m}. \tag{AI.2}$$

It should be considered to be the equilibrium density matrix.

In the isotropic Fermi liquid the f-function has the form*

$$f = \eta(\boldsymbol{p}, \boldsymbol{p}') + (\boldsymbol{\sigma\sigma}')\zeta(\boldsymbol{p}, \boldsymbol{p}'). \tag{AI.3}$$

There is a certain connection between $b(\boldsymbol{p})$ in formula (AI.1) and function ζ in (AI.3). In order to find it, let us see how the electron energy varies upon rotation of the vector $\boldsymbol{m}$ by an angle $\delta\theta$. Here, $\delta\boldsymbol{m} = [\delta\boldsymbol{\theta m}]$ and, according to (AI.1), we have

$$\delta\varepsilon = -b[\boldsymbol{m\sigma}]\delta\boldsymbol{\theta}. \tag{AI.4}$$

* In principle, there may be terms proportional to $\boldsymbol{m}(\boldsymbol{\sigma} + \boldsymbol{\sigma}')$ and $(\boldsymbol{m\sigma})(\boldsymbol{m\sigma}')$, but it can be shown that they are unimportant in further calculations and are therefore omitted here.

But, on the other hand, as the value of m varies, the equilibrium distribution function n_0 varies as well:

$$\delta n_0(\boldsymbol{p}, \boldsymbol{\sigma}) = \tfrac{1}{2}(n^+ - n^-)[\boldsymbol{m\sigma}]\delta\theta$$

and the energy also varies with it:

$$\delta\varepsilon = \mathrm{Sp}_{\sigma'} \int f\delta n_0 \frac{\mathrm{d}^3 p}{(2\pi\hbar)^3}$$
$$= \mathrm{Sp}_{\sigma'} \int \tfrac{1}{2} f(n^+ - n^-)[\boldsymbol{m\sigma}']\delta\theta \frac{\mathrm{d}^3 p}{(2\pi\hbar)^3}. \tag{AI.5}$$

Equating (AI.4) and (AI.5) at an arbitrary $\delta\theta$, we obtain

$$-b[\boldsymbol{m\sigma}] = \mathrm{Sp}_{\sigma'} \int \tfrac{1}{2} f(n^+ - n^-)[\boldsymbol{m\sigma}'] \frac{\mathrm{d}^3 p}{(2\pi\hbar)^3}.$$

Substituting (AI.3) into the above expression, we find

$$b(\boldsymbol{p}) = -\int \zeta(\boldsymbol{p}, \boldsymbol{p}')(n^+ - n^-) \frac{\mathrm{d}^3 p}{(2\pi\hbar)^3}. \tag{AI.6}$$

From formula (AI.6) we can find the condition for the generation of ferromagnetism. Let us consider the case of an infinitestimal b. Then, $n^+ - n^-$ may be expanded in b as

$$n^+ - n^- \approx -2 \frac{\partial n_{\mathrm{F}}(\varepsilon_0)}{\partial \varepsilon_0} b.$$

Substitution of the above relation into (AI.6) leads to

$$b(\boldsymbol{p}) = -2 \int \zeta(\boldsymbol{p}, \boldsymbol{p}') b(\boldsymbol{p}') \frac{\mathrm{d}^3 p}{(2\pi\hbar)^3 v'}, \tag{AI.7}$$

where the integration runs over the Fermi surface. In an isotropic liquid b depends only on $|\boldsymbol{p}|$. Assuming $|\boldsymbol{p}| = p_0$, we obtain in (AI.7) the function ζ, which depends only on the angle $\theta = (\widehat{\boldsymbol{p}, \boldsymbol{p}'})$. From (AI.7) we find

$$Z_0 = \nu(\mu) \int \zeta(0) \frac{\mathrm{d}\Omega}{4\pi} = -1,$$

which is the same value obtained from the condition of the infinite paramagnetic susceptibility (see eq. 13.16). Thus, for ferromagnetism to exist it is necessary that

$$Z_0 < -1. \tag{AI.8}$$

In the presence of ferromagnetism there arises a new type of spin waves. In order to obtain their spectrum, we use a kinetic equation in the form (13.25), which in this case reads

$$\frac{\partial n}{\partial t} + \frac{\partial \varepsilon}{\partial \boldsymbol{p}} \frac{\partial n}{\partial \boldsymbol{r}} - \frac{\partial \varepsilon}{\partial \boldsymbol{r}} \frac{\partial n}{\partial \boldsymbol{p}} + \frac{\mathrm{i}}{\hbar}[\varepsilon, n] = 0. \tag{AI.9}$$

where $[\varepsilon, n] = \varepsilon n - n\varepsilon$. We will look for the distribution function in the form $n = n_0 + \delta n$, where n_0 is the equilibrium function (AI.2) and δn depends on the coordinates and time as $\exp[\mathrm{i}(\boldsymbol{kr} - \omega t)]$.

Suppose we have δn as the sum of two terms, one depending on the spin, the other independent of it:

$$\delta n = \nu(\boldsymbol{p}) + \boldsymbol{\nu}(\boldsymbol{p})\,\boldsymbol{\sigma}. \tag{AI.10}$$

We substitute the distribution function $n_0 + \delta n$ into eq. (AI.9), taking into account the variation of the energy with the distribution function. Retaining the terms linear in δn, we obtain a system of equations for ν and $\boldsymbol{\nu}$.

This system of equations splits into two systems, the first of which corresponding to oscillations in the electron density and to the associated oscillations in the spin projection onto the direction of the magnetic field, $\nu_z = \boldsymbol{\nu m}$. However, since the oscillations of the electron density are accompanied by a change in charge density, it is necessary to take into account the electric field that arises in this case. This immediately leads to frequencies ω of order $\mu/\hbar$ - just as in section 13.5 - and for this reason such oscillations need not be considered here. It is the second system of equations, corresponding to the oscillations of the transverse components of the total spin, we are interested in.

Note that in nonferromagnetic metals, for which the energy and n_0 do not include terms with $\boldsymbol{m}$, the oscillations in the quantity $\boldsymbol{\nu}$ are not associated with the oscillations in the density, and therefore, in principle, there may arise spin waves with a longitudinal polarization and a linear dependence of ω on k (section 13.6).

We introduce, instead of the components ν_x and ν_y, the quantities $\nu_\pm = \nu_x \pm \mathrm{i}\nu_y$. Then, from the system of equations for ν_x and ν_y we obtain for ν_+,

$$\begin{aligned}&-\omega\nu_+ + \boldsymbol{kv}\nu_+ + [\boldsymbol{k}(\boldsymbol{v}-\boldsymbol{u})\,\delta(\varepsilon_0 - b - \mu) + \boldsymbol{k}(\boldsymbol{v}+\boldsymbol{u})\,\delta(\varepsilon_0 + b - \mu)]\\ &\quad\times\int \zeta(\boldsymbol{p},\boldsymbol{p}')\,\nu_+(\boldsymbol{p}')\frac{\mathrm{d}^3p'}{(2\pi\hbar)^3} - \frac{2b}{h}\nu_+ - \frac{2}{\hbar}(n_+ - n_-)\\ &\quad\times\int \zeta(\boldsymbol{p},\boldsymbol{p}')\,\nu_+(\boldsymbol{p}')\frac{\mathrm{d}^3p'}{(2\pi\hbar)^3},\end{aligned} \tag{AI.11}$$

where $\boldsymbol{v} = \partial\varepsilon_0/\partial\boldsymbol{p}$ and $\boldsymbol{u} = \partial b/\partial\boldsymbol{p}$.

The equation for ν_- differs in sign for the last two terms and can formally be derived from (AI.11) by using the transformations $\omega \to -\omega$ and $\boldsymbol{k} \to -\boldsymbol{k}$.

We will solve eq. (AI.11) by means of successive approximations. At $\boldsymbol{k} = 0$ we have

$$\omega^{(0)}\nu^{(0)} + \frac{2b}{h}\nu^{(0)} + \frac{2}{\hbar}(n_+ - n_-)\int \zeta(\boldsymbol{p},\boldsymbol{p}')\,\nu^{(0)}(\boldsymbol{p}')\frac{\mathrm{d}^3p'}{(2\pi\hbar)^3} = 0. \tag{AI.12}$$

Let us integrate this equation over $\mathrm{d}^3\boldsymbol{p}$. In the integral of the last term we replace the variables $\boldsymbol{p} \rightleftarrows \boldsymbol{p}'$. Following this, condition (AI.6) leads to the cancellation of the integrals of the last two terms. As a result, we have $\omega^{(0)}=0$ and for $\nu^{(0)}$ we obtain (using eq. AI.6)

$$\nu^{(0)}=A(n^+-n^-),$$

where A is a constant. Thus, the variation of the distribution function in this order is proportional to $(n^+-n^-)\sigma_x$ or $(n^+-n^-)\sigma_y$. Comparing with formula (AI.2), we see that in this case the oscillations reduce to the rotation of the total magnetic moment.

In the next approximation with respect to k we have

$$A\omega^{(1)}(n^+-n^-)+\frac{2b}{\hbar}\nu^{(1)}+\frac{2}{\hbar}(n^+-n^-)\int \zeta(\boldsymbol{p},\boldsymbol{p}')\,\nu^{(1)}(\boldsymbol{p}')\frac{\mathrm{d}^3\boldsymbol{p}'}{(2\pi\hbar)^3}$$
$$=A(\boldsymbol{kv})(n^+-n^-)-Ab[\boldsymbol{k}(\boldsymbol{v}-\boldsymbol{u})\,\delta(\varepsilon_0-b-\mu)+\boldsymbol{k}(\boldsymbol{v}+\boldsymbol{u})\,\delta(\varepsilon_0+b-\mu)], \tag{AI.13}$$

where again we have used relation (AI.6).

Let us integrate this equation over $\mathrm{d}^3\boldsymbol{p}$, taking account of (AI.6) and of the circumstance that the functions ε_0 and b must be even with respect to $\boldsymbol{p}$ and, hence, $\boldsymbol{u}$ and $\boldsymbol{v}$ are odd functions. Here we find that $\omega^{(1)}=0$, i.e., ω is at least of second order in $\boldsymbol{k}$.

According to eq. (AI.13), the variation of the distribution function consists of three parts. One of them is nonzero in the interval where the difference n^+-n^- does not vanish. The other two are different from zero only on the corresponding Fermi surfaces and are proportional to $\delta(\varepsilon_0-b-\mu)$ and $\delta(\varepsilon_0+b-\mu)$, respectively. We denote the quantity $\nu^{(1)}/A$ by ν_1 (it no longer depends on A) and write ν_1 in the form

$$\nu_1=B(\boldsymbol{p})\,(n^+-n^-)-\tfrac{1}{2}\hbar[\boldsymbol{k}(\boldsymbol{v}-\boldsymbol{u})\,\delta(\varepsilon_0-b-\mu)$$
$$+\boldsymbol{k}(\boldsymbol{v}+\boldsymbol{u})\,\delta(\varepsilon_0+b-\mu). \tag{AI.14}$$

Substituting this expression into (AI.13), we arrive at the equation for the function $B(\boldsymbol{p})$:

$$\frac{2b}{\hbar}B(\boldsymbol{p})+\frac{2}{\hbar}\int \zeta(\boldsymbol{p},\boldsymbol{p}')\,B(\boldsymbol{p}')\,(n^+-n^-)\frac{\mathrm{d}^3\boldsymbol{p}'}{(2\pi\hbar)^3}$$
$$=(\boldsymbol{kv})+\int\frac{\mathrm{d}S'}{(2\pi\hbar)^3}\zeta(\boldsymbol{p},\boldsymbol{p}')\,\boldsymbol{k}(\boldsymbol{v}'-\boldsymbol{u}')\frac{1}{|\boldsymbol{v}'-\boldsymbol{u}'|}\bigg|_{\varepsilon_0-b=\mu}$$
$$+\int\frac{\mathrm{d}S'}{(2\pi\hbar)^3}\zeta(\boldsymbol{p},\boldsymbol{p}')\,\boldsymbol{k}(\boldsymbol{v}'+\boldsymbol{u}')\frac{1}{|\boldsymbol{v}'+\boldsymbol{u}'|}\bigg|_{\varepsilon_0+b=\mu}. \tag{AI.15}$$

In the general case of the arbitrary function ζ this equation cannot be solved for $B(\boldsymbol{p})$.

In order to find the relation between ω and $\boldsymbol{k}$, we write the equation for $\nu^{(2)}$ and integrate it over $\mathrm{d}^3\boldsymbol{p}$. This yields

$$\omega^{(2)}=\left\{\int\int \boldsymbol{k}\boldsymbol{v}\nu_1 \frac{\mathrm{d}^3\boldsymbol{p}}{(2\pi\hbar)^3}+\int\int [\boldsymbol{k}(\boldsymbol{v}-\boldsymbol{u})\,\delta(\varepsilon_0-b-\mu)+\boldsymbol{k}(\boldsymbol{v}+\boldsymbol{u})\,\delta(\varepsilon_0+b-\mu)]\right.$$
$$\left.\times\zeta(\boldsymbol{p},\boldsymbol{p}')\,\nu_1(\boldsymbol{p}')\frac{\mathrm{d}^3\boldsymbol{p}\,\mathrm{d}^3\boldsymbol{p}'}{(2\pi\hbar)^6}\right\}$$
$$\times\left[\int\int (n^+-n^-)\frac{\mathrm{d}^3\boldsymbol{p}}{(2\pi\hbar)^3}\right]^{-1}.$$

We substitute expression (AI.13) into this formula and rearrange it with the aid of eq. (AI.15) for B. This yields

$$\omega^{(2)}=\left\{\int\int \boldsymbol{k}\boldsymbol{v}(n^+-n^-)\,B(\boldsymbol{p})\frac{\mathrm{d}^3\boldsymbol{p}}{(2\pi\hbar)^3}\right.$$
$$-\int\frac{\mathrm{d}S}{(2\pi\hbar)^3}\,bB(\boldsymbol{p})\,\boldsymbol{k}(\boldsymbol{v}-\boldsymbol{u})\frac{1}{|\boldsymbol{v}-\boldsymbol{u}|}\bigg|_{\varepsilon_0-b=\mu}$$
$$\left.-\int\frac{\mathrm{d}S}{(2\pi\hbar)^3}\,bB(\boldsymbol{p})\,\boldsymbol{k}(\boldsymbol{v}+\boldsymbol{u})\frac{1}{|\boldsymbol{v}+\boldsymbol{u}|}\bigg|_{\varepsilon_0+b=\mu}\right\}$$
$$\times\left[\int\int (n^+-n^-)\frac{\mathrm{d}^3\boldsymbol{p}}{(2\pi\hbar)^3}\right]^{-1}. \tag{AI.16}$$

Since, according to (AI.15), function B is of first order in $\boldsymbol{k}$ and odd in $\boldsymbol{p}$, it is clear that ω is a quadratic function of $\boldsymbol{k}$.

Equations (AI.15) and (AI.16) define $\omega(\boldsymbol{k})$ completely. In the simplest case $\zeta=\text{const.}$ eq. (AI.15) may be solved. Since $\boldsymbol{B}$ is an odd function of the quasimomentum and n^+ and n^- are even, all the integrals in eq. (AI.15) vanish. As a result, we obtain $B=\hbar\boldsymbol{k}\boldsymbol{v}/(2b)$. Substituting this into (AI.16) and taking into account that, in accordance with relation (AI.16) at $\zeta=\text{const.}$ function $b(\boldsymbol{p})$ is a constant and $\boldsymbol{u}=\partial b/\partial\boldsymbol{p}=0$, we arrive at

$$\omega=\left[\int\int \hbar(\boldsymbol{k}\boldsymbol{v})^2(n^+-n^-)\frac{\mathrm{d}^3\boldsymbol{p}}{(2\pi\hbar)^3}-\hbar b\int(\boldsymbol{k}\boldsymbol{v})^2\frac{\mathrm{d}S}{(2\pi\hbar)^3 v}\bigg|_{\varepsilon_0=\mu+b}\right.$$
$$\left.-\hbar b\int(\boldsymbol{k}\boldsymbol{v})^2\frac{\mathrm{d}S}{(2\pi\hbar)^3 v}\bigg|_{\varepsilon_0=\mu-b}\right]$$
$$\times\left[2b\int(n^+-n^-)\frac{\mathrm{d}^3\boldsymbol{p}}{(2\pi\hbar)^3}\right]^{-1}. \tag{AI.17}$$

The estimate for ω can be carried out using the model in which $\varepsilon_0=p^2/2m$). Here it turns out that $\omega\sim\hbar k^2v^2/b$ (we assume $b\lessdot\mu$).

Thus, transverse spin waves with a frequency ω proportional to k^2 can propagate in a ferromagnetic metal. Quasiparticles with an energy $\hbar\omega=Ck^2$ correspond to these spin waves. They are sometimes called magnons.

Appendix

II

Second-order phase transitions

The theory of phase transitions has been described in a large number of monographs and textbooks, and it would be simpler to refer the reader to the literature. In particular, we can recommend the books written by Landau and Lifshitz (1976) and by Patashinskii and Pokrovskii (1982). However, for the completeness of presentation we will describe the basic aspects of this theory.

The concept of second-order phase transitions was proposed by Landau in 1937 (Landau 1937). According to his idea, these are transitions in which the state of the substance varies continuously and the symmetry is changed in a jumpwise manner. Examples are the transitions of magnetic substances from the paramagnetic to the ferro- or antiferromagnetic state, the transition of a metal from the normal to the superconducting state, certain structural transitions of crystals, in particular ferroelectric transitions.

In accordance with the Landau theory, in all cases of second-order phase transition we can define a certain parameter which characterizes the transition and which is called the order parameter. This parameter is equal to zero on one side of the transition point and is nonzero on the other side (i.e. in the other phase). For instance, in a ferromagnetic transition the order parameter may be the spontaneous magnetization vector $\boldsymbol{M}$. In this case, the order parameter has three components. In some structural transitions the probability of the arrangement of atoms in the lattice varies. In such a case the difference in the probabilities of various positions may serve as the order parameter, which is then a scalar. With the superconducting transition the order parameter is a complex scalar quantity, namely the coherent wave function of the Cooper pairs present in the Bose condensate. Here the order parameter has two real components. There are also transitions with a larger number of components of the order parameter.

In a second-order phase transition the order parameter varies continuously. Nevertheless, at the transition point the order parameter has a singularity because once it has vanished it remains to be equal to zero. From this it is rather obvious that the thermodynamic quantities must also exhibit singularities at the transition point. This means that, although the order parameter becomes small in the vicinity of the transition point, the thermodynamic potential, in general, cannot be expanded into a series in powers of the order parameter. In spite of this, we will assume first that the singularity is weak (i.e. it manifests itself in the far expansion terms), or it occurs in a very narrow vicinity of the phase transition point, which will not be considered here; we may then expand the thermodynamic potential into a series in powers of the order parameter. This expansion forms the basis of the Landau theory (Landau 1937a, b).

In what follows we will consider, for the sake of simplicity, the scalar order parameter φ. We also introduce a certain "external field" h, which causes the appearance of a finite φ. In this case, the thermodynamic potential in the neighborhood of the transition point may be written in the form

$$\Phi = \Phi_0 + \Phi_1\varphi + \Phi_2\varphi^2 + \Phi_3\varphi^3 + \Phi_4\varphi^4 + \cdots - Vh\varphi. \tag{AII.1}$$

The last term describes the energy associated with the external field (similar to $-\boldsymbol{MH}$, see Appendix III); V is the volume of the system.

The coefficients in this series are functions of pressure p and temperature T. It should be remarked that in the equilibrium state Φ depends only on p and T. In this case, in formula (AII.1) φ is not an independent variable and is determined from the minimum condition for Φ.

It can be shown (see Landau and Lifshitz 1976) that from symmetry considerations in all cases of phase transitions, without exceptions, $\Phi_1 \equiv 0$. This is quite evident in the examples given here. Since Φ is a real scalar function, it cannot, in general, contain odd powers of the vector or of the complex function.

Thus, at small φ the key role is played by the term with φ^2. If the coefficient $\Phi_2 > 0$, then $\varphi = 0$ corresponds to the minimum of Φ. Hence, in order for $\varphi \neq 0$ to correspond to the minimum of the thermodynamic potential below the transition point, the coefficient Φ_2 must change sign at the transition point. Thus, we obtain the condition

$$\Phi_2^{(c)}(p, T) = 0. \tag{AII.2}$$

In the examples given above, all the terms odd in φ are equal to zero and, hence, $\Phi_3 = 0$. If this is so, then condition (AII.2) is the only condition to deal with. It describes the transition curve $p(T)$. However, in contrast to the coefficient Φ_1, the symmetry requirements in certain cases permit the finiteness of Φ_2. If condition (AII.2) is fulfilled, then at the transition point we must have $\Phi_3 = 0$ and $\Phi_4 > 0$, because the equilibrium value of the order parameter at this point must be $\varphi = 0$. In such a case we have two equations for two variables, i.e., the strictly definite values of p_c and T_c. Recent investigations have shown that this situation actually corresponds to the critical point. For example, if we deal with a gas–liquid transition, then for φ we may take $\varphi = \rho - \rho_c$, where ρ is the density and $\rho_c = \rho(p_c, T_c)$ is the critical density. Evidently, the third-order term in φ must generally speaking not vanish. This situation is not examined here; we confine ourselves to true second-order phase transitions, i.e., to the cases where $\Phi_3 \equiv 0$.

For definiteness we fix the pressure and consider only the transition with respect to temperature. Assuming the absence of a singularity, we expand all the coefficients into series in $T - T_c$, because the theory refers only to the vicinity of the transition point. According to condition (AII.2), the expansion of Φ_2 must start with a first-order term. We write it in the form

$$\Phi_2 \approx \tfrac{1}{2}\alpha\tau V, \tag{AII.3}$$

where we have introduced $\tau = (T - T_c)/T_c$. Since $\Phi_2 > 0$ at $T > T_c$, we have $\alpha > 0$. At the transition point the determining role is played by the term with φ^4. In order for $\varphi = 0$ to correspond to the minimum of Φ, it is necessary that $\Phi_4^{(c)} > 0$. In view of this, we may replace Φ_4 by $\Phi_4(T_c) > 0$. We redefine this coefficient as

$$\Phi_4 \approx \Phi_4(T_c) \equiv \tfrac{1}{4}bV. \tag{AII.4}$$

Thus, instead of (AII.1) we obtain

$$\Phi = \Phi_0 + \tfrac{1}{2}\alpha V\tau\varphi^2 + \tfrac{1}{4}bV\varphi^4 - h\varphi V. \tag{AII.5}$$

Consider first the case $h = 0$. Taking the derivative with respect to φ we obtain the extremum condition

$$(\alpha\tau + b\varphi^2)\varphi = 0.$$

This equation has the solution $\varphi = 0$, which corresponds to the minimum if $T > T_c$. In the case $T < T_c$ the solution $\varphi = 0$ corresponds to the maximum of Φ, and the other root, i.e.

$$\varphi_0 = \left(\frac{\alpha|\tau|}{b}\right)^{1/2} \tag{AII.6}$$

corresponds to the minimum of Φ.

Knowing Φ, we can find the behavior of the entropy and the heat capacity. Since $(\partial\Phi/\partial\varphi)_T = 0$ (recalling that $p = \text{const.}$),

$$S = -\frac{\partial\Phi}{\partial T} = -\left(\frac{\partial\Phi}{\partial T}\right)_\varphi - \left(\frac{\partial\Phi}{\partial\varphi}\right)_T \frac{d\varphi}{dT} = -\left(\frac{\partial\Phi}{\partial T}\right)_\varphi$$

$$= S_0 - \frac{\alpha V}{2T_c}\varphi^2 = S_0 + \frac{\alpha^2 V\tau}{2bT_c}. \tag{AII.7}$$

At $T = T_c$ ($\tau = 0$) we have $S = S_0$, i.e., the entropy is continuous and, hence, the latent heat of transition $q = T(S_1 - S_2)$ is equal to zero. The heat capacity is given by

$$C_p = \frac{T}{V}\left(\frac{\partial S}{\partial T}\right)_p = C_{p0} + \frac{\alpha^2}{2bT_c} \tag{AII.8}$$

(where we have replaced $T \approx T_c$). Since $C_p = C_{p0}$ at $T > T_c$, it follows from formula (AII.8) that the heat capacity experiences a jump at $T = T_c$.

Note that the value of $\varphi = 0$ below the transition point corresponds to the maximum of Φ and the root of (AII.6) above the transition point ($\tau > 0$) is absent at all. This means that, in contrast to first-order phase transitions, in second-order phase transitions metastable phases (of the type of an overheated liquid or overcooled vapor) are absent. In experiments this criterion is often the simplest method of determination of the nature of the transition.

Now let $h \neq 0$. The minimum of Φ is given by

$$\alpha\tau\varphi + b\varphi^3 = h. \tag{AII.9}$$

If $\tau = 0$ [recall that $\tau = (T - T_c)/T_c$], then

$$\varphi = \left(\frac{h}{b}\right)^{1/3}.$$

With $h \to 0$ we can determine the susceptibility $\chi = (\partial\varphi/\partial h)_{h\to 0}$. From (AII.7) we find

$$\frac{\partial\varphi}{\partial h}(\alpha\tau + 3b\varphi^2) = 1. \tag{AII.10}$$

At $\tau > 0$ we substitute $\varphi = 0$ into the term in parenthesis and obtain

$$\chi = \frac{1}{\alpha\tau}. \tag{AII.11}$$

But if $\tau < 0$, we substitute the value of φ from (AII.6) into (AII.10), which gives

$$\chi = \frac{1}{2\alpha|\tau|}. \tag{AII.12}$$

Thus, in the vicinity of the transition point power dependences of the physical quantities on τ and h arise:

$$\varphi(\tau, h = 0) \propto |\tau|^{\beta},$$

$$\varphi(\tau = 0, h) \propto h^{1/\delta},$$

$$C_p(\tau, 0) \propto |\tau|^{-\alpha},$$

$$\chi(\tau, 0) \propto |\tau|^{-\gamma}, \tag{AII.13}$$

where in accordance with the Landau theory $\beta = \frac{1}{2}$, $\delta = 3$, $\alpha = 0$, and $\gamma = 1$. In experiments power dependences of the type (AII.13) are really observed. However, the exponents (they are called critical exponents or critical indices) do not always coincide with the predictions of the Landau theory.

What makes the simple Landau theory incorrect? The cause can be suggested by the similarity between a second-order phase transition and the critical point. As is known, the vicinity of the critical point in a liquid is characterized by the so-called critical opalascence. This implies a drastic increase in light scattering caused by an increase in density fluctuations. An analogous phenomenon is observed for a second-order phase transition: a drastic increase in diffuse scattering of neutrons in a magnetic metal in the vicinity of the ferromagnetic transition. This phenomenon is also associated with an increase in fluctuations. This can be understood if it is taken into account that in the vicinity of the phase transition point, just as near the critical point, the phases differ little from one another, and therefore the fluctuations in the order parameter are associated with a small increase in energy.

In order to determine the probability of fluctuations, we assume that in a certain volume V of the equilibrium system with a total volume $V+V_0$ (V_0 is the volume of the medium) a fluctuation arises with an energy ΔE. However, since the volume V is not isolated and is surrounded by the medium, it follows that the energy of the medium is also changed. The total change in energy is given by

$$\Delta E_{\text{total}} = \Delta E + \Delta E_{\text{med}} = \Delta E - p_0 \Delta V_0 + T_0 \Delta S_0,$$

where the subscript "0" refers to the medium. Since the total volume is given, it follows that

$$\Delta V + \Delta V_0 = 0.$$

At the same time, since we are dealing with an equilibrium system, the entropy must not increase as a result of fluctuations, i.e.

$$\Delta S + \Delta S_0 = 0.$$

The pressure and temperature are constant along the entire system. From this follows

$$\begin{aligned}\Delta E_{\text{total}} &= \Delta E + p_0 \Delta V - T_0 \Delta S \\ &= \Delta(E + p_0 V - T_0 S) = \Delta\Phi,\end{aligned}$$

where $\Delta\Phi$ is the change of the thermodynamic potential of the subsystem with a volume V. Thus, we come to the conclusion that the fluctuation probability is

$$W \propto \exp\left(-\frac{\Delta E_{\text{total}}}{T}\right) = \exp\left(-\frac{\Delta\Phi}{T}\right). \tag{AII.14}$$

In order to establish the form of $\Delta\Phi$, we assume that the key role will be played by small fluctuations of the order parameter, which vary smoothly in space (they are called long-wavelength fiuctuations). In order to take into account the possibility of such spatial variations, the term $\frac{1}{2}c(\nabla\varphi)^2$ must be added to the thermodynamic potential of the unit volume. Since the parameter φ becomes inhomogeneous, Φ is written in the form of an integral:

$$\Phi[\varphi] = \Phi_0 + \int \left[\tfrac{1}{2}c(\nabla\varphi)^2 + \tfrac{1}{2}\alpha\tau\varphi^2 + \tfrac{1}{4}b\varphi^4 - h\varphi\right] \mathrm{d}V. \tag{AII.15}$$

This is known as the Landau functional: $\Phi[\varphi]$ is determined by the function $\varphi(r)$ rather than by the number φ.

Let us assume that not only the equilibrium value $\varphi = \varphi_0$ is small but that the fluctuations φ are also small as compared to φ_0:

$$\Delta\varphi \ll \varphi_0.$$

In this case we may expand (AII.15) into a series in powers of $\Delta\varphi$. Let us write $\Delta\varphi$ in the form of the Fourier integral:

$$\Delta\varphi = V^{-1/2}\sum_k \varphi_k \, e^{ikr}. \tag{AII.16}$$

Since $\Delta\varphi$ is real, it follows that $\varphi_{-k} = \varphi_k^*$. Substituting $\varphi = \varphi_0 + \Delta\varphi$ into (AII.15) and separating the part due to fluctuations, we obtain

$$\Delta\Phi = \sum_k (ck^2 + \alpha\tau + 3\varphi_0^2 b)|\varphi_k|^2. \tag{AII.17}$$

Here we assume that $\Delta\varphi$ does not contain a constant part, i.e. $\boldsymbol{k} \neq 0$ (the constant part is incorporated into φ_0). The summation in (AII.17) runs over the half-sphere of the directions $\boldsymbol{k}$.

If (AII.17) is substituted into the fluctuation probability (AII.14), W splits into a product of separate factors with different $\boldsymbol{k}$. Hence, the fluctuations with different $\boldsymbol{k}$ or different wavelengths are statistically independent in the approximation under consideration.

Let us find the fluctuational contribution to the heat capacity. To this end, we have to determine the fluctuational part of the thermodynamic potential:

$$\Phi_{\text{fl}} = -T \ln \sum \exp\left(-\frac{\Delta\Phi}{T}\right),$$

where the sum runs over all permissible fluctuations, i.e., over all complex amplitudes φ_k. Substituting (AII.17) in the above equation, we get

$$\Phi_{\text{fl}} = -T \ln \Pi_k \int \exp\left[-\frac{(ck^2 + \alpha\tau + 3\varphi_0^2 b)|\varphi_k|^2}{T}\right] \mathrm{d}\,\mathrm{Re}\,\varphi \;\mathrm{d}\,\mathrm{Im}\,\varphi$$

$$= -T \sum_k \ln \int_0^\infty \exp\left[-\frac{(ck^2 + \alpha\tau + 3\varphi_0^2 b)|\varphi_k|^2}{T}\right] 2\pi|\varphi_k| \,\mathrm{d}\,|\varphi_k|.$$

Here we have passed to the integration over $|\varphi_k|$ and the phase of the complex φ_k. Calculating the integral, we obtain

$$\Phi_{\text{fl}} = -T \sum_k \ln\left[\frac{\pi T}{(ck^2 + \alpha\tau + 3\varphi_0^2 b)}\right].$$

In order to find the heat capacity, we have to differentiate Φ_{fl} twice with respect to temperature. Evidently, the term obtained by differentiation of $\tau = (T - T_c)/T_c$ will diverge most strongly:

$$C_{\text{fl}} = -\frac{T}{V}\frac{\partial^2 \Phi_{\text{fl}}}{\partial T^2} \approx -\frac{1}{VT_c}\frac{\partial^2 \Phi_{\text{fl}}}{\partial \tau^2}$$

$$= \frac{\alpha^2}{V}\sum_k \frac{1}{(ck^2 + \alpha\tau + 3\varphi_0^2 b)^2}. \tag{AII.18}$$

Passing from the summation to the integration:

$$\sum_{k} \rightarrow V \int d^3 k/(2\pi)^3 \rightarrow V \int k^2\, dk(2\pi)^2$$

(recall that we integrate over a half-sphere), we obtain

$$C_{\text{fl}} = \frac{\alpha^2}{16\pi c^{3/2}} \frac{1}{(\alpha\tau + 3\varphi_0^2 b)^{1/2}}. \tag{AII.19}$$

The derivation given here is called the Ornstein–Zernike theory (Landau and Lifshitz 1976).

From expression (AII.19) it follows that if we substitute $\varphi_0 = 0$ for $\tau > 0$ or φ_0 for $\tau < 0$ according to (AII.6), we obtain

$$C_{\text{fl}} \propto |\tau|^{-1/2}.$$

In the above derivation it was assumed that the fluctuations are small and have a long wavelength. That they are long-wavelength fluctuations is confirmed by the fact that the region where k is small is the most significant in the integral over k (for the heat capacity!). As regards the condition of the smallness of fluctuations, the corresponding criterion is deduced if we compare the fluctuational correction to heat capacity with a jump in heat capacity obtained according to the Landau theory. Expression (AII.19) must be small compared to the addition to C_{p0} in (AII.8). Recalling that the theory under discussion applies only at $|\tau| \ll 1$, we obtain (Levanyuk 1959; Ginzburg 1960)

$$1 \gg |\tau| \gg \frac{b^2 T_c^2}{c^3 \alpha}. \tag{AII.20}$$

From this it is seen that if the quantity that limits $|\tau|$ from below is small, there is a region where the Landau theory is valid. But if it is of the order of unity, the Landau theory does not apply at any temperature. We shall soon see that the lower limit of $|\tau|$ never exceeds unity in order of magnitude.

The question of the validity of the Landau theory must be analyzed for each particular transition separately. Here, in different cases we obtain different results. For example, for the superconducting transition the lower limit of $|\tau|$ is always very small* and the Landau theory is practically always valid. For the transition of liquid He^4 (the helium isotope with an atomic weight of 4) to the superfluid state, the Landau theory is not valid, and for the superfluid transition of liquid He^3, on the contrary, the Landau theory applies practically at all temperatures ($|\tau| \ll 1$). For magnetic transitions, either ferromagnetic or antiferromagnetic, the right-hand side in (AII.20) is of the order of 0.1–0.2, i.e., the Landau theory is approximately meaningful, although deviations arise in the immediate (but quite observable) vicinity of the transition.

* It is larger in the high-temperature ceramics.

Criterion (AII.20) is expressed in terms of the parameters of the Landau functional (AII.15), rather than in terms of the microscopic characteristics of the problem, i.e., in terms of the properties of the particles that make up the system and of their interactions. Therefore, it is sometimes difficult to apply it to a real system. This difficulty can be surmounted. Let us consider the system at a temperature much lower than T_c. Although the Landau functional is not applicable for a quantitative description, it may be used for estimates. The characteristic energy of the unit volume can be defined as the product of the energy of one particle, which is naturally considered to be of order T_c, and the density of the particles, i.e. $N/V \sim r_0^{-3}$, where r_0 is the average distance between the particles. On the other hand, according to (AII.15), the same quantity is of order $b\varphi^4$ or $\alpha\varphi^2$ (assuming $\tau \sim 1$). According to (AII.6) $\varphi^2 \sim \alpha/b$, from which follows $r_0^3 \sim T_c b/\alpha^2$.

In any system the motion of particles is correlated at certain distances. The correlation radius r_c may be of order r_0, but it may also be much larger. This correlation is reflected in the term with $c(\nabla\varphi)^2$ in (AII.15). Strictly speaking, it is correct only at a large correlation radius, when φ varies slowly in space; otherwise, one has to take into account higher derivatives. However, we will use this term for estimates. By order of magnitude, $c(\nabla\varphi)^2 \sim c\varphi^2/r_c^2$. Comparing it with $\alpha\tau\varphi^2$, we find (at $\tau \sim 1$) $r_c \sim (c/\alpha)^{1/2}$. Let us now look at the ratio r_0^6/r_c^6:

$$\left(\frac{r_0}{r_c}\right)^6 \sim \frac{T_c^2}{\alpha c^3}. \tag{AII.21}$$

The right-hand side coincides with the right-hand side of (AII.20). From this it follows that we must know the ratio of the correlation radius or the correlation length to the interatomic distance at a temperature much lower than T_c. If this ratio is considerably larger than unity, the Landau theory has a region of validity; if it is of the order of unity, the Landau theory is not applicable at all. Evidently, the correlation length cannot be smaller than the interatomic distance, i.e., the right-hand side in (AII.20) may be either small or of the order of unity.

If $|\tau| \ll b^2T_c^2/(\alpha c^3)$ (this region is called critical), then the fluctuations become large and therefore neither the Landau theory nor the Ornstein–Zernike theory, which is based on the idea of small fluctuations, is valid. The hypothesis of gauge invariance or "scaling" has been proposed (Patashinskii and Pokrovskii 1964, 1966, Kadanoff 1966) as the solution of the problem.

In the preceding paragraph we have discussed the correlation length r_c. If it is determined near T_c by equating the first two terms of the Landau functional $c(\nabla\varphi)^2 \sim c\varphi^2/r_c^2 \sim \alpha\tau\varphi^2$), then it is found to be of the order of

$$r_c \sim \left(\frac{c}{\alpha\tau}\right)^{1/2}. \tag{AII.22}$$

In fact, in the critical region this formula does not hold, but it would be natural to suppose that its basic qualitative consequence is retained in the exact theory:

$$r_c \to \infty \text{ with } \tau \to 0. \tag{AII.23}$$

In the critical region the contribution of fluctuations to the thermodynamic quantities becomes dominant. But the correlation length, which governs the characteristic size of fluctuations, increases with decreasing τ. In view of this, in the critical region the properties of a substance are determined not by the interaction of individual particles but by the interaction of volumes with a characteristic linear size of order r_c. But all this is based nevertheless on the interaction of individual particles.

The relationship between these statements is established by the scaling hypothesis. According to this hypothesis, in going from the interaction of individual particles to the interaction of volumes containing several particles the most significant part of the interaction will retain its form to within a constant factor dependent on the ratio of the scales. Hence, it becomes possible to go from one scale to the other, i.e., to measure the length by any scale. Here, however, all the physical relations, say $\varphi(\tau, 0)$, $\varphi(0, h)$ or $\chi(\tau)$ must be independent of the choice of the scale. The thermodynamic potential of the entire system must also remain constant.

This requirement can be realized as follows. We have to assume that when the scale is changed, each physical quantity is transformed according to the law

$$A(\lambda \boldsymbol{r}) = \lambda^{-\Delta_A} A(\boldsymbol{r}), \qquad \text{(AII.24)}$$

that is, it is multiplied by $\lambda^{-\Delta_A}$, where Δ_A is a certain characteristic number called the anomalous dimensionality of the quantity A. The anomalous dimensionality must not coincide at all with the power of the length in physical dimensionality. All physical relationships must have the form of power laws of the type (AII.13). Indeed, according to (AII.24),

$$\varphi(\lambda \boldsymbol{r}) = \lambda^{-\Delta_\varphi} \varphi(\boldsymbol{r}), \qquad \tau(\lambda \boldsymbol{r}) = \lambda^{-\Delta_\tau} \tau(\boldsymbol{r}).$$

In order for $\varphi(\lambda \boldsymbol{r})$ to be connected with $\tau(\lambda \boldsymbol{r})$ by the same relation that connects $\varphi(\boldsymbol{r})$ with $\tau(\boldsymbol{r})$, the factors that depend on λ must cancel out. This is possible with a power relation of the type (AII.13), where

$$\beta = \frac{\Delta_\varphi}{\Delta_\tau}. \qquad \text{(AII.25)}$$

In this way we establish first the very fact of the power dependence $\varphi(\tau)$ (eq. AII.13) and, second, we obtain the relationship between the critical index β and the anomalous dimensionalities of Δ_φ and Δ_τ.

It should be noted that the concept of scale invariance is as yet a hypothesis. Actually, in going from the interaction of individual particles to the interaction of the volumes we obtain a set of terms of complicated form and the statement that the key role is played by an interaction similar to the interaction between individual particles cannot be proved. For example, for magnetics in going from the interaction of the spins of individual atoms to the interaction of the volumes the result is not expressed as the interaction of the total spins of these volumes.

On the other hand, physical relations of the type (AII.13), if they are correct, must directly follow from the microscopic theory, which is based on the initial interaction of individual particles. Such a program has been realized for several models in two-dimensional space, i.e. in a plane; the results are consistent with (AII.13). But so far this has not been accomplished for any (even the simplest) three-dimensional model, not to mention real physical systems. Nevertheless, the undoubtful success of the scaling hypothesis in the interpretation of experimental data has led to its general recognition.

Having made this digression, we continue to describe the consequences of the scaling hypothesis. For the sake of generality, we shall consider not the three-dimensional but the d-dimensional space. This will prove useful in what follows. The last term in the Landau functional (AII.15) must be regarded as the definition of the field h responsible for the appearance of a finite order parameter at $\tau = 0$. Therefore, this term must retain its form in a scaling transformation. For this to be true, it is necessary that $\lambda^{-\Delta_h} \cdot \lambda^{-\Delta_\varphi} \cdot \lambda^{d} = 1$, i.e.

$$\Delta_\varphi + \Delta_h = d. \tag{AII.26}$$

The magnetic susceptibility is $\chi = (\partial\varphi/\partial h)_{h=0}$. Hence, its anomalous dimensionality is $\Delta_\varphi - \Delta_h$. According to the last formula of (AII.13) for $\chi(\tau)$ we see that $-\gamma$ plays the same role as β in the formula for $\varphi(\tau)$. Hence, $-\gamma = (\Delta_\varphi - \Delta_h)/\Delta_\tau$, i.e.

$$\gamma = \frac{\Delta_h - \Delta_\varphi}{\Delta_\tau}. \tag{AII.27}$$

The second formula of (AII.13) leads in an analogous way to

$$\delta = \frac{\Delta_h}{\Delta_\varphi}. \tag{AII.28}$$

The subscript α refers to the heat capacity of the unit volume. Hence, $C = -V_d^{-1} T_c^{-1}\, \partial\Phi[\varphi]/\partial\tau^2$ (V_d is the volume in d-dimensional space). Since Φ is invariant, it follows that $\Delta_c = -2\Delta_\tau + d$. According to the third formula of (AII.13),

$$\alpha = 2 - \frac{d}{\Delta_\tau}. \tag{AII.29}$$

Instead of the index Δ_φ it is customary to introduce the index η in accordance with

$$\Delta_\varphi = \tfrac{1}{2}(d - 2 + \eta). \tag{AII.30}$$

The origin of this definition is as follows. If the Landau functional had not contained a term with φ^4, the fluctuations of the individual Fourier components φ_k would always have been statistically independent, just as in the case of small fluctuations. Here the dimensionality of Δ_φ could have been determined from the invariance of the first term in (AII.15): $2\Delta_\varphi + 2 - d = 0$, i.e. $\Delta_\varphi = \frac{1}{2}(d-2)$. The presence of the interaction of fluctuations with different $\boldsymbol{k}$ leads to the appearance of some anomalous dimensionality for the coefficient c. This is taken into account by introducing the index term η in (AII.30).

Finally, let us define the critical index ν for the correlation length:

$$r_c(\tau) \propto \tau^{-\nu}. \tag{AII.31}$$

When the scale is changed, r_c is simply multiplied by λ. Hence,

$$\nu = \Delta_\tau^{-1}. \tag{AII.32}$$

We exclude the "nonphysical" indices from eqs. (AII.25)-(AII.32), retaining only α, β, γ, δ, ν, and η. It is not difficult to check the fulfillment of the following formulas:

$$\alpha + 2\beta + \gamma = 2, \qquad \beta + \gamma = \beta\delta,$$
$$(2-\eta)\nu = \gamma, \qquad 2 - \alpha = d\nu. \tag{AII.33}$$

Thus, 4 relations are imposed on the 6 indices, i.e., only two of them are independent. We should note the nontriviality of this conclusion. The six indices introduced here are physically independent and are determined from quite different experiments. This makes it possible to test the scaling-invariance hypothesis. It has been strikingly well confirmed in all the transitions that have been investigated.

As regards the two independent indices, they may be determined on the basis of the experimental fact that the indices α and η are small. Assuming α, $\eta = 0$, we obtain from (AII.33) for the three-dimensional case:

$$\nu = \tfrac{2}{3}, \qquad \gamma = \tfrac{4}{3}, \qquad \beta = \tfrac{1}{3}, \qquad \delta = 5. \tag{AII.34}$$

These numbers give approximate values of the indices for any transitions in three-dimensional space. A more exact calculation with account taken of α and η leads to the dependence of the indices on the number of components of the order parameter. Various phase transitions having an order parameter with the same number of components have identical critical indices and are said to belong to the same universality class.

An exact calculation of the critical indices is a complicated task and falls beyond the scope of the present book. We only indicate here what may be used for the derivation concerned. We substitute into (AII.33) the indices from the Landau theory: $\alpha = 0$, $\beta = \frac{1}{2}$, $\gamma = 1$, $\delta = 3$, $\eta = 0$, and $\nu = \frac{1}{2}$ [the last one follows from eq. (AII.22)]. The first three relations in (AII.33) are fulfilled and the last one is satisfied only at $d = 4$. The dimensionality $d = 4$ is said to be critical for the Landau theory. Formally, the Landau theory refers to any number of dimensions. At $d > 4$ the fluctuational correction is always small, and the Landau theory is exact (although there are no physical objects with $d > 4$). But if $d < 4$, the fluctuational correction in the critical region is large.

In the case of $d = 4$ there arises a special, so-called logarithmic situation. The main fluctuational corrections diverge as $\ln|\tau|$ rather than by a power law. An example is the earlier evaluated integral (AII.18) for the first fluctuational correction to the heat capacity. At $d=4$,

$$\sum_k \to V_4 \int k^3\,\mathrm{d}k \quad \text{and} \quad C_{fl} \propto \ln|\tau|.$$

With the logarithmic situation the singular parts of all the quantities can be calculated exactly. They are expressed in powers of $\ln|\tau|$ rather than in powers of τ. It is interesting to note that phase transitions really exist in certain ferroelectric and antiferromagnetic substances, which simulate the four-dimensional situation (Larkin and Khmel'nitskii 1969).

The indices for the ordinary three-dimensional case have been calculated by Wilson and Fisher (1972), who extended the theory to a space with a non-integer number of dimensions and found the expansions of the indices in $\varepsilon = 4-d$ [the true expansion parameter is $\varepsilon(n+2)/(n+8)$, where n is the number of components of the order parameter]. The results have been found to be in satisfactory agreement with experiment (see Patashinskii and Pokrovskii 1982).

Appendix

III

Thermodynamics in a magnetic field

We start with the Maxwell equations

$$\operatorname{rot} \boldsymbol{H} = \frac{4\pi}{c} j + \frac{1}{c}\frac{\partial D}{\partial t}, \tag{AIII.1}$$

$$\operatorname{rot} \boldsymbol{E} = -\frac{1}{c}\frac{\partial \boldsymbol{B}}{\partial t}, \tag{AIII.2}$$

where $\boldsymbol{E}$ is the electric field $\boldsymbol{D}$ is the electrical induction $\boldsymbol{H}$ is the magnetic field and $\boldsymbol{B}$ is the magnetic induction. We multiply eq. (AIII.1) by $\boldsymbol{E}$ and subtract from it eq. (AIII.2), which is multiplied by $\boldsymbol{H}$. This leads to

$$\boldsymbol{E} \operatorname{rot} \boldsymbol{H} - \boldsymbol{H} \operatorname{rot} \boldsymbol{E} = \frac{4\pi}{c} \boldsymbol{jE} + \frac{1}{c}\left(\frac{\boldsymbol{E}\partial D}{\partial t} + \frac{\boldsymbol{H}\partial \boldsymbol{B}}{\partial t}\right).$$

The expresion on the left-hand side is equal to $-\operatorname{div}[\boldsymbol{EH}]$.

The relation obtained is now written in the form

$$-\boldsymbol{jE} = \frac{c}{4\pi} \operatorname{div}[\boldsymbol{EH}] + \frac{1}{4\pi}\left(\frac{\boldsymbol{E}\partial \boldsymbol{D}}{\partial t} + \frac{\boldsymbol{H}\partial \boldsymbol{B}}{\partial t}\right). \tag{AIII.3}$$

Let us now integrate over the volume of the system and over a small time interval δt:

$$-\int \boldsymbol{jE}\, \mathrm{d}V \delta t = \frac{c}{4\pi} \oint [\boldsymbol{EH}]\, \mathrm{d}\boldsymbol{S} + \frac{1}{4\pi}(\boldsymbol{E}\delta \boldsymbol{D} + \boldsymbol{H}\delta \boldsymbol{B}) \tag{AIII.4}$$

(here we have used Gauss' theorem $\int \operatorname{div} \boldsymbol{a}\, \mathrm{d}V = \oint \boldsymbol{a}\, \mathrm{d}\boldsymbol{S}$). Relations (AIII.3) and (AIII.4) are different forms of the Poynting theorem, which expresses the law of energy conservation in electrodynamics. If we write $\boldsymbol{j} = n_e e\boldsymbol{v}$, where n_e is the electron density and e is the charge, then $\boldsymbol{jE} = n_e \cdot e\boldsymbol{E} \cdot \boldsymbol{v}$. But $e\boldsymbol{E}$ is the force acting on electrons and, hence, $\boldsymbol{jE}\delta t = n_e \cdot e\boldsymbol{E} \cdot \delta \boldsymbol{r}$ is the work done by the field over the system of electrons. When it is taken with opposite sign, this quantity expresses a change in the energy of the system. This change is made up of two parts. The first term on the right-hand side expresses the energy flux through the surface, and the second the change in the field energy.

We are interested only in the second term for a case where there is only a magnetic field, i.e. at $\boldsymbol{E}, \boldsymbol{D}=0$. In accordance with the result obtained, the free-energy change in a magnetic field is given by

$$dF = dF_0 + \frac{1}{4\pi}\boldsymbol{H}\,d\boldsymbol{B} = -S\,dT - p\,dV + \frac{1}{4\pi}\boldsymbol{H}\,d\boldsymbol{B}. \tag{AIII.5}$$

Thus, the free energy is a thermodynamic potential, which depends on the variables T, V and $\boldsymbol{B}$. This is not convenient because $\boldsymbol{B}$ is the averaged value of the microscopic field in the specimen. In experiments we cannot record this quantity and we can specify the external field; therefore, we have to use the corresponding thermodynamic potential. To this end, we subtract the quantity $\boldsymbol{HB}/4\pi$ from F. As a result, we obtain

$$dF_H = d\left(F - \frac{\boldsymbol{HB}}{4\pi}\right) = dF_0 - \frac{1}{4\pi}\boldsymbol{B}\,d\boldsymbol{H}. \tag{AIII.6}$$

Use is often made of the energy $F'_H = F_H + H^2/8\pi$, for which

$$dF'_H = dF_0 - \frac{1}{4\pi}(\boldsymbol{B}-\boldsymbol{H})\,d\boldsymbol{H} = dF_0 - \boldsymbol{M}\,d\boldsymbol{H}, \tag{AIII.7}$$

where $\boldsymbol{M}$ is the magnetization connected with $\boldsymbol{B}$ and $\boldsymbol{H}$ by the relation

$$\boldsymbol{B} = \boldsymbol{H} + 4\pi\boldsymbol{M}. \tag{AIII.8}$$

The free energies F_H and F'_H depend on the Maxwell field $\boldsymbol{H}$. This field coincides with the external field for cylindrical geometry, i.e., for a case where a cylindrical specimen of an arbitrary cross-section is placed in a longitudinal magnetic field. If there is a linear relation between $\boldsymbol{B}$ and $\boldsymbol{H}$, $\boldsymbol{B} = \mu\boldsymbol{H}$, $\boldsymbol{M} = \chi\boldsymbol{H}$, $\mu = 1 + 4\pi\chi$, where μ is the magnetic permeability and χ is the magnetic susceptibility, then relations (AIII.5), (AIII.6) and (AIII.7) can be integrated, which gives

$$F = F_0 + \frac{B^2}{8\pi\mu}, \tag{AIII.9}$$

$$F_H = F_0 - \frac{\mu H^2}{8\pi}, \tag{AIII.10}$$

$$F'_H = F_0 - \tfrac{1}{2}\chi H^2 = F_0 - \tfrac{1}{2}\boldsymbol{MH}. \tag{AIII.11}$$

However, we do not always have to deal with cylindrical geometry. With a different geometry the Maxwell field $\boldsymbol{H}$ has no clear-cut physical meaning. In actual setting up experiments a specimen of arbitrary shape is placed in a homogeneous magnetic field created by external current sources. The field in the vicinity of the specimen is distorted. However, at a large distance from the specimen the field is the same as in the absence of the specimen. It is this field that can be specified. Without giving the derivation, we write here only the result (see the derivation in the work by Landau and Lifshitz 1982). The convenient thermodynamic potential in this case is the analog of the free energy F'_H, but for the entire object under study rather than for the unit volume. We have the following relation for it:

$$\mathcal{F} = \mathcal{F}_0 - \tfrac{1}{2}\mathcal{M}\boldsymbol{H}_0, \tag{AIII.12}$$

where $\mathcal{F}_0 = F_0 V$ is the free energy in the absence of the field; $\boldsymbol{H}_0$ is the field at infinity, and

$$\mathcal{M} = \int \mathbf{M}\, \mathrm{d}V \tag{AIII.13}$$

is the total magnetic moment of the specimen.

Expression (AIII.12) describes the total free energy; in particular, it includes the energy change associated with the distortion of the field around the specimen.

In many sections of the present book we consider a magnetic specimen of ellipsoidal form placed in an external magnetic field, which is homogeneous at infinity. In such a case it can be shown that the Maxwell field inside the specimen $\boldsymbol{H}_\mathrm{i}$ is homogeneous, parallel to the external field at infinity and is given by

$$\boldsymbol{H}_\mathrm{i} = \boldsymbol{H}_0 - 4\pi n \boldsymbol{M}, \tag{AIII.14}$$

where $\boldsymbol{M}$ is the magnetization of the specimen and n is the so-called demagnetization factor, which depends only on the geometry of the specimen, i.e., on the ratio of the semi-axes of the ellipsoid and on its orientation with respect to the external field. The magnetic induction is given by

$$\boldsymbol{B} = \boldsymbol{H}_\mathrm{i} + 4\pi \boldsymbol{M}. \tag{AIII.15}$$

It is $\boldsymbol{B}$ which determines the average field in the specimen.

The field outside the specimen is deduced from the solution of the Maxwell equation in vacuum,

$$\mathrm{rot}\, \boldsymbol{H} = 0, \qquad \mathrm{div}\, \boldsymbol{H} = 0, \tag{AIII.16}$$

and from the corresponding boundary conditions,

$$H_{\mathrm{vac}\parallel} = H_{\mathrm{i}\parallel}, \qquad H_{\mathrm{vac}\perp} = B_{\mathrm{vac}\perp} = B_\perp. \tag{AIII.17}$$

The quantities on the right-hand side are the limits taken from the "inside".

Let us consider, for example, the problem of the sphere. Let the magnetic permeability by μ. Then, the equation for the field H_i inside the sphere is rot $\boldsymbol{H}_i = 0$, div $\boldsymbol{B} = 0$ or, which is the same thing, div $\boldsymbol{H}_i = 0$ ($\boldsymbol{B} = \mu \boldsymbol{H}_i$). Hence, both outside and inside the sphere we may introduce the potential φ and write

$$\boldsymbol{H} = \nabla \varphi, \qquad \Delta\varphi = 0. \tag{AIII.18}$$

In spherical coordinates with account taken of the symmetry of the problem we have

$$\frac{\partial^2 \varphi}{\partial r^2} + \frac{2}{r}\frac{\partial \varphi}{\partial r} + \frac{1}{r^2 \sin\theta}\frac{\partial}{\partial \theta}\sin\theta \frac{\partial \varphi}{\partial \theta} = 0. \tag{AIII.19}$$

The boundary conditions are given in formulas (AIII.17) and $\varphi = H_0 z = H_0 r \cos\theta$ for $r \to \infty$. It is not difficult to see that the solution has the form

$$\varphi = \begin{cases} (H_0 r + A r^{-3}) \cos\theta, & r > R, \\ H_i r \cos\theta (\text{i.e. } H_i = H_{iz} = \text{const}), & r < R, \end{cases}$$

where R is the radius of the sphere and A and H_i are constants. Substitution into the boundary conditions (AIII.17) gives

$$H_i = \frac{3H_0}{2+\mu}, \qquad A = H_0 R^3 \frac{1-\mu}{2+\mu}. \tag{AIII.20}$$

Introducing $M = (B - H_i)/4\pi = (\mu - 1)H_i/4\pi$ and writing H_i in the form (AIII.14), we find that $n = \frac{1}{3}$. In an analogous way we can find for n the following values: $n = 0$ for a cylinder in a longitudinal field, $n = \frac{1}{2}$ for a cylinder in a transverse field, $n = 1$ for a disc in a field normal to its plane.

References

Abrahams, E., and I.O. Kulik, 1977, Preprint (FTINT, Kharkov).
Abrahams, E., P.W. Anderson, D.C. Licciardello and T.V. Ramakrishnan, 1979, Phys. Rev. Lett. **42**, 673.
Abrikosov, A.A., 1952, Dokl. Akad. Nauk SSSR **86**, 489.
Abrikosov, A.A., 1957, Zh. Eksp. & Teor. Fiz. **32**, 1442.
Abrikosov, A.A., 1960, Zh. Eksp. & Teor. Fiz. **39**, 1797.
Abrikosov, A.A., 1961, Zh. Eksp. & Teor. Fiz. **41**, 569.
Abrikosov, A.A., 1964a, Zh. Eksp. & Teor. Fiz. **46**, 1464.
Abrikosov, A.A., 1964b, Zh. Eksp. & Teor. Fiz. **47**, 720.
Abrikosov, A.A., 1965, Physics **2**, 21.
Abrikosov, A.A., 1969, Zh. Eksp. & Teor. Fiz. **56**, 1391.
Abrikosov, A.A., 1978, Pis'ma v Zh. Eksp. & Teor. Fiz. **27**, 235; J. Less-Common Met. **62**, 451.
Abrikosov, A.A., 1981, Solid State Commun. **37**, 997.
Abrikosov, A.A., and A.I. Buzdin, 1988, Pis'ma v Zh. Eksp. & Teor. Fiz. **47**, 204.
Abrikosov, A.A., and I.E. Dzyaloshinskii, 1958, Zh. Eksp. & Teor. Fiz. **35**, 771 [1959, Sov. Phys.-JETP **8**, 535].
Abrikosov, A.A., and L.A. Fal'kovskii, 1961, Zh. Eksp. & Teor. Fiz. **40**, 262.
Abrikosov, A.A., and L.A. Fal'kovskii, 1962, Zh. Eksp. & Teor. Fiz. **43**, 1090.
Abrikosov, A.A., and L.A. Fal'kovskii, 1987, Pis'ma v Zh. Eksp. & Teor. Fiz. **46**, 236.
Abrikosov, A.A., and V.M. Genkin, 1973, Zh. Eksp. & Teor. Fiz. **65**, 842.
Abrikosov, A.A., and L.P. Gor'kov, 1958, Zh. Eksp. & Teor. Fiz. **35**, 1558 [Sov. Phys.-JETP **8**, 1090].
Abrikosov, A.A., and L.P. Gor'kov, 1959, Zh. Eksp. & Teor. Fiz. **36**, 319 [Sov. Phys.-JETP **9**, 220].
Abrikosov, A.A., and L.P. Gor'kov, 1960a, Zh. Eksp. & Teor. Fiz. **39**, 178.
Abrikosov, A.A., and L.P. Gor'kov, 1960b, Zh. Eksp. & Teor. Fiz. **39**, 480.
Abrikosov, A.A., and L.P. Gor'kov, 1962a, Zh. Eksp. & Teor. Fiz. **42**, 1089.
Abrikosov, A.A., and L.P. Gor'kov, 1962b, Zh. Eksp. & Teor. Fiz. **43**, 2230.
Abrikosov, A.A., and I.M. Khalatnikov, 1958, Usp. Fiz. Nauk. **95**, 551.
Abrikosov, A.A., and A.V. Pantsulaya, 1986, Fiz. Tverd. Tela **28**, 2140.
Abrikosov, A.A., and I.A. Ryzhkin, 1978, Adv. Phys. **27**, 147.
Abrikosov, A.A., L.P. Gor'kov and I.M. Khalatnikov, 1958, Zh. Eksp. & Teor. Fiz. **35**, 265 [Sov. Phys.-JETP **8**, 182].
Abrikosov, A.A., L.P. Gor'kov and I.M. Khalatnikov, 1959, Zh. Eksp. & Teor. Fiz. **37**, 187 [Sov. Phys.-JETP **10**, 132].
Abrikosov, A.A., L.P. Gor'kov and I.E. Dzyaloshinskii, 1962, Methods of Quantum Field theory in Statistical Physics (GIFML, Moscow) [1963,Quantum Field Theoretical Methods in Statistical Physics (Prentice Hall, Englewood Cliffs, NJ; Pergamon Press, London)].
Adams, E.N., and T.D. Holstein, 1959, J. Phys. & Chem. Solids **10**, 254.
Aharonov, Y., and D. Bohm, 1959, Phys. Rev. **115**, 485.
Akhiezer, A.I., M.I. Kaganov and G.Ya. Lyubarskii, 1957, Zh. Eksp. & Teor. Fiz. **32**, 837.
Akhmedov, S.Sh., V.R. Karasik and A.I. Rusinov, 1969, Zh. Eksp. & Teor. Fiz. **56**, 444.
Alekseevskii, N.E., and Yu.P. Gaidukov, 1956, Zh. Eksp. & Teor. Fiz. **31**, 947.

Alekseevskii, N.E., and Yu.P. Gaidukov, 1957, Zh. Eksp. & Teor. Fiz. **32**, 1589.

Alekseevskii, N.E., and Yu.P. Gaidukov, 1958, Zh. Eksp. & Teor. Fiz. **35**, 554.

Alekseevskii, N.E., Yu.P. Gaidukov, I.M. Lifshitz and V.G. Peschanskii, 1960, Zh. Eksp. & Teor. Fiz. **39**, 1201.

Alexandrov, A.S., J. Ranninger and S. Robaszkiewicz, 1986, Phys. Rev. B **33**, 4526.

Alfeev, V.N., 1979, Superconductors, Semiconductors and Paraelectrics in Cryoelectronics (Soviet Radio Publishers, Moscow). In Russian.

Allen, V., and V. Mitrovic, 1982, Solid State Commun. **37**, 1.

Allender, D., J. Bray and J. Bardeen, 1973, Phys. Rev. **7**, 1020; **8**, 4433.

Al'tshuler, B.L., 1983, Dissertation (LIYaF Akad. Nauk SSSR).

Al'tshuler, B.L., 1985, Pis'ma v Zh. Eksp. & Teor. Fiz. **41**, 530.

Al'tshuler, B.L., and A.G. Aronov, 1979, Zh. Eksp. & Teor. Fiz. **77**, 2028.

Al'tshuler, B.L., and D.E. Khmel'nitskii, 1985, Pis'ma v Zh. Eksp. & Teor. Fiz. **42**, 291.

Al'tshuler, B.L., D.E. Khmel'nitsky, A.I. Larkin and P.A. Lee, 1980, Phys. Rev. B **20**, 5142.

Al'tshuler, B.L., A.G. Aronov and V.Z. Spivak, 1981, Pis'ma v Zh. Eksp. & Teor. Fiz. **33**, 101.

Al'tshuler, B.L., A.G. Aronov and D.E. Khmel'nitsky, 1982, J. Phys. C **15**, 7369.

Ambegoakar, V., and A. Baratoff, 1963, Phys. Rev. Lett. **10**, 486; **11**, 104.

Anderson, P.W., 1958, Phys. Rev. **109**, 1492.

Anderson, P.W., 1962, Phys. Rev. Lett. **9**, 209.

Anderson, P.W., 1963, in: Proc. 8th Int. Conf. on Low-Temperature Physics, London.

Anderson, P.W., 1966, Rev. Mod. Phys. **38**, 298.

Anderson, P.W., 1987, Science **235**, 1196.

Anderson, P.W., and Y.B. Kim, 1964, Rev. Mod. Phys. **36**, 39.

Anderson, P.W., and H. Suhl, 1959, Phys. Rev. **116**, 898.

Andreev, A.F., 1964, Zh. Eksp. & Teor. Fiz. **46**, 185, 1823; **47**, 2222.

Andreev, A.F., 1965, Zh. Eksp. & Teor. Fiz. **49**, 655.

Andreev, A.F., 1966, Zh. Eksp. & Teor. Fiz. **51**, 1510.

Andreev, A.F., 1968, Zh. Eksp. & Teor. Fiz. **54**, 1510.

Andreev, A.F., 1987, Pis'ma v Zh. Eksp. & Teor. Fiz. **46**, 463.

Aronov, A.G., Yu.M. Gal'perin, V.L. Gurevich and V.I. Kozub, 1981, Adv. Phys. **30**, 539.

Aslamazov, L.G., and M.V. Fistul', 1979, Pis'ma v Zh. Eksp. & Teor. Fiz. **30**, 233.

Aslamazov, L.G., and M.V. Fistul', 1981, Zh. Eksp. & Teor. Fiz. **81**, 382.

Aslamazov, L.G., and M.V. Fistul', 1982, Zh. Eksp. & Teor. Fiz. **83**, 1170.

Aslamazov, L.G., and M.V. Fistul', 1984, Zh. Eksp. & Teor. Fiz. **86**, 1516.

Aslamazov, L.G., and A.I. Larkin, 1968a, Zh. Eksp. & Teor. Fiz. **55**, 1477.

Aslamazov, L.G., and A.I. Larkin, 1968b, Phys. Lett. A **26**, 238.

Aslamazov, L.G., and A.I. Larkin, 1969, Pis'ma v Zh. Eksp. & Teor. Fiz. **9**, 150.

Aslamazov, L.G., and A.I. Larkin, 1978, Zh. Eksp. & Teor. Fiz. **74**, 2184.

Austin, B.J., Heine and L.J. Sham, 1962, Phys. Rev. **127**, 276.

Azbel', M.Ya., 1958, Zh. Eksp. & Teor. Fiz. **34**, 968, 1158.

Azbel', M.Ya., 1960, Zh. Eksp. & Teor. Fiz. **39**, 14.

Azbel', M.Ya., 1967, Zh. Eksp. & Teor. Fiz. **53**, 1751.

Azbel', M.Ya., and E.A. Kaner, 1956, Zh. Eksp. & Teor. Fiz. **32**, 896.

Azbel', M.Ya., V.I. Gerasimenko and I.M. Lifshitz, 1956, Zh. Eksp. & Teor. Fiz. **31**, 357.

Azbel', M.Ya., V.I. Gerasimenko and I.M. Lifshitz, 1957a, Zh. Eksp. & Teor. Fiz. **32**, 1212.

Azbel', M.Ya., M.I. Kaganov and I.M. Lifshitz, 1957b, Zh. Eksp. & Teor. Fiz. **32**, 1188.

Balatskii, A.V., L.I. Burlachkov and L.P. Gor'kov, 1986, Zh. Eksp. & Teor. Fiz. **90**, 1478.

Bardeen, J., 1950, Phys. Rev. **79**, 167; **80**, 567.

Bardeen, J., 1951a, Phys. Rev. **81**, 829, 1070.

Bardeen, J., 1951b, Phys. Rev. **82**, 987.

Bardeen, J., L.N. Cooper and J.R. Schrieffer, 1957, Phys. Rev. **106**, 162; **108**, 1175.

Barone, A., and G. Paterno, 1982, Physics and Applications of the Josephson Effect (Wiley, New York).

Bassani, F., and V. Celli, 1959, Nuovo Cim. **11**, 805.

Bazhenov, A.V., A.V. Gorbunov, H.V. Klassen, S.F. Kondakov, I.V. Kukushkin, O.V. Misochko, V.B. Timofeev, L.I. Chernyshova and B.N. Shepel', 1987, Pis'ma v Zh. Eksp. & Teor. Fiz. **46**, 35.

Bean, C.P., and J.B. Livingston, 1964, Phys. Rev. Lett. **12**, 14.

Berezinskii, V.L., 1973, Zh. Eksp. & Teor. Fiz. **65**, 1251.

Bergmann, G., 1982, Phys. Rev. Lett. **48**, 1046; Phys. Rev. B **25**, 2937.

Biondi, M.A., and M.P. Garfunkel, 1959, Phys. Rev. **116**, 853.

Blount, E.I., 1962, Phys. Rev. **126**, 1636.

Bogoliubov, N.N., 1958, Zh. Eksp. & Teor. Fiz. **34**, 58; Nuovo Cim. **7**, 794.

Bogoliubov, N.N., V.V. Tolmachev and D.V. Shirkov, 1958, A New Method in the Theory of Superconductivity (Nauka, Moscow) [1959 (Consultants Bureau, New York)].

Bommel, H., 1955, Phys. Rev. **100**, 758.

Brandt, N.B., and V.V. Moshchalkov, 1984, Adv. Phys. **33**, 373.

Brandt, N.B., and V.A. Ventsel', 1958, Zh. Eksp. & Teor. Fiz. **35**, 1081.

Brandt, N.B., S.V. Demishev, A.A. Dmitriev and V.B. Moshchalkov, 1981, Solid State Commun. **37**, 643.

Brovman, E.G., and Yu.M. Kagan, 1974, Usp. Fiz. Nauk **112**, 369.

Brust, D., 1964, Phys. Rev. **134**, 1337.

Bulaevskii, L.N., A.A. Sobyanin and D.I. Khomskii, 1984, Zh. Eksp. & Teor. Fiz. **87**, 1490.

Bulaevskii, L.N., A.I. Buzdin, M.L. Kulik and S.V. Panjukov, 1985, Adv. Phys. **34**, 175.

Buzdin, A.I., and L.N. Bulaevskii, 1986, Usp. Fiz. Nauk **149**, 45.

Buzdin, A.I., and A.S. Mikhailov, 1986, Zh. Eksp. & Teor. Fiz. **90**, 294.

Bychkov, Yu.A., L.E. Gurevich and G.M. Nedlin, 1959, Zh. Eksp. & Teor. Fiz. **37**, 534.

Caroli, C., and J. Matricon, 1965, Phys. Kondens. Mater. **3**, 380.

Caroli, C., P.G. De Gennes and J. Matricon, 1965, J. Phys. Lett. (France) **9**, 307.

Clogston, A.M., 1962, Phys. Rev. Lett. **9**, 266.

Cohen, M.H., and L.M. Falicov, 1961, Phys. Rev. Lett. **7**, 231.

Cohen, M.H., and V. Heine, 1961, Phys. Rev. **122**, 1821.

Condon, J.H., 1966, Phys. Rev. **145**, 526.

Condon, J.H., and R.E. Walstedt, 1968, Phys. Rev. Lett. **21**, 612.

Cooper, L.N., 1956, Phys. Rev. **104**, 1189.

Corak, W.S., and C.W. Satterthwaite, 1954, Phys. Rev. **99**, 1660.

Corak, W.S., B.B. Goodman, C.B. Satterthwaite and A. Wexler, 1954, Phys. Rev. **96**, 1442.

Corak, W.S., B.B. Goodman, C.B. Satterthwaite and A. Wexler, 1956, Phys. Rev. **102**, 656.

Cribier, D., B. Jacrot, L.M. Rao and B. Farnoux, 1964, Phys. Lett. **9**, 106.

De Gennes, P., 1966, Superconductivity of Metals and Alloys (Benjamin, New York).

De Haas, W.J., and P.M. van Alphen, 1930, Leiden Commun. A **212**, 215; Proc. Amsterdam Acad. Sci. **33**, 1106.

Deaver, D.S., and W.M. Fairbank, 1961, Phys. Rev. Lett. **7**, 43.

Dingle, R.B., 1952, Proc. R. Soc. London Ser. A **211**, 517.

Dmitrenko, I.M., I.K. Yanson and V.M. Svistunov, 1965, Pis'ma v Zh. Eksp. & Teor. Fiz. **2**, 17.

Dolgopolov, V.T., 1975, Zh. Eksp. & Teor. Fiz. **68**, 355.

Dolgopolov, V.T., 1980, Usp. Fiz. Nauk **130**, 241.

Doll, R., and M. Näbauer, 1961, Phys. Rev. Lett. **7**, 51.

Dreizin, Yu.A., and A.M. Dykhne, 1972, Zh. Eksp. & Teor. Fiz. **63**, 242.

Dyson, F.J., 1955, Phys. Rev. **98**, 349.

Dzyaloshinskii, I.E., and E.I. Katz, 1968, Zh. Eksp. & Teor. Fiz. **55**, 338.

Eck, R.E., D.J. Scalapino and B.N. Taylor, 1964, Phys. Rev. Lett. **13**, 15.

Efetov, K.B., 1983, Adv. Phys. **32**, 53.

Eliashberg, G.M., 1970, Pis'ma v Zh. Eksp. & Teor. Fiz. **11**, 186.

Eliashberg, G.M., 1987, Pis'ma v Zh. Eksp. & Teor. Fiz. **46**, 94.

Essmann, U., and H. Träuble, 1967, Phys. Lett. A **24**, 526; Phys. Status Solidi **20**, 95.

Falicov, L.M., and S. Golin, 1965, Phys. Rev. A **137**, 871.

Fal'kovskii, L.A., 1971, Zh. Eksp. & Teor. Fiz. **60**, 838.
Fal'kovskii, L.A., 1979, J. Low-Temp. Phys. **36**, 713.
Fal'kovskii, L.A., 1981, Zh. Eksp. & Teor. Fiz. **80**, 775.
Fal'kovskii, L.A., 1983, Adv. Phys. **32**, 753.
Fang, M.M., V.G. Kogan, D.K. Finnemore, et al., 1987, preprint (Ames Laboratory).
Ferrell, R., and R. Prange, 1963, Phys. Rev. Lett. **10**, 479.
Feynman, R.P., 1955, in: Progress in Low-Temperature Physics, Vol. 1, ed. D.F. Brewer (North-Holland, Amsterdam) ch. 11.
Fiske, M.D., 1964, Rev. Mod. Phys. **36**, 221.
Friedel, J., 1952, Philos. Mag. **43**, 153.
Friedel, J., 1954, Philos. Mag. Suppl. **3**, 446.
Friedel, J., 1958, Nuovo Cimento Suppl. **7**, 287.
Frölich, H., 1950, Phys. Rev. **79**, 845.
Fukuyama, H., and K. Yosida, 1987, Jpn. J. Appl. Phys. **26**, L160.
Fulde, P., and R.A. Ferrell, 1964, Phys. Rev. A **135**, 550.
Gal'perin, Yu.M., and V.I. Kozub, 1975, Fiz. Tverd. Tela **17**, 2222.
Gal'perin, Yu.M., V.D. Kagan and V.I. Kozub, 1972, Zh. Eksp. & Teor. Fiz. **62**, 1521.
Gal'perin, Yu.M., V.L. Gurevich and V.I. Kozub, 1973, Pis'ma v Zh. Eksp. & Teor. Fiz. **17**, 687.
Gal'perin, Yu.M., V.L. Gurevich and V.I. Kozub, 1974, Zh. Eksp. & Teor. Fiz. **66**, 1387.
Gal'perin, Yu.M., V.L. Gurevich and V.I. Kozub, 1979, Usp. Fiz. Nauk **128**, 107.
Gamble, F.R., R.A. Klemm, F.J. Di Salvo and T.H. Geballe, 1976, Science **168**, 568.
Gantmakher, V.F., 1962, Zh. Eksp. & Teor. Fiz. **43**, 345.
Gantmakher, V.F., and E.A. Kaner, 1963, Zh. Eksp. & Teor. Fiz. **45**, 1430.
Gantmakher, V.F., and Yu.V. Sharvin, 1965, Zh. Eksp. & Teor. Fiz. **48**, 1077.
Geilikman, B.T., 1958, Zh. Eksp. & Teor. Fiz. **34**, 1042.
Gershenson, M.E., and V.N. Gubankov, 1982, Solid State Commun. **41**, 33.
Giaever, I., 1960a, Phys. Rev. Lett. **5**, 147.
Giaever, I., 1960b, Phys. Rev. Lett. **5**, 464.
Ginzburg, V.L., 1952, Dokl. Akad. Nauk SSSR **83**, 385.
Ginzburg, V.L., 1960, Fiz. Tverd. Tela **2**, 2034.
Ginzburg, V.L., 1964, Zh. Eksp. & Teor. Fiz. **47**, 2318; Phys. Lett. **13**, 10.
Ginzburg, V.L., 1968, Usp. Fiz. Nauk **95**, 91.
Ginzburg, V.L., 1970, Usp. Fiz. Nauk **101**, 185.
Ginzburg, V.L., 1976, Usp. Fiz. Nauk **118**, 315.
Ginzburg, V.L., and D.A. Kirzhnits, eds, 1977, The Problem of High-Temperature Superconductivity (Nauka, Moscow). In Russian.
Ginzburg, V.L., and L.D. Landau, 1950, Zh. Eksp. & Teor. Fiz. **20**, 1064.
Glover, R.E., 1967, Phys. Lett. A **25**, 542.
Gor'kov, L.P., 1958, Zh. Eksp. & Teor. Fiz. **34**, 735.
Gor'kov, L.P., 1959a, Zh. Eksp. & Teor. Fiz. **36**, 1918.
Gor'kov, L.P., 1959b, Zh. Eksp. & Teor. Fiz. **37**, 833.
Gor'kov, L.P., 1959c, Zh. Eksp. & Teor. Fiz. **37**, 1407.
Gor'kov, L.P., 1963, Zh. Eksp. & Teor. Fiz. **44**, 767.
Gor'kov, L.P., A.I. Larkin and D.E. Khmel'nitsky, 1979, Pis'ma v Zh. Eksp. & Teor. Fiz. **30**, 248.
Grimes, C.C., and S. Shapiro, 1968, Phys. Rev. **169**, 397.
Gurevich, V.L., 1958, Zh. Eksp. & Teor. Fiz. **35**, 669.
Gurevich, V.L., 1959, Zh. Eksp. & Teor. Fiz. **37**, 71.
Gurevich, V.L., V.G. Skobov and Yu.A. Firsov, 1961, Zh. Eksp. & Teor. Fiz. **40**, 786.
Gurvitch, M., 1983, Phys. Rev. B **28**, 544.
Gurzhi, R.N., 1964, Zh. Eksp. & Teor. Fiz. **47**, 1415.
Gurzhi, R.N., and A.I. Kopeliovich, 1973, Zh. Eksp. & Teor. Fiz. **64**, 380.
Halperin, B.I., and T.M. Rice, 1968, Rev. Mod. Phys. **40**, 755.

Harris, R., L.J. Lewis and M.J. Zuckermann, 1983, J. Phys. F **13**, 2323.
Harrison, W.A., 1959, Phys. Rev. **116**, 555.
Harrison, W.A., 1960, Phys. Rev. **118**, 1182.
Heine, V., and I. Abarenkov, 1964, Philos. Mag. **9**, 451.
Heine, V., and I. Abarenkov, 1965, Philos. Mag. **12**, 529.
Helfand, E., and N.R. Werthamer, 1966, Phys. Rev. **147**, 288.
Herring, C., 1940, Phys. Rev. **57**, 1169.
Hess, H.F., K. De Conde, T.F. Rosenbaum, and G.A. Thomas, 1982, Phys. Rev. B **25**, 5578.
Hikami, S., A.I. Larkin and Y. Nagaoka, 1980, Prog. Theor. Phys. **63**, 707.
Hohenberg, P.C., 1963, Zh. Eksp. & Teor. Fiz. **45**, 1208.
Inderhees, S.E., M.B. Salamon, N. Goldenfeld, et al., 1987, preprint (University of Illinois).
Ivlev, B.I., and N.B. Kopnin, 1978, Pis'ma v Zh. Eksp. & Teor. Fiz. **28**, 640.
Ivlev, B.I., and N.B. Kopnin, 1984, Adv. Phys. **33**, 47.
Ivlev, V.I., and G.M. Eliashberg, 1971, Pis'ma v Zh. Eksp. & Teor. Fiz. **13**, 464.
Jaccarino, V., and M. Peter, 1962, Phys. Rev. Lett. **9**, 290.
Johnson, W.J., 1968, Ph.D. Thesis (University of Wisconsin).
Josephson, B.D., 1962, Phys. Rev. Lett. **1**, 251.
Josephson, B.D., 1964, Rev. Mod. Phys. **36**, 216.
Josephson, B.D., 1965, Adv. Phys. **14**, 419.
Justi, E., 1940, Phys. Zs. **41**, 563.
Kadanoff, L.P., 1966, Physics **2**, 263.
Kagan, Yu., and A.P. Zhernov, 1971, Zh. Eksp. & Teor. Fiz. **60**, 1832.
Kamerlingh Onnes, H., 1911, Commun. Phys. Lab. Univ. Leiden Nos. 119b, 120b, 122b, 124c.
Kamerlingh Onnes, H., 1914, Commun. Phys. Lab. Univ. Leiden No. 139f.
Kaner, E.A., 1958, Dokl. Akad. Nauk SSSR **119**, 471.
Kaner, E.A., 1963, Zh. Eksp. & Teor. Fiz. **44**, 1036.
Kaner, E.A., and V.G. Skobov, 1963, Zh. Eksp. & Teor. Fiz. **45**, 610.
Kaner, E.A., and V.G. Skobov, 1968, Adv. Phys. **17**, 605.
Kaner, E.A., V.G. Peschanskii and I.A. Privorotskii, 1961, Zh. Eksp. & Teor. Fiz. **40**, 214.
Kapitza, P.L., 1929, Proc. R. Soc. London **123**, 292.
Kapitza, P.L., 1938, Dokl. Akad. Nauk. SSSR **18**, 21; Nature **141**, 74.
Kasuya, T., 1956, Prog. Theor. Phys. **16**, 45.
Khaikin, M.S., 1960, Zh. Eksp. & Teor. Fiz. **39**, 212.
Khaikin, M.S., 1961, Zh. Eksp. & Teor. Fiz. **41**, 1773.
Khaikin, M.S., 1968, Usp. Fiz. Nauk **96**, 409.
Khaikin, M.S., and V.S. Edel'man, 1965, Zh. Eksp. & Teor. Fiz. **49**, 1695.
Khaikin, M.S., and I.N. Khlyustikov, 1981, Pis'ma v Zh. Eksp. & Teor. Fiz. **33**, 167; **34**, 207.
Khaikin, M.S., and I.N. Khlyustikov, 1983, Pis'ma v Zh. Eksp. & Teor. Fiz. **38**, 191.
Khaikin, M.S., L.A. Fal'kovskii, V.S. Edel'man and R.T. Mina, 1963, Zh. Eksp. & Teor. Fiz. **45**, 1704.
Khlyustikov, I.N., and A.I. Buzdin, 1987, Adv. Phys. **36**, 271.
Kleiner, W.H., L.M. Roth and S.H. Autler, 1964, Phys. Rev. A **133**, 1226.
Knight, W.D., 1949, Phys. Rev. **76**, 1259.
Kondo, J., 1964, Prog. Theor. Phys. **32**, 37.
Kondo, J., 1965, Prog. Theor. Phys. **34**, 372.
Konstantinov, O.V., and V.I. Perel', 1960, Zh. Eksp. & Teor. Fiz. **38**, 161.
Kozub, V., 1974, private communication.
Kozub, V., 1985, Thesis (Ioffe Physico-Technical Institute, Leningrad).
Kramer, B., G. Bergmann and Y. Bruynseraede, eds, 1984, Localization, Interaction and Transport Phenomena, Springer Series in Solid-State Sciences 61.
Krylov, I.P., and Yu.V. Sharvin, 1970, Pis'ma Zh. Eksp. & Teor. Fiz. **12**, 102.
Krylov, I.P., and Yu.V. Sharvin, 1973, Zh. Eksp. & Teor. Fiz. **64**, 946.
Kulik, I.O., 1965, Pis'ma Zh. Eksp. & Teor. Fiz. **2**, 134.

Kulik, I.O., 1968a, Zh. Eksp. & Teor. Fiz. **55**, 218.
Kulik, I.O., 1968b, Zh. Eksp. & Teor. Fiz. **55**, 889.
Kulik, I.O., 1984, Kharkov: Preprint (FTINT).
Kulik, I.O., and A.G. Pedan, 1980, Zh. Eksp. & Teor. Fiz. **79**, 1469.
Kulik, I.O., and A.G. Pedan, 1982, Fiz. Nizk. Temp. **8**, 236.
Kulik, I.O., and A.G. Pedan, 1983, Fiz. Nizk. Temp. **9**, 256.
Laeng, R., and L. Rinderer, 1972, Cryogenics **12**, 315.
Landau, L.D., 1930, Z. Phys. **64**, 629.
Landau, L.D., 1937a, Zh. Eksp. & Teor. Fiz. **7**, 19; Phys. Z. der Sowjet Union **11**, 26.
Landau, L.D., 1937b, Phys. Z. der Sowjet Union **11**, 129.
Landau, L.D., 1941a, Zh. Eksp. & Teor. Fiz. **11**, 581.
Landau, L.D., 1941b, J. Phys. (USSR) **5**, 71.
Landau, L.D., 1944, Zh. Eksp. & Teor. Fiz. **11**, 592.
Landau, L.D., 1947, J. Phys. (USSR) **11**, 91.
Landau, L.D., 1956, Zh. Eksp. & Teor. Fiz. **30**, 1058 (Sov. Phys.-JETP **3**, 920).
Landau, L.D., 1957, Zh. Eksp. & Teor. Fiz. **32**, 59 (Sov Phys.-JETP. **5**, 101).
Landau, L.D., and E.M. Lifshitz, 1974, Quantum Mechanics (Nauka, Moscow) [1977 (Pergamon Press, London)].
Landau, L.D., and E.M. Lifshitz, 1976, Statistical Physics (Nauka, Moscow) [1979 (Pergamon Press, London)].
Landau, L.D., and E.M. Lifshitz, 1982, Electrodynamics of Continuous Media (Nauka, Moscow) [1960 (Pergamon Press, London)].
Langenberg, D., D. Scalapino, B. Taylor and R. Eck, 1965, Phys. Rev. Lett. **15**, 294.
Langer, J.S., and V. Ambegoakar, 1967, Phys. Rev. **164**, 498.
Larkin, A.I., and D.E. Khmel'nitskii, 1969, Zh. Eksp. & Teor. Fiz. **56**, 2087.
Larkin, A.I., and D.E. Khmel'nitskii, 1982, Usp. Fiz. Nauk **136**, 536.
Larkin, A.I., and D.E. Khmel'nitskii, 1986, Zh. Eksp. & Teor. Fiz. **91**, 1815.
Larkin, A.I., and Yu.N. Ovchinnikov, 1964, Zh. Eksp. & Teor. Fiz. **47**, 1136.
Leggett, A.J., 1975, Rev. Mod. Phys. **47**, 33.
Levanyuk, A.P., 1959, Zh. Eksp. & Teor. Fiz. **36**, 810.
Lifshitz, E.M., and L.P. Pitaevskii, 1978, Statistical Physics. Part II (Nauka, Moscow).
Lifshitz, E.M., and L.P. Pitaevskii, 1979, Physical Kinetics (Nauka, Moscow).
Lifshitz, I.M., 1960, Zh. Eksp. & Teor. Fiz. **38**, 1569.
Lifshitz, I.M., and A.M. Kosevich, 1955, Zh. Eksp. & Teor. Fiz. **29**, 730.
Lifshitz, I.M., and V.G. Peschanskii, 1958, Zh. Eksp. & Teor. Fiz. **35**, 1251.
Lifshitz, I.M., M.Ya. Azbel and M.I. Kaganov, 1956, Zh. Eksp. & Teor. Fiz. **31**, 63.
Lifshitz, I.M., M.Ya. Azbel and M.I. Kaganov, 1957, Sov. Phys.-JETP **4**, 41.
Little, W.A., 1964, Phys. Rev. A **134**, 1416.
Little, W.A., 1967, Phys. Rev. **156**, 398.
Little, W.A., and R.D. Parks, 1962, Phys. Rev. Lett. **9**, 9.
Little, W.A., and R.D. Parks, 1964, Phys. Rev. A **133**, 97.
London, F., 1936, Physica **3**, 450.
London, F., 1937, Une conception nouvelle de la supraconductibilité, (Paris).
London, F., 1950, Superfluids, Vol. 1, New York.
London, F., and H. London, 1935, Proc. R. Soc. London Ser. A **149**, 71; Physica **2**, 341.
London, H., 1940, Proc. R. Soc. London Ser. A **176**, 522.
Luttinger, J.M., and J.C. Ward, 1960a, Phys. Rev. **118**, 1417.
Luttinger, J.M., and J.C. Ward, 1960b, Phys. Rev. **119**, 1153.
Lyons, K.B., S.H. Liou, M. Hong, H.S. Chen, J. Kuo and T.J. Negran, 1987, Phys. Rev. B **36**, 5592.
Maki, K., 1968, Prog. Theor. Phys. **39**, 897; **40**, 193.
Maki, K., 1969, in: Superconductivity, ed. R.D. Parks (Dekker, New York) ch. 18.
Mattis, D.C., and J. Bardeen, 1958, Phys. Rev. **111**, 412.

Maxwell, E., 1950, Phys. Rev. **78**, 477.
McCumber, D.E., 1968, J. Appl. Phys. **39**, 2503, 3113.
McMillan, M.L., 1968, Phys. Rev. **267**, 331.
Meissner, H., 1958, Phys. Rev. **109**, 686; ibid. **117**, 672.
Meissner, H., 1959, Phys. Rev. Lett. **2**, 458.
Meissner, W., and R. Ochsenfeld, 1933, Naturwiss. **21**, 787.
Meshkovskii, A.G., and A.I. Shal'nikov, 1947, Zh. Eksp. & Teor. Fiz. **17**, 851.
Meyer, J., and G.V. Minnigerode, 1972, Phys. Rev. Lett. A **38**, 529.
Mihaly, L.M., and A. Zawadowsky, 1978, J. Phys. Lett. **39**, L483.
Morris, R.C., R.V. Coleman and R. Bhandari, 1972, Phys. Rev. B **5**, 895.
Mott, N.F., 1961, Philos. Mag. **6**, 287.
Mott, N.F., 1974, Metal-Insulator Transitions (Taylor and Francis, London).
Mott, N.F., and E.A. Davis, 1971, Electronic Processes in Non-crystalline Materials (Clarendon Press, Oxford).
Mott, N.F., and W.D. Twose, 1961, Adv. Phys. **10**, 107.
Nee, T.W., and R.E. Prange, 1967, Phys. Rev. Lett. A **25**, 583.
Nicol, J., S. Shapiro and P.H. Smith, 1960, Phys. Rev. Lett. **5**, 461.
Nozières, P., 1974, J. Low-Temp. Phys. **17**, 13.
Nozières, P., and A. Blandin, 1980, J. Phys. (France) **41**, 193.
Onsager, L., 1949, Nuovo Cim. Suppl. **6**, 249.
Onsager, L., 1952, Philos. Mag. **43**, 1006.
Patashinskii, A.Z., and V.L. Pokrovskii, 1964, Zh. Eksp. & Teor. Fiz. **46**, 944.
Patashinskii, A.Z., and V.L. Pokrovskii, 1966, Zh. Eksp. & Teor. Fiz. **50**, 439.
Patashinskii, A.Z., and V.L. Pokrovskii, 1982, Fluctuational Theory of Phase Transitions (Nauka, Moscow).
Pearl, J., 1964, J. Appl. Phys. Lett. **5**, 65.
Peierls, R.E., 1933, Z. Phys. **81**, 186.
Peierls, R.E., 1936, Proc. R. Soc. London **155**, 613.
Peierls, R.E., 1955, Quantum Theory of Solids (Oxford Univ. Press, London).
Phillips, J.C., and L. Kleinman, 1959, Phys. Rev. **116**, 287, 880.
Pippard, A.B., 1953, Proc. R. Soc. London **216**, 547.
Pippard, A.B., 1955, Philos. Trans. R. Soc. London A **248**, 97.
Pippard, A.B., 1957, Philos. Mag. **2**, 1147.
Pippard, A.B., 1963, Proc. R. Soc. London Ser. A **272**, 192.
Platzman, P.M., and P.A. Wolff, 1967, Phys. Rev. Lett. **18**, 280.
Pokrovskii, V.L., 1961, Zh. Eksp. & Teor. Fiz. **40**, 641.
Pomeranchuk, I.Ya., 1958, Zh. Eksp. & Teor. Fiz. **35**, 992.
Privorotskii, I.A., 1967, Zh. Eksp. & Teor. Fiz. **52**, 1755.
Reuter, G.E., and E.H. Sondheimer, 1948, Proc. R. Soc. London Ser. A **195**, 336.
Reynolds, C.A., B. Serin, W.H. Wright and L.B. Nesbitt, 1950, Phys. Rev. **78**, 487.
Rice, T.M., 1965, Phys. Rev. A **140**, 1889.
Rowell, J.M., 1963, Phys. Rev. Lett. **11**, 200.
Ruderman, M.A., and C. Kittel, 1954, Phys. Rev. **96**, 99.
Saint-James, D., 1965, Phys. Rev. Lett. **16**, 218.
Saint-James, D., and P.G. De Gennes, 1963, Phys. Rev. Lett. **7**, 306.
Saint-James, D., G. Sarma and E. Thomas, 1969, Type II Superconductors (Oxford).
Sampsell, J.B., and J.C. Garland, 1976, Phys. Rev. B **13**, 583.
Schneider, T., 1969, Helv. Phys. Acta **42**, 957.
Schultz, S., and G. Dunifer, 1967, Phys. Rev. Lett. **18**, 283.
Schütter, H.B., M. Jarrel and D.J. Scalapino, 1987, Phys. Rev. Lett. **58**, 1147.
Shapira, Y., and B. Lax, 1965, Phys. Rev. A **138**, 1191.
Shapiro, S., 1963, Phys. Rev. Lett. **11**, 80.

Shapiro, S., A.R. Janus and S. Holly, 1964, Rev. Mod. Phys. **36**, 223.
Sharvin, D. Yu., and Yu.V. Sharvin, 1981, Pis'ma Zh. Eksp. & Teor. Fiz. **34**, 285.
Sharvin, Yu.V., 1957, Zh. Eksp. & Teor. Fiz. **33**, 1341.
Sharvin, Yu.V., 1984, Physica B **126**, 288.
Sharvin, Yu.V., and B.M. Balashova, 1956, Zh. Eksp. & Teor. Fiz. **31**, 40.
Shaw, R.W., 1968, Phys. Rev. **174**, 769.
Shoenberg, D., 1939, Proc. R. Soc. London Ser. A **170**, 341.
Shoenberg, D., 1952, Superconductivity (Cambridge).
Shoenberg, D., 1962, Philos. Trans. R. Soc. London A **255**, 85.
Shubnikov, L.V., and W.J. de Haas, 1930, Leiden Commun. **207**, 210; Proc. Amsterdam Acad. Sci. **33**, 418.
Silin, V.P., 1958, Zh. Eksp. & Teor. Fiz. **35**, 1243 (Sov. Phys.-JETP **8**, 870).
Silsbee, F.B., 1916, J. Wash. Acad Sci. **6**, 597.
Skobov, V.G., 1962, Dissertation, IFP Akad. Nauk SSSR, Moscow.
Skocpol, W.J., M.R. Beasley and M. Tinkham, 1974, J. Low-Temp. Phys. **16**, 145.
Sondheimer, E.H., 1950, Phys. Rev. **80**, 401.
Stewart, W.C., 1968, Appl. Phys. Rev. Lett. **12**, 277.
Stone, A.D., and Y. Imry, 1986, Phys. Rev. Lett. **56**, 189.
Suhl, H., 1965, Phys. Rev. A **138**, 515; Physics **2**, 39.
Swihart, J.C., 1961, J. Appl. Phys. **32**, 461.
Thompson, R.S., 1970, Phys. Rev. B **1**, 327.
Thouless, D.J., 1977, Phys. Rev. Lett. **39**, 1167.
Tsvelick, A.M., and P.B. Wiegmann, 1983, Adv. Phys. **32**, 453.
Varlamov, A.A., and A.V. Pantsulaya, 1985, Zh. Eksp. & Teor. Fiz. **89**, 2188.
Volodin, A.P., M.S. Khaikin and V.S. Edel'man, 1973, Zh. Eksp. & Teor. Fiz. **65**, 2105.
Werthamer, N.R., 1963, Phys. Rev. **132**, 2440.
Wheatley, J.C., 1975, Rev. Mod. Phys. **47**, 415.
Wilson, K., 1974, in: Noble Symposium - 24. Collective Properties of Physical Systems (Academic Press, New York).
Wilson, K., 1975, Rev. Mod. Phys. **47**, 773.
Wilson, K.G., and M.E. Fisher, 1972, Phys. Rev. Lett. **28**, 240.
Wyatt, A.F., V.M. Dmitriev, W.S. Moore and F.W. Sheard, 1966, Phys. Rev. Lett. **16**, 1166.
Yanson, I.K., V.M. Svistunov and I.M. Dmitrenko, 1965, Zh. Eksp. & Teor. Fiz. **48**, 976.
Yegorov, V.S., and A.N. Fedorov, 1982, Pis'ma Zh. Eksp. & Teor. Fiz. **35**, 375.
Yegorov, V.S., M.Yu. Lavrenyuk, N.Ya. Minina and A.M. Savin, 1984, Pis'ma Zh. Eksp. & Teor. Fiz. **40**, 25 .
Yosida, K., 1957, Phys. Rev. **106**, 893.
Zaitsev, R.O., 1965, Zh. Eksp. & Teor. Fiz. **48**, 664, 1759.
Zavaritskii, N.V., 1951, Dokl. Akad. Nauk SSSR **78**, 665.
Zavaritskii, N.V., 1952, Dokl. Akad. Nauk SSSR **86**, 501.
Zavaritskii, N.V., 1965, Pis'ma v Zh. Eksp. & Teor. Fiz. **2**, 168.
Zavaritskii, N.V., 1974, Pis'ma v Zh. Eksp. & Teor. Fiz. **19**, 205.
Zernov, V.B., and Yu.V. Sharvin, 1959, Zh. Eksp. & Teor. Fiz. **36**, 1038.
Zhernov, A.P., and Yu. Kagan, 1978, Fiz. Tverd. Tela **20**, 3306.
Ziesche, P., and G. Lehmann, 1983, Ergebnisse in der Electronentheorie der Metalle (Akademie-Verlag, Berlin).
Ziman, J., 1971, Calculation of Bloch Functions (Academic Press, New York).
Zvezdin, A.K., and D.I. Khomskii, 1987, Pis'ma v Zh. Eksp. & Teor. Fiz. **46**, 102.

Suggested reading

Alfeev, V.N., 1979, Superconductors, Semiconductors and Paraelectrics in Cryoelectronics (Soviet Radio Publisher, Moscow). In Russian.

Al'tov, V.A., V.V. Zenkevich, M.G. Kremlev and V.V. Sychev, 1984, Stabilization of Superconducting Magnetic Systems, Energoatomizdat. In Russian.

Ashcroft, N.W., and N.D. Mermin, 1976, Solid State Physics (Holt, Rinehart and Winston, New York).

Aslamazov, L.G., and V.N. Gubankov, 1982, Weak Superconductivity (Znanic, Moscow). In Russian.

Barone, A., and G. Paterno, 1982, Physics of Applications of the Josephson Effect (Wiley, New York).

Bogoliubov, N.N., V.V. Tolmachev and D.V. Shirkov, 1958, A New Method in the Theory of Superconductivity (Nauka, Moscow) [1959 (Consultants Bureau, New York)].

Brandt, N.B., and S.M. Chudinov, 1973, The Electronic Structure of Metals (Moscow State Univ. Press, Moscow). In Russian.

Brandt, N.B., and S.M. Chudinov, 1983, Experimental Methods of Investigation of the Energy Spectra of Electrons and Phonons in Metals (Moscow Univ. Press, Moscow). In Russian.

Bremer, J., 1962, Superconductive Devices (New York).

Buckel, W., 1972, Supraleitung Grundlagen Anwendungen (Weinheim).

Callaway, J., 1964, Energy Band Theory (New York).

De Gennes, P., 1966, Superconductivity of Metals and Alloys (New York).

Gantmakher, V.F., and I.B. Levinson, 1984, Scattering of Current Carriers in Metals and Semiconductors (Nauka, Moscow). In Russian.

Geilikman, B.T., and V.Z. Kresin, 1972, Kinetic and Nonstationary Phenomena in Superconductors (Nauka, Moscow). In Russian.

Ginzburg, V.L., 1946, Superconductivity (Nauka, Moscow). In Russian.

Ginzburg, V.L., and D.A. Kirzhnits, eds, 1977, The Problem of High-Temperature Superconductivity (Nauka, Moscow). In Russian.

Harrison, W.A., 1966, Pseudopotentials in the Theory of Metals (New York).

Harrison, W.A., 1980, Electronic Structure and the Properties of Solids (Freeman, San Francisco).

Jones, H., 1960, The Theory of the Brillouin Zone and Electronic States in Crystals (North-Holland, New York).

Kaganov, M.I., and V.S. Edel'man, eds, 1985, Conduction Electrons (Nauka, Moscow). In Russian.

Kittel, C., 1971, Introduction to Solid State Physics, 4th Ed. (Wiley, New York).

Kulik, I.O., and I.K. Yanson, 1970, The Josephson Effect in Superconducting Tunnel Structures (Nauka, Moscow). In Russian.

Liftshitz, I.M., M.Ya. Azbel and M.I. Kaganov, 1971, Electronic Theory of Metals (Nauka, Moscow). In Russian.

Likharev, K.K., 1985, Introduction to the Dynamics of Josephson Transitions (Nauka, Moscow). In Russian.

Lynton, E.A., 1969, Superconductivity, 3rd Ed. (London).

Mints, R.G., and A.L. Rakhmanov, 1984, Instabilities in Superconductors (Nauka, Moscow). In Russian.

Mott, N.F., 1974, Metal-Insulator Transitions (Taylor and Francis, London).

Mott, N.F., and E.A. Davis, 1971, Electronic Processes in Non-Crystalline Materials (Clarendon Press, Oxford).

Parks, R.D., ed., 1969, Superconductivity (New York).
Patashinskii, A.Z., and V.L. Pokrovskii, 1982, The Fluctuation Theory of Phase Transitions (Nauka, Moscow). In Russian.
Peierls, R.E., 1955, Quantum Theory of Solids (Oxford Univ. Press, London).
Pines, D., and Ph. Nozières, 1966, The Theory of Quantum Liquids (New York).
Rose-Innes, A.C., and E.H. Rhoderick, 1977, Introduction to Superconductivity, 2nd Ed. (Pergamon Press, New York).
Saint-James, D., G. Sarma and E. Thomas, 1969, Type II Superconductors (Oxford).
Schrieffer, J.R., 1964, Theory of Superconductivity (Benjamin, New York).
Shmidt, V.V., 1982, Introduction to the Physics of Superconductors (Nauka, Moscow). In Russian.
Shoenberg, D., 1952, Superconductivity (Cambridge).
Solymar, L., 1972, Superconductivity Tunnelling and Applications (London).
Tinkham, M., 1975, Introduction to Superconductivity (McGraw-Hill, New York).
Vonsovskii, S.V., and M.I. Katsnelson, 1983, Quantum Physics of Solids (Nauka, Moscow). In Russian.
Vonsovskii, S.V., Yu.A. Izyumov and E.Z. Kurmaev, 1977, Superconductivity of Transition Metals, Their Alloys and Compounds (Nauka, Moscow). In Russian.
Yastrebov, L.I., and A.A. Kannelson, 1981, Foundations of the One-Electron Theory of Solids (Nauka, Moscow). In Russian.
Ziesche, P., and G. Lehmann, eds, 1983, Ergebnisse in der Electronentheorie der Metalle (Akademie-Verlag, Berlin).
Ziman, J., ed., Physics of Metals. Electrons, Vol. 1.
Ziman, J., 1960, Electrons and Phonons (Oxford).
Ziman, J., 1964, Principles of the Theory of Solids (Cambridge Univ. Press, London).
Ziman, J., 1971, Calculation of Bloch Functions (Academic Press, New York).

*Author index**

*Page numbers in italics refer to the reference section.

Subject index